STUDY GUIDE AND SOLUTIONS MANUAL
TO ACCOMPANY

ORGANIC CHEMISTRY

SIXTH EDITION

STUDY GUIDE
AND
SOLUTIONS MANUAL
TO ACCOMPANY

ORGANIC
CHEMISTRY
SIXTH EDITION

T. W. GRAHAM SOLOMONS
University of South Florida

JOHN WILEY & SONS, INC.
New York   Chichester   Brisbane   Toronto   Singapore

Acquisitions Editor	Nedah Rose
Associate Editor	Joan Kalkut
Marketing Manager	Catherine Faduska
Senior Production Editor	Suzanne Magida
Cover Designer	Madelyn Lesure
Associate Manufacturing Manager	Marsheela Evans

This book was set in 10/13 Times Roman by York Production Services and printed and bound by Hamilton Printing. The cover was printed by Phoenix Color.

Recognizing the importance of preserving what has been written, it is a policy of John Wiley & Sons, Inc., to have books of enduring value published in the United States printed on acid-free paper, and we exert out best efforts to that end.

ISBN 0-471-12108-8

Printed in the United States of America

10 9 8 7 6

TO THE STUDENT

Contrary to what you may have heard, organic chemistry does not have to be a difficult course. It will be a rigorous course, and it will offer a challenge. You will learn more in it than in almost any course you will take, and what you learn will have a special relevance to the world around you. However, because organic chemistry can be approached in a logical and systematic way, you will find that with the right study habits, mastering organic chemistry can be a deeply satisfying experience.

Here, then, are some suggestions about how to study:

1. ***Keep up with your work from day to day—never let yourself get behind.*** Organic chemistry is a course in which one idea almost always builds on another that has gone before. It is essential, therefore, that you keep up, or better yet, be a little ahead of your instructor. You should always try, in your class preparation, to stay one day ahead of your instructor's lectures. The lecture, then, will be much more helpful because you will already have some understanding of the assigned material. The lecture will clarify and expand ideas that are already familiar ones.

2. ***Study material in small units, and be sure that you understand each new section before you go on to the next.*** Again, because of the cumulative nature of organic chemistry, your studying will be much more effective if you take each new idea as it comes to try to understand it completely before you move on to the new concept. One way to check your progress is to work each of the in-chapter problems when you come to it. These problems have been written just for this purpose and are designed to help you decide whether or not you understand the material that has just been explained. If you can work the in-chapter problem, then you should go on; if you cannot, then you should go back and study the preceding material again.

3. ***Work all of the assigned problems.*** Work your problems in a notebook and bring this book with you when you go to see your instructor for extra help.

4. ***Write when you study.*** Write the reactions, mechanisms, structures, and so on, over and over again. Organic chemistry is best assimilated through the fingertips by writing, and not through the eyes by simply looking, or by highlighting material in the text, or by referring to flash cards. There is a good reason for this. Organic structures, mechanisms, and reactions are complex. If you simply examine them, you may think you understand them thoroughly but that will be a misperception. The reaction, mechanism, may make sense to you in a certain way, but you need a deeper understanding than this. You need to know the material so thoroughly that you can explain it to someone else. This level of understanding comes to most of us (those of us without photographic memories) through writing. Only by writing the reaction mechanisms do we

pay sufficient attention to their details, such as which atoms are connected to which atoms, which bonds break in a reaction, and which bonds form, and the three-dimensional aspects of the structures. When we write reactions and mechanisms, connections are made in our brains that provide the long-term memory needed for success in organic chemistry. I can guarantee that your grade in the course will be directly proportional to the number of pages of paper that you fill with your own writing in studying during the term.

5. *Use the answers to the problems in the study guide in the proper way.* Refer to the answers only in two circumstances: (a) When you have finished a problem, use the study guide to check your answer. (b) When, after making a real effort to solve the problem, you find that you are completely stuck, then look at the answer for a clue, and go back to work out the problem on your own. The value of a problem is in solving it. If you simply read the problem and look up the answer, you will deprive yourself of an important way to learn.

6. *Use molecular models when you study.* Because of the three-dimensional nature of most organic molecules, molecular models can be an invaluable aid to your understanding of them. Buy yourself an inexpensive molecular model set, and use it when you need to see the three-dimensional aspect of the particular topic. Appendix D of this study guide provides a set of highly useful molecular model exercises.

ACKNOWLEDGMENTS

I am especially grateful to Keith Buszek of Kansas State University for reviewing all of the answers to the new problems contained in the previous edition. I am also grateful to Ronald Starkey of the University of Wisconsin for providing the Molecular Model Exercises contained in Appendix B.

I also thank those who have made many helpful suggestions for earlier versions of this study guide: These include: George R. Jurch, George R. Wenzinger, and J. E. Fernandez at the University of South Florida; Darell Berlin, Oklahoma State University; John Mangravite, West Chester State College; J. G. Traynham, Louisiana State University; and Desmond M. S. Wheeler, University of Nebraska.

I am very greatful to Professor John Thompson of Texas A&M University—Kingsville for completing the arduous task of creating all of the structures found in this Study Guide.

Once again, I am indebted to Robert G. Johnson of Xavier University for his help. I thank him for his reviews, for contributing problems and answers, and for proofreading all of this edition.

CONTENTS

1 CARBON COMPOUNDS AND CHEMICAL BONDS

SOLUTIONS TO PROBLEMS

Another Approach to Writing Lewis Structures

When we write Lewis structures using this method, we assemble the molecule or ion from the constituent atoms showing only the valence electrons (i.e., the electrons of the outermost shell). By having the atoms share electrons, we try to give each atom the electronic structure of a noble gas. For example, we give hydrogen atoms two electrons because this gives them the structure of helium. We give carbon, nitrogen, oxygen, and fluorine atoms eight electrons because this gives them the electronic structure of neon. The number of valence electrons of an atom can be obtained from the periodic table because it is equal to the group number of the atom. Carbon, for example, is in group IVA and has four valence electrons; fluorine, in group VIIA, has seven; hydrogen, in group IA, has one. As an illustration, let us write the Lewis structure for CH_3F. In the example below, we will at first show a hydrogen's electron as x, carbon's electrons as o's, and fluorine's electrons as dots.

Example A

$$3\ H^{\times},\ \overset{o}{\underset{o}{o}}C^{o},\ \text{and}\ \cdot\overset{..}{\underset{..}{F}}\vdots\ \text{are assembled as}$$

$$\begin{array}{c} H \\ \overset{\times o}{\underset{\times o}{H^{o}_{\times}C^{o}_{o}F}}\vdots \\ H \end{array} \quad \text{or} \quad \begin{array}{c} H \\ H\vdots\overset{..}{\underset{..}{C}}\vdots\overset{..}{\underset{..}{F}}\vdots \\ H \end{array}$$

If the structure is an ion, we add or subtract electrons to give it the proper charge. As an example, consider the chlorate ion, ClO_3^-.

Example B

$$\overset{..}{\underset{..}{Cl}}\cdot,\ \text{and}\ \overset{oo}{\underset{oo}{o}O\overset{}{\underset{}{o}}}\ \text{and an extra electron x are assembled as}$$

$$\left[\begin{array}{c} \overset{oo}{\underset{oo}{o}O\overset{}{o}} \\ \overset{oo}{\underset{oo}{o}O}\vdots Cl_{\times}\overset{oo}{\underset{oo}{O}o} \end{array} \right]^{-} \quad \text{or} \quad \left[\begin{array}{c} \vdots\overset{..}{\underset{..}{O}}\vdots \\ \vdots\overset{..}{\underset{..}{O}}\vdots Cl\vdots\overset{..}{\underset{..}{O}}\vdots \end{array} \right]^{-}$$

1.1 (a) H—F̈:   (d) H—Ö—N̈＝Ö: (g) H—Ö—P—Ö—H with ＝O above and Ö—H below

(b) :F̈—F̈:   (e) H—Ö—S̈—Ö—H with ＝O above   (h) H—Ö—C—Ö—H with ＝O above

(c) H—C—F̈: with H above and below   (f) [H—B—H with H above and below]⁻   (i) H—C≡N:

1.2 (a) :Ö＝N̈—Ö: (c) ⁻:C≡N: (e) H—Ö—C—Ö:⁻ with ＝O above

(b) H—N̈: with H below   (d) H—Ö—S—Ö: with ＝O above and ＝O below   (f) H—C≡C:⁻

1.3 (a) H—C—C⁺ with H H above and H H below   (c) H—C—Ö:⁻ with ＝O above   (e) H—C—N⁺—H with H H above and H H below

(b) H—Ö⁺—H with H below   (d) H—C—H with :Ö:⁻ above and H below   (f) H—Ö⁺—H and H—C—H with H below

1.4 (a) H—C with Ö· above and Ö:⁻ below   ⟷   H—C with Ö:⁻ above and Ö· below

(b) and (c) Since the two resonance structures are equivalent, each should make an equal contribution to the overall hybrid. The C–O bonds should therefore, be of equal length (they should be of bond order 1.5), and each oxygen atom should bear a 0.5 negative charge.

1.5 (a) In its ground state, the valence electrons of carbon might be disposed as shown in the following figure.

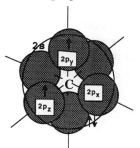

The electronic configuration of a ground state carbon atom. The *p* orbitals are designated $2p_x$, $2p_y$, and $2p_z$ to indicate their respective orientations along the *x*, *y*, and *z* axes. The assignment of the unpaired electrons to the $2p_y$ and $2p_x$ orbitals is arbitrary. They could also have been placed in the $2p_x$ and $2p_z$ or $2p_y$ and $2p_z$ orbitals. (To have placed them both in the same orbital would not have been correct, however, for this would have violated Hund's rule.) (Section 1.11.)

The formation of the covalent bonds of methane *from individual atoms* requires that the carbon atom overlap its orbitals containing *single electrons* with 1*s* orbitals of hydrogen atoms (which also contain a single electron). If a ground state carbon atom were to combine with hydrogen atoms in this way, the result would be that depicted below. *Only two carbon-hydrogen bonds would be formed, and these would be at right angles to each other.*

The hypothetical formation of CH_2 from a carbon atom in its ground state.

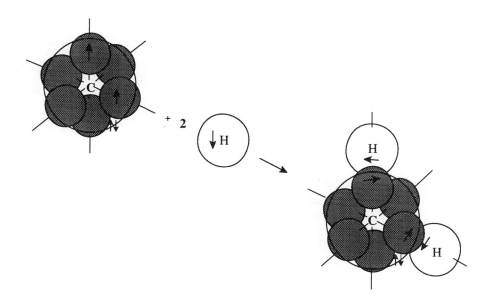

(b) An excited-state carbon atom might combine with four hydrogen atoms as shown in the figure above.

The promotion of an electron from the 2*s* orbital to the $2p_z$ orbital requires energy. The amount of energy required has been determined and is equal to 96 kcal mol^{-1}. This expenditure of energy can be rationalized by arguing that the energy released when two additional covalent bonds form would more than compensate for that required to excite the electron. No doubt this is true, but it solves only one problem. The problems that cannot be solved by using an excited-state carbon as a basis for a model of methane are the problems of the carbon-hydrogen bond angles and the apparent equivalence of all four carbon-hydrogen bonds. Three of the hydrogens—those overlapping their $1s$ orbitals with the three p orbitals—would, in this model, be at angles of 90° with respect to each other; the fourth hydrogen, the one overlapping its $1s$ orbital with the $2s$ orbital of carbon, would be at some other angle, probably as far from the other bonds as the confines of the molecule would allow. Basing our model of methane on this excited state of carbon gives us a carbon that is tetravalent *but one that is not tetrahedral,* and it predicts a structure for methane in which one carbon-hydrogen bond differs from the other three.

The hypothetical formation of CH_4 from an excited-state carbon atom.

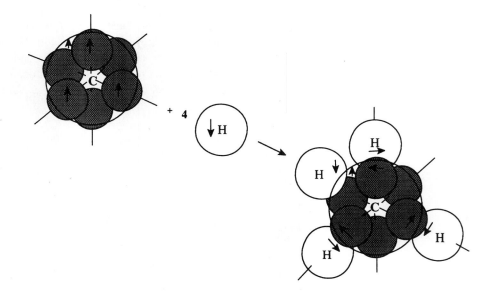

1.6 (a) It should be monovalent because only one orbital $(2p)$ contains a single electron; the $2s$ orbital is filled. (b) The two p orbitals lie at 90° to one another; the resulting bonds would also lie at 90° to each other. Thus, BH_3, based on an excited state of boron, would have the following structure. The angles of 135° result by dividing (360–90) by 2.

1.7 If the geometry of the carbon atom were square planar, two isomeric compounds with the formula CH_2Cl_2 should exist.

H-----Cl   H-----Cl
   C   and   C
H-----Cl   Cl-----H

The fact that only one form of CH_2Cl_2 has ever been detected supports a tetrahedral geometry for carbon, because with a tetrahedral geometry only one compound with the formula CH_2Cl_2 is possible. All of the following structures are just different orientations of the same molecule.

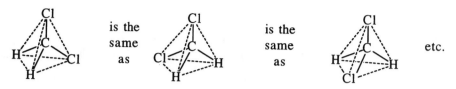

is the same as   is the same as   etc.

Make models to convince yourself of this fact.

1.8   (a) There are four bonding pairs. The geometry is tetrahedral.

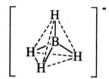

(b) There are two bonding pairs about the central atom. The geometry is linear.

$$:\!\ddot{F}\!-\!Be\!-\!\ddot{F}\!:$$

(c) There are four bonding pairs. The geometry is tetrahedral.

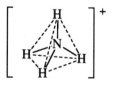

(d) There are two bonding pairs and two nonbonding pairs. The geometry is angular.

(e) There are three bonding pairs. The geometry is trigonal planar.

(f) There are four bonding pairs around the central atom. The geometry is tetrahedral.

(g) There are four bonding pairs around the central atom. The geometry is tetrahadral.

(h) There are three bonding pairs and one nonbonding pair around the central atom. The geometry is trigonal pyramidal.

$$\left[\ \overset{\cdot\cdot}{\underset{Cl}{\overset{\displaystyle Cl}{\diagup}}} \overset{\displaystyle C}{\diagdown} _{Cl} \ \right]^{-}$$

1.9 (a) $\underset{H}{\overset{H}{\diagdown}} C = C \underset{H}{\overset{H}{\diagup}}$ 120°, 120° , trigonal planar at each carbon atom.

(b) H—C≡C—H 180° linear

(c) H—C≡N: 180° linear

1.10 (a) H—F or $\overset{\delta+ \ \ \delta-}{H—F}$ (c) Br—Br $\mu = 0$

(b) I—Br or $\overset{\delta+ \ \ \delta-}{I—Br}$ (d) F—F $\mu = 0$

1.11 VSEPR theory predicts a planar structure for BF_3.

$$\underset{F \ \ \mu = 0 \ \ F}{\overset{F}{\underset{\diagup \overset{\displaystyle B}{} \diagdown}{}}}$$

The vector sum of the bond moments of a trigonal planar structure would be zero, resulting in a prediction of $\mu = 0$ for BF_3. This correlates with the experimental observation and confirms the prediction of VSEPR theory.

1.12   The shape of $CCl_2{=}CCl_2$ (below is such that the vector sum of all of the C—Cl bond moments is zero.

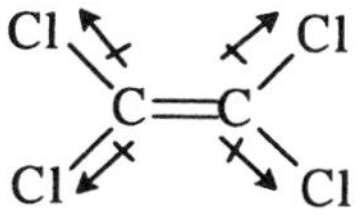

1.13   The fact that SO_2 has a dipole moment indicates that the molecule is angular, not linear.

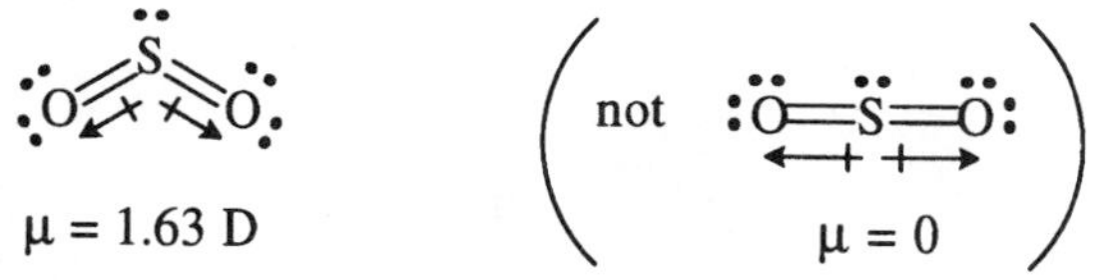

$\mu = 1.63\ D$ $\left(\text{not}\quad \mu = 0\right)$

An angular shape is what we would expect from VSEPR theory, too.

The fact that CO_2 has no dipole moment indicates that its shape is linear, not angular.

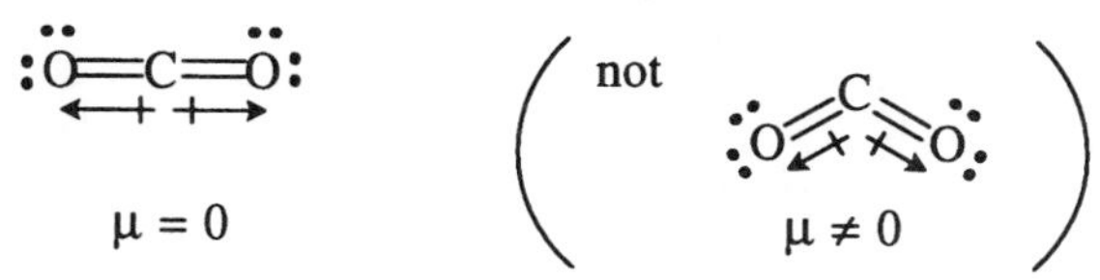

$\mu = 0$ $\left(\text{not}\quad \mu \neq 0\right)$

Again, this is what VSEPR theory predicts.

1.14

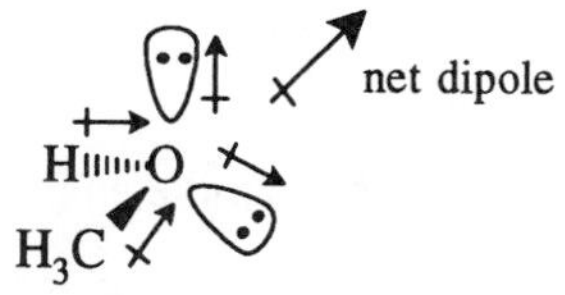

net dipole

1.15   In $CFCl_3$ the large C–F bond moment opposes the C–Cl moments. Because hydrogen is much less electronegative than fluorine no such opposing effect occurs in $CHCl_3$; therefore, it has a larger net dipole moment.

Smaller net
dipole moment

Larger net
dipole moment

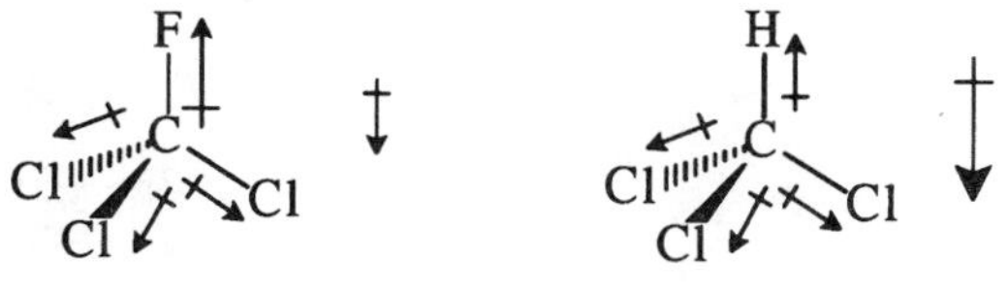

1.16

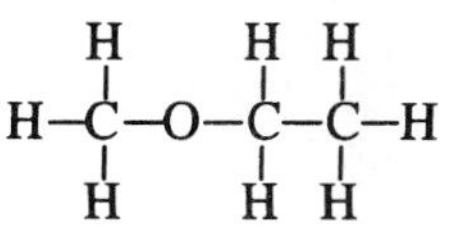

1.17   (a)

(b)

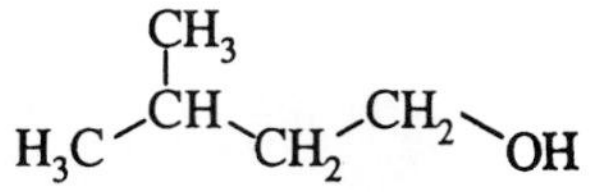

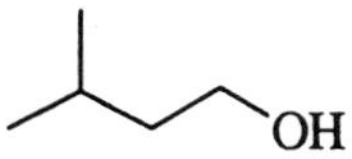

(c)

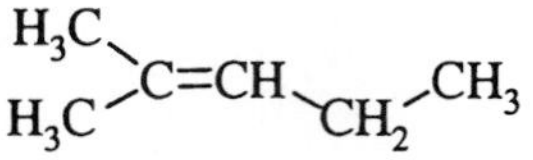

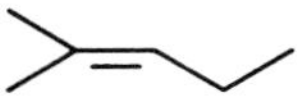

(d)

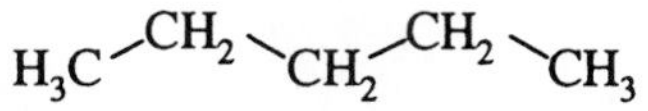

(e)

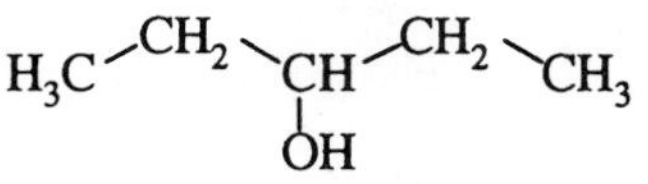

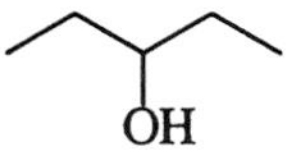

(f)

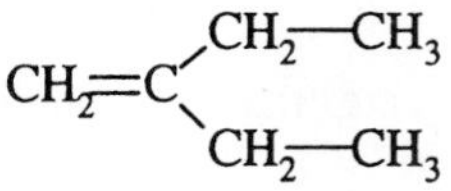

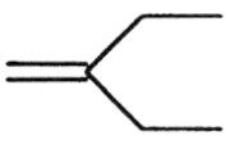

(g)

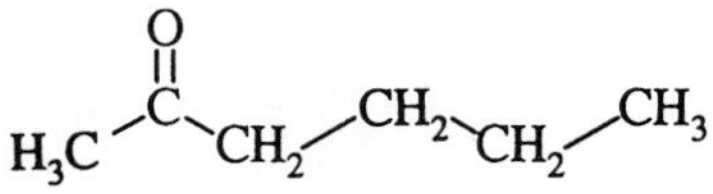

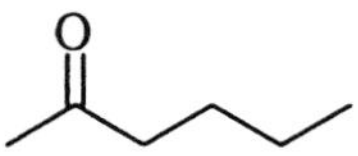

(h)

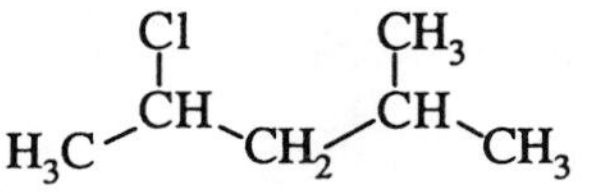

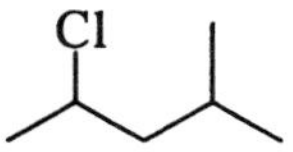

1.18 (a) and (d) are constitutional isomers with the molecular formula C_5H_{12}.

(b) and (e) are constitutional isomers with the molecular formula $C_5H_{12}O$.

(c) and (f) are constitutional isomers with the molecular formula C_6H_{12}.

1.19 (a)

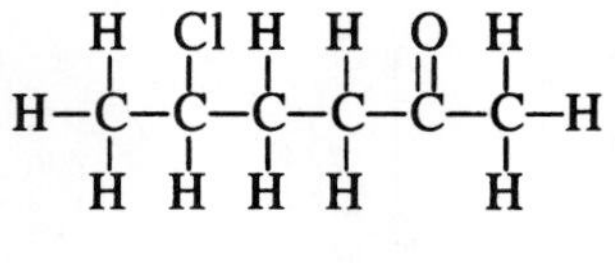

(b)

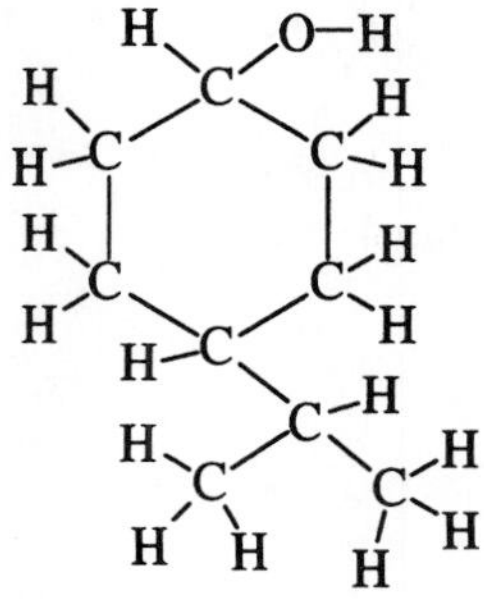

(c)

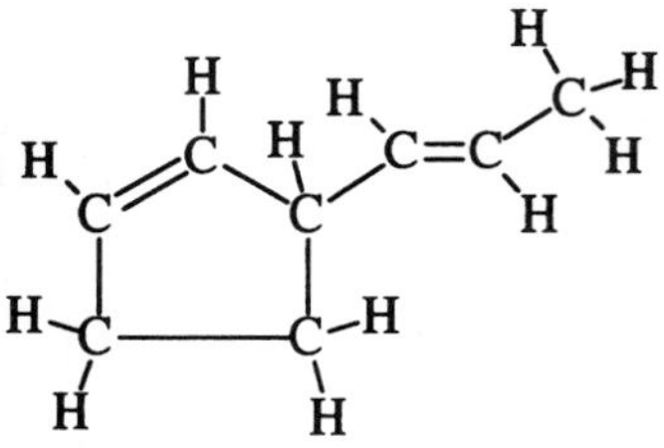

1.20 (a)

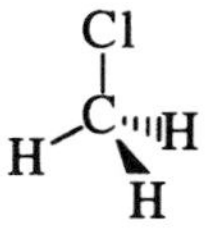

(Note that the **Cl** atom and the three **H** atoms may be written at any of the four positions.)

(b)

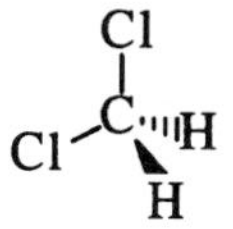

or 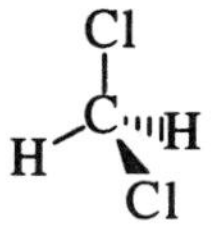 and so on

(c)

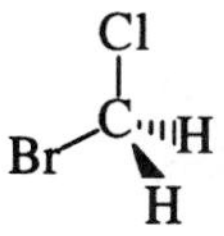

and others

(d)

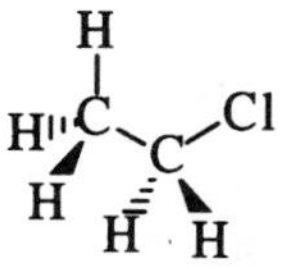

and others

1.21 (a) Na^+ has the electronic configuration, $1s^2 2s^2 2p^6$, of Ne.

(b) Cl^- has the electronic configuration, $1s^2 2s^2 2p^6 3s^2 3p^6$, of Ar.

(c) F^+ and (h) Br^+ do not have the electronic configuration of a noble gas.

(d) H^- has the electronic configuration, $1s^2$, of He.

(e) Ca^{2+} has the electronic configuration, $1s^2 2s^2 2p^6 3s^2 3p^6$ of Ar.

(f) S^{2-} has the electronic configuration, $1s^2 2s^2 2p^6 3s^2 3p^6$ of Ar.

(g) O^{2-} has the electronic configuration, $1s^2 2s^2 2p^6$ of Ne.

1.22 (a) 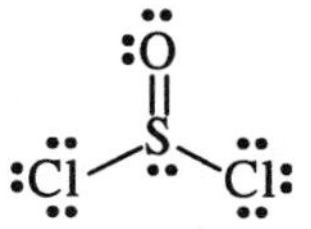 (b) 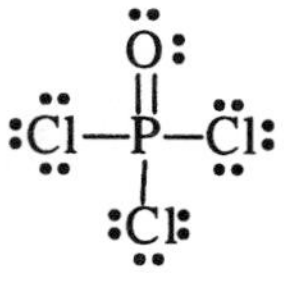 (c) 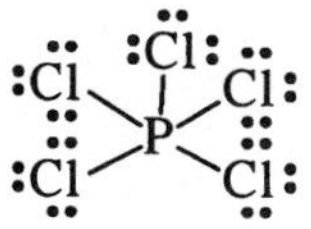 (d)

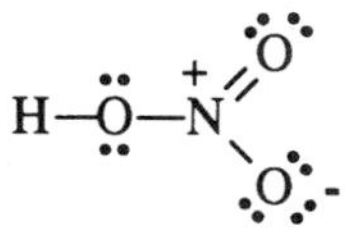

1.23 (a) 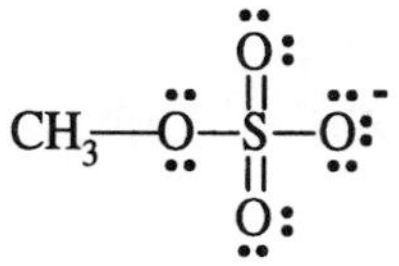 (c)

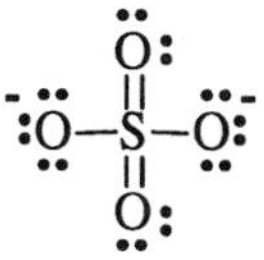

(b) 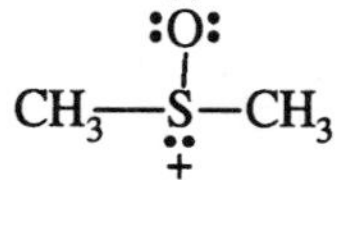 (d)

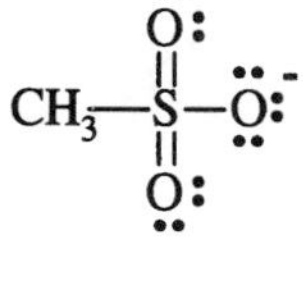

1.24 (a) $(CH_3)_2CHCH_2OH$ (c)

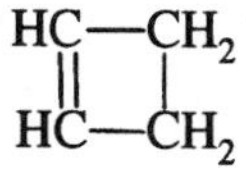

(b) 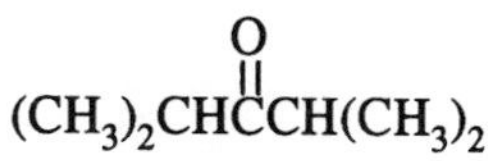 (d) $(CH_3)_2CHCH_2CH_2OH$

1.25 (a) $C_4H_{10}O$
(b) $C_7H_{14}O$
(c) C_4H_6
(d) $C_5H_{12}O$

1.26 (a) Different compounds, not isomeric
(b) Constitutional isomers
(c) Same compound
(d) Same compound
(e) Same compound
(f) Constitutional isomers
(g) Different compounds, not isomeric
(h) Same compound
(i) Different compounds, not isomeric
(j) Same compound
(k) Constitutional isomers
(l) Different compounds, not isomeric
(m) Same compound
(n) Same compound
(o) Same compound
(p) Constitutional isomers

1.27 (a)
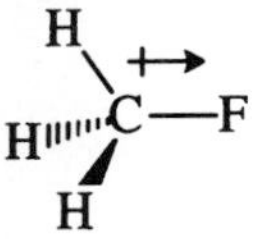

(b)
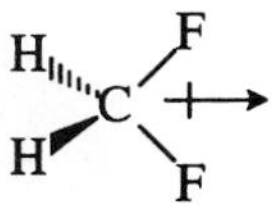

(c)
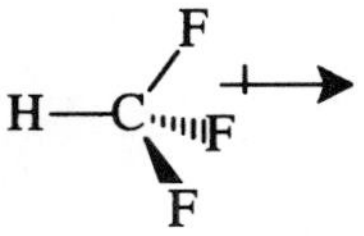

(d)
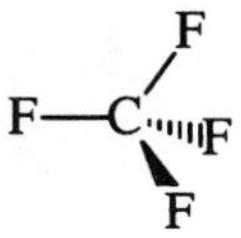

No dipole moment

(e)
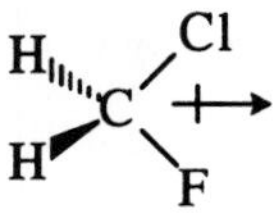

(f)
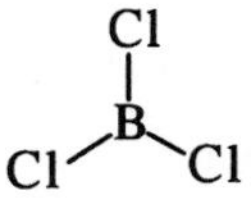

No dipole moment

(g) F—Be—F No dipole moment

(h)
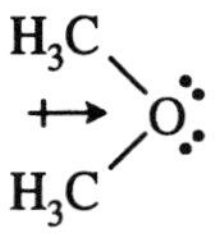

(i)
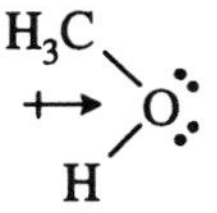

(j)
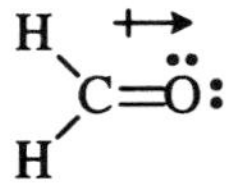

1.28 (a)
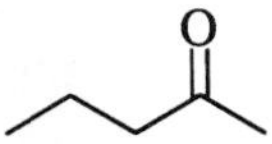

(b)
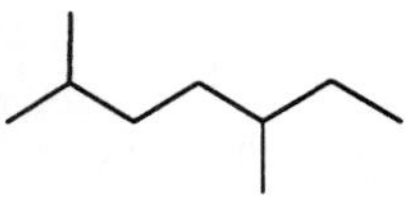

(c)
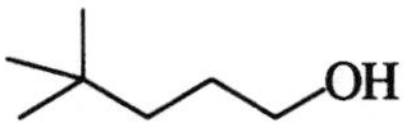

(d)
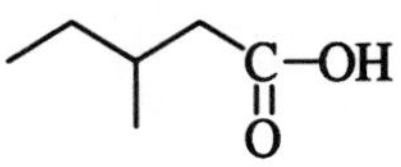

(e)
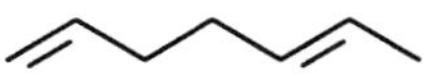

(f)
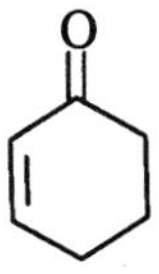

1.29 (a)

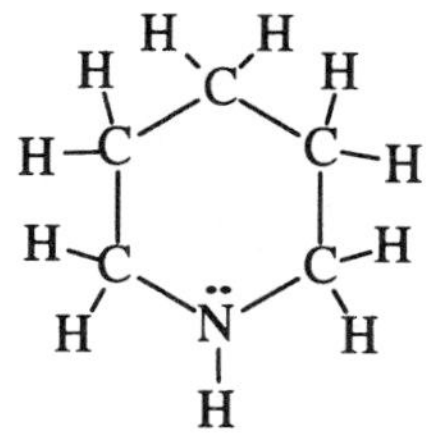

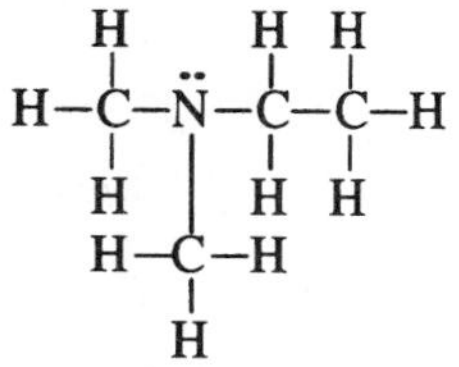

(b)

(c)

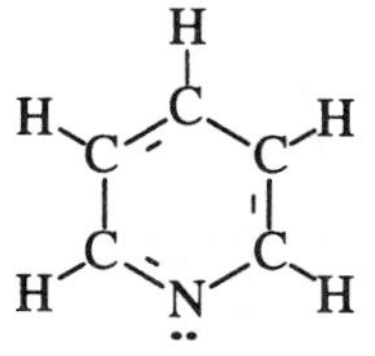

(d)

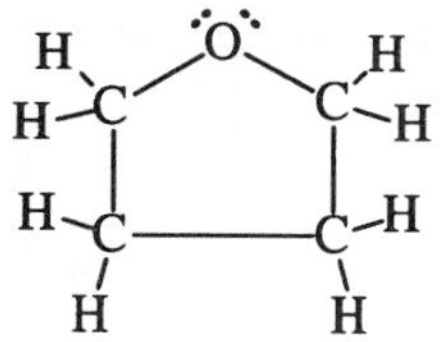

1.30 $CH_2=CHCH_2CH_3$ $CH_3CH=CHCH_3$ $CH_2=CCH_3$
                                                              CH_3

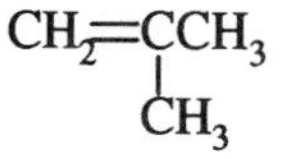

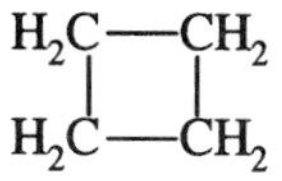

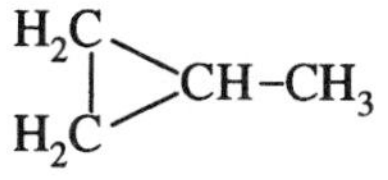

$H_2C—CH_2$ H_2C
$H_2C—CH_2$ $$$>CH—CH_3$
                         H_2C

1.31 $H-\overset{\overset{\displaystyle H}{|}}{\underset{\underset{\displaystyle H}{|}}{C}}-\overset{+}{N}\overset{\ddot{O}\cdot}{\underset{\ddot{O}:^-}{}}$ (⟷ $H-\overset{\overset{\displaystyle H}{|}}{\underset{\underset{\displaystyle H}{|}}{C}}-\overset{+}{N}\overset{\ddot{O}:^-}{\underset{\ddot{O}\cdot}{}}$)

$H-\ddot{O}-\overset{\overset{\displaystyle :\ddot{O}:}{\|}}{C}-\ddot{N}-H$
                          H

$H-\overset{\overset{\displaystyle H}{|}}{\underset{\underset{\displaystyle H}{|}}{C}}-\ddot{O}-\ddot{N}=\ddot{O}:$

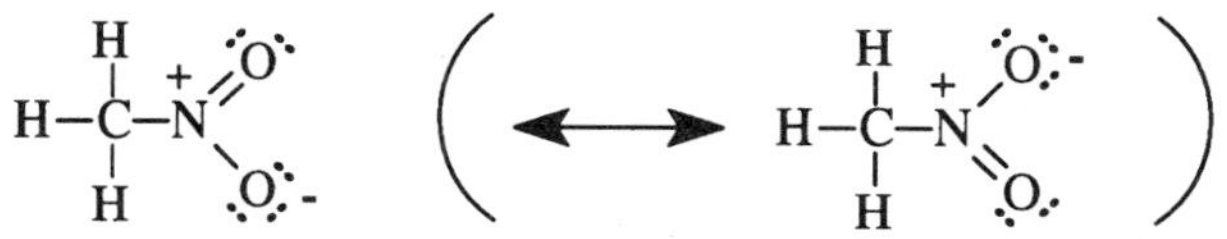

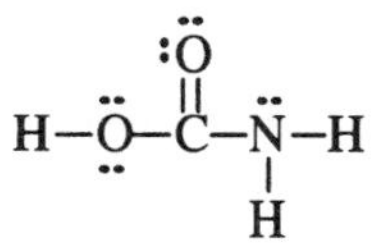

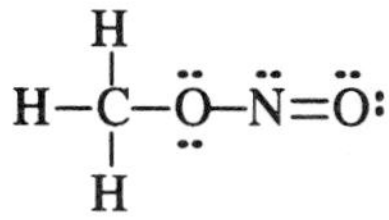

(Other structures are possible.)

1.32   A carbon-chlorine bond is longer than a carbon-fluorine bond because chlorine is a larger atom than fluorine. Thus in $\overset{\delta+}{C}H_3\overset{\delta-}{-Cl}$ the distance, d, that separates the charges is greater than in $\overset{\delta+}{C}H_3\overset{\delta-}{-F}$. The greater value of d for CH_3Cl more than compensates for the smaller value of e and thus the dipole moment ($e \times d$) is larger.

1.33   (a) While the structures differ in the position of their electrons, they also differ in the positions of their nuclei and thus *they are not resonance structures*. (In cyanic acid the hydrogen nucleus is bonded to oxygen; in isocyanic acid it is bonded to nitrogen.)

(b) The anion obtained from either acid is a resonance hybrid of the following structures: $:\ddot{O}-C\equiv N:$ ⟷ $:\ddot{O}=C=\ddot{N}:^-$

1.34

$$H-\overset{\displaystyle H}{\underset{\displaystyle H}{C}}$$

(a) A + charge. ($F = 4 - \% = +1$)

(b) A + charge. (It is called a methyl cation.)

(c) Trigonal planar, that is,

(d) sp^2

1.35

$$H-\overset{\displaystyle H}{\underset{\displaystyle H}{C}}:$$

(a) A − charge. ($F = 4 - \% - 2 = -1$)

(b) A − charge. (It is called a methyl anion.)

(c) Trigonal pyramidal, that is

(d) sp^3

1.36

$$H-\overset{\displaystyle H}{\underset{\displaystyle H}{C}}\cdot$$

(a) No formal charge. ($F = 4 - \% - 1 = 0$)

(b) No charge.

(c) sp^2, that is,

1.37   (a) and (b)

(c) Because the two resonance structures are equivalent, they should make equal contributions to the hybrid and, therefore, the bonds should be the same length.

(d) Yes. We consider the central atom to have two groups or units of bonding electrons and one unshared pair.

(e) Measurement of the dipole moment. If ozone has a dipole moment, which it does, then we conclude that the molecule is angular. If it had no dipole moment, we would have to conclude that the molecule is linear. In a (hypothetical) linear molecule the bond moments would cancel, that is,

1.38

Structures **A** and **C** are equivalent and, therefore, make equal contributions to the hybrid. The bonds of the hybrid, therefore, have the same length.

1.39   (a)

(b) $(CH_3)_2NH$ $CH_3CH_2NH_2$

(c) $(CH_3)_3N$ $CH_3CH_2NHCH_3$ $CH_3CH_2CH_2NH_2$ CH_3CHCH_3
                                                                              |
                                                                             NH_2

(d)

1.40   (a) Dimethyl ether: There are four electron pairs around the central oxygen: two bonding pairs and two nonbonding pairs. We would expect sp^3 hybridization of the oxygen with a bond angle of approximately 109.5° between the methyl groups.

(b) Trimethylamine: There are four electron pairs around the central nitrogen: three bonding pairs and one nonbonding pair. We would expect sp^3 hybridization of the nitrogen with a bond angle of approximately 109.5° between the methyl groups.

$$H_3C^{\text{\tiny\textbf{w}}}N\diagdown CH_3$$
$$H_3C$$

(c) Trimethylboron: There are only three bonding electron pairs around the central boron. We would expect sp^2 hybridization of the boron with a bond angle of 120° between the methyl groups.

$$CH_3$$
$$|$$
$$B$$
$$H_3C \diagup \diagdown CH_3$$

(d) Dimethylberyllium: There are only two bonding electron pairs around the central beryllium atom. We would expect sp hybridization of the beryllium atom with a bond angle of 180° between the methyl groups.

$$H_3C-Be-CH_3$$

1.41   Without one (or more) polar bonds, a molecule cannot possess a dipole moment and, therefore, it cannot be polar. If the bonds are directed so that the bond moments cancel, however, the molecule will not be polar even though it has polar bonds.

1.42   (a) $\overset{..}{:}\overset{+}{O}=\overset{+}{N}=\overset{..}{O}:$

(b) Linear

(c) Carbon dioxide

QUIZ

1.1   Which of the following is a valid Lewis dot formula for the nitrite ion ($NO_2{}^-$)?

(a) $:\overset{..}{\underset{..}{O}}-\overset{..}{N}=\overset{..}{O}:$ (b) $:\overset{..}{O}=\overset{..}{N}-\overset{..}{\underset{..}{O}}:$ (c) $:\overset{..}{\underset{..}{O}}-\overset{..}{N}=\overset{..}{O}:$ (d) Two of
these

(e) None of the above

1.2   What is the hybridization state of the boron atom in BF_3?

   (a) s (b) p (c) sp (d) sp^2 (e) sp^3

1.3 BF_3 reacts with NH_3 to produce a compound, $F-\overset{F\ \ H}{\underset{F\ \ H}{B-N}}-H$. The hybridization state of **B** is

   (a) s (b) p (c) sp (d) sp^2 (e) sp^3

1.4   The formal charge on N in the compound given in Problem 1.3 is

   (a) -2 (b) -1 (c) 0   (d) $+1$ (e) $+2$

1.5   The correct bond-line formula of the compound whose condensed formula is $CH_3CHClCH_2CH(CH_3)CH(CH_3)_2$ is

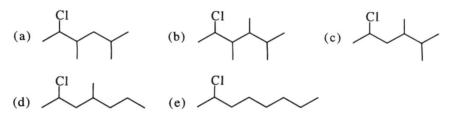

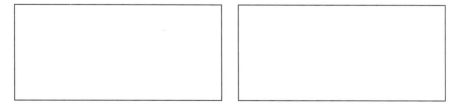

1.6   Write another resonance structure for the formate ion.

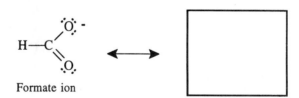

Formate ion

1.7   In the boxes below write condensed structural formulas for constitutional isomers of $CH_3(CH_2)_3CH_3$.

1.8 Write a three-dimensional formula for a constitutional isomer of compound **A** given below. Complete the partial structure shown.

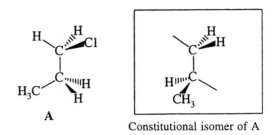

A

Constitutional isomer of A

1.9 Consider the molecule $(CH_3)_3B$ and give the following:

(a) Hybridization state of boron

(b) Hybridization state of carbon atoms

(c) Formal charge on boron

(d) Orientation of groups around boron

(e) Dipole moment of $(CH_3)_3B$

1.10 Give the formal charge on oxygen in each compound.

(a) $CH_3-\overset{\cdot\cdot}{\underset{\underset{CH_3}{|}}{O}}-CH_3$

(b) $CH_3-\overset{\underset{CH_3}{|}}{C}=\overset{\cdot\cdot}{O}:$

(c) $CH_3-\overset{\overset{H}{|}}{\underset{\underset{H}{|}}{C}}-\overset{\cdot\cdot}{\underset{\cdot\cdot}{O}}:$

1.11 Write another resonance structure in which all of the atoms have a formal charge of zero.

$$H-\overset{\overset{\overset{\cdot\cdot}{O}:}{|}}{C}=\overset{+}{\underset{\underset{H}{|}}{N}}-H \longleftrightarrow$$

1.12 Indicate the direction of the net dipole moment of the following molecule.

$$\overset{H_3C}{\underset{H_3C}{>}}C\overset{Cl}{\underset{F}{<}}$$

2 REPRESENTATIVE CARBON COMPOUNDS

SOLUTIONS TO PROBLEMS

2.1 (a) Cis-trans isomers are not possible.

(b)
$$
\begin{array}{ccc}
\underset{H}{\overset{H_3C}{>}} C = C \underset{H}{\overset{CH_3}{<}} & \text{and} & \underset{H}{\overset{H_3C}{>}} C = C \underset{CH_3}{\overset{H}{<}}
\end{array}
$$

(c) Cis-trans isomers are not possible.

(d)
$$
\begin{array}{ccc}
\underset{H}{\overset{CH_3CH_2}{>}} C = C \underset{H}{\overset{Cl}{<}} & \text{and} & \underset{H}{\overset{CH_3CH_2}{>}} C = C \underset{Cl}{\overset{H}{<}}
\end{array}
$$

2.2 (a)
$$
\underset{H}{\overset{F}{>}} C = C \underset{H}{\overset{F}{<}} \quad \uparrow \text{net dipole moment}
$$

(b)
$$
\underset{F}{\overset{H}{>}} C = C \underset{H}{\overset{F}{<}} \quad \mu = 0
$$

(c)
$$
\underset{H}{\overset{H}{>}} C = C \underset{F}{\overset{F}{<}} \quad \longleftrightarrow \text{net dipole moment}
$$

(d)
$$
\underset{F}{\overset{F}{>}} C = C \underset{F}{\overset{F}{<}} \quad \mu = 0
$$

2.3 (a)
$$
\underset{H}{\overset{H}{>}} C = C \underset{Br}{\overset{Br}{<}} \qquad \underbrace{\underset{Br}{\overset{H}{>}} C = C \underset{Br}{\overset{H}{<}} \quad \underset{Br}{\overset{H}{>}} C = C \underset{H}{\overset{Br}{<}}}_{\text{Cis-trans isomers}}
$$
cis   trans

(b)
$$
\underset{Br}{\overset{Br}{>}} C = C \underset{Cl}{\overset{Cl}{<}} \qquad \underbrace{\underset{Cl}{\overset{Br}{>}} C = C \underset{Cl}{\overset{Br}{<}} \quad \underset{Br}{\overset{Cl}{>}} C = C \underset{Cl}{\overset{Br}{<}}}_{\text{Cis-trans isomers}}
$$
cis   trans

2.4 (a) $CH_3CH_2CH_2CH_2Br$ and CH_3CHCH_2Br
$\overset{|}{C}H_3$

(b) $CH_3\overset{|}{C}HCH_2CH_3$
$\underset{Br}{}$

(c) $CH_3-\overset{\overset{\displaystyle CH_3}{|}}{\underset{\underset{\displaystyle CH_3}{|}}{C}}-Br$

2.5 (a) CH_3CH_2F

(b) $CH_3\overset{|}{C}HCH_3$ or $(CH_3)_2CHCl$
$\underset{Cl}{}$

(c) Propyl bromide

(d) Isopropyl fluoride

(e) Phenyl iodide

2.6 (a) $CH_3CH_2CH_2CH_2OH$ and $CH_3\overset{|}{C}HCH_2OH$
$\overset{|}{C}H_3$

(b) $CH_3\overset{|}{C}HCH_2CH_3$
$\underset{OH}{}$

(c) $CH_3-\overset{\overset{\displaystyle CH_3}{|}}{\underset{\underset{\displaystyle CH_3}{|}}{C}}-OH$

2.7 (a) $CH_3CH_2CH_2OH$

(b) $CH_3\overset{|}{C}HCH_3$ or $CH_3\overset{|}{C}H$ -OH
$\underset{OH}{}$ $\underset{CH_3}{}$

2.8 (a) $CH_3CH_2-O-CH_2CH_3$

(b) $CH_3CH_2-O-CH_2CH_2CH_3$

(c) $CH_3CH_2-O-\overset{|}{C}HCH_3$
$\underset{CH_3}{}$

(d) Methyl propyl ether

(e) Diisopropyl ether

(f) Methyl phenyl ether

2.9 (a) $CH_3CH_2CH_2NH_2$

(b) $CH_3-\overset{|}{N}-CH_3$ or $(CH_3)_3N$
$\underset{CH_3}{}$

(c) $CH_3-\overset{\overset{\displaystyle }{|}}{N}-CH_2CH_3$
$\overset{\diagup CH \diagdown}{H_3C \qquad CH_3}$

(d) Isopropylpropylamine

(e) Tripropylamine

(f) Methylphenylamine

(g) Dimethylphenylamine

2.10 (a) (a) only   (b) (d,f)   (c) (b, c, e, g)

2.11 (a) $CH_3-\overset{\overset{\displaystyle CH_3}{|}}{\underset{\underset{\displaystyle CH_3}{|}}{\ddot{N}}}:$ + $H-\ddot{\underset{\cdot\cdot}{C}l}:$ $\longrightarrow$ $CH_3-\overset{\overset{\displaystyle CH_3}{|}}{\underset{\underset{\displaystyle CH_3}{|}}{\overset{+}{N}}}-H$ + $:\ddot{\underset{\cdot\cdot}{C}l}:^{-}$

(b) sp^3

2.12 (a) $CH_3CH_2CH_2CH_2OH$ would boil higher because its molecules can form hydrogen bonds to each other through the $-\overset{..}{\underset{..}{O}}-H$ group.

(b) $CH_3CH_2NHCH_3$ would boil higher because its molecules can form hydrogen bonds to each other through the $-\overset{..}{N}-H$ group.

(c) $HOCH_2CH_2CH_2OH$ because by having two $-\overset{..}{\underset{..}{O}}-H$ groups, it can form more hydrogen bonds.

2.13 Cyclopropane would have the higher melting point because its cyclic structure gives it a rigid compact shape that would permit stronger crystal lattice forces.

2.14 (a) Ketone (b) Alkyne (c) Alcohol (d) Aldehyde

(e) Alcohol (f) Alkene

2.15 (a) Three carbon-carbon double bonds (alkene) and a 2° alcohol

(b) Phenyl, carboxylic acid, amide, ester, and a 1° amine

(c) Phenyl and a 1° amine

(d) Carbon-carbon double bond and a 2° alcohol

(e) Phenyl, ester, and a 3° amine

(f) Carbon-carbon double bond and an aldehyde

(g) Carbon-carbon double bond and 2 ester groups

2.16

$CH_3CH_2CH_2CH_2Br$	$CH_3CH_2\overset{\underset{\textstyle Br}{\mid}}{C}HCH_3$	$CH_3\overset{\underset{\textstyle CH_3}{\mid}}{C}HCH_2Br$	$CH_3-\overset{\overset{\textstyle CH_3}{\mid}}{\underset{\underset{\textstyle Br}{\mid}}{C}}-CH_3$
1° Alkyl halide	2° Alkyl halide	1° Alkyl halide	3° Alkyl halide

2.17

$CH_3CH_2CH_2CH_2OH$	$CH_3CH_2\overset{\underset{\textstyle OH}{\mid}}{C}HCH_3$	$CH_3\overset{\underset{\textstyle CH_3}{\mid}}{C}HCH_2OH$	$CH_3-\overset{\overset{\textstyle CH_3}{\mid}}{\underset{\underset{\textstyle OH}{\mid}}{C}}-CH_3$
1° Alcohol	2° Alcohol	1° Alcohol	3° Alcohol

$CH_3OCH_2CH_2CH_3$	$CH_3O\overset{\underset{\textstyle CH_3}{\mid}}{C}HCH_3$	$CH_3CH_2OCH_2CH_3$
Ether	Ether	Ether

2.18 Any four of the following:

$$CH_3\overset{\overset{\displaystyle O}{\|}}{C}CH_3 \qquad CH_3CH_2\overset{\overset{\displaystyle O}{\|}}{C}H \qquad \begin{array}{c} H_2C—CH_2 \\ H_2C—O \end{array} \qquad H_3C—HC\underset{}{\overset{\displaystyle O}{\diagdown\!\diagup}}CH_2$$

Ketone   Aldehyde   Ether   Ether

$$CH_2{=}CHCH_2OH \qquad CH_2{=}CH–O–CH_3 \qquad \begin{array}{c} H_2C \\ | \\ H_2C \end{array}\!\!\!>\!CHOH$$

Alkene, alcohol   Alkene, ether   Alcohol

2.19 (a) 1°   (b) 2°   (c) 3°   (d) 3°   (e) 2°

2.20 (a) 2°   (b) 1°   (c) 3°   (d) 2°   (e) 2°   (f) 3°

2.21 (a) $CH_3OCH_2CH_2CH_3$ CH_3OCHCH_3 $CH_3CH_2OCH_2CH_3$
 $\underset{CH_3}{|}$

(b) $\begin{array}{c} H_2C \\ | \\ H_2C \end{array}\!\!\!>\!CH{-}CH_2OH$ $CH_2{=}CHCH_2CH_2OH$ $CH_3CH{=}CHCH_2OH$

(c) $\begin{array}{c} H_2C \\ | \\ H_2C \end{array}\!\!\!>\!CH{-}OH$ (d) $\begin{array}{c} H_2C \\ | \\ H_2C \end{array}\!\!\!>\!\overset{\overset{\displaystyle OH}{|}}{C}{-}CH_3$

(e) $H{-}\overset{\overset{\displaystyle O}{\|}}{C}{-}O{-}CH_2CH_3$ $CH_3{-}\overset{\overset{\displaystyle O}{\|}}{C}{-}O{-}CH_3$

(f) $CH_3CH_2CH_2CH_2CH_2Br$ $\underset{CH_3}{CH_3CHCH_2CH_2Br}$ $\underset{CH_3}{CH_3CH_2CHCH_2Br}$

 $CH_3{-}\overset{\overset{\displaystyle CH_3}{|}}{\underset{\underset{\displaystyle CH_3}{|}}{C}}{-}CH_2{-}Br$

(g) $\underset{Br}{CH_3CHCH_2CH_2CH_3}$ $\underset{Br}{CH_3CH_2CHCH_2CH_3}$ $\underset{Br}{CH_3CH\overset{\overset{\displaystyle CH_3}{|}}{C}HCH_3}$

(h) $CH_3CH_2{-}\overset{\overset{\displaystyle CH_3}{|}}{\underset{\underset{\displaystyle CH_3}{|}}{C}}{-}Br$

(i) $CH_3CH_2CH_2CH_2\overset{\overset{\displaystyle O}{\|}}{C}H$ $\underset{CH_3}{CH_3CHCH_2\overset{\overset{\displaystyle O}{\|}}{C}H}$ $\underset{CH_3}{CH_3CH_2CH\overset{\overset{\displaystyle O}{\|}}{C}H}$

(j) $CH_3\overset{\overset{O}{\|}}{C}CH_2CH_2CH_3$ $CH_3CH_2\overset{\overset{O}{\|}}{C}CH_2CH_3$ $CH_3\overset{}{\underset{\underset{CH_3}{|}}{CH}}\overset{\overset{O}{\|}}{C}CH_3$

(k) $CH_3CH_2CH_2NH_2$ $CH_3\overset{}{\underset{\underset{NH_2}{}}{\overset{\overset{CH_3}{|}}{CH}}}$

(l) $CH_3CH_2NHCH_3$ (m) $(CH_3)_3N$

(n) $H-\overset{\overset{O}{\|}}{C}-NHCH_3$ $CH_3-\overset{\overset{O}{\|}}{C}-NH_2$

2.22   (a) $CH_3CH_2CH_2OH$ because its molecules can form hydrogen bonds to each other through its $-\overset{\cdot\cdot}{\underset{\cdot\cdot}{O}}-H$ group.

(b) $HOCH_2CH_2OH$ because with two $\overset{\cdot\cdot}{\underset{\cdot\cdot}{O}}-H$ groups, its molecules can form more hydrogen bonds with each other.

(c)   ∿∿OH   because its molecules can form hydrogen bonds to each other.

(d)   ▷—OH   [Same reason as (c)].

(e)   ⬡NH   because its molecules can form hydrogen bonds to each other through its $-\overset{}{\underset{|}{\overset{\cdot\cdot}{N}}}-H$ group.

(f) $\overset{F}{\diagdown}\diagdown=\diagup\overset{F}{\diagup}$ because its molecules will have a larger dipole moment. (The trans compound will have $\mu = 0$.)

(g)   ∿⌒$\overset{\overset{O}{\|}}{C}$OH   [Same reason as (c)].

(h)   Nonane, because of its larger molecular weight and larger size, will have larger van der Waals attractions.

(i)   ⟩=O   because its carbonyl group is far more polar than the double bond of   ⟩= .

2.23   (a) $CH_3CH_2\overset{\overset{O}{\|}}{C}NH_2$ $CH_3\overset{\overset{O}{\|}}{C}\overset{}{\underset{\underset{H}{|}}{N}}CH_3$ $H\overset{\overset{O}{\|}}{C}\overset{}{\underset{\underset{H}{|}}{N}}CH_2CH_3$ $H-\overset{\overset{O}{\|}}{C}-\overset{}{\underset{\underset{CH_3}{|}}{N}}-CH_3$

(b) The last one given above [i.e., $\overset{\overset{O}{\|}}{H C}N(CH_3)_3$] because it does not have a hydrogen that is covalently bonded to nitrogen and, therefore, its molecules cannot form hydrogen bonds to each other. The other molecules all have a hydrogen covalently bonded to nitrogen and, therefore, hydrogen-bond formation is possible. With the first molecule, for example, hydrogen bonds could form in the following way:

$$CH_3CH_2C\overset{O----H-\overset{\overset{H}{|}}{N}}{\underset{\overset{\bullet\bullet}{N}-H----O}{}}CCH_2CH_3$$

2.24   An ester group, $-\overset{|}{\underset{\underset{(CH_2)_n}{}}{C}}-\overset{\overset{O}{\|}}{C}-O-\overset{|}{C}-$

2.25   The attractive forces between hydrogen fluoride molecules are the very strong dipole–dipole attractions that we call *hydrogen bonds*. (The partial positive charge of a hydrogen fluoride molecule is relatively exposed because it resides on the hydrogen nucleus. By contrast, the positive charge of an ethyl fluoride molecule is buried in the ethyl group and is shielded by the surrounding electrons. Thus the positive end of one hydrogen fluoride molecule can approach the negative end of another hydrogen fluoride molecule much more closely, with the result that the attractive force between them is much stronger.)

2.26   (a) and (b) are polar and hence are able to dissolve ionic compounds. (c) and (d) are nonpolar and will not dissolve ionic compounds.

QUIZ

2.1   Which of the following pairs of compounds is *not* a pair of constitutional isomers?

(a) $CH_3-O-CH=CH_2$ and $CH_3CH_2\overset{\overset{O}{\|}}{C}H$

(b) ⬠   and $CH_3CH=CHCH_2CH_3$

(c) $CH_3\overset{\overset{O}{\|}}{C}-OH$ and $HO-CH_2\overset{\overset{O}{\|}}{C}H$

(d) $CH_3CH_2C\equiv CH$ and $CH_3CH=C=CH_2$

(e) $CH_3\underset{\overset{|}{CH_3}}{C}HCH(CH_3)_2$ and $(CH_3)_2CHCH(CH_3)_2$

2.2 Which of the answers to Problem 2.1 contains an ether group?

2.3 Which of the following pairs of structures represents a pair of isomers?

(a)
$$\begin{array}{ccc} H_3C & & CH_3 \\ & C=C & \\ H_3C & & H \end{array}$$
and
$$\begin{array}{ccc} H_3C & & H \\ & C=C & \\ H_3C & & CH_3 \end{array}$$

(b) $CH_3C\equiv CCH_3$ and $CH_3CH_2C\equiv CH$

(c)
$$\begin{array}{ccc} H & F \\ C-C & Cl \\ Cl & H \\ F & \end{array}$$
and
$$\begin{array}{ccc} H & H \\ C-C & Cl \\ Cl & F \\ F & \end{array}$$

(d)
$$\begin{array}{c} CH_3 \\ | \\ CH_3CH_2CHCH_2CH_3 \end{array}$$
and
$$\begin{array}{c} CH_3CH_2CHCH_3 \\ | \\ CH_2CH_3 \end{array}$$

(e) More than one of these pairs are isomers.

2.4 Give a structural formula for each of the following:

(a) A tertiary alcohol with the formula $C_5H_{12}O$

(b) An *N,N*–disubstituted amide with the formula C_4H_9NO

(c) The alkene isomer of $C_2H_2Cl_2$ that has no dipole moment

(d) An ester with the formula $C_2H_4O_2$

(e) The isomer of $C_2H_2Cl_2$ that cannot show cis-trans isomerism

(f) The isomer of C_3H_8O that would have the lowest boiling point

(g) The isomer of $C_4H_{11}N$ that would have the lowest boiling point

2.5   Write the bond-line formula for a constitutional isomer of the compound shown below that does not contain a double bond.

$CH_3CH_2CH{=}CH_2$

2.6   Circle the compound in each pair that would have the higher boiling point.

(a) $CH_3CH_2CH_2OH$ or CH_3CH_2CHO

(b) 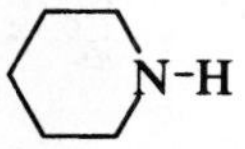 ⬡N–H   or   ⬠N–CH$_3$

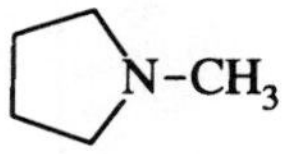

(c) $CH_3\overset{\text{O}}{\overset{\|}{C}}OCH_3$ or $CH_3CH_2\overset{\text{O}}{\overset{\|}{C}}OH$

(d) $CH_3OCH_2CH_2OH$ or $CH_3OCH_2OCH_3$

(e) $CH_3CH_2\overset{\text{O}}{\overset{\|}{C}}NHCH_3$ or $CH_3\overset{\text{O}}{\overset{\|}{C}}N(CH_3)_2$

2.7   Give an acceptable name for each of the following:

(a) 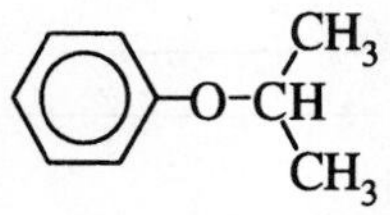 ⬡–O–CH$\overset{\displaystyle CH_3}{\underset{\displaystyle CH_3}{}}$

(b) $CH_3CH_2{-}N{-}C_6H_5$
        $\underset{\displaystyle CH_3}{|}$

(c) CH_3CHNH_2
        $\underset{\displaystyle CH_3}{|}$

3

AN INTRODUCTION TO
ORGANIC REACTIONS: ACIDS AND BASES

SOLUTIONS TO PROBLEMS

3.1 (a) $CH_3-\overset{..}{\underset{..}{O}}-H$ + $\overset{\overset{..}{:}\overset{..}{F}:}{\underset{:\overset{..}{F}:}{B-\overset{..}{F}:}}$ $\longrightarrow$ $CH_3-\overset{+}{\underset{H}{\overset{..}{O}}}-\overset{\overset{:\overset{..}{F}:}{|}}{\underset{:\overset{..}{F}:}{B}}{-}\overset{..}{F}:$

(b) $CH_3-\overset{..}{\underset{..}{Cl}}:$ + $\overset{\overset{:\overset{..}{Cl}:}{|}}{\underset{:\overset{..}{Cl}:}{Al}-\overset{..}{Cl}:}$ $\longrightarrow$ $CH_3-\overset{+}{\overset{..}{Cl}}-\overset{\overset{:\overset{..}{Cl}:}{|}}{\underset{:\overset{..}{Cl}:}{Al}}{-}\overset{-}{\overset{..}{Cl}}:$

(c) $CH_3-\overset{..}{\underset{..}{O}}-CH_3$ + $\overset{\overset{:\overset{..}{F}:}{|}}{\underset{:\overset{..}{F}:}{B}-\overset{..}{F}:}$ $\longrightarrow$ $CH_3-\overset{+}{\underset{CH_3}{\overset{..}{O}}}-\overset{\overset{:\overset{..}{F}:}{|}}{\underset{:\overset{..}{F}:}{B}}{-}\overset{-}{\overset{..}{F}}:$

3.2 (a) Lewis base   (b) Lewis acid

    (c) Lewis base   (d) Lewis base

    (e) Lewis acid   (f) Lewis base

3.3 $CH_3-\overset{\overset{H}{|}}{\underset{CH_3}{N}}:$ + $\overset{\overset{:\overset{..}{F}:}{|}}{\underset{:\overset{..}{F}:}{B}-\overset{..}{F}:}$ $\longrightarrow$ $CH_3-\overset{\overset{H}{|}}{\underset{CH_3}{\overset{+}{N}}}-\overset{\overset{:\overset{..}{F}:}{|}}{\underset{:\overset{..}{F}:}{B}}{-}\overset{..}{F}:$

       Lewis base   Lewis acid

3.4 (a) $K_a = \dfrac{[H_3O^+]\,[HCO_2^-]}{[HCO_2H]} = 1.78 \times 10^{-4}$

Let $x = [H_3O^+] = [HCO_2^-]$ at equilibrium
then, $0.1 - x = [HCO_2H]$ at equilibrium
but, since the K_a is very small, x will be very small and $0.1 - x \approx 0.1$
Therefore,

$$\frac{(x)\,(x)}{0.1} = 1.78 \times 10^{-4}$$
$$x^2 = 1.78 \times 10^{-5}$$
$$x = 0.0042 = [H_3O^+] = [HCO_2^-]$$

(b)   % Ionized $= \dfrac{[H_3O^+]}{0.1} \times 100$ or $\dfrac{[HCO_2^-]}{0.1} \times 100$

$$= \dfrac{.0042}{0.1} \times 100 = 4.2\%$$

3.5 (a) $pK_a = -\log 10^{-7} = -(-7) = 7$

(b) $pK_a = -\log 5.0 = -0.699$

(c)   Since the acid with a $pK_a = 5$ has a larger K_a, it is the stronger acid.

3.6 When H_3O^+ acts as an acid in aqueous solution, the equation is

$$H_3O^+ + H_2O \rightleftharpoons H_2O + H_3O^+$$

and K_a is

$$K_a = \dfrac{[H_2O][H_3O^+]}{[H_3O^+]} = [H_2O]$$

The molar concentration of H_2O in pure H_2O, that is, $[H_2O] = 55.5$; therefore, $K_a = 55.5$

The pK_a is

$$pK_a = -\log 55.5 = -1.74$$

3.7 The pK_a of the methylaminium ion is equal to 10.6 (Section 3.5C). Since the pK_a of the anilinium ion is equal to 4.6, the anilinium ion is a stronger acid than the methylaminium ion, and aniline ($C_6H_5NH_2$) is a weaker base than methylamine (CH_3NH_2).

3.8 (a)   Negative. Because the atoms are constrained to one molecule in the product, they have to become more ordered.

(b)   Approximately zero.

(c)   Positive. Because the atoms are in two separate product molecules, they become more disordered.

3.9 (a)   If $K_{eq} = 1$
then,
$$\log K_{eq} = 0 = \dfrac{-\Delta G^\circ}{2.303RT}$$
$$\Delta G^\circ = 0$$

(b)   If $K_{eq} = 10$
then,
$$\log K_{eq} = 1 = \dfrac{-\Delta G^\circ}{2.303RT}$$
$$\Delta G^\circ = -(2.303)(0.001987 \text{ kcal mol}^{-1} \text{ K}^{-1})(298 \text{ K}) = -1.36 \text{ kcal mol}^{-1}$$

(c) $\Delta G^\circ = \Delta H^\circ - T\Delta S^\circ$

$\Delta G^\circ = \Delta H^\circ = -1.36 \text{ kcal mol}^{-1}$ if $\Delta S^\circ = 0$

3.10 Structures **A** and **B** make equal contributions to the overall hybrid. This means that the carbon-oxygen bonds should be the same length and that the oxygens should bear equal positive charges.

A   **B**

3.11 (a) $CHCl_2CO_2H$ would be the stronger acid because the electron-withdrawing inductive effect of *two chlorine atoms* would make its hydroxyl proton more positive. The electron-withdrawing effect of the two chlorine atoms would also stabilize the dichloroacetate ion more effectively by dispersing its negative charge more extensively.

(b) CCl_3CO_2H would be the stronger acid for reasons similar to those given in (a), except here there are three versus two electron-withdrawing chlorine atoms involved.

(c) CH_2FCO_2H would be the stronger acid because the electron-withdrawing effect of a fluorine atom is greater than that of a bromine atom (fluorine is more electronegative).

(d) CH_2FCO_2H is the stronger acid because the fluorine atom is nearer the carboxyl group and is, therefore, better able to exert its electron-withdrawing inductive effect. (*Remember:* Inductive effects weaken steadily as the distance between the substituent and the group increases.)

3.12 All compounds containing oxygen and most compounds containing nitrogen will have an unshared electron pair on their oxygen or nitrogen atom. These compounds can, therefore, act as bases and accept a proton from concentrated sulfuric acid. When they accept a proton, these compounds become either oxonium ions or ammonium ions, and having become ionic, they are soluble in the polar medium of sulfuric acid. The only nitrogen compounds that do not have an electron pair on their nitrogen atom are quaternary ammonium compounds, and these, already being ionic, also dissolve in the polar medium of concentrated sulfuric acid.

3.13 (a)

3.13 (b)

(c)

$$H-\overset{\overset{\displaystyle |}{\underset{\displaystyle |}{N}}}{H}-H \quad + \quad :CH_2CH_3 \quad \xrightarrow{\text{hexane}} \quad :NH_2^- \quad + \quad CH_3CH_3$$

Stronger acid	Stronger base	Weaker base	Weaker acid
$pK_a = 38$	(from CH_3CH_2Li)		$pK_a = 50$

(d)

$$H-\overset{\overset{\displaystyle +}{\underset{\displaystyle |}{\overset{\displaystyle H}{N}}}}{\underset{\displaystyle H}{}}-H \quad + \quad :\overset{..}{N}H_2^- \quad \xrightarrow{\text{liq. } NH_3} \quad :NH_3 \quad + \quad :NH_3$$

Stronger acid	Stronger base	Weaker base	Weaker acid
$pK_a = 9.2$	(from $NaNH_2$)		$pK_a = 38$
(from NH_4Cl)			

(e)

$$H-\overset{..}{\underset{..}{O}}-H \quad + \quad :\overset{..}{\underset{..}{O}}C(CH_3)_3 \quad \xrightarrow{H_2O} \quad H-\overset{..}{\underset{..}{O}}:^- \quad + \quad HOC(CH_3)_3$$

Stronger acid	Stronger base	Weaker base	Weaker acid
$pK_a = 15.7$			$pK_a = 17$

(f) No appreciable acid-base reaction would occur because NaOH is not a strong enough base to remove a proton from $(CH_3)_3COH$.

3.14 (a) $HC{\equiv}CH \ + \ NaH \ \xrightarrow{\text{hexane}} \ HC{\equiv}CNa \ + \ H_2$

(b) $HC{\equiv}CNa \ + \ D_2O \ \xrightarrow{\text{hexane}} \ HC{\equiv}CD \ + \ NaOD$

(c) $CH_3CH_2Li \ + \ D_2O \ \xrightarrow{\text{hexane}} \ CH_3CH_2D \ + \ LiOD$

(d) $CH_3CH_2OH \ + \ NaH \ \xrightarrow{\text{hexane}} \ CH_3CH_2ONa \ + \ H_2$

(e) $CH_3CH_2ONa \ + \ T_2O \ \xrightarrow{\text{hexane}} \ CH_3CH_2OT \ + \ NaOT$

(f) $CH_3CH_2CH_2Li \ + \ D_2O \ \xrightarrow{\text{hexane}} \ CH_3CH_2CH_2D \ + \ LiOD$

3.15 (a) $:\overset{..}{N}H_2^-$ (the amide ion) (d) $H-C{\equiv}C:^-$ (the ethynide ion)

(b) $H-\overset{..}{\underset{..}{O}}:^-$ (the hydroxide ion) (e) $CH_3\overset{..}{\underset{..}{O}}:^-$ (the methoxide ion)

(c) $:H^-$ (the hydride ion) (f) H_2O (water)

3.16 $:\overset{..}{N}H_2^- \ > \ :H^- \ > \ H-C{\equiv}C:^- \ > \ CH_3\overset{..}{\underset{..}{O}}:^- \ \approx \ H-\overset{..}{\underset{..}{O}}:^- \ > \ H_2O$

3.17 (a) H_2SO_4 (d) NH_3

(b) H_3O^+ (e) CH_3CH_3

(c) $CH_3NH_3^+$ (f) CH_3CO_2H

3.18 $H_2SO_4 > H_3O^+ > CH_3CO_2H > CH_3NH_3^+ > NH_3 > CH_3CH_3$

3.19   (a)

$$CH_3CH_2-\ddot{\underset{..}{Cl}}: \ + \ AlCl_3 \longrightarrow CH_3CH_2-\overset{+}{\ddot{Cl}}-\overset{:\ddot{Cl}:}{\underset{:\ddot{Cl}:}{\overset{|}{\underset{|}{Al}}}}-\ddot{\underset{..}{Cl}}:$$

Lewis base   Lewis acid

(b)

$$CH_3-\overset{..}{\underset{..}{O}}H \ + \ BF_3 \longrightarrow CH_3-\overset{+}{\underset{H}{\ddot{O}}}-\overset{:\ddot{F}:}{\underset{:\ddot{F}:}{\overset{|}{\underset{|}{B}}}}-\ddot{\underset{..}{F}}:$$

Lewis base   Lewis acid

(c)

$$CH_3-\overset{CH_3}{\underset{CH_3}{\overset{|}{\underset{|}{C}}}}{}^+ \ + \ H_2\ddot{O}: \longrightarrow CH_3-\overset{CH_3}{\underset{CH_3}{\overset{|}{\underset{|}{C}}}}-\overset{..}{O}H_2{}^+$$

Lewis acid   Lewis base

3.20   (a)

$$CH_3-\overset{..}{\underset{..}{O}}H \ + \ H-\ddot{\underset{..}{I}}: \longrightarrow CH_3-\overset{+}{\underset{H}{\ddot{O}}}-H \ + \ :\ddot{\underset{..}{I}}:^-$$

(b)

$$CH_3-\overset{..}{N}H_2 \ + \ H-\ddot{\underset{..}{Cl}}: \longrightarrow CH_3-\overset{+}{\underset{H}{\overset{|}{\underset{|}{N}}}}-H \ + \ :\ddot{\underset{..}{Cl}}:^-$$

(c)

$$\underset{H}{\overset{H}{>}}C=C\underset{H}{\overset{H}{<}} \ + \ H-\ddot{\underset{..}{F}}: \longrightarrow H-\overset{H}{\underset{+}{\overset{|}{\underset{|}{C}}}}-\overset{H}{\underset{H}{\overset{|}{\underset{|}{C}}}}-H \ + \ :\ddot{\underset{..}{F}}:^-$$

3.21   Because the proton attached to the highly electronegative oxygen atom of CH_3OH is much more acidic than the protons attached to the much less electronegative carbon atom.

3.22 $$CH_3CH_2-\overset{..}{\underset{..}{O}}-H \ + \ ^-:C\equiv C-H \ \xrightarrow{\text{liq. NH}_3} \ CH_3CH_2-\overset{..}{\underset{..}{O}}:^- \ + \ H-C\equiv C-H$$

3.23   (a) $pK_a = -\log 1.77 \times 10^{-4} = 4 \ -0.248 = 3.752$

(b) $K_a = 10^{-13}$

3.24   (a)   HB is the stronger acid because it has the smaller pK_a.

(b)   Yes. Since A^- is the stronger base and HB is the stronger acid, the following acid-base reaction will take place.

$$A:^- \ + \ H-B \longrightarrow A-H \ + \ B:^-$$

Stronger base   Stronger acid   Weaker acid   Weaker base

$pK_a = 10$ $pK_a = 20$

3.25 (a) $CH_3CH_2-\overset{\overset{\text{O}}{\|}}{C}-\ddot{\ddot{O}}-H + :\ddot{O}-H \longrightarrow CH_3CH_2-\overset{\overset{\text{O}}{\|}}{C}-\ddot{O}: + H-\ddot{O}-H$

(b) $C_6H_5-\overset{\overset{:\ddot{O}:}{\|}}{\underset{\underset{:\ddot{O}:}{\|}}{S}}-\ddot{O}-H + :\ddot{O}-H \longrightarrow C_6H_5-\overset{\overset{:\ddot{O}:}{\|}}{\underset{\underset{:\ddot{O}:}{\|}}{S}}-\ddot{O}: + H-\ddot{O}-H$

(c) No appreciable acid-base reaction takes place because CH_3CH_2ONa is too weak a base to remove a proton from ethyne.

(d) $H-C\equiv C-H + {}^-:CH_2CH_3 \xrightarrow{\text{hexane}} H-C\equiv C:^- + CH_3CH_3$
(from LiCH$_2$CH$_3$)

(e) $CH_3-CH_2-\ddot{O}-H + {}^-:CH_2CH_3 \xrightarrow{\text{hexane}} CH_3-CH_2-\ddot{O}:^- + CH_3CH_3$
(from LiCH$_2$CH$_3$)

3.26 (a) $CH_3CH_2CH_2Br + 2 Li \xrightarrow{\text{ether}} CH_3CH_2CH_2Li + LiBr$

then

$CH_3CH_2CH_2Li + D_2O \longrightarrow CH_3CH_2CH_2D + LiOD$

(b) $CH_3-\overset{\overset{CH_3}{|}}{CH}-Br + 2 Li \xrightarrow{\text{ether}} CH_3-\overset{\overset{CH_3}{|}}{CH}-Li + LiBr$

then

$CH_3-\overset{\overset{CH_3}{|}}{CH}-Li + D_2O \longrightarrow CH_3-\overset{\overset{CH_3}{|}}{CH}-D + LiOD$

(c) $CH_3-\overset{\overset{CH_3}{|}}{\underset{\underset{CH_3}{|}}{C}}-Br + 2 Li \xrightarrow{\text{ether}} CH_3-\overset{\overset{CH_3}{|}}{\underset{\underset{CH_3}{|}}{C}}-Li + LiBr$

then

$CH_3-\overset{\overset{CH_3}{|}}{\underset{\underset{CH_3}{|}}{C}}-Li + D_2O \longrightarrow CH_3-\overset{\overset{CH_3}{|}}{\underset{\underset{CH_3}{|}}{C}}-D + LiOD$

3.27 (a) $CH_3-C\equiv C-H + NaNH_2 \xrightarrow{\text{ether}} CH_3-C\equiv C:^- Na^+ + NH_3$

then

$CH_3-C\equiv C:^- Na^+ + T_2O \longrightarrow CH_3-C\equiv C-T + NaOT$

(b) $CH_3-\overset{\underset{\underset{CH_3}{|}}{}}{CH}-O-H + NaH \longrightarrow CH_3-\overset{\underset{\underset{CH_3}{|}}{}}{CH}-ONa + H_2$

then

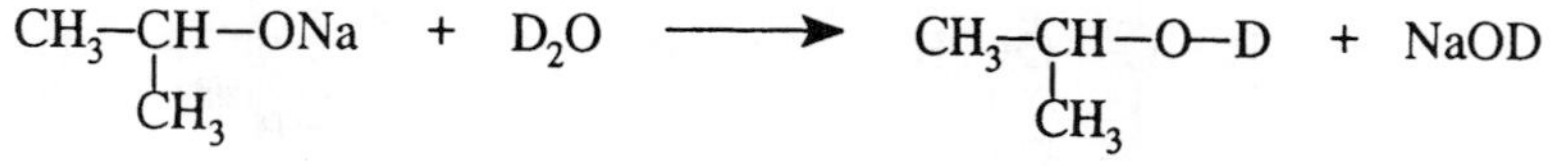

$$CH_3-CH-ONa \ + \ D_2O \ \longrightarrow \ CH_3-CH-O-D \ + \ NaOD$$
$$\quad\quad\ |\quad\quad\quad\quad\quad\quad\quad\quad\quad\quad\quad\quad\ |$$
$$\quad\quad CH_3 \quad\quad\quad\quad\quad\quad\quad\quad\quad\quad CH_3$$

(c) $CH_3CH_2CH_2OH \ + \ NaH \ \longrightarrow \ CH_3CH_2CH_2ONa \ + \ H_2$

then

$$CH_3CH_2CH_2ONa \ + \ D_2O \ \longrightarrow \ CH_3CH_2CH_2OD \ + \ NaOD$$

3.28 (a) $CH_3CH_2OH \ > \ CH_3CH_2NH_2 \ > \ CH_3CH_2CH_3$

Oxygen is more electronegative than nitrogen, which is more electronegative than carbon. The O-H bond is most polarized, the N-H bond is next, and the C-H bond is least polarized.

(b) $CH_3CH_2O^- \ < \ CH_3CH_2NH^- \ < \ CH_3CH_2CH_2^-$

The weaker the acid, the stronger the conjugate base.

3.29 (a) $CH_3C{\equiv}CH \ > \ CH_3CH{=}CH_2 \ > \ CH_3CH_2CH_3$

(b) $CH_3CHClCO_2H \ > \ CH_3CH_2CO_2H \ > \ CH_3CH_2CH_2OH$

(c) $CH_3CH_2OH_2^+ \ > \ CH_3CH_2OH \ > \ CH_3OCH_3$

3.30 (a) $CH_3NH_3^+ \ < \ CH_3NH_2 \ < \ CH_3NH^-$

(b) $CH_3O^- \ < \ CH_3NH^- \ < \ CH_3CH_2^-$

(c) $CH_3C{\equiv}C^- \ < \ CH_3CH{=}CH^- \ < \ CH_3CH_2CH_2^-$

3.31 The acidic hydrogens must be attached to oxygen atoms. In H_3PO_3, one hydrogen is bonded to a phosphorus atom:

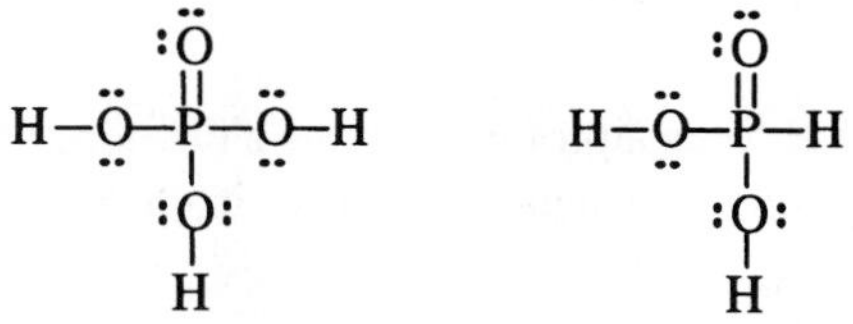

3.32 (a)

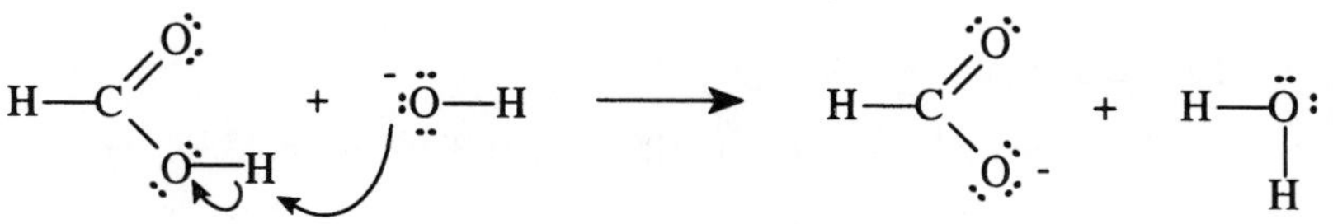

(b)

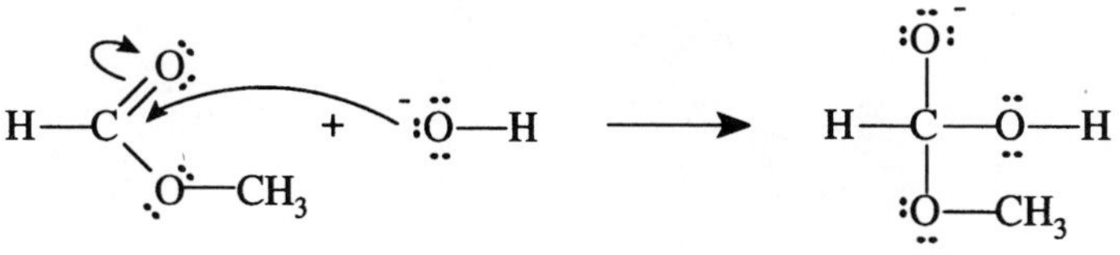

(c)   [reaction structures]

(d) $H{-}\ddot{O}:^{-}$ + $CH_3{-}\ddot{I}:$ $\longrightarrow$ $H{-}\ddot{O}{-}CH_3$ + $:\ddot{I}:^{-}$

(e)   [reaction structures] $CH_2{=}C$ (with CH_3 groups) + $:\ddot{Cl}:^{-}$

3.33   (a)   Assume that the acidic and basic groups of glycine in its two forms have acidities and basicities similar to those of acetic acid and methylamine. Then consider the equilibrium between the two forms:

[structural diagram of glycine equilibrium]

Stronger base   Stronger acid   ⇌   Weaker acid   Weaker base

We see that the ionic form contains the groups that are the weaker acid and weaker base. The equilibrium, therefore, will favor this form.

(b)   The high melting point shows that the ionic structure better represents glycine.

3.34   (a)   The second carboxyl group of malonic acid acts as an electron-withdrawing group and stabilizes the conjugate base formed (i.e., $HO_2CCH_2CO_2^-$) when malonic acid loses a proton. [Any factor that stabilizes the conjugate base of an acid always increases the strength of the acid (Section 3.9c).] An important factor here may be an entropy effect as explained in Section 3.8.

(b)   When $^-O_2CCH_2CO_2H$ loses a proton, it forms a dianion, $^-O_2CCH_2CO_2^-$. This dianion is destabilized by having two negative charges in close proximity.

3.35   HB is the stronger acid.

3.36 $\Delta G^\circ = \Delta H^\circ - T\Delta S^\circ$

$= 1.5\ \text{kcal mol}^{-1} - (298\ \text{K})(0.0020\ \text{kcal mol}^{-1}\text{K}^{-1})$

$= 0.9\ \text{kcal mol}^{-1}$

$\log K_{eq} = \log K_a = -\text{p}K_a = -\dfrac{\Delta G^\circ}{2.303RT}$

$\text{p}K_a = \dfrac{\Delta G^\circ}{2.303RT}$

$\text{p}K_a = \dfrac{0.9\ \text{kcal mol}^{-1}}{(2.303)(0.001987\ \text{kcal mol}^{-1}\text{K}^{-1})(298\ \text{K})}$

$\text{p}K_a = 0.66$

3.37   The dianion is a hybrid of the following resonance structures:

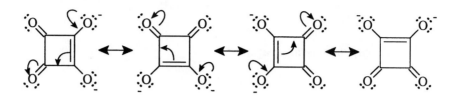

If we mentally fashion a hybrid of these structures, we see that each carbon-carbon bond is a single bond in three structures and a double bond in one. Each carbon-oxygen bond is a double bond in two structures and a single bond in two structures. Therefore, we would expect all of the carbon-carbon bonds to be equivalent and of the same length, and exactly the same can be said for the carbon-oxygen bonds.

QUIZ

3.1   Which of the following is the strongest acid?

(a) $CH_3CH_2CO_2H$ (b) CH_3CH_3 (c) CH_3CH_2OH (d) $CH_2=CH_2$

3.2   Which of the following is the strongest base?

(a) CH_3ONa (b) $NaNH_2$ (c) CH_3CH_2Li (d) $NaOH$ (e) CH_3CO_2Na

3.3   Dissolving $NaNH_2$ in water will give:

(a)   A solution containing solvated Na^+ and NH_2^- ions.

(b)   A solution containing solvated Na^+ ions, OH^- ions, and NH_3.

(c) NH_3 and metallic Na.

(d)   Solvated Na^+ ions and hydrogen gas.

(e)   None of the above.

3.4   Which base is strong enough to convert $(CH_3)_3COH$ into $(CH_3)_3CONa$ in a reaction that goes to completion?

(a) $NaNH_2$ (b) CH_3CH_2Na (c) $NaOH$ (d) CH_3CO_2Na

(e)   More than one of the above.

3.5   Which would be the strongest acid?

(a) $CH_3CH_2CH_2CO_2H$ (b) $CH_3CH_2CHFCO_2H$ (c) $CH_3CHFCH_2CO_2H$

(d) $CH_2FCH_2CH_2CO_2H$ (e) $CH_3CH_2CH_2CH_2OH$

3.6   Which would be the weakest base?

(a) CH_3CO_2Na (b) CF_3CO_2Na (c) CHF_2CO_2Na (d) CH_2FCO_2Na

3.7   What acid-base reaction (if any) would occur when NaF is dissolved in H_2SO_4?

3.8   The pK_a of $CH_3NH_3^+$ equals 10.6; the pK_a of $(CH_3)_2NH_2^+$ equals 10.7. Which is the stronger base, CH_3NH_2 or $(CH_3)_2NH$?

3.9   Supply the missing reagents.

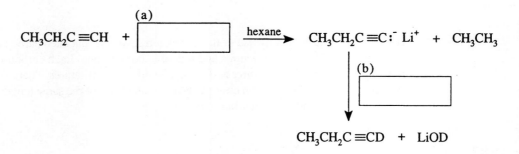

3.10   Supply the missing intermediates and reagents.

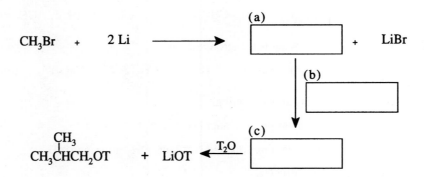

4

ALKANES AND CYCLOALKANES: CONFORMATIONS OF MOLECULES

SOLUTIONS TO PROBLEMS

4.1

Heptane ✓   2-Methylhexane ✓   3-Methylhexane ✓

2,2-Dimethylpentane ✓   3,3-Dimethylpentane ✓   2,3-Dimethylpentane ✓

2,4-Dimethylpentane ✓   3-Ethylpentane ✓   2,2,3-Trimethylbutane

4.2 (a) $CH_3CH_2CH_2CH_2CH_2$ — $CH_3CH_2CH_2CHCH_3$ $CH_3CH_2CHCH_2CH_3$

          1-Pentyl   2-Pentyl   3-Pentyl

      $CH_3CH_2CH(CH_3)CH_2$— $CH_3CH(CH_3)CH_2CH_2$—

         2-Methyl-1-butyl   3-Methyl-1-butyl

       $CH_3CHCH(CH_3)CH_3$ $CH_3C(CH_3)CH_2CH_3$

        3-Methyl-2-butyl   2-Methyl-2-butyl

(b)   See the answer to Problem 4.1 for the names of C_7H_{16} isomers.

4.3 (a) $CH_3CH_2CH_2CH_2Cl$
1-Chlorobutane

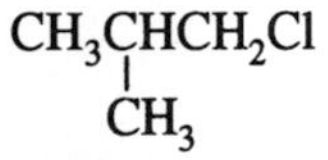

CH_3CHCH_2Cl
       $|$
      CH_3
1-Chloro-2-methylpropane

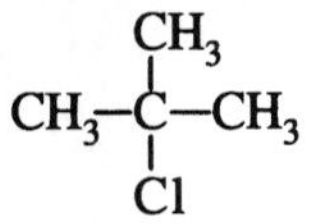

$CH_3CH_2CHCH_3$
          $|$
          Cl
2-Chlorobutane

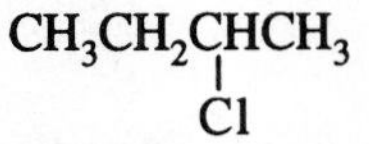

      CH_3
       $|$
CH_3-C-CH_3
       $|$
      Cl
2-Chloro-2-methylpropane

(b) $CH_3CH_2CH_2CH_2CH_2Br$
1-Bromopentane

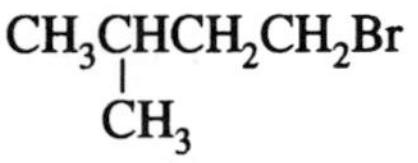

$CH_3CHCH_2CH_2Br$
       $|$
      CH_3
1-Bromo-3-methylbutane

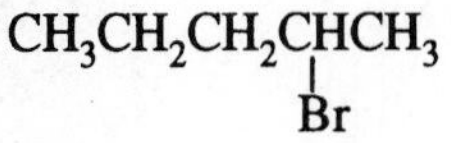

$CH_3CH_2CH_2CHCH_3$
              $|$
              Br
2-Bromopentane

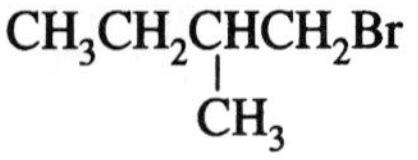

$CH_3CH_2CHCH_2Br$
          $|$
         CH_3
1-Bromo-2-methylbutane

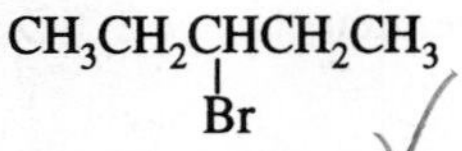

$CH_3CH_2CHCH_2CH_3$
          $|$
          Br
3-Bromopentane

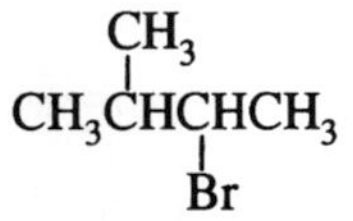

       CH_3
        $|$
$CH_3CHCHCH_3$
          $|$
          Br
2-Bromo-3-methylbutane

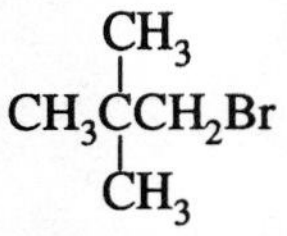

      CH_3
       $|$
CH_3CCH_2Br
       $|$
      CH_3
1-Bromo-2,2-dimethylpropane

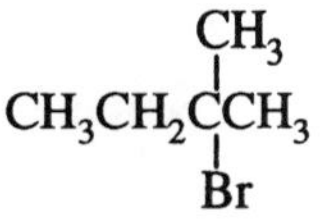

      CH_3
       $|$
$CH_3CH_2CCH_3$
       $|$
      Br
2-Bromo-2-methylbutane

4.4 (a) $CH_3CH_2CH_2CH_2OH$
1-Butanol

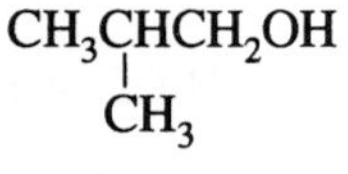

CH_3CHCH_2OH
       $|$
      CH_3
2-Methyl-1-propanol

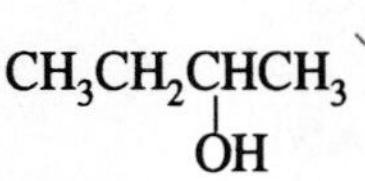

$CH_3CH_2CHCH_3$
          $|$
          OH
2-Butanol

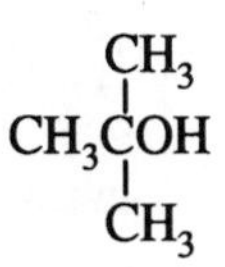

      CH_3
       $|$
CH_3COH
       $|$
      CH_3
2-Methyl-2-propanol

(b) $CH_3CH_2CH_2CH_2CH_2OH$

1-Pentanol

$CH_3CH_2CH_2CHCH_3$
                     OH

2-Pentanol

$CH_3CH_2CHCH_2CH_3$
               OH

3-Pentanol

         CH_3

CH_3CCH_2OH
       CH_3

2,2-Dimethyl-1-propanol

$CH_3CHCH_2CH_2OH$
     CH_3

3-Methyl-1-butanol

$CH_3CH_2CHCH_2OH$
         CH_3

2-Methyl-1-butanol

     CH_3

$CH_3CHCHCH_3$
         OH

3-Methyl-2-butanol

        CH_3

$CH_3CH_2CCH_3$
         OH

2-Methyl-2-butanol

4.5   (a)   1-(1-Methylethyl)-2-(1,1-dimethylethyl)cyclopentane or
1-*tert*-butyl-2-isopropylcyclopentane

    (b)   1-Methyl-2-(2-methylpropyl)cyclohexane or
1-isobutyl-2-methylcyclohexane

    (c)   Butylcyclohexane

    (d)   1-Chloro-2,4-dimethylcyclohexane

    (e)   2-Chlorocyclopentanol

    (f)   3-(1,1-Dimethylethyl)cyclohexanol or 3-*tert*-butylcyclohexanol

4.6   (a)   2-Chlorobicyclo[1.1.0]butane

    (b)   Bicyclo[3.2.1]octane

    (c)   Bicyclo[2.1.1]hexane

    (d)   9-Chlorobicyclo[3.3.1]nonane

    (e)   2-Methylbicyclo[2.2.2]octane

    (f) 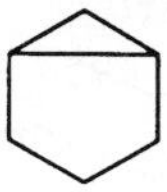   Bicyclo[3.1.0]hexane or 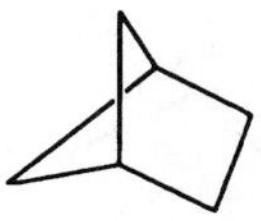   bicyclo[2.1.1]hexane

4.7

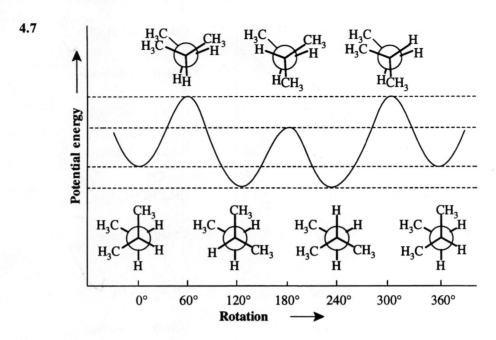

4.8

$$\log K_{eq} = \frac{\Delta G°}{-2.303RT} = \frac{-1800 \text{ cal}}{(-2.303)(1.987 \text{ cal K}^{-1})(298 \text{ K})} = 1.32$$

$$K_{eq} = 20.89$$

Let e = amount of equatorial form

and a = amount of axial form

then, $K_{eq} = \frac{e}{a} = 20.89$

$$e = 20.89a$$

$$\%e = \frac{20.89a}{a + 20.89a} \times 100 = 95.4\%$$

4.9 (a)

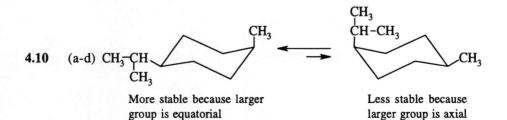

4.10 (a-d)

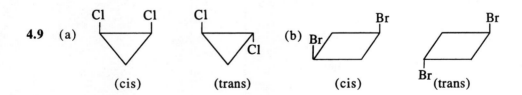

More stable because larger group is equatorial

Less stable because larger group is axial

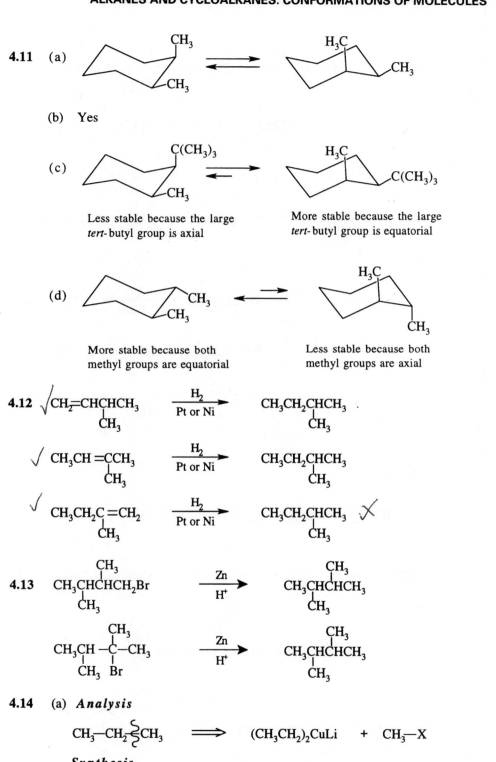

4.11 (a)

(b)   Yes

(c)

Less stable because the large
tert-butyl group is axial

More stable because the large
tert-butyl group is equatorial

(d)

More stable because both
methyl groups are equatorial

Less stable because both
methyl groups are axial

4.12 $CH_2=CHCHCH_3$ over $\overset{H_2}{\underset{Pt \text{ or } Ni}{\longrightarrow}}$ $CH_3CH_2CHCH_3$
with CH_3 substituents

$CH_3CH=CCH_3$ over $\overset{H_2}{\underset{Pt \text{ or } Ni}{\longrightarrow}}$ $CH_3CH_2CHCH_3$
with CH_3 substituents

$CH_3CH_2C=CH_2$ over $\overset{H_2}{\underset{Pt \text{ or } Ni}{\longrightarrow}}$ $CH_3CH_2CHCH_3$ ✗
with CH_3 substituents

4.13 $CH_3CHCHCH_2Br$ (with CH_3, CH_3 substituents) $\overset{Zn}{\underset{H^+}{\longrightarrow}}$ $CH_3CHCHCH_3$ (with CH_3, CH_3 substituents)

$CH_3CH-C-CH_3$ (with CH_3 top, CH_3 Br bottom) $\overset{Zn}{\underset{H^+}{\longrightarrow}}$ $CH_3CHCHCH_3$ (with CH_3, CH_3 substituents)

4.14   (a) *Analysis*

$CH_3-CH_2\overset{\xi}{\xi}CH_3$ $\implies$ $(CH_3CH_2)_2CuLi$ + CH_3-X

Synthesis

CH_3CH_2Br $\overset{Li}{\underset{\substack{\text{diethyl ether} \\ (-LiBr)}}{\longrightarrow}}$ CH_3CH_2Li $\overset{CuI}{\underset{(-LiI)}{\longrightarrow}}$ $(CH_3CH_2)_2CuLi$

$(CH_3CH_2)_2CuLi$ + CH_3I $\longrightarrow$ $CH_3CH_2CH_3$ + CH_3CH_2Cu + LiI

(b) *Analysis*

$$CH_3-CH_2\overset{\zeta}{\underset{\zeta}{}}CH_2-CH_3 \implies (CH_3CH_2)_2CuLi + CH_3CH_2-X$$

Synthesis

$$(CH_3CH_2)_2CuLi + CH_3CH_2I \longrightarrow CH_3CH_2CH_2CH_3 + CH_3CH_2Cu + LiI$$
[from (a)]

(c) *Analysis*

$$\underset{\substack{\\ CH_3}}{CH_3-CH-CH_2}\overset{\zeta}{\underset{\zeta}{}}CH_3 \implies \left(\underset{CH_3}{CH_3\overset{|}{C}HCH_2}\right)_2 CuLi + CH_3-X$$

Synthesis

$$\underset{CH_3}{CH_3\overset{|}{C}HCH_2Br} \xrightarrow[\substack{\text{diethyl ether} \\ (-LiBr)}]{Li} \underset{CH_3}{CH_3\overset{|}{C}HCH_2Li} \xrightarrow[(-LiI)]{CuI} \left(\underset{CH_3}{CH_3\overset{|}{C}HCH_2}\right)_2 CuLi$$

$$\xrightarrow{CH_3I} \underset{CH_3}{CH_3\overset{|}{C}HCH_2CH_3} + \underset{CH_3}{CH_3\overset{|}{C}HCH_2Cu} + LiI$$

(d) *Analysis*

$$\underset{CH_3}{CH_3\overset{|}{C}HCH_2CH_2}\overset{\zeta}{\underset{\zeta}{}}\underset{CH_3}{CH_2CH_2\overset{|}{C}HCH_3} \implies \left(\underset{CH_3}{CH_3\overset{|}{C}HCH_2CH_2}\right)_2 CuLi$$
$$+ \underset{CH_3}{CH_3\overset{|}{C}HCH_2CH_2-X}$$

Synthesis

$$\underset{CH_3}{CH_3\overset{|}{C}HCH_2CH_2I} \xrightarrow[\substack{\text{diethyl ether} \\ (-LiI)}]{Li} \underset{CH_3}{CH_3\overset{|}{C}HCH_2CH_2Li} \xrightarrow[(-LiI)]{CuI}$$

$$\left(\underset{CH_3}{CH_3\overset{|}{C}HCH_2CH_2}\right)_2 CuLi \xrightarrow{\underset{CH_3}{CH_3\overset{|}{C}HCH_2CH_2I}} \underset{CH_3}{CH_3\overset{|}{C}HCH_2CH_2CH_2CH_2\overset{|}{C}HCH_3}$$

$$+ \underset{CH_3}{CH_3\overset{|}{C}HCH_2CH_2Cu} + LiI$$

Other syntheses are possible in each part except (a).

(e) *Analysis*

$$\implies (CH_3CH_2)_2CuLi +$$

Synthesis

$$(CH_3CH_2)_2CuLi +$$ $$\longrightarrow$$ $$+ CH_3CH_2Cu + LiBr$$

(f) *Analysis*

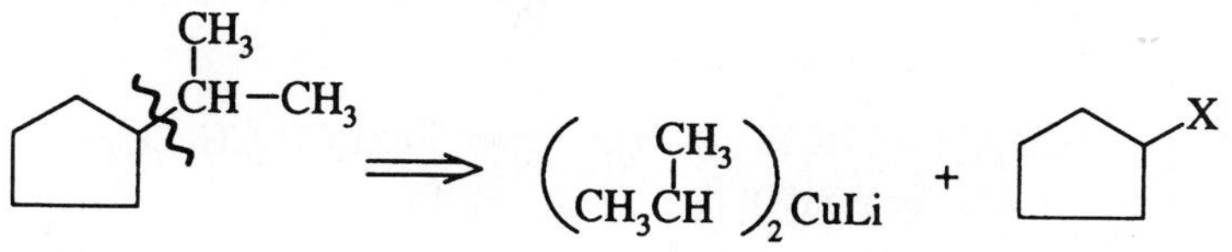

Synthesis

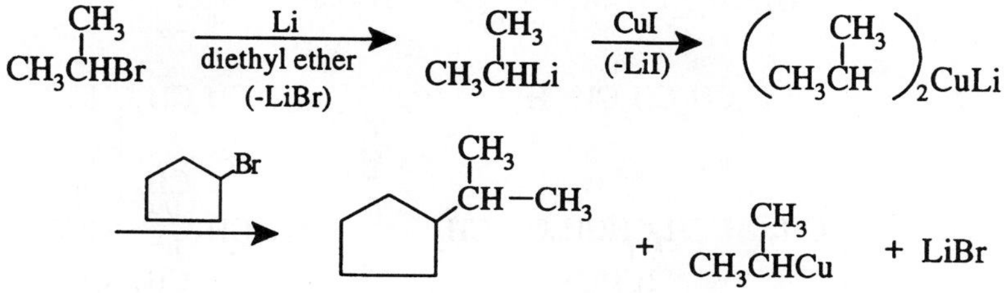

(g) *Analysis*

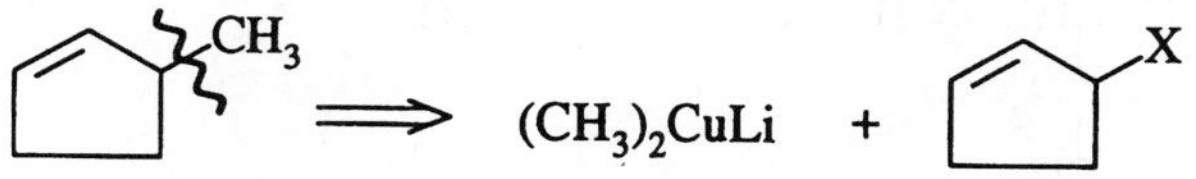

Synthesis

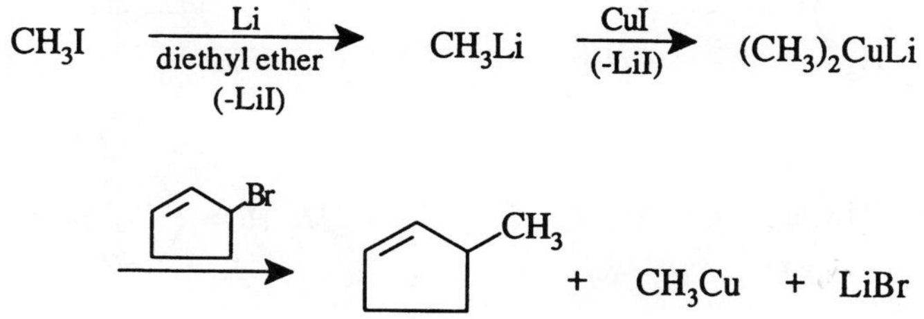

$$CH_3I \xrightarrow[\substack{\text{diethyl ether} \\ (-LiI)}]{Li} CH_3Li \xrightarrow[(-LiI)]{CuI} (CH_3)_2CuLi$$

4.15 (a) $CH_3CH_2Br + (CH_3CH_2)_2CuLi \longrightarrow CH_3CH_2CH_2CH_3$

(b) $CH_3CH_2CH_2Br + (CH_3)_2CuLi \longrightarrow CH_3CH_2CH_2CH_3$

(c) $CH_3CH_2CH_2CH_2Br \xrightarrow[H^+]{Zn} CH_3CH_2CH_2CH_3$

or

$$CH_3CH_2\underset{\underset{Br}{|}}{C}HCH_3 \xrightarrow[H^+]{Zn} CH_3CH_2CH_2CH_3$$

(d) $CH_3CH_2CH=CH_2$ $\xrightarrow[\text{Ni or Pt}]{H_2}$ $CH_3CH_2CH_2CH_3$

or

$CH_3CH=CHCH_3$ $\xrightarrow[\text{Ni or Pt}]{H_2}$ $CH_3CH_2CH_2CH_3$
(cis- or trans-)

4.16 (a) $[CH_3(CH_2)_3CH_2]_2CuLi$

(b) $CH_3(CH_2)_3CH_2Br$ $\xrightarrow[\text{ether}]{Li}$ $CH_3(CH_2)_3CH_2Li$ $\xrightarrow{CuI}$ $[CH_3(CH_2)_3CH_2]_2CuLi$

4.17 (a) $\underset{\underset{Cl}{|}}{ClCH_2CH_2CH_2CHCH_3}$

(b) $\underset{\underset{Br}{|}}{CH_3CH_2CHCH_3}$

(c) $\underset{\underset{\underset{CH_3}{|}}{CHCH_3}}{CH_3CH_2CH_2CHCH_2CH_2CH_3}$

(d) $\underset{\underset{CH_3}{|}}{CH_3}\overset{\overset{CH_3}{|}}{\underset{}{C}}-\underset{\underset{CH_3}{|}}{CHCH_2CH_3}$

(e) $\underset{\underset{\underset{CH_3}{|}}{CH_2}}{CH_3CHCHCH_2CH_2CH_3}$ with CH_3 top

(f)
Cl Cl
(cyclopentane with two Cl on one carbon)

(g)
H   H
(cyclopropane)
H_3C CH_3

(h)
H CH_3
(cyclopropane)
H_3C H

(i) $\underset{\underset{OH}{|}\quad\underset{CH_3}{|}}{CH_3CHCH_2CHCH_3}$

(j) $HO-$ (cyclohexane) $-\text{''''}CH_2\underset{\overset{|}{CH_3}}{CH}CH_3$

or

$HO-$ (cyclohexane) $-CH_2CH(CH_3)_2$

(k) (cyclopropane)$-CH_2CH_2CH_2CHCH_2CH_3$ (with cyclopropane substituent)

(l) $CH_3-\overset{\overset{CH_3}{|}}{\underset{\underset{CH_3}{|}}{C}}-CH_2OH$

(m) (bicyclic structure)

(n) (norbornane structure)

(o) (bicyclopentyl structure)

4.18 (a) 3,3,4-Trimethylhexane

(b) 2,2-Dimethyl-1-butanol

(c) 3,5,7-Trimethylnonane

(d) 3-Methyl-4-heptanol

(e) 2-Bromobicyclo[3.3.1]nonane

(f) 2,5-Dibromo-4-ethyloctane

(g) Cyclobutylcyclopentane

(h) 7-Chlorobicyclo[2.2.1]heptane

4.19 The two secondary carbon atoms in *sec*-butyl alcohol are equivalent; however, there are three five-carbon alcohols (pentyl alcohols) that contain a secondary carbon atom.

4.20 (a)

$$CH_3-\underset{\underset{H_3C}{|}}{\overset{\overset{H_3C}{|}}{C}}-\underset{\underset{CH_3}{|}}{\overset{\overset{CH_3}{|}}{C}}-CH_3$$

2,2,3,3 -Tetramethylbutane

(b)

Cyclohexane

(c)

1,1-Dimethylcyclobutane

(d)

Bicyclo [2.2.2] octane

4.21 Each of the desired alkenes must have the same carbon skeleton as 2-methylbutane,

$$C-\overset{\overset{C}{|}}{C}-C-C;$$ they are therefore

$$CH_2{=}\overset{\overset{CH_3}{|}}{C}CH_2CH_3$$

$$CH_3\overset{\overset{CH_3}{|}}{C}{=}CHCH_3 \Bigg\} + H_2 \xrightarrow[C_2H_5OH]{Ni} CH_3\overset{\overset{CH_3}{|}}{C}HCH_2CH_3$$

$$CH_3\overset{\overset{CH_3}{|}}{C}HCH{=}CH_2$$

4.22 Only one isomer of C_6H_{14} can be produced from five isomeric hexyl chlorides $(C_6H_{13}Cl)$.

The alkane is 2-methylpentane, $CH_3\overset{\overset{CH_3}{|}}{C}HCH_2CH_2CH_3$. The five alkyl chlorides are

$$ClCH_2\overset{\overset{CH_3}{|}}{C}HCH_2CH_2CH_3 \qquad CH_3\overset{\overset{CH_3}{|}}{C}ClCH_2CH_2CH_3 \qquad CH_3\overset{\overset{CH_3}{|}}{C}HCHClCH_2CH_3$$

$$CH_3\overset{\overset{CH_3}{|}}{C}HCH_2CHClCH_3 \quad \text{and} \quad CH_3\overset{\overset{CH_3}{|}}{C}HCH_2CH_2CH_2Cl$$

4.23

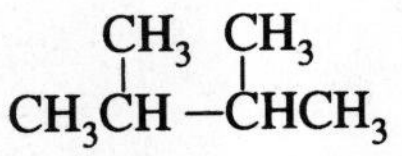

$$\underset{\substack{|\quad\;\;|\\ CH_3CH-CHCH_3}}{CH_3\;\;CH_3}$$ 2,3-Dimethylbutane

From two alkyl chlorides

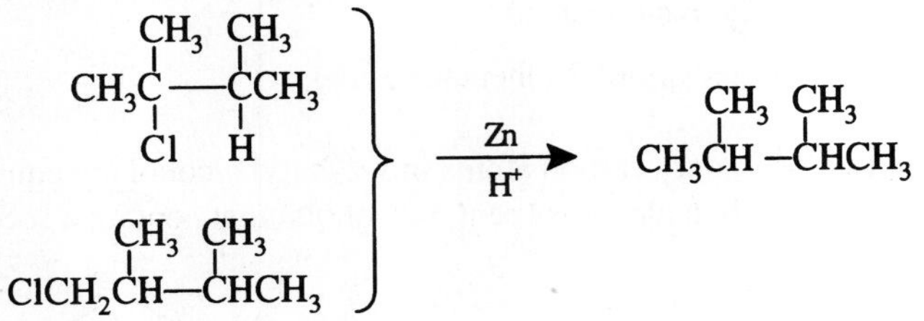

$$\left.\begin{array}{c} \underset{\substack{|\quad\;\;|\\ Cl\quad\;\;H}}{\overset{CH_3\;\;CH_3}{CH_3C-CCH_3}} \\[2em] \underset{\substack{|\quad\;\;|\\ ClCH_2CH-CHCH_3}}{CH_3\;\;CH_3} \end{array}\right\} \xrightarrow[H^+]{Zn} \underset{\substack{|\quad\;\;|\\ CH_3CH-CHCH_3}}{CH_3\;\;CH_3}$$

From two alkenes

$$\left.\begin{array}{c} \underset{\substack{|\\ H}}{\overset{CH_3\;\;CH_3}{CH_2{=}C-CCH_3}} \\[2em] \overset{CH_3\;\;CH_3}{CH_3C{=}CCH_3} \end{array}\right\} \xrightarrow[Ni]{H_2} \underset{\substack{|\quad\;\;|\\ CH_3CH-CHCH_3}}{CH_3\;\;CH_3}$$

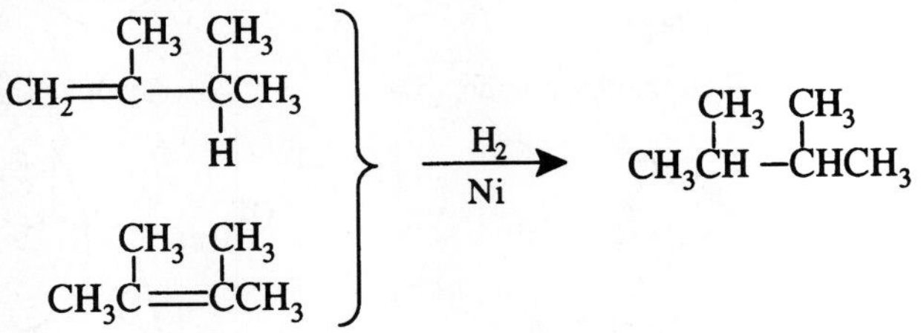

4.24

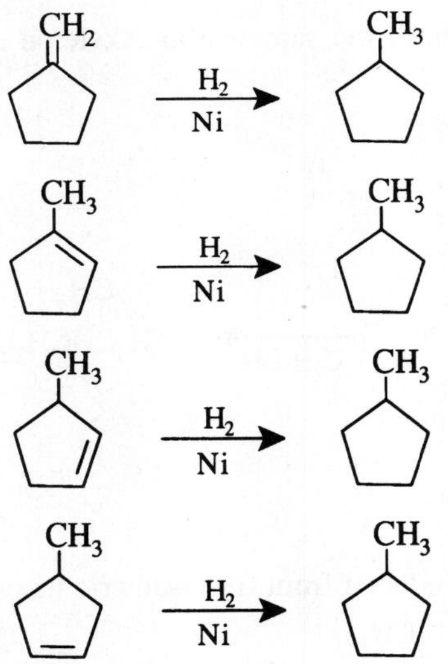

4.25 $(CH_3)_3CCH_3$ is the most stable isomer (i.e., it is the isomer with the lowest potential energy) because it evolves the least amount of heat on a molar basis when subjected to complete combustion.

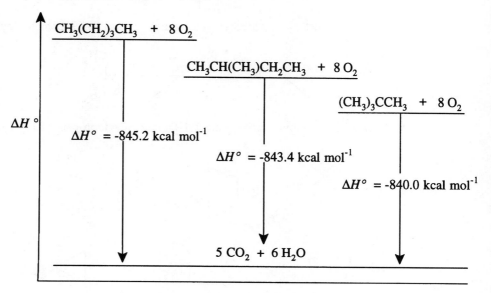

$$CH_3(CH_2)_3CH_3 \ + \ 8\,O_2$$

$$CH_3CH(CH_3)CH_2CH_3 \ + \ 8\,O_2$$

$$(CH_3)_3CCH_3 \ + \ 8\,O_2$$

$\Delta H\,°$

$\Delta H° = -845.2$ kcal mol^{-1}

$\Delta H° = -843.4$ kcal mol^{-1}

$\Delta H° = -840.0$ kcal mol^{-1}

$$5\,CO_2 \ + \ 6\,H_2O$$

4.26   A homologous series is one in which each member of the series differs from the one preceding it by a constant amount, usually a CH_2 group. A homologous series of alkyl halides would be the following:

CH_3X
CH_3CH_2X
$CH_3(CH_2)_2X$
$CH_3(CH_2)_3X$
$CH_3(CH_2)_4X$
etc.

4.27

This conformation is *less stable* because 1,3-diaxial interactions with the large *tert*-butyl group cause considerable repulsion.

This conformation is *more stable* because 1,3-diaxial interactions with the smaller methyl group are less repulsive.

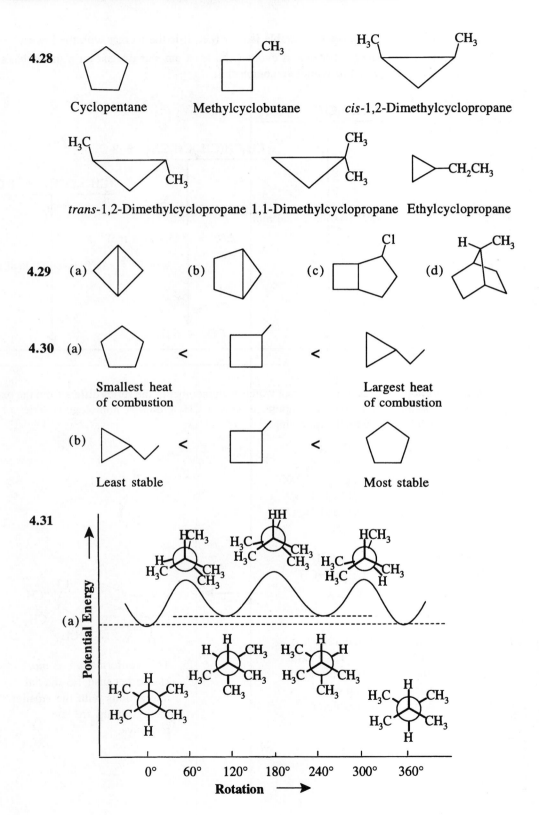

4.28

Cyclopentane   Methylcyclobutane   *cis*-1,2-Dimethylcyclopropane

trans-1,2-Dimethylcyclopropane   1,1-Dimethylcyclopropane   Ethylcyclopropane

4.29   (a)   (b)   (c)   (d)

4.30   (a)   <   <

Smallest heat   Largest heat
of combustion   of combustion

(b)   <   <

Least stable   Most stable

4.31

(a)   Potential Energy

0°   60°   120°   180°   240°   300°   360°

Rotation ⟶

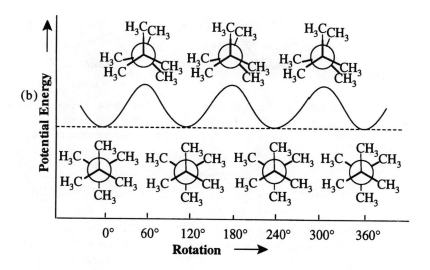

4.32 (a) Pentane would boil higher because its chain is unbranched. Chain-branching lowers the boiling point.

(b) Heptane would boil higher because it has the larger molecular weight and would, because of its larger surface area, have larger van der Waals attractions.

(c) 2-Chloropropane because it is more polar and because it has a larger molecular weight.

(d) 1-Propanol would boil higher because its molecules would be associated with each other through hydrogen-bond formation.

(e) Propanone (CH_3COCH_3) would boil higher because its molecules are more polar.

4.33 *trans*-1,2-Dimethylcyclopropane would be more stable because there is less crowding between its methyl groups.

Less stable   More stable

4.34 (a)

trans   cis

More stable because both methyl groups are equatorial in the most stable conformation

Less stable because one methyl group must be axial and so has the larger heat of combustion

(b)

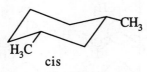 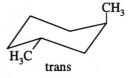

More stable because both methyl
groups are equatorial in the most
stable conformation

Less stable because one methyl
group must be axial and so has
the larger heat of combustion

(c)

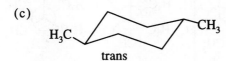 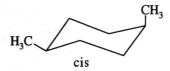

More stable because both methyl
groups are equatorial in the most
stable conformation

Less stable because one methyl
group must be axial and so has
the larger heat of combustion

4.35   (a)

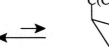

More stable conformation because
both alkyl groups are equatorial

(b)

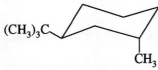

More stable because larger group
is equatorial

(c)

More stable conformation because
both alkyl groups are equatorial

(d)

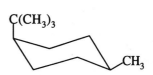

More stable because larger group
is equatorial

4.36   If the cyclobutane ring were planar, the C—Br bond moments would exactly cancel in the trans isomer. The fact that *trans*-1,3-dibromocyclobutane has a dipole moment shows the ring is not planar.

Planar form
μ = 0

Bent form
μ ≠ 0

4.37   (a) $A = \left(\begin{array}{c} H_3C \\ CHCH_2CH_2 \\ H_3C \end{array} \right)_2 CuLi$ (b)   B = ⬠—Br

(c)   C = ⬡—Br   (d)   D = CH_3CH –$\overset{\overset{\displaystyle CH_3}{|}}{C}HCH_2Br$ with $\underset{|}{CH_3}$

(e)   E = CH_3–$\overset{\overset{\displaystyle CH_3}{|}}{\underset{\underset{\displaystyle CH_3}{|}}{C}}$–$CH_2Br$ (f)   F = H_2/Ni or H_2/Pt

4.38   (a) [cyclohexane with CH₃ groups]   (b) From Table 4.10 we find that this is *cis*-1,2-dimethylcyclohexane.

(c) Since catalytic hydrogenation produces the cis isomer, both hydrogen atoms must have added from the same side of the double bond. (As we shall see in Section 7.6A, this type of addition is called a syn addition.

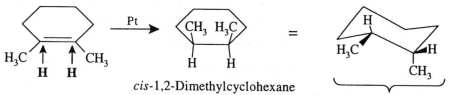

cis-1,2-Dimethylcyclohexane

The cis isomer is produced when both hydrogen atoms add from the same side.

4.39   (a) From Table 4.10 we find that this is *trans*-1,2-dichlorocyclohexane.

(b) Since the product is the trans isomer, we can conclude that the chlorine atoms have added from opposite sides of the double bond.

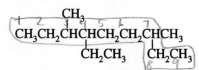

trans-1,2-Dichlorocyclohexane

The trans isomer is produced when the chlorine atoms add from opposite sides of the double bond.

4.40 If *trans*-1,3-di-*tert*-butylcyclohexane were to adopt a chair conformation, one *tert*-butyl group would have to be axial. It is, therefore, more energetically favorable for the molecule to adopt a twist boat conformation.

QUIZ

4.1 Consider the properties of the following compounds:

NAME	FORMULA	BOILING POINT (°C)	MOLECULAR WEIGHT
Ethane	CH_3CH_3	−88.2	30
Fluoromethane	CH_3F	−78.6	34
Methanol	CH_3OH	+64.7	32

Select the answer that explains why methanol boils so much higher than ethane or fluoromethane, even though they all have nearly equal molecular weights.

(a) Ion-ion forces between molecules.

(b) Weak dipole-dipole forces between molecules.

(c) Hydrogen bonding between molecules.

(d) van der Waals forces between molecules.

(e) Covalent bonding between molecules.

4.2 Select the correct name of the compound whose structure is

$$CH_3CH_2CHCHCH_2CH_2CHCH_3$$

(a) 2,5-Diethyl-6-methyloctane

(b) 4,7-Diethyl-3-methyloctane

(c)   4-Ethyl-3,7-dimethylnonane

(d)   6-Ethyl-3,7-dimethylnonane

(e)   More than one of the above

4.3   Select the correct name of the compound whose structure is $CH_3\overset{\overset{\displaystyle CH_3}{|}}{C}HCH_2Cl$

(a)   Butyl chloride

(b)   Isobutyl chloride

(c)   *sec*-Butyl chloride

(d)   *tert*-Butyl chloride

(e)   More than one of the above

4.4   The structure shown in Problem 4.2 has:

(a)   1°, 2°, and 3° carbon atoms

(b)   1° and 2° carbon atoms only

(c)   1° and 3° carbon atoms only

(d)   2° and 3° carbon atoms only

(e)   None of the above

4.5   How many isomers are possible for C_3H_7Br?

(a)   1   (b)   2   (c)   3   (d)   4   (e)   5

4.6   Which isomer of 1,3-dimethylcyclohexane is more stable?

(a)   cis   (b)   trans   (c)   Both are equally stable

(d)   Impossible to tell

(e)   More than one of the above

4.7   Which is the lowest energy conformation of *trans*- 1,4-dimethylcyclohexane?

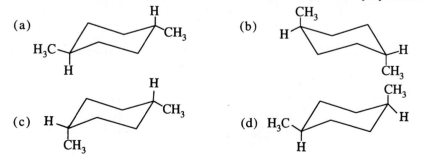

(e)   More than one of the above

4.8   Supply the missing structures

(a)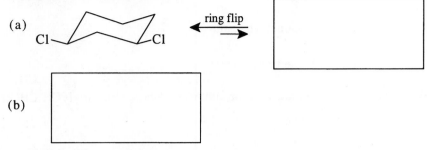

(b)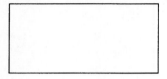

2-Bromobicyclo[2.2.1]heptane

(c)   Newman projection for a gauche form of 1,2-dibromoethane

4.9   Supply the missing reagents:

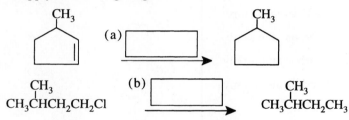

4.10   Complete the following synthesis:

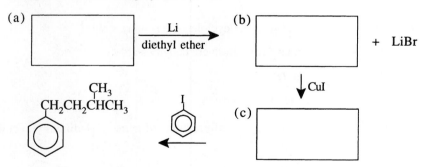

4.11   The most stable conformation of *trans*-1-isopropyl-3-methylcyclohexane:

5 STEREOCHEMISTRY: CHIRAL MOLECULES

SOLUTIONS TO PROBLEMS

5.1 (a) Achiral   (c) Chiral   (e) Chiral   (g) Chiral   (i) Chiral

(b) Achiral   (d) Chiral   (f) Chiral   (h) Achiral

5.2 (a) Yes   (b) No   (c) No

5.3 (a) They are the same molecule. (b) They are enantiomers.

5.4 (a), (b), and (f) do not have stereocenters.

(c)

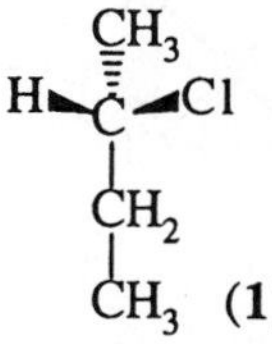

$$H\text{---}\underset{\underset{\underset{CH_3}{|}}{CH_2}}{\overset{\overset{CH_3}{\|}}{C}}\text{---}Cl \quad (1) \qquad Cl\text{---}\underset{\underset{\underset{CH_3}{|}}{CH_2}}{\overset{\overset{CH_3}{\|}}{C}}\text{---}H \quad (2)$$

(d)

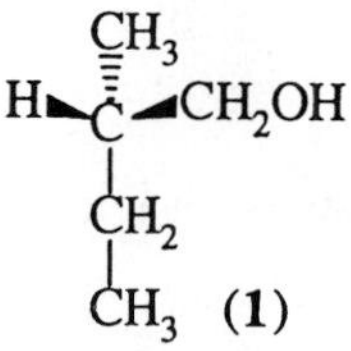

$$H\text{---}\underset{\underset{\underset{CH_3}{|}}{CH_2}}{\overset{\overset{CH_3}{\|}}{C}}\text{---}CH_2OH \quad (1) \qquad HOCH_2\text{---}\underset{\underset{\underset{CH_3}{|}}{CH_2}}{\overset{\overset{CH_3}{\|}}{C}}\text{---}H \quad (2)$$

(e)

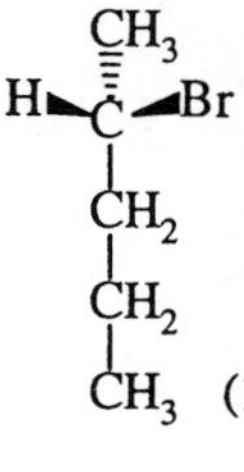

$$H\text{---}\underset{\underset{\underset{\underset{CH_3}{|}}{CH_2}}{CH_2}}{\overset{\overset{CH_3}{\|}}{C}}\text{---}Br \quad (1) \qquad Br\text{---}\underset{\underset{\underset{\underset{CH_3}{|}}{CH_2}}{CH_2}}{\overset{\overset{CH_3}{\|}}{C}}\text{---}H \quad (2)$$

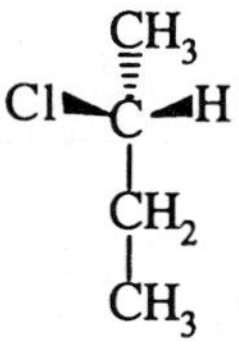

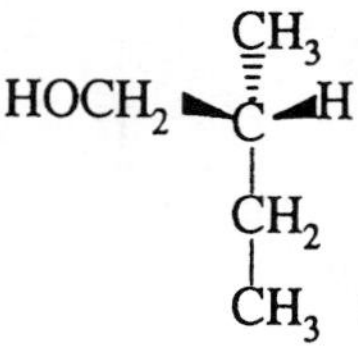

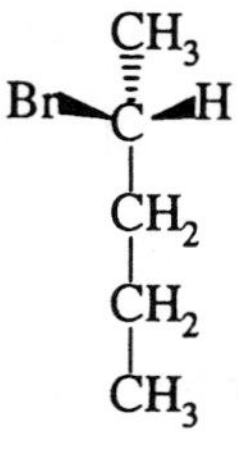

(g)

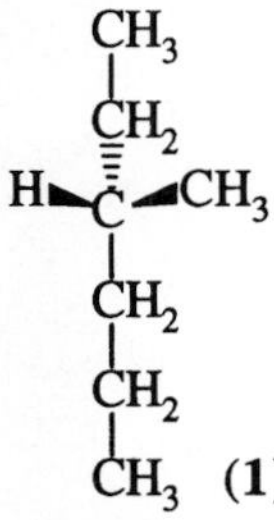

CH$_3$
CH$_2$
H—C—CH$_3$
CH$_2$
CH$_2$
CH$_3$ **(1)**

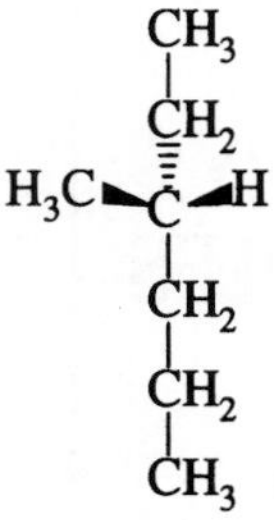

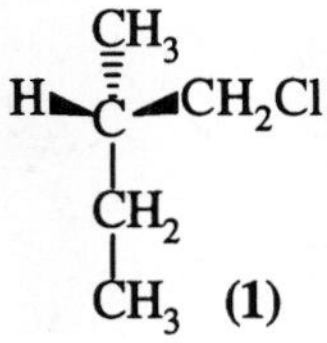

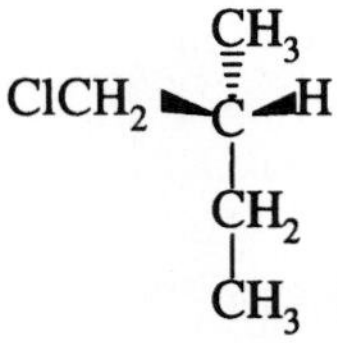

CH$_3$
CH$_2$
H$_3$C—C—H
CH$_2$
CH$_2$
CH$_3$ **(2)**

(h)

CH$_3$
H—C—CH$_2$Cl
CH$_2$
CH$_3$ **(1)**

CH$_3$
ClCH$_2$—C—H
CH$_2$
CH$_3$ **(2)**

5.5 (a)

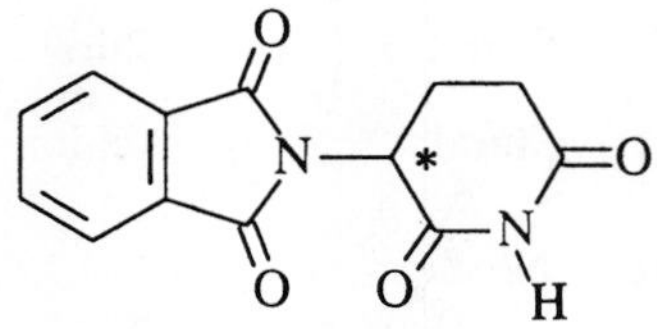

Thalidomide

(b)

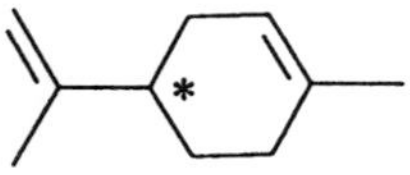

Limonene

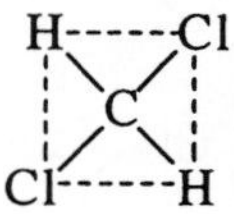

5.6 (a) Two in each instance.

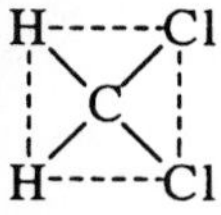

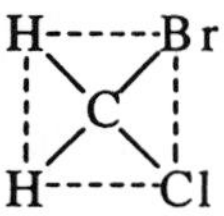

H-----Cl
C
H-----Cl

and

H-----Cl
C
Cl-----H

or

H-----Br
C
H-----Cl

and

H-----Br
C
Cl-----H

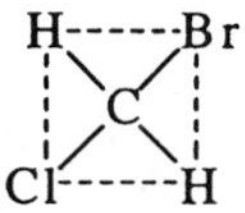

(b) Only one in each instance.

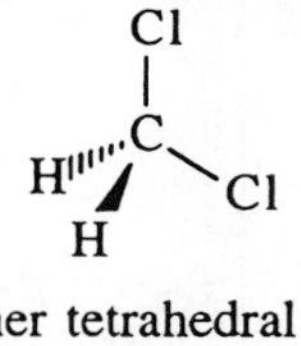

Cl
C
H''' Cl
H

or

Br
C
H''' Cl
H

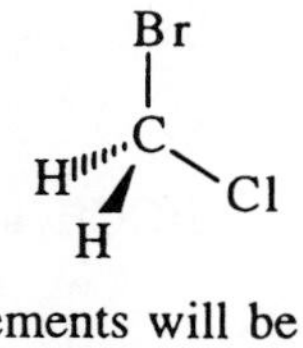

Any other tetrahedral arrangements will be superposable on one or the other of the above.

(c) Three:

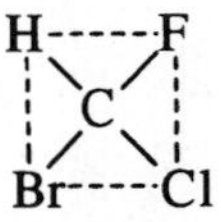

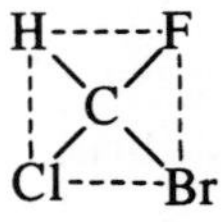

H-----F
C
Br-----Cl

H-----F
C
Cl-----Br

and

H-----Br
C
Cl-----F

(b) Two:

F
C
H''' Br
Cl

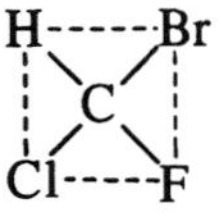

and

F
C
Br '''H
Cl

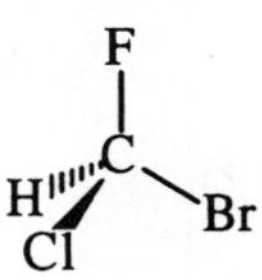

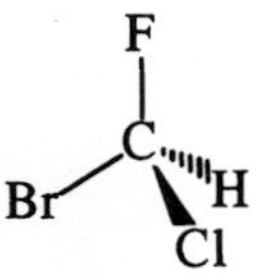

5.7 The following items possess a plane of symmetry, and are, therefore, achiral.

(a) A screwdriver

(b) A baseball bat (ignoring any writing on it)

(h) A hammer

5.8 In each instance below, the plane defined by the page is a plane of symmetry.

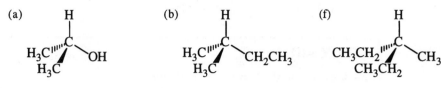

(a)

(b)

(f)

5.9

(S)   (R)

5.10 (c) (1) is (S)
(2) is (R)

(d) (1) is (S)
(2) is (R)

(e) (1) is (S)
(2) is (R)

(g) (1) is (S)
(2) is (R)

(h) (1) is (S)
(2) is (R)

5.11 (a) $-Cl$ > $-SH$ > $-OH$ > $-H$

(b) $-CH_2Br$ > $-CH_2Cl$ > $-CH_2OH$ > $-CH_3$

(c) $-OH$ > $-CHO$ > $-CH_3$ > $-H$

(d) $-C(CH_3)_3$ > $-CH=CH_2$ > $-CH(CH_3)_2$ > $-H$

(e) $-OCH_3$ > $-N(CH_3)_2$ > $-CH_3$ > $-H$

5.12 (a) (S)   (b) (R)   (c) (S)

5.13 (a) Enantiomers

(b) Two molecules of the same compound

(c) Enantiomers

5.14

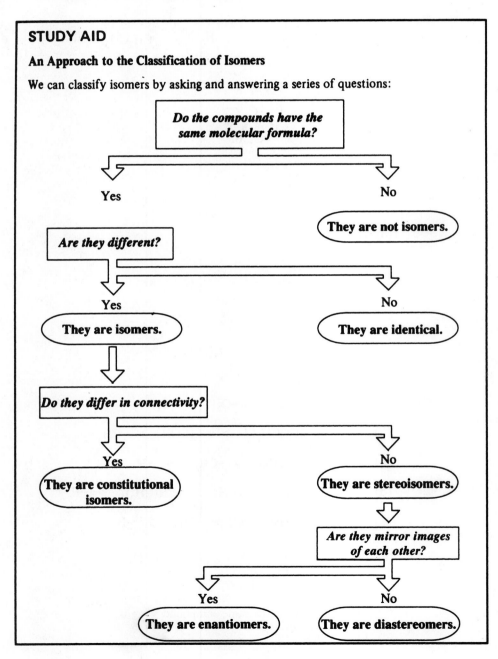

(*S*)-(+)-Carvone   (*R*)-(−)-Carvone

STUDY AID

An Approach to the Classification of Isomers

We can classify isomers by asking and answering a series of questions:

Do the compounds have the same molecular formula?

Yes

No

They are not isomers.

Are they different?

Yes

No

They are isomers.

They are identical.

Do they differ in connectivity?

Yes

No

They are constitutional isomers.

They are stereoisomers.

Are they mirror images of each other?

Yes

No

They are enantiomers.

They are diastereomers.

5.15 (a) Enant. Excess $=\dfrac{\text{observed rotation}}{\text{specific rotation of pure enantiomer}}\times 100$

$=\dfrac{+1.151°}{+5.756°}\times 100$

$= 20.00\%$

(b) since the (R) enantiomer (see Section 5.7C) is +, the (R) enantiomer is present in excess.

5.16 (a)

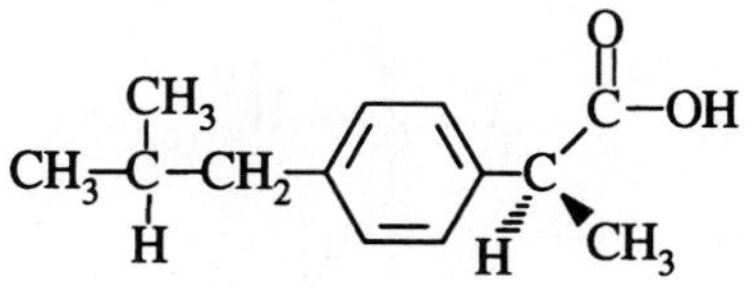

(S)-Ibuprofen

(b)

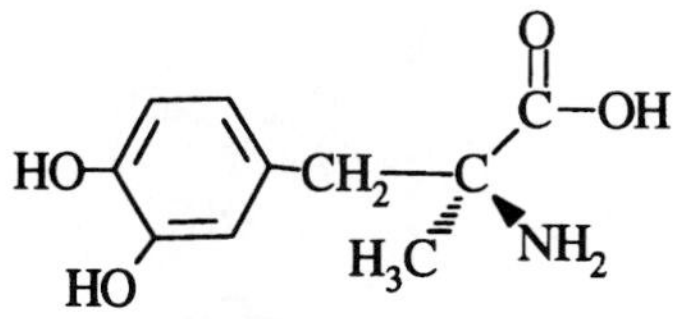

(S)-Methyldopa

(c)

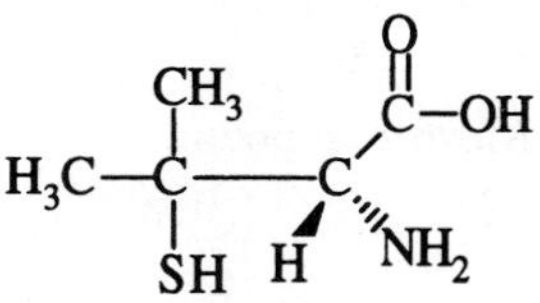

(S)-Penicillamine

5.17 (a) Diastereomers.

(b) Diastereomers in each instance.

(c) No, diastereomers have different melting points.

(d) No, diastereomers have different boiling points.

(e) No, diastereomers have different vapor pressures.

(f) Yes, because they have different vapor pressures.

5.18 (a) It would be optically active.

(b) It would be optically active.

(c) No, because it is a meso compound.

(d) No, because it would be a racemic form.

5.19 (a) Represents **A**

(b) Represents **C**

(c) Represents **B**

5.20 (a)

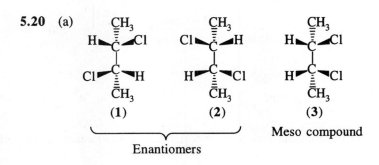

(1)   (2)   (3)

Meso compound

Enantiomers

(b)

CH_3
H—C—OH
CH_2
HO—C—H
CH_3
(1)

CH_3
HO—C—H
CH_2
H—C—OH
CH_3
(2)

CH_3
H—C—OH
CH_2
H—C—OH
CH_3
(3)

Meso compound

Enantiomers

(c)

CH_2Cl
H—C—F
F—C—H
CH_2Cl
(1)

CH_2Cl
F—C—H
H—C—F
CH_2Cl
(2)

CH_2Cl
H—C—F
H—C—F
CH_2Cl
(3)

Meso compound

Enantiomers

(d)

CH_3
H—C—OH
CH_2
Cl—C—H
CH_3
(1)

CH_3
HO—C—H
CH_2
H—C—Cl
CH_3
(2)

CH_3
H—C—OH
CH_2
H—C—Cl
CH_3
(3)

CH_3
HO—C—H
CH_2
Cl—C—H
CH_3
(4)

Enantiomers   Enantiomers

(e)

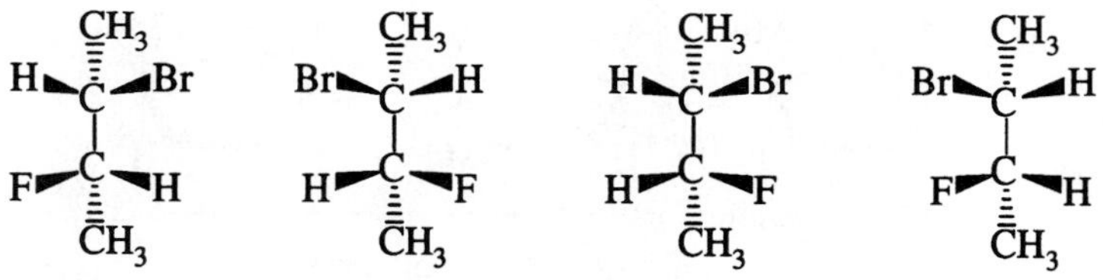

| (1) | (2) | (3) | (4) |

Enantiomers   Enantiomers

5.21 **B** is (2*S*,3*S*)-2,3-Dibromobutane
C is (2*R*,3*S*)-2,3-Dibromobutane

5.22 (a) (**1**) is (2*S*,3*S*)-2,3-Dichlorobutane
(**2**) is (2*R*,3*R*)-2,3-Dichlorobutane
(**3**) is (2*R*,3*S*)-2,3-Dichlorobutane

(b) (**1**) is (2*S*,4*S*)-2,4-Pentanediol
(**2**) is (2*R*,4*R*)-2,4-Pentanediol
(**3**) is (2*R*,4*S*)-2,4-Pentanediol

(c) (**1**) is (2*R*,3*R*)-1,4-Dichloro-2,3-difluorobutane
(**2**) is (2*S*,3*S*)-1,4-Dichloro-2,3-difluorobutane
(**3**) is (2*R*,3*S*)-1,4-Dichloro-2,3-difluorobutane

(d) (**1**) is (2*S*,4*S*)-4-Chloro-2-pentanol
(**2**) is (2*R*,4*R*)-4-Chloro-2-pentanol
(**3**) is (2*S*,4*R*)-4-Chloro-2-pentanol
(**4**) is (2*R*,4*S*)-4-Chloro-2-pentanol

(e) (**1**) is (2*S*,3*S*)-2-Bromo-3-fluorobutane
(**2**) is (2*R*,3*R*)-2-Bromo-3-fluorobutane
(**3**) is (2*S*,3*R*)-2-Bromo-3-fluorobutane
(**4**) is (2*R*,3*S*)-2-Bromo-3-fluorobutane

5.23

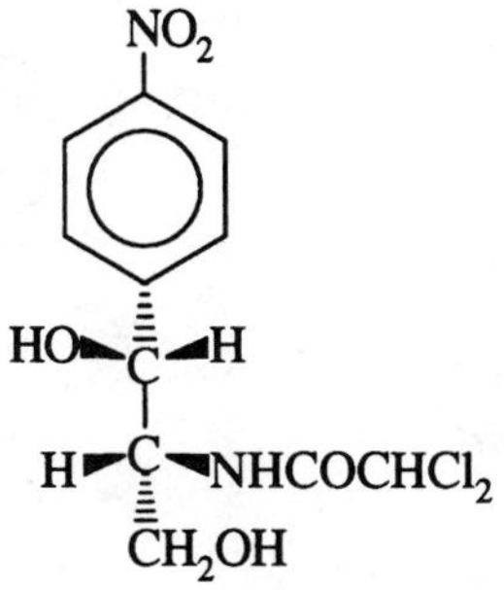

Chloramphenicol

5.24 (a) No   (b) Yes   (c) No   (d) No   (e) Diastereomers

(f) Diastereomers

5.25

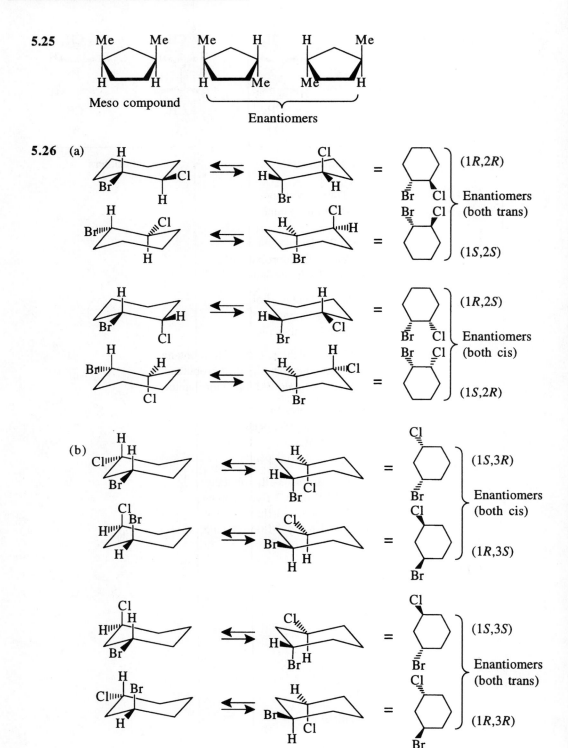

Meso compound

Enantiomers

(c)

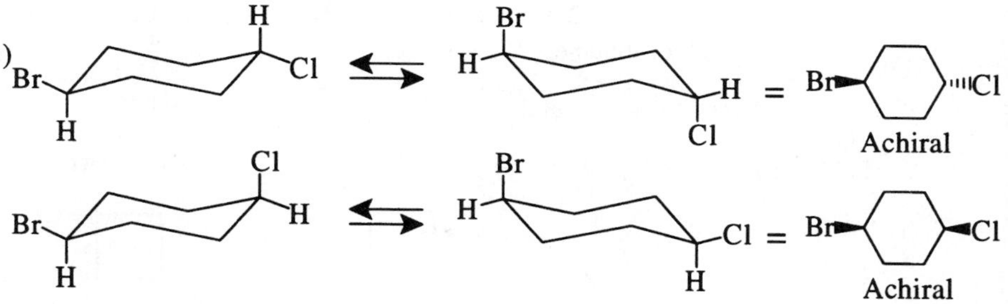
Br
H
Cl
Br
H
Cl
Achiral
Cl
Br
H
Br
H
Cl
Achiral

5.27 See Problem 5.26. The molecules in (c) are achiral, so they have no (*R*-*S*) designation.

5.28

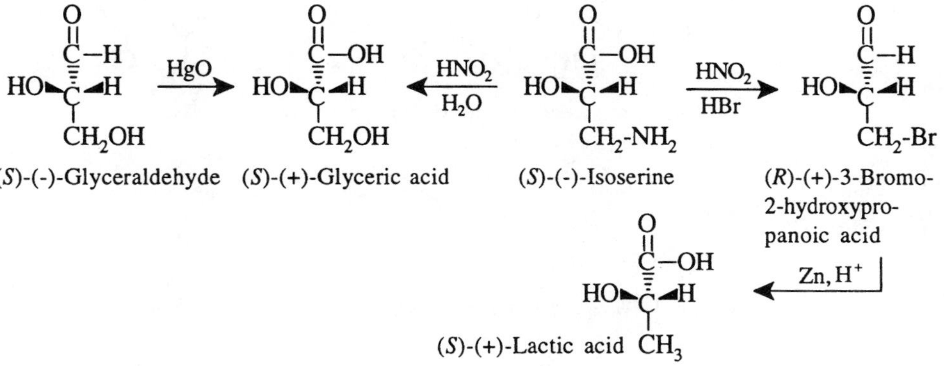
HgO
HNO2
H2O
HNO2
HBr
Zn, H+
(S)-(-)-Glyceraldehyde
(S)-(+)-Glyceric acid
(S)-(-)-Isoserine
(R)-(+)-3-Bromo-2-hydroxypropanoic acid
(S)-(+)-Lactic acid

5.29

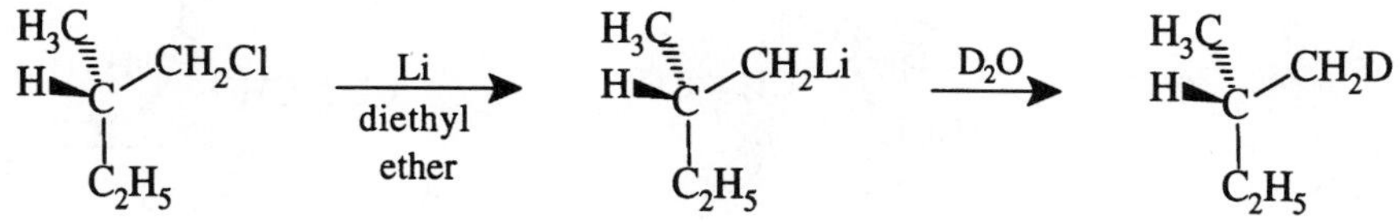
Li
diethyl
ether
D2O

(*S*)-(+)-1-Chloro-2-methylbutane (*R*)-1-Deuterio-2-methylbutane

5.30 (a), (b), (f) and (g) only

5.31 (a) Seven. Consider Table 4.2 and notice that all of the alkanes there (through six carbons) are achiral.

(b) (*R*)- and (*S*)-3-Methylhexane and (*R*)- and (*S*)-2,3-dimethylpentane.

5.32 (a) and (b)

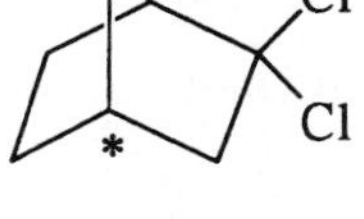
Cl
Cl

(c) Four

(d) Because a trans arrangement of the one carbon bridge is structurally impossible. Such a molecule would have too much strain.

5.33   (a) **A** is (*R*,*S*)-2,3-dichlorobutane; **B** is (*S*,*S*)-2,3-dichlorobutane; **C** is (*R*,*R*)-2,3-dichlorobutane.

(b) **A**

5.34   (a) 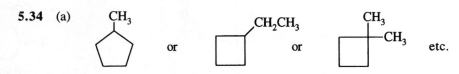   etc.

(b)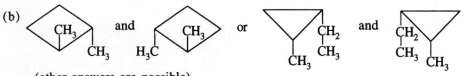

(other answers are possible)

(c) ![structure] and ![structure]

(other answers are possible)

(d) ![structure] and ![structure]

(e) ![structure] and ![structure]

(other answers are possible)

5.35   (a) Same   (j) Enantiomers that are interconvertible by a ring flip

(b) Enantiomers   (k) Diastereomers

(c) Diastereomers   (l) Same

(d) Same   (m) Diastereomers

(e) Same   (n) Constitutional isomers

(f) Constitutional isomers (o) Diastereomers

(g) Diastereomers   (p) Same

(h) Enantiomers   (q) Same

(i) Same

5.36 All of these molecules are expected to be planar. Their stereochemistry is identical to that of the corresponding chloroethenes. (a) can exist as cis and trans isomers. Only one compound exists in the case of (b) and (c).

5.37 (a)

(1)   (2)   (3)   (4)

(b) (3) and (4) are chiral and are enantiomers of each other.

(c) Three fractions: a fraction containing (1), a fraction containing (2), and a fraction containing (3) and (4) [because, being enantiomers, (3) and (4) would have the same vapor pressure].

(d) None

5.38 (a)

(b) No, they are not superposable.

(c) No, and they are, therefore, enantiomers of each other.

(d)

(e) No, they are not superposable.

(f) Yes, and they are, therefore, just different conformations of the same molecule.

5.39 (a)

(b) Yes, and therefore *trans*-1.4-diethylcyclohexane is achiral.
(c) No, they are different orientations of the same molecule.
(d) Yes, *cis*-1,4-diethylcyclohexane is a stereoisomer (a diastereomer) of *trans*-1.4-diethylcyclohexane.

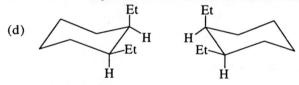

cis-1,4-Diethylcyclohexane

(e) No, it, too, is superposable on its mirror image. (Notice, too, that the plane of the page constitutes a plane of symmetry for both *cis*-1,4-diethylcyclohexane and for *trans*-1.4-diethylcyclohexane as we have drawn them.)

5.40 *trans*-1,3-Diethylcyclohexane can exist in the following enantiomeric forms.

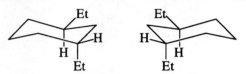

trans-1,3-Diethylcyclohexane enantiomers

cis-1,3-Diethylcyclohexane consists of achiral molecules because they have a plane of symmetry. [The plane of the page (below) is a plane of symmetry.]

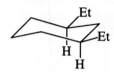

cis-1,3-Diethylcyclohexane

5.41 (a) Since it is optically active and not resolvable, it must be the meso form:

CO$_2$H
H—C—OH
H—C—OH
CO$_2$H
(meso)

(b)

CO$_2$H
H—C—OH
HO—C—H
CO$_2$H

CO$_2$H
HO—C—H
H—C—OH
CO$_2$H

(c) No   (d) A racemic modification

5.42

H$_3$C
H—C—CH$_2$Cl
C$_2$H$_5$
$\xrightarrow[\text{ether}]{\text{Li} \atop \text{diethyl}}$
H$_3$C
H—C—CH$_2$Li
C$_2$H$_5$
$\xrightarrow{\text{CuI}}$
$\left(\begin{array}{c} \text{H}_3\text{C} \\ \text{H—C—CH}_2 \\ \text{C}_2\text{H}_5 \end{array} \right)_2 \text{CuLi}$

(b) Yes, because no bonds to the stereocenter are broken.

(c) (3S,6S)-3,6-Dimethyloctane, because no bonds to either stereocenter are broken.

(d) Yes, the molecules are chiral and the product is not a racemate.

(e) (3R,6S)-3,6-Dimethyloctane.

(f) No, this is a meso compound.

(g) Start with (R)-(-)-1-chloro-2-methylbutane and convert it to a lithium dialkylcuprate. Then allow this lithium dialkylcuprate to react with (R)-(-)-1-chloro-2-methylbutane.

QUIZ

5.1 Describe the relationship between the two structures shown.

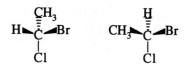

(a) Enantiomers   (b) Diastereomers   (c) Constitutional isomers

(d) Conformations   (e) Two molecules of the same compound

5.2 Which of the following molecule(s) possess(es) a plane of symmetry?

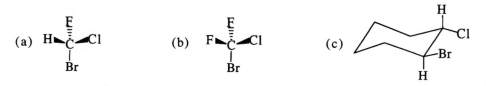

(d) More than one of these   (e) None of these

5.3 Give the (R-S) designation of the structure shown.

$$HO-\overset{\overset{O}{\|}}{C}\longrightarrow\overset{\overset{CH_3}{\vdots}}{\underset{Cl}{C}}\longrightarrow H$$

(a) (R)   (b) (S)   (c) Neither, because this molecule has no stereocenter

(d) Impossible to tell

5.4 Select the words that best describe the following structure:

$$H\longrightarrow\overset{\overset{CH_3}{\vdots}}{C}\longrightarrow Cl$$
$$H\longrightarrow\underset{\underset{CH_3}{\vdots}}{C}\longrightarrow Cl$$

(a) Chiral   (b) Meso form   (c) Achiral   (d) Has a plane of symmetry

(e) More than one of these

5.5 Select the words that best describe what happens to the optical rotation of the alkene shown when it is hydrogenated to the alkane according to the following equation:

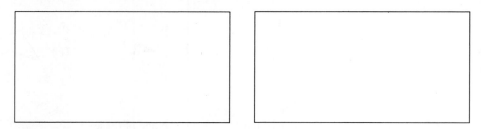

(a) Increases   (b) Drops to zero   (c) Changes sign

(d) Stays the same   (e) Impossible to predict

5.6 There are two compounds with the formula C_7H_{16} that are capable of existing as enantiomers. Write three-dimensional formulas for the (*S*) isomer of each.

5.7 Compound A is optically active and is the (*S*) isomer.

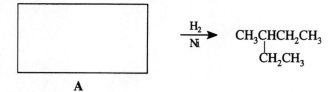

A

5.8 Give a three-dimensional formula for the product **B**.

(*S*)-1-Bromo-2-methylbutane + $(CH_3CH_2)_2CuLi$ —ether→

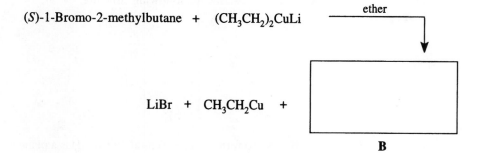

LiBr + CH_3CH_2Cu +

B

6

IONIC REACTIONS—
NUCLEOPHILIC SUBSTITUTION AND
ELIMINATION REACTIONS
OF ALKYL HALIDES

SOLUTIONS TO PROBLEMS

6.1 (a) $CH_3-\ddot{I}:$ + $CH_3CH_2-\ddot{O}:^-$ $\longrightarrow$ $CH_3-\ddot{O}-CH_2CH_3$ + $:\ddot{I}:^-$

Substrate   Nucleophile   Leaving group

(b) $:\ddot{I}:^-$ + $CH_3CH_2-\ddot{B}r:$ $\longrightarrow$ $CH_3CH_2-\ddot{I}:$ + $:\ddot{B}r:^-$

Nucleo-phile   Substrate   Leaving group

(c) $CH_3\ddot{O}H$ $(CH_3)_3C-\ddot{C}l:$ $\longrightarrow$ $(CH_3)_3C-\ddot{O}-CH_3$ + $:\ddot{C}l:^-$ + H^+

Nucleo-phile   Substrate   Leaving group

(d) $CH_3CH_2CH_2-\ddot{B}r:$ + $^-:C\equiv N:$ $\longrightarrow$ $CH_3CH_2CH_2-CN$ + $:\ddot{B}r:^-$

Substrate   Nucleo-phile   Leaving group

(e) $C_6H_5CH_2-\ddot{B}r:$ + $:NH_3$ $\longrightarrow$ $C_6H_5CH_2-\ddot{N}H_2$ + $:\ddot{B}r:^-$ + H^+

Substrate   Nucleo-phile   Leaving group

6.2

6.3 (a)   We know that when a secondary alkyl halide reacts with hydroxide ion by substitution, the reaction occurs with *inversion of configuration* because the reaction is S_N2. If we know that the configuration of (−)-2-butanol (from Section 5.7C) is that shown here, then we can conclude that (+)-2-chlorobutane has the opposite configuration.

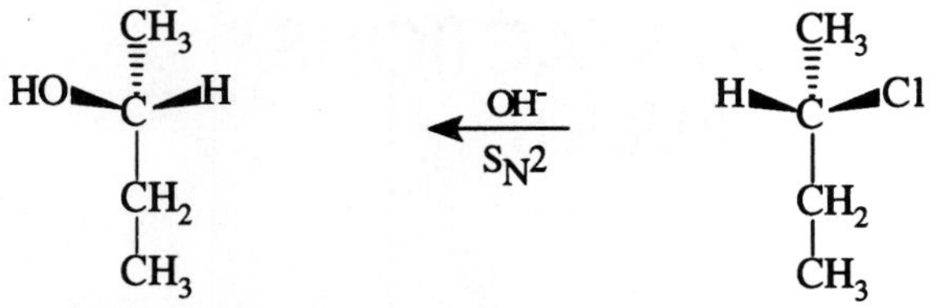

(*R*)-(-)-2-Butanol
$[\alpha]_D^{25°} = -13.52°$

(*S*)-(+)-2-Chlorobutane
$[\alpha]_D^{25°} = +36.00°$

(b)   Again the reaction is S_N2. Because we now know the configuration of (+)-2-chlorobutane to be (*S*) [cf., part (a)], we can conclude that the configuration of (−)-2-iodobutane is (*R*).

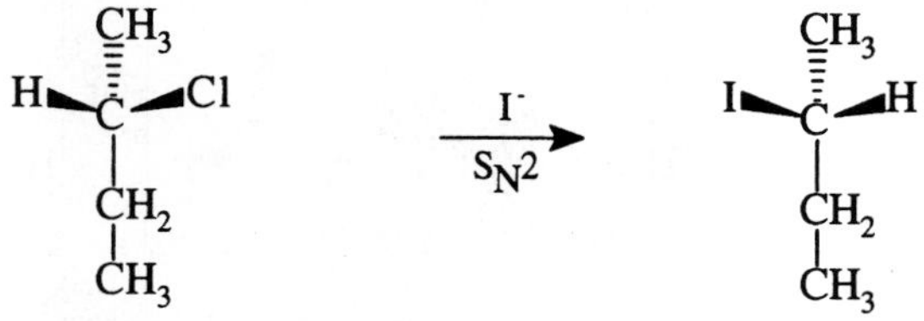

(*S*)-(+)-2-Chlorobutane   (*R*)-(-)-2-Iodobutane

(+)-2-Iodobutane has the (*S*) configuration

6.4   (a)   (b)

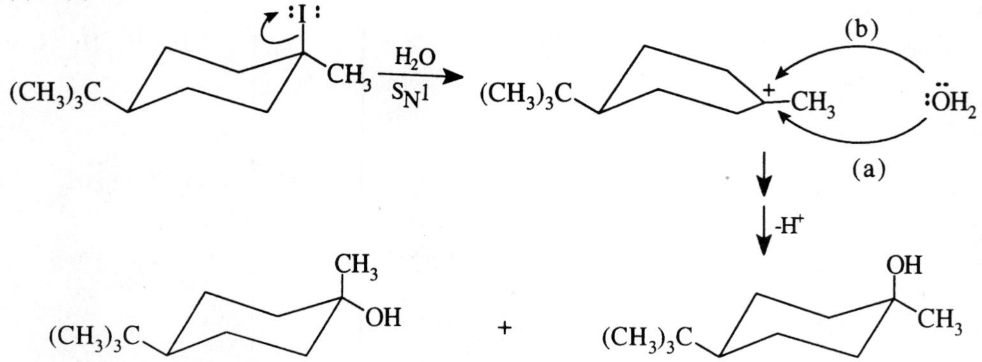

By path (a)   +   By path (b)

6.5

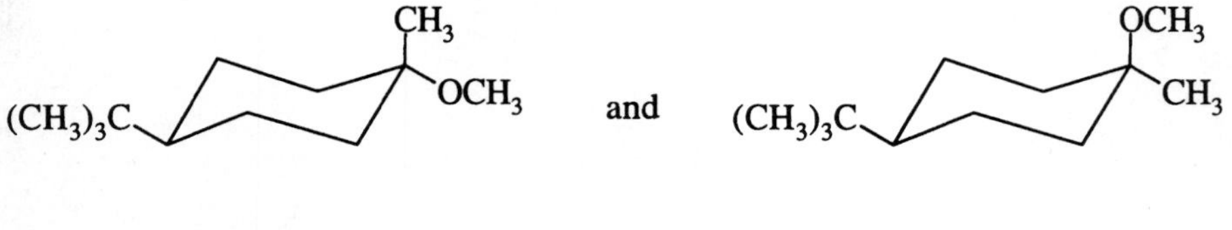

$(CH_3)_3C$ [structure with OCH₃]   and $(CH_3)_3C$ [structure with OCH₃]

6.6 (a) Being primary halides, the reactions are most likely to be S_N2, with the nucleophile in each instance being a molecule of the solvent (i.e., a molecule of ethanol).

(b) Steric hindrance provided by the substituent or substituents on the carbon β to the carbon bearing the leaving group. With each addition of a methyl group at the β carbon (below), the number of pathways open to the attacking nucleophile becomes fewer.

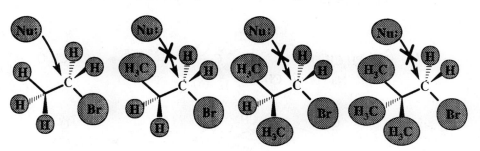

6.7 **Protic solvents** are those that have an H bonded to an oxygen or nitrogen (or to another strongly electronegative atom). Therefore, the protic solvents are formic acid, $H\overset{\overset{\displaystyle O}{\|}}{C}OH$;

formamide, $H\overset{\overset{\displaystyle O}{\|}}{C}NH_2$; ammonia, NH_3, and ethylene glycol, $HOCH_2CH_2OH$.

Aprotic solvents lack an H bonded to a strongly electronegative element. Aprotic solvents in this list are acetone, $CH_3\overset{\overset{\displaystyle O}{\|}}{C}CH_3$; acetonitrile $CH_3C\equiv N$; sulfur dioxide, SO_2; and trimethylamine, $N(CH_3)_3$.

6.8 The reaction is an S_N2 reaction. In the polar aprotic solvent (DMF), the nucleophile (CN^-) will be relatively unencumbered by solvent molecules, and, therefore, it will be more reactive than in ethanol. As a result, the reaction will occur faster in *N,N*-dimethylformamide.

6.9 (a) CH_3O^-

(b) H_2S

(c) $(CH_3)_3P$

6.10 (a) Increasing the percentage of water in the mixture increases the polarity of the solvent. (Water is more polar than methanol.) Increasing the polarity of the solvent increases the rate of the solvolysis because separated charges develop in the transition state. The more polar the solvent, the more the transition state is stabilized (Section 6.15D).

(b) In an S_N2 reaction of this type, the charge becomes dispersed in the transition state:

$$I^- + CH_3CH_2{-}Cl \longrightarrow \left[\overset{\delta-}{I}{-----}\overset{\overset{\displaystyle CH_3}{|}}{\underset{}{C}H_2}{----}\overset{\delta-}{Cl} \right] \longrightarrow ICH_2CH_3 + Cl^-$$

Reactants   Transition state

Charge is concentrated   *Charge is dispersed*

Increasing the polarity of the solvent increases the stabilization of the reactant I$^-$ more than the stabilization of the transition state, and thereby increases the free energy of activation, thus decreasing the rate of reaction.

6.11 $CH_3OSO_2CF_3$ > CH_3I > CH_3Br > CH_3Cl > CH_3F > $^{14}CH_3OH$

(Most reactive)   (Least reactive)

6.12 (a)

(b)

(c)

(d)

6.13 (a) $CH_3CH_2CH_2Br$ + $NaOH$ $\longrightarrow$ $CH_3CH_2CH_2OH$ + $NaBr$

(b) $CH_3CH_2CH_2Br$ + NaI $\longrightarrow$ $CH_3CH_2CH_2I$ + $NaBr$

(c) $CH_3CH_2CH_2Br$ + CH_3CH_2ONa $\longrightarrow$ $CH_3CH_2CH_2OCH_2CH_3$ + $NaBr$

(d) $CH_3CH_2CH_2Br$ + CH_3SNa $\longrightarrow$ $CH_3CH_2CH_2SCH_3$ + $NaBr$

(e) $CH_3CH_2CH_2Br$ + $CH_3\overset{O}{\overset{\|}{C}}-ONa$ $\longrightarrow$ $CH_3CH_2CH_2O\overset{O}{\overset{\|}{C}}CH_3$ + $NaBr$

(f) $CH_3CH_2CH_2Br$ + NaN_3 $\longrightarrow$ $CH_3CH_2CH_2N_3$ + $NaBr$

(g) $CH_3CH_2CH_2Br$ + $:N(CH_3)_3$ $\longrightarrow$ $CH_3CH_2CH_2-\overset{CH_3}{\underset{CH_3}{N^{+}}}CH_3$ Br$^-$

(h) $CH_3CH_2CH_2Br$ + $NaCN$ $\longrightarrow$ $CH_3CH_2CH_2CN$ + $NaBr$

(i) $CH_3CH_2CH_2Br$ + $NaSH$ $\longrightarrow$ $CH_3CH_2CH_2SH$ + $NaBr$

6.14 (a) 1-Bromopropane would react more rapidly because, being a primary halide, it is less hindered.

(b) 1-Iodobutane, because iodide ion is a better leaving group than chloride ion.

(c) 1-Chlorobutane, because the carbon bearing the leaving group is less hindered than in 1-chloro-2-methylpropane.

(d) 1-Chloro-3-methylbutane, because the carbon bearing the leaving group is less hindered than in 1-chloro-2-methylbutane.

(e) 1-Chlorohexane because it is a primary halide. Phenyl halides are unreactive in S_N2 reactions.

6.15 (a) Reaction (1) because ethoxide ion is a stronger nucleophile than ethanol.

(b) Reaction (2) because the ethyl sulfide ion is a stronger nucleophile than the ethoxide ion in a protic solvent. (Because sulfur is larger than oxygen, the ethyl sulfide ion is less solvated and it is more polarizable.)

(c) Reaction (2) because triphenylphosphine $[(C_6H_5)_3P]$ is a stronger nucleophile than triphenylamine. (Phosphorus atoms are larger than nitrogen atoms.)

(d) Reaction (2) because in an S_N2 reaction the rate depends on the concentration of the substrate and the nucleophile. In reaction (2) the concentration of the nucleophile is twice that of the reaction (1).

6.16 (a) Reaction (2) because bromide ion is a better leaving group than chloride ion.

(b) Reaction (1) because water is a more polar solvent than methanol, and S_N1 reactions take place faster in more polar solvents.

(c) Reaction (2) because the concentration of the substrate is twice that of reaction (1). (The major reaction would be E2. However, the problem asks us to consider that small portion of the overall reaction that proceeds by an S_N1 pathway.)

(d) Both reactions would take place at the same rate because, being S_N1 reactions, they are independent of the concentration of the nucleophile.
(The major reaction would be E2. However, the problem asks us to consider that small portion of the overall reaction that proceeds by an S_N1 pathway.)

(e) Reaction (1) because the substrate is a tertiary halide. Phenyl halides are unreactive in S_N1 reactions.

6.17 Possible methods are given here.

(a) $CH_3Cl \xrightarrow[\substack{CH_3OH \\ S_N2}]{I^-} CH_3I$

(b) $CH_3CH_2Cl \xrightarrow[\substack{CH_3OH \\ S_N2}]{I^-} CH_3CH_2I$

(c) $CH_3Cl \xrightarrow[\substack{CH_3OH/H_2O \\ S_N2}]{OH^-} CH_3OH$

(d) $CH_3CH_2Cl \xrightarrow[\substack{CH_3OH/H_2O \\ S_N2}]{OH^-} CH_3CH_2OH$

(e) $CH_3Cl \xrightarrow[\substack{CH_3OH \\ S_N2}]{SH^-} CH_3SH$

(f) $CH_3CH_2Cl \xrightarrow[\substack{CH_3OH \\ S_N2}]{SH^-} CH_3CH_2SH$

(g) $CH_3Cl \xrightarrow[DMF]{CN^-} CH_3CN$

(h) $CH_3CH_2Cl \xrightarrow[DMF]{CN^-} CH_3CH_2CN$

(i) $CH_3OH \xrightarrow[(-H_2)]{NaH} CH_3ONa \xrightarrow[CH_3OH]{CH_3I} CH_3OCH_3$

(j) $CH_3CH_2OH \xrightarrow[(-H_2)]{NaH} CH_3CH_2ONa \xrightarrow{CH_3I} CH_3CH_2OCH_3$

(k)

6.18 (a) The reaction will not take place because the leaving group would have to be a methyl anion, a very powerful base, and a very poor leaving group.

(b) The reaction will not take place because the leaving group would have to be a hydride ion, a very powerful base, and a very poor leaving group.

(c) The reaction will not take place because the leaving group would have to be a carbanion, a very powerful base, and a very poor leaving group.

(d) The reaction will not take place by an S_N2 mechanism because the substrate is a tertiary halide, and is, therefore, not susceptible to S_N2 attack because of the steric hindrance. (A very small amount of S_N1 reaction may take place, but the main reaction will be E2 to produce an alkene.)

(e) The reaction will not take place because the leaving group would have to be a CH_3O^- ion, a strong base, and a very poor leaving group.

(f) The reaction will not take place because the first reaction that would take place would be an acid-base reaction that would convert the ammonia to an ammonium ion. An ammonium ion, because it lacks an electron pair, is not nucleophilic.

$$NH_3 + CH_3\overset{+}{O}H_2 \rightleftarrows NH_4^+ + CH_3OH$$

6.19   The better yield will be obtained by using the secondary halide, 1-bromo-1-phenylethane, because the desired reaction is E2. Using the primary halide will result in substantial S_N2 reaction as well, producing the alcohol instead of the desired alkene.

6.20   Reaction (2) would give the better yield because the desired reaction is an S_N2 reaction, and the substrate is a methyl halide. Use of reaction (1) would, because the substrate is a secondary halide, result in considerable elimination by an E2 pathway.

6.21   (a)   The major product would be $CH_3CH_2CH_2CH_2CH_2OCH_2CH_3$ (by an S_N2 mechanism) because the substrate is primary and the nucleophile-base is not hindered. Some $CH_3CH_2CH_2CH=CH_2$ would be produced by an E2 mechanism.

(b)   The major product would be $CH_3CH_2CH_2CH=CH_2$ (by an E2 mechanism), even though the substrate is primary because the base is a hindered strong base. Some $CH_3CH_2CH_2CH_2CH_2OC(CH_3)_3$ would be produced by an S_N2 mechanism.

(c)   For all practical purposes, $(CH_3)_2C=CH_2$ (by an E2 mechanism) would be the only product because the substrate is tertiary and the base is strong.

(d)   Same answer as (c) above.

(e)   (formed by an S_N2 mechanism) would, for all practical purposes, be the only product. Iodide ion is a very weak base and a good nucleophile.

(f)   Because the substrate is tertiary and the only nucleophile is the solvent, the mechanism is E1. The two products that follow would be formed.

(g) $CH_3CH=CHCH_2CH_3$ (by an E2 mechanism) would be the major product because the substrate is secondary and the base/nucleophile is a strong base. Some of the ether, $CH_3CH_2CH(OCH_3)CH_2CH_3$, would be formed by an S_N2 pathway.

(h)   The major product would be $CH_3CH_2CH(O_2CCH_3)CH_2CH_3$ (by an S_N2 mechanism) because the acetate ion is a weak base. Some $CH_3CH=CHCH_2CH_2CH_3$ might be formed by an E2 pathway.

(i) $CH_3CH=CHCH_3$ and $CH_2=CHCH_2CH_3$ (by E2) would be major products, and (S)-$CH_3CH(OH)CH_2CH_3$ (by S_N2) would be the minor product.

(j) $(\pm)$-$CH_3CH_2\underset{\underset{OCH_3}{|}}{C}(CH_3)CH_2CH_2CH_3$ (by S_N1) would be the major product.

$CH_3CH_2\underset{\underset{CH_3}{|}}{C}=CHCH_2CH_3$, $CH_3CH=\underset{\underset{CH_3}{|}}{C}CH_2CH_2CH_3$, and

$CH_3CH_2\overset{\overset{CH_2}{||}}{C}CH_2CH_2CH_3$ (by E1) would be minor products.

(k) (R)-$CH_3CHIC_6H_{13}$ (by S_N2) would be the only product.

6.22 (a), (b), and (c) are all S_N2 reactions and, therefore, proceed with inversion of configuration. The products are

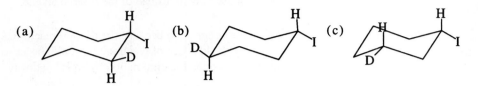

(a) (b) (c)

(d) is an S_N1 reaction. The carbocation that forms can react with either nucleophile (H_2O or CH_3OH) from either the top or bottom side of the molecule. Four substitution products (below) would be obtained. (Considerable elimination by an E1 path would also occur.)

and

and

6.23 Isobutyl bromide is more sterically hindered than ethyl bromide because of the methyl groups on the β carbon atom.

$$H_3C \overset{\beta}{\underset{H_3C}{\diagdown}} \overset{\alpha}{\underset{}{C}} - CH_2 - Br \qquad H \overset{}{\underset{H}{\diagdown}} C - CH_2 - Br$$

Isobutyl bromide Ethyl bromide

This steric hindrance causes isobutyl bromide to react more slowly in S_N2 reactions and to give relatively more elimination (by an E2 path) when a strong base is used.

6.24 (a) S_N2 because the substrate is a 1° halide.

(b) Rate = k [CH_3CH_2Cl][I^-]

$= 5 \times 10^{-5}\,\text{L mol}^{-1}\text{s}^{-1} \times 0.1\,\text{mol L}^{-1} \times 0.1\,\text{mol L}^{-1}$

Rate = $5 \times 10^{-7}\,\text{mol L}^{-1}\text{s}^{-1}$

(c) $1 \times 10^{-6}\,\text{mol L}^{-1}\text{s}^{-1}$

(d) $1 \times 10^{-6}\,\text{mol L}^{-1}\text{s}^{-1}$

(e) $2 \times 10^{-6}\,\text{mol L}^{-1}\text{s}^{-1}$

6.25 (a) CH_3NH^- because it is the stronger base.

(b) CH_3O^- because it is the stronger base.

(c) CH_3SH because sulfur atoms are larger and more polarizable than oxygen atoms.

(d) $(C_6H_5)_3P$ because phosphorus atoms are larger and more polarizable than nitrogen atoms.

(e) H_2O because it is the stronger base.

(f) NH_3 because it is the stronger base.

(g) HS^- because it is the stronger base.

(h) OH^- because it is the stronger base.

6.26 (a) $HOCH_2CH_2Br$ + OH^-

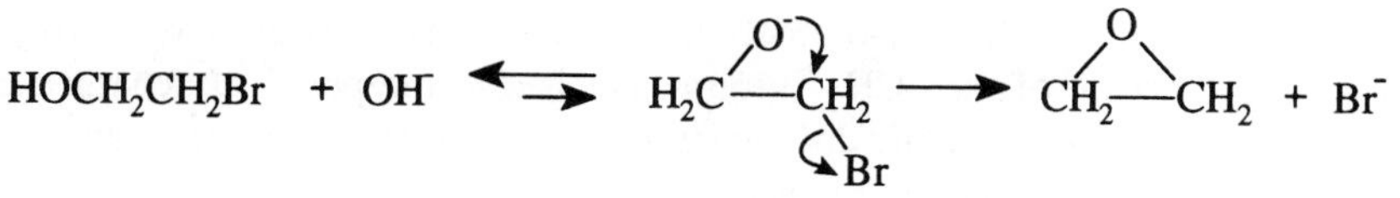

(b)

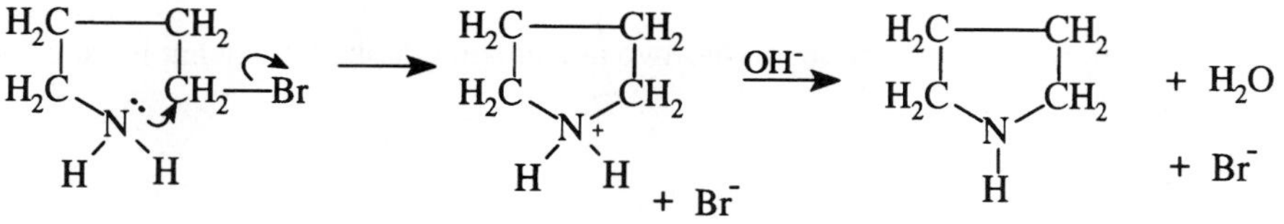

6.27   Iodide ion is a good nucleophile and a good leaving group; it can rapidly convert an alkyl chloride or alkyl bromide into an alkyl iodide, and the alkyl iodide can then react rapidly with another nucleophile. With methyl bromide in water, for example, the following reaction can take place:

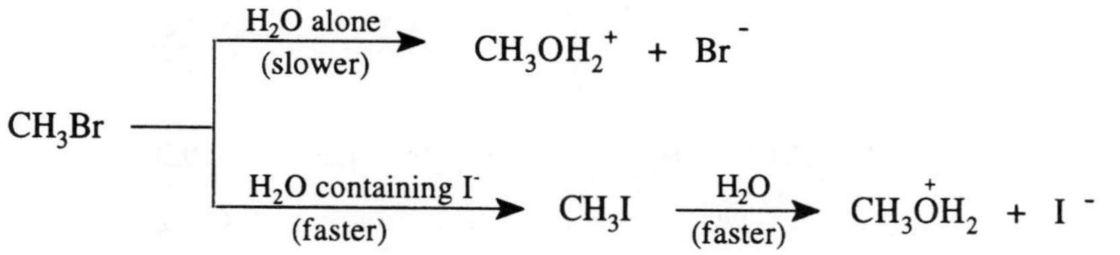

6.28   *tert*-Butyl alcohol and *tert*-butyl methyl ether are formed via an S_N1 mechanism. The rate of the reaction is independent of the concentration of methoxide ion (from sodium methoxide). This, however, is only one reaction that causes *tert*-butyl bromide to disappear. A competing reaction that also causes *tert*-butyl bromide to disappear is an E2 reaction in which methoxide ion reacts with *tert*-butyl bromide. This reaction is dependent on the concentration of methoxide ion; therefore, increasing the methoxide ion concentration causes an increase in the rate of disappearance of *tert*-butyl bromide.

6.29 (a)   You should use a strong base, such as RO^-, at a higher temperature to bring about an E2 reaction.

(b)   Here we want an S_N1 reaction. We use ethanol as the solvent *and as the nucleophile,* and we carry out the reaction at a low temperature so that elimination will be minimized.

6.30   1-Bromobicyclo[2.2.1]heptane is unreactive in an S_N2 reaction because it is a tertiary halide and its ring structure makes the backside of the carbon bearing the leaving group completely inaccessible to attack by a nucleophile.

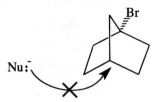

1-Bromobicyclo[2.2.1]heptane is unreactive in an S_N1 reaction because the ring structure makes it impossible for the carbocation that must be formed to assume the required trigonal planar geometry around the positively charged carbon. Any carbocation formed from 1-bromobicyclo[2.2.1]heptane would have a trigonal pyramidal arrangement of the $-CH_2-$ groups attached to the positively charged carbon (make a model). Such a structure does not allow stabilization of the carbocation by overlap of sp^3 orbitals from the alkyl groups (see Fig. 6.10).

6.31   The cyanide ion has two nucleophilic atoms; it is what is called an ambident nucleophile.

$$^-:C{\equiv}N:$$

It can react with a substrate using either atom, although the carbon atom is more nucleophilic.

$$Br{-}CH_2CH_3 \ + \ ^-:C{\equiv}N: \ \longrightarrow \ CH_3CH_2{-}C{\equiv}N:$$

$$^-:C{\equiv}N: \ + \ CH_3CH_2{-}Br \ \longrightarrow \ CH_3CH_2{-}\ddot{N}{=}C:$$

6.32   (a) $\underset{\overset{|}{OH}}{CH_3CH_2CHCH_3} \ \xrightarrow[\text{ether }(-H_2)]{\text{NaH}} \ \underset{\overset{|}{O^-Na^+}}{CH_3CH_2CHCH_3}$

$\underset{\overset{|}{OCH_2CH_2CH_2CH_3}}{CH_3CH_2CHCH_3} \ \xleftarrow{\text{(-NaBr)}} \ CH_3CH_2CH_2CH_2{-}Br$

(b) $(CH_3)_3CSH \ \xrightarrow[\substack{\text{ether} \\ (-H_2)}]{\text{NaH}} \ (CH_3)_3CS^-Na^+ \ \xrightarrow[\text{(-NaBr)}]{CH_3CH_2{-}Br} \ (CH_3)_3C{-}S{-}CH_2CH_3$

(c) $\underset{\overset{|}{CH_3}}{\overset{\overset{CH_3}{|}}{CH_3{-}C{-}CH_2OH}} \ \xrightarrow[\substack{\text{ether} \\ (-H_2)}]{\text{NaH}} \ \underset{\overset{|}{CH_3}}{\overset{\overset{CH_3}{|}}{CH_3{-}C{-}CH_2O^-Na^+}} \ \xrightarrow[\text{(-NaI)}]{CH_3I}$

$$\underset{\overset{|}{CH_3}}{\overset{\overset{CH_3}{|}}{CH_3{-}C{-}CH_2{-}OCH_3}}$$

(d) C_6H_5OH $\xrightarrow[\text{ether}\ (-H_2)]{\text{NaH}}$ $C_6H_5O^-Na^+$ $\xrightarrow[(-NaI)]{CH_3I}$ $C_6H_5OCH_3$

(e) $C_6H_5CH_2-Br + Na^{+-}:C\equiv N: \xrightarrow[(-NaBr)]{EtOH} C_6H_5CH_2-C\equiv N:$

(f) $CH_3\overset{O}{\overset{\|}{C}}-O^-Na^+ + C_6H_5CH_2-Br \xrightarrow[(-NaBr)]{CH_3CO_2H} C_6H_5CH_2O\overset{O}{\overset{\|}{C}}CH_3$

(g) 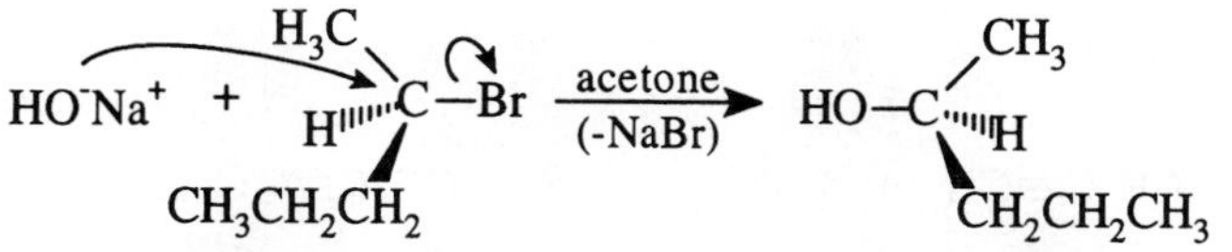 $HO^-Na^+ +$

$\xrightarrow[(-NaBr)]{\text{acetone}}$

(R)-2-Bromopentane   (S)-2-Pentanol

(h) 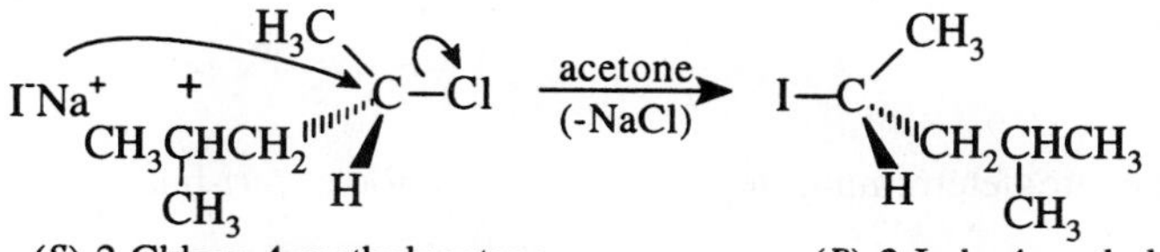 $I^-Na^+ +$

$\xrightarrow[(-NaCl)]{\text{acetone}}$

(S)-2-Chloro-4-methylpentane   (R)-2-Iodo-4-methylpentane

(i) $(CH_3)_3CCHCH_3$ $\xrightarrow[\text{EtOH}\ (-NaBr)]{EtONa^+}$ $(CH_3)_3CCH=CH_2 + EtOH$
    $\quad\quad\quad\ \ |$
    $\quad\quad\quad\ \ Br$

(j) 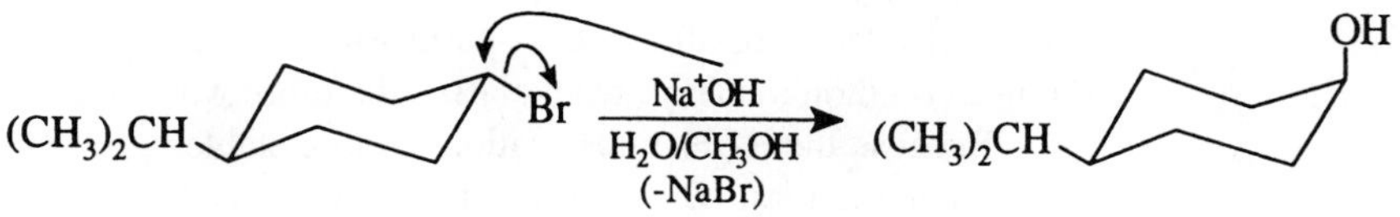 $(CH_3)_2CH$

$\xrightarrow[H_2O/CH_3OH\ (-NaBr)]{Na^+OH^-}$ $(CH_3)_2CH$

(k) 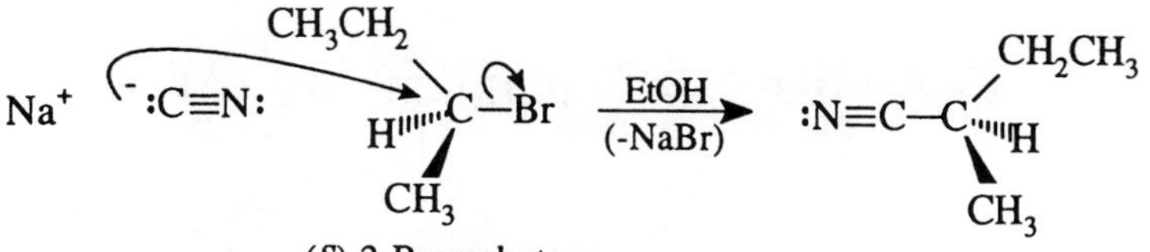 Na^+ $^-:C\equiv N:$

$\xrightarrow[(-NaBr)]{EtOH}$

(S)-2-Bromobutane

(l) 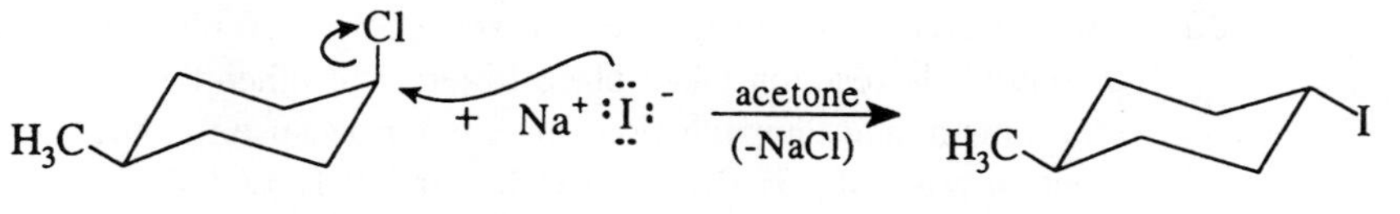 H_3C

$+ Na^+:\overset{..}{\underset{..}{I}}:^-$ $\xrightarrow[(-NaCl)]{\text{acetone}}$ H_3C

6.33 (a)

(Formation of this product depends on the fact that bromide ion is a much better leaving group than fluoride ion.)

(b) $CH_3CH_2CHCH_2CH_2CH_2-I$
    $\quad\quad\quad\ |$
    $\quad\quad\quad\ Cl$

(Formation of this product depends on the greater reactivity of 1° substrates in S_N2 reactions.)

(c) $S\begin{smallmatrix} CH_2CH_2 \\ \diagdown \\ CH_2CH_2 \end{smallmatrix}S$ (Here two S_N2 reactions produce a cyclic molecule.)

(d) $Cl-CH_2CH_2CH_2CH_2OH$ + NaH $\xrightarrow[Et_2O]{-H_2}$ $Cl-CH_2CH_2CH_2CH_2O^-$

(e) $CH_3C\equiv CH$ + $NaNH_2$ $\longrightarrow$ $CH_3C\equiv C:^- Na^+$ $\xrightarrow{CH_3-I}$

$\downarrow$ (-NaI)

$CH_3C\equiv CCH_3$

6.34 The rate-determining step in the S_N1 reaction of *tert*-butyl bromide is the following:

$(CH_3)_3C-Br$ $\underset{x}{\overset{slow}{\rightleftarrows}}$ $(CH_3)_3C^+$ + Br^-

$\downarrow H_2O$

$(CH_3)_3COH_2^+$

$(CH_3)_3C^+$ is so unstable that it reacts almost immediately with one of the surrounding water molecules, and, for all practical purposes, no reverse reaction with Br^- takes place. Adding a common ion (Br^- from NaBr), therefore, has no effect on the rate.

Because the $(C_6H_5)_2CH^+$ cation is more stable, a reversible first step occurs and adding a common ion (Br^-) slows the overall reaction by increasing the rate at which $(C_6H_5)_2CH^+$ is converted back to $(C_6H_5)_2CHBr$.

$(C_6H_5)_2CHBr$ $\rightleftarrows$ $(C_6H_5)_2CH^+$ + Br^-

$\downarrow H_2O$

$(C_6H_5)_2CHOH_2^+$

6.35 Two different mechanisms are involved. $(CH_3)_3CBr$ reacts by an S_N1 mechanism, and apparently this reaction takes place faster. The other three alkyl halids react by an S_N2 mechanism, and their reactions are slower because the nucleophile (H_2O) is weak. The reaction rates of CH_3Br, CH_3CH_2Br, and $(CH_3)_2CHBr$ are affected by the steric hindrance, and thus their order of reactivity is $CH_3Br > CH_3CH_2Br > (CH_3)_2CHBr$.

6.36 The nitrite ion is an *ambident nucleophile;* that is, it is an ion with two nucleophilic sites. The equivalent oxygen atoms and the nitrogen atom are nucleophilic.

Nucleophilic site

$:\ddot{O}=\ddot{N}-\ddot{O}:^-$

Nucleophilic site

6.37 (a) The transition state has the form:

$$\overset{\delta^+}{Nu}\text{----}R\text{-----}\overset{\delta^-}{L}$$

in which charges are developing. The more polar the solvent, the better it can solvate the transition state, thus lowering the free energy of activation and increasing the reaction rate.

(b) The transition state has the form:

$$\overset{\delta^+}{R}\text{-----------}\overset{\delta^+}{L}$$

in which the charge is becoming dispersed. A polar solvent is less able to solvate this transition state than it is to solvate the reactant. The free energy of activation, therefore, will become somewhat larger as the solvent polarity increases, and the rate will be slower.

6.38 (a) $Cl\text{-}CH_2\overset{\overset{\displaystyle CH_3}{|}}{\underset{\underset{\displaystyle CH_3}{|}}{C}}\text{-}CH_2\text{-}CH_2\text{-}I$

(b) $HO\text{-}\overset{\overset{\displaystyle CH_3}{|}}{\underset{\underset{\displaystyle CH_3}{|}}{C}}\text{-}CH_2\text{-}CH_2\text{-}Cl$ + some alkene

6.39 (a) In an S_N1 reaction the carbocation intermediate reacts rapidly with any nucleophile it encounters in a Lewis acid–Lewis base reaction. In the case of the S_N2 reaction, the leaving group departs only when "pushed out" by the attacking nucleophile, and some nucleophiles are better than others.

(b) CN^- is a much better nucleophile than ethanol and hence the nitrile is formed in the S_N2 reaction of $CH_3CH_2CH_2CH_2Cl$. In the case of $(CH_3)_3CCl$ the *tert*-butyl cation reacts chiefly with the nucleophile present in higher concentration, here the ethanol solvent.

6.40

	$\Delta H°$, kcal mol^{-1}	
$(CH_3)_3C\text{-}Cl \longrightarrow (CH_3)_3C\cdot + Cl\cdot$	+78	Homolytic bond dissoc. energy
$(CH_3)_3C\cdot \longrightarrow (CH_3)_3C^+ + e^-$	+171	Ionization potential
$Cl\cdot + e^- \longrightarrow Cl^-$	−79	Electron affinity
$(CH_3)_3C\text{-}Cl \longrightarrow (CH_3)_3C^+ + Cl^-$	+170	Heterolytic bond dissoc. energy

6.41 (a) The entropy term is slightly favorable. (The enthalpy term is highly unfavorable.)

(b) $\Delta G° = \Delta H° - T\Delta S°$

$$= 6.36 \text{ kcal mol}^{-1} - (298)(0.00115 \text{ kcal mol}^{-1})$$

$$= 6.02 \text{ kcal mol}^{-1}$$

(c) $\log K_{eq} = \dfrac{-\Delta G°}{2.303RT}$

$= \dfrac{-6.02 \text{ kcal mol}^{-1}}{(2.303)(0.001987 \text{ kcal mol}^{-1} \text{ K}^{-1})(298 \text{ K})}$

$= -4.4145$

$K_{eq} = 10^{-4.4145} = 3.84 \times 10^{-5}$

(d) The equilibrium is very much more favorable in aqueous solution because solvation of the products (ethanol, hydronium ions, and chloride ions) takes place and thereby stabilizes them.

QUIZ

6.1 Which set of conditions would you use to obtain the best yield in the reaction shown?

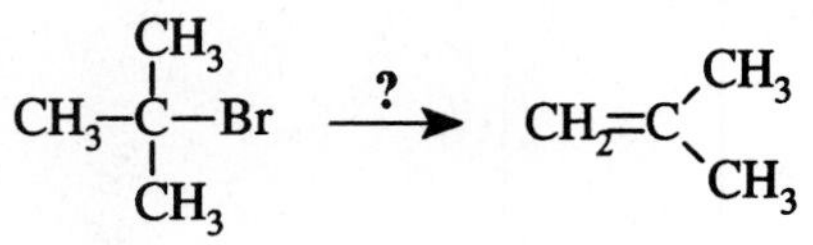

$$CH_3-\underset{\underset{CH_3}{|}}{\overset{\overset{CH_3}{|}}{C}}-Br \xrightarrow{?} CH_2=C\overset{CH_3}{\underset{CH_3}{\diagdown}}$$

(a) H_2O, heat   (b) CH_3CH_2ONa/CH_3CH_2OH, heat   (c) Heat alone

(d) H_2SO_4 (e) None of the above

6.2 Which of the following reactions would give the best yield?

(a) $CH_3ONa + (CH_3)_2CHBr \longrightarrow CH_3OCH(CH_3)_2$

(b) $(CH_3)_2CHONa + CH_3Br \longrightarrow CH_3OCH(CH_3)_2$

(c) $CH_3OH + (CH_3)_2CHBr \xrightarrow{\text{heat}} CH_3OCH(CH_3)_2$

6.3 A kinetic study yielded the following reaction rate data:

Experiment Number	Initial Concentrations		Initial Rate of Disappearance of R—Br and Formation of R—OH
	$[OH^-]$	$[R-Br]$	
1	0.50	0.50	1.00
2	0.50	0.25	0.50
3	0.25	0.25	0.25

Which of the following statements best describe this reaction?

(a) The reaction is second order.   (b) The reaction is first order.

(c) The reaction is S_N1.   (d) Increasing the concentration of OH^- has no effect on the rate.

(e) More than one of the above.

6.4   There are four compounds with the formula C_4H_9Br. List them in order of decreasing reactivity in an S_N2 reaction.

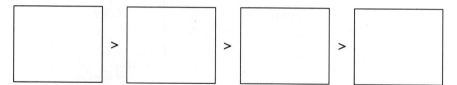

6.5   Supply the missing reactants, reagents, intermediates, or products.

A (C_4H_9Br) $\xrightarrow[\text{CH}_3\text{CO}_2\text{H}]{\text{CH}_3\overset{\overset{\text{O}}{\|}}{\text{C}}\text{ONa}}$ $\underset{H_3C}{\overset{CH_3CH_2}{\underset{H''''}{\diagup}}}C-O-\overset{\overset{O}{\|}}{C}CH_3$

$CH_3CH_2CH_2CH_2Br$ $\xrightarrow[\text{(CH}_3)_3\text{COH}]{\text{(CH}_3)_3\text{CO}^-\text{K}^+}$ $CH_3CH_2CH_2CH_2OC(CH_3)_3$ +   [ ] **B** (Minor product)

$\underset{\overset{|}{CH_2}}{\overset{CH_3}{\underset{\overset{|}{CH_3}}{H\diagdown C\diagup CH_2Cl}}}$ $\xrightarrow[\text{CH}_3\text{OH}]{\text{CH}_3\text{O}^-\text{Na}^+}$ [ ] **C**   + $\underset{\overset{|}{CH_2}}{\overset{CH_3}{\underset{\overset{|}{CH_3}}{C=CH_2}}}$

$H_3C\diagdown\diagup\diagdown\diagup\diagup^{Br}$ $\xrightarrow[25°C]{\text{Na}^+\text{CN}^-}$ [ ] **D** Major product   +   (cyclohexene with CH_3)   Minor product

6.6   Which S_N2 reaction will occur most rapidly. (Assume the concentrations and temperatures are all the same.)

(a) $CH_3O^- + CH_3CH_2-F \longrightarrow CH_3CH_2OCH_3 + F^-$

(b) $CH_3O^- + CH_3CH_2-I \longrightarrow CH_3CH_2OCH_3 + I^-$

(c) $CH_3O^- + CH_3CH_2-Cl \longrightarrow CH_3CH_2OCH_3 + Cl^-$

(d) $CH_3O^- + CH_3CH_2-Br \longrightarrow CH_3CH_2OCH_3 + Br^-$

6.7 Give three-dimensional formulas for the missing compounds.

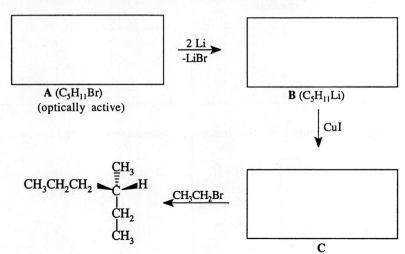

A (C$_5$H$_{11}$Br)
(optically active)

$\xrightarrow[\text{-LiBr}]{\text{2 Li}}$

B (C$_5$H$_{11}$Li)

$\downarrow$ CuI

$$\text{CH}_3\text{CH}_2\text{CH}_2 \overset{\overset{\displaystyle\text{CH}_3}{|}}{\underset{\underset{\displaystyle\text{CH}_3}{|}}{\underset{\displaystyle\text{CH}_2}{C}}} \text{H}$$

$\xleftarrow{\text{CH}_3\text{CH}_2\text{Br}}$

C

SPECIAL TOPIC
A Biological Nucleophilic Substitution Reaction:
Biological Methylation

A.1 (a) 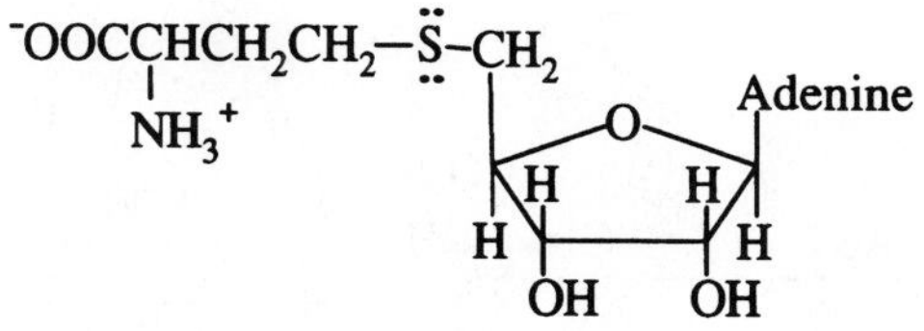 $^-OOCCHCH_2CH_2-\overset{..}{\underset{..}{S}}-CH_2$

$\qquad\qquad\underset{NH_3^+}{|}$

Adenine

H   H

H   H

OH   OH

(b) 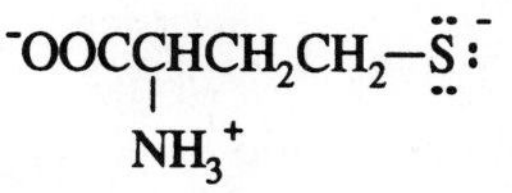 $^-OOCCHCH_2CH_2-\overset{..}{\underset{..}{S}}\,{}^-$

$\qquad\underset{NH_3^+}{|}$

(c)   The leaving group (a) is a weaker base than (b); therefore (a) is the better leaving group. The reaction with methionine would be much slower than the reaction with *S*-adenosylmethionine.

SPECIAL TOPIC
Neighboring Group Participation in Nucleophilic Substitution Reactions— A Deeper Look

SOLUTIONS TO PROBLEMS

B.1 (a) and (b)

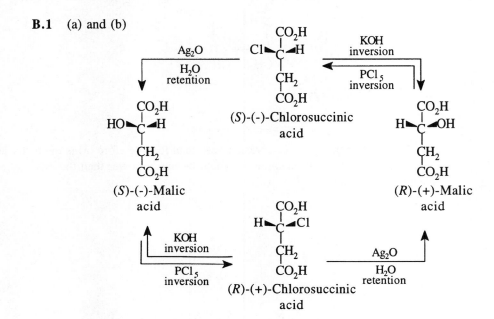

(c) The reaction takes place with retention of configuration.

(d)

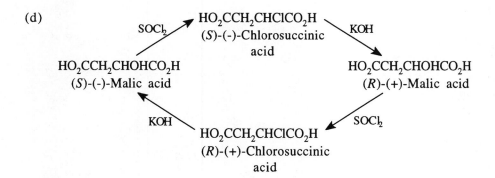

7

ALKENES AND ALKYNES I: PROPERTIES AND SYNTHESIS

SOLUTIONS TO PROBLEMS

7.1 (a) *trans*-3-Heptene

(b) 2,5-Dimethyl-2-octene

(c) 4-Ethyl-2-methyl-1-hexene

(d) 3,5-Dimethylcyclohexene

(e) 4-Methyl-4-penten-2-ol

(f) 2-Chloro-3-methyl-3-cyclohexen-1-ol

7.2 (a)  (b) 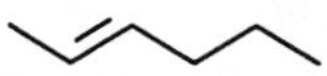 (c)

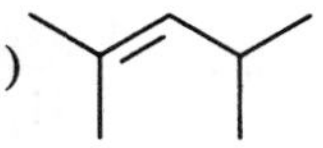

(d) 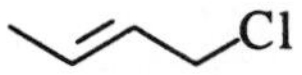 (e) 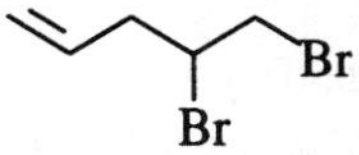 (f)

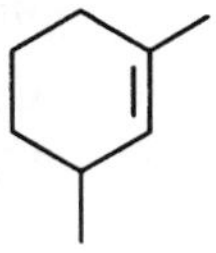

(g) 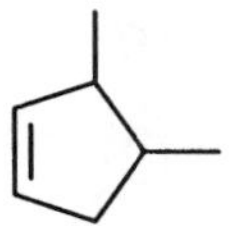 (h) 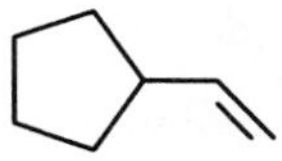 (i)

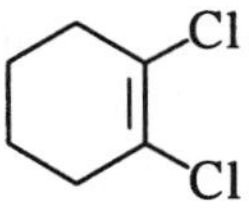

(j)

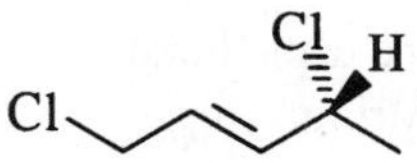

7.3 (a) (*E*)-1-Bromo-1-chloro-1-pentene

(b) (*E*)-2-Bromo-1-chloro-1-iodo-1-butene

(c) (*Z*)-3,5-Dimethyl-2-hexene

(d) (*Z*)-1-Chloro-1-iodo-2-methyl-1-butene

(e) (*Z*,4*S*)-3,4-Dimethyl-2-hexene

(f) (*Z*,3*S*)-1-Bromo-2-chloro-3-methyl-1-hexene

7.4

(R)-3-Methyl-1-pentyne   (S)-3-Methyl-1-pentyne

7.5   (a) C_6H_{14} = formula of alkane
$\underline{C_6H_{12}}$ = formula of 2-hexene
    H_2 = difference = 1 pair of hydrogen atoms

Index of hydrogen deficiency = 1

(b) C_6H_{14} = formula of alkane
$\underline{C_6H_{12}}$ = formula of methylcyclopentane
    H_2 = difference = 1 pair of hydrogen atoms

Index of hydrogen deficiency = 1

(c) No, all isomers of C_6H_{12}, for example, have the same index of hydrogen deficiency.

(d) No

(e) C_6H_{14} = formula of alkane
$\underline{C_6H_{10}}$ = formula of 2-hexyne
    H_4 = difference = 2 pairs of hydrogen atoms

Index of hydrogen deficiency = 2

(f) $C_{10}H_{22}$ (alkane)
$\underline{C_{10}H_{16}}$ (compound)
    H_6 = difference = 3 pairs of hydrogen atoms

Index of hydrogen deficiency = 3

The structural possibilities are thus

3 double bonds
1 double bond and one triple bond
2 double bonds and 1 ring
1 double bond and 2 rings
3 rings
1 triple bond and one ring

7.6   (a) $C_{15}H_{32}$ = formula of alkane
$\underline{C_{15}H_{24}}$ = formula of zingiberene
    H_8 = difference = 4 pairs of hydrogen atoms

Index of hydrogen deficiency = 4

(b) Since 1 mol of zingiberene absorbs 3 mol of hydrogen, one molecule of zingiberene must contain three double bonds. (We are told that molecules of zingiberene do not contain any triple bonds.)

(c) If a molecule of zingiberene has three double bonds and an index of hydrogen deficiency equal to 4, it must have one ring. (The structural formula for zingiberene can be found in Problem 23.2.)

7.7 (a), (b)

$$CH_2{=}\overset{\overset{\displaystyle CH_3}{|}}{C}CH_2CH_3 \quad \xrightarrow[\text{Pt}]{H_2} \quad CH_3\overset{\overset{\displaystyle CH_3}{|}}{C}HCH_2CH_3 \qquad \Delta H° = \text{-28.5 kcal mol}^{-1}$$

2-Methyl-1-butene
(disubstituted)

$$CH_3\overset{\overset{\displaystyle CH_3}{|}}{C}HCH{=}CH_2 \quad \xrightarrow[\text{Pt}]{H_2} \quad CH_3\overset{\overset{\displaystyle CH_3}{|}}{C}HCH_2CH_3 \qquad \Delta H° = \text{-30.3 kcal mol}^{-1}$$

3-Methyl-1-butene
(monosubstituted)

$$CH_3\overset{\overset{\displaystyle CH_3}{|}}{C}{=}CHCH_3 \quad \xrightarrow[\text{Pt}]{H_2} \quad CH_3\overset{\overset{\displaystyle CH_3}{|}}{C}HCH_2CH_3 \qquad \Delta H° = \text{-26.9 kcal mol}^{-1}$$

2-Methyl-2-butene
(trisubstituted)

(c) Yes, because hydrogenation converts each alkene into the same product.

(d) $CH_3\overset{\overset{\displaystyle CH_3}{|}}{C}{=}CHCH_3$ > $CH_2{=}\overset{\overset{\displaystyle CH_3}{|}}{C}CH_2CH_3$ > $CH_3\overset{\overset{\displaystyle CH_3}{|}}{C}HCH{=}CH_2$

    (trisubstituted)   (disubstituted)   (monosubstituted)

Notice that this predicted order of stability is confirmed by the heats of hydrogenation. 2-Methyl-2-butene evolves the least heat; therefore, it is the most stable. 3-Methyl-1-butene evolves the most heat; therefore, it is the least stable.

(e) $CH_2{=}CHCH_2CH_2CH_3$

1-Pentene

$$\underset{H}{\overset{H_3C}{\diagdown}}C{=}C\underset{H}{\overset{CH_2CH_3}{\diagup}}$$

cis-2-Pentene

$$\underset{H}{\overset{H_3C}{\diagdown}}C{=}C\underset{CH_2CH_3}{\overset{H}{\diagup}}$$

trans-2-Pentene

(f) Heats of combustion, because complete combustion would convert all of the alkenes to the same products. (All of the alkenes have the formula C_5H_{10}.)

$$C_5H_{10} + 7\,1/2\,O_2 \longrightarrow 5\,CO_2 + 5\,H_2O$$

7.8 (a) 2,3-Dimethyl-2-butene would be the more stable because the double bond is tetrasubstituted. 2-Methyl-2-pentene has a trisubstituted double bond.

(b) trans-3-Hexene would be the more stable because alkenes with trans double bonds are more stable than those with cis double bonds.

(c) cis-3-Hexene would be more stable because its double bond is disubstituted. The double bond of 1-hexene is monosubstituted.

(d) 2-Methyl-2-pentene would be the more stable because its double bond is trisubstituted. The double bond of trans-2-hexene is disubstituted.

7.9 The relative stabilities of the pairs of alkenes in parts (b) and (c) in Problem 7.8 could be determined by measuring heats of hydrogenation, because in each instance the two alkenes would yield the same product. Heats of combustion could be used to determine the relative stabilities of the alkene pairs in parts (a) and (d) [and also those in parts (b) and (c)] because on complete combustion the alkenes produce the same number of molar equivalents of CO_2 and H_2O.

7.10 (a)

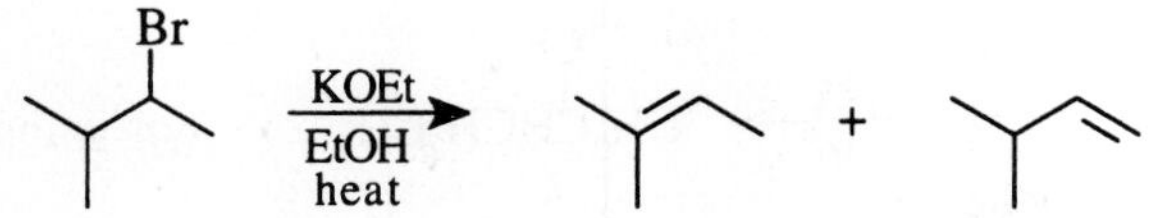

(trisubstituted,   (monosubstituted,
more stable)   less stable)
Major product Minor product

(b)

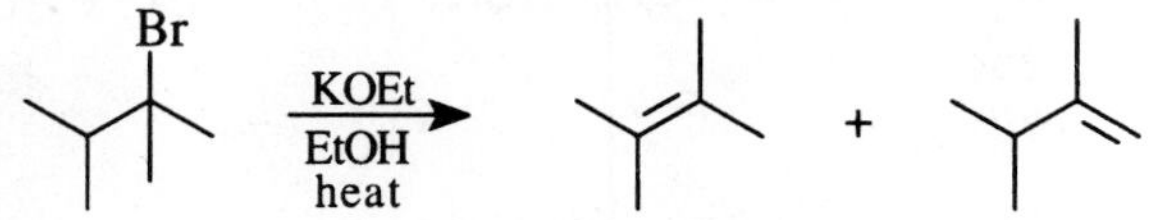

(tetrasubstituted,   (disubstituted,
more stable)   less stable)
Major product Minor product

7.11 An anti periplanar transition state allows the molecule to assume the more stable staggered conformation,

whereas a syn periplanar transition state requires the molecule to assume the less stable eclipsed conformation:

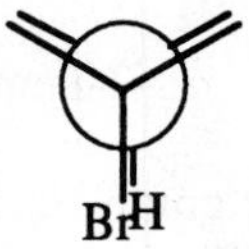

7.12 *cis*-1-Bromo-4-*tert*-butylcyclohexane can assume an anti periplanar transition state in which the bulky *tert*-butyl group is equatorial:

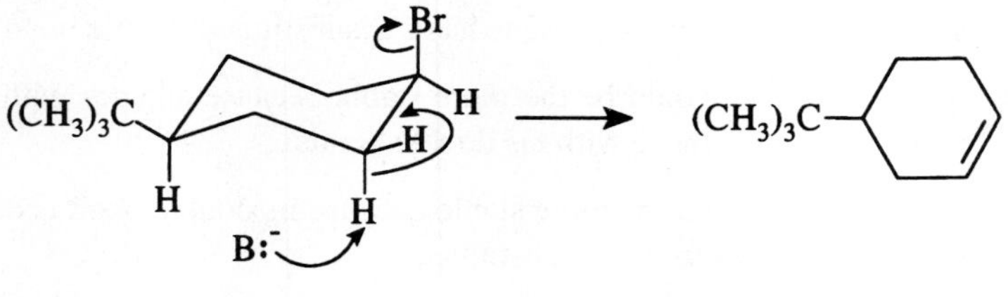

The conformation (above), because it is relatively stable, is assumed by most of the molecules present, and, therefore, the reaction is rapid.

On the other hand, for *trans*-1-bromo-4-*tert*-butylcyclohexane to assume an anti periplanar transition state, the molecule must assume a conformation in which the large *tert*-butyl group is axial:

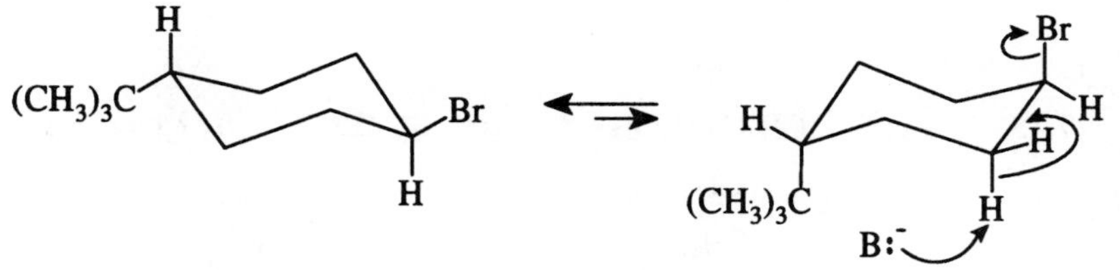

Such a conformation is of high energy; therefore, very few molecules assume this conformation. The reaction, consequently, is very slow.

7.13 (a) Anti periplanar elimination can occur in two ways with the cis isomer.

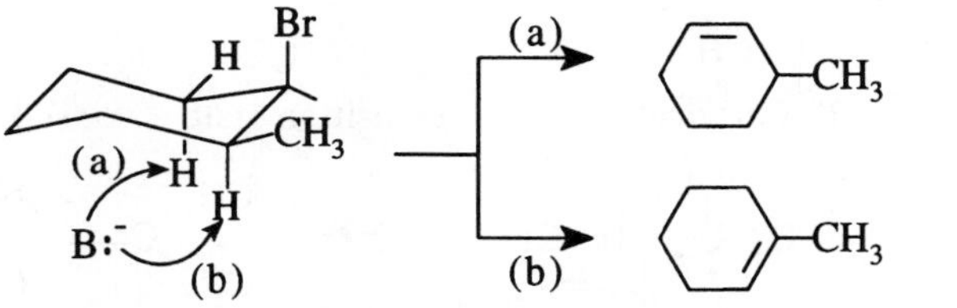

cis-1-Bromo-2-methylcyclohexane   (major products)

(b) Anti periplanar elimination can occur in only one way with the trans isomer.

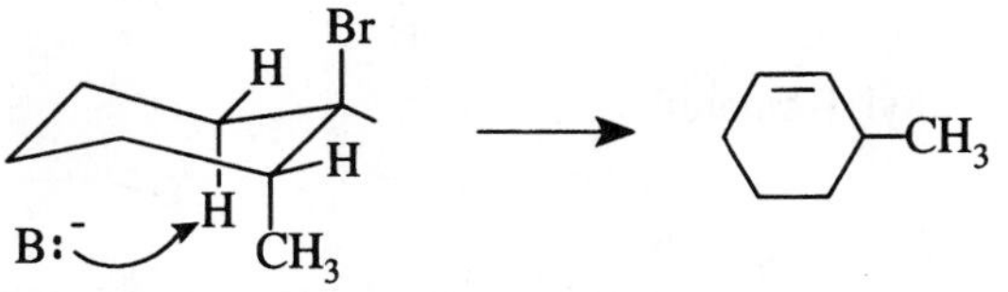

trans-1-Bromo-2-methylcyclohexane

7.14 (a) (1) $CH_3-CH-\overset{\cdot\cdot}{O}H \;+\; H-\overset{\cdot\cdot}{\overset{+}{O}}-H \;\rightleftharpoons\; CH_3-CH-\overset{H}{\overset{|}{\overset{+}{O}}}-H \;+\; H_2O$
$\;\; \overset{|}{CH_3} \overset{|}{H}$

(2) $CH_3-CH-\overset{+}{\overset{|}{O}}-H \;\rightleftharpoons\; CH_3-CH^+ \;+\; H_2O$
$\;\; \overset{|}{CH_3} \overset{|}{CH_3}$

(3) $CH_3-\overset{+}{CH}-CH_2-H \;+\; :\overset{\cdot\cdot}{O}H_2 \;\rightleftharpoons\; CH_3-CH=CH_2 \;+\; H_3O^+$

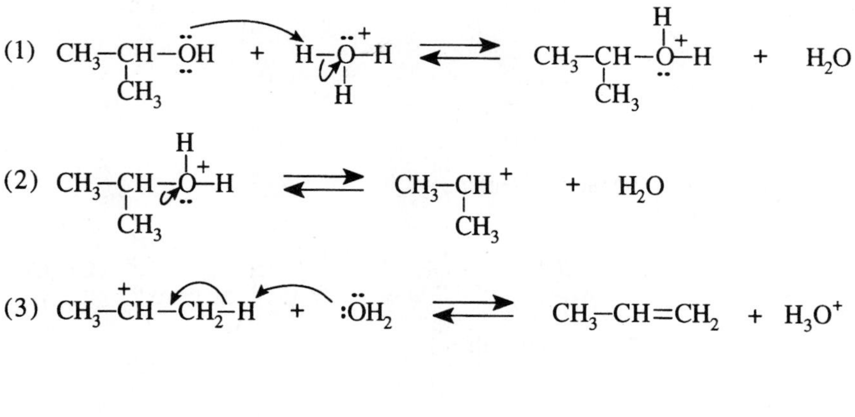

(b) By donating a proton to the $-OH$ group of the alcohol in step (1), the acid allows the loss of a relatively stable, weakly basic, leaving group (H_2O) in step (2). In the absence of an acid, the leaving group would have to be the strongly basic OH^- ion, and such steps almost never occur.

7.15

$$(1) \quad CH_3\overset{\overset{\displaystyle CH_3}{|}}{\underset{\underset{\displaystyle CH_3}{|}}{C}}CH_2OH \; + \; H^+ \; \rightleftharpoons \; CH_3\overset{\overset{\displaystyle CH_3}{|}}{\underset{\underset{\displaystyle CH_3}{|}}{C}}CH_2\overset{+}{O}H_2$$

$$(2) \quad CH_3\overset{\overset{\displaystyle CH_3}{|}}{\underset{\underset{\displaystyle CH_3}{|}}{C}}CH_2\overset{+}{O}H_2 \; \rightleftharpoons \; CH_3\overset{\overset{\displaystyle CH_3}{|}}{\underset{\underset{\displaystyle CH_3}{|}}{\overset{+}{C}}}CH_2 \; + \; H_2O$$

1° Carbocation

$$(3) \quad CH_3\overset{\overset{\displaystyle CH_3}{|}}{\underset{\underset{\displaystyle CH_3}{|}}{\overset{+}{C}}}CH_2 \; \longrightarrow \; \left[CH_3\overset{\overset{\displaystyle CH_3}{|}}{\underset{\underset{\displaystyle \overset{+}{C}H_3}{}}{C}}\text{---}CH_2 \right] \; \longrightarrow \; CH_3\overset{\overset{\displaystyle CH_3}{|}}{\underset{}{C}}\text{---}\overset{+}{C}H_2\text{--}CH_3$$

1° Carbocation   Transition state   3° Carbocation

$$(4) \quad CH_3\text{--}\overset{\overset{\displaystyle CH_3}{|}}{\underset{\underset{\displaystyle H}{+}}{C}}\text{--}CH\text{--}CH_3 \; \longrightarrow \; \overset{CH_3}{\underset{CH_3}{>}}C=C\overset{H}{\underset{CH_3}{<}} \; + \; H_3O^+$$
$$\qquad\qquad\qquad\qquad :OH_2$$

2-Methyl-2-butene

7.16

$$CH_3CH_2\underset{\underset{\displaystyle CH_3}{|}}{C}HCH_2\text{--}\overset{..}{\underset{..}{O}}H \; + \; H\text{--}\overset{+}{\underset{\underset{\displaystyle H}{|}}{O}}\text{--}H \; \underset{(+H_2O)}{\overset{(-H_2O)}{\rightleftharpoons}} \; CH_3CH_2\underset{\underset{\displaystyle CH_3}{|}}{C}HCH_2\text{--}\overset{+}{O}H_2 \; \underset{(+H_2O)}{\overset{(-H_2O)}{\rightarrow}}$$

2-Methyl-1-butanol

$$CH_3CH_2\text{--}\overset{\overset{\displaystyle H}{|}}{\underset{\underset{\displaystyle CH_3}{|}}{C}}\text{--}\overset{+}{C}H_2 \; \overset{\text{hydride}}{\underset{\text{shift}}{\longrightarrow}} \; CH_3\text{--}\overset{\overset{\displaystyle H}{|}}{C}H\text{--}\overset{+}{\underset{\underset{\displaystyle CH_3}{|}}{C}}\text{--}CH_3 \; \overset{:\overset{..}{O}H_2}{\rightleftharpoons}$$

1° Cation   3° Cation

$$CH_3CH=\underset{\underset{\displaystyle CH_3}{|}}{C}\text{--}CH_3 \; + \; H_3O^+$$

2-Methyl-2-butene

$$CH_3\underset{\underset{\displaystyle CH_3}{|}}{C}HCH_2CH_2\text{--}\overset{..}{\underset{..}{O}}H \; + \; H\text{--}\overset{+}{\underset{\underset{\displaystyle H}{|}}{O}}\text{--}H \; \underset{(+H_2O)}{\overset{(-H_2O)}{\rightleftharpoons}} \; CH_3\underset{\underset{\displaystyle CH_3}{|}}{C}HCH_2CH_2\text{--}\overset{+}{O}H_2 \; \underset{(+H_2O)}{\overset{(-H_2O)}{\rightarrow}}$$

3-Methyl-1-butanol

$$CH_3CH-CH-CH_2^+ \xrightarrow[\text{shift}]{\text{hydride}} CH_3-C-CH-CH_3 \quad \rightleftarrows$$
$$\overset{|}{CH_3} \qquad\qquad\qquad\qquad \overset{|}{CH_3}$$

$$CH_3C=CH-CH_3 \quad + \quad H_3O^+$$
$$\overset{|}{CH_3}$$
2-Methyl-2-butene

7.17 HO

Isoborneol

$$\xrightarrow[(-H_2O)]{H^+}$$

Camphene

7.18 (a) $CH_3\overset{\overset{O}{\|}}{C}CH_3 \xrightarrow[0°C]{PCl_5} CH_3CCl_2CH_3 \xrightarrow[\substack{\text{mineral oil, heat} \\ (2)\ H^+}]{(1)\ 3\ NaNH_2} CH_3C\equiv CH$

(b) $CH_3CH_2CHBr_2 \xrightarrow[\substack{\text{mineral oil, heat} \\ (2)\ H^+}]{(1)\ 3\ NaNH_2} CH_3C\equiv CH$

(c) $CH_3CHBrCH_2Br \xrightarrow{\text{[same as (b)]}} CH_3C\equiv CH$

(d) $CH_3CH=CH_2 \xrightarrow[CCl_4]{Br_2} CH_3\underset{\underset{Br}{|}}{C}HCH_2Br \xrightarrow{\text{[same as (b)]}} CH_3C\equiv CH$

7.19 (a) $CH_3CH=CH_2 + NaNH_2 \longrightarrow$ No reaction

(b) $CH_3C\equiv C-H \quad + \quad Na^+\ :\ddot{N}H_2 \longrightarrow CH_3C\equiv C:^-\ Na^+ \quad + \quad :NH_3$

Stronger   Stronger   Weaker   Weaker
acid   base   base   acid

(c) $CH_3CH_2CH_3 + NaNH_2 \longrightarrow$ No reaction

(d) $CH_3C\equiv C:^- \quad + \quad H-\ddot{O}CH_2CH_3 \longrightarrow CH_3C\equiv CH \quad + \quad :\ddot{O}CH_2CH_3$

Stronger   Stronger   Weaker   Weaker
base   acid   acid   base

(e) $CH_3C\equiv C:^- \quad + \quad H-\overset{+}{N}H_3 \longrightarrow CH_3C\equiv CH \quad + \quad :NH_3$

Stronger   Stronger acid   Weaker   Weaker
base   acid   base

7.20

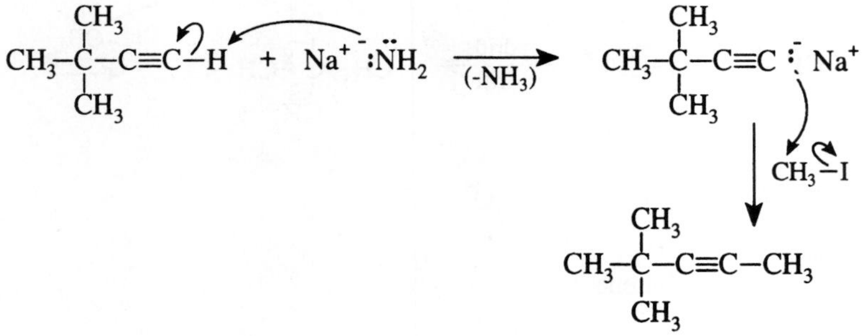

$$CH_3\text{-}\overset{\displaystyle CH_3}{\underset{\displaystyle CH_3}{\overset{|}{\underset{|}{C}}}}\text{-}C\equiv C\text{-}H \;+\; Na^+ \;:\!\ddot{N}H_2 \;\xrightarrow[(-NH_3)]{}\; CH_3\text{-}\overset{\displaystyle CH_3}{\underset{\displaystyle CH_3}{\overset{|}{\underset{|}{C}}}}\text{-}C\equiv C\!:\; Na^+$$

$$CH_3\!-\!I$$

$$CH_3\text{-}\overset{\displaystyle CH_3}{\underset{\displaystyle CH_3}{\overset{|}{\underset{|}{C}}}}\text{-}C\equiv C\text{-}CH_3$$

7.21   (a) We designate the position of the double bond by using the *lower* number of the two numbers of the doubly bonded carbon atoms, and the chain is numbered from the end nearer the double bond. The correct name is *trans*-2-pentene

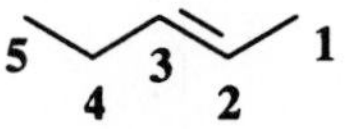

not

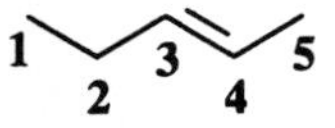

(b) We must choose the longest chain for the base name. The correct name is 2-methyl-propene.

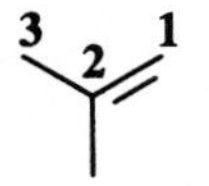

(c) We use the lower number of the two doubly bonded carbon atoms to designate the position of the double bond. The correct name is 1-methylcyclohexene.

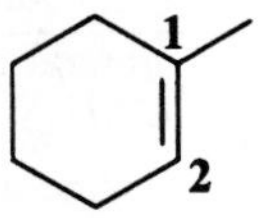

(d) We must number the ring starting with the double bond in the direction that gives the substituent the lower number. The correct name is 3-methylcyclobutene.

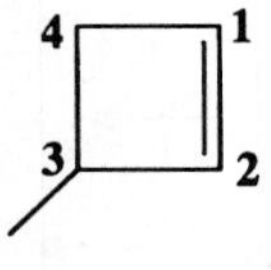

not

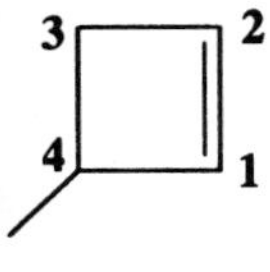

(e) We number in the way that gives the double bond *and the substituent* the lower number. The correct name is (Z)-2-chloro-2-butene.

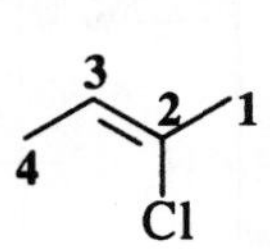

not

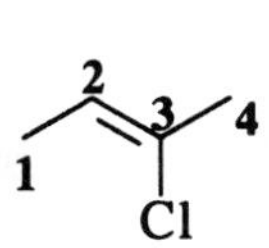

(f) We number the ring starting with the double bond so as to give the substituents the lower numbers. The correct name is 3,4-dichlorocyclohexene.

not

7.22 (a)   (b)   (c)

(d)   (e)   (f) $CH_2{=}CHCBr_3$

(g)   (h)   (i)

(j) [structure]   (k) $CH_3{-}C{\equiv}C{-}$[structure]   (l) [structure]

7.23 (a)

[structure]
(Z,4R)-4-Bromo-2-hexene

[structure]
(Z,4S)-4-Bromo-2-hexene

[structure]
(E,4R)-4-Bromo-2-hexene

[structure]
(E,4S)-4-Bromo-2-hexene

(b)

[structure]
(3R,4Z)-3-Chloro-1,4-hexadiene

[structure]
(3S,4Z)-3-Chloro-1,4-hexadiene

[structure]
(3R,4E)-3-Chloro-1,4-hexadiene

[structure]
(3S,4E)-3-Chloro-1,4-hexadiene

(c)

(2E,4R)-2,4-Dichloro-2-pentene

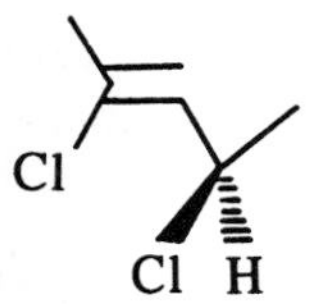

(2Z,4R)-2,4-Dichloro-2-pentene

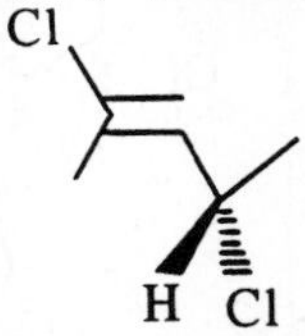

(2E,4S)-2,4-Dichloro-2-pentene

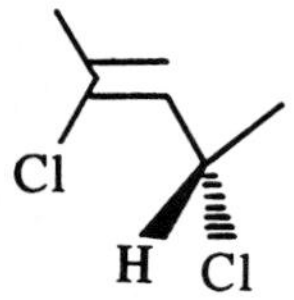

(2Z,4S)-2,4-Dichloro-2-pentene

(d)

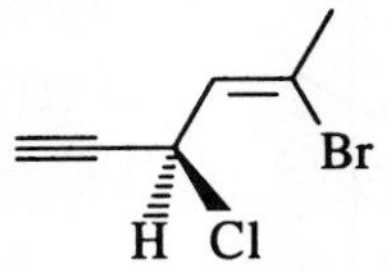

(3R,4Z)-5-Bromo-3-chloro-4-
hexen-1-yne

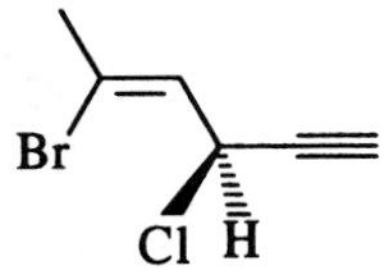

(3S,4Z)-5-Bromo-3-chloro-4-
hexen-1-yne

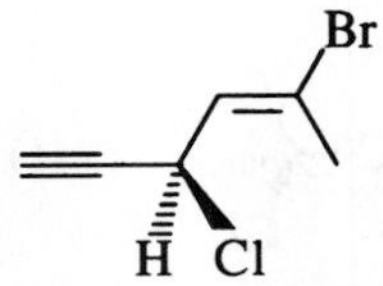

(3R,4E)-5-Bromo-3-chloro-4-
hexen-1-yne

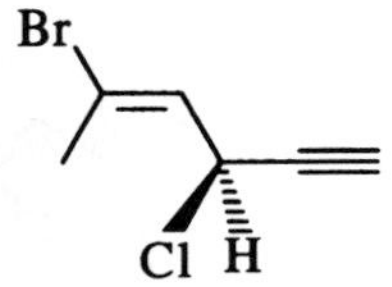

(3S,4E)-5-Bromo-3-chloro-4-
hexen-1-yne

7.24   (a)   *(E)*-3,5-Dimethyl-2-hexene

(b)   4-Chloro-3-methylcyclopentene

(c)   6-Methyl-3-heptyne

(d)   1-*sec*-Butyl-2-methylcyclohexene of 1-methyl-2(1-methylpropyl)cyclohexene

(e)   *(Z,5R)*-5-Chloro-3-hepten-6-yne

(f)   2-Pentyl-1-heptene

7.25   (a) $CH_3CH_2CH_2Cl$ $\xrightarrow[\text{(CH}_3)_3\text{COH}]{\text{(CH}_3)_3\text{COK}}$ $CH_3CH=CH_2$

(b) $CH_3\underset{\underset{Cl}{|}}{C}HCH_3$ $\xrightarrow[\text{CH}_3\text{CH}_2\text{OH}]{\text{CH}_3\text{CH}_2\text{ONa}}$ $CH_3CH=CH_2$

(c) $CH_3CH_2CH_2OH$ $\xrightarrow{\text{H}^+, \text{ heat}}$ $CH_3CH=CH_2$

(d) $CH_3\underset{\underset{OH}{|}}{C}HCH_3$ $\xrightarrow{\text{H}^+, \text{ heat}}$ $CH_3CH=CH_2$

(e) $CH_3\underset{\underset{Br}{|}}{C}HCH_2Br$ $\xrightarrow[\text{or NaI/acetone}]{\text{Zn, CH}_3\text{CO}_2\text{H}}$ $CH_3CH=CH_2$

(f) $CH_3C\equiv CH$ $\xrightarrow[\text{Ni}_2\text{B (P-2)}]{\text{H}_2}$ $CH_3CH=CH_2$

7.26 (a)

$$\xrightarrow[\text{CH}_3\text{CH}_2\text{OH}]{\text{CH}_3\text{CH}_2\text{ONa}}$$

(b)

$$\xrightarrow[\text{CH}_3\text{CO}_2\text{H}]{\text{Zn}}$$

(c)

$$\xrightarrow[]{\text{H}^+,\ \text{heat}}$$

7.27 (a) $HC\equiv CH$ $\xrightarrow[\text{liq. NH}_3]{\text{NaNH}_2}$ $HC\equiv C:^- Na^+$ $\xrightarrow[(-\text{NaI})]{\text{CH}_3-\text{I}}$ $HC\equiv C-CH_3$

(b) $HC\equiv CH$ $\xrightarrow[\text{liq. NH}_3]{\text{NaNH}_2}$ $HC\equiv C:^- Na^+$ $\xrightarrow[(-\text{NaBr})]{\text{CH}_3\text{CH}_2-\text{Br}}$ $HC\equiv C-CH_2CH_3$

(c) $CH_3C\equiv CH$ [from(a)] $\xrightarrow[\text{liq. NH}_3]{\text{NaNH}_2}$ $CH_3C\equiv C:^- Na^+$ $\xrightarrow[(-\text{NaI})]{\text{CH}_3-\text{I}}$ $CH_3-C\equiv C-CH_3$

(d) $CH_3-C\equiv C-CH_3$ [from(c)] $\xrightarrow[\text{Ni}_2\text{B (P-2)}]{\text{H}_2}$ $\underset{H}{\overset{CH_3}{\diagdown}}C=C\underset{H}{\overset{CH_3}{\diagup}}$

(e) $CH_3-C\equiv C-CH_3$ [from(c)] $\xrightarrow[(2)\ \text{NH}_4\text{Cl}]{(1)\ \text{Li, CH}_3\text{CH}_2\text{NH}_2}$ $\underset{H}{\overset{CH_3}{\diagdown}}C=C\underset{CH_3}{\overset{H}{\diagup}}$

(f) $HC\equiv C:^- Na^+$ [from(a)] $\xrightarrow[(-\text{NaBr})]{\text{CH}_3\text{CH}_2\text{CH}_2-\text{Br}}$ $HC\equiv C-CH_2CH_2CH_3$

(g) $CH_3CH_2CH_2C\equiv CH$ [from(f)] $\xrightarrow[\text{liq. NH}_3]{\text{NaNH}_2}$ $CH_3CH_2CH_2C\equiv C:^- Na^+$ $\xrightarrow[(-\text{NaI})]{\text{CH}_3-\text{I}}$

$$CH_3CH_2CH_2C\equiv CCH_3$$

(h) $CH_3CH_2CH_2C\equiv CCH_3$ [from(g)] $\xrightarrow[\text{Ni}_2\text{B (P-2)}]{\text{H}_2}$ $\underset{H}{\overset{CH_3CH_2CH_2}{\diagdown}}C=C\underset{H}{\overset{CH_3}{\diagup}}$

(i) $CH_3CH_2CH_2C\equiv CCH_3$ [from(g)] $\xrightarrow[(2)\ \text{NH}_4\text{Cl}]{(1)\ \text{Li, CH}_3\text{CH}_2\text{NH}_2}$ $\underset{H}{\overset{CH_3CH_2CH_2}{\diagdown}}C=C\underset{CH_3}{\overset{H}{\diagup}}$

(j) $HC\equiv C:^- Na^+$ [from(a)] $\xrightarrow[(-\text{NaBr})]{\text{CH}_3\text{CH}_2-\text{Br}}$ $HC\equiv CCH_2CH_3$ $\xrightarrow[\text{liq. NH}_3]{\text{NaNH}_2}$ $CH_3CH_2C\equiv C:^- Na^+$

$$\xrightarrow[]{\text{CH}_3\text{CH}_2-\text{Br}} CH_3CH_2C\equiv CCH_2CH_3$$

(k) $CH_3CH_2C\equiv C\colon^- Na^+ \xrightarrow{D_2O} CH_3CH_2C\equiv CD$
[from(j)]

(l) $CH_3C\equiv CCH_3 \xrightarrow[Ni_2B \ (P-2)]{D_2}$
$$\underset{D}{\overset{CH_3}{\diagdown}}C=C\underset{D}{\overset{CH_3}{\diagup}}$$
[from(c)]

7.28 We notice that the deuterium atoms are cis to each other, and we conclude, therefore, that we need to choose a method that will cause a syn addition of deuterium. One way would be to use D_2 and a metal catalyst (Section 7.6)

7.29 Dehydration of *trans*-2-methylcyclohexanol proceeds through the formation of a carbocation (through an E1 reaction of the protonated alcohol) and leads preferentially to the more stable alkene. 1-Methylcyclohexene (below) is more stable than 3-methylcyclohexene (the minor product of the dehydration) because its double bond is more highly substituted.

(major)	(minor)
Trisubstituted	Disubstituted
double bond	double bond

Dehydrohalogenation of *trans*-1-bromo-2-methylcyclohexane is an E2 reaction and must proceed through an anti periplanar transition state. Such a transition state is possible only for the elimination leading to 3-methylcyclohexene (cf. Problem 7.13).

3-Methylcyclohexene

7.30 (a) $C_6H_5-\underset{\underset{Br}{|}}{\overset{\overset{Br}{|}}{C}}-CH_3 \xrightarrow[\text{mineral oil, heat}]{3\,NaNH_2} C_6H_5-C\equiv C\colon^- Na^+ \xrightarrow{NH_4Cl} C_6H_5-C\equiv CH$

(b) $C_6H_5CH_2-CHBr_2 \xrightarrow[\text{mineral oil, heat}]{3\,NaNH_2} C_6H_5C\equiv C\colon^- Na^+ \xrightarrow{NH_4Cl} C_6H_5-C\equiv CH$

(c) $C_6H_5CH=CH_2$ $\xrightarrow[CCl_4]{Br_2}$ $C_6H_5-\underset{\underset{Br}{|}}{C}H-\underset{\underset{Br}{|}}{C}H_2$ $\xrightarrow[\text{mineral oil, heat}]{3\ NaNH_2}$ $C_6H_5C\equiv C\overset{..}{:}{}^-Na^+$

$\xrightarrow{NH_4Cl}$ $C_6H_5-C\equiv CH$

(d) $C_6H_5-\overset{\overset{O}{\|}}{C}-CH_3$ $\xrightarrow{PCl_5}$ $C_6H_5-\underset{\underset{Cl}{|}}{\overset{\overset{Cl}{|}}{C}}-CH_3$ $\xrightarrow[\text{mineral oil, heat}]{3\ NaNH_2}$ $C_6H_5C\equiv C\overset{..}{:}{}^-Na^+$

$\xrightarrow{NH_4Cl}$ $C_6H_5-C\equiv CH$

7.31 Cyclobutane is less stable than any of the butene isomers.

7.32 1-Pentene, 806.7 kcal mol^{-1}

cis-2-Pentene, 805.2 kcal mol^{-1}

trans-2-Pentene, 804.2 kcal mol^{-1}

2-Methyl-1-butene, 803.4 kcal mol^{-1}

2-Methyl-2-butene, 801.8 kcal mol^{-1}

7.33 1-Pentanol > 1-pentyne > 1-pentene > pentane
(See Section 3.7 for the explanation.)

7.34 (a) $CH_3\underset{\underset{Br}{|}}{\overset{\overset{CH_3}{|}}{C}}HCHCH_3$ $\xrightarrow[\substack{EtOH,\\ heat}]{NaOEt}$ $\underset{H}{\overset{CH_3}{\diagdown}}C=C\underset{\diagdown CH_3}{\diagup CH_3}$ (major) $+$ $CH_2=CH\overset{\overset{CH_3}{|}}{C}HCH_3$ (minor)

(b) $CH_3-\underset{\underset{CH_3}{|}}{\overset{\overset{CH_3}{|}}{C}}-\underset{\underset{Br}{|}}{C}H-CH_3$ $\xrightarrow[\substack{EtOH,\\ heat}]{NaOEt}$ $CH_3-\underset{\underset{CH_3}{|}}{\overset{\overset{CH_3}{|}}{C}}-CH=CH_2$

(c) $CH_3CH_2\underset{\underset{Br}{|}}{\overset{\overset{CH_3}{|}}{C}}CH_2CH_3$ $\xrightarrow[\substack{EtOH,\\ heat}]{NaOEt}$ $CH_3CH=\overset{\overset{CH_3}{|}}{C}CH_2CH_3$ (major) $+$ $CH_2=C(CH_2CH_3)_2$ (minor)

(d) (cyclohexane with Br and CH_3) $\xrightarrow[\substack{EtOH,\\ heat}]{NaOEt}$ (cyclohexene)−CH$_3$ (major) $+$ (cyclohexane)=CH$_2$ (minor)

(e) (bicyclic structure with Br, H, CH_2CH_3) $\xrightarrow[\substack{EtOH,\\ heat}]{NaOEt}$ (cyclohexene−CH$_2$CH$_3$) (major) $+$ (cyclohexene−CH$_2$CH$_3$) (minor)

(f) (bicyclic structure with Br, H, CH_2CH_3) $\xrightarrow[\substack{EtOH,\\ heat}]{NaOEt}$ (cyclohexene−CH$_2$CH$_3$)

7.35 (a) $CH_3CHCHCH_3$ (with CH_3 on C3 and Br on C2) $\xrightarrow[\substack{t\text{-BuOH,}\\ \text{heat}}]{KOt\text{-Bu}}$ $CH_2{=}CHCHCH_3$ (with CH_3) **(major)** + $CH_3CH{=}CCH_3$ (with CH_3) **(minor)**

(b) $CH_3CCH_2CH_2$ (with two CH_3 and Br) $\xrightarrow[\substack{t\text{-BuOH,}\\ \text{heat}}]{KOt\text{-Bu}}$ $CH_3CCH{=}CH_2$ (with two CH_3) **(only)**

(c) $CH_3CH_2CCH_2CH_3$ (with CH_3 and Br) $\xrightarrow[\substack{t\text{-BuOH,}\\ \text{heat}}]{KOt\text{-Bu}}$ $CH_3CH_2CCH_2CH_3$ (with CH_2 double bond) **(major)** + $CH_3CH{=}CCH_2CH_3$ (with CH_3) **(minor)**

(d) $CH_3CCH_2CHCH_3$ (with two CH_3 and Br) $\xrightarrow[\substack{t\text{-BuOH,}\\ \text{heat}}]{KOt\text{-Bu}}$ $CH_3CCH_2CH{=}CH_2$ (with two CH_3) **(major)** + $CH_3CCH{=}CHCH_3$ (with two CH_3) **(minor)**

(e) (cyclohexane with Br and CH_3) $\xrightarrow[\substack{t\text{-BuOH,}\\ \text{heat}}]{KOt\text{-Bu}}$ (cyclohexane with $=CH_2$) **(major)** + (cyclohexene with $-CH_3$) **(minor)**

7.36 (a) $CH_3CH_2CH_2CH_2CH_2Br$ $\xrightarrow[(CH_3)_3COH]{(CH_3)_3COK}$ $CH_3CH_2CH_2CH{=}CH_2$

(b) $CH_3CHCH_2CH_2Br$ (with CH_3) $\xrightarrow[(CH_3)_3COH]{(CH_3)_3COK}$ $CH_3CHCH{=}CH_2$ (with CH_3)

(c) $CH_3CHCHCH_2Br$ (with two CH_3) $\xrightarrow[(CH_3)_3COH]{(CH_3)_3COK}$ $CH_3CHC{=}CH_2$ (with two CH_3)

(d) H_3C-(cyclohexane)$-Br$ $\xrightarrow[(CH_3)_3COH]{(CH_3)_3COK}$ H_3C-(cyclohexene)

(e) (cyclopentane with Br and H_3C) $\xrightarrow[CH_3CH_2OH]{CH_3CH_2ONa}$ H_3C-(cyclopentene)

7.37

$CH_3CCH_2CH_3$ (with CH_3 and OH) **3°** $>$ $CH_3CHCHCH_3$ (with CH_3 and OH) **2°** $>$ $CH_3CH_2CH_2CH_2CH_2OH$ **1°**

7.38 (a)

(b)

(c)

(d)

(e)

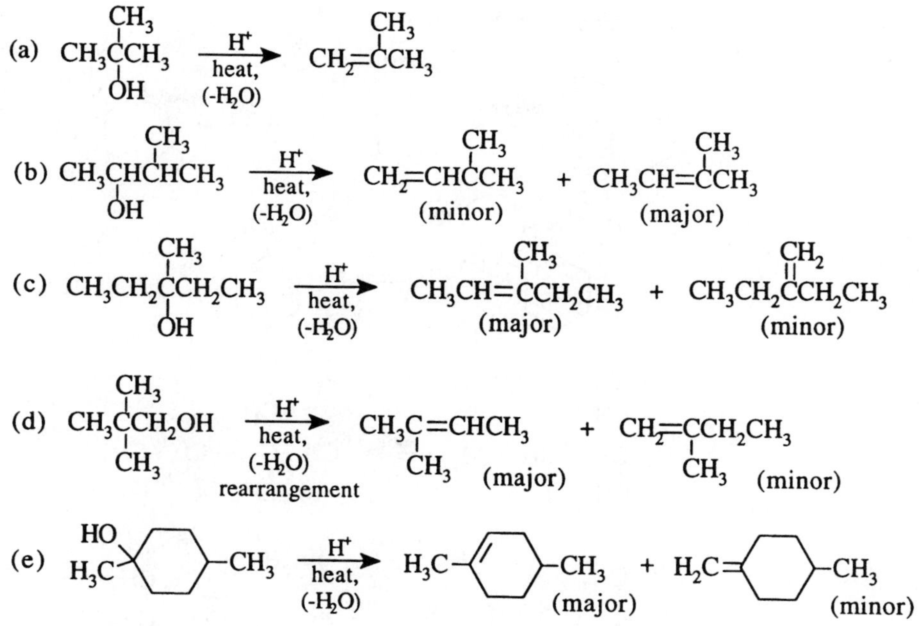

7.39 The alkene cannot be formed because the double bond in the product is too highly strained. Recall that the atoms at each carbon of a double bond prefer to be in the same plane.

7.40 Only the deuterium atom can assume the anti periplanar orientation necessary for an E2 reaction to occur.

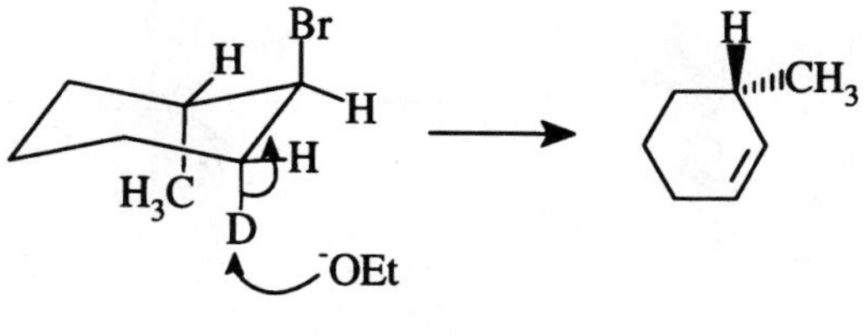

7.41 (a) A hydride shift occurs.

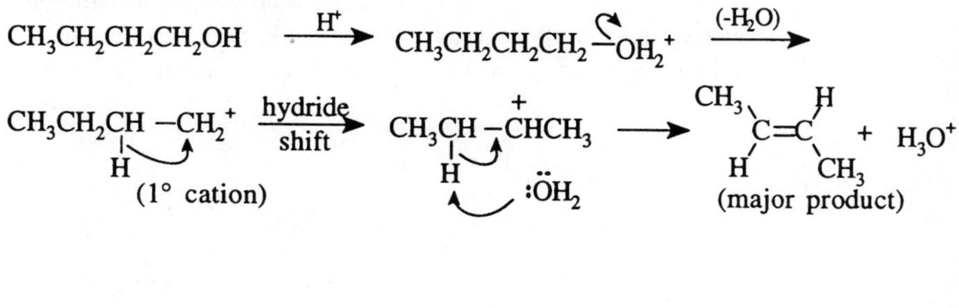

(b) A methanide shift occurs.

(c) A methanide shift occurs.

(d) The required anti periplanar transtition state leads only to (E) alkene:

(1S,2R)

7.42 In the first step, cholesterol reacts with bromine to form the *vic*-dibromide. This product is then purified by crystallization, and then treatment with zinc in ethanol converts the pure *vic*-dibromide back to cholesterol. (Recrystallization of the *vic*-dibromide is especially easy because it has a higher melting point than cholesterol.)

Crude cholesterol   Careful recrystallization to purify

$$\xrightarrow[\text{(-ZnBr}_2)]{\text{Zn, EtOH}}$$

HO   Pure cholesterol

7.43   (a)   Caryophyllene has the same molecular formula as zingiberene (Problem 7.6); thus it, too, has an index of hydrogen deficiency equal to 4. That 1 mol of caryophyllene absorbs 2 mol of hydrogen on catalytic hydrogenation indicates the presence of two double bonds per molecule.

(b)   Caryophyllene molecules must also have two rings. (See Problem 23.2 for the structure of caryophyllene.)

7.44   (a) $C_{30}H_{62}$ = formula of alkane
$\underline{C_{30}H_{50}}$ = formula of squalene
     H_{12} = difference = 6 pairs of hydrogen atoms

Index of hydrogen deficiency = 6

(b)   Molecules of squalene contain six double bonds.

(c)   Squalene molecules contain no rings. (See Problem 23.2 for the structural formula of squalene.)

7.45   (a)   We are given (Section 7.9A) the following heats of hydrogenation:
     cis-2-Butene + H_2 $\xrightarrow{\text{Pt}}$ butane $\Delta H° = -28.6 \text{ kcal mol}^{-1}$
     $trans$-2-Butene + H_2 $\xrightarrow{\text{Pt}}$ butane $\Delta H° = -27.6 \text{ kcal mol}^{-1}$

Thus, for
     cis-2-Butene $\longrightarrow$ $trans$-2-butene $\Delta H° = -1.0 \text{ kcal mol}^{-1}$

(b)   Converting cis-2-butene into $trans$-2-butene involves breaking the π bond. Therefore, we would expect the energy of activation to be at least as large as the π-bond strength, that is, at least 63 kcal mol^{-1}.

(c)

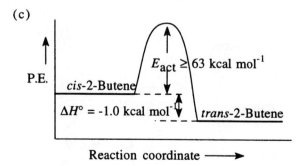

Reaction coordinate $\longrightarrow$

7.46 (a)

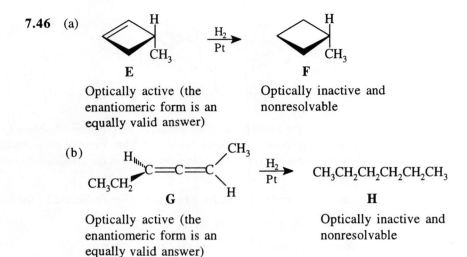

E

Optically active (the
enantiomeric form is an
equally valid answer)

F

Optically inactive and
nonresolvable

(b)

$$H_{\prime\prime\prime}\!\!\!\diagdown\!\!C\!\!=\!\!C\!\!=\!\!C\diagup\!\!\!\overset{CH_3}{\underset{H}{}}\quad\overset{H_2}{\underset{Pt}{\longrightarrow}}\quad CH_3CH_2CH_2CH_2CH_2CH_3$$

G

Optically active (the
enantiomeric form is an
equally valid answer)

H

Optically inactive and
nonresolvable

7.47 That **I** and **J** rotate plane-polarized light in the same direction tells us that **I** and **J** are not
enantiomers of each other. Thus, the following are possible structures for **I, J,** and **K.** (The
enantiomers of **I, J,** and **K** would form another set of structures, and other answers are
possible as well.)

$$CH_3CH\diagdown\!\!\!\overset{CH_3}{\underset{CH=CH_2}{\overset{CH_3}{\underset{|}{C}}}}\!\!\!\diagup H \qquad \overset{H_2}{\underset{Pt}{\longrightarrow}}$$

I
Optically active

$$CH_2\!\!=\!\!C\diagdown\!\!\!\overset{CH_3}{\underset{CH_2CH_3}{\overset{CH_3}{\underset{|}{C}}}}\!\!\!\diagup H \qquad \overset{H_2}{\underset{Pt}{\longrightarrow}}$$

J
Optically active

$$CH_3CH\diagdown\!\!\!\overset{CH_3}{\underset{CH_2CH_3}{\overset{CH_3}{\underset{|}{C}}}}\!\!\!\diagup H$$

K
Optically
active

7.48 The following are possible structures:

$$\overset{H_3C}{\underset{H}{}}\diagdown\!\!C\!\!=\!\!C\diagup\!\!\!\overset{CH_3}{\underset{\underset{CH_3}{CHCH_3}}{}}\qquad\overset{H_2}{\underset{Pt}{\longrightarrow}}$$

L

$$\overset{CH_3}{\underset{H}{}}\diagdown\!\!C\!\!=\!\!C\diagup\!\!\!\overset{\overset{CH_3}{|}{CHCH_3}}{\underset{CH_3}{}}\qquad\overset{H_2}{\underset{Pt}{\longrightarrow}}$$

M

$$CH_3CH_2\overset{\overset{CH_3}{|}}{C}HCH(CH_3)_2$$

N
Optically inactive
but resolvable

(other answers are possible as well)

7.49   (a)   With either the (1R,2R)- or the (1S,2S)-1,2-dibromo-1,2-diphenylethane, only one conformation will allow an anti periplanar arrangement of the H- and Br-. In either case, the elimination leads only to (Z)-1-bromo-1,2-diphenylethene:

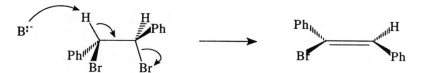

(1R,2R)-1,2-Dibromo-1,2-diphenylethane   (Z)-1-Bromo-1,2-diphenylethene
(anti periplanar orientation of H- and -Br)

(1S,2S)-1,2-Dibromo-1,2-diphenylethane   (Z)-1-Bromo-1,2-diphenylethene
(anti periplanar orientation of H- and -Br)

(b)   With (1R,2S)-1,2-dibromo-1,2-diphenylethane, only one conformation will allow an anti periplanar arrangement of the H- and Br-. In this case, the elimination leads only to (E)-1-bromo-1,2-diphenylethene:

(1R,2S)-1,2-Dibromo-1,2-diphenylethane   (E)-1-Bromo-1,2-diphenylethene
(anti periplanar orientation of H- and -Br)

(c)   With (1R,2S)-1,2-dibromo-1,2-diphenylethane, only one conformation will allow an anti periplanar arrangement of both bromine atoms. In this case, the elimination leads only to (E)-1,2-diphenylethene:

(1R,2S)-1,2-Dibromo-1,2-diphenylethane   (E)-1,2-Diphenylethene
(anti periplanar orientation of both -Br atoms)

QUIZ

7.1   Which conditions/reagents would you emply to obtain the best yields in the following reaction?

$$CH_3CH_2\overset{|}{\underset{Br}{C}}HCH_3 \xrightarrow{\quad ? \quad} CH_3CH_2CH=CH_2$$

(a) H_2O/heat

(c) $(CH_3)_3COK/(CH_3)_3COH$, heat

(b) CH_3CH_2ONa/CH_3CH_2OH, heat

(d) Reaction cannot occur as shown

7.2   Which of the following names is incorrect?

(a)   1-Butene   (b)   *trans*-2-Butene   (c)   (Z)-2-Chloro-2-pentene

(d)   1,1-Dimethylcyclopentene   (e)   Cyclohexene

7.3   Select the major product of the reaction

$$CH_3CH_2\overset{\overset{\displaystyle CH_3}{|}}{\underset{\underset{\displaystyle Br}{|}}{C}}-CH(CH_3)_2 \xrightarrow[C_2H_5OH]{C_2H_5ONa} \quad ?$$

(a) $CH_3CH_2\overset{\overset{\displaystyle CH_3}{|}}{C}=C(CH_3)_2$

(b) $CH_3CH_2\overset{\overset{\displaystyle CH_2}{||}}{C}-CH(CH_3)_2$

(c) $CH_3CH=\overset{\overset{\displaystyle CH_3}{|}}{C}-CH(CH_3)_2$

(d) $CH_2=CH-\overset{\overset{\displaystyle CH_3}{|}}{C}H-CH(CH_3)_2$

(e) $CH_3CH_2\overset{\overset{\displaystyle CH_3}{|}}{\underset{\underset{\displaystyle OC_2H_5}{|}}{C}}-CH(CH_3)_2$

7.4   Supply the missing reagents.

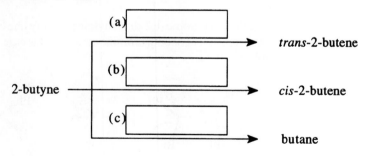

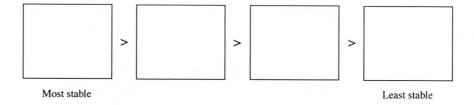

7.5 Arrange the following alkenes in order of decreasing stability.

1-Pentene, *cis*-2-pentene, *trans*-2-pentene, 2-methyl-2-butene

	>		>		>	

Most stable   Least stable

7.6 Complete the following synthesis.

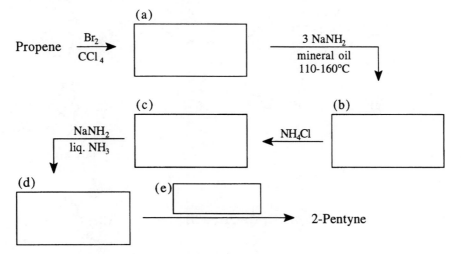

8
ALKENES AND ALKYNES II: ADDITION REACTIONS

SOLUTIONS TO PROBLEMS

8.1 CH$_3$CHCH$_2$I
　　　|
　　　Br　2-Bromo-1-iodopropane

8.2 (a) CH$_3$CH$_2$C=CH$_2$ + $\overset{\delta^+}{H}$—$\overset{\delta^-}{Br}$ ⟶ CH$_3$CH$_2\overset{+}{C}$—CH$_3$ $\xrightarrow{Br^-}$ CH$_3$CH$_2$CCH$_3$
　　　　　|　　　　　　　　　　　　　　　　　|　　　　　　　　　|
　　　　　CH$_3$　　　　　　　　　　　　　　CH$_3$　　　　Br／CH$_3$

(b) $\overset{CH_3}{|}$
　　CH$_3$C=CHCH$_3$ + $\overset{\delta^+}{I}$—$\overset{\delta^-}{Cl}$ ⟶ CH$_3\overset{CH_3}{\underset{+}{C}}$—$\overset{I}{C}HCH_3$
　　　　　　　　　　　　　　　　　　　　　　　　　　↘ Cl$^-$

　　　　　　　　　　　　　　　　　　　　CH$_3$ I
　　　　　　　　　　　　　　　　CH$_3$C—CHCH$_3$
　　　　　　　　　　　　　　　　　　|
　　　　　　　　　　　　　　　　　Cl

(c) (structure) + $\overset{\delta^+}{H}$—$\overset{\delta^-}{I}$ ⟶ (structure) $\xrightarrow{I^-}$ (structure)

8.3 (a) $\overset{CH_3}{|}$
　　CH$_3$—CH—CH=CH$_2$ + $\overset{\delta^+}{H}$—$\overset{\delta^-}{Cl}$ ⟶ CH$_3\overset{CH_3}{\underset{}{C}}$—$\overset{+}{C}$H—CH$_3$
　　　　　　　　　　　　　　　　　　　　　　　　　　H　2° Carbocation

| 1,2-hydride shift

CH$_3$　　　　　　　　　　　　　CH$_3$
　|　　　　　　　　　　　　　　　|
CH$_3$C—CH$_2$—CH$_3$ ⟵$_{Cl^-}$ CH$_3$C—CH$_2$—CH$_3$
　|　　　　　　　　　　　　　　+
　Cl　　　　　　　　　　　　　　3° Carbocation
2-Chloro-2-methylbutane
(from rearranged cation)

CH$_3$
　+|
CH$_3$C—CH—CH$_3$ $\xrightarrow{Cl^-}$ CH$_3$C—CH—CH$_3$
　|　　　　　　　　　　　　　|　　|
　H　　　　　　　　　　　　　H　Cl
Unrearranged 2° cation　　　　CH$_3$ Cl
　　　　　　　　　　　　　2-Chloro-3-methylbutane
　　　　　　　　　　　　　(from unrearranged cation)

(b) $CH_3\overset{\underset{\displaystyle CH_3}{|}}{\underset{\underset{\displaystyle CH_3}{|}}{C}}-CH=CH_2$ + $\overset{\delta^+}{H}-\overset{\delta^-}{Cl}$ $\longrightarrow$ $CH_3\overset{\underset{\displaystyle CH_3}{|}}{\underset{\underset{\displaystyle CH_3}{|}}{\overset{+}{C}}}-\overset{}{C}H-CH_3$

CH_3 2° Carbocation

1,2-methanide shift

Cl^-

$H_3C-\overset{\underset{\displaystyle H_3C}{|}}{\underset{\underset{\displaystyle }{}}{C}}-\overset{\underset{\displaystyle Cl}{|}}{C}H-CH_3$

3-Chloro-2,2-dimethylbutane
(from unrearranged carbocation)

$CH_3-\overset{+}{C}-\overset{\underset{\displaystyle CH_3}{|}}{\underset{\underset{\displaystyle CH_3}{|}}{C}}H-CH_3$

Cl^-

$CH_3-\overset{\underset{\displaystyle Cl}{|}}{C}-\overset{\underset{\displaystyle CH_3}{|}}{C}H-CH_3$

2-Chloro-2,3-dimethylbutane
(from rearranged carbocation)

8.4 $CH_2{=}CH_2$ + H_2SO_4 $\longrightarrow$ $CH_3CH_2OSO_3H$ $\xrightarrow[\text{heat}]{H_2O}$ CH_3CH_2OH + H_2SO_4

8.5   (a) Use a high concentration of water because we want the cation produced to react with water. And use a strong acid whose conjugate base is a very weak nucleophile. (For this reason we would not use HI, HBr, or HCl.) An excellent method, therefore, is to use dilute sulfuric acid.

$CH_2{=}CH_2$ + $H-\overset{\underset{\displaystyle H}{|}}{\overset{+}{\underset{}{O}}}-H$ $\rightleftarrows$ $CH_3-\overset{+}{C}H_2$ + $\overset{\underset{\displaystyle H}{|}}{:\ddot{O}-H}$ $\rightleftarrows$

(from dilute H_2SO_4)

$CH_3-CH_2-\overset{\underset{\displaystyle H}{|}}{\overset{+}{\underset{}{O}}}H$ + $\overset{\underset{\displaystyle H}{|}}{:\ddot{O}-H}$ $\rightleftarrows$ $CH_3-CH_2-\ddot{O}-H$ + $H-\overset{\underset{\displaystyle H}{|}}{\overset{+}{\underset{}{O}}}-H$

(b) Use a low concentration of water (i.e., use concentrated H_2SO_4) and use a higher temperature to encourage elimination.

(c) 2-Propanol (isopropyl alcohol) would be the product because a 2° carbocation would be formed as the intermediate.

$CH_3CH{=}CH_2$ + $H-\overset{\underset{\displaystyle H}{|}}{\overset{+}{\underset{}{O}}}-H$ $\rightleftarrows$ $CH_3-\overset{+}{C}H-CH_3$ + $\overset{\underset{\displaystyle H}{|}}{:\ddot{O}-H}$ $\rightleftarrows$

$CH_3-\overset{\underset{\displaystyle CH_3}{|}}{C}H-CH_3 + :\overset{\underset{\displaystyle H}{|}}{\overset{+}{\underset{}{O}}}-H$ $\overset{:\ddot{O}H}{\underset{H}{|}}$ $\rightleftarrows$ $CH_3-\overset{\underset{\displaystyle OH}{|}}{C}H-CH_3$ + H_3O^+

8.6 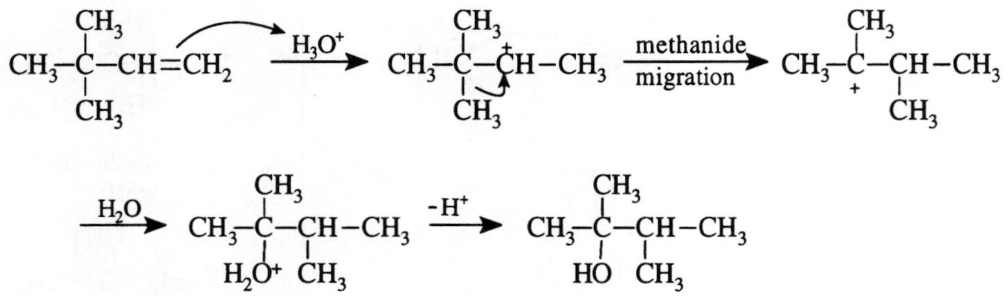

8.7 The order reflects the relative ease with which these alkenes accept a proton and form a carbocation. $(CH_3)_2C{=}CH_2$ reacts fastest because it leads to a tertiary cation,

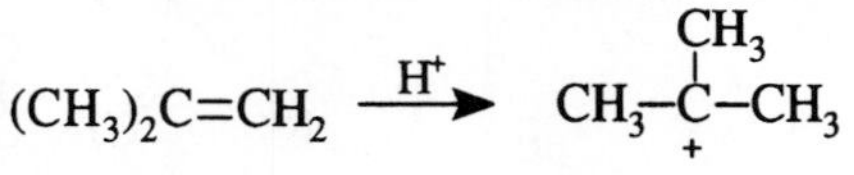
3° Carbocation

$CH_3CH{=}CH_2$ leads to a secondary cation,

$$CH_3CH{=}CH_2 \xrightarrow{H^+} CH_3\overset{+}{C}H\text{-}CH_3 \quad \text{2° Carbocation}$$

and $CH_2{=}CH_2$ reacts most slowly because it leads to a primary carbocation.

$$CH_2{=}CH_2 \xrightarrow{H^+} CH_3{-}CH_2^+ \quad \text{1° Carbocation}$$

Recall that formation of the cation is the rate-determining step in acid-catalyzed hydration and that the order of stabilities of carbocations is the following:

$$3° > 2° > 1° > {}^+CH_3$$

8.8 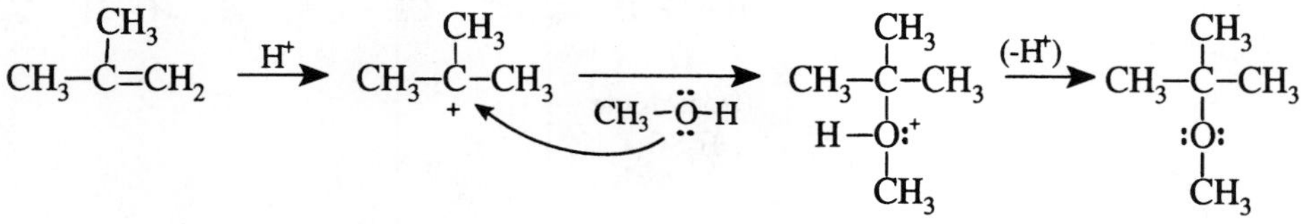

8.9 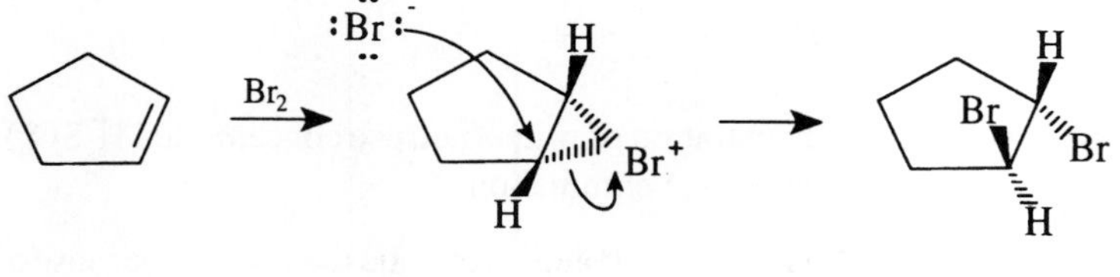

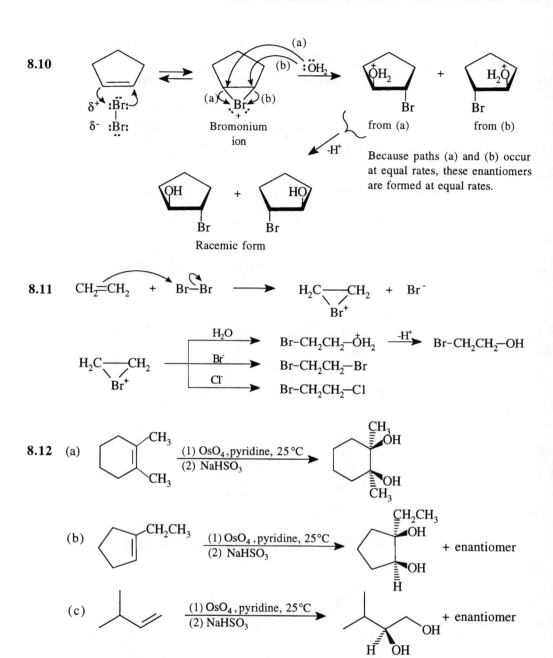

8.10

Bromonium
ion

from (a)   from (b)

-H⁺

Because paths (a) and (b) occur
at equal rates, these enantiomers
are formed at equal rates.

Racemic form

8.11 $CH_2{=}CH_2$ + Br—Br ⟶ $H_2C{-\!\!-}CH_2$ + Br⁻
                                                    Br⁺

$H_2C{-\!\!-}CH_2$ ⟶ $\xrightarrow{H_2O}$ Br–CH₂CH₂–$\overset{+}{O}H_2$ $\xrightarrow{-H^+}$ Br–CH₂CH₂–OH
        Br⁺ $\xrightarrow{Br^-}$ Br–CH₂CH₂–Br
                 $\xrightarrow{Cl^-}$ Br–CH₂CH₂–Cl

8.12 (a)
$\xrightarrow[\text{(2) NaHSO}_3]{\text{(1) OsO}_4\text{,pyridine, 25°C}}$

(b)
$\xrightarrow[\text{(2) NaHSO}_3]{\text{(1) OsO}_4\text{,pyridine, 25°C}}$ + enantiomer

(c)
$\xrightarrow[\text{(2) NaHSO}_3]{\text{(1) OsO}_4\text{,pyridine, 25°C}}$ + enantiomer

8.13 (a) Syn-hydroxylation at either face of (Z)- or cis-alkene leads to the meso compound (2R,3S)-2,3-butanediol.

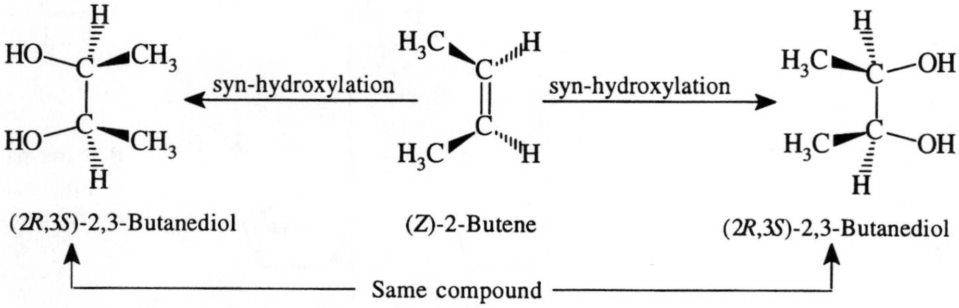

| (2R,3S)-2,3-Butanediol | (Z)-2-Butene | (2R,3S)-2,3-Butanediol |

Same compound

(b) Syn-hydroxylation at one face of the (E)- or trans-alkene leads to the (2R,3R)-enantiomer; at the other face, which is equally likely, it leads to the (2S,3S)-enantiomer.

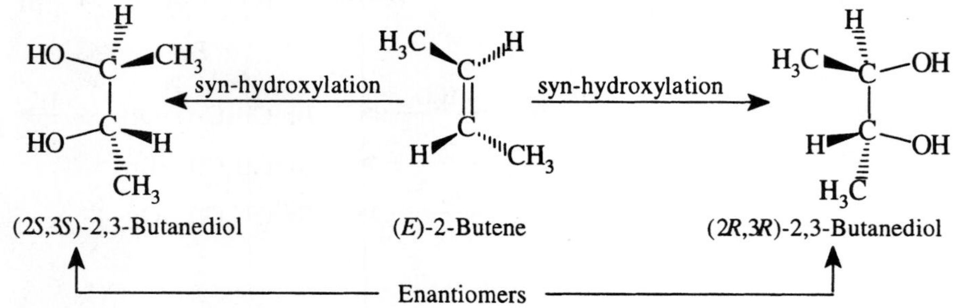

| (2S,3S)-2,3-Butanediol | (E)-2-Butene | (2R,3R)-2,3-Butanediol |

Enantiomers

8.14 (a)

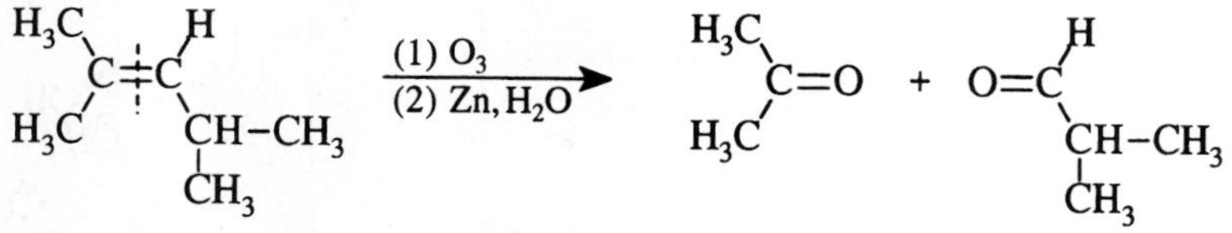

(b) $CH_3CH_2CH\!\!=\!\!\!\!=\!\!CHCH_2CH_3$ $\xrightarrow[\text{(2) Zn,H}_2\text{O}]{\text{(1) O}_3}$ 2 $CH_3CH_2CH\!\!=\!\!O$

(c)

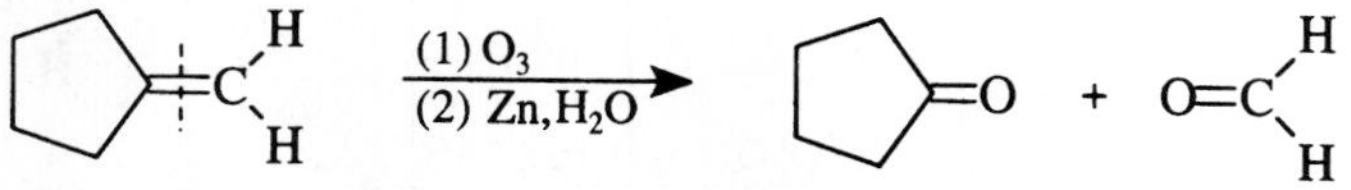

8.15 Ordinary alkenes *are* more reactive toward electrophilic reagents. But the alkenes obtained from the addition of an electrophilic reagent to an alkyne have at least one electronegative atom (Cl, Br, etc.) attached to a carbon atom of the double bond.

or

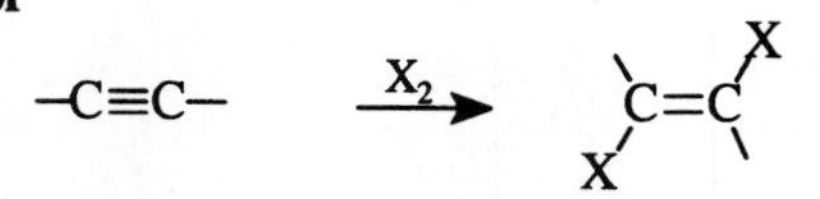

These alkenes are less reactive than alkynes toward electrophilic addition because the electronegative group makes the double bond "electron poor."

8.16 The problem as given in the text is in error.

Compounds **A** and **B** have the molecular formula C_8H_{14}, **not** C_8H_{12}. However, they can be distinguished from each other:

$$\textbf{A} = CH_3CH_2CH_2C{\equiv}CCH_2CH_2CH_3 \xrightarrow[Pt]{H_2} CH_3(CH_2)_6CH_3$$

$$\xrightarrow[(2)H_2O]{(1)\,O_3} 2\ CH_3CH_2CH_2CO_2H$$

$$\textbf{B} = CH_3CH_2CH_2CH_2CH_2CH_2C{\equiv}CH \xrightarrow[Pt]{H_2} CH_3(CH_2)_6CH_3$$

$$\xrightarrow[NaCN]{Ag(NH_3)_2OH} CH_3CH_2CH_2CH_2CH_2CH_2C{\equiv}CAg{\downarrow}$$

Compound **C** does have the formula C_8H_{12}; it is not an isomer of **A** and **B**.

This is the chemistry described in the question:

8.17 By converting the 3-hexyne to *cis*-3-hexene using H_2/Ni_2B (P-2).

Then, addition of bromine to *cis*-3-hexene will yield $(3R,4R)$, and $(3S,4S)$-3,4-dibromo-hexane as a racemic form.

8.18 (a) $CH_3CH_2\underset{\underset{I}{|}}{C}HCH_3$ (b) $CH_3CH_2CH_2CH_3$ (c) $CH_3CH_2\underset{\underset{OH}{|}}{C}HCH_3$

(d) $CH_3CH_2\underset{\underset{OSO_3H}{|}}{C}HCH_3$ (e) $CH_3CH_2\underset{\underset{OH}{|}}{C}HCH_3$ (f) $CH_3CH_2\underset{\underset{Br}{|}}{C}HCH_3$

(g) $CH_3CH_2\underset{\underset{Br}{|}}{C}HCH_2Br$ (h) $CH_3CH_2CH{=}CH_2$ (i) $CH_3CH_2\underset{\underset{OH}{|}}{C}HCH_2Br$

(j) $CH_3CH_2\underset{\underset{Cl}{|}}{C}HCH_3$ (k) $CH_3CH_2\underset{\underset{OH}{|}}{C}HCH_2OH$ (l) $CH_3CH_2\overset{O}{\overset{||}{C}}H \;+\; H\overset{O}{\overset{||}{C}}H$

(m) $CH_3CH_2\underset{\underset{OH}{|}}{C}HCH_2OH$ (n) $CH_3CH_2CO_2H \;+\; CO_2$

8.19 (a)   (b)   (c)

(d)   (e)   (f)

(g) + enantiomer   (h)   (i) + enantiomer

(j)   (k)   (l) $H\overset{O}{\overset{||}{C}}(CH_2)_4\overset{O}{\overset{||}{C}}H$

(m)   (n) $HO\overset{O}{\overset{||}{C}}(CH_2)_4\overset{O}{\overset{||}{C}}OH$

8.20 (a) $CH_3CH_2\underset{\underset{Br}{|}}{C}{=}CHBr$ (b) $CH_3CH_2\underset{\underset{Br}{|}}{C}{=}CH_2$ (c) $CH_3CH_2CBr_2CH_3$

(d) $CH_3CH_2CH_2CH_3$ (e) $CH_3CH_2CH{=}CH_2$ (f) $CH_3CH_2C{\equiv}CCH_3$

(g) $CH_3CH_2C{\equiv}CH$ and $CH_2{=}\underset{\underset{CH_3}{|}}{C}CH_3$ [An E2 reaction would take place when $CH_3CH_2C{\equiv}CNa$ is treated with $(CH_3)_3CBr$.]

(h) $CH_3CH_2C{\equiv}CAg$ (i) $CH_3CH_2C{\equiv}CCu$

8.21 (a) $CH_3\underset{\underset{Br}{|}}{C}{=}CHCH_3$ (b) $CH_3CBr_2CH_2CH_3$ (c)

(d) $CH_3CBr_2CBr_2CH_3$ (e)
$$\underset{H}{\overset{H_3C}{}}C=C\underset{H}{\overset{CH_3}{}}$$
(f)
$$\underset{H}{\overset{H_3C}{}}C=C\underset{CH_3}{\overset{Cl}{}}$$

(g)
$$\underset{H}{\overset{H_3C}{}}C=C\underset{CH_3}{\overset{H}{}}$$
(h) $CH_3CH_2CH_2CH_3$ (i) No reaction

(j) $CH_3CH_2CH_2CH_3$ (k) CH_3CO_2H (2 molar equivalents)

(l) CH_3CO_2H (2 molar equivalents)   (m) No reaction

8.22 (a) $CH_3CH_2CH=CH_2 \xrightarrow[CCl_4]{Br_2} CH_3CH_2\underset{\underset{Br}{|}}{C}H-\underset{\underset{Br}{|}}{C}H_2 \xrightarrow[\substack{\text{mineral oil,}\\ \text{heat}}]{3\,NaNH_2}$

$CH_3CH_2C\equiv CNa \xrightarrow{NH_4Cl} CH_3CH_2C\equiv CH$

(b) $CH_3CH_2CH_2CH_2Cl \xrightarrow[t\text{-BuOH, heat}]{KOt\text{-Bu}} CH_3CH_2CH=CH_2$ Then as in (a)

(c) $CH_3CH_2CH=CHCl \xrightarrow[\substack{\text{mineral oil,}\\ \text{heat}}]{2\,NaNH_2} CH_3CH_2C\equiv CNa \xrightarrow{NH_4Cl} CH_3CH_2C\equiv CH$

(d) $CH_3CH_2CH_2\underset{\underset{Cl}{|}}{C}H-Cl \xrightarrow[\substack{\text{mineral oil,}\\ \text{heat}}]{3\,NaNH_2} CH_3CH_2C\equiv CNa \xrightarrow{NH_4Cl} CH_3CH_2C\equiv CH$

(e) $H-C\equiv C-H \xrightarrow[\text{liq.}NH_3]{NaNH_2} H-C\equiv CNa \xrightarrow{CH_3CH_2\text{-}Br} CH_3CH_2C\equiv CH$

8.23 (a) $CH_3\overset{\overset{CH_3}{|}}{C}=CH_2 \xrightarrow{H_3O^+, H_2O} CH_3\underset{\underset{OH}{|}}{\overset{\overset{CH_3}{|}}{C}}CH_3$

(b) $CH_3\overset{\overset{CH_3}{|}}{C}=CH_2 \xrightarrow{HCl} CH_3\underset{\underset{Cl}{|}}{\overset{\overset{CH_3}{|}}{C}}CH_3$

(c) $CH_3\overset{\overset{CH_3}{|}}{C}=CH_2 \xrightarrow[\text{(no peroxides)}]{HBr} CH_3\underset{\underset{Br}{|}}{\overset{\overset{CH_3}{|}}{C}}CH_3$

(d) $CH_3\overset{\overset{CH_3}{|}}{C}=CH_2 \xrightarrow{HF} CH_3\underset{\underset{F}{|}}{\overset{\overset{CH_3}{|}}{C}}CH_3$

(e) $CH_3\overset{\overset{CH_3}{|}}{C}=CH_2 \xrightarrow{Cl_2, H_2O} CH_3\underset{\underset{OH}{|}}{\overset{\overset{CH_3}{|}}{C}}CH_2Cl$

8.24   (a) $C_{10}H_{22}$ (saturated alkane)

$\underline{C_{10}H_{16}}$ (formula of myrcene)

H_6 = 3 pairs of hydrogen atoms

Index of hydrogen deficiency (IHD) = 3

(b) Myrcene contains no rings because complete hydrogenation gives $C_{10}H_{22}$, which corresponds to an alkane.

(c) That myrcene absorbs three molar equivalents of H_2 on hydrogenation indicates that it contains three double bonds.

(d) Three structures are possible; however, only one gives 2,6-dimethyloctane on complete hydrogenation. Myrcene is therefore

$$
\begin{array}{cc}
CH_3 & CH_2 \\
| & || \\
CH_3C{=}CHCH_2CH_2CCH{=}CH_2
\end{array}
$$

(e) $O{=}CHCH_2CH_2\overset{\overset{\displaystyle O}{||}}{C}CH{=}O$

8.25 $CH_3CH{=}CH_2$ + $H{-}Cl$ $\longrightarrow$ $CH_3{-}\overset{+}{C}H{-}CH_3$ $\xrightarrow{\quad H\ddot{O}{-}CH_2CH_3 \quad}$

$$
\begin{array}{ccc}
CH_3CHCH_3 & & CH_3CHCH_3 \\
| & \xleftarrow{\ -H^+\ } & | \\
OCH_2CH_3 & & \overset{+}{:}\underset{\underset{\displaystyle H}{|}}{O}{-}CH_2CH_3
\end{array}
$$

8.26   The rate-determining step in each reaction is the formation of a carbocation when the alkene accepts a proton from HI. When 2-methylpropene reacts, it forms a 3° carbocation (the most stable); therefore, it reacts fastest. When ethene reacts, it forms a 1° carbocation (the least stable); therefore, it reacts the slowest.

$$
\begin{array}{ccccc}
CH_3 & & CH_3 & & CH_3 \\
| & \xrightarrow[\text{fastest}]{H{-}I} & | & \xrightarrow{\ I^- \ } & | \\
CH_3{-}C{=}CH_2 & & CH_3{-}\underset{+}{C}{-}CH_3 & & CH_3\underset{|}{C}CH_3 \\
& & 3°\ \text{Cation} & & I
\end{array}
$$

$$
CH_3CH{=}CH_2 \xrightarrow{\ H{-}I\ } \underset{2°\ \text{Cation}}{CH_3{-}\overset{+}{C}H{-}CH_3} \xrightarrow{\ I^-\ } \underset{I}{CH_3CHCH_3}
$$

$$
CH_2{=}CH_2 \xrightarrow[\text{slowest}]{H{-}I} \underset{1°\ \text{Cation}}{CH_3{-}CH_2{}^+} \xrightarrow{\ I^-\ } CH_3CH_2I
$$

8.27 $CH_3\underset{\underset{\displaystyle CH_3}{|}}{C}HCH_2CH_2CH_2\underset{\underset{\displaystyle CH_3}{|}}{C}HCH_2CH_2CH_2\underset{\underset{\displaystyle CH_3}{|}}{C}HCH_2CH_3$

2,6,10-Trimethyldodecane

8.28

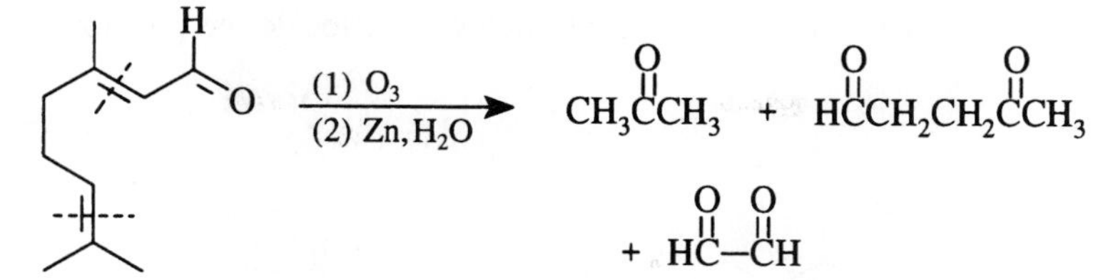

$$\text{CH}_3\overset{O}{\underset{}{\text{C}}}\text{CH}_3 \;+\; \text{H}\overset{O}{\underset{}{\text{C}}}\text{CH}_2\text{CH}_2\overset{O}{\underset{}{\text{C}}}\text{CH}_3$$

$$+\; \text{H}\overset{O}{\underset{}{\text{C}}}-\overset{O}{\underset{}{\text{C}}}\text{H}$$

8.29

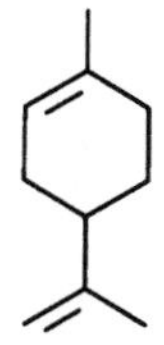

Limonene

8.30

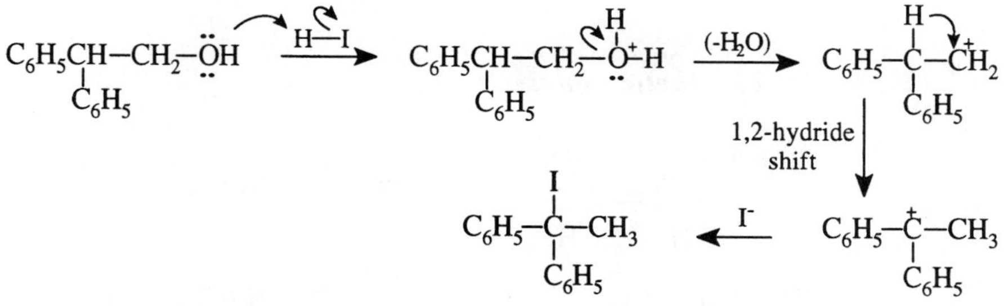

1,2-hydride shift

8.31

2° Cation

3° Cation

1,2-methanide shift

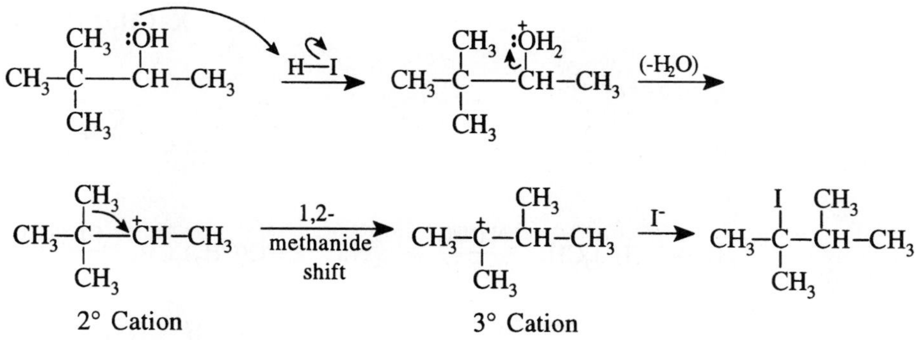

8.32 (a) The hydrogenation experiment discloses the carbon skeleton of the pheromone.

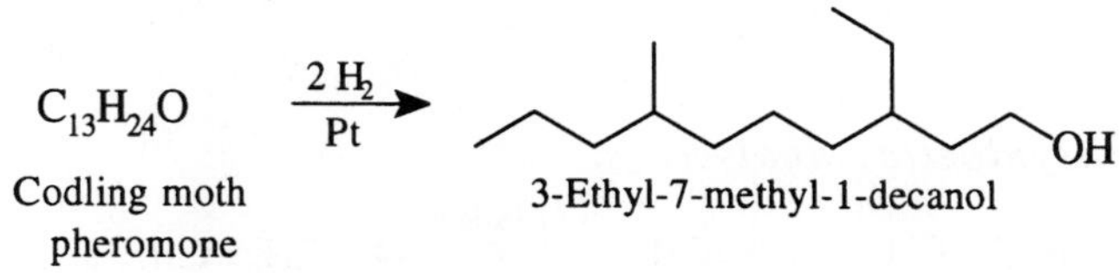

$$\text{C}_{13}\text{H}_{24}\text{O} \xrightarrow[\text{Pt}]{2\ \text{H}_2}$$

Codling moth pheromone

3-Ethyl-7-methyl-1-decanol

The ozonolysis experiment allows us to locate the position of the double bonds.

$$\xrightarrow[\text{(2) Zn,H}_2\text{O}]{\text{(1) O}_3}$$

(b)

8.33 Retrosynthetic analysis

$$CH_3(CH_2)_{12} \backslash C{=}C / (CH_2)_7CH_3 \Longrightarrow CH_3(CH_2)_{12}C{\equiv}C(CH_2)_7CH_3$$

Syn addition of H$_2$

$$CH_3(CH_2)_{12}{+}C{\equiv}C{+}(CH_2)_7CH_3 \Longrightarrow CH_3(CH_2)_{11}CH_2X \; + \; HC{\equiv}CH \; + $$
$$X{-}CH_2(CH_2)_6CH_3$$

Synthesis

$$HC{\equiv}CH \xrightarrow[\text{liq. NH}_3]{\text{NaNH}_2} HC{\equiv}C\overset{..}{:}Na^+ \xrightarrow[(-\text{NaBr})]{CH_3(CH_2)_6CH_2{-}Br}$$

$$HC{\equiv}C(CH_2)_7CH_3 \xrightarrow[\text{liq.NH}_3]{\text{NaNH}_2} Na^+\overset{..}{:}C{\equiv}C(CH_2)_7CH_3 \xrightarrow[(-\text{NaBr})]{CH_3(CH_2)_{11}CH_2{-}Br}$$

$$CH_3(CH_2)_{12}C{\equiv}C(CH_2)_7CH_3 \xrightarrow[\text{Ni}_2\text{B (P-2)}]{\text{H}_2} CH_3(CH_2)_{12}\backslash C{=}C / (CH_2)_7CH_3$$

Muscalure

8.34 Retrosynthetic analysis

$$CH_3CBr_2CH_2CH_2CH_2CH_2CH_3 \xrightarrow[\text{nikov}\atop\text{addition}]{\text{Markov-}} CH_2{=}CBrCH_2CH_2CH_2CH_2CH_3 \; + \; HBr$$

$$\xrightarrow[\text{nikov}\atop\text{addition}]{\text{Markov-}} HC{\equiv}CCH_2CH_2CH_2CH_2CH_3 \Longrightarrow HC{\equiv}CH \; + \; BrCH_2CH_2CH_2CH_2CH_3$$
$$+ \; HBr$$

Synthesis

$$HC{\equiv}CH \xrightarrow[\text{liq.NH}_3]{\text{NaNH}_2} HC{\equiv}C\overset{..}{:}Na^+ \xrightarrow[(-\text{NaBr})]{CH_3CH_2CH_2CH_2CH_2{-}Br}$$

$$HC \equiv CCH_2CH_2CH_2CH_2CH_3 \xrightarrow[\text{CH}_3\text{COBr/alumina}]{\text{"HBr"}} \underset{H}{\overset{Br}{HC=CCH_2CH_2CH_2CH_2CH_3}}$$

$$\xrightarrow[\text{CH}_3\text{COBr/alumina}]{\text{"HBr"}} CH_3CBr_2CH_2CH_2CH_2CH_2CH_3$$

8.35 Syn hydrogenation of the triple bond is required. So use H_2 and Ni_2B(P-2) or H_2 and Lindlar's catalyst.

8.36 (a)

and enantiomer through syn addition

(b)

and enantiomer through syn addition

(c)

and enantiomer through anti addition

(d)

and enantiomer through anti addition

8.37 (a) (2S,3R)- [the enantiomer is (2R,3S)-]

(b) (2S,3S)- [the enantiomer is (2R,3R)-]

(c) (2 S,3 R)- [the enantiomer is (2 R,3 S)-]

(d) (2 S,3 S)- [the enantiomer is (2 R,3 R)-]

8.38

Bromonium ion

The bromonium ion reacts with a chloride ion to produce the *trans*-1-bromo-2-chlorocyclohexane enantiomers.

8.39 (a) (1) $CH_3(CH_2)_3CH_3 \xrightarrow[\text{CCl}_4,\text{ dark}]{\text{Br}_2}$ No reaction

(2) $CH_3(CH_2)_2C \equiv CH \xrightarrow[\text{CCl}_4,\text{ dark}]{\text{Br}_2} CH_3(CH_2)_2CBr = CHBr$
(excess) $+$
$CH_3(CH_2)_2CBr_2CHBr_2$

Rapid decolorization of the reddish-brown solution of Br_2 in CCl_4 would occur in (2) if the 1-pentyne were present in excess. (Other tests, i.e., cold concd. H_2SO_4 or cold dil. $KMnO_4$ are also possible.)

(b) (1) $CH_3(CH_2)_3CH_3$ $\xrightarrow[\text{KMnO}_4]{\text{cold, dil.}}$ No reaction

(2) $CH_3(CH_2)_2CH{=}CH_2$ $\xrightarrow[\text{KMnO}_4]{\text{cold, dil.}}$ $CH_3(CH_2)_2\underset{\underset{OH}{|}}{CH}CH_2OH$ + MnO_2

The purple color of $KMnO_4$ would disappear in (2) and a brown precipitate of MnO_2 would appear. (Other tests, i.e., Br_2/CCl_4 or cold concd. H_2SO_4 are possible.)

(c) (1) $CH_3(CH_2)_2CH{=}CH_2$ $\xrightarrow{\text{Ag(NH}_3)_2\text{OH}}$ No reaction

(2) $CH_3(CH_2)_2C{\equiv}CH$ $\xrightarrow{\text{Ag(NH}_3)_2\text{OH}}$ $CH_3(CH_2)_2C{\equiv}CAg\downarrow$

In (2) a precipitate would appear.

(d) (1) $CH_3(CH_2)_3CH_3$ $\xrightarrow[\text{EtOH}]{\text{AgNO}_3}$ No reaction

(2) $CH_3(CH_2)_3CH_2Br$ $\xrightarrow[\text{EtOH}]{\text{AgNO}_3}$ $CH_3(CH_2)_2CH{=}CH_2$
$+$
$CH_3(CH_2)_3CH_2OEt$ + $AgBr\downarrow$

In (2) a precipitate would appear.

(e) (1) $CH_3C{\equiv}CCH_2CH_3$ $\xrightarrow{\text{Ag(NH}_3)_2\text{OH}}$ No reaction

(2) $CH_3(CH_2)_2C{\equiv}CH$ $\xrightarrow{\text{Ag(NH}_3)_2\text{OH}}$ $CH_3(CH_2)_2C{\equiv}CAg\downarrow$

In (2) a precipitate would appear.

(f) (1) $CH_3(CH_2)_2CH{=}CH_2$ $\xrightarrow[\text{CCl}_4,\text{ dark}]{\text{Br}_2}$ $CH_3(CH_2)_2CHBrCH_2Br$
(excess)

(2) $CH_3(CH_2)_3CH_2OH$ $\xrightarrow[\text{CCl}_4,\text{ dark}]{\text{Br}_2}$ No reaction

In (1) rapid decolorization of the bromine solution would occur.

(g) (1) $CH_3(CH_2)_3CH_3$ $\xrightarrow[\text{H}_2\text{SO}_4]{\text{cold, concd.}}$ No reaction, insoluble

(2) $CH_3(CH_2)_3CH_2OH$ $\xrightarrow[\text{H}_2\text{SO}_4]{\text{cold, concd.}}$ $CH_3(CH_2)_3CH_2OH_2{}^+\,HSO_4{}^-$
+ other products

In (2) the organic compound would dissolve.

(h) (1) $CH_3CH_2CH = CHCH_2Br$ $\xrightarrow[CCl_4, \ dark]{Br_2}$ $CH_3CH_2\overset{\overset{\displaystyle Br}{|}}{C}H\overset{\overset{}{|}}{\underset{\underset{\displaystyle Br}{|}}{C}}HCH_2Br$
      (excess)

     (2) $CH_3(CH_2)_3CH_2Br$ $\xrightarrow[CCl_4, \ dark]{Br_2}$ No reaction

In (1) rapid decolorization of the bromine solution would occur.

(i) (1) $CH_3(CH_2)_3CH_2OH$ $\xrightarrow[CCl_4, \ dark]{Br_2}$ No reaction

     (2) $CH_3CH_2CH = CHCH_2OH$ $\xrightarrow[CCl_4, \ dark]{Br_2}$ $CH_3CH_2\overset{\overset{\displaystyle Br}{|}}{C}H\overset{}{\underset{\underset{\displaystyle Br}{|}}{C}}HCH_2OH$
      (excess)

In (2) rapid decolorization of the bromine solution would occur.

8.40 Because of the electron-withdrawing nature of chlorine, the electron density at the double bond is greatly reduced and attack by the electrophilic bromine does not occur.

8.41 The index of hydrogen deficiency of **A**, **B**, and **C** is two.

$$C_6H_{14}$$
$$C_6H_{10}$$
$$H_4 \ = \ 2 \ \text{pairs of hydrogen atoms}$$

This result suggests the presence of a triple bond, two double bonds, a double bond and a ring, or two rings. The fact that **A**, **B**, and **C** all decolorize Br_2/CCl_4 and dissolve in concd. H_2SO_4 suggests they all have a carbon-carbon multiple bond.

A must be a terminal alkyne, because

$$\textbf{A} \ \xrightarrow{Ag(NH_3)_2OH} \ RC{\equiv}CAg{\downarrow}$$

Since **A** gives hexane on catalytic hydrogenation, **A** must be 1-hexyne.

$$\underset{\textbf{A}}{CH_3(CH_2)_3C{\equiv}CH} \ \xrightarrow[Pt]{2\,H_2} \ CH_3(CH_2)_4CH_3$$

This is confirmed by the oxidation experiment

$$CH_3(CH_2)_3C{\equiv}CH \ \xrightarrow[\text{(2) } H_3O^+]{\text{(1) } KMnO_4, \ OH^-, \ heat} \ CH_3(CH_2)_3CO_2H \ + \ CO_2{\uparrow}$$

The oxidation experiment shows that **B** is 3-hexyne.

$$\underset{\textbf{B}}{CH_3CH_2C{\equiv}CCH_2CH_3} \ \xrightarrow[\text{(2) } H_3O^+]{\text{(1) } KMnO_4, \ OH^-, \ heat} \ 2 \ CH_3CH_2CO_2H$$

Oxidation of **C** shows that it is cyclohexene.

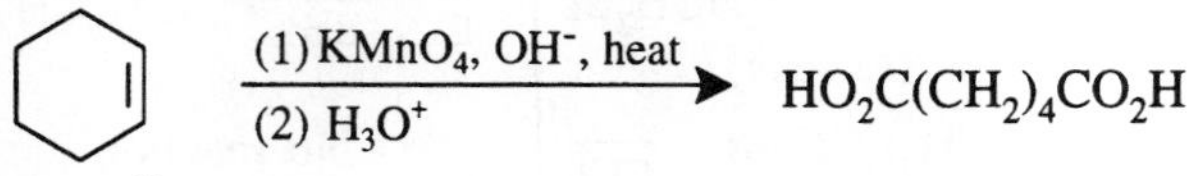

$$\text{(1) KMnO}_4, \text{OH}^-, \text{heat} \atop \text{(2) H}_3\text{O}^+} \longrightarrow \text{HO}_2\text{C(CH}_2)_4\text{CO}_2\text{H}$$

8.42 (a) Four

(b)

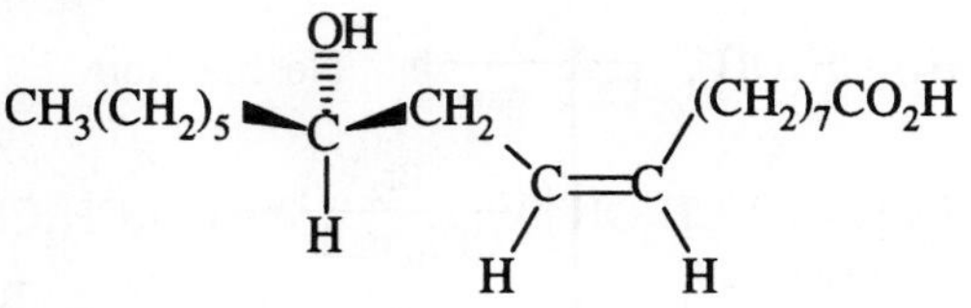

+ enantiomer

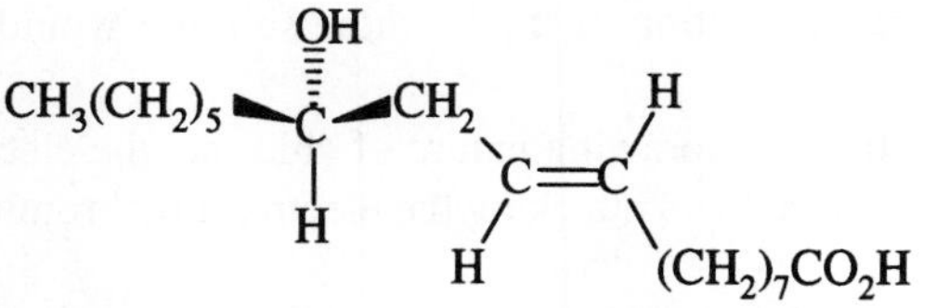

+ enantiomer

8.43 Hydroxylations by KMnO_4 are syn hydroxylations (cf. Section 8.9). Thus, maleic acid must be the *cis*-dicarboxylic acid:

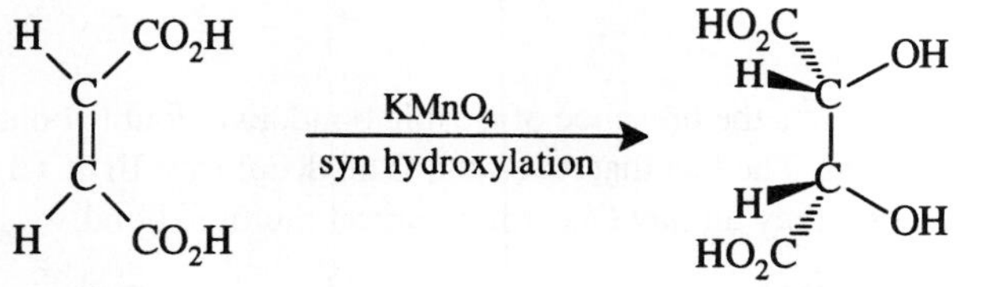

Maleic acid   *meso*-Tartaric acid

Fumaric acid must be the *trans*-dicarboxylic acid:

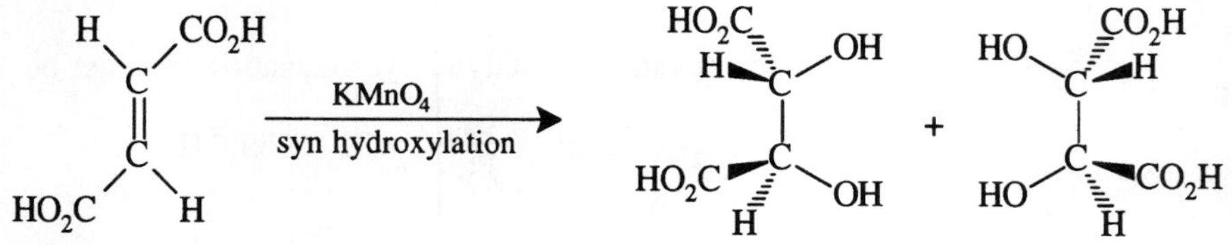

Fumaric acid   (±)-Tartaric acid

8.44 (a) The addition of bromine is an anti addition. Thus, fumaric acid yields a meso compound.

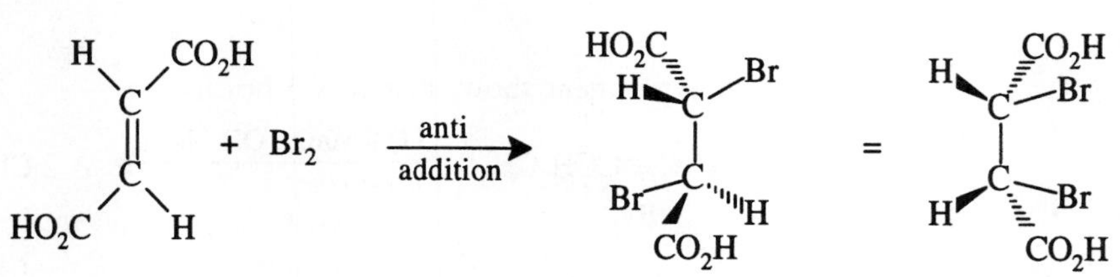

A meso compound

(b) Maleic acid adds bromine to yield a racemic modification.

8.45

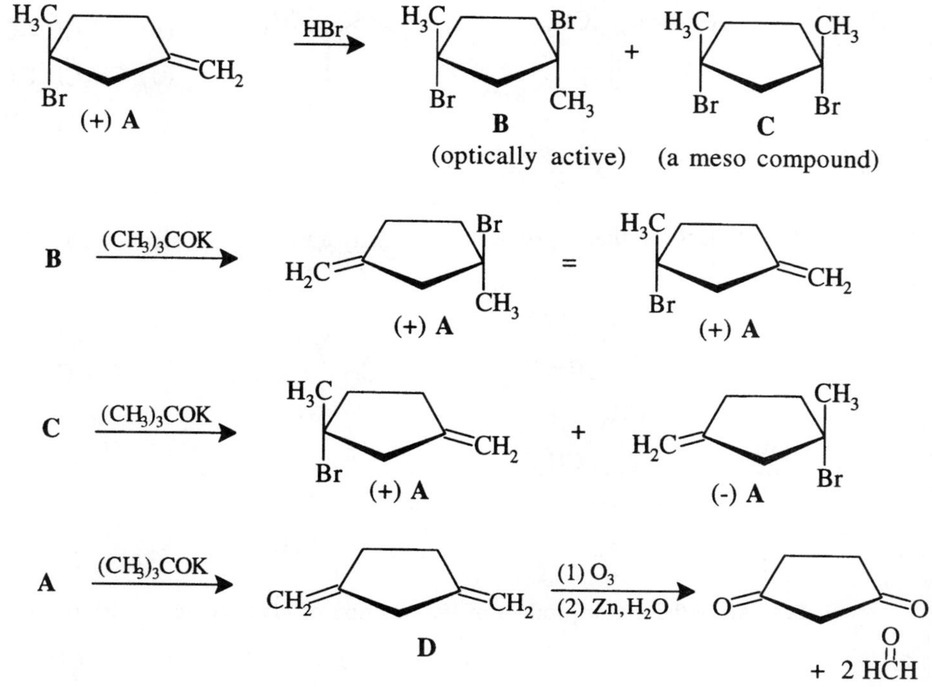

8.46

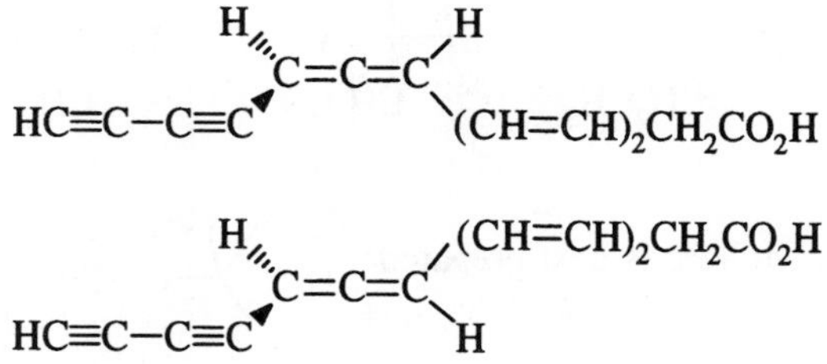

8.47

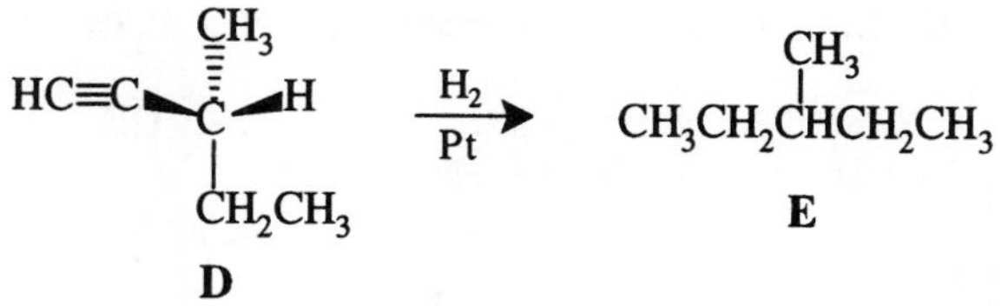

D

Optically active
(the other enantiomer is
an equally valid answer)

E

Optically inactive
(nonresolvable)

QUIZ

8.1   A hydrocarbon whose molecular formula is C_7H_{12}, on catalytic hydrogenation (excess H_2/Pt), yields C_7H_{16}. The original hydrocarbon adds bromine and also reacts with $Ag(NH_3)_2{}^+OH^-$ to give a precipitate. Which of the following is a plausible choice of structure for the original hydrocarbon?

(a) 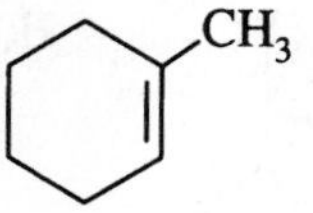 (b) 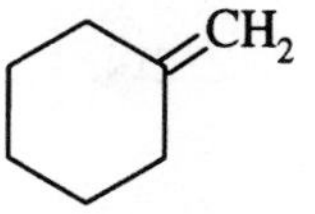 (c) $CH_3CH=CHCH=CHCH_2CH_3$

(d) $CH_3CH_2CH_2C\equiv CCH_2CH_3$

(e) $CH_3CH_2CH_2CH_2CH_2C\equiv CH$

8.2 Select the major product of the dehydration of the alcohol, $CH_3\overset{\underset{\displaystyle CH_3}{|}}{\underset{\underset{\displaystyle CH_3}{|}}{C}}-\overset{\overset{\displaystyle OH}{|}}{C}HCH_3$

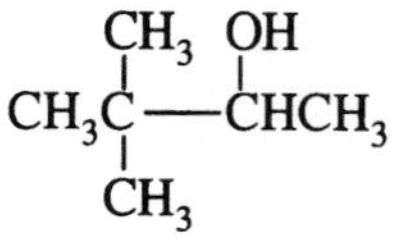

(a) $CH_3\overset{\overset{\displaystyle CH_3}{|}}{\underset{\underset{\displaystyle CH_3}{|}}{C}}-CH=CH_2$
(b) $CH_3\overset{\overset{\displaystyle CH_3}{|}}{\underset{\underset{\displaystyle CH_3}{|}}{C}}=CHCH_3$
(c) $CH_3\overset{\overset{\displaystyle CH_3}{|}}{C}=\overset{\overset{\displaystyle }{}}{\underset{\underset{\displaystyle CH_3}{|}}{C}}CH_3$

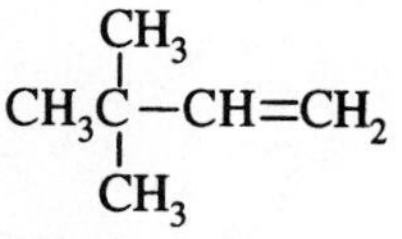

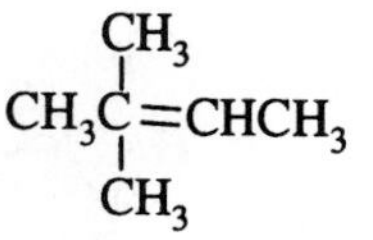

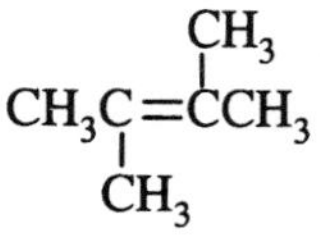

(d) 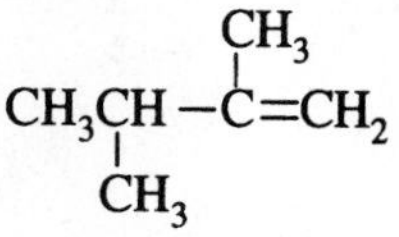 $CH_3\overset{}{\underset{\underset{\displaystyle CH_3}{|}}{C}}H-\overset{\overset{\displaystyle CH_3}{|}}{C}=CH_2$
(e) 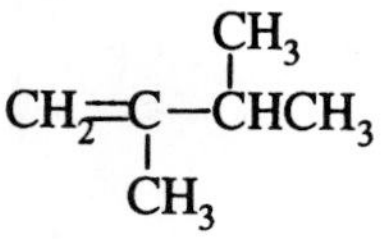 $CH_2=\overset{}{C}-\overset{\overset{\displaystyle CH_3}{|}}{\underset{\underset{\displaystyle CH_3}{|}}{C}}HCH_3$

8.3 Give the major product of the reaction of *cis*-2-pentene with bromine.

(a)

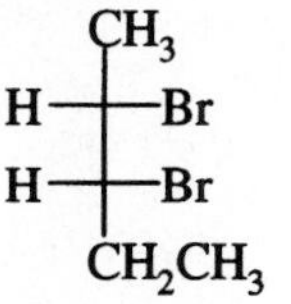

CH₃
H—Br
H—Br
CH₂CH₃

(b)

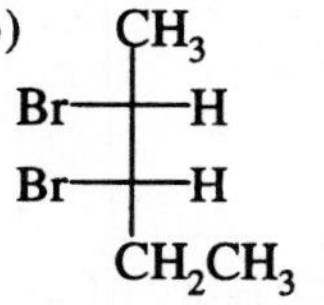

CH₃
Br—H
Br—H
CH₂CH₃

(c)

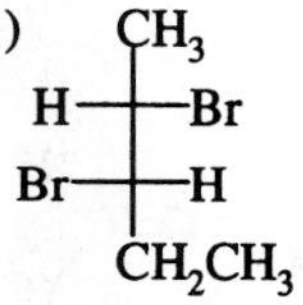

CH₃
H—Br
Br—H
CH₂CH₃

(d)

CH₃
Br—H
H—Br
CH₂CH₃

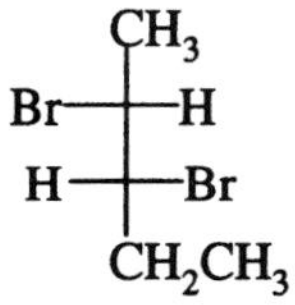

(e) A racemic mixture of (c) and (d)

8.4 The compound shown here is best prepared by which sequence of reactions?

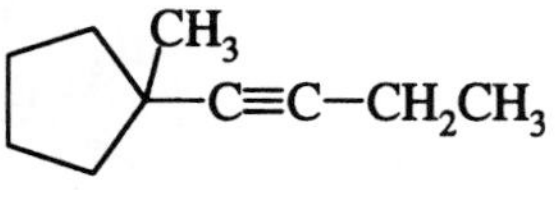

(a) 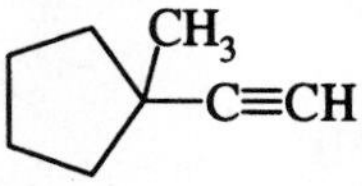 $-C\equiv CH$ + $NaNH_2$ $\longrightarrow$ then CH_3CH_2Br $\longrightarrow$ product

(b) $CH_3CH_2C\equiv CH$ + $NaNH_2$ $\longrightarrow$ then $\longrightarrow$ product

(c) 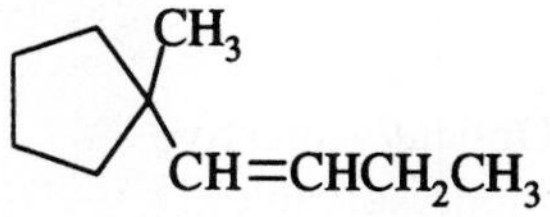 + H_2 $\xrightarrow{Pt}$ product

(d) 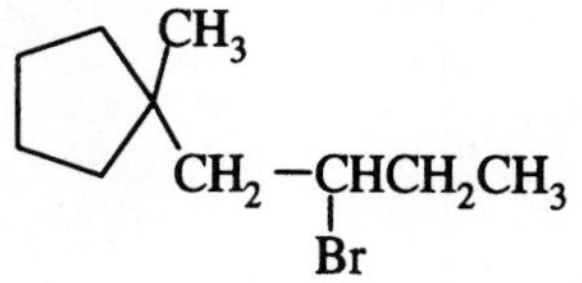 $\xrightarrow[C_2H_5OH]{NaOC_2H_5}$ product

8.5 A compound whose formula is C_6H_{10} (Compound **A**) reacts with H_2/Pt in excess to give a product C_6H_{12}, which does not decolorize Br_2/CCl_4. Compound **A** does not give any visible reaction with $Ag(NH_3)_2^+OH^-$.

Ozonolysis of **A** gives 1 mol of $\overset{\overset{\text{O}}{\|}}{\text{HCH}}$ and 1 mol of 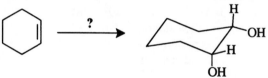 =O . Give the structure of **A**.

(a) <!-- cyclohexene structure -->

(b) $CH_3CH_2CH_2C \equiv CCH_3$

(c) $CH_3CH_2CH_2CH_2C \equiv CH$

(d) <!-- methylenecyclopentane structure --> =CH_2

(e) $CH_2\!\!=\!\!CHCH_2CH_2CH\!=\!CH_2$

8.6 Compound **B** (C_5H_{10}) does not dissolve in cold, concentrated H_2SO_4. What is **B**?

(a) $CH_2\!\!=\!\!CHCH_2CH_2CH_3$ (b) $CH_3CH\!=\!CHCH_2CH_3$

(c) <!-- cyclopentane structure -->   (d) <!-- cyclopentene structure -->

8.7 Which reaction sequence converts cyclohexene to *cis*-1,2-dihydroxycyclohexane? That is,

<!-- cyclohexene ──?──> cis-1,2-dihydroxycyclohexane structure -->

(a) Cold, dilute, aqueous $KMnO_4$, OH^- (b) (1) O_3 (2) Zn/H_2O

(c) (1) OsO_4 (2) $NaHSO_3$ (d) (1) $R\overset{\overset{\text{O}}{\|}}{C}\text{-OOH}$ (2) H_3O^+/H_2O

(e) More than one of these

8.8 Which of the following sequences leads to the best synthesis of the compound $CH_3CH_2C \equiv CH$? (Assume that the quantities of reagents are sufficient to carry out the desired reaction.)

(a) $CH_3CH_2CH\!=\!CH_2$ $\xrightarrow{Br_2}$ $\xrightarrow[H_2O]{NaOH}$

(b) $CH_3CH_2CH\!=\!CH_2$ $\xrightarrow{Br_2}$ $\xrightarrow{NaNH_2}$

(c) $CH_3CH_2CH_2CHBr_2$ $\xrightarrow{H_2SO_4}$

(d) $CH_3CH_2CH_2CH_3$ $\xrightarrow[light]{Br_2}$ $\xrightarrow{NaNH_2}$

(e) $CH_3CH_2CH\!=\!CH_2$ $\xrightarrow{O_3}$ $\xrightarrow{Zn, H_2O}$

9 RADICAL REACTIONS

SOLUTIONS TO PROBLEMS

9.1 (a)   H–H   +   F–F $\longrightarrow$ 2 H–F

             ($DH° = 104$)   ($DH° = 38$) $2(DH° = 136)$

                 +142 kcal mol^{-1} − 272 kcal mol^{-1}

                   is required for   is evolved in

                   bond cleavage   bond formation

$$\Delta H° = +142 - 272$$
$$= -130 \text{ kcal mol}^{-1}$$

(b)   CH$_3$–H   +   F–F $\longrightarrow$ CH$_3$–F   +   H–F

           ($DH° = 104$)   ($DH° = 38$)   ($DH° = 108$)   ($DH° = 136$)

               +142 kcal mol^{-1} − 244 kcal mol^{-1}

               is required for   is evolved in

               bond cleavage   bond formation

$$\Delta H° = +142 - 244$$
$$= -102 \text{ kcal mol}^{-1}$$

(c)   CH$_3$–H   +   Cl–Cl $\longrightarrow$ CH$_3$–Cl   +   H–Cl

           ($DH° = 104$)   ($DH° = 58$)   ($DH° = 83.5$)   ($DH° = 103$)

               +162 kcal mol^{-1} − 186.5 kcal mol^{-1}

               is required for   is evolved in

               bond cleavage   bond formation

$$\Delta H° = +162 - 186.5$$
$$= -24.5 \text{ kcal mol}^{-1}$$

(d)   CH$_3$–H   +   Br–Br $\longrightarrow$ CH$_3$–Br   +   H–Br

           ($DH° = 104$)   ($DH° = 46$)   ($DH° = 70$)   ($DH° = 87.5$)

               + 150 kcal mol^{-1} − 157.5 kcal mol^{-1}

               is required for   is evolved in

               bond cleavage   bond formation

$$\Delta H° = +150 - 157.5$$
$$= 7.5 \text{ kcal mol}^{-1}$$

(e) CH$_3$–H   +   I–I $\longrightarrow$ CH$_3$–I   +   H–I
   (DH° = 104)   (DH° = 36)   (DH° = 56)   (DH° = 71)

   +140 kcal mol^{-1} − 127 kcal mol^{-1}
   is required for   is evolved in
   bond cleavage   bond formation

$$\Delta H^\circ = +140 - 127$$
$$= +13 \text{ kcal mol}^{-1}$$

(f) CH$_3$CH$_2$–H   +   Cl–Cl $\longrightarrow$ CH$_3$CH$_2$–Cl   +   H–Cl
   (DH° = 98)   (DH° = 58)   (DH° = 81.5)   (DH° = 103)

   +156 kcal mol^{-1} − 184.5 kcal mol^{-1}
   is required for   is evolved in
   bond cleavage   bond formation

$$\Delta H^\circ = +156 - 184.5$$
$$= -28.5 \text{ kcal mol}^{-1}$$

(g) (CH$_3$)$_2$CH–H   +   Cl–Cl $\longrightarrow$ (CH$_3$)$_2$CH–Cl   +   H–Cl
   (DH° = 94.5)   (DH° = 58)   (DH° = 81)   (DH° = 103)

   +152.5 kcal mol^{-1} − 184 kcal mol^{-1}
   is required for   is evolved in
   bond cleavage   bond formation

$$\Delta H^\circ = +152.5 - 184$$
$$= -31.5 \text{ kcal mol}^{-1}$$

(h) (CH$_3$)$_3$C–H   +   Cl–Cl $\longrightarrow$ (CH$_3$)$_3$C–Cl   +   H–Cl
   (DH° = 91)   (DH° = 58)   (DH° = 78.5)   (DH° = 103)

   +149 kcal mol^{-1} − 181.5 kcal mol^{-1}
   is required for   is evolved in
   bond cleavage   bond formation

$$\Delta H^\circ = +149 - 181.5$$
$$= -32.5 \text{ kcal mol}^{-1}$$

9.2   (a)

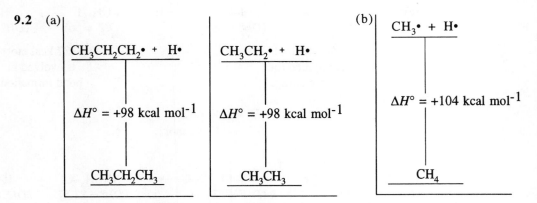

Each 1° radical has the same (98 kcal mol^{-1}) greater potential energy than the alkane from which each was formed. Therefore, the two 1° radicals have the same stability.

The methyl radical has greater potential energy than the 1° radicals, therefore the methyl radical is less stable.

9.3 $CH_3CH_2\overset{\displaystyle CH_3}{\underset{\displaystyle CH_3}{C}}\cdot$ > $CH_3CH_2\overset{\displaystyle CH_3}{\underset{\displaystyle}{CH}}\cdot$ > $(CH_3)_2CHCH_2\cdot$ > $CH_3\cdot$

   3°   >   **2°**   >   **1°**   >   **Methyl**

9.4   The compounds all have different boiling points. They could, therefore, be separated by careful fractional distillation. Or, because the compounds have different vapor pressures, they could be easily separated by gas-liquid chromatography.

9.5   A small amount of ethane is formed by the combination of two methyl radicals:

$$2CH_3\cdot \longrightarrow CH_3 : CH_3$$

This ethane then reacts with chlorine in a substitution reaction (see Section 9.6) to form chloroethane.

   The significance of this observation is that it is evidence for the proposal that the combination of methyl radicals is one of the chain-terminating steps in the chlorination of methane.

9.6   The use of a large excess of chlorine allows all of the chlorinated methanes (CH_3Cl, CH_2Cl_2, and $CHCl_3$) to react with chlorine.

9.7   *Chain Initiation*

   Step 1   F–F $\longrightarrow$ $2F\cdot$ $\Delta H° = +38$ kcal mol^{-1}
   ($DH° = 38$)

Chain Propagation

Step 2   CH$_3$–H + F· $\longrightarrow$ CH$_3$· + H–F $\Delta H° = -32$ kcal mol^{-1}
   $(DH° = 104)$ $(DH° = 136)$

Step 3   CH$_3$· + F–F $\longrightarrow$ CH$_3$–F + F· $\Delta H° = -70$ kcal mol^{-1}
             $(DH° = 38)$ $(DH° = 108)$

Chain Termination

CH$_3$· + F· $\longrightarrow$ CH$_3$–F $\Delta H° = -108$ kcal mol^{-1}
                   $(DH° = 108)$

CH$_3$· + CH$_3$· $\longrightarrow$ CH$_3$–CH$_3$ $\Delta H° = -88$ kcal mol^{-1}
                           $(DH° = 88)$

F· + F· $\longrightarrow$ F–F $\Delta H° = -38$ kcal mol^{-1}
                   $(DH° = 38)$

9.8   CH$_3$–H + F̶· $\longrightarrow$ C̶H$_3$· + H–F $\Delta H° = -32$ kcal mol^{-1}

C̶H$_3$· + F–F $\longrightarrow$ CH$_3$–F + F̶· $\Delta H° = -70$ kcal mol^{-1}
───
CH$_3$–H + F–F $\longrightarrow$ CH$_3$–F + H–F $\Delta H° = -102$ kcal mol^{-1}

9.9   (a) Reactions (3), (5), and (6) should have $E_{act} = 0$ because these are gas-phase reactions in which small radicals combine to form molecules.

(b) Reactions (1), (2), and (4) should have $E_{act} > 0$ because in them covalent bonds are broken.

(c) Reactions (1) and (2) should have $E_{act} = \Delta H°$ because in them bonds are broken homolytically but no bonds are formed.

9.10   (a)

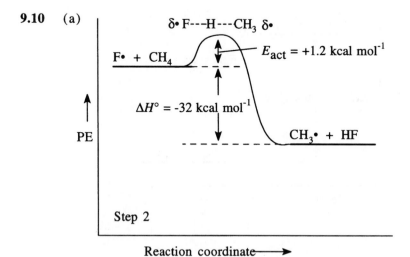

(b)

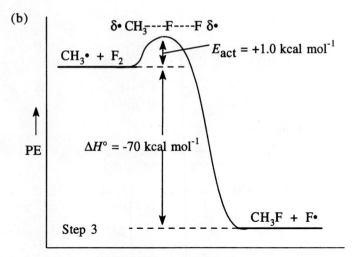

(c)

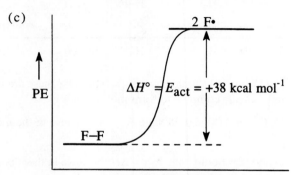

(d)

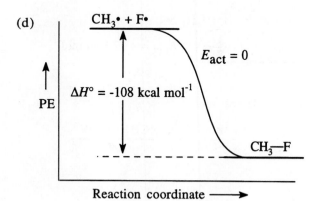

(e)   Notice that this is the reverse of Step (2) in part (a)

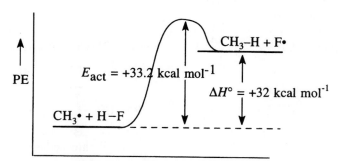

9.11   (a) CH_3CH_2-H + $Cl\cdot$ $\longrightarrow$ $CH_3CH_2\cdot$ + $H-Cl$
            $(DH° = 98)$ $(DH° = 103)$

   $\Delta H° = -103 + 98 = -5$ kcal mol^{-1}

(b)

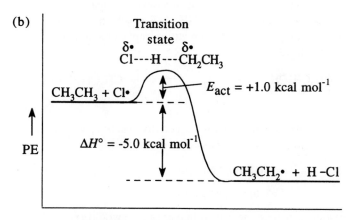

(c)   The hydrogen abstraction step for ethane,

   CH_3CH_2-H + $Cl\cdot$ $\longrightarrow$ $CH_3CH_2\cdot$ +   HCl $(E_{act} = 1.0$ kcal mol$^{-1})$

   has a much lower energy of activation than the corresponding step for methane:

   CH_3-H + $Cl\cdot$ $\longrightarrow$ $CH_3\cdot$ +   HCl $(E_{act} = 3.8$ kcal mol$^{-1})$

   Therefore, ethyl radicals form much more rapidly in the mixture than methyl
   radicals, and this, leads to the more rapid formation of ethyl chloride.

9.12

$$Cl_2 \xrightarrow[\text{or heat}]{h\nu} 2\ Cl\bullet$$

$$
\text{Step 2a}\quad Cl\bullet\ +\ \underset{\overset{|}{H}}{\overset{\overset{Cl}{|}}{H:\!C\text{–}CH_3}}\ \longrightarrow\ H:\!Cl\ +\ \underset{\overset{|}{H}}{\overset{\overset{Cl}{|}}{\bullet C\text{–}CH_3}}
$$

$$
\text{Step 3a}\quad \underset{\overset{|}{H}}{\overset{\overset{Cl}{|}}{CH_3\!C\bullet}}\ +\ Cl:\!Cl\ \longrightarrow\ \underset{\overset{|}{H}}{\overset{\overset{Cl}{|}}{CH_3\text{–}C\text{–}Cl}}\ +\ Cl\bullet
$$
$$\text{1,1-dichloroethane}$$

Step 2b $Cl\bullet\ +\ H\text{–}CH_2CH_2Cl\ \longrightarrow\ H:\!Cl\ +\ \bullet CH_2CH_2Cl$

Step 3b $\bullet CH_2CH_2Cl\ +\ Cl:\!Cl\ \longrightarrow\ ClCH_2CH_2Cl\ +\ Cl\bullet$
$$\text{1,2-dichloroethane}$$

9.13 (a) There are a total of eight hydrogen atoms in propane. There are six equivalent 1° hydrogen atoms, replacement of any one of which leads to propyl chloride, and there are two equivalent 2° hydrogen atoms, replacement of any one of which leads to isopropyl chloride.

$$CH_3CH_2CH_3\ +\ Cl_2\ \longrightarrow\ CH_3CH_2CH_2Cl\ +\ \underset{\overset{|}{Cl}}{CH_3CHCH_3}$$

If all the hydrogen atoms were equally reactive, we would expect to obtain 75% propyl chloride and 25% isopropyl chloride:

$$\%\text{ Propyl chloride}\ =\ \%\ \times\ 100\ =\ 75\%$$
$$\%\text{ Isopropyl chloride}\ =\ \tfrac{2}{8}\ \times\ 100\ =\ 25\%$$

(b) Reasoning in the same way as in part (a), we would expect 90% isobutyl chloride and 10% *tert*-butyl chloride, if the hydrogen atoms were equally reactive.

$$\underset{\overset{|}{CH_3}}{\overset{\overset{CH_3}{|}}{CH_3\text{–}C\text{–}H}}\ +\ Cl_2\ \longrightarrow\ (CH_3)_2CHCH_2Cl\ +\ (CH_3)_3CCl$$

$$\%\text{ Isobutyl chloride}\ =\ \tfrac{9}{10}\ \times\ 100\ =\ 90\%$$
$$\%\ \textit{tert}\text{-Butyl chloride}\ =\ \tfrac{1}{10}\ \times\ 100\ =\ 10\%$$

(c) In the case of propane (see Section 9.6), we actually get more than twice as much isopropyl chloride (55%) than we would expect if the 1° and 2° hydrogen atoms were equally reactive (25%). Clearly, then, 2° hydrogen atoms are more than twice as reactive as 1° hydrogen atoms.

In the case of isobutane, we get almost four times as much *tert*-butyl chloride (37%) as we would get (10%) if the 1° and 3° hydrogen atoms were equally reactive. The order of reactivity of the hydrogens then must be

$$3°\ >\ 2°\ >\ 1°$$

9.14 The hydrogen atoms of these molecules are all equivalent. Replacing any one of them yields the same product.

$$\triangle \ + \ Cl_2 \ \xrightarrow{light} \ \triangle\!\!-Cl \quad \text{(+ more highly chlorinated products)}$$

$$\square \ + \ Cl_2 \ \xrightarrow{light} \ \square\!\!-Cl \quad \text{(+ more highly chlorinated products)}$$

We can minimize the amounts of more highly chlorinated products formed by using a large excess of the cyclopropane or cyclobutane. (And we can recover the unreacted cyclopropane or cyclobutane after the reaction is over.)

9.15 (a) [cyclopentane structure]

(b) $\begin{array}{c} H_3C \quad CH_3 \\ | \qquad | \\ CH_3-C-C-CH_3 \\ | \qquad | \\ H_3C \quad CH_3 \end{array}$

(c) $\begin{array}{c} CH_3 \\ | \\ CH_3-C-CH_3 \\ | \\ CH_3 \end{array}$

9.16 At lower temperatures, isomer distribution accurately reflects the inherent reactivities of the hydrogens of the alkanes. As the temperature is raised, chlorine atoms become more reactive, and hence less discriminating. If the temperature is high enough, hydrogens are replaced by chlorine on a purely statistical basis.

9.17 (a)

[Structure of 2-chloropentane] $\xrightarrow[\text{light}]{Cl_2}$ [structure] (2S,4S)-2,4-Dichloropentane + [structure] (2R,4S)-2,4-Dichloropentane

(b) They are diastereomers. (They are stereoisomers, but they are not mirror images of each other.)

(c) No, (2R,4S)-2,4-dichloropentane is achiral because it is a meso compound. (It has a plane of symmetry passing through C3.)

(d) No, the achiral meso compound would not be optically active.

(e) Yes, by fractional distillation or by gas-liquid chromatography. (Diastereomers have different physical properties. Therefore, the two isomers would have different vapor pressures.)

(f and g) In addition to the (2S,3S)-2,4-dichloropentane, (2S,3R)-2,4-dichloropentane, (2S,4S)-2,4-dichloropentane, and (2R,4S)-2,4-dichloropentane isomers described previously, we would also get:

(optically active)   (optically inactive)   (optically active)

9.18   (a) Seven (see below)

(c)   None of the fractions would show optical activity. Fractions (b), (d), and (f) are racemic forms; all others contain achiral molecules.

9.19   (a)   No, the only fractions that would contain chiral molecules (as enantiomers) would be those containing 1-chloro-2-methylbutane and the one containing 2-chloro-3-methylbutane. These fractions would not show optical activity, however, because they would contain racemic forms of the enantiomers.

(b)   Yes, the fractions containing 1-chloro-2-methylbutane and the one containing 2-chloro-3-methylbutane.

9.20   *Chain-Initiating Step*

$$Cl_2 \xrightarrow[\text{light}]{\text{heat},h\nu} 2\,Cl\cdot$$

Chain-Propagating Steps

$$CH_3CH_2CH_2 \cdot + HCl \xrightarrow{Cl_2} CH_3CH_2CH_2Cl + Cl \cdot$$

$$CH_3CH_2CH_2 \xleftarrow{Cl \cdot}$$

$$CH_3CH \cdot \underset{CH_3}{\overset{}{|}} + HCl \xrightarrow{Cl_2} CH_3CHCl \underset{CH_3}{\overset{}{|}} + Cl \cdot$$

9.21 (a) Three

$$CH_3CH_2CH_2CH_3 \xrightarrow{Cl_2} CH_3CH_2CH_2CH_2Cl + \underset{\mathbf{II}}{\begin{array}{c} CH_3 \\ H\text{---}C\text{---}Cl \\ CH_2 \\ CH_3 \end{array}} + \underset{\mathbf{III}}{\begin{array}{c} CH_3 \\ Cl\text{---}C\text{---}H \\ CH_2 \\ CH_3 \end{array}}$$

$$\mathbf{I}$$

Enantiomers as a
racemic form

(b) Only two: one fraction containing I, and another fraction containing the enantiomers II and III as a racemic form. (The enantiomers, having the same boiling points, would distill in the same fraction.)

(c) Both of them.

(d) The fraction containing the enantiomers.

9.22 (a) Five

$$\underset{A}{\begin{array}{c} CH_3 \\ Cl\text{---}C\text{---}H \\ CH_2 \\ CH_3 \end{array}} \xrightarrow{Cl_2} \underset{A}{\begin{array}{c} CH_2Cl \\ Cl\text{---}C\text{---}H \\ CH_2 \\ CH_3 \end{array}} + \underset{B}{\begin{array}{c} CH_3 \\ CCl_2 \\ CH_2 \\ CH_3 \end{array}} + \underset{C}{\begin{array}{c} CH_3 \\ Cl\text{---}C\text{---}H \\ Cl\text{---}C\text{---}H \\ CH_3 \end{array}} + \underset{D}{\begin{array}{c} CH_3 \\ Cl\text{---}C\text{---}H \\ H\text{---}C\text{---}Cl \\ CH_3 \end{array}} + \underset{E}{\begin{array}{c} CH_3 \\ Cl\text{---}C\text{---}H \\ CH_2 \\ CH_2Cl \end{array}}$$

(b) Five. None of the fractions would be a racemic form.

(c) The fractions containing **A, D,** and **E.** The fraction containing **B** and **C** would be optically inactive. (**B** contains no stereocenter and **C** is a meso compound.)

9.23 (a) Oxygen-oxygen bonds are especially weak, that is,

$$\text{HO---OH} \qquad DH° = 51 \text{ kcal mol}^{-1}$$
$$\text{CH}_3\text{CH}_2\text{O---OCH}_3 \qquad DH° = 44 \text{ kcal mol}^{-1}$$

This means that a peroxide will dissociate into radicals at a relatively low temperature.

$$\text{RO---OR} \xrightarrow{100\text{-}200°C} 2 \text{ RO} \cdot$$

Oxygen-hydrogen single bonds, on the other hand, are very strong. (For HO-H, $DH° = 119$ kcal mol^{-1}.) This means that reactions like the following will be highly exothermic.

$$RO• + R-H \longrightarrow RO-H + R•$$

(b) *Step 1* $(CH_3)_3CO-OC(CH_3)_3 \xrightarrow{\text{heat}} 2 \ (CH_3)_3CO•$ $\left.\begin{array}{c} \\ \\ \end{array}\right\}$ Chain Initiation

Step 2 $(CH_3)_3CO• + R-H \longrightarrow (CH_3)_3COH + R•$

Step 3 $R• + Cl-Cl \longrightarrow R-Cl + Cl•$ $\left.\begin{array}{c} \\ \\ \end{array}\right\}$ Chain Propagation

Step 4 $Cl• + R-H \longrightarrow H-Cl + R•$

9.24
$$\underset{(3°)}{CH_3\overset{CH_3}{\underset{•}{C}}CH_2CH_3} > \underset{(2°)}{CH_3\overset{CH_3}{\underset{•}{CH}}CHCH_3} > \underset{(1°)}{CH_3\overset{CH_3}{CH}CH_2CH_2•} \sim \underset{(1°)}{CH_3\overset{CH_2•}{CH}CH_2CH_3}$$

9.25   (1) $Cl_2 \longrightarrow 2 \ Cl•$ $\Delta H° = +58$ kcal mol^{-1}

(2) $Cl• + CH_4 \longrightarrow CH_3Cl + H•$ $\Delta H° = +20.5$ kcal mol^{-1}

(3) $H• + Cl_2 \longrightarrow HCl + Cl•$ $\Delta H° = -45$ kcal mol^{-1}

This mechanism is highly unlikely to compete with one given in Section 9.4 because step (2) of this mechanism is highly endothermic ($\Delta H° = +20.5$ kcal mol^{-1}). This means that the energy of activation for this step, a chain-propagating step, will have to be larger than $+20.5$ kcal mol^{-1}. Notice in Section 9.5B that neither of the chain-propagating steps for the mechanism given in Section 9.4 has an energy of activation greater than $+3.8$ kcal mol^{-1}. The alternative mechanism given in this problem, consequently, will proceed at a rate that is very much slower than the one given in Section 9.4 and will, for all practical purposes, not compete with it at all.

9.26   **(a)**

(b) $CH_3-\overset{\overset{CH_3}{|}}{\underset{\underset{CH_3}{|}}{C}}-CH_3 \xrightarrow[\substack{\text{heat,}\\\text{light}}]{Br_2} CH_3-\overset{\overset{CH_3}{|}}{\underset{\underset{CH_3}{|}}{C}}-CH_2Br \xrightarrow[\substack{\text{diethyl}\\\text{ether}}]{Li} CH_3-\overset{\overset{CH_3}{|}}{\underset{\underset{CH_3}{|}}{C}}-CH_2Li \xrightarrow{CuI}$

$[(CH_3)_3CCH_2]_2CuLi + CH_3Br \longrightarrow CH_3-\overset{\overset{CH_3}{|}}{\underset{\underset{CH_3}{|}}{C}}-CH_2-CH_3$

$\overset{}{\underset{CH_4}{\underset{\uparrow Br_2 \,\, \substack{\text{heat,}\\\text{light}}}{}}}$

(c) $CH_3CH_2\overset{\curvearrowleft}{-Br} + Na^+ I^- \xrightarrow[\text{acetone}]{S_N2} CH_3CH_2-I$
[from part (a)]

(d) $CH_3CH_2\overset{\curvearrowleft}{-Br} + Na^+ OH^- \xrightarrow[\text{acetone}]{S_N2} CH_3CH_2-OH \xrightarrow[(-H_2)]{Na}$
[from part (a)]

$CH_3CH_2-O^-Na^+ + CH_3CH_2\overset{\curvearrowleft}{-Br} \xrightarrow{S_N2} CH_3CH_2-O-CH_2CH_3$
[from part (a)]

(e) $\xrightarrow[E2]{(CH_3)_3CO^-K^+}$
[from part (a)]

(f) $CH_3-\overset{\overset{CH_3}{|}}{\underset{\underset{CH_3}{|}}{C}}-H \xrightarrow[\substack{\text{heat,}\\\text{light}}]{Br_2} CH_3-\overset{\overset{CH_3}{|}}{\underset{\underset{CH_3}{|}}{C}}-Br \xrightarrow[\substack{\text{diethyl}\\\text{ether}}]{Li} CH_3-\overset{\overset{CH_3}{|}}{\underset{\underset{CH_3}{|}}{C}}-Li \xrightarrow{T_2O} CH_3-\overset{\overset{CH_3}{|}}{\underset{\underset{CH_3}{|}}{C}}-T$

(g) $CH_3CH_2\overset{\curvearrowleft}{-Br} + Na^+ N_3^- \xrightarrow[\text{acetone}]{S_N2} CH_3CH_2-N_3$
[from part (a)]

9.27 Chlorine atoms are electronegative and abstract hydrogens according to their "electron richness." The electronegative fluorine atom reduces electron density in proportion to its proximity to the different CH_2 groups. The CH_3 group is the least reactive site because the bond dissociation energy for a CH_3 group is greater than that for a CH_2 group.

Use the single-bond dissociation energies of Table 9.1:

Table 9.1   Single-bond homolytic dissociation energies $DH°$ at 25°C

Compound	kcal mol^{-1}	kJ mol^{-1}	Compound	kcal mol^{-1}	kJ mol^{-1}
		A : B $\longrightarrow$ A· + B·			
H–H	104	435	$(CH_3)_2CH–H$	94.5	395
D–D	106	444	$(CH_3)_2CH–F$	105	439
F–F	38	159	$(CH_3)_2CH–Cl$	81	339
Cl–Cl	58	243	$(CH_3)_2CH–Br$	68	285
Br–Br	46	192	$(CH_3)_2CH–I$	53	222
I–I	36	151	$(CH_3)_2CH–OH$	92	385
H–F	136	569	$(CH_3)_2CH–OCH_3$	80.5	337
H–Cl	103	431	$(CH_3)_2CHCH_2–H$	98	410
H–Br	87.5	366	$(CH_3)_3C–H$	91	381
H–I	71	297	$(CH_3)_3C–Cl$	78.5	328
$CH_3–H$	104	435	$(CH_3)_3C–Br$	63	264
$CH_3–F$	108	452	$(CH_3)_3C–I$	49.5	207
$CH_3–Cl$	83.5	349	$(CH_3)_3C–OH$	90.5	379
$CH_3–Br$	70	293	$(CH_3)_3C–OCH_3$	78	326
$CH_3–I$	56	234	$C_6H_5CH_2–H$	85	356
$CH_3–OH$	91.5	383	$CH_2=CHCH_2–H$	85	356
$CH_3–OCH_3$	80	335	$CH_2=CH–H$	108	452
$CH_3CH_2–H$	98	410	$C_6H_5–H$	110	460
$CH_3CH_2–F$	106	444	$HC≡C–H$	125	523
$CH_3CH_2–Cl$	81.5	341	$CH_3–CH_3$	88	368
$CH_3CH_2–Br$	69	289	$CH_3CH_2–CH_3$	85	356
$CH_3CH_2–I$	53.5	224	$CH_3CH_2CH_2–CH_3$	85	356
$CH_3CH_2–OH$	91.5	383	$CH_3CH_2–CH_2CH_3$	82	343
$CH_3CH_2–OCH_3$	80	335	$(CH_3)_2CH–CH_3$	84	351
			$(CH_3)_3C–CH_3$	80	335
$CH_3CH_2CH_2–H$	98	410	HO–H	119	498
$CH_3CH_2CH_2–F$	106	444	HOO–H	90	377
$CH_3CH_2CH_2–Cl$	81.5	341	HO–OH	51	213
$CH_3CH_2CH_2–Br$	69	289	$CH_3CH_2O–OCH_3$	44	184
$CH_3CH_2CH_2–I$	53.5	224	$CH_3CH_2O–H$	103	431
$CH_3CH_2CH_2–OH$	91.5	383			
$CH_3CH_2CH_2–OCH_3$	80	335	$CH_3\overset{\displaystyle O}{\overset{\|}{C}}–H$	87	364

QUIZ

9.1 On the basis of Table 9.1, what is the order of decreasing stability of the radicals, $HC{\equiv}C\cdot$ $CH_2{=}CH\cdot$ $CH_2{=}CHCH_2\cdot$?

(a) $HC{\equiv}C\cdot > CH_2{=}CH\cdot > CH_2{=}CHCH_2\cdot$

(b) $CH_2{=}CH\cdot > HC{\equiv}C\cdot > CH_2{=}CHCH_2\cdot$

(c) $CH_2{=}CHCH_2\cdot > HC{\equiv}C\cdot > CH_2{=}CH\cdot$

(d) $CH_2{=}CHCH_2\cdot > CH_2{=}CH\cdot > HC{\equiv}C\cdot$

(e) $CH_2{=}CH\cdot > CH_2{=}CHCH_2\cdot > HC{\equiv}C\cdot$

9.2 In the radical chlorination of methane, one propagation step is shown as

$$Cl\cdot + CH_4 \longrightarrow HCl + \cdot CH_3$$

Why do we eliminate the possibility that this step goes as shown below?

$$Cl\cdot + CH_4 \longrightarrow CH_3Cl + H\cdot$$

(a) Because in the next propagation step, $H\cdot$ would have to react with Cl_2 to form $Cl\cdot$ and HCl; this reaction is not feasible.

(b) Because this alternative step has a more endothermic $\Delta H°$ than the first.

(c) Because free hydrogen atoms cannot exist.

(d) Because this alternative step is not consistent with the high photochemical efficiency of this reaction.

9.3 Pure (S)-$CH_3CH_2CHBrCH_3$ is subjected to monobromination to form several isomers of $C_4H_8Br_2$. Which of the following is not produced?

(a)   (b)   (c)

(d) $CH_3CH_2CBr_2CH_3$ (e) (R)-$CH_3CH_2CHBrCH_2Br$

9.4 Using the data of Table 9.1, calculate the heat of reaction, $\Delta H°$, of the reaction,

$$CH_3CH_3 + Br_2 \longrightarrow CH_3CH_2Br + HBr$$

(a) 12.5 kcal mol^{-1} (b) -12.5 kcal mol^{-1} (c) 300.5 kcal mol^{-1}

(d) -300.5 kcal mol^{-1} (e) -58.5 kcal mol^{-1}

9.5   Which gas-phase reaction would have $E_{act} = 0$?

   (a) $CH_3^\bullet + (CH_3)_3C-H \longrightarrow CH_4 + (CH_3)_3C^\bullet$

   (b) $CH_3^\bullet + CH_3CH_3 \longrightarrow CH_4 + CH_3CH_2^\bullet$

   (c) $CH_3CH_2^\bullet + CH_3CH_2^\bullet \longrightarrow CH_3CH_2CH_2CH_3$

   (d) $Br^\bullet + H-Cl \longrightarrow H-Br + Cl^\bullet$

   (e) $Br^\bullet + H-I \longrightarrow H-Br + I^\bullet$

9.6   What is the most stable radical that would be formed in the following reaction?

$$Cl^\bullet + CH_3CH_2\overset{\overset{\textstyle CH_3}{|}}{C}HCH_3 \longrightarrow \boxed{} + HCl$$

9.7   The reaction of 2-methylbutane with chlorine would yield a total of _____ different monochloro products (including stereoisomers).

9.8   For which reaction would the transition state most resemble the products?

   (a) $CH_4 + F^\bullet \longrightarrow CH_3^\bullet + HF$

   (b) $CH_4 + Cl^\bullet \longrightarrow CH_3^\bullet + HCl$

   (c) $CH_4 + Br^\bullet \longrightarrow CH_3^\bullet + HBr$

   (d) $CH_4 + I^\bullet \longrightarrow CH_3^\bullet + HI$

C

SPECIAL TOPIC
Chain-Growth Polymers

C.1 Head to tail polymerization leads to a more stable radical on the growing polymer chain. In head to tail coupling, the radical is 2° (actually 2° benzylic, and as we shall see in Section 15.12A this makes it even more stable). In head to head coupling, the radical is 1°.

C.2 (a)

$$R\cdot \quad + \quad CH_2{=}CH{-}OCH_3 \quad \longrightarrow \quad R{-}CH_2{-}CH\cdot{-}OCH_3 \quad \xrightarrow{\ CH_2=CH{-}OCH_3\ }$$

(from initiator)   Monomer

$$R{-}CH_2{-}\underset{OCH_3}{CH}{-}CH_2{-}\underset{OCH_3}{CH}\cdot \quad \xrightarrow{\ CH_2=CH{-}OCH_3\ } \quad \text{etc.}$$

(b)

$$R\cdot \quad + \quad CH_2{=}CCl_2 \quad \longrightarrow \quad R{-}CH_2{-}CCl_2\cdot \quad \xrightarrow{\ CH_2=CCl_2\ }$$

(from initiator)   Monomer

$$R{-}CH_2{-}CCl_2{-}CH_2{-}CCl_2\cdot \quad \xrightarrow{\ CH_2=CCl_2\ } \quad \text{etc.}$$

C.3 In the cationic polymerization of isobutylene (see text), the growing polymer chain has a stable 3° carbocation at the end. In the cationic polymerization of ethene, for example, the intermediates would be much less stable 1° cations.

$$H^+ \quad + \quad CH_2{=}CH_2 \quad \longrightarrow \quad CH_3CH_2^+ \quad \xrightarrow{\ CH_2=CH_2\ } \quad CH_3CH_2CH_2CH_2^+ \quad \xrightarrow{\ etc.\ }$$

1° Carbocation

With vinyl chloride, the cations at the end of the growing chain would be destabilized by electron-withdrawing groups.

$$H^+ \quad + \quad CH_2{=}\underset{Cl}{CH} \quad \longrightarrow \quad CH_3\underset{Cl}{CH^+} \quad \xrightarrow{\ CH_2=CHCl\ } \quad CH_3\underset{Cl}{CH}CH_2\underset{Cl}{CH^+} \quad \xrightarrow{\ etc.\ }$$

$$H^+ \quad + \quad CH_2{=}\underset{CN}{CH} \quad \longrightarrow \quad CH_3\underset{CN}{CH^+} \quad \xrightarrow{\ CH_2=CHCN\ } \quad CH_3\underset{CN}{CH}CH_2\underset{CN}{CH^+} \quad \xrightarrow{\ etc.\ }$$

C.4

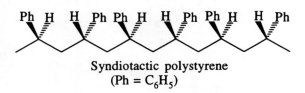

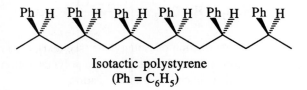

C.5 (a)

Ph H Ph H H Ph H Ph Ph H H Ph

Atactic polystyrene
(Ph = C_6H_5)

Ph H H Ph Ph H H Ph Ph H H Ph

Syndiotactic polystyrene
(Ph = C_6H_5)

Ph H Ph H Ph H Ph H Ph H Ph H

Isotactic polystyrene
(Ph = C_6H_5)

(b) The solution of isotactic polystyrene.

D

SPECIAL TOPIC
Divalent Carbon Compounds: Carbenes

D.1 (a)

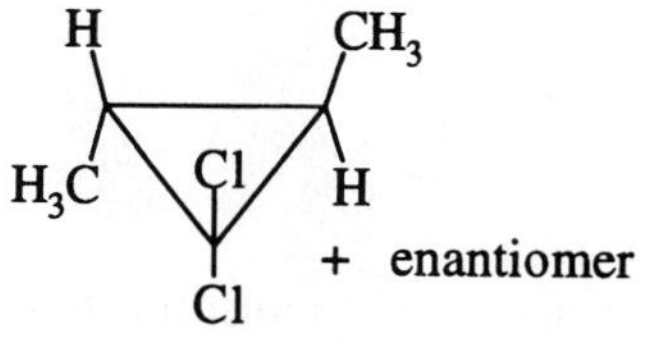

+ enantiomer

(b)

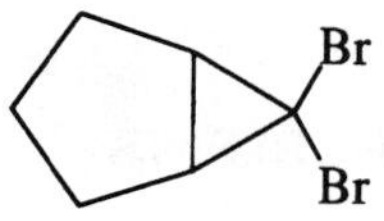

(c)

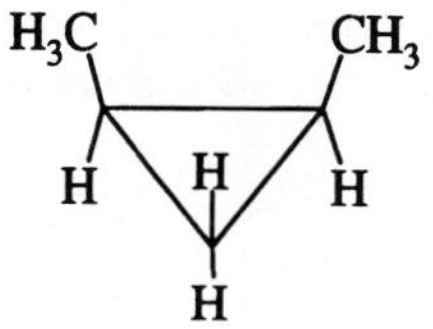

D.2

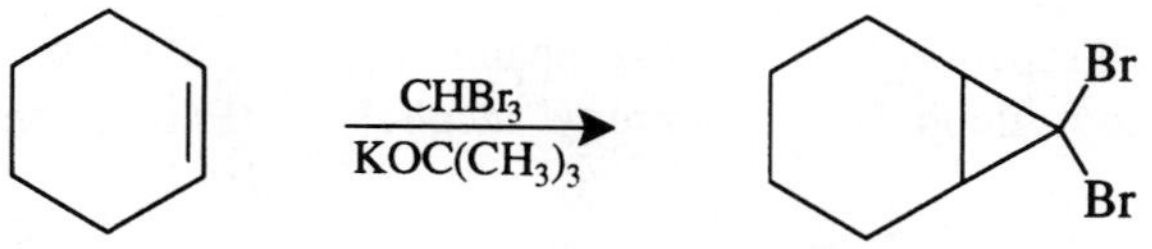

$$\xrightarrow[\text{KOC(CH}_3)_3]{\text{CHBr}_3}$$

D.3

$$\xrightarrow[\text{Zn(Cu)}]{\text{CH}_3\text{CHI}_2}$$

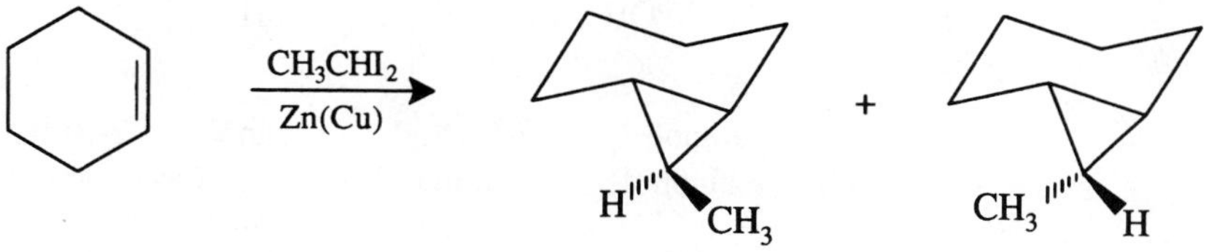

10 ALCOHOLS AND ETHERS

SOLUTIONS TO PROBLEMS

10.1 These names mix two systems of nomenclature (radicofunctional and substitutive; see Section 4.3F). The proper names are: isopropyl alcohol (radicofunctional) or 2-propanol (substitutive), and *tert*-butyl alcohol (radicofunctional) and 2-methyl-2-propanol (substitutive). Names with mixed systems of nomenclature should not be used.

10.2 (a)

1-Propanol   2-Propanol   Methoxyethane
(Propyl alcohol)   (Isopropyl alcohol)   (Ethyl methyl ether)

(b)

1-Butanol   2-Methyl-1-propanol   2-Butanol   2-Methyl-2-propanol
(Butyl alcohol)   (Isobutyl alcohol)   (*sec*-Butyl alcohol)   (*tert*-Butyl alcohol)

1-Methoxypropane   Ethoxyethane   2-Methoxypropane
(Methyl propyl ether)   (Diethyl ether)   (Isopropyl methyl ether)

10.3 The presence of two —OH groups in each molecule of 1,2-propanediol and 1,3-propanediol allows their molecules to form more hydrogen bonds. Greater hydrogen-bond formation means that the molecules of 1,2-propanediol and 1,3-propanediol are more highly associated, and, consequently, their boiling points are higher.

10.4 (a) CH_3CH_2OH (b) $CH_3\underset{\underset{\displaystyle OH}{|}}{C}HCH_3$ (c) $CH_3\underset{\underset{\displaystyle CH_3}{|}}{\overset{\overset{\displaystyle CH_3}{|}}{C}}{-}OH$ (d) $CH_3\underset{\underset{\displaystyle OH}{|}}{\overset{\overset{\displaystyle CH_3}{|}}{C}}CH_2CH_3$

10.5 A rearrangement takes place.

$$CH_3-\underset{\underset{CH_3}{|}}{\overset{\overset{CH_3}{|}}{C}}-CH{=}CH_2 \ + \ H{-}\overset{+}{\underset{\underset{H}{|}}{O}}{-}H \longrightarrow CH_3-\underset{\underset{CH_3}{|}}{\overset{\overset{CH_3}{|}}{C}}{-}\overset{+}{C}H{-}CH_3 \xrightarrow[\text{shift}]{\text{1,2-methanide}}$$

$$CH_3-\overset{+}{\underset{\underset{CH_3\ CH_3}{|\ \ \ |}}{C}}{-}CH{-}CH_3 \xrightarrow{:\ddot{O}H_2} CH_3-\underset{\underset{CH_3\ CH_3}{|\ \ \ |}}{\overset{\overset{\overset{H}{|}}{\overset{+}{O}{-}H}}{C}}{-}CH{-}CH_3 \xrightarrow{:\ddot{O}H_2}$$

$$CH_3-\underset{\underset{CH_3\ CH_3}{|\ \ \ |}}{\overset{\overset{OH}{|}}{C}}{-}CH{-}CH_3$$

2,3-Dimethyl-2-butanol
(major product)

10.6 (a) $CH_3\underset{\underset{CH_3}{|}}{\overset{\overset{CH_3}{|}}{C}}{=}CH_2 \xrightarrow[\text{THF-H}_2\text{O}]{\text{Hg(OAc)}_2} CH_3{-}\underset{\underset{OH}{|}}{\overset{\overset{CH_3}{|}}{C}}{-}CH_2HgOAc \xrightarrow[\text{OH}^-]{\text{NaBH}_4} CH_3{-}\underset{\underset{OH}{|}}{\overset{\overset{CH_3}{|}}{C}}{-}CH_3$

(b) $CH_3CH{=}CH_2 \xrightarrow[\text{THF-H}_2\text{O}]{\text{Hg(OAc)}_2} CH_3\underset{\underset{OH}{|}}{CH}CH_2HgOAc \xrightarrow[\text{OH}^-]{\text{NaBH}_4} CH_3\underset{\underset{OH}{|}}{CH}CH_3$

(c) $CH_3\underset{\underset{CH_3}{|}}{C}{=}CHCH_3 \xrightarrow[\text{THF-H}_2\text{O}]{\text{Hg(OAc)}_2} CH_3\underset{\underset{CH_3}{|}}{\overset{\overset{HO}{|}}{C}}{-}\overset{\overset{HgOAc}{|}}{C}HCH_3 \xrightarrow[\text{OH}^-]{\text{NaBH}_4} CH_3\underset{\underset{CH_3}{|}}{\overset{\overset{OH}{|}}{C}}CH_2CH_3$

10.7 (a)

$$\overset{\diagdown}{\underset{\diagup}{C}}{=}\overset{\diagup}{\underset{\diagdown}{C}} \ + \ {}^+HgO\overset{\overset{O}{\|}}{C}CF_3 \longrightarrow \ {-}\overset{|}{\underset{\delta_+}{C}}{\cdots}\underset{\underset{\delta_+}{HgO\overset{\overset{O}{\|}}{C}CF_3}}{\overset{|}{C}}{-} \xrightarrow{R{-}\ddot{O}{-}H} {-}\overset{\overset{R\overset{+}{O}H}{|}}{C}{-}\underset{\underset{HgO\overset{\overset{O}{\|}}{C}CF_3}{|}}{\overset{|}{C}}{-}$$

$$\xrightarrow{-H^+} {-}\overset{\overset{RO}{|}}{\underset{|}{C}}{-}\underset{\underset{HgO\overset{\overset{O}{\|}}{C}CF_3}{|}}{\overset{|}{C}}{-}$$

(b) $CH_3{-}\underset{\underset{CH_3}{|}}{C}{=}CH_2 \xrightarrow[\substack{\text{THF-CH}_3\text{OH}\\\text{solvomercuration}}]{\text{Hg(O\overset{\overset{O}{\|}}{C}CF}_3)_2} CH_3{-}\underset{\underset{OCH_3}{|}}{\overset{\overset{CH_3}{|}}{C}}{-}CH_2HgO\overset{\overset{O}{\|}}{C}CF_3 \xrightarrow[\text{demercuration}]{\text{NaBH}_4/\text{OH}^-}$

$$CH_3\underset{\underset{OCH_3}{|}}{\overset{\overset{CH_3}{|}}{C}}CH_3 \ + \ Hg \ + \ CF_3COO^-$$

10.8 (a) $3\ CH_3CH_2CH=CH_2 \xrightarrow{THF:BH_3} (CH_3CH_2CH_2CH_2)_3B$

(b) $3\ CH_3-\underset{\underset{CH_3}{|}}{C}=CH_2 \xrightarrow{THF:BH_3} (CH_3\underset{\underset{CH_3}{|}}{C}HCH_2)_3B$

(c) $3\ CH_3CH=CHCH_3 \xrightarrow{THF:BH_3} (CH_3CH_2\underset{\underset{CH_3}{|}}{C}H)_3B$

(d) 3 $\xrightarrow[\substack{\text{syn addition} \\ \text{anti Markovnikov}}]{THF:BH_3}$ + enantiomer

10.9 $CH_3\underset{\underset{CH_3}{|}}{C}=CHCH_3 \xrightarrow{THF:BH_3}$

Disiamylborane

10.10 (a) $CH_3CH_2CH_2CH=CH_2 \xrightarrow[\text{(2) } H_2O_2,\ OH]{\text{(1) } THF:BH_3} CH_3CH_2CH_2CH_2CH_2OH$
1-Pentene

(b) $CH_3CH_2CH_2\underset{\underset{CH_3}{|}}{C}=CH_2 \xrightarrow[\text{(2) } H_2O_2,\ OH]{\text{(1) } THF:BH_3} CH_3CH_2CH_2\underset{\underset{CH_3}{|}}{C}HCH_2OH$
2-Methyl-1-pentene

(c) $CH_3CH_2\underset{\underset{CH_3}{|}}{C}=CHCH_3 \xrightarrow[\text{(2) } H_2O_2,\ OH]{\text{(1) } THF:BH_3} CH_3CH_2\underset{\underset{CH_3}{|}}{C}H\overset{\overset{OH}{|}}{C}HCH_3$
3-Methyl-2-pentene

(d) $CH_3\underset{\underset{CH_3}{|}}{C}=CHCH_2CH_3 \xrightarrow[\text{(2) } H_2O_2,\ OH]{\text{(1) } THF:BH_3} CH_3\underset{\underset{CH_3}{|}}{C}H\overset{\overset{OH}{|}}{C}HCH_2CH_3$
2-Methyl-2-pentene

(e) —CH_3 $\xrightarrow[\text{(2) } H_2O_2,\ OH]{\text{(1) } THF:BH_3}$ + enantiomer
1-Methylcyclobutene

10.11 (a) $3\ CH_3\underset{\underset{CH_3}{|}}{C}HCH=CH_2 \xrightarrow{THF:BH_3}$ $B \xrightarrow{CH_3CO_2D}$

$3\ CH_3\underset{\underset{CH_3}{|}}{C}HCH_2CH_2D$

(b) $CH_3C=CHCH_3$ $\xrightarrow{THF:BH_3}$ $CH_3CH-\overset{\overset{\displaystyle -B-}{|}}{CH}-CH_3$ $\xrightarrow{CH_3CO_2D}$

with CH_3 substituent, CH_3 substituent

$CH_3\underset{\underset{\displaystyle CH_3}{|}}{CH}CHDCH_3$

(c) $-CH_3$ $\xrightarrow{THF:BH_3}$ $\overset{CH_3}{\underset{H}{\overset{H}{}}}$ (+ enantiomer) $\xrightarrow{CH_3CO_2D}$

$\overset{B-}{|}$

$\overset{CH_3}{\underset{H}{\overset{H}{}}}$ (+ enantiomer)

D

(d) $\overset{CH_3}{}$ $\xrightarrow{THF:BD_3}$ $\overset{D}{}$ $-CH_3$ $\xrightarrow{CH_3CO_2T}$

$-B-$ (+ enantiomer)

H

$\overset{D}{}$ $-CH_3$

$-T$ (+ enantiomer)

H

10.12 (a) CH_3CH_2OH + $NaNH_2$ $\longrightarrow$ CH_3CH_2ONa + NH_3

Stronger acid · Stronger base · Weaker base · Weaker acid

(b) CH_3CH_2OH + $HC\equiv CNa$ $\longrightarrow$ CH_3CH_2ONa + $HC\equiv CH$

(c) CH_3CH_2OH + CH_3CH_2Li $\longrightarrow$ CH_3CH_2OLi + CH_3CH_3

10.13   Use an alcohol containing labeled oxygen. If all of the label appears in the sulfonate ester, then one can conclude that the alcohol C–O bond does not break during the reaction:

$$R-{}^{18}O-H \quad + \quad R'-SO_2Cl \quad \xrightarrow[(-HCl)]{base} \quad R-{}^{18}O-SO_2-R'$$

10.14 (a) $H_3C-\bigcirc-SO_2OH$ $\xrightarrow[(-POCl_3 -HCl)]{PCl_5}$ $H_3C-\bigcirc-SO_2Cl$

$\xrightarrow[base\ (-HCl)]{CH_3OH}$ $H_3C-\bigcirc-SO_2OCH_3$

(b) $CH_3SO_2OH \xrightarrow[\text{(- POCl}_3\text{-HCl)}]{PCl_5} CH_3SO_2Cl \xrightarrow[\text{base (-HCl)}]{(CH_3)_2CHCH_2OH}$

$CH_3SO_2OCH_2CH(CH_3)_2$

(c) $CH_3SO_2Cl \xrightarrow[\text{base (-HCl)}]{(CH_3)_3COH} CH_3SO_2OC(CH_3)_3$

10.15   (a)

$$\underset{\substack{\text{(R)-2-Butanol}}}{\overset{\substack{H_3C \\ H_5C_2}}{\underset{}{H\text{''''}C-OH}}} + \text{TsCl} \xrightarrow[\text{(-HCl)}]{\text{retention}} \overset{\substack{H_3C \\ H_5C_2}}{\underset{}{H\text{''''}C-OTs}}$$

(b) $HO^- + \overset{\substack{H_3C \\ H_5C_2}}{H\text{''''}C-OTs} \xrightarrow[S_N2]{\text{inversion}} HO-\overset{\substack{CH_3 \\ C_2H_5}}{C\text{''''}H} + OTs^-$

(c)

cis-4-Methyl-
cyclohexanol

$\xrightarrow[\text{retention}]{\text{TsCl}}$

$\xrightarrow[\text{inversion}]{Cl^-}$

trans-1-Chloro-4-
methylcyclohexane

10.16   (a) Tertiary alcohols react faster than secondary alcohols because they form more stable carbocations; that is, 3° rather than 2°:

$$CH_3-\underset{\substack{| \\ CH_3}}{\overset{\substack{CH_3 \\ |}}{C}}-\overset{\substack{+ \\ |}}{\underset{\substack{| \\ H}}{O}}-H \rightleftharpoons CH_3-\underset{\substack{| \\ CH_3}}{\overset{\substack{CH_3 \\ |}}{C}}+ + H_2O$$

$$\xrightarrow{Cl^-} CH_3-\underset{\substack{| \\ CH_3}}{\overset{\substack{CH_3 \\ |}}{C}}-Cl$$

(b) CH_3OH reacts faster than 1° alcohols because it offers less hindrance to S_N2 attack. (Recall that CH_3OH and 1° alcohols must react through an S_N2 mechanism.)

10.17

$$\underset{\substack{| \\ OH}}{CH_3\overset{\substack{CH_3 \\ |}}{C}HCHCH_3} + HBr \rightleftharpoons \underset{\substack{+ \\ OH_2}}{CH_3\overset{\substack{CH_3 \\ |}}{C}HCHCH_3} + Br^-$$

$$\Big\updownarrow \text{-H}_2O$$

$$\underset{\substack{| \\ Br}}{CH_3\overset{\substack{CH_3 \\ |}}{C}CH_2CH_3} \xleftarrow{Br^-} \underset{\substack{+}}{CH_3\overset{\substack{CH_3 \\ |}}{C}CH_2CH_3} \xleftarrow{} \underset{\substack{+}}{CH_3\overset{\substack{CH_3 \\ |}}{C}HCHCH_3}$$

10.18 (a)

$$CH_3-\underset{\underset{CH_3}{|}}{\overset{\overset{CH_3}{|}}{C}}-OH \xrightarrow{H^+} CH_3-\underset{\underset{CH_3}{|}}{\overset{\overset{CH_3}{|}}{C}}-\overset{+}{O}H_2 \xrightarrow{-H_2O} CH_3-\underset{\underset{CH_3}{|}}{\overset{\overset{CH_3}{|}}{C}}+ \xrightarrow[(1° \text{ only})]{R-OH}$$

$$CH_3-\underset{\underset{CH_3}{|}}{\overset{\overset{CH_3}{|}}{C}}-\overset{+}{\underset{H}{O}}-R \xrightarrow{-H^+} CH_3-\underset{\underset{CH_3}{|}}{\overset{\overset{CH_3}{|}}{C}}-O-R$$

This reaction succeeds because a 3° carbocation is much more stable than a 1° carbocation. Consequently, mixing the 1° alcohol and H_2SO_4 does not lead to formation of appreciable amounts of a 1° carbocation. However, when the 3° alcohol is added, it is rapidly converted to a 3° carbocation, which then reacts with the 1° alcohol that is present in the mixture.

10.19 (a) (1)

$$CH_3\overset{\overset{CH_3}{|}}{\underset{}{CHO^-}} Na^+ + CH_3-L \longrightarrow CH_3\overset{\overset{CH_3}{|}}{\underset{}{CHO}}-CH_3 + L^- + Na^+$$
$$(L = X, OSO_2R, \text{ or } OSO_2OR)$$

(2)

$$CH_3O^- + CH_3-\overset{\overset{CH_3}{|}}{\underset{}{CH}}-L \longrightarrow CH_3O-\overset{\overset{CH_3}{|}}{\underset{}{CHCH_3}} + L^-$$
$$(L = X, OSO_2R, \text{ or } OSO_2OR)$$

(b) Both methods involve S_N2 reactions. Therefore, method (1) is better because substitution takes place at an unhindered methyl carbon atom. In method (2) where substitution must take place at a relatively hindered secondary carbon atom, the reaction would be accompanied by considerable elimination.

10.20 Reaction of the alcohol with K and then of the resulting salt with C_2H_5Br does not break bonds to the stereocenter, and these reactions therefore occur with retention of configuration at the stereocenter.

Reaction of the tosylate, $C_6H_5CH_2\underset{\underset{OTs}{|}}{CHCH_3}$, with C_2H_5OH in K_2CO_3 solution, however,

is an S_N2 reaction that takes place at the stereocenter and thus it occurs with inversion at the stereocenter.

10.21

$$Cl-CH_2\overset{\overset{CH_2-CH_2}{|\quad\quad|}}{\underset{\underset{OH}{|}}{CH_2}} \rightleftharpoons \overset{OH^-}{} Cl-CH_2\overset{\overset{CH_2-CH_2}{|\quad\quad|}}{\underset{\underset{:\ddot{O}:^-}{}}{CH_2}} \longrightarrow \overset{\overset{CH_2-CH_2}{|\quad\quad|}}{\underset{\underset{O}{\diagdown\diagup}}{CH_2\quad CH_2}} + Cl^-$$
$$+ H_2O$$

10.22 (a) $HO^- + HOCH_2-CH_2-Cl \rightleftarrows H_2O + {}^-\ddot{O}-CH_2-CH_2 \overset{\curvearrowleft}{\overset{\frown}{C}l} \longrightarrow$

$$H_2C\overset{O}{\overset{\diagup\diagdown}{-}}CH_2 + Cl^-$$

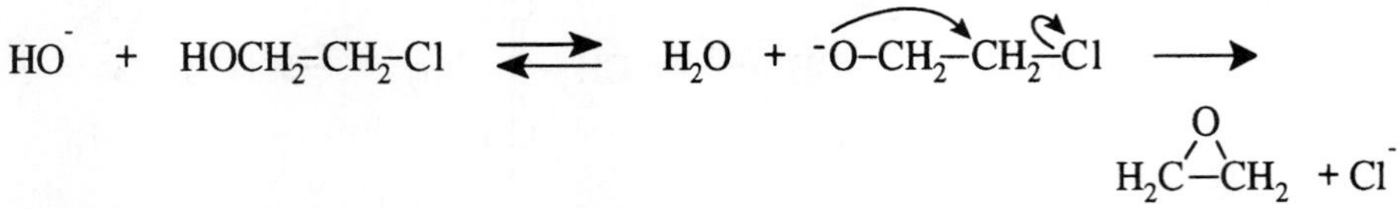

(b) The $-\ddot{\ddot{O}}{:}^-$ group must displace the Cl⁻ from the backside,

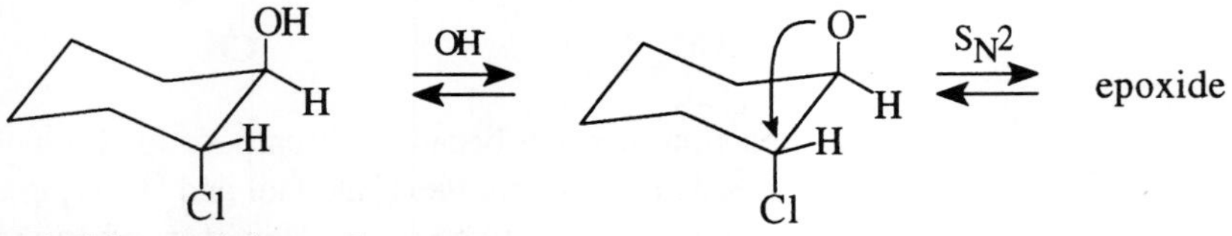

trans-2-Chlorocyclohexanol

Backside attack is not possible with the cis isomer (below); therefore, it does not form an epoxide.

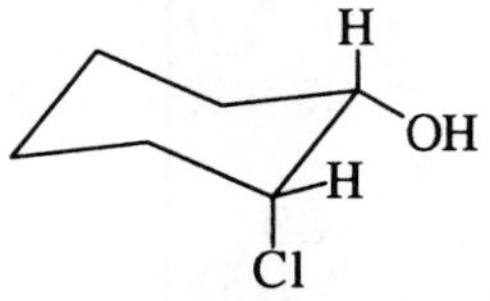

cis-2-Chlorocyclohexanol

10.23 (a)

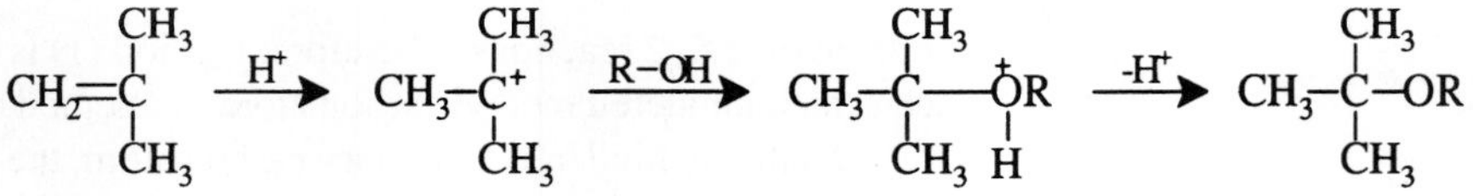

(b) The *tert*-butyl group is easily removed because, in acid, it is easily converted to a relatively stable, tertiary carbocation.

(c)

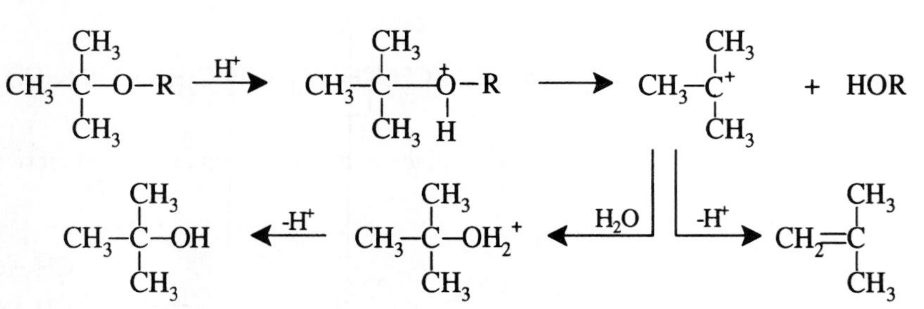

10.24 (a)

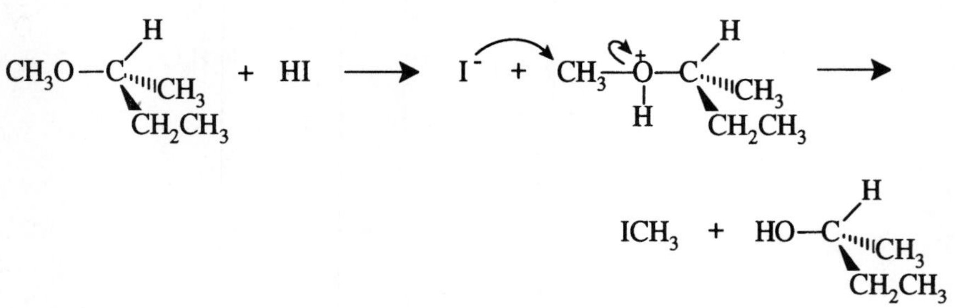

S_N2 attack of I^- occurs at the methyl carbon atom because it is less hindered; therefore, the bond between the *sec*-butyl group and the oxygen is not broken.

(b) CH_3–O–$C(CH_3)_3$ + HI $\longrightarrow$ CH_3–$\overset{+}{\underset{H}{O}}$–$C(CH_3)_3$ + $\overset{-}{I}$

CH_3OH + CH_3–$\overset{CH_3}{\underset{CH_3}{C}}$+ $\xrightarrow{I^-}$ CH_3–$\overset{CH_3}{\underset{CH_3}{C}}$—I

In this reaction the much more stable *tert*-butyl cation is produced. It then combines with I^- to form *tert*-butyl iodide.

10.25 (a) $H_2C\overset{\diagdown\diagup}{\underset{O}{—}}CH_2$ $\xrightarrow{H^+}$ $H_2C\overset{\diagdown\diagup}{\underset{\overset{+}{O}\,\underset{H}{|}}{—}}CH_2$ $\xrightarrow{CH_3–\overset{..}{\overset{..}{O}}H}$ $\overset{H}{\overset{|}{\underset{}{+\overset{}{O}-CH_3}}}$ $\underset{\underset{OH}{|}}{CH_2–CH_2}$ $\longrightarrow$ $HOCH_2CH_2OCH_3$
Methyl cellosolve

(b) An analogous reaction yields ethyl cellosolve, $HOCH_2CH_2OCH_2CH_3$.

(c) $H_2C\overset{\diagdown\diagup}{\underset{O}{—}}CH_2$ $\xrightarrow{I^-}$ $\underset{\underset{O^-}{|}}{\overset{\overset{I}{|}}{CH_2CH_2}}$ $\xrightarrow{H_2O}$ $HOCH_2CH_2I$ + OH^-

(d) $H_2C\overset{\diagdown\diagup}{\underset{O}{—}}CH_2$ $\xrightarrow{:NH_3}$ $\underset{\underset{O^-}{|}}{\overset{\overset{+}{NH_3}}{CH_2CH_2}}$ $\longrightarrow$ $HOCH_2CH_2NH_2$

(e) $H_2C\overset{\diagdown\diagup}{\underset{O}{—}}CH_2$ $\xrightarrow{CH_3O^-}$ $\underset{\underset{O^-}{|}}{\overset{\overset{OCH_3}{|}}{CH_2CH_2}}$ $\xrightarrow{CH_3OH}$ $HOCH_2CH_2OCH_3$ + CH_3O^-

10.26 The reaction is an S_N2 reaction, and thus nucleophilic attack takes place much more rapidly at the primary carbon atom than at the more hindered secondary carbon atom.

$CH_3\overset{CH_3}{\overset{|}{C}}\overset{\diagdown\diagup}{\underset{O}{—}}CH_2$ + CH_3O^- $\xrightarrow[CH_3OH]{fast}$ $CH_3\overset{CH_3}{\underset{\underset{OH}{|}}{\overset{|}{C}}}CH_2OCH_3$ Major product

$CH_3\overset{CH_3}{\overset{|}{C}}\overset{\diagdown\diagup}{\underset{O}{—}}CH_2$ + CH_3O^- $\xrightarrow[CH_3OH]{slow}$ $CH_3\overset{CH_3}{\underset{\underset{OCH_3}{|}}{\overset{|}{C}}}CH_2OH$ Minor product

10.27 Ethoxide ion attacks the epoxide ring at the primary carbon because it is less hindered, and the following reactions take place.

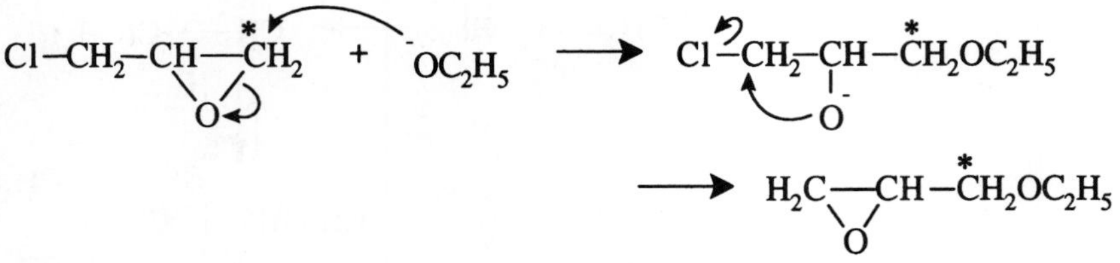

$$Cl-CH_2-CH \overset{*}{\underset{\underset{\text{O}}{|}}{-CH_2}} \ + \ ^-OC_2H_5 \longrightarrow Cl-CH_2-\underset{\underset{\text{O}^-}{|}}{CH}-\overset{*}{CH_2}OC_2H_5$$

$$\longrightarrow H_2C\underset{\underset{\text{O}}{\diagdown\diagup}}{-CH}-\overset{*}{CH_2}OC_2H_5$$

10.28

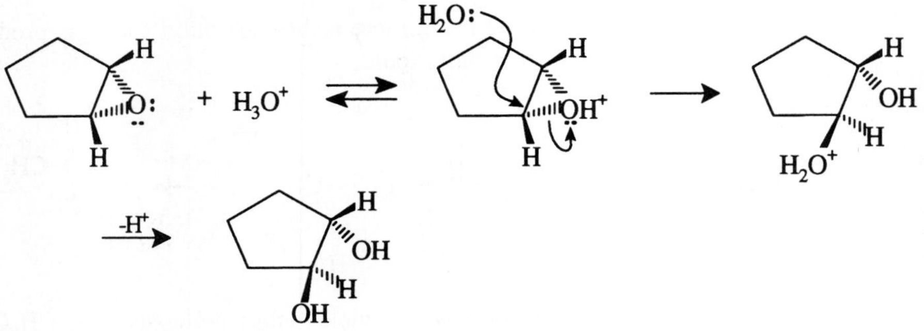

10.29

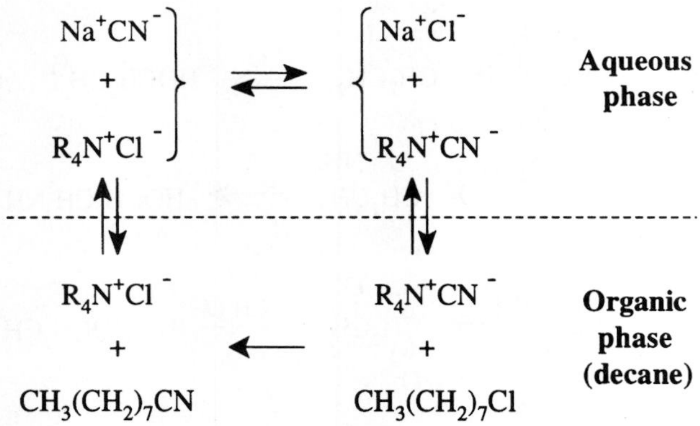

10.30   (a)

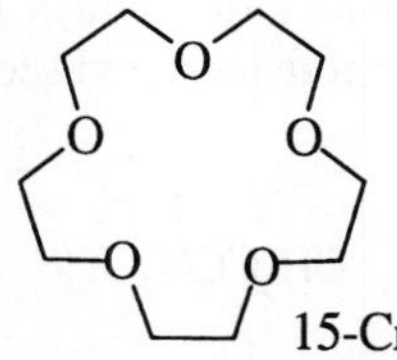

15-Crown-5

(b)

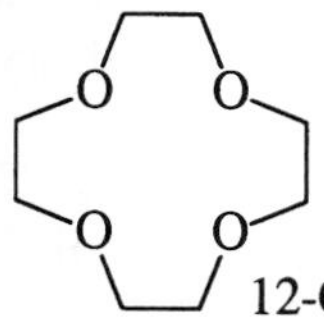

12-Crown-4

10.31   (a)   3,3-Dimethyl-1-butanol

   (b)   4-Penten-2-ol

   (c)   2-Methyl-1,4-butanediol

   (d)   2-Phenylethanol

   (e)   1-Methyl-2-cyclopenten-1-ol

   (f)   *cis*-3-Methylcyclohexanol

10.32 (a)
$$\underset{H}{\overset{H_3C}{}}C=C\underset{H}{\overset{CH_2OH}{}}$$

(b)
$$\underset{HOCH_2}{\overset{HOCH_2CH_2}{}}C\underset{\cdots\cdots H}{\overset{OH}{}}$$

(c)

(d)
$$\overset{OH}{\underset{\square}{}}\!\!-CH_2CH_3$$

(e) $CH_3CH_2C\equiv CCHCH_2OH$
$$\quad\quad\quad\quad\quad\quad\underset{Cl}{|}$$

(f) $H_2C\!-\!CH_2$
$$H_2C\underset{O}{}CH_2$$

(g) $CH_3CHCH_2CH_2CH_3$
$$\quad\underset{OCH_2CH_3}{|}$$

(h) $CH_3CH_2\!-\!O\!-\!\bigcirc$

(i) $\underset{CH_3CH}{\overset{CH_3}{|}}\!-\!O\!-\!\underset{CHCH_3}{\overset{CH_3}{|}}$

(j) $CH_3CH_2\!-\!O\!-\!CH_2CH_2OH$

10.33 (a) $CH_3CH_2CH=CH_2 \xrightarrow[\text{(hydroboration)}]{\text{THF: BH}_3} (CH_3CH_2CH_2CH_2)_3B$

$$\xrightarrow[\text{(oxidation)}]{H_2O_2/OH^-} CH_3CH_2CH_2CH_2OH$$

(b) $CH_3CH_2CH_2CH_2Cl \xrightarrow{OH^-} CH_3CH_2CH_2CH_2OH$

(c) $CH_3CH_2CHCH_3 \xrightarrow{(CH_3)_3COK/(CH_3)_3COH} CH_3CH_2CH=CH_2$
$$\quad\quad\quad\underset{Cl}{|}$$

$$\xrightarrow[\text{ROOR}]{HBr} CH_3CH_2CH_2CH_2Br \xrightarrow{OH^-} CH_3CH_2CH_2CH_2OH$$

(d) $CH_3CH_2C\equiv CH \xrightarrow[\text{Ni}_2\text{B (P-2)}]{H_2} CH_3CH_2CH=CH_2$

$$\xrightarrow{\text{[as in (a)]}} CH_3CH_2CH_2CH_2OH$$

10.34 (a) $3\ CH_3CH_2CHCH_3 + PBr_3 \longrightarrow 3\ CH_3CH_2CHCH_3 + H_3PO_3$
$$\quad\quad\quad\quad\underset{OH}{|}\quad\quad\quad\quad\quad\quad\quad\quad\quad\underset{Br}{|}$$

(b) $CH_3CH_2CH_2CH_2OH \xrightarrow{PBr_3} CH_3CH_2CH_2CH_2Br \xrightarrow{(CH_3)_3COK}$

$$CH_3CH_2CH=CH_2 \xrightarrow[\text{(no peroxides)}]{HBr} CH_3CH_2CHCH_3$$
$$\quad\quad\quad\quad\quad\quad\quad\quad\quad\quad\quad\quad\quad\quad\underset{Br}{|}$$

(c)   See (b) above.

(d) $CH_3CH_2C{\equiv}CH$ $\xrightarrow[\text{Ni}_2\text{B (P-2)}]{\text{H}_2}$ $CH_3CH_2CH{=}CH_2$ $\xrightarrow[\text{(no peroxides)}]{\text{HBr}}$

$$CH_3CH_2\underset{\underset{Br}{|}}{C}HCH_3$$

10.35 (a) [cyclohexanol] OH + $SOCl_2$ $\longrightarrow$ [cyclohexyl chloride] Cl + SO_2 + HCl

(b) [cyclohexene] + HCl $\longrightarrow$ [chlorocyclohexane] Cl

(c) [1-methylcyclohexene] CH_3 $\xrightarrow[\text{(no peroxides)}]{\text{HBr}}$ [1-bromo-1-methylcyclohexane] $\overset{Br}{\underset{}{}}CH_3$

(d) [1-methylcyclohexene] CH_3 $\xrightarrow[\text{(2) H}_2\text{O}_2,\text{OH}^-]{\text{(1) THF: BH}_3}$ [trans-2-methylcyclohexanol with H, ''''CH$_3$, ''''H, OH] + enantiomer

(e) [1-bromo-1-methylcyclohexane] $\overset{Br}{\underset{}{}}CH_3$ $\xrightarrow[\text{t-BuOH}]{\text{t-BuOK}}$ [methylenecyclohexane] CH_2 $\xrightarrow[\text{(2) H}_2\text{O}_2,\text{OH}]{\text{(1) THF: BH}_3}$ [cyclohexylmethanol] CH_2OH

10.36 (a) $CH_3CH_2CH_2CH_2ONa$ Sodium butoxide

(b) $CH_3CH_2CH_2CH_2OCH_2CH_2CH_3$ Butyl propyl ether

(c) $CH_3SO_2OCH_2CH_2CH_2CH_3$ Butyl mesylate

(d) $H_3C{-}\langle\bigcirc\rangle{-}SO_2OCH_2CH_2CH_2CH_3$ Butyl tosylate

(e) $CH_3OCH_2CH_2CH_2CH_3$ 1-Methoxybutane

(f) $CH_3CH_2CH_2CH_2I$ Butyl iodide

(g) $CH_3CH_2CH_2CH_2Cl$ Butyl chloride

(h) same as (g)

(i) $CH_3CH_2CH_2CH_2OCH_2CH_2CH_2CH_3$ Dibutyl ether

(j) $CH_3CH_2CH_2CH_2Br$ Butyl bromide

10.37 (a) $\overset{\overset{\text{CH}_3}{|}}{\text{CH}_3\text{CH}_2\text{CHONa}}$ Sodium *sec*-butoxide

(b) $\overset{\overset{\text{CH}_3}{|}}{\text{CH}_3\text{CH}_2\text{CHOCH}_2\text{CH}_2\text{CH}_3}$ *sec*-Butyl propyl ether

(c) $\overset{\overset{\text{CH}_3}{|}}{\text{CH}_3\text{SO}_2\text{OCHCH}_2\text{CH}_3}$ *sec*-Butyl mesylate

(d) $\text{H}_3\text{C}-$⬡$-\text{SO}_2\text{OCHCH}_2\text{CH}_3$ *sec*-Butyl tosylate $\overset{\text{CH}_3}{|}$

(e) $\overset{\overset{\text{CH}_3}{|}}{\text{CH}_3\text{OCHCH}_2\text{CH}_3}$ 2-Methoxybutane

(f) $\underset{\underset{\text{I}}{|}}{\text{CH}_3\text{CHCH}_2\text{CH}_3}$ *sec*-Butyl iodide

(g) $\underset{\underset{\text{Cl}}{|}}{\text{CH}_3\text{CHCH}_2\text{CH}_3}$ *sec*-Butyl chloride

(h)   same as (g)

(i) $\underset{\text{H}}{\overset{\text{H}_3\text{C}}{}}\text{C}=\text{C}\underset{\text{CH}_3}{\overset{\text{H}}{}}$ mainly, *trans*-2-Butene

(j) $\underset{\underset{\text{Br}}{|}}{\text{CH}_3\text{CHCH}_2\text{CH}_3}$ *sec*-Butyl bromide

10.38 (a) $\text{CH}_3\text{Br} + \text{CH}_3\text{CH}_2\text{Br}$ (c) $\text{Br}-\text{CH}_2\text{CH}_2\text{CH}_2\text{CH}_2-\text{Br}$

(b) $\overset{\overset{\text{CH}_3}{|}}{\underset{\underset{\text{CH}_3}{|}}{\text{CH}_3\text{C}-\text{Br}}} + \text{CH}_3\text{CH}_2\text{Br}$ (d) $\text{Br}-\text{CH}_2\text{CH}_2-\text{Br}$ (2 molar equivalent

10.39

+ H_2O 3° Carbocation
is more stable

10.40 (a)

$$CH_3\text{-}\underset{\underset{CH_3}{|}}{\overset{\overset{CH_3}{|}}{C}}\text{-}CH=CH_2 \;+\; THF\!:\!BH_3 \longrightarrow \left(CH_3\text{-}\underset{\underset{CH_3}{|}}{\overset{\overset{CH_3}{|}}{C}}\text{-}CH_2\text{-}CH_2\right)_{\!3}\!B$$

$$\xrightarrow[\;]{H_2O_2,\,OH,\,H_2O}\; CH_3\text{-}\underset{\underset{CH_3}{|}}{\overset{\overset{CH_3}{|}}{C}}\text{-}CH_2\text{-}CH_2OH$$

(b) $CH_3CH_2CH_2CH_2CH=CH_2 \xrightarrow[\text{(2) } H_2O_2,\,OH,\,H_2O]{\text{(1) THF: }BH_3} CH_3CH_2CH_2CH_2CH_2CH_2OH$

(c) Ph–CH=CH$_2$ $\xrightarrow[\text{(2) } H_2O_2,\,OH,\,H_2O]{\text{(1) THF: }BH_3}$ Ph–CH$_2$CH$_2$OH

(d) 1-methylcyclopentene $\xrightarrow[\text{(2) } H_2O_2,\,OH,\,H_2O]{\text{(1) THF: }BH_3}$ *trans*-2-methylcyclopentanol (H, ""CH$_3$, OH, H) + enantiomer

10.41 (a) cyclohexane ring with H, T, CH$_3$, H   (b) ring with H, D, CH$_3$, D   (c) ring with H, OH, CH$_3$, D

10.42 (a)

$$CH_3\underset{\underset{H}{|}}{\overset{\overset{CH_3}{|}}{C}}CH_3 \xrightarrow[\text{heat},\,h\nu]{Br_2} CH_3\underset{\underset{Br}{|}}{\overset{\overset{CH_3}{|}}{C}}CH_3$$

(b)

$$CH_3\underset{\underset{Br}{|}}{\overset{\overset{CH_3}{|}}{C}}CH_3 \xrightarrow[CH_3CH_2OH]{CH_3CH_2ONa} CH_3\overset{\overset{CH_3}{|}}{C}=CH_2$$

(c)

$$CH_3\overset{\overset{CH_3}{|}}{C}=CH_2 \xrightarrow[\text{peroxides}]{HBr} CH_3\overset{\overset{CH_3}{|}}{C}HCH_2Br$$

(d)

$$CH_3\overset{\overset{CH_3}{|}}{C}HCH_2Br \xrightarrow[\text{acetone}]{KI} CH_3\overset{\overset{CH_3}{|}}{C}HCH_2I$$

(e)

$$CH_3\overset{\overset{CH_3}{|}}{C}HCH_2Br \xrightarrow[H_2O]{OH} CH_3\overset{\overset{CH_3}{|}}{C}HCH_2OH$$

or

$$CH_3\overset{\overset{CH_3}{|}}{C}=CH_2 \xrightarrow[\text{(2) } H_2O_2,\,OH]{\text{(1) }(BH_3)_2} CH_3\overset{\overset{CH_3}{|}}{C}HCH_2OH$$

(f)

$$CH_3\overset{\overset{CH_3}{|}}{C}=CH_2 \xrightarrow[\text{heat}]{H_3O^+,\,H_2O} CH_3\underset{\underset{OH}{|}}{\overset{\overset{CH_3}{|}}{C}}CH_3$$

(g) $CH_3\overset{\overset{\displaystyle CH_3}{|}}{C}HCH_2Br$ $\xrightarrow[CH_3OH]{CH_3ONa}$ $CH_3\overset{\overset{\displaystyle CH_3}{|}}{C}HCH_2OCH_3$

(h) $CH_3\overset{\overset{\displaystyle CH_3}{|}}{C}HCH_2Br$ $\xrightarrow[\underset{CH_3\overset{\overset{\displaystyle O}{||}}{C}OH}{}]{CH_3\overset{\overset{\displaystyle O}{||}}{C}ONa}$ $CH_3\overset{\overset{\displaystyle CH_3}{|}}{C}HCH_2O\overset{\overset{\displaystyle O}{||}}{C}CH_3$

(i) $CH_3\overset{\overset{\displaystyle CH_3}{|}}{C}HCH_2Br$ $\xrightarrow{NaCN}$ $CH_3\overset{\overset{\displaystyle CH_3}{|}}{C}HCH_2CN$

(j) $CH_3\overset{\overset{\displaystyle CH_3}{|}}{C}HCH_2Br$ $\xrightarrow{CH_3SNa}$ $CH_3\overset{\overset{\displaystyle CH_3}{|}}{C}HCH_2SCH_3$

or $CH_3\overset{\overset{\displaystyle CH_3}{|}}{C}=CH_2$ $\xrightarrow[peroxides]{CH_3SH}$ $CH_3\overset{\overset{\displaystyle CH_3}{|}}{C}HCH_2SCH_3$

(k) $CH_3\overset{\overset{\displaystyle CH_3}{|}}{C}=CH_2$ $\xrightarrow[peroxides]{CBr_4}$ $CH_3\overset{\overset{\displaystyle CH_3}{|}}{\underset{\underset{\displaystyle Br}{|}}{C}}CH_2CBr_3$

10.43 (a)

(b) The trans product because the Cl⁻ attacks anti to the epoxide and an inversion of configuration occurs.

10.44 $HC{\equiv}CH$ $\xrightarrow[liq.\ NH_3]{NaNH_2}$ $HC{\equiv}CNa$ $\xrightarrow{CH_3\overset{\overset{\displaystyle CH_3}{|}}{C}H(CH_2)_4Br}$ $CH_3\overset{\overset{\displaystyle CH_3}{|}}{C}H(CH_2)_4C{\equiv}CH$

$A\ (C_9H_{16})$

$\xrightarrow[liq.\ NH_3]{NaNH_2}$ $CH_3\overset{\overset{\displaystyle CH_3}{|}}{C}H(CH_2)_4C{\equiv}CNa$ $\xrightarrow{CH_3(CH_2)_9Br}$ $CH_3\overset{\overset{\displaystyle CH_3}{|}}{C}H(CH_2)_4C{\equiv}C(CH_2)_9CH_3$

$B\ (C_9H_{15}Na)$ $\qquad\qquad\qquad\qquad$ $C\ (C_{19}H_{36})$

$$\xrightarrow[\text{Ni}_2\text{B (P-2)}]{\text{H}_2} \quad \underset{\underset{H \quad\quad H}{\underset{\diagdown}{C=C}}}{CH_3CH(CH_2)_4 \diagdown \quad \diagup (CH_2)_9CH_3} \overset{CH_3}{} \quad \textbf{D} \ (C_{19}H_{38}) \quad \xrightarrow{C_6H_5CO_3H}$$

$$\underset{\underset{H \quad O \quad H}{\underset{\diagdown}{C-C}}}{CH_3CH(CH_2)_4 \diagdown \quad \diagup (CH_2)_9CH_3} \overset{CH_3}{} \quad \textbf{E} \ (C_{19}H_{38}O)$$

10.45 (a) $\underset{}{\overset{CH_3}{CH_3\overset{|}{C}=CH_2}} \quad \xrightarrow[\text{(2) H}_2\text{O}_2,\ \text{OH}]{\text{(1) THF: BH}_3} \quad \underset{}{\overset{CH_3}{CH_3\overset{|}{C}HCH_2OH}}$

(b) $\underset{}{\overset{CH_3}{CH_3\overset{|}{C}=CH_2}} \quad \xrightarrow[\text{(2) CH}_3\text{CO}_2\text{T}]{\text{(1) THF: BH}_3} \quad \underset{}{\overset{CH_3}{CH_3\overset{|}{C}HCH_2T}}$

(c) $\underset{}{\overset{CH_3}{CH_3\overset{|}{C}=CH_2}} \quad \xrightarrow{\text{THF: BD}_3} \quad \underset{\underset{D}{|}}{\overset{CH_3}{CH_3\overset{|}{C}-CH_2-B\diagup}} \quad \xrightarrow{\text{CH}_3\text{CO}_2\text{T}} \quad \underset{\underset{D}{|}}{\overset{CH_3}{CH_3\overset{|}{C}CH_2T}}$

(d) $\underset{}{\overset{CH_3}{CH_3\overset{|}{C}HCH_2OH}} \quad \xrightarrow{\text{Na}} \quad \underset{}{\overset{CH_3}{CH_3\overset{|}{C}HCH_2ONa}} \quad \xrightarrow{\text{CH}_3\text{CH}_2\text{Br}}$

$$\underset{}{\overset{CH_3}{CH_3\overset{|}{C}HCH_2-O-CH_2CH_3}}$$

10.46 (a) $CH_3CH_2CH_2CH=CH_2 \quad \xrightarrow{\text{Hg(OAc)}_2} \quad \underset{\underset{OH}{|}}{CH_3CH_2CH_2CHCH_2HgOAc}$

$$\xrightarrow[\text{OH}]{\text{NaBH}_4} \quad \underset{\underset{OH}{|}}{CH_3CH_2CH_2CHCH_3}$$

(b) ⬠$-CH=CH_2 \quad \xrightarrow[\text{(2) NaBH}_4/\text{OH}^-]{\text{(1) Hg(OAc)}_2} \quad$ ⬠$-\underset{\underset{OH}{|}}{CHCH_3}$

(c) $\underset{}{\overset{CH_3}{CH_3CH=\overset{|}{C}CH_2CH_3}} \quad \xrightarrow[\text{(2) NaBH}_4/\text{OH}^-]{\text{(1) Hg(OAc)}_2} \quad \underset{\underset{OH}{|}}{\overset{CH_3}{CH_3CH_2\overset{|}{C}CH_2CH_3}}$

(d) ⬠$=CHCH_3 \quad \xrightarrow[\text{(2) NaBH}_4/\text{OH}^-]{\text{(1) Hg(OAc)}_2} \quad$ ⬠$\overset{CH_2CH_3}{\underset{OH}{}}$

10.47 (a) A = + enantiomer   B = + enantiomer

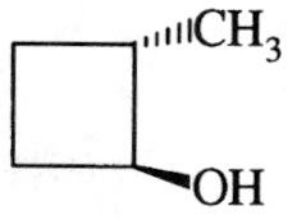

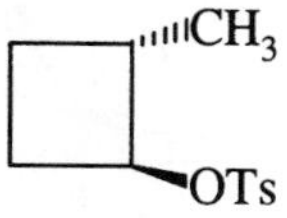

C = + enantiomer

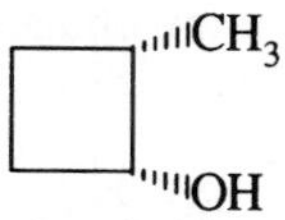

(b) Diastereomers

(c) D = + enantiomer

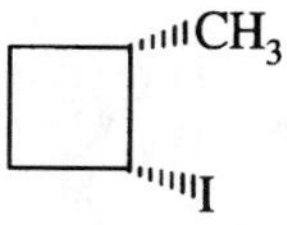

(d) E =

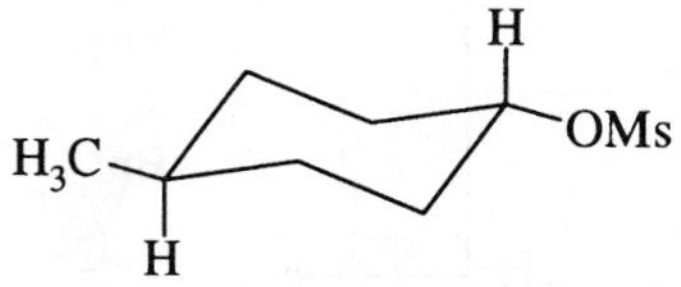

F =

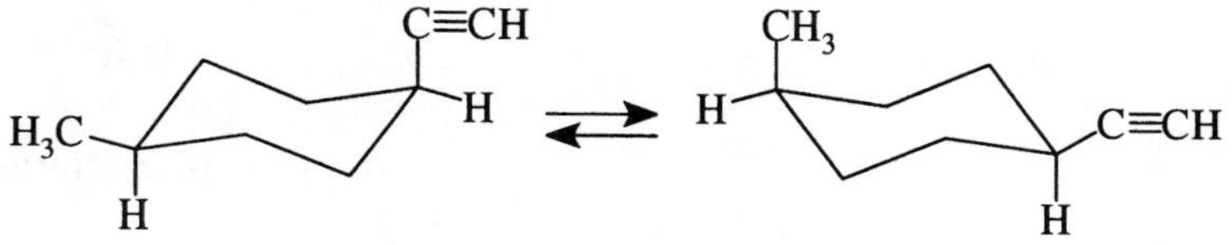

(e) H 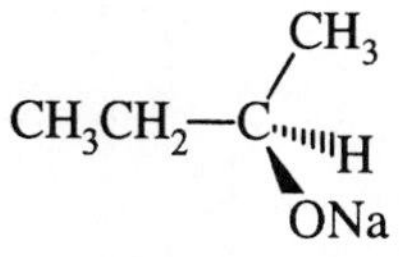 =   J =

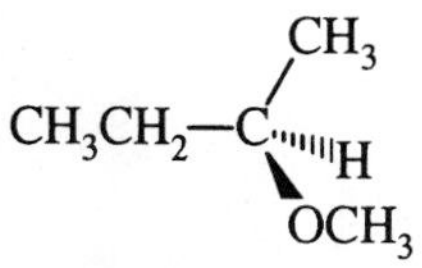

(f) K 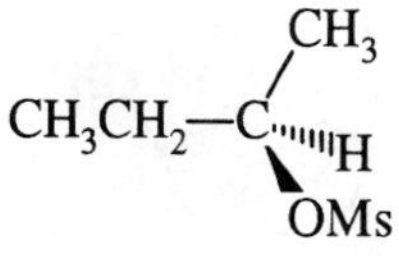 =   L =

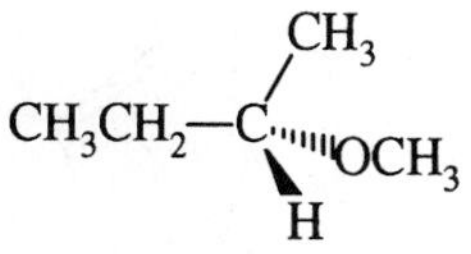

(g) Enantiomers

10.48 The reactions proceed through the formation of bromonium ions identical to those formed in the bromination of *trans*- and *cis*-2-butene (see Section 8.6A).

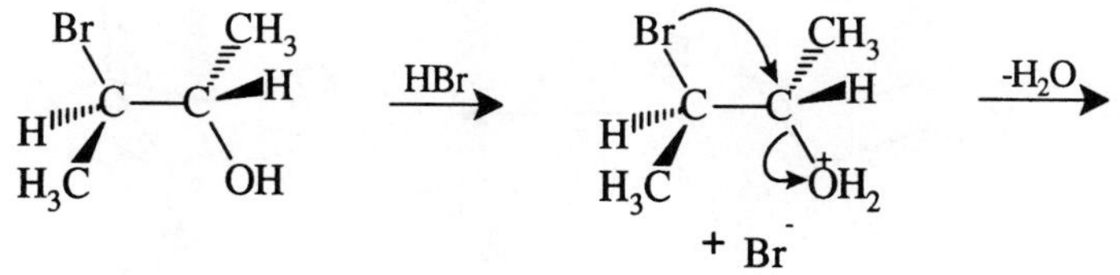

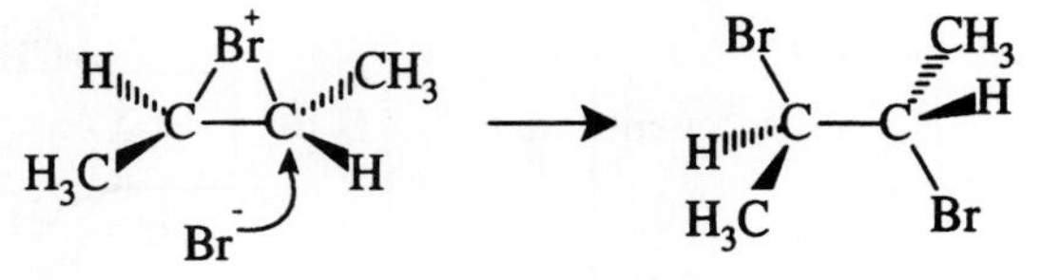

(Attack at the other carbon atom of the bromonium ion gives the same product.)

meso-2,3-Dibromobutane

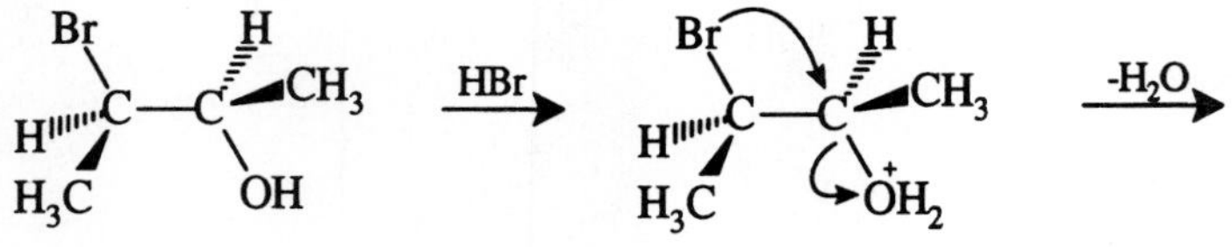

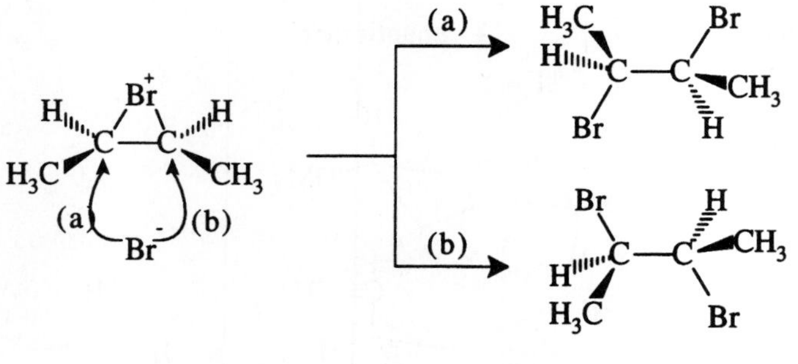

(±)-2,3-Dibromobutane

QUIZ

10.1   Which set of reagents would effect the conversion,

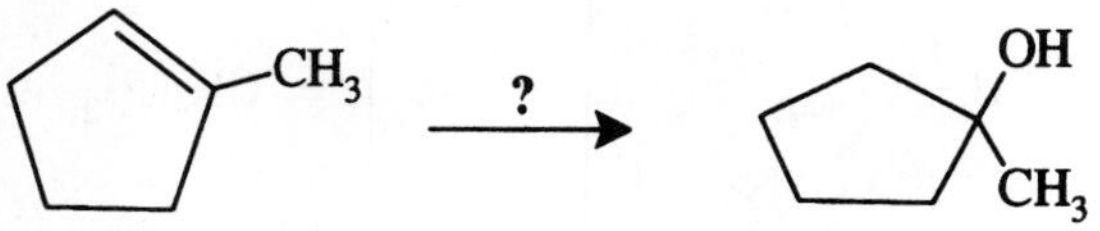

(a) BH_3:THF, then H_2O_2/OH^- (b) $H_2O/Hg\,(OAc)_2THF$, then $NaBH_4/OH^-$

(c) H_3O^+, H_2O, heat   (d) More than one of these   (e) None of these

10.2   Which of the reagents in item 10.1 would effect the conversion,

+ enantiomer

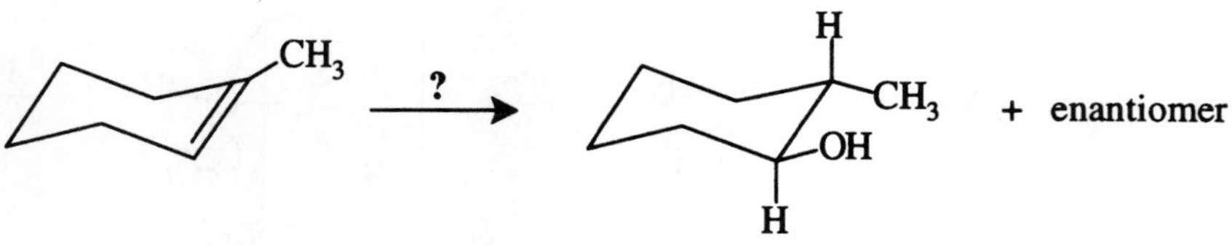

10.3 The following compounds have identical molecular weights. Which would have the lowest boiling point?

(a) 1-Butanol

(b) 2-Butanol

(c) 2-Methyl-1-propanol

(d) 1,1-Dimethylethanol

(e) 1-Methoxypropane

10.4 Complete the following synthesis:

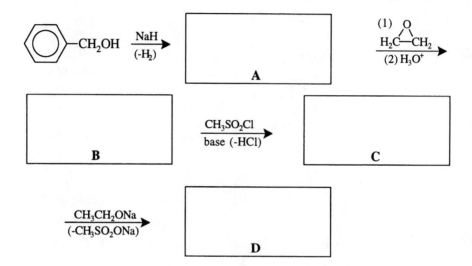

11 ALCOHOLS FROM CARBONYL COMPOUNDS: OXIDATION–REDUCTION AND ORGANOMETALLIC COMPOUNDS

SOLUTIONS TO PROBLEMS

11.1 (a)

$$H-\underset{\underset{H}{|}}{\overset{\overset{H}{|}}{C}}-O-H$$

3 **H** = -3
<u>1 **O** = +1</u>
Total = -2   = oxidation state of C

$$H-\overset{\overset{O}{\|}}{C}-O-H$$

1 **H** = -1
<u>3 **O** = +3</u>
Total = +2   = oxidation state of C

$$H-\overset{\overset{O}{\|}}{C}-H$$

2 **H** = -2
<u>2 **O** = +2</u>
Total = 0   = oxidation state of C

(b) CH_4 CH_3OH $H-\overset{\overset{O}{\|}}{C}-H$ $H-\overset{\overset{O}{\|}}{C}-O-H$ CO_2
       -4   -2   0   +2   +4

(c)   A change from -2 to 0

(d)   An oxidation, since the oxidation state increases

(e)   A reduction from +6 to +3

11.2 (a)

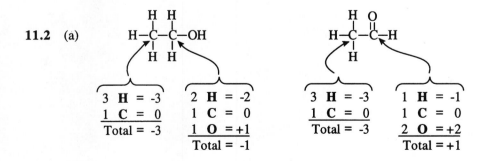

3 **H** = -3
<u>1 **C** = 0</u>
Total = -3

2 **H** = -2
1 **C** = 0
<u>1 **O** = +1</u>
Total = -1

3 **H** = -3
<u>1 **C** = 0</u>
Total = -3

1 **H** = -1
1 **C** = 0
<u>2 **O** = +2</u>
Total = +1

(b) Only the carbon atom of the $-CH_2OH$ group of ethanol undergoes a change in oxidation state. The oxidation state of the carbon atom in the CH_3- group remains unchanged.

(c)

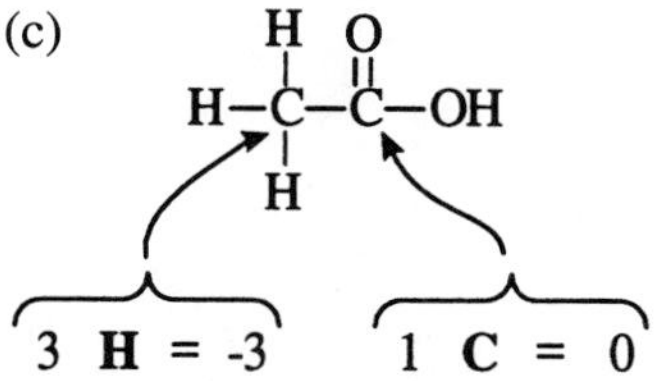

$$\begin{array}{cc} & H & O \\ & | & \| \\ H-&C-&C-OH \\ & | \\ & H \end{array}$$

The oxygen-bearing carbon atom increases its oxidation state from +1 (in acetaldehyde) to +3 (in acetic acid).

$$
\begin{array}{cc}
3\ H = -3 & 1\ C = 0 \\
1\ C = 0 & 3\ O = +3 \\
\hline
\text{Total} = -3 & \text{Total} = +3
\end{array}
$$

11.3   (a) If we consider the hydrogenation of ethene as an example, we find that the oxidation state of carbon decreases. Thus, because the reaction involves the *addition* of *hydrogen,* it is both an *addition reaction* and a *reduction.*

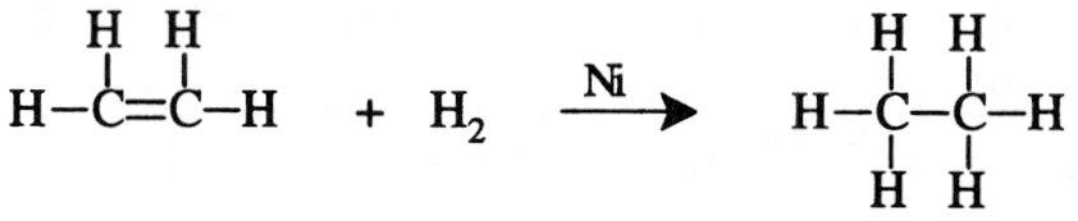

$$\begin{array}{ccc} H & H & \\ | & | & \\ H-C=C-H & + H_2 & \xrightarrow{\ Ni\ } \end{array} \quad \begin{array}{c} H\ \ H \\ |\ \ | \\ H-C-C-H \\ |\ \ | \\ H\ \ H \end{array}$$

$$
\begin{array}{cc}
2\ H = -2 & 3\ H = -3 \\
2\ C = 0 & 1\ C = 0 \\
\hline
\text{Total} = -2 & \text{Total} = -3
\end{array}
$$

(b) The hydrogenation of acetaldehyde is not only an addition reaction, but it is also a *reduction* because the carbon atom of the C=O group goes from a + 1 to a − 1 oxidation state. The reverse reaction (the *dehydrogenation* of ethanol) is not only an *elimination* reaction, but also an *oxidation.*

Ion-Electron Half-Reaction Method for Balancing Organic Oxidation–Reduction Equations

Only two simple rules are needed:

Rule 1 Electrons (e^-) together with protons (H$^+$) are arbitrarily considered the reducing agents in the half-reaction for the reduction of the oxidizing agent. Ion charges are balanced by *adding electrons to the left-hand side.* (If the reaction is run in neutral or basic solution, add an equal number of OH$^-$ ions to both sides of the balanced half-reaction to neutralize the H$^+$, and show the resulting H$^+$ + OH$^-$ as H_2O.)

Rule 2 Water (H_2O) is arbitrarily taken as the formal source of oxygen for the oxidation of the organic compound, producing *product, protons,* and *electrons* on the right-hand side. (Again, use OH$^-$ to neutralize H$^+$ in the *balanced* half-reaction in neutral or basic media.)

EXAMPLE 1

Write a balanced equation for the oxidation of RCH_2OH to RCO_2H by $Cr_2O_7^{2-}$ in acid solution.

Reduction half-reaction:

$$Cr_2O_7^{2-} + H^+ + e^- \longrightarrow 2Cr^{3+} + 7H_2O$$

Balancing atoms and charges:

$$Cr_2O_7^{2-} + 14H^+ + 6e^- = 2Cr^{3+} + 7H_2O$$

Oxidation half-reaction:

$$RCH_2OH + H_2O = RCO_2H + 4H^+ + 4e^-$$

The least common multiple of a 6-electron uptake in the reduction step and a 4-electron loss in the oxidation step is 12, so we multiply the first half-reaction by 2 and the second by 3, and add:

$$3RCH_2OH + 3H_2O + 2Cr_2O_7^{2-} + 28H^+ = 3RCO_2H + 12H^+ + 4Cr^{3+} + 14H_2O$$

Canceling common terms, we get:

$$3RCH_2OH + 2Cr_2O_7^{2-} + 16H^+ = 3RCO_2H + 4Cr^{3+} + 11H_2O$$

This shows that the oxidation of 3 mol of a primary alcohol to a carboxylic acid requires 2 mol of dichromate.

EXAMPLE 2

Write a balanced equation for the oxidation of styrene to benzoate ion and carbonate ion by MnO_4^- in alkaline solution.

Reduction:

$$MnO_4^- + 4H^+ + 3e^- = MnO_2 + 2H_2O \text{ (in acid)}$$

Since this reaction is carried out in basic solution, we must add 4 OH^- to neutralize the 4H^+ on the left side, and, of course, 4 OH^- to the right side to maintain a balanced equation.

$$MnO_4^- + 4H^+ + 4OH^- + 3e^- = MnO_2 + 2H_2O + 4OH^-$$

or, $MnO_4^- + 2H_2O + 3e^- = MnO_2 + 4OH^-$

Oxidation:

$$ArCH{=}CH_2 + 5H_2O = ArCO_2^- + CO_3^{2-} + 13H^+ + 10e^-$$

We add 13 OH^- to each side to neutralize the H^+ on the right side,

$$ArCH{=}CH_2 + 5H_2O + 13OH^- = ArCO_2^- + CO_3^{2-} + 13H_2O + 10e^-$$

The least common multiple is 30, so we multiply the reduction half-reaction by 10 and the oxidation half-reaction by 3 and add:

$$3ArCH{=}CH_2 + 39\,OH^- + 10MnO_4^- + 20H_2O = 3ArCO_2^- + 3CO_3^{2-} +$$
$$24H_2O + 10MnO_2 + 40\,OH^-$$

Canceling:

$$3ArCH{=}CH_2 + 10MnO_4^- = 3ArCO_2^- + 3CO_3^{2-} + 4H_2O + 10MnO_2 + OH^-$$

SAMPLE PROBLEMS

Using the ion–electron half-reaction method, write balanced equations for the following oxidation reactions.

(a) Cyclohexene + $MnO_4^- + H^+$ $\xrightarrow{\text{(hot)}}$ $HO_2C(CH_2)_4CO_2H + Mn^{2+} + H_2O$

(b) Cyclopentene + $MnO_4^- + H_2O$ $\xrightarrow{\text{(cold)}}$ cis-1,2-cyclopentanediol + $MnO_2 + OH^-$

(c) Cyclopentanol + HNO_3 $\xrightarrow{\text{(hot)}}$ $HO_2C(CH_2)_3CO_2H + NO_2 + H_2O$

(d) 1,2,3-Cyclohexanetriol + HIO_4 $\xrightarrow{\text{(cold)}}$ $OCH(CH_2)_3CHO + HCO_2H + HIO_3$

SOLUTIONS TO SAMPLE PROBLEMS

(a) Reduction:

$$MnO_4^- + 8\,H^+ + 5\,e^- = Mn^{2+} + 4\,H_2O$$

Oxidation:

The least common multiple is 40:

$$8\,MnO_4^- + 64\,H^+ + 40\,e^- = 8\,Mn^{2+} + 32\,H_2O$$

Adding and canceling:

(b) Reduction:

$$MnO_4^- + 2\,H_2O + 3\,e^- = MnO_2 + 4\,OH^-$$

Oxidation:

$$\text{cyclopentene} + 2\,OH^- = \text{cyclopentane-1,2-diol} + 2\,e^-$$

The least common multiple is 6:

$$2\,MnO_4^- + 4\,H_2O + 6\,e^- = 2\,MnO_2 + 8\,OH^-$$

$$3\,(\text{cyclopentene}) + 6\,OH^- = 3\,(\text{cyclopentane-1,2-diol}) + 6\,e^-$$

Adding and cancelling:

$$3\,(\text{cyclopentene}) + 2\,MnO_4^- + 4\,H_2O = 3\,(\text{cyclopentane-1,2-diol}) + 2\,MnO_2 + 2\,OH^-$$

(c) Reduction:

$$HNO_3 + H^+ + e^- = NO_2 + H_2O$$

Oxidation:

$$\text{cyclopentanol} + 3\,H_2O = \text{glutaric acid (diCO}_2\text{H)} + 8\,H^+ + 8\,e^-$$

The least common multiple is 8:

$$8\,HNO_3 + 8\,H^+ + 8\,e^- = 8\,NO_2 + 8\,H_2O$$

$$\text{cyclopentanol} + 3\,H_2O = \text{diCO}_2\text{H} + 8\,H^+ + 8\,e^-$$

Adding and cancelling:

$$\text{cyclopentanol} + 8\,HNO_3 = \text{diCO}_2\text{H} + 8\,NO_2 + 5\,H_2O$$

(d) Reduction:

$$HIO_4 + 2\,H^+ + 2\,e^- = HIO_3 + H_2O$$

Oxidation:

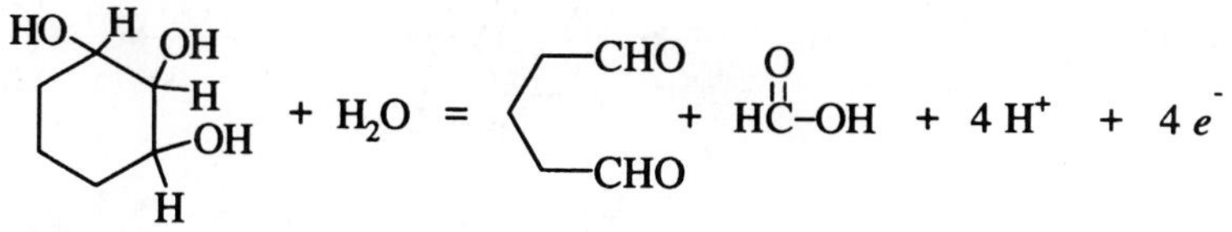

The least common multiple is 4:

$$2\,HIO_4 + 4\,H^+ + 4\,e^- = 2\,HIO_3 + 2\,H_2O$$

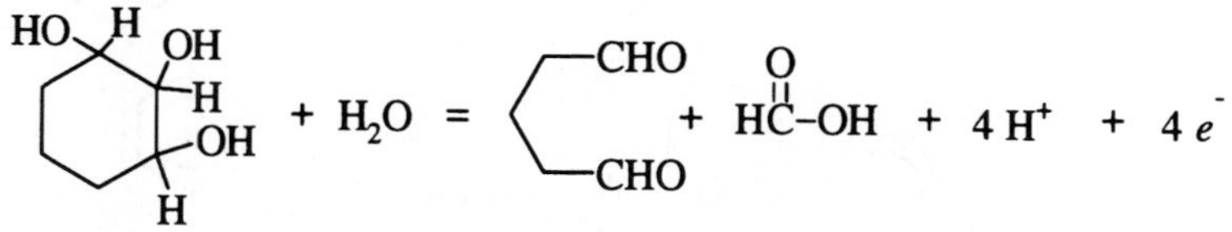

Adding and cancelling:

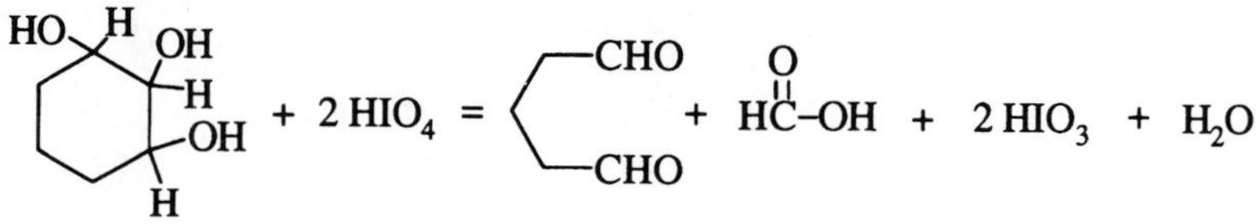

11.4 (a) $LiAlH_4$

(b) $LiAlH_4$

(c) $NaBH_4$

11.5 (a) 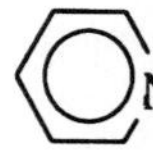 $NH^+ \; CrO_3Cl$ (PCC)/CH_2Cl_2

(b) $KMnO_4$, OH^-, H_2O, heat; then H_3O^+

(c) H_2CrO_4/acetone

(d) (1) O_3 (2) Zn, H_2O

11.6 (a)

$$\overset{\delta^-\;\delta^+}{C_6H_5{:}Li} + H{-}OH \longrightarrow C_6H_6 + LiOH$$

Stronger base	Stronger acid	Weaker acid	Weaker base

(b)

$$\overset{\delta^-\;\delta^+}{C_6H_5{:}Li} + H{-}OEt \longrightarrow C_6H_6 + LiOEt$$

Stronger base	Stronger acid	Weaker acid	Weaker base

11.7

$$\text{Ph-Br} \xrightarrow[\text{Et}_2\text{O}]{\text{Mg}} \text{Ph} \overset{\delta^-}{\underset{}{}} \overset{\delta^+}{:} \text{MgBr} \xrightarrow{\text{D-OD}} \text{Ph-D}$$

11.8

$$\text{Ph-MgBr} + \underset{\text{Ph}}{\overset{\text{Cl}}{\underset{}{\text{C}}}}=\text{O} \longrightarrow \left[\underset{\text{Ph}}{\overset{\text{Ph}}{\underset{}{\text{C}}}} \overset{\text{Cl}}{\underset{}{}} \text{O-MgBr} \right] \xrightarrow{\text{-MgBrCl}}$$

$$\left[\underset{\text{Ph}}{\overset{\text{Ph}}{\underset{}{\text{C}}}}=\text{O} \right] \xrightarrow{\text{Ph-MgBr}} \underset{\text{Ph}}{\overset{\text{Ph}}{\underset{\text{Ph}}{\text{C}}}} \text{-O-MgBr} \xrightarrow{\text{H}_3\text{O}^+} \underset{\text{Ph}}{\overset{\text{Ph}}{\underset{\text{Ph}}{\text{C}}}} \text{-OH}$$

11.9 (a) (1) $\underset{}{\overset{\text{O}}{\underset{}{\text{CH}_3\text{CCH}_2\text{CH}_3}}} + \text{CH}_3\text{MgI} \xrightarrow[\text{(2) H}_3\text{O}^+]{\text{(1) ether}} \underset{\text{CH}_3}{\overset{\text{OH}}{\text{CH}_3\text{CCH}_2\text{CH}_3}}$

(2) $\underset{}{\overset{\text{O}}{\underset{}{\text{CH}_3\text{CCH}_3}}} + \text{CH}_3\text{CH}_2\text{MgBr} \xrightarrow[\text{(2) H}_3\text{O}^+]{\text{(1) ether}} \underset{\text{CH}_3}{\overset{\text{OH}}{\text{CH}_3\text{CCH}_2\text{CH}_3}}$

(3) $\underset{}{\overset{\text{O}}{\underset{}{\text{CH}_3\text{CH}_2\text{COCH}_2\text{CH}_3}}} + 2\text{ CH}_3\text{MgI} \xrightarrow[\text{(2) H}_3\text{O}^+]{\text{(1) ether}} \underset{\text{CH}_3}{\overset{\text{OH}}{\text{CH}_3\text{CCH}_2\text{CH}_3}}$

(b) (1) $\underset{}{\overset{\text{O}}{\underset{}{\text{CH}_3\text{CH}_2\text{CCH}_2\text{CH}_3}}} + \text{CH}_3\text{MgI} \xrightarrow[\text{(2) H}_3\text{O}^+]{\text{(1) ether}} \underset{\text{CH}_3}{\overset{\text{OH}}{\text{CH}_3\text{CH}_2\text{CCH}_2\text{CH}_3}}$

(2) $\underset{}{\overset{\text{O}}{\underset{}{\text{CH}_3\text{CH}_2\text{CCH}_3}}} + \text{CH}_3\text{CH}_2\text{MgBr} \xrightarrow[\text{(2) H}_3\text{O}^+]{\text{(1) ether}} \underset{\text{CH}_3}{\overset{\text{OH}}{\text{CH}_3\text{CH}_2\text{CCH}_2\text{CH}_3}}$

(3) $\underset{}{\overset{\text{O}}{\underset{}{\text{CH}_3\text{COCH}_3}}} + 2\text{ CH}_3\text{CH}_2\text{MgBr} \xrightarrow[\text{(2) H}_3\text{O}^+]{\text{(1) ether}} \underset{\text{CH}_3}{\overset{\text{OH}}{\text{CH}_3\text{CH}_2\text{CCH}_2\text{CH}_3}}$

(c) (1) $CH_3\overset{O}{\overset{\|}{C}}H$ + $BrMgCHCH_2CH_3$ $\xrightarrow[\text{(2) } H_3O^+]{\text{(1) ether}}$ $CH_3\overset{OH}{\overset{|}{C}}HCHCH_2CH_3$
$\qquad\qquad\qquad\qquad\quad \overset{|}{C}H_2 \qquad\qquad\qquad\qquad\qquad\qquad\qquad \overset{|}{C}H_2$
$\qquad\qquad\qquad\qquad\quad \overset{|}{C}H_3 \qquad\qquad\qquad\qquad\qquad\qquad\qquad \overset{|}{C}H_3$

(2) CH_3MgI + $H\overset{O}{\overset{\|}{C}}CHCH_2CH_3$ $\xrightarrow[\text{(2) } H_3O^+]{\text{(1) ether}}$ $CH_3\overset{OH}{\overset{|}{C}}HCHCH_2CH_3$
$\qquad\qquad\qquad\qquad\qquad \overset{|}{C}H_2 \qquad\qquad\qquad\qquad\qquad\qquad\qquad \overset{|}{C}H_2$
$\qquad\qquad\qquad\qquad\qquad \overset{|}{C}H_3 \qquad\qquad\qquad\qquad\qquad\qquad\qquad \overset{|}{C}H_3$

(d) (1) $C_6H_5\overset{O}{\overset{\|}{C}}CH_3$ + $CH_3CH_2CH_2MgBr$ $\xrightarrow[\text{(2) } H_3O^+]{\text{(1) ether}}$ $C_6H_5\overset{OH}{\underset{\overset{|}{C}H_3}{\overset{|}{C}}}CH_2CH_2CH_3$

(2) $C_6H_5\overset{O}{\overset{\|}{C}}CH_2CH_2CH_3$ + CH_3MgI $\xrightarrow[\text{(2) } H_3O^+]{\text{(1) ether}}$ $C_6H_5\overset{OH}{\underset{\overset{|}{C}H_3}{\overset{|}{C}}}CH_2CH_2CH_3$

(3) $CH_3\overset{O}{\overset{\|}{C}}CH_2CH_2CH_3$ + C_6H_5MgBr $\xrightarrow[\text{(2) } H_3O^+]{\text{(1) ether}}$ $C_6H_5\overset{OH}{\underset{\overset{|}{C}H_3}{\overset{|}{C}}}CH_2CH_2CH_3$

(e) (1) $C_6H_5\overset{O}{\overset{\|}{C}}OEt$ + $2\ C_6H_5MgBr$ $\xrightarrow[\text{(2) } H_3O^+]{\text{(1) ether}}$ $(C_6H_5)_3COH$

(2) $C_6H_5\overset{O}{\overset{\|}{C}}C_6H_5$ + C_6H_5MgBr $\xrightarrow[\text{(2) } H_3O^+]{\text{(1) ether}}$ $(C_6H_5)_3COH$

11.10 (a) $CH_3CH_2CH_2OH$ $\xrightarrow[CH_2Cl_2]{PCC}$ $CH_3CH_2\overset{O}{\overset{\|}{C}}H$ $\xrightarrow[\text{ether}]{C_6H_5MgBr}$ $CH_3CH_2\overset{OMgBr}{\overset{|}{C}}HC_6H_5$

$\qquad\qquad\qquad\qquad\qquad\qquad\qquad\qquad\qquad\qquad\qquad \xrightarrow[H_2O]{H_3O^+}$ $CH_3CH_2\overset{OH}{\overset{|}{C}}HC_6H_5$

(b) C_6H_5MgBr $\xrightarrow[\text{(2) } H_3O^+]{\text{(1) } H\overset{O}{\overset{\|}{C}}H, \text{ ether}}$ $C_6H_5CH_2OH$ $\xrightarrow[CH_2Cl_2]{PCC}$ $C_6H_5\overset{O}{\overset{\|}{C}}H$

(c) $CH_3CH_2\overset{\overset{\displaystyle O}{\|}}{C}OCH_3$ $\xrightarrow[\text{[from part (a)]}]{\overset{\displaystyle 2\ C_6H_5MgBr}{\text{ether}}}$ $C_6H_5\overset{\overset{\displaystyle OMgBr}{|}}{\underset{\underset{\displaystyle C_6H_5}{|}}{C}}CH_2CH_3$ $\xrightarrow[H_2O]{H_3O^+}$ $C_6H_5\overset{\overset{\displaystyle OH}{|}}{\underset{\underset{\displaystyle C_6H_5}{|}}{C}}CH_2CH_3$

(d) $CH_3\overset{\overset{\displaystyle }{\underset{\underset{\displaystyle CH_3}{|}}{C}}}HCH_2OH$ $\xrightarrow[CH_2Cl_2]{PCC}$ $CH_3\overset{\overset{\displaystyle }{\underset{\underset{\displaystyle CH_3}{|}}{C}}}H\overset{\overset{\displaystyle O}{\|}}{C}H$ $\xrightarrow[\text{ether}]{C_6H_5MgBr}$ $CH_3\overset{\overset{\displaystyle }{\underset{\underset{\displaystyle CH_3}{|}}{C}}}H\overset{\overset{\displaystyle OMgBr}{|}}{C}HC_6H_5$

$\xrightarrow[H_2O]{H_3O^+}$ $CH_3\overset{\overset{\displaystyle }{\underset{\underset{\displaystyle CH_3}{|}}{C}}}H\overset{\overset{\displaystyle OH}{|}}{C}HC_6H_5$

11.11 (a) $(CH_3)_2CHCH_2OH$ + $(CH_3)_2C=CH_2$

(b) $(CH_3)_2CHCH_2CN$ (c) $(CH_3)_2C=CH_2$

(d) $CH_3\overset{\overset{\displaystyle }{\underset{\underset{\displaystyle CH_3}{|}}{C}}}HCH_2OCH_3$ + $(CH_3)_2C=CH_2$

(e) $(CH_3)_2CHCH_2-\overset{\overset{\displaystyle OH}{|}}{\underset{\underset{\displaystyle CH_3}{|}}{C}}-CH_3$ (f) $(CH_3)_2CHCH_2\overset{\overset{\displaystyle OH}{|}}{C}HCH_3$

(g) $(CH_3)_2CHCH_2\overset{\overset{\displaystyle OH}{|}}{\underset{\underset{\displaystyle CH_3}{|}}{C}}CH_2CH(CH_3)_2$ (h) $(CH_3)_2CHCH_2CH_2CH_2OH$

(i) $(CH_3)_2CHCH_2CH_2OH$ (j) $(CH_3)_2CHCH_3$

(k) $(CH_3)_2CHCH_3$ + $CH_3C\equiv CLi$

11.12 (a) CH_3CH_3 (b) CH_3CH_2D (c) $C_6H_5\overset{\overset{\displaystyle OH}{|}}{C}HCH_2CH_3$

(d) $C_6H_5-\overset{\overset{\displaystyle OH}{|}}{\underset{\underset{\displaystyle CH_2CH_3}{|}}{C}}-C_6H_5$ (e) $C_6H_5-\overset{\overset{\displaystyle OH}{|}}{\underset{\underset{\displaystyle CH_2CH_3}{|}}{C}}-CH_2CH_3$ (f) $C_6H_5-\overset{\overset{\displaystyle OH}{|}}{\underset{\underset{\displaystyle CH_3}{|}}{C}}-CH_2CH_3$

(g) CH_3CH_3 + $CH_3CH_2C\equiv C-\overset{\overset{\displaystyle OH}{|}}{C}HCH_3$

(h) CH_3CH_3 + ⬠—$MgBr$

11.13 (a) $(CH_3)_2CHCHCH_2CH_2CH_3$
                                   |
                                   OH

(b) $(CH_3)_2CHCCH_2CH_2CH_3$
                       |
                       OH (above)
                       |
                       CH_3 (below)

(c) $CH_3CH_2CH_3$ + $CH_3CH_2CH_2C \equiv C-\overset{\overset{OH}{|}}{\underset{\underset{CH_3}{|}}{C}}-CH_3$ (d) $CH_3CH_2CH_3$

(e) $CH_3CH_2CH_2CH_2CH =CH_2$ (f) $CH_3CH_2CH_2-$ <cyclopentane ring>

(g) $\underset{H}{\overset{CH_3CH_2CH_2}{\diagdown}}C=C\underset{H}{\overset{CH_3}{\diagup}}$ (h) $CH_3CH_2CH_2CH_3$

Note: This variation of the Corey-Posner, Whitesides-House synthesis is stereo-specific.

(i) $CH_3CH_2CH_2D$

11.14 (a) $LiAlH_4$ (d) (1) $KMnO_4$, OH^-, heat,   (2) H_3O^+

(b) $NaBH_4$ (e) PCC/CH_2Cl_2

(c) $LiAlH_4$

11.15 (a) $3 (CH_3)_2CHOH$ + PBr_3 $\longrightarrow$ $(CH_3)_2CHBr$ + H_3PO_3

$(CH_3)_2CHBr$ + Mg $\xrightarrow{ether}$ $(CH_3)_2CHMgBr$

$(CH_3)_2CHMgBr$ + $CH_3\overset{O}{\overset{||}{C}}H$ $\xrightarrow[(2) H_3O^+]{(1) \text{ether}}$ $(CH_3)_2CHCHCH_3$ (with OH on middle C)

(b) $(CH_3)_2CHMgBr$ + $H\overset{O}{\overset{||}{C}}H$ $\xrightarrow[(2) H_3O^+]{(1) \text{ether}}$ $(CH_3)_2CHCH_2OH$
    [from part (a)]

(c) $(CH_3)_2CHMgBr$ + $H_2C\overset{O}{\overset{\diagup\diagdown}{\!-\!}}CH_2$ $\xrightarrow[(2) H_3O^+]{(1) \text{ether}}$ $(CH_3)_2CHCH_2CH_2OH$
    [from part (a)]

$(CH_3)_2CHCH_2CH_2Cl$ $\xleftarrow{SOCl_2}$

(d) $(CH_3)_2CHMgBr$ + $H\overset{O}{\overset{||}{C}}CH(CH_3)_2$ $\xrightarrow[(2) H_3O^+]{(1) \text{ether}}$ $(CH_3)_2CHCHCH(CH_3)_2$ (with OH)
    [from part (a)]

(e) $(CH_3)_2CHMgBr$ + D_2O $\longrightarrow$ $(CH_3)_2CHD$
[from part (a)]

(f) 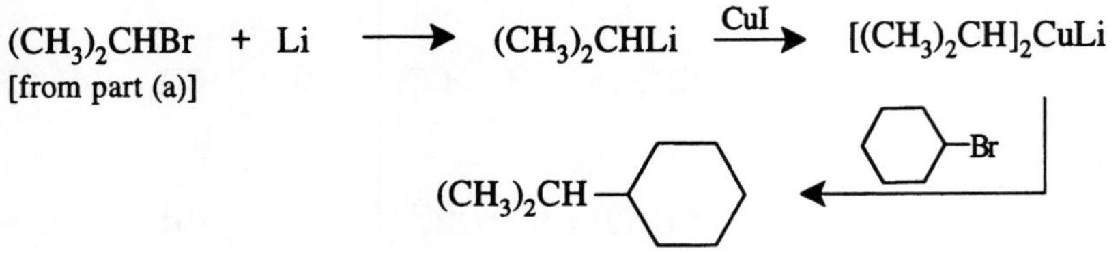 $(CH_3)_2CHBr$ + Li $\longrightarrow$ $(CH_3)_2CHLi$ $\xrightarrow{CuI}$ $[(CH_3)_2CH]_2CuLi$
[from part (a)]

11.16 (a) 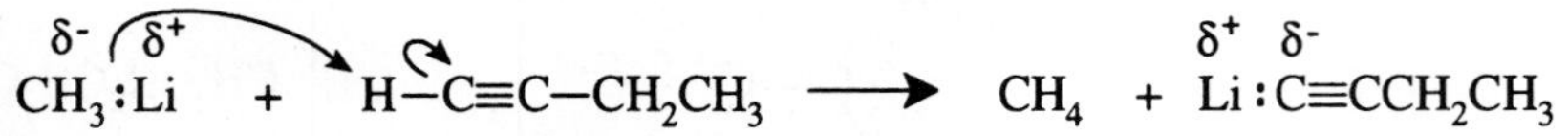 $\overset{\delta^-}{CH_3}:Li$ + $\overset{\delta^+}{H}-C\equiv C-CH_2CH_3$ $\longrightarrow$ CH_4 + $\overset{\delta^+}{Li}:\overset{\delta^-}{C}\equiv CCH_2CH_3$

(b) 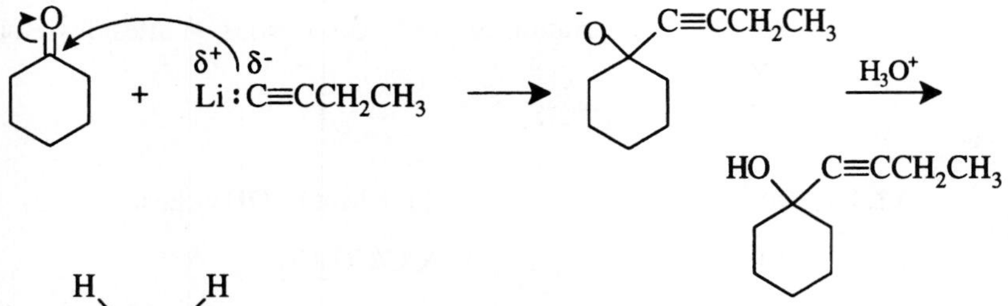   + $\overset{\delta^+}{Li}:\overset{\delta^-}{C}\equiv CCH_2CH_3$ $\longrightarrow$ $\xrightarrow{H_3O^+}$

(c)

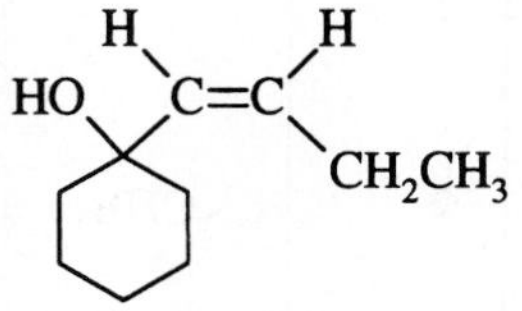

(d) 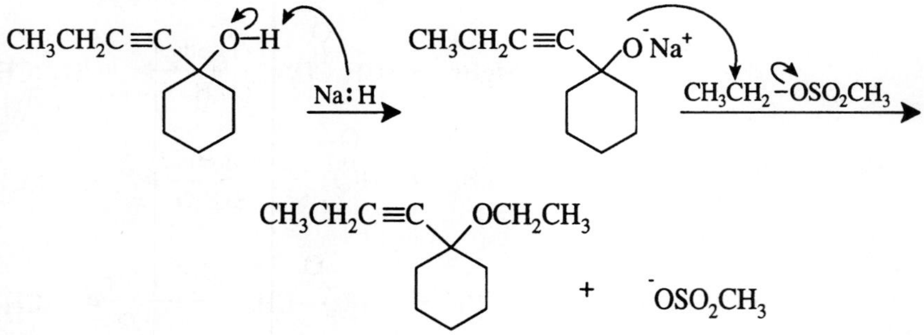 $\xrightarrow{Na:H}$ $\xrightarrow{CH_3CH_2-OSO_2CH_3}$

+ $^-OSO_2CH_3$

(e) $CH_3CH_2\underset{\underset{OH}{|}}{C}HCH_3$

(f) $CH_3CH_2\underset{\underset{OSO_2CH_3}{|}}{C}HCH_3$

(g) $CH_3CH_2CHCH_3$ + Na^+ $\overset{\overset{O}{\|}}{OCCH_3}$ $\longrightarrow$ $CH_3CH_2CHCH_3$
$\quad\quad\quad \overset{|}{OSO_2CH_3}$ $\quad\quad\quad\quad\quad\quad\quad\quad\quad\quad\quad\quad \overset{|}{O}\underset{\overset{\|}{O}}{CCH_3}$

$\quad\quad\quad\quad\quad\quad\quad\quad\quad\quad\quad\quad\quad\quad\quad\quad\quad\quad\quad$ + Na^+ $^-OSO_2CH_3$

(h) $CH_3CH_2\underset{\overset{|}{OH}}{C}HCH_3$ + CH_3CH_2OH

11.17 (a) $CH_3CH_2CH_2CH_2CH_2OH$ $\xrightarrow{PBr_3}$ $CH_3CH_2CH_2CH_2CH_2Br$

(b) $CH_3CH_2CH_2CH_2CH_2Br$ $\xrightarrow[\text{(CH}_3)_3\text{COH}]{\text{(CH}_3)_3\text{COK}}$ $CH_3CH_2CH_2CH=CH_2$
$\quad$ [from (a)]

(c) $CH_3CH_2CH_2CH=CH_2$ $\xrightarrow[\text{(2) NaBH}_4\text{/OH}^-]{\text{(1) Hg(OAc)}_2\text{THF·H}_2\text{O}}$ $CH_3CH_2CH_2\underset{\overset{|}{OH}}{C}HCH_3$
$\quad$ [from (b)]

(d) $CH_3CH_2CH_2\underset{\overset{|}{OH}}{C}HCH_3$ $\xrightarrow[\text{acetone, H}_2\text{O}]{\text{H}_2\text{CrO}_4}$ $CH_3CH_2CH_2\underset{\overset{\|}{O}}{C}CH_3$
$\quad$ [from (c)]

(e) $CH_3CH_2CH_2\underset{\overset{|}{OH}}{C}HCH_3$ $\xrightarrow{PBr_3}$ $CH_3CH_2CH_2\underset{\overset{|}{Br}}{C}HCH_3$
$\quad$ [from (c)]

(f) $CH_3CH_2CH_2CH_2CH_2Br$ $\xrightarrow[\text{Et}_2\text{O}]{\text{Mg}}$ $CH_3CH_2CH_2CH_2CH_2MgBr$ $\xrightarrow[\text{(2) H}_3\text{O}^+]{\text{(1) HCH}\overset{O}{\|}}$
$\quad$ [from (a)]

$\quad\quad\quad\quad\quad\quad\quad\quad\quad\quad\quad\quad\quad\quad\quad\quad\quad\quad\quad$ $CH_3CH_2CH_2CH_2CH_2CH_2OH$

(g) $CH_3CH_2CH_2CH_2CH_2MgBr$ $\xrightarrow[\text{(2) H}_3\text{O}^+]{\text{(1) H}_2\text{C}\overset{O}{-}\text{CH}_2}$ $CH_3CH_2CH_2CH_2CH_2CH_2CH_2OH$
$\quad$ [from (f)]

(h) $CH_3CH_2CH_2CH_2CH_2OH$ $\xrightarrow[\text{CH}_2\text{Cl}_2]{\text{PCC}}$ $CH_3CH_2CH_2CH_2\overset{\overset{O}{\|}}{C}H$

(i) $CH_3CH_2CH_2\underset{\overset{|}{OH}}{C}HCH_3$ $\xrightarrow[\text{acetone, H}_2\text{O}]{\text{H}_2\text{CrO}_4}$ $CH_3CH_2CH_2\underset{\overset{\|}{O}}{C}CH_3$
$\quad$ [from (c)]

(j) $CH_3CH_2CH_2CH_2CH_2OH$ $\xrightarrow[\text{(2) H}_3\text{O}^+]{\text{(1) KMnO}_4\text{, OH}^-\text{, heat}}$ $CH_3CH_2CH_2CH_2\overset{\overset{O}{\|}}{C}OH$

(k) (1) $CH_3CH_2CH_2CH_2CH_2OH$ $\xrightarrow[140° \text{ C}]{\text{H}_2\text{SO}_4}$ $(CH_3CH_2CH_2CH_2CH_2)_2O$

$\quad$ (2) $CH_3CH_2CH_2CH_2CH_2OH$ $\xrightarrow[(-\text{H}_2)]{\text{NaH}}$ $CH_3CH_2CH_2CH_2CH_2ONa$

$\quad\quad$ $\xrightarrow[\text{[from (a)]}]{CH_3CH_2CH_2CH_2CH_2Br}$ $(CH_3CH_2CH_2CH_2CH_2)_2O$

(l) $CH_3CH_2CH_2CH=CH_2$ $\xrightarrow[CCl_4]{Br_2}$ $CH_3CH_2CH_2\underset{\underset{Br}{|}}{CH}CH_2Br$ $\xrightarrow[heat]{3\ NaNH_2}$

[from (b)]

$CH_3CH_2CH_2C\equiv CNa$ $\xrightarrow{H^+}$ $CH_3CH_2CH_2C\equiv CH$

(m) $CH_3CH_2CH_2C\equiv CH$ $\xrightarrow{HBr}$ $CH_3CH_2CH_2\underset{\underset{Br}{|}}{C}=CH_2$

[from (l)]

(n) $CH_3CH_2CH_2CH_2CH_2Br$ $\xrightarrow[Et_2O]{Li}$ $CH_3CH_2CH_2CH_2CH_2Li$

[from (a)]

(o) $CH_3CH_2CH_2CH_2CH_2Li$ $\xrightarrow{Cu\,I}$ $(CH_3CH_2CH_2CH_2CH_2)_2CuLi$

[from (n)]

$\xrightarrow[\text{[from (a)]}]{CH_3CH_2CH_2CH_2CH_2Br}$ $CH_3(CH_2)_8CH_3$

(p) $CH_3CH_2CH_2CH_2CH_2MgBr$ $\xrightarrow[(2)\ H_3O^+]{(1)CH_3(CH_2)_2\overset{O}{\overset{||}{C}}CH_3\ \text{[from (d)]}}$

[from (f)]

$CH_3(CH_2)_4\underset{\underset{CH_3}{|}}{\overset{\overset{OH}{|}}{C}}(CH_2)_2CH_3$

11.18 (a) $C_6H_5C\equiv CH$ $\xrightarrow[ether\ (-CH_4)]{CH_3MgBr}$ $C_6H_5C\equiv CMgBr$ $\xrightarrow[(2)\ H_3O^+]{(1)CH_3\overset{O}{\overset{||}{C}}CH_3}$

$C_6H_5C\equiv CC(OH)(CH_3)_2$

(b) $C_6H_5COCH_3$ $\xrightarrow[OH^-]{NaBH_4}$ $C_6H_5\underset{\underset{OH}{|}}{CH}CH_3$

(c) $C_6H_5C\equiv CH$ $\xrightarrow[Ni_2B\ (P\text{-}2)]{H_2}$ $C_6H_5CH=CH_2$

(d) $C_6H_5CH=CH_2$ $\xrightarrow[(2)H_2O_2,\ OH^-]{(1)\ THF:\ BH_3}$ $C_6H_5CH_2CH_2OH$

(e) $C_6H_5CH_2CH_2OH$ $\xrightarrow[(2)\ Mg,\ Et_2O]{(1)\ PBr_3}$ $C_6H_5CH_2CH_2MgBr$ $\xrightarrow[(2)\ H_3O^+]{(1)\ H_2C\overset{O}{\overset{\diagdown\,\diagup}{-}}CH_2}$

$C_6H_5CH_2CH_2CH_2CH_2OH$

(f) $C_6H_5CH_2CH_2OH$ $\xrightarrow[(-H_2)]{NaH}$ $C_6H_5CH_2CH_2ONa$ $\xrightarrow{CH_3I}$

$C_6H_5CH_2CH_2OCH_3$

11.19 (a) *Partial Analysis*

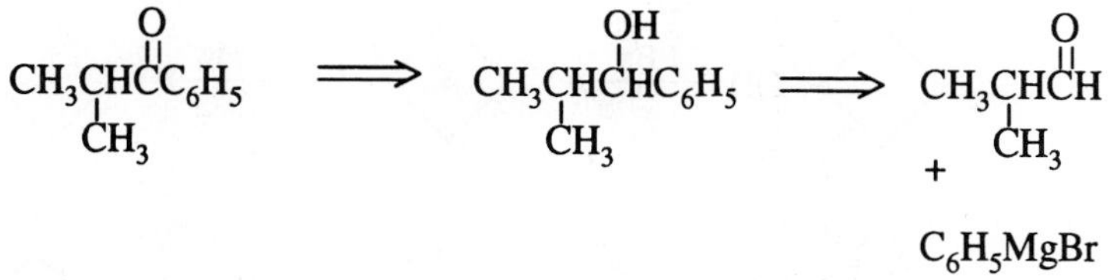

$$CH_3CHCC_6H_5 \Longrightarrow CH_3CHCHC_6H_5 \Longrightarrow CH_3CHCH$$

(with O above the C=O carbonyl, CH₃ below the first CH; middle structure has OH above, CH₃ below; right structure has O on carbonyl, CH₃ below)

$$+$$

$$C_6H_5MgBr$$

Synthesis

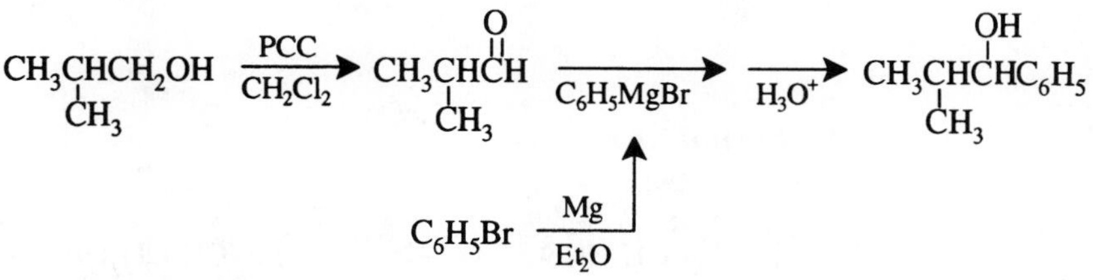

$$CH_3CHCH_2OH \xrightarrow[CH_2Cl_2]{PCC} CH_3CHCH \xrightarrow{C_6H_5MgBr} \xrightarrow{H_3O^+} CH_3CHCHC_6H_5$$

(left: CH₃ below first CH; middle: O on carbonyl, CH₃ below; right: OH above, CH₃ below)

$$C_6H_5Br \xrightarrow[Et_2O]{Mg}$$

(b) *Partial Analysis*

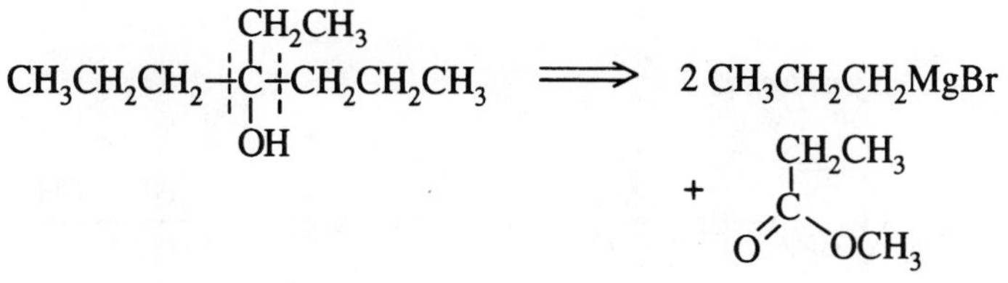

$$CH_3CH_2CH_2{-}\overset{\overset{\displaystyle CH_2CH_3}{|}}{\underset{\underset{\displaystyle OH}{|}}{C}}{-}CH_2CH_2CH_3 \Longrightarrow 2\ CH_3CH_2CH_2MgBr$$

$$+$$

(ester structure: CH₂CH₃ attached to C, with O double bond and OCH₃)

Synthesis

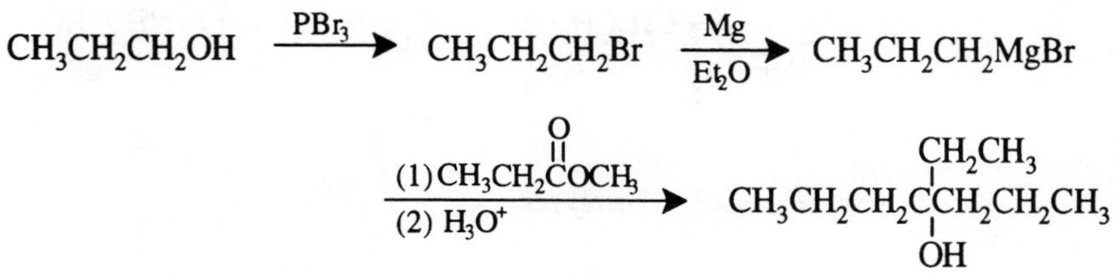

$$CH_3CH_2CH_2OH \xrightarrow{PBr_3} CH_3CH_2CH_2Br \xrightarrow[Et_2O]{Mg} CH_3CH_2CH_2MgBr$$

$$\xrightarrow[\text{(2) } H_3O^+]{\text{(1) } CH_3CH_2COCH_3} CH_3CH_2CH_2\overset{\overset{\displaystyle CH_2CH_3}{|}}{\underset{\underset{\displaystyle OH}{|}}{C}}CH_2CH_2CH_3$$

(reagent (1) has O on carbonyl of CH₃CH₂COCH₃)

(c) *Partial Analysis*

(cyclobutane ring)$-\overset{\overset{\displaystyle CH_3}{|}}{\underset{\underset{\displaystyle OH}{|}}{CH}}CHCH_3 \Longrightarrow$ (cyclobutane ring)$-MgBr + HCCHCH_3$

(right product: CH₃ above, O double bond below the HC)

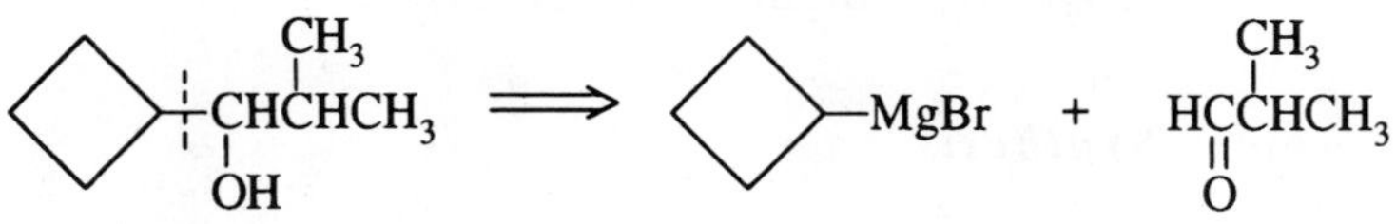

Synthesis

$$\text{cyclobutyl-OH} \xrightarrow{\text{PBr}_3} \text{cyclobutyl-Br} \xrightarrow[\text{Et}_2\text{O}]{\text{Mg}} \text{cyclobutyl-MgBr}$$

$$\xrightarrow[\text{[from (a)]}]{\underset{\text{CH}_3\text{CHCHO}}{\overset{\text{CH}_3}{}}} \text{cyclobutyl-}\underset{\overset{|}{\text{OMgBr}}}{\overset{\overset{\text{CH}_3}{|}}{\text{CHCHCH}_3}} \xrightarrow{\text{H}_3\text{O}^+} \text{cyclobutyl-}\underset{\overset{|}{\text{OH}}}{\overset{\overset{\text{CH}_3}{|}}{\text{CHCHCH}_3}}$$

(d) *Partial Analysis*

$$\text{Ph-CH}_2\text{CHO} \implies \text{Ph-CH}_2\text{CH}_2\text{OH} \implies \text{C}_6\text{H}_5\text{MgBr}$$

$$+ \quad \text{H}_2\text{C}\overset{O}{\underset{}{-}}\text{CH}_2$$

Synthesis

$$\text{Ph-Br} \xrightarrow[\text{Et}_2\text{O}]{\text{Mg}} \text{Ph-MgBr} \xrightarrow[\text{(2) H}_3\text{O}^+]{\text{(1) H}_2\text{C}\overset{O}{-}\text{CH}_2}$$

$$\text{Ph-CH}_2\text{CH}_2\text{OH} \xrightarrow[\text{CH}_2\text{Cl}_2]{\text{PCC}} \text{Ph-CH}_2\text{CHO}$$

(e) *Partial Analysis*

$$(\text{CH}_3)_2\text{CHCH}_2\text{CH}_2\text{CO}_2\text{H} \implies (\text{CH}_3)_2\text{CHCH}_2 \dotplus \text{CH}_2\text{CH}_2\text{OH} \implies$$

$$(\text{CH}_3)_2\text{CHCH}_2\text{MgBr} \quad + \quad \text{H}_2\text{C}\overset{O}{-}\text{CH}_2$$

Synthesis

$$(\text{CH}_3)_2\text{CHCH}_2\text{OH} \xrightarrow{\text{PBr}_3} (\text{CH}_3)_2\text{CHCH}_2\text{Br} \xrightarrow[\text{Et}_2\text{O}]{\text{Mg}} (\text{CH}_3)_2\text{CHCH}_2\text{MgBr}$$

$$\xrightarrow[\text{(2) H}_3\text{O}^+]{\text{(1) H}_2\text{C}\overset{O}{-}\text{CH}_2} (\text{CH}_3)_2\text{CHCH}_2\text{CH}_2\text{CH}_2\text{OH} \xrightarrow[\text{(2) H}_3\text{O}^+]{\text{(1) KMnO}_4,\text{ OH}^-,\text{ heat}}$$

$$(\text{CH}_3)_2\text{CHCH}_2\text{CH}_2\text{CO}_2\text{H}$$

(f) *Partial   Analysis*

$$\square\!\!-\!\!\overset{\text{OH}}{\underset{\vdots}{\text{C}}}\!\!-\!\!CH_2CH_2CH_3 \implies \square\!\!=\!\!O + CH_3CH_2CH_2MgBr$$

Synthesis

$$\square\!\!-\!\!OH \xrightarrow{H_2CrO_4} \square\!\!=\!\!O \xrightarrow[\text{(2) } H_3O^+]{\substack{\text{(1) } CH_3CH_2CH_2MgBr \\ \text{[from (b)]}}} \square\!\!-\!\!\overset{\text{OH}}{\underset{}{\text{C}}}\!\!-\!\!CH_2CH_2CH_3$$

(g) *Partial   Analysis*

$$CH_3CH_2CH_2\underset{\underset{O}{\|}}{C}CH_2CH(CH_3)_2 \implies CH_3CH_2CH_2\underset{\underset{OH}{|}}{CH}\!\!\!\vdots\!\!\!CH_2CH(CH_3)_2 \implies$$

$$CH_3CH_2CH_2\overset{\overset{O}{\|}}{CH} + BrMgCH_2CH(CH_3)_2$$

Synthesis

$$CH_3CH_2CH_2CH_2OH \xrightarrow[CH_2Cl_2]{PCC} CH_3CH_2CH_2\overset{\overset{O}{\|}}{CH} \xrightarrow[\text{(2) } H_3O^+]{\substack{\text{(1)}(CH_3)_2CHCH_2MgBr \\ \text{[from (e)]}}}$$

$$CH_3CH_2CH_2\underset{\underset{OH}{|}}{CH}CH_2CH(CH_3)_2 \xrightarrow{H_2CrO_4} CH_3CH_2CH_2\underset{\underset{O}{\|}}{C}CH_2CH(CH_3)_2$$

(h) *Partial   Analysis*

$$CH_3CH_2\underset{\underset{C_6H_5}{|}}{\overset{\overset{Br}{|}}{C}}CH_2CH_3 \implies CH_3CH_2\underset{\underset{C_6H_5}{|}}{\overset{\overset{OH}{|}}{C}}\!\!\!\vdots\!\!\!CH_2CH_3 \implies$$

$$CH_3CH_2MgBr + CH_3CH_2\underset{\underset{O}{\|}}{C}C_6H_5 \implies CH_3CH_2\overset{\overset{OH}{|}}{CH}\!\!-\!\!C_6H_5 \implies$$

$$CH_3CH_2\overset{\overset{O}{\|}}{CH} + C_6H_5MgBr$$

Synthesis

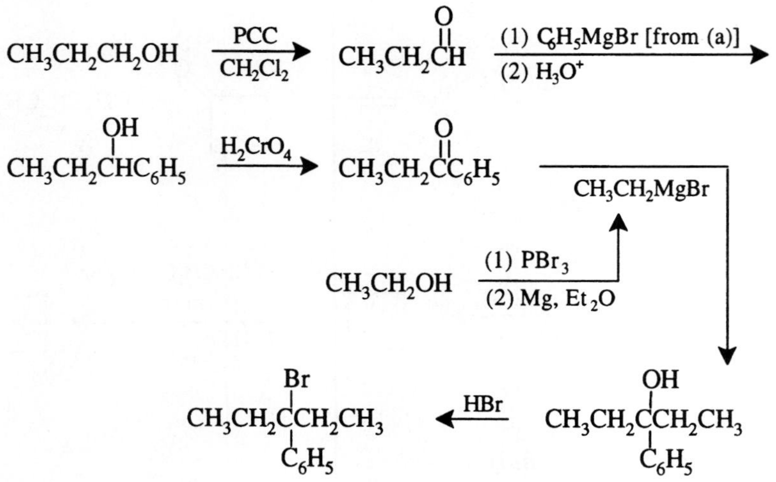

$$CH_3CH_2CH_2OH \xrightarrow[CH_2Cl_2]{PCC} CH_3CH_2\overset{\displaystyle O}{\overset{\|}{C}}H \xrightarrow[(2)\ H_3O^+]{(1)\ C_6H_5MgBr\ [from\ (a)]}$$

$$CH_3CH_2\overset{\displaystyle OH}{\underset{|}{C}}HC_6H_5 \xrightarrow{H_2CrO_4} CH_3CH_2\overset{\displaystyle O}{\overset{\|}{C}}C_6H_5$$

$$CH_3CH_2MgBr$$

$$CH_3CH_2OH \xrightarrow[(2)\ Mg,\ Et_2O]{(1)\ PBr_3} CH_3CH_2MgBr$$

$$CH_3CH_2\overset{\displaystyle Br}{\underset{\underset{\displaystyle C_6H_5}{|}}{\overset{|}{C}}}CH_2CH_3 \xleftarrow{HBr} CH_3CH_2\overset{\displaystyle OH}{\underset{\underset{\displaystyle C_6H_5}{|}}{\overset{|}{C}}}CH_2CH_3$$

11.20 *Analysis*

$$C_6H_5Br + Mg \qquad CH_2{=}CHCH_2CH_3$$

$$+ R\overset{\displaystyle O}{\overset{\|}{C}}OOH$$

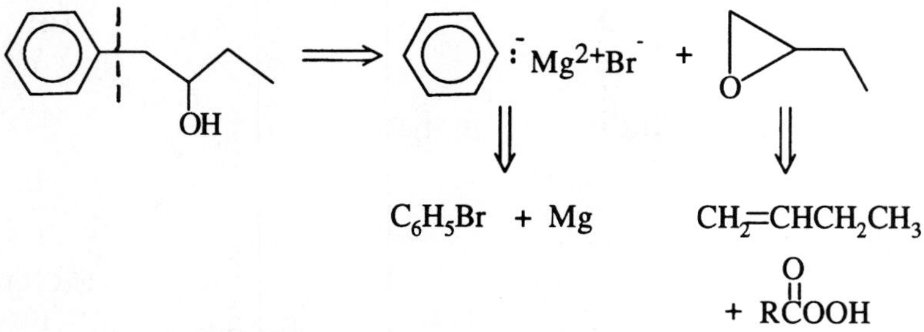

Synthesis

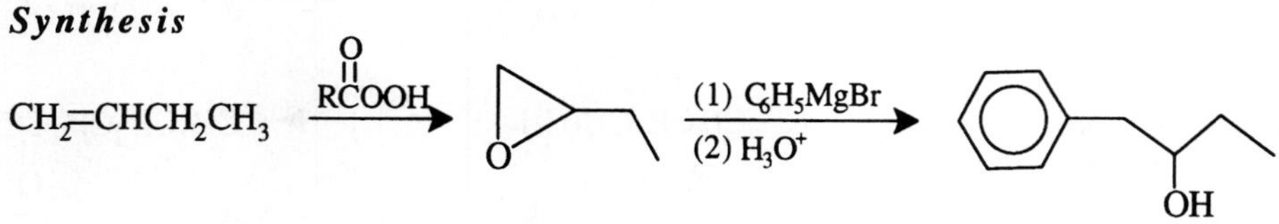

11.21 The starting compound is a cyclic ester. Addition of two molar equivalents of CH_3MgI will (after acidification) furnish the desired product.

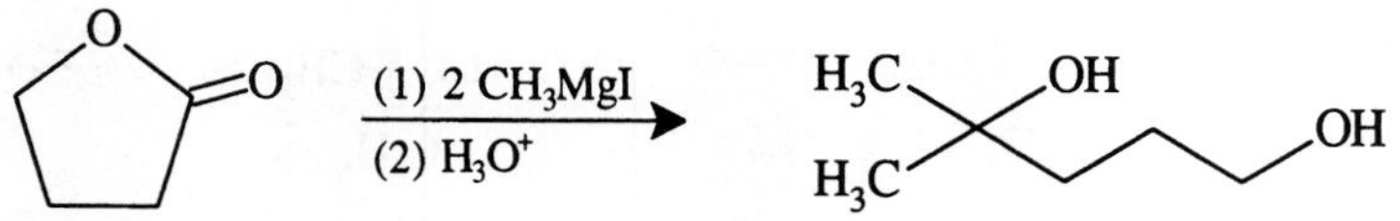

11.22 *Analysis*

$$CH_3-CH_2-\underset{\underset{OH}{|}}{\overset{\overset{CH_3}{|}}{C}}\text{┼}C\equiv CH \implies CH_3-CH_2-\underset{\underset{O}{\|}}{\overset{\overset{CH_3}{|}}{C}} \quad + \quad Na^+ \; \overset{\bullet\bullet}{:}C\equiv CH$$

Synthesis

$$HC\equiv CH \xrightarrow[\text{liq. NH}_3]{\text{NaNH}_2} HC\equiv C\overset{\bullet\bullet}{:}{}^-Na^+ \xrightarrow[\text{(2) H}_3O^+]{\text{(1) CH}_3\text{CH}_2\overset{\overset{O}{\|}}{C}CH_3} CH_3-CH_2-\underset{\underset{OH}{|}}{\overset{\overset{CH_3}{|}}{C}}-C\equiv CH$$

11.23 The three- and four-membered rings are strained, and so they open on reaction with RMgX or RLi. THF possesses an essentially unstrained ring and hence is far more resistant to attack by an organometallic compound.

11.24 (a) $RMgX \; + \; C_2H_5-O-\overset{\overset{O}{\|}}{C}-O-C_2H_5 \longrightarrow \left[C_2H_5-O-\underset{\underset{R}{|}}{\overset{\overset{OMgX}{|}}{C}}-O-C_2H_5 \right]$

$$\xrightarrow{-C_2H_5OMgX} \left[R-\overset{\overset{O}{\|}}{C}-O-C_2H_5 \right] \xrightarrow{RMgX} \left[R-\underset{\underset{R}{|}}{\overset{\overset{OMgX}{|}}{C}}-O-C_2H_5 \right] \xrightarrow{-C_2H_5OMgX}$$

$$\left[R-\overset{\overset{O}{\|}}{C}-R \right] \xrightarrow{RMgX} R-\underset{\underset{R}{|}}{\overset{\overset{OMgX}{|}}{C}}-R \xrightarrow{H_2O} R-\underset{\underset{R}{|}}{\overset{\overset{OH}{|}}{C}}-R$$

(b) $RMgX \; + \; H-\overset{\overset{O}{\|}}{C}-O-C_2H_5 \longrightarrow \left[H-\underset{\underset{R}{|}}{\overset{\overset{OMgX}{|}}{C}}-O-C_2H_5 \right] \xrightarrow{-C_2H_5OMgX}$

$$\left[H-\overset{\overset{O}{\|}}{C}-R \right] \xrightarrow{RMgX} H-\underset{\underset{R}{|}}{\overset{\overset{OMgX}{|}}{C}}-R \xrightarrow{H_2O} H-\underset{\underset{R}{|}}{\overset{\overset{OH}{|}}{C}}-R$$

QUIZ

11.1 Which of the following could be employed to transform ethanol into $CH_3CH_2CH_2OH$?

(a) Ethanol + HBr, then Mg/diethyl ether, then H_3O^+

(b) Ethanol + HBr, then Mg/diethyl ether, then $H\overset{\overset{O}{\|}}{C}H$, then H_3O^+

(c) Ethanol + $H_2SO_4/140°$ C

(d) Ethanol + Na, then $H\overset{\overset{O}{\|}}{C}H$, then H_3O^+

(e) Ethanol + $H_2SO_4/180°$ C, then $H_2C\overset{O}{\diagup\!\!\diagdown}CH_2$

11.2   The principal product(s) formed when *1 mol* of methylmagnesium iodide reacts with 1 mol of $CH_3\overset{\overset{}{\underset{\|}{}}O}{C}CH_2CH_2OH$.

(a) $CH_4 + CH_3\overset{\overset{}{\underset{\|}{}}O}{C}CH_2CH_2OMgI$

(b) $CH_3\overset{\overset{OMgI}{\|}}{\underset{\underset{CH_3}{\|}}{C}}CH_2CH_2OH$

(c) $CH_3\overset{\overset{}{\underset{\|}{}}O}{C}CH_2CH_2OCH_3$

(d) $CH_3\overset{\overset{CH_3}{\|}}{\underset{\underset{OH}{\|}}{C}}CH_2CH_2OCH_3$

(e) None of the above

11.3   Supply the missing reagents.

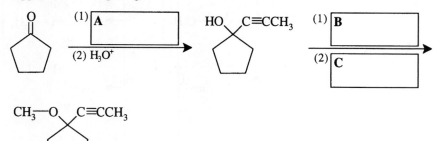

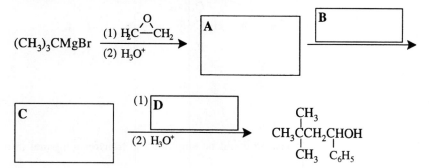

11.4   Supply the missing reagents and intermediates.

11.5   Supply the missing starting compound.

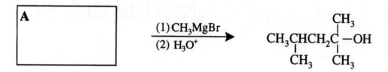

12 CONJUGATED UNSATURATED SYSTEMS

SOLUTIONS TO PROBLEMS

12.1 (a) $^{14}CH_2=CHCH_2X$ and $CH_2=CH\overset{14}{\text{—}}CH_2X$

(b) The reaction proceeds through the resonance-stabilized radical.

$$^{14}CH_2=CH-\overset{\bullet}{C}H_2 \longleftrightarrow {}^{14}\overset{\bullet}{C}H_2-CH=CH_2 \quad \text{or} \quad {}^{14}\overset{\delta\bullet}{CH_2}\text{---}CH\overset{\delta\bullet}{\text{---}}CH_2$$

Thus, attack on X_2 can occur by the carbon atom at either end of the chain since these atoms are equivalent.

(c) 50:50 because attack at the two ends of the chain is equally probable.

12.2 (a)

$$CH_3-\underset{+}{CH}\overset{2}{\underset{\underset{\textbf{D}}{CH}}{\diagup}}\underset{1}{\diagdown}CH_2 \longleftrightarrow CH_3-\overset{4}{\underset{\textbf{E}}{CH}}\overset{2}{\underset{\underset{}{CH}}{\diagup}}\underset{+}{\overset{1}{CH_2}} \quad \text{or} \quad CH_3-\underset{\delta^+}{\overset{3}{CH}}\overset{2}{\underset{\underset{\textbf{F}}{CH}}{\diagup}}\underset{\delta^+}{\overset{1}{CH_2}}$$

(b) We know that the allylic cation is almost as stable as a tertiary carbocation. Here we find not only the resonance stabilization of an allylic cation but also the additional stabilization that arises from contributor **D** in which the plus charge is on a secondary carbon atom.

(c) $CH_3-\overset{Cl}{\overset{|}{CH}}-CH=CH_2$ and $CH_3-CH=CH-CH_2-Cl$, because the Cl will attack the chain at the two positive centers shown in structure **F**.

12.3 (a) $CH_2=\overset{\overset{CH_3}{|}}{C}-\overset{\bullet}{C}H_2 \longleftrightarrow {}^{\bullet}CH_2-\overset{\overset{CH_3}{|}}{C}=CH_2$

(b) $CH_2=CH-\underset{+}{CH}-CH=CH_2 \longleftrightarrow {}^{+}CH_2-CH=CH-CH=CH_2 \longleftrightarrow$

$$CH_2=CH-CH=CH-\overset{+}{C}H_2$$

(c)

(d)

(e) $CH_3CH=CH-CH=\overset{+}{\underset{\cdot\cdot}{O}}H$ ⟷ $CH_3CH=CH-\overset{+}{C}H-\underset{\cdot\cdot}{\ddot{O}}H$ ⟷

$CH_3\overset{+}{C}H-CH=CH-\ddot{O}H$

(f) $CH_2=CH-\underset{\cdot\cdot}{\overset{+}{\ddot{B}r}}:$ ⟷ $:\!\overset{-}{C}H_2-CH=\overset{+}{\ddot{B}r}:$

(g)

(h) $:\!\overset{-}{C}H_2-\overset{\overset{\displaystyle\ddot{O}:}{\|}}{C}-CH_3$ ⟷ $CH_2\!=\!\overset{\overset{\displaystyle:\!\ddot{O}:^{-}}{\|}}{C}-CH_3$

(i) $CH_3-\underset{\cdot\cdot}{\overset{\cdot\cdot}{S}}-\overset{+}{C}H_2$ ⟷ $CH_3-\overset{+}{\underset{\cdot\cdot}{S}}=CH_2$

(j) $CH_3-\overset{+}{N}\!\!\begin{smallmatrix}\nearrow\ddot{O}\cdot\\[2pt]\searrow\ddot{O}:^{-}\end{smallmatrix}$ ⟷ $CH_3-\overset{+}{N}\!\!\begin{smallmatrix}\nearrow\ddot{O}:^{-}\\[2pt]\searrow\ddot{O}\cdot\end{smallmatrix}$ ⟷ $CH_3-\overset{2+}{N}\!\!\begin{smallmatrix}\nearrow\ddot{O}:^{-}\\[2pt]\searrow\ddot{O}:^{-}\end{smallmatrix}$
(minor)

12.4   (a) $CH_3CH_2\overset{\overset{\displaystyle CH_3}{|}}{\underset{+}{C}}-CH=CH_2$ because the positive charge is on a tertiary carbon atom rather than a primary one (rule 8).

(b)   because the positive charge is on a secondary carbon atom rather than a primary one (rule 8).

(c) $CH_2\!=\!\overset{+}{N}(CH_3)_2$ because all atoms have a complete octet (rule 8b), and there are more covalent bonds (rule 8a).

(d) $CH_3-\overset{\overset{\displaystyle O}{\|}}{C}-OH$ because it has no charge separation (rule 8c).

(e) $CH_2\!=\!CH\overset{\displaystyle\bullet}{C}HCH=CH_2$ because the radical is on a secondary carbon atom rather than a primary one (rule 8).

(f) $:\!NH_2-C\equiv N:$ because it has no charge separation (rule 8c).

12.5 In resonance product structures, the positions of the nuclei must remain the same for all structures (rule 2). The keto and enol forms shown differ not only in the positions of their electrons, but also in the position of one of the hydrogen atoms. In the enol form, it is attached to an oxygen atom; in the keto form, it has been moved so that it is attached to a carbon atom.

12.6 (a) *cis*-1,3-Pentadiene, *trans,trans*-2,4-hexadiene, *cis,trans*-2,4-hexadiene, and 1,3-cyclohexadiene are conjugated dienes.

(b) 1,4-Cyclohexadiene is an isolated diene.

(c) 1-Penten-4-yne is an isolated enyne.

12.7 (a) Recall that 1,2 and 1,4 addition refer to the conjugated system itself and not the entire carbon chain. $CH_3CH_2\underset{\underset{Cl}{|}}{CH}CH=CHCH_3$ and $CH_3CH_2CH=CH\underset{\underset{Cl}{|}}{CH}CH_3$

(b) The most stable cation is a hybrid of equivalent forms:

$$CH_3\overset{+}{C}HCH=CHCH_3 \longleftrightarrow CH_3CH=CH\overset{+}{C}HCH_3$$

Thus, 1,4 and 1,2 addition yield the same product.

$$CH_3\underset{\underset{Cl}{|}}{CH}CH=CHCH_3$$

12.8 Addition of the proton gives the resonance hybrid.

(a) $CH_3-\overset{+}{C}H-CH=CH_2 \longleftrightarrow CH_3-CH=CH-\overset{+}{C}H_2$

$\quad\quad\quad$ **I** $\quad\quad\quad\quad\quad\quad\quad\quad\quad$ **II**

The inductive effect of the methyl group in **I** stabilizes the positive charge on the adjacent carbon. Such stabilization of the positive charge does not occur in **II**. Because **I** contributes more heavily to the resonance hybrid than does **II**, C2 bears a greater positive charge and reacts faster with the bromide ion.

(b) In the 1,4-addition product, the double bond is more highly substituted than in the 1,2-addition product; hence it is the more stable alkene.

12.9 (a), (c)

(b) π-Electron interaction occurs here.

Endo adduct

12.10 (a)

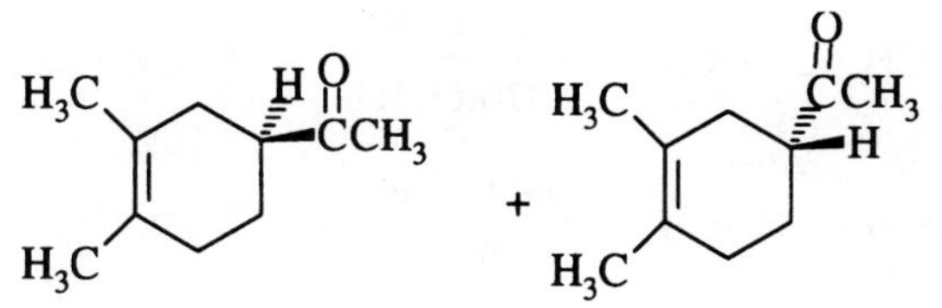

$+$

(b)

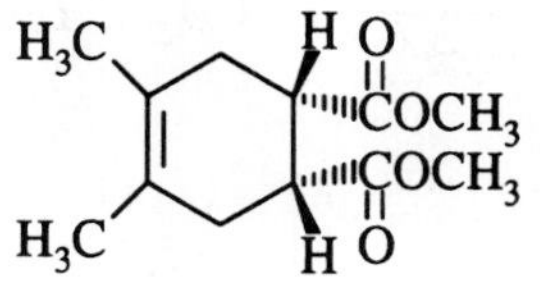

(c)

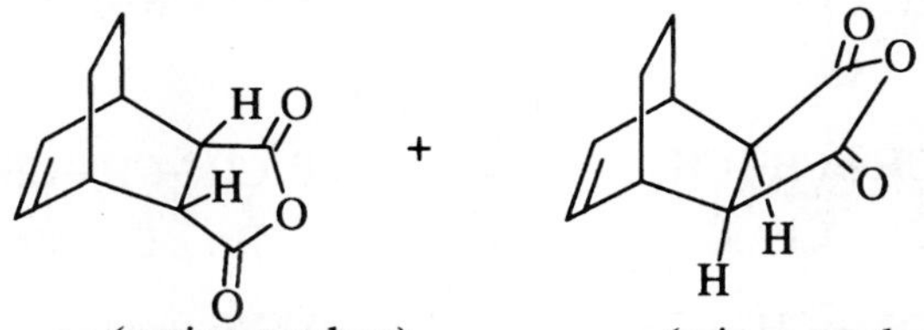

(major product)   (minor product)

12.11 Use the trans diester because the stereochemistry is retained in the adduct.

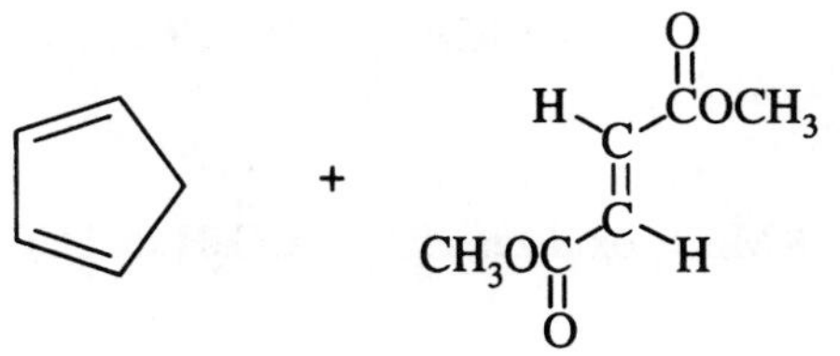

12.12

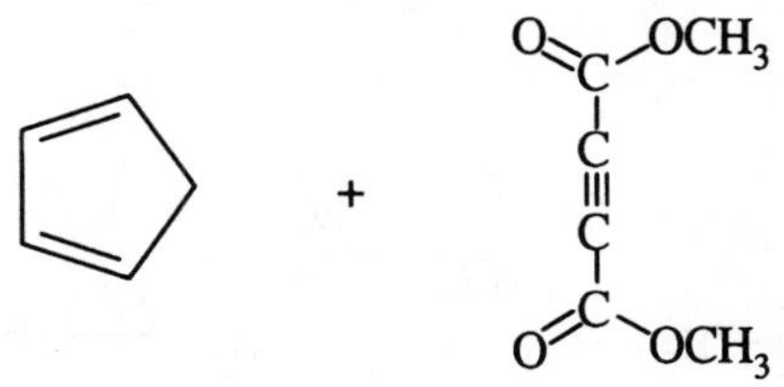

12.13 (a) $BrCH_2CH_2CH_2CH_2Br$ $\xrightarrow[\text{(CH}_3)_3\text{COH}]{\text{(CH}_3)_3\text{COK}}$ $CH_2{=}CH{-}CH{=}CH_2$

(b) $HOCH_2CH_2CH_2CH_2OH$ $\xrightarrow[\text{heat}]{\text{concd. H}_2\text{SO}_4}$ $CH_2{=}CH{-}CH{=}CH_2$

(c) $CH_2{=}CH{-}CH_2CH_2{-}OH$ $\xrightarrow[\text{heat}]{\text{concd. H}_2\text{SO}_4}$ $CH_2{=}CH{-}CH{=}CH_2$

(d) $CH_2{=}CH{-}CH_2CH_2{-}Cl$ $\xrightarrow[\text{(CH}_3)_3\text{COH}]{\text{(CH}_3)_3\text{COK}}$ $CH_2{=}CH{-}CH{=}CH_2$

(e) $CH_2=CH-CH-CH_3$ $\xrightarrow[\text{(CH}_3\text{)}_3\text{COH}]{\text{(CH}_3\text{)}_3\text{COK}}$ $CH_2=CH-CH=CH_2$
            |
            Cl

(f) $CH_2=CH-CH-CH_3$ $\xrightarrow[\text{heat}]{\text{concd. H}_2\text{SO}_4}$ $CH_2=CH-CH=CH_2$
            |
            OH

(g) $CH_2=CH-C\equiv CH$ $+$ H_2 $\xrightarrow{\text{Ni}_2\text{B (P-2)}}$ $CH_2=CH-CH=CH_2$

12.14 $CH_2=C\!\!-\!\!-\!\!C=CH_2$
                        | |
                        CH_3 CH_3

12.15   (a) $Cl-CH_2CHCH=CH_2$ $+$ $Cl-CH_2-CH=CH-CH_2-Cl$
                                |
                                Cl

(b) $CH_2-CH-CH-CH_2$ (c) $CH_2-CH-CH-CH_2$
        | | | | | | | |
        Cl   Cl   Cl   Cl   Br   Br   Br   Br

(d) $CH_3-CH_2-CH_2-CH_3$ (e)   No reaction

(f) $Cl-CH_2-CH-CH=CH_2$ $+$ $Cl-CH_2-CH=CH-CH_2-OH$
                      |
                      OH

$[+ClCH_2CHCH=CH_2$ $+$ $ClCH_2CH=CHCH_2Cl$ $]$
                |
                Cl

(g) $4\ CO_2$ (*Note:* $KMnO_4$ oxidizes HO_2C-CO_2H to $2\ CO_2$)

(h) $CH_3-CH-CH=CH_2$ $+$ $CH_3-CH=CH-CH_2-OH$
                  |
                  OH

12.16   (a) $CH_2=CH-CH_2-CH_3$ $+$ (NBS) $\xrightarrow{\text{CCl}_4}$ $CH_2=CH-\overset{\displaystyle Br}{\underset{\displaystyle |}{CH}}-CH_3$

$\left(+ CH_2-CH=CH-CH_3 \atop\quad | \atop\quad Br \right)$

$\xrightarrow[\text{(CH}_3\text{)}_3\text{COH}]{\text{(CH}_3\text{)}_3\text{COK}}$ $CH_2=CH-CH=CH_2$

Note: In the second step, both allylic halides undergo elimination of **HBr** to yield 1,3-butadiene; therefore, separating the mixture produced in the first step is unnecessary. The $BrCH_2CH=CHCH_3$ undergoes a 1,4 elimination (the opposite of a 1,4 addition).

(b) $CH_2=CH-CH_2-CH_2-CH_3$ + NBS $\xrightarrow{CCl_4}$ $CH_2=CH-\overset{\overset{\displaystyle Br}{|}}{CH}-CH_2-CH_3$

$\left(+ \underset{\underset{\displaystyle Br}{|}}{CH_2}-CH=CH-CH_2-CH_3 \right)$ $\xrightarrow[\text{(CH}_3)_3\text{COH}]{\text{(CH}_3)_3\text{COK}}$ $CH_2=CH-CH=CH-CH_3$

Here again both products undergo elimination of **HBr** to yield 1,3-pentadiene.

(c) $CH_3CH_2CH_2CH_2OH$ $\xrightarrow[\text{heat}]{\text{concd. H}_2\text{SO}_4}$ $CH_3CH=CHCH_3$ $\qquad$ [as in (a)]

$(+ CH_3CH_2CH=CH_2)$

$\underset{\underset{\displaystyle Br}{|}}{CH_2}-CH=CH-\underset{\underset{\displaystyle Br}{|}}{CH_2}$ $\xleftarrow[\text{heat}]{Br_2}$ $CH_2=CH-CH=CH_2$ $\longleftarrow$

(d) $CH_3-CH=CH-CH_3$ + NBS $\xrightarrow{CCl_4}$ $CH_3-CH=CH-CH_2Br$

$+ CH_2=CH-CHBr-CH_3$

(e) ⬠ $+ Br_2$ $\xrightarrow[\text{heat}]{\text{light}}$ ⬠(Br) $\xrightarrow[\text{(CH}_3)_3\text{COH}]{\text{(CH}_3)_3\text{COK}}$ ⬠ $\xrightarrow[\text{CCl}_4]{\text{NBS}}$ ⬠(Br)

(excess)

(f) ⬠(Br) $\xrightarrow[\text{(CH}_3)_3\text{COH}]{\text{(CH}_3)_3\text{COK}}$ ⬠ $\left(\text{same as } ⬠ \right)$

12.17 $R-\overset{..}{\underset{..}{O}}-\overset{..}{\underset{..}{O}}-R$ $\xrightarrow[\text{or light}]{\text{heat}}$ $2\ R-\overset{..}{\underset{..}{O}}•$

$R-\overset{..}{O}•$ + $H-\overset{..}{\underset{..}{Br}}:$ $\longrightarrow$ $R-\overset{..}{\underset{..}{O}}-H$ + $•\overset{..}{\underset{..}{Br}}:$

$CH_2=CH-CH=CH_2$ + $•\overset{..}{\underset{..}{Br}}:$ $\longrightarrow$ $\left[CH_2=CH-\overset{•}{C}H-\underset{\underset{\displaystyle Br}{|}}{CH_2} \longleftrightarrow \overset{•}{C}H_2-CH=CH-\underset{\underset{\displaystyle Br}{|}}{CH_2} \right]$

$\xrightarrow{\text{HBr}}$ $CH_2=CH-\underset{\underset{\displaystyle H}{|}}{CH}-\underset{\underset{\displaystyle Br}{|}}{CH_2}$ + $\underset{\underset{\displaystyle H}{|}}{CH_2}-CH=CH-\underset{\underset{\displaystyle Br}{|}}{CH_2}$ + $•\overset{..}{\underset{..}{Br}}:$

(cis and trans)

12.18 (a) $Ag(NH_3)_2OH$ gives a precipitate with 1-butyne only.

(b) 1,3-Butadiene decolorizes bromine solution; butane does not.

(c) H_2SO_4 dissolves $CH_2=CHCH_2CH_2OH$. Butane does not dissolve.

(d) $AgNO_3$ in C_2H_5OH gives a AgBr precipitate with $CH_2=CHCH_2CH_2Br$. No reaction with 1,3-butadiene.

(e) $AgNO_3$ in C_2H_5OH gives a AgBr precipitate with $BrCH_2CH=CHCH_2Br$ (it is an allylic bromide), but not with $CH_3C=C\text{-}CH_3$ (a vinylic bromide).

$$\underset{\overset{|}{Br}\ \ \overset{|}{Br}}{}$$

12.19 (a) Because a highly resonance-stabilized radical is formed:

$$CH_2=CH-\overset{\bullet}{C}H-CH=CH_2 \longleftrightarrow CH_2=CH-CH=CH-\overset{\bullet}{C}H_2 \longleftrightarrow$$
$$\overset{\bullet}{C}H_2-CH=CH-CH=CH_2$$

(b) Because the carbanion is more stable:

$$CH_2=CH-\overset{\overset{\bullet\bullet}{-}}{C}H-CH=CH_2 \longleftrightarrow CH_2=CH-CH=CH-\overset{\overset{\bullet\bullet}{-}}{C}H_2 \longleftrightarrow$$
$$\overset{\overset{\bullet\bullet}{-}}{C}H_2-CH=CH-CH=CH_2$$

That is, we can write more reasonance structures of nearly equal energies.

12.20

$$CH_2=\overset{\overset{\textstyle CH_3}{|}}{C}-CH=CH_2 \xrightarrow{\ H^+\ }$$

$$\left[CH_3-\overset{\overset{\textstyle CH_3}{|}}{\underset{+}{C}}-CH=CH_2 \longleftrightarrow CH_3-\overset{\overset{\textstyle CH_3}{|}}{C}=CH-\underset{+}{C}H_2 \right]\ \mathbf{I}$$

$$\left[CH_2=\overset{\overset{\textstyle CH_3}{|}}{\underset{+}{C}}-CH-CH_3 \longleftrightarrow \underset{+}{C}H_2-\overset{\overset{\textstyle CH_3}{|}}{C}=CH-CH_3 \right]\ \mathbf{II}$$

The resonance hybrid, **I,** has the positive charge, in part, on the tertiary carbon atom; in **II,** the positive charge is on primary and secondary carbon atoms only. Therefore, hybrid **I** is more stable and will be the intermediate carbocation. A 1,4 addition to **I** gives

$$CH_3-\overset{\overset{\textstyle CH_3}{|}}{C}=CH-CH_2-Cl$$

12.21 (a)   (b)

(c)

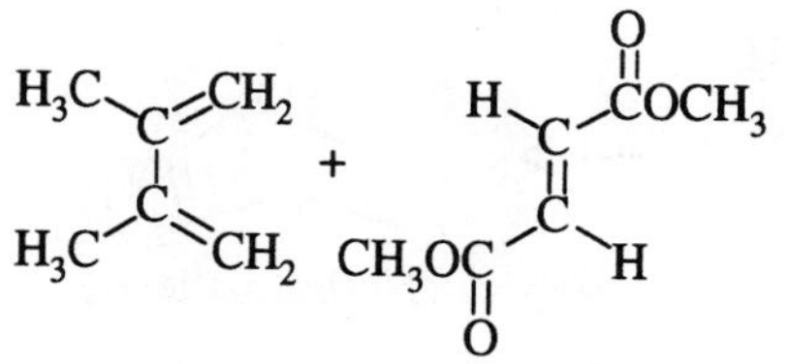

(d)

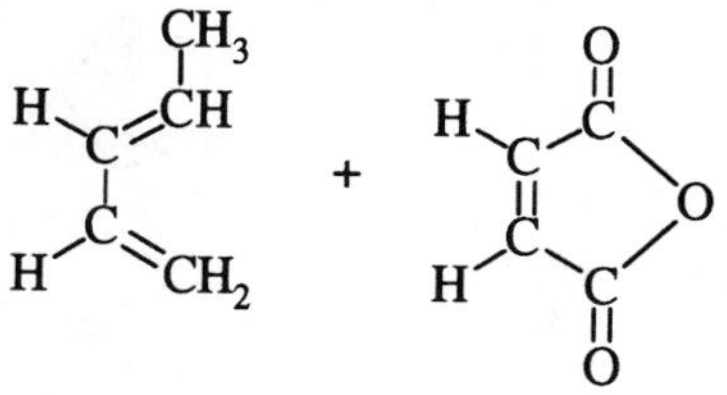

(e)

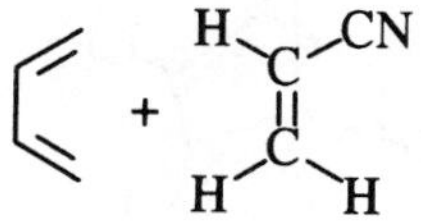

(f)

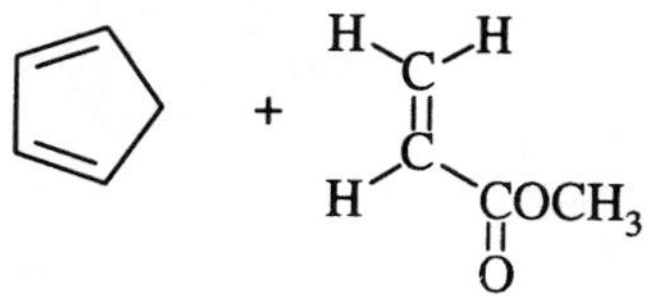

12.22 Neither compound can assume the s-cis conformation. 1,3-Butadiyne is linear, and

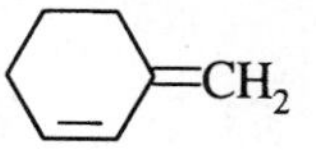

is held in an s-trans conformation by the requirements of the ring.

12.23 (a) 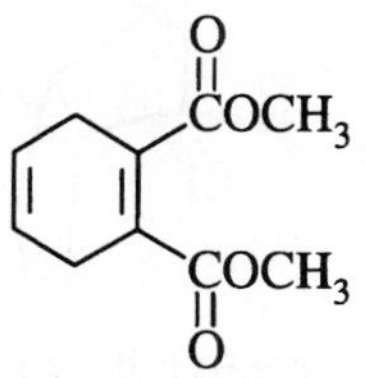   (b)

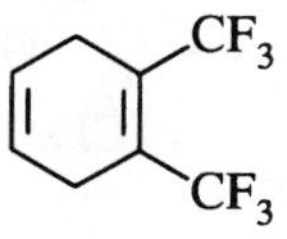

12.24

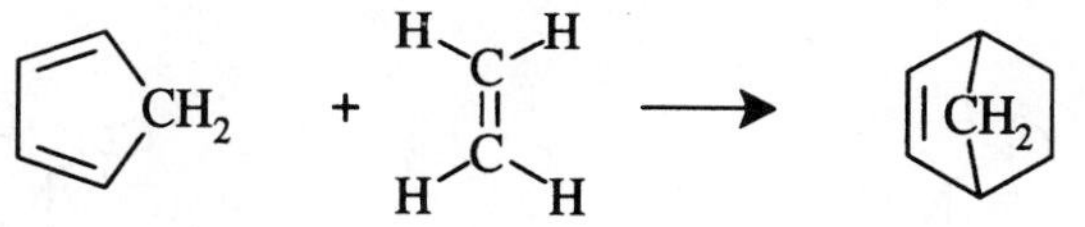

12.25 The endo adduct is less stable than the exo, but is produced at a faster rate at 25°C. At 90°C the Diels-Alder reaction becomes reversible; an equilibrium is established, and the more stable exo adduct predominates.

12.26

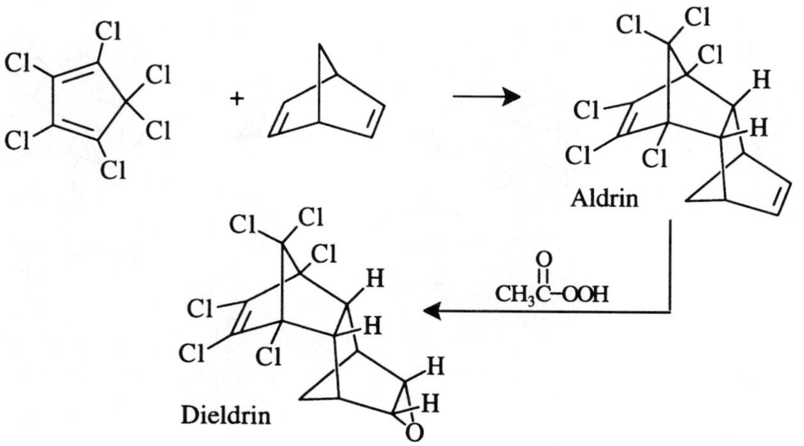

Aldrin

CH_3C-OOH

Dieldrin

12.27 (a)

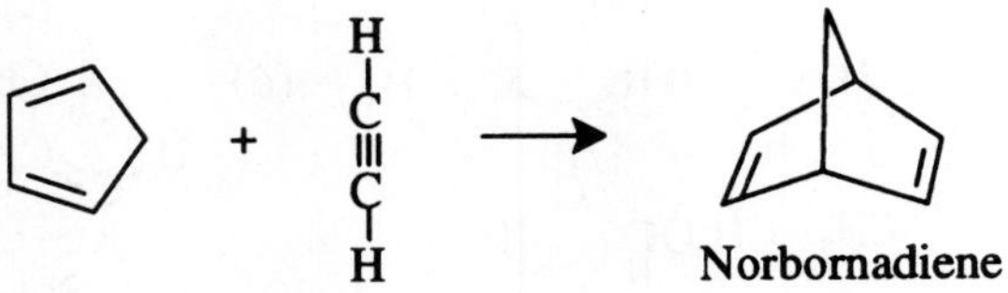

Norbornadiene

(b)

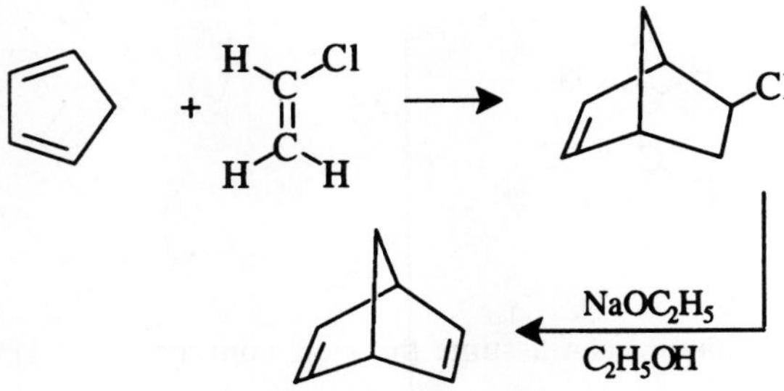

12.28

Chlordan

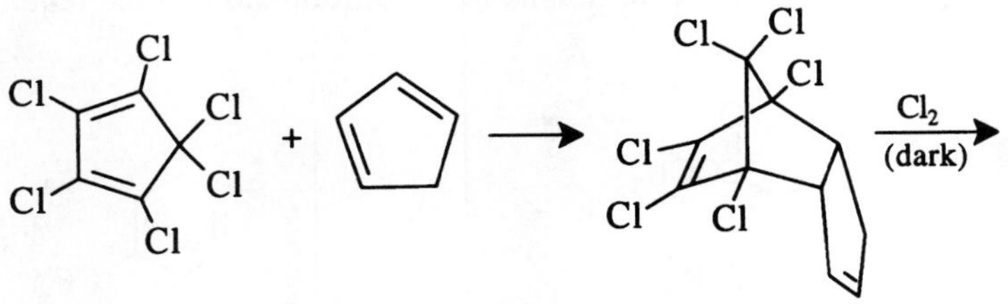

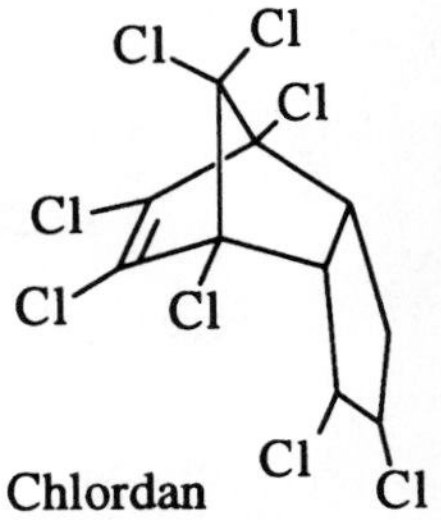

Note: The other double bond is less reactive because of the presence of the two chlorine substituents.

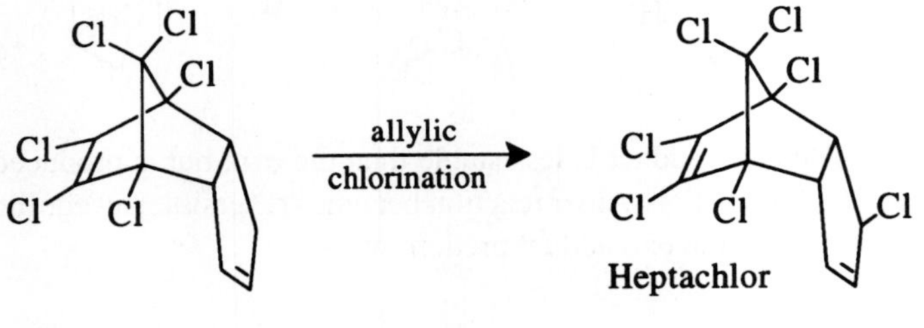

allylic chlorination

Heptachlor

12.29

Isodrin

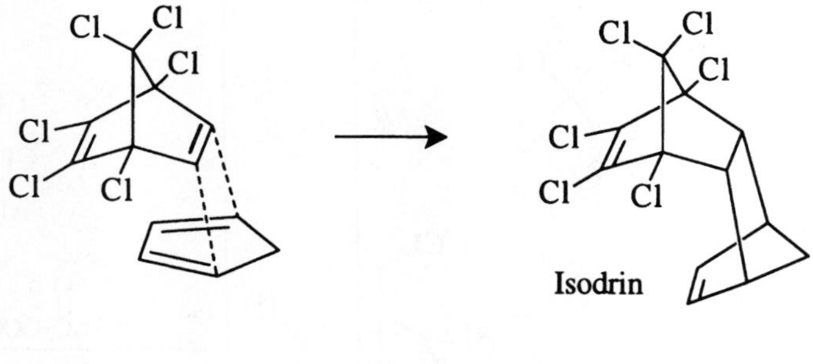

12.30   Protonation of the alcohol and loss of water lead to an allylic cation that can react with a chloride ion at either C1 or C3.

$$CH_3CH=CHCH_2OH \xrightarrow{H^+} CH_3CH=CHCH_2-\overset{\overset{\displaystyle H}{|}}{\underset{+}{O}}-H \xrightarrow{-H_2O}$$

$$CH_3CH=CH\overset{+}{C}H_2 \longleftrightarrow CH_3\overset{+}{C}HCH=CH_2 \xrightarrow{Cl^-}$$

$$CH_3CH=CHCH_2Cl + CH_3\underset{\underset{\displaystyle Cl}{|}}{C}HCH=CH_2$$

12.31   (1) $CH_2=CH-CH=CH_2 + Cl_2 \longrightarrow ClCH_2-\overset{+}{C}H-CH=CH_2$

$$\updownarrow$$

$$ClCH_2-CH=CH-\overset{+}{C}H_2$$

$$\underbrace{}$$

$$ClCH_2-\overset{\delta^+}{CH}\text{==}\overset{}{CH}\text{==}\overset{\delta^+}{CH_2}$$

(2) $ClCH_2-\overset{\delta^+}{CH}\text{==}CH\text{==}\overset{\delta^+}{CH_2} \xrightarrow[(-H^+)]{CH_3OH} ClCH_2-\underset{\underset{\displaystyle OCH_3}{|}}{C}H-CH=CH_2$

$$+ ClCH_2-CH=CH-CH_2OCH_3$$

12.32   A six-membered ring cannot accommodate a triple bond because of the strain that would be introduced.

Too highly strained

12.33   The products are $CH_3CH_2\underset{\underset{\displaystyle Br}{|}}{C}HCH=CH_2$ and $CH_3CH_2CH=CHCH_2Br$ (cis and trans). They are formed from an allylic radical in the following way:

$$Br_2 \longrightarrow 2 Br\cdot \quad \text{(from NBS)}$$

$$Br\cdot \ + \ CH_3CH_2CH_2CH=CH_2 \ \longrightarrow \ CH_3CH_2\overset{\cdot}{C}HCH=CH_2$$

$$\updownarrow \qquad\qquad + \ HBr$$

$$CH_3CH_2CH=CH\overset{\cdot}{C}H_2$$

$$CH_3CH_2\overset{\delta\cdot}{CH}=CH=\overset{\delta\cdot}{CH_2} \ + \ Br_2 \ \longrightarrow \ CH_3CH_2\underset{\underset{Br}{|}}{C}HCH=CH_2$$

$$+ \ Br\cdot$$

$$+ \ CH_3CH_2CH=CHCH_2Br$$
$$(cis \ and \ trans)$$

12.34 (a) The same carbocation (a resonance hybrid) is produced in the dissociation step:

$$\underset{\underset{|}{CH_3}}{CH_3C}=CHCH_2Cl$$

$$\overset{Ag^+}{\searrow}$$

$$\underset{\underset{Cl}{|}}{CH_3}\underset{\underset{|}{C}}{C}-CH=CH_2 \quad \overset{Ag^+}{\nearrow}$$

$$\underset{\underset{+}{}}{CH_3}-\underset{\underset{|}{CH_3}}{C}-CH=CH_2 \ \longleftrightarrow \ CH_3-\underset{\underset{|}{CH_3}}{C}=CH-\underset{+}{CH_2} \ + \ AgCl$$

$$\mathbf{I} \qquad\qquad\qquad \mathbf{II}$$

$$\downarrow H_2O$$

$$CH_3-\underset{\underset{OH}{|}}{\overset{\overset{CH_3}{|}}{C}}-CH=CH_2 \ + \ CH_3-\overset{\overset{CH_3}{|}}{C}=CH-CH_2OH$$

$$(85\%) \qquad\qquad\qquad (15\%)$$

(b) Structure **I** contributes more than **II** to the resonance hybrid of the carbocation (rule 8). Therefore, the hybrid carbocation has a larger partial positive charge on the tertiary carbon atom than on the primary carbon atom. Reaction of the carbocation with water will therefore occur more frequently at the tertiary carbon atom.

12.35 (a) Propyne.   (b) Base (: B$^-$) removes a proton, leaving the anion whose resonance structures are shown:

$$CH_2{=}C{=}CH_2 \ + \ :B^- \ \rightleftharpoons \ H{:}B \ + \ \overset{H}{\underset{H}{}}C{=}C{=}\overset{..-}{\underset{H}{C}} \ \longleftrightarrow \ \overset{H}{\underset{H}{}}\overset{..-}{C}-C{\equiv}C-H$$

$$\mathbf{I} \qquad\qquad\qquad\qquad \mathbf{II}$$

Reaction with H:B may then occur at the CH$_2$ carbanion. The overall reaction is

$$CH_2{=}C{=}CH_2 \ + \ :B^- \ \rightleftharpoons \ [CH_2{=}C{=}\overset{..-}{CH} \ \longleftrightarrow \ \overset{..-}{CH_2}-C{\equiv}CH \ ] \ + \ H{:}B$$

$$\updownarrow$$

$$CH_3-C{\equiv}CH \ + \ :B^-$$

12.36 The first crystalline solid is the Diels-Alder adduct below, mp 125°C,

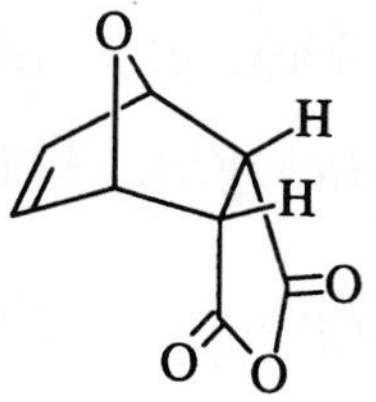

On melting, this adduct undergoes a reverse Diels-Alder reaction, yielding furan (which vaporizes) and maleic anhydride, mp 56°C,

Furan

Maleic
anhydride
(mp 56° C)

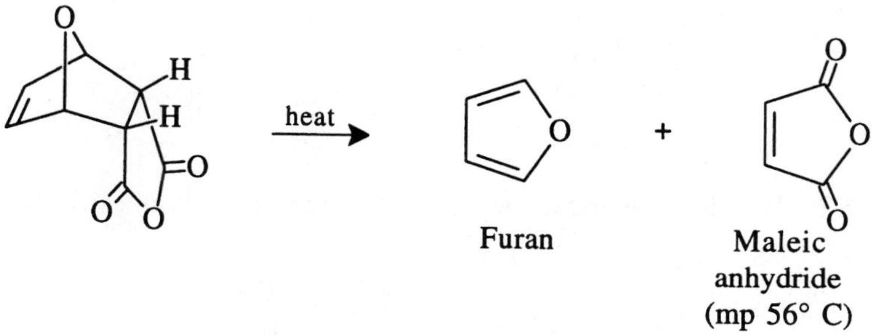

12.37 (a)

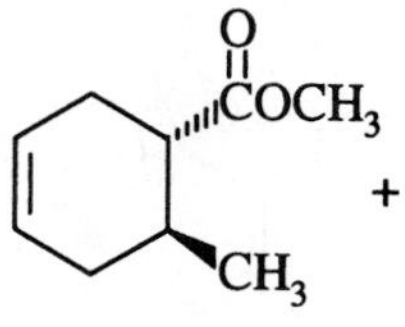

+ enantiomer

(b)

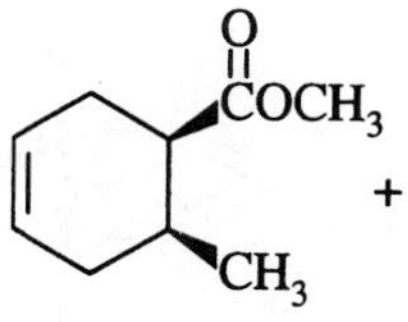

+ enantiomer

(c)

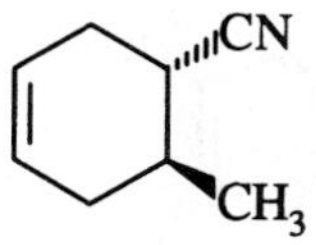

+ enantiomer

(d)

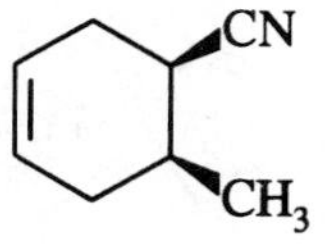

+ enantiomer

12.38 The product formed when butyl bromide undergoes elimination is 1-butene, a simple monosubstituted alkene. When 4-bromo-1-butene undergoes elimination, the product is 1,3-butadiene, a conjugated diene, and therefore, a more stable product. The transition states leading to the products reflect the relative stabilities of the products. Since the transition state leading to 1,3-butadiene has the lower free energy of activation of the two, the elimination reaction of 4-bromo-1-butene will occur more rapidly.

12.39 The diene portion of the molecule is locked into an s-trans conformation. It cannot, therefore, achieve the s-cis conformation necessary for a Diels-Alder reaction.

QUIZ

12.1 Give the 1,4-addition product of the following reaction:

$$CH_3CH=CHCH=CHCH_3 + HCl \longrightarrow ?$$

(a) CH₃CH=CHC=CHCH₃
        |
        Cl

(b) CH₃CH₂CHCH=CHCH₃
          |
          Cl

(c) CH₂CH=CHCH=CHCH₃
  |
  Cl

(d) CH₃CH₂CH=CHCHCH₃
            |
            Cl

(e) CH₃CH₂CHCHCH₂CH₃
        | |
        Cl Cl

12.2 Which diene and dienophile could be used to synthesize the following compound?

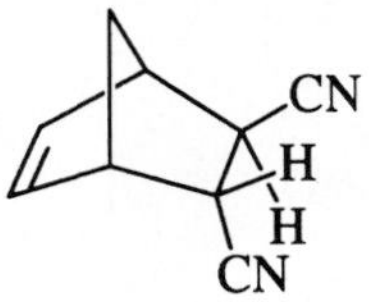

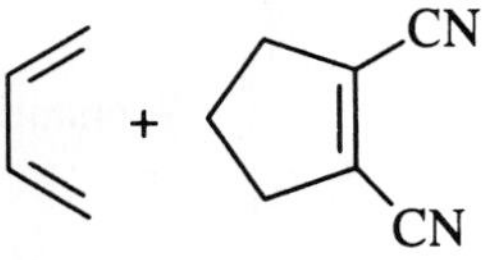

12.3 Which reagent(s) could be used to carry out the following reaction?

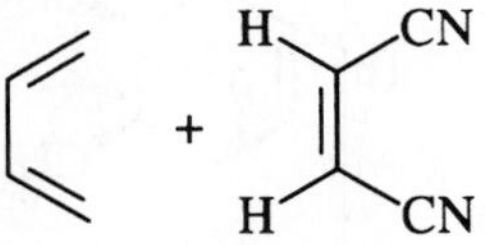

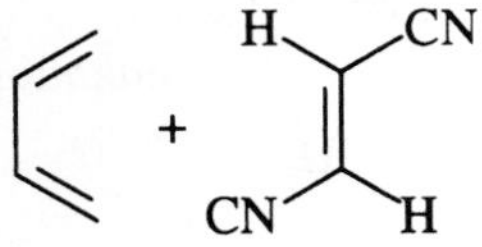

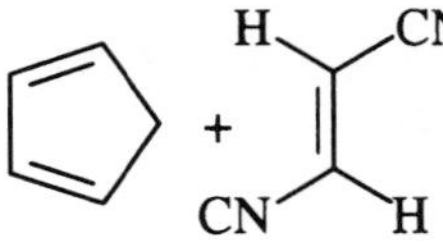

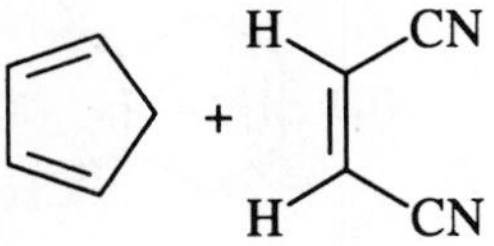

(a) NBS/CCl₄ (NBS = succinimide structure with NBr)

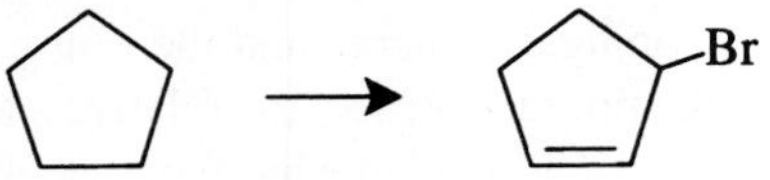

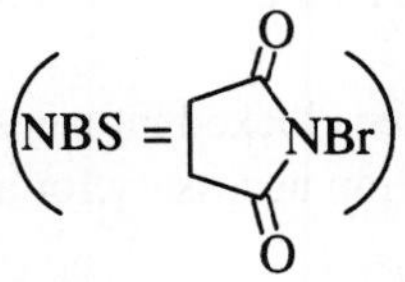

(b) NBS/CCl₄, then Br₂/hv

(c) Br₂/hv, then (CH₃)₃COK/(CH₃)₃COH, then NBS/CCl₄

(d) (CH₃)₃COK/(CH₃)₃COH, then NBS/CCl₄

12.4 Which of the following structures does not contribute to the hybrid for the carbocation formed when 4-chloro-2-pentene ionizes in an S_N1 reaction?

(a) $CH_3CH=CH\overset{+}{C}HCH_3$ (b) $CH_3\overset{+}{C}HCH=CHCH_3$ (c) $CH_3\overset{+}{C}HCH_2CH=CH_2$

(d) All of these contribute to the resonance hybrid.

12.5 Which of the following resonance structures accounts at least in part for the lack of S_N2 reactivity of vinyl chloride?

(a) $CH_2{=}CH-\overset{\cdot\cdot}{\underset{\cdot\cdot}{C}}l{:}$ (b) $\overset{\cdot\cdot}{C}H_2-CH={\overset{\cdot\cdot}{C}}l{:}^+$ (c) Neither   (d) Both

ANSWERS TO
FIRST REVIEW PROBLEM SET

1 (a)

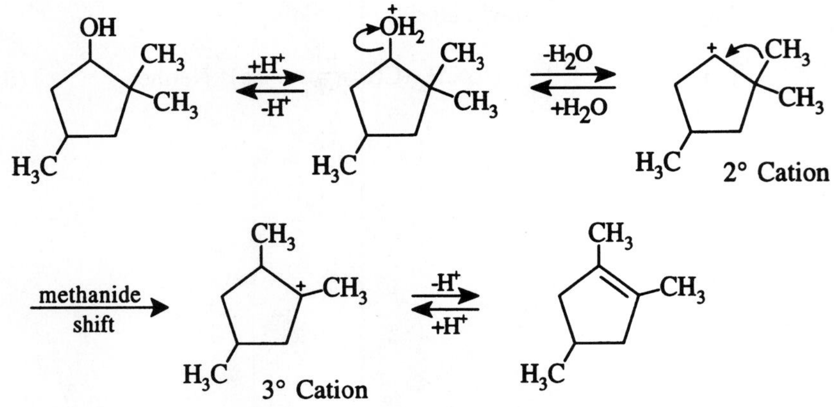

2° Cation

methanide shift

3° Cation

(b)

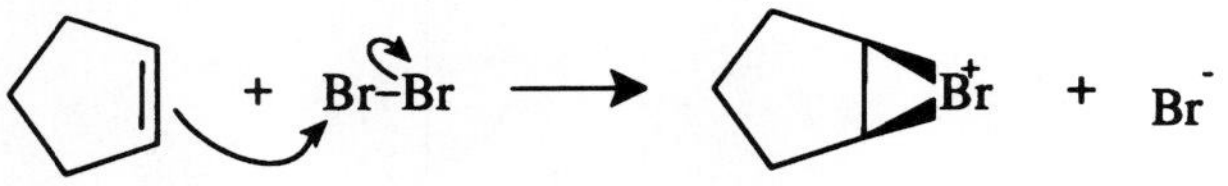

then,

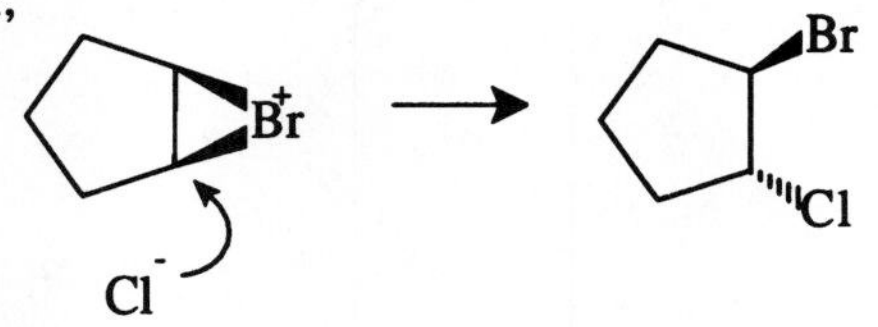

(c) The enantiomer of the product given would be formed in an equimolar amount via the following reaction:

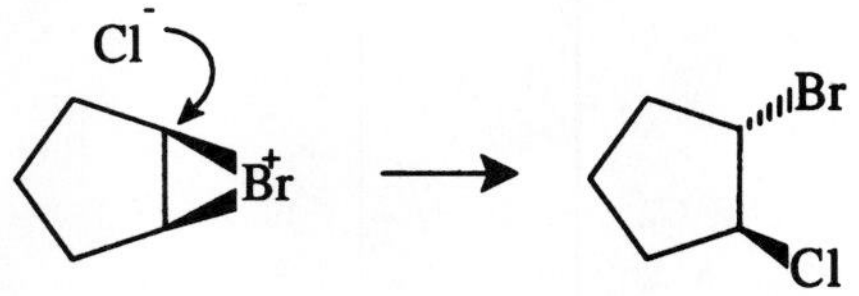

The *trans*-1, 2-dibromocyclopentane would be formed as a racemic form via the reaction of the bromonium ion with a bromide ion:

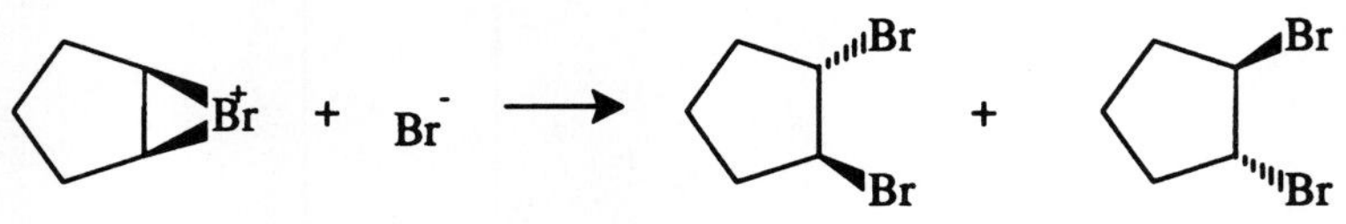

Racemic *trans*-1,2-dibromocyclopentane

And, *trans*-2-bromocyclopentanol (the bromohydrin) would be formed (as a racemic form) via the reaction of the bromonium ion with water.

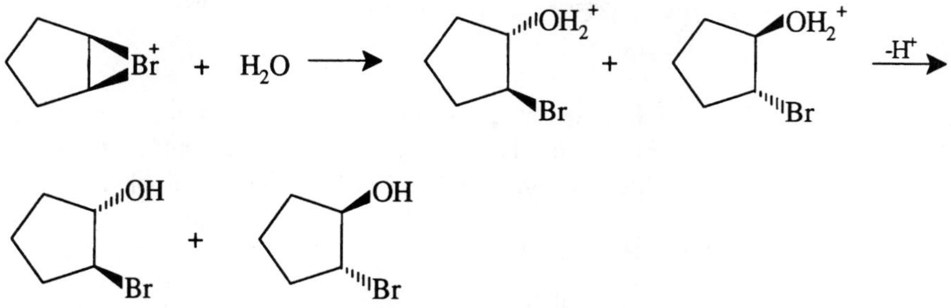

Racemic *trans*-2-bromocyclopentanol

2

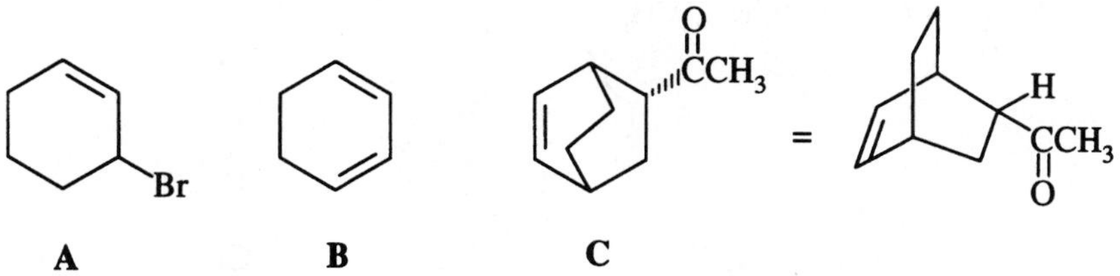

A   **B**   **C**

A is formed by an allylic bromination. **B** is formed by an E2 elimination. **C** is formed via a Diels-Alder reaction that yields predominantly the endo product. Ozonolysis of the double bond then yields the product in which all three substituents are on the same side of the cyclohexane ring.

3   All of these differences can be explained by the contribution to the $CH_2=CHCl$ molecule made by the following **A** and **B** resonance structures.

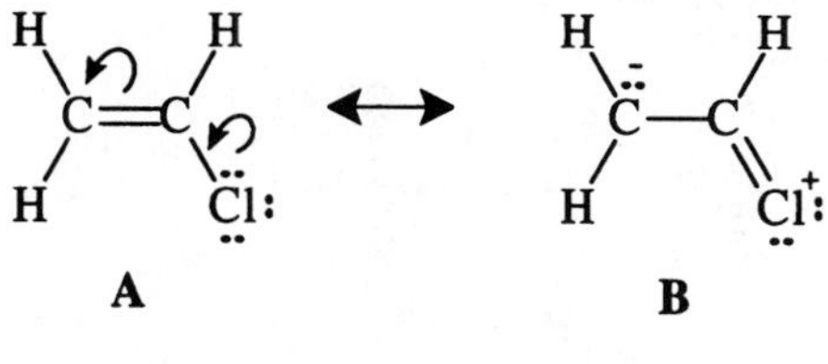

A   **B**

(a)   Because of the contribution made to the hybrid by **B**, the C–Cl bond of $CH_2=CH–Cl$ has some double-bond character and is, therefore, shorter than the "pure" single bond of $CH_3CH_2–Cl$.

(b)   The contribution made to the hybrid by **B** imparts some single-bond character to the carbon-carbon double bond of $CH_2=CH–Cl$, causing it to be longer than the "pure" double bond of $CH_2=CH_2$.

(c)   Electronegativity differences would cause a carbon-chlorine bond to be polarized as follows:

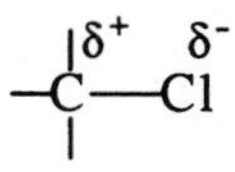

And this effect accounts, almost entirely, for the dipole moment of CH_3CH_2Cl.

$$\overset{\delta^+}{}\overset{\delta^-}{}$$
$$CH_3CH_2\!\!-\!\!Cl \qquad \qquad \mu = 2.05\ D$$
$$\overset{\longrightarrow}{+}$$

With $CH_2{=}CH{-}Cl$, however, the resonance contribution of **B** tends to oppose the polarization of the C–Cl bond caused by electronegativity differences. That is, the resonance effect partially cancels the electronegativity effect, causing the dipole moment to be smaller.

4 $A = CH_3(CH_2)_{11}CH_2C{\equiv}CH$

$B = CH_3(CH_2)_{11}CH_2C{\equiv}CNa$

$C = CH_3(CH_2)_{11}CH_2C{\equiv}CCH_2(CH_2)_6CH_3$

Muscalure $= CH_3(CH_2)_{11}CH_2$ $CH_2(CH_2)_6CH_3$

5

(*E*)-2,3-Diphenyl-2-butene $\qquad\qquad$ (*Z*)-2,3-Diphenyl-2-butene

Because catalytic hydrogenation is a syn addition, catalytic hydrogenation of the (*Z*) isomer would yield a meso compound.

(*Z*) $\qquad\qquad$ (by addition $\qquad$ (by addition at
$\qquad\qquad\qquad$ at one face) $\qquad$ the other face)
$\qquad\qquad\qquad$ A meso compound

Syn addition of hydrogen to the (*E*) isomer would yield a racemic form:

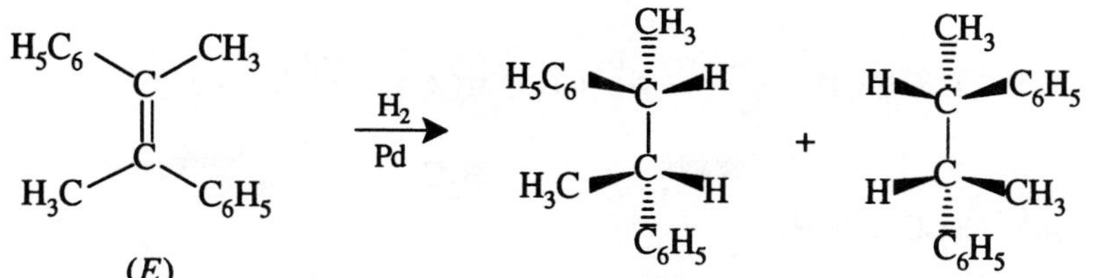

(E)

(by addition
at one face)

(by addition at
the other face)

Enantiomers - a racemic form

6   From the molecular formula of **A** and of its hydrogenation product **B,** we can conclude
that **A** has two rings and a double bond. (**B** has two rings.)

From the product of strong oxidation with $KMnO_4$ and its stereochemistry (i.e.,
compound **C**), we can deduce the structure of **A**.

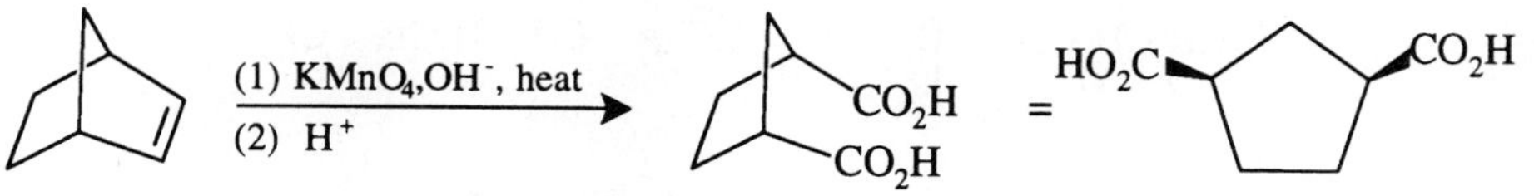

A

meso-1,3-Cyclopentane-
dicarboxylic acid

Compound **B** is bicyclo[2.2.1]heptane and **C** is a glycol.

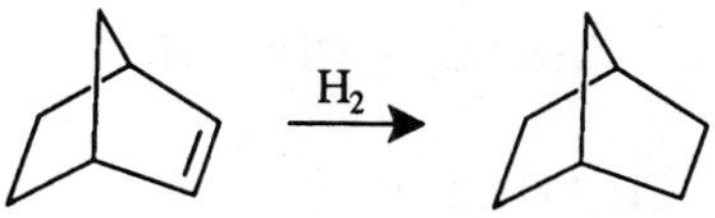

A   **B**

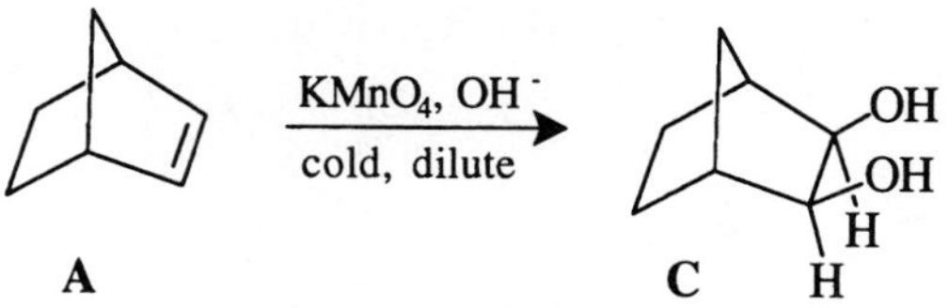

A   **C**

Notice that **C** is also a meso compound.

7 (a) $CH_3C\equiv CH$ $\xrightarrow[\text{liq. NH}_3]{\text{NaNH}_2}$ $CH_3C\equiv CNa$ $\xrightarrow{\text{CH}_3\text{I}}$ $CH_3C\equiv CCH_3$

(b) $CH_3C\equiv CCH_3$ $\xrightarrow{\underset{\text{Ni}_2\text{B (P-2)}}{\text{H}_2}}$

[from (a)]

$$\underset{H}{\overset{H_3C}{>}}C=C\underset{H}{\overset{CH_3}{<}}$$

(c) $CH_3C\equiv CCH_3$ $\xrightarrow[\text{NH}_3]{\text{Li}}$

[from (a)]

$$\underset{H}{\overset{H_3C}{>}}C=C\underset{CH_3}{\overset{H}{<}}$$

(d) $CH_3CH=CHCH_3$ $\xrightarrow{\text{THF:BH}_3}$ $\underset{\overset{|}{B-}\atop|}{CH_3CH_2CHCH_3}$ $\xrightarrow[160°C]{\text{heat}}$

[from (b) or (c)]

$CH_3CH_2CH_2CH_2-\underset{|}{\overset{|}{B}}-$ $\xrightarrow[160°C]{\text{1-decene}}$ $CH_3CH_2CH=CH_2$

(e) $CH_3CH_2CH=CH_2$ $\xrightarrow[\text{CCl}_4]{\text{NBS}}$ $\left.\begin{array}{c}\underset{\overset{|}{Br}}{CH_3CHCH_2}=CH_2\\+\\CH_3CH=CHCH_2Br\end{array}\right\}$ $\dfrac{(CH_3)_3COK}{(CH_3)_3COH}$

[from (d)]

$CH_2=CH-CH=CH_2$

(f) $CH_3CH_2CH=CH_2$ $\xrightarrow[\text{ROOR}]{\text{HBr}}$ $CH_3CH_2CH_2CH_2Br$

[from (d)]

(g) $CH_3CH=CHCH_3$ $\xrightarrow[\text{no peroxides}]{\text{HBr}}$ $\underset{\overset{|}{Br}}{CH_3CH_2CHCH_3}$

[from (b) or (c)]

or

$CH_3CH_2CH=CH_2$ $\xrightarrow[\text{no peroxides}]{\text{HBr}}$ $\underset{\overset{|}{Br}}{CH_3CH_2CHCH_3}$

[from (d)]

(h) $$\underset{H}{\overset{H_3C}{>}}C=C\underset{CH_3}{\overset{H}{<}}$$ $\xrightarrow[\text{(anti addition)}]{\underset{\text{CCl}_4}{\text{Br}_2}}$

[from (c)]

(cf. Section 8.7)

$$\begin{array}{c} CH_3 \\ Br\underset{\cdots}{\overset{}{\underset{|}{C}}}H \\ Br\overset{}{\underset{|}{C}}H \\ \bar{C}H_3 \end{array}$$

(2R, 3S)

A meso compound

(i)

$$\underset{\text{[from (b)]}}{\underset{\substack{H \\ \\}}{\overset{\substack{H_3C \\ \\}}{C}}}=\underset{\substack{H \\ \\}}{\overset{\substack{CH_3 \\ \\}}{C}}$$

$\xrightarrow[\substack{CCl_4 \\ (\text{anti addition})}]{Br_2}$

(2R, 3R)   (2S, 3S)

A racemic form

(j)

$$\underset{\text{[from (b)]}}{\underset{\substack{H \\ \\}}{\overset{\substack{H_3C \\ \\}}{C}}}=\underset{\substack{H \\ \\}}{\overset{\substack{CH_3 \\ \\}}{C}}$$

$\xrightarrow[\substack{(2)\ NaHSO_3 \\ (\text{syn addition})}]{(1)\ OsO_4}$

(cf. Section 8.9)

or

$$\underset{\text{[from (c)]}}{\underset{\substack{H \\ \\}}{\overset{\substack{H_3C \\ \\}}{C}}}=\underset{\substack{CH_3 \\ \\}}{\overset{\substack{H \\ \\}}{C}}$$

$\xrightarrow[\substack{(2)\ H_3O^+,\ heat \\ (\text{anti addition})}]{(1)\ RCOOH}$

(cf. Section 10.20)

(k) $CH_3C{\equiv}CCH_3$ $\xrightarrow[CH_3CO_2H]{HBr,Br^-}$ (cf. Section 8.13)

8 $\underset{\substack{| \\ CH_3}}{CH_3CHCH_2CH_3}$ $\xrightarrow[hv,\ heat]{Br_2}$ $\underset{\substack{| \\ CH_3}}{CH_3\overset{\substack{Br \\ |}}{C}CH_2CH_3}$ (cf. Section 9.6)

(a) $\underset{\substack{| \\ CH_3}}{CH_3\overset{\substack{Br \\ |}}{C}CH_2CH_3}$ $\xrightarrow[\substack{CH_3CH_2OH \\ heat}]{CH_3CH_2ONa}$ $\underset{\substack{| \\ CH_3}}{CH_3C{=}CHCH_3}$

(b) $\underset{\substack{| \\ CH_3}}{CH_3C{=}CHCH_3}$ $\xrightarrow[H_2O]{H_3O^+}$ $\underset{\substack{| \\ CH_3}}{CH_3\overset{\substack{OH \\ |}}{C}CH_2CH_3}$

[from (a)]

(c) $\underset{\substack{| \\ CH_3}}{CH_3C{=}CHCH_3}$ $\xrightarrow[\substack{(2)H_2O_2,\ OH^-}]{(1)\ THF:\ BH_3}$ $\underset{\substack{| \\ CH_3}}{CH_3CH\overset{\substack{OH \\ |}}{C}HCH_3}$

[from (a)]

(d) 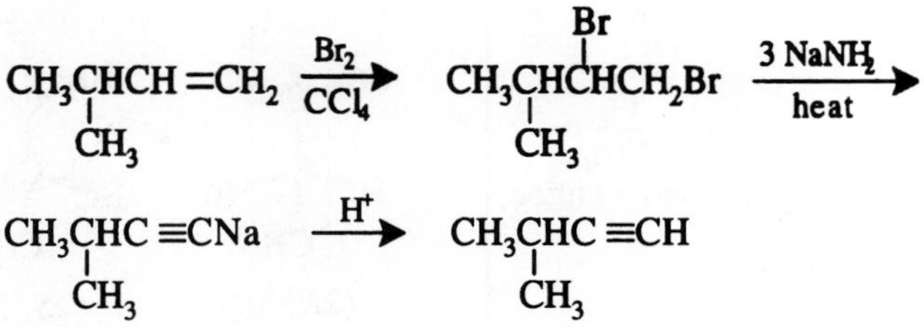 $CH_3CHCH=CH_2$ $\xrightarrow{Br_2}$ $CH_3CHCHCH_2Br$ $\xrightarrow[\text{heat}]{3\ NaNH_2}$
        | CCl$_4$ |
        CH$_3$ Br (top), CH$_3$

$CH_3CHC\equiv CNa$ $\xrightarrow{H^+}$ $CH_3CHC\equiv CH$
        | |
        CH$_3$ CH$_3$

(e) 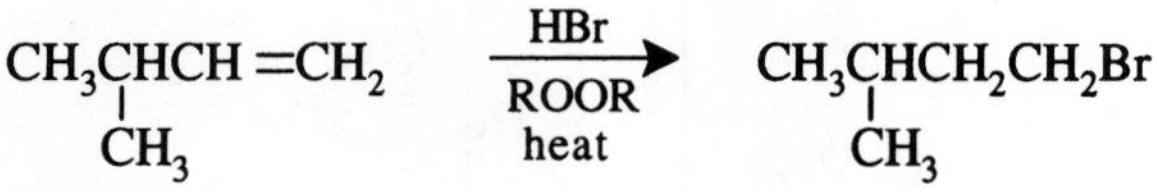 $CH_3CHCH=CH_2$ $\xrightarrow[\text{heat}]{\substack{HBr \\ ROOR}}$ $CH_3CHCH_2CH_2Br$
        | |
        CH$_3$ CH$_3$

(f) 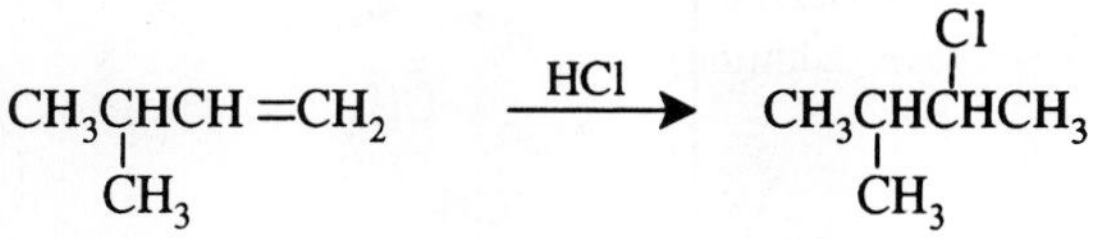 $CH_3CHCH=CH_2$ $\xrightarrow{HCl}$ $CH_3CHCHCH_3$
        | | |
        CH$_3$ CH$_3$ Cl

(g) 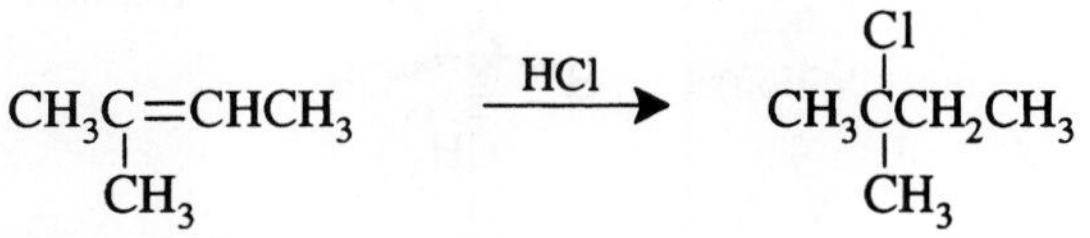 $CH_3C=CHCH_3$ $\xrightarrow{HCl}$ $CH_3CCH_2CH_3$
        | |
        CH$_3$ CH$_3$, Cl
    [from (a)]

(h) 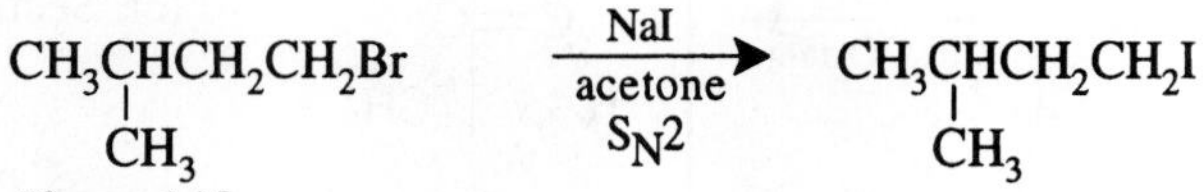 $CH_3CHCH_2CH_2Br$ $\xrightarrow[S_N2]{\substack{NaI \\ acetone}}$ $CH_3CHCH_2CH_2I$
        | |
        CH$_3$ CH$_3$
    [from (e)]

(i) 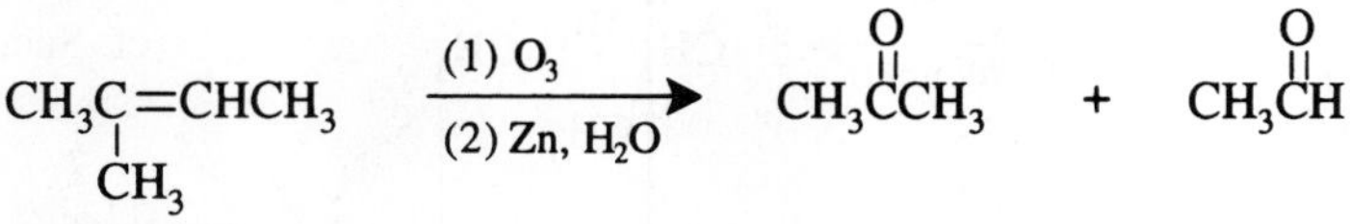 $CH_3C=CHCH_3$ $\xrightarrow[(2)\ Zn,\ H_2O]{(1)\ O_3}$ $CH_3\overset{O}{\overset{\|}{C}}CH_3$ $+$ $CH_3\overset{O}{\overset{\|}{C}}H$
        |
        CH$_3$
    [from (a)]

(j) 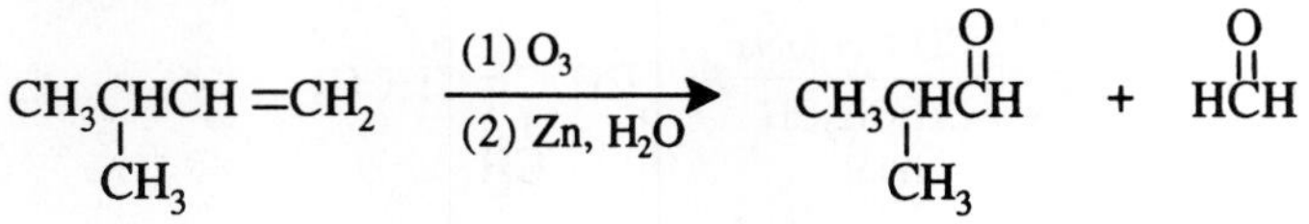 $CH_3CHCH=CH_2$ $\xrightarrow[(2)\ Zn,\ H_2O]{(1)\ O_3}$ $CH_3CH\overset{O}{\overset{\|}{C}}H$ $+$ $H\overset{O}{\overset{\|}{C}}H$
        | |
        CH$_3$ CH$_3$

9   $CH_3\overset{CH_3}{\underset{CH_3}{C}}CH_2CH_3$ $\xrightarrow[hv,\text{heat}]{Cl_2}$ $CH_3\overset{CH_2Cl}{\underset{CH_3}{C}}CH_2CH_3$ $+$ $CH_3\overset{CH_3}{\underset{CH_3}{C}}CHClCH_3$

    A   **B**   **C**

$+$ $CH_3\overset{CH_3}{\underset{CH_3}{C}}CH_2CH_2Cl$

    D

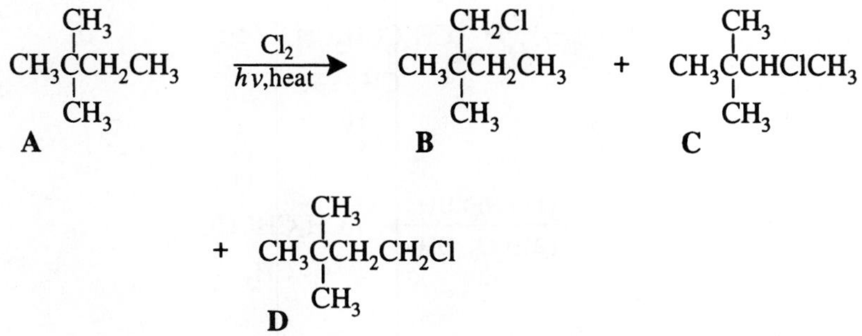

B cannot undergo dehydrohalogenation because it has no β hydrogen; however, **C** and **D** can, as shown next.

$$
\begin{array}{c}
\underset{\underset{CH_3}{|}}{\overset{\overset{CH_3}{|}}{CH_3CCHClCH_3}} \\
\mathbf{C}
\end{array}
$$

$$
\begin{array}{c}
\underset{\underset{CH_3}{|}}{\overset{\overset{CH_3}{|}}{CH_3CCH_2CH_2Cl}} \\
\mathbf{D}
\end{array}
$$

$$\xrightarrow[\text{(CH}_3)_3\text{COH}]{\text{(CH}_3)_3\text{COK}}$$

$$
\underset{\underset{CH_3}{|}}{\overset{\overset{CH_3}{|}}{CH_3CCH=CH_2}} \quad \xrightarrow[\text{Pt}]{\text{H}_2} \quad \mathbf{A}
$$

$$\mathbf{E}$$

$$
\underset{\underset{CH_3}{|}}{\overset{\overset{CH_3}{|}}{CH_3CCH=CH_2}} \quad \xrightarrow{\text{HCl}} \quad \left[\underset{\underset{CH_3}{|}}{\overset{\overset{CH_3}{|}}{CH_3C-\overset{+}{C}HCH_3}} \right] \quad \longrightarrow \quad \left[\underset{CH_3\overset{+}{C}-CHCH_3}{\overset{\overset{CH_3}{|}}{}} \right]
$$

$$\mathbf{E} \qquad\qquad + Cl^- \qquad\qquad\qquad + Cl^-$$

$$
\longrightarrow \quad \underset{\underset{CH_3}{|}}{\overset{\overset{Cl\ \ CH_3}{|\ \ \ |}}{CH_3C-CHCH_3}} \quad \xrightarrow[\text{CH}_3\text{CO}_2\text{H}]{\text{Zn}} \quad \underset{\underset{CH_3}{|}}{\overset{\overset{CH_3}{|}}{CH_3CHCHCH_3}}
$$

$$\mathbf{F} \qquad\qquad\qquad \mathbf{G}$$

10 $CH_3C\equiv CCH_3$

$$\xrightarrow{\text{H}_2,\ \text{Pt}} \quad CH_3CH_2CH_2CH_3$$

$$\xrightarrow{\text{Ag(NH}_3)_2\text{OH}} \quad \text{No reaction}$$

$\mathbf{A}$

$$\downarrow \text{H}_2 \ | \ \text{Ni}_2\text{B (P-2)}$$

$$\xrightarrow[\text{(2)NaHSO}_3]{\text{(1) OsO}_4}$$

(syn hydroxylation)

$\mathbf{B}$

$\mathbf{C}$

(a meso compound)

11   The eliminations are anti eliminations, requiring an anti periplanar arrangement of the bromine atoms.

meso-2,3-Dibromobutane   *trans*-2-Butene   + IBr

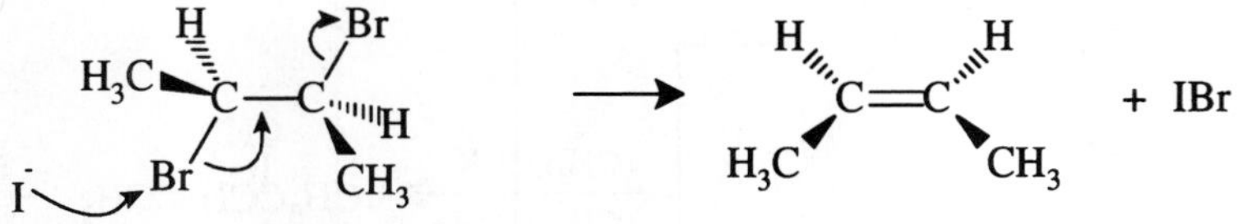

(2S,3S)-2,3-Dibromobutane   cis-2-Butene   + IBr

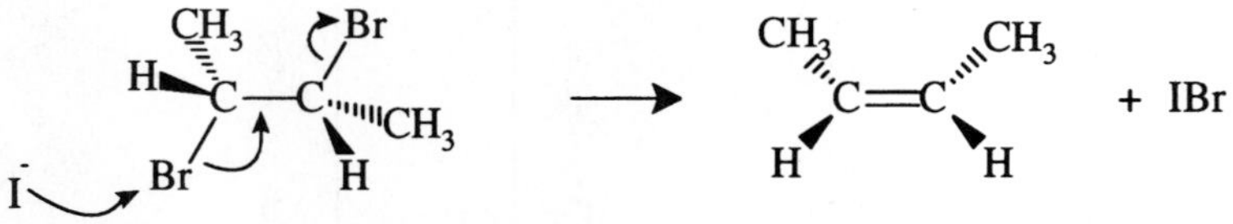

(2R,3R)-2,3-Dibromobutane   cis-2-Butene   + IBr

12   The eliminations are anti eliminations, requiring an anti periplanar arrangement of the
−H and −Br.

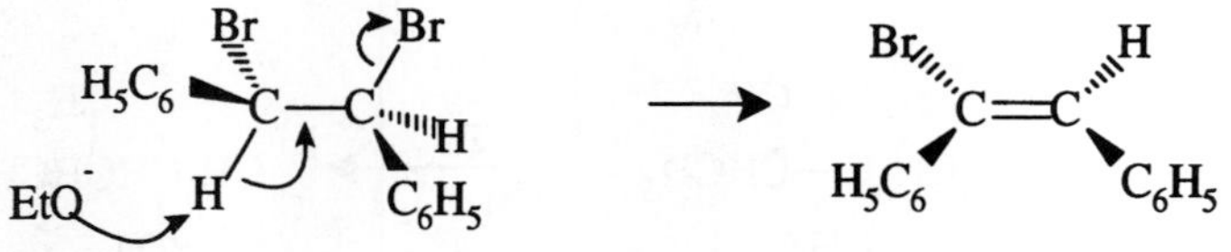

meso-1,2-Dibromo-
1,2-diphenylethane

(E) -1-Bromo-1,2-
diphenylethene

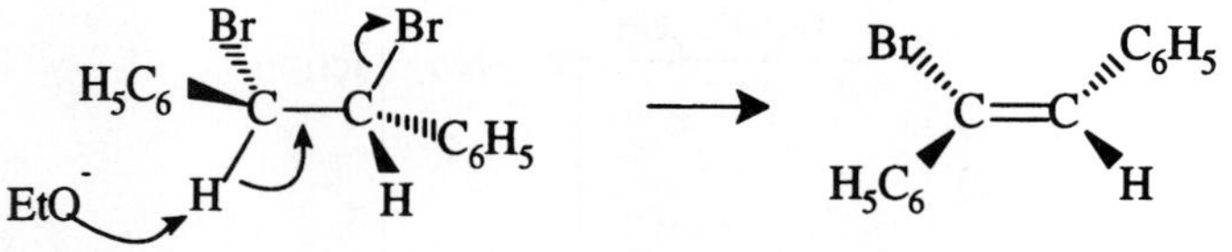

(2R,3R)-1,2-Dibromo-
1,2-diphenylethane

(Z) -1-Bromo-1,2-
diphenylethene

(2S,3S)-1,2-Dibromo-1,2-diphenylethane will also give (Z)-1-bromo-1,2-diphenylethene
in an anti elimination.

13   In all the following structures, notice that the large *tert*-butyl group is equatorial.

(a) 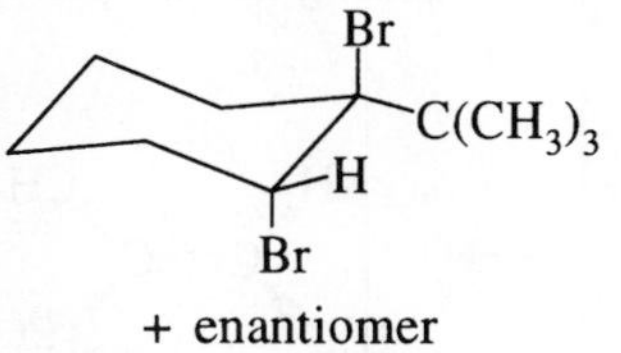   Br ... C(CH₃)₃   H   Br   (bromine addition is anti; cf.
Section 8.7)

+ enantiomer
as a racemic form

(b)

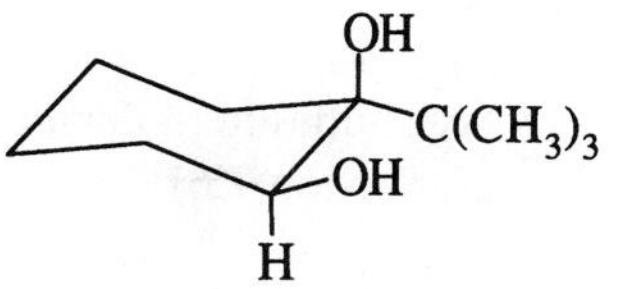

+ enantiomer
as a racemic form

(syn hydroxylation; cf. Section 8.9)

(c)

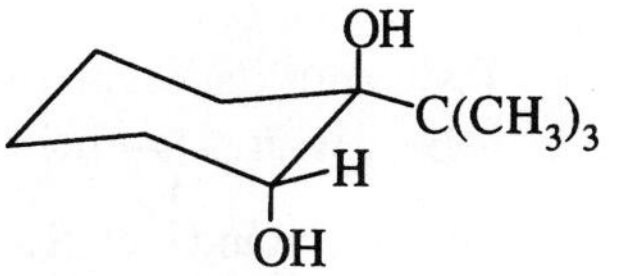

+ enantiomer
as a racemic form

(anti hydroxylation; cf. Section 10.20)

(d)

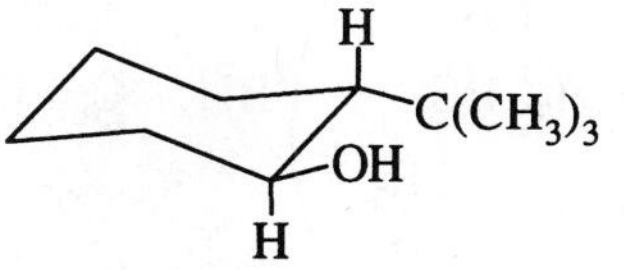

+ enantiomer
as a racemic form

(syn and anti Markovnikov addition of −H and −OH; cf. Section 10.7)

(e)

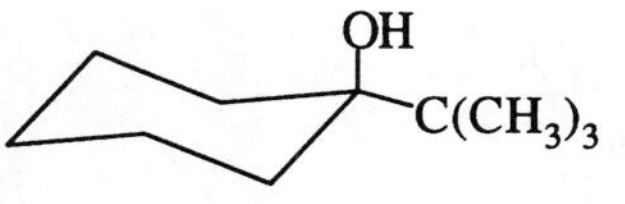

(Markovnikov addition of -H and -OH cf. Section 10.4A)

(f)

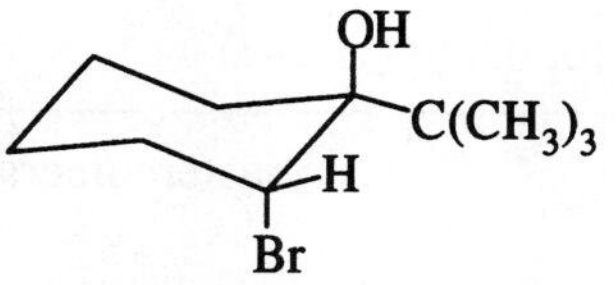

+ enantiomer
as a racemic form

(anti addition of −Br and −OH, with −Br and −OH placement resulting from the more stable partial carbocation in the intermediate bromonium ion; cf. Section 8.8)

(g)

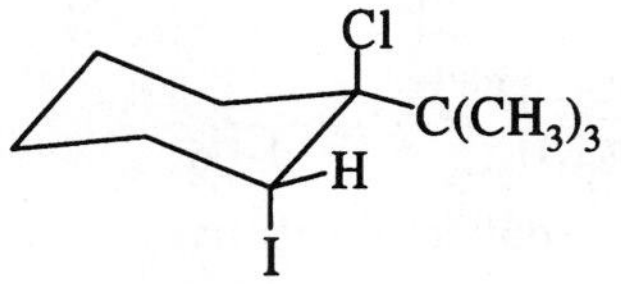

+ enantiomer
as a racemic form

(anti addition of −I and −Cl, following Markovnikov's rule; cf. Section 8.2)

(h) $HC(CH_2)_4CC(CH_3)_3$
       (with two C=O groups shown above)

(i) 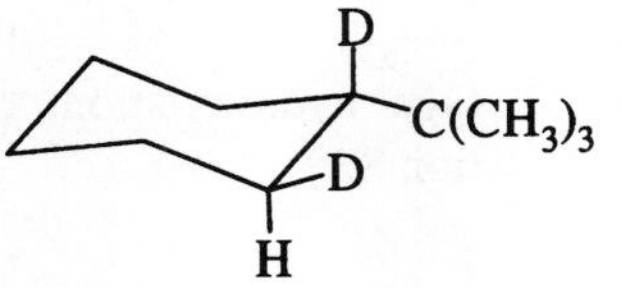 (syn addition of deuterium, cf. Section 7.6)

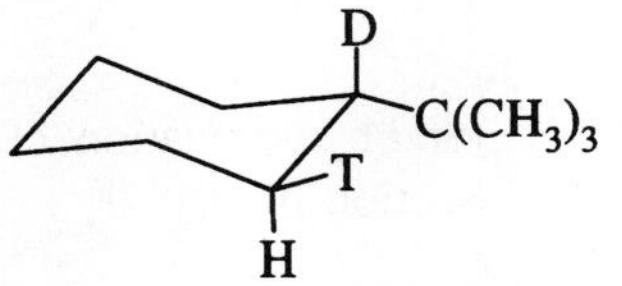

+ enantiomer
as a racemic form

(j) (syn, anti Markovnikov addition of −D and
−B− , with −B− being replaced by -T
where it stands; cf. Section 10.7)

+ enantiomer
as a racemic form

14 $A = CH_3\overset{\overset{\displaystyle CH_3}{|}}{C}=CHCH_2CH_3$ $B = \left(CH_3\overset{\overset{\displaystyle CH_3}{|}}{C}HCH\overset{\overset{\displaystyle}{\underset{\underset{\displaystyle CH_3}{|}}{CH_2}}}{|}BH\right)_2$ $C = CH_3\overset{\overset{\displaystyle CH_3}{|}}{C}H\overset{\underset{\displaystyle OH}{|}}{C}HCH_2CH_3$

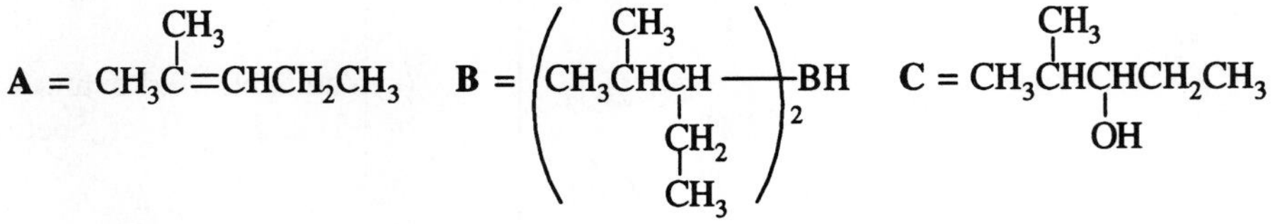

15 (a) The following products are diastereomers. They would have different boiling points and would be in separate fractions. Each fraction would be optically active.

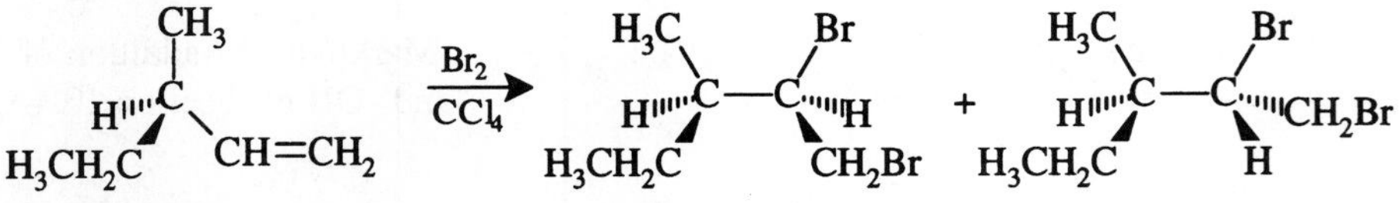

(R)-3-Methyl-1-pentene   (optically active)   (optically active)

Diastereomers

(b) Only one product is formed. It is achiral, and, therefore, it would not be optically active.

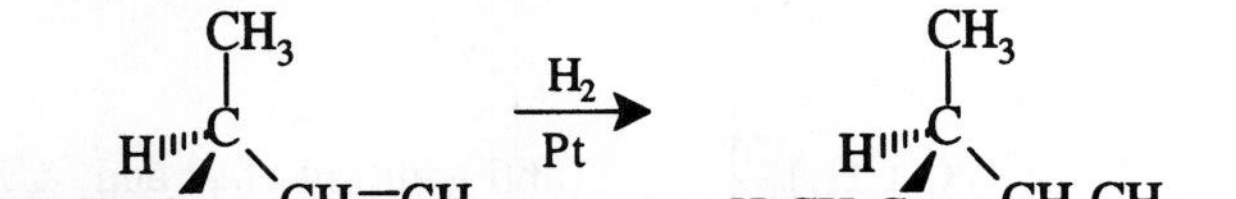

(optically inactive)

(c) Two diastereomeric products are formed. Two fractions would be obtained. Each fraction would be optically active.

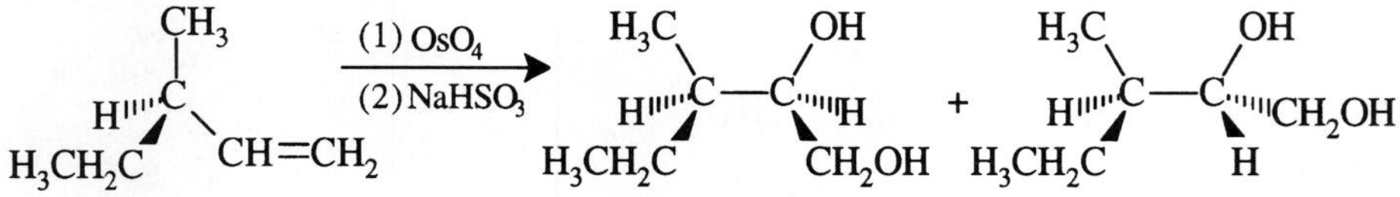

(optically active)   (optically active)

Diastereomers

(d) One optically active compound is produced.

$$\underset{\underset{H_3CH_2C}{}}{\overset{\overset{CH_3}{|}}{\overset{H^{\text{\tiny III}}\!\!-C}{}}}\!\!\!\diagdown_{CH=CH_2} \quad \xrightarrow[\text{(2) } H_2O_2, OH^-]{\text{(1) THF:BH}_3} \quad \underset{\underset{H_3CH_2C}{}}{\overset{\overset{CH_3}{|}}{\overset{H^{\text{\tiny III}}\!\!-C}{}}}\!\!\!\diagdown_{CH_2CH_2OH}$$

(optically active)

(e) Two diastereomeric products are formed. Two fractions would be obtained. Each fraction would be optically active.

$$\underset{\underset{H_3CH_2C}{}}{\overset{\overset{CH_3}{|}}{\overset{H^{\text{\tiny III}}\!\!-C}{}}}\!\!\!\diagdown_{CH=CH_2} \quad \xrightarrow[\text{(2) NaBH}_4, OH^-]{\text{(1) Hg(OAc)}_2, \text{THF-H}_2O}$$

$$\underset{H_3CH_2C}{\overset{H_3C}{\diagdown}}C\!\!-\!\!C{\overset{OH}{\diagup}}_{\underset{CH_3}{H}} \quad + \quad \underset{H_3CH_2C}{\overset{H_3C}{\diagdown}}C\!\!-\!\!C{\overset{OH}{\diagup}}_{\underset{H}{CH_3}}$$

(optically active)   (optically active)

Diastereomers

(f) Two diastereomeric products are formed. Two fractions would be obtained. Each fraction would be optically active.

$$\underset{\underset{H_3CH_2C}{}}{\overset{\overset{CH_3}{|}}{\overset{H^{\text{\tiny III}}\!\!-C}{}}}\!\!\!\diagdown_{CH=CH_2} \quad \xrightarrow[\text{(2) } H_3O^+, H_2O]{\text{(1)}}$$

$$\left[\underset{\text{COO}^-}{\overset{\overset{O}{\overset{||}{\overset{}{}}}}{\overset{CO_2H}{\bigcirc}}} \right]_2 Mg^{2+}$$

$$\underset{H_3CH_2C}{\overset{H_3C}{\diagdown}}C\!\!-\!\!C{\overset{OH}{\diagup}}_{\underset{CH_2OH}{H}} \quad + \quad \underset{H_3CH_2C}{\overset{H_3C}{\diagdown}}C\!\!-\!\!C{\overset{OH}{\diagup}}_{\underset{H}{CH_2OH}}$$

(optically active)   (optically active)

Diastereomers

16

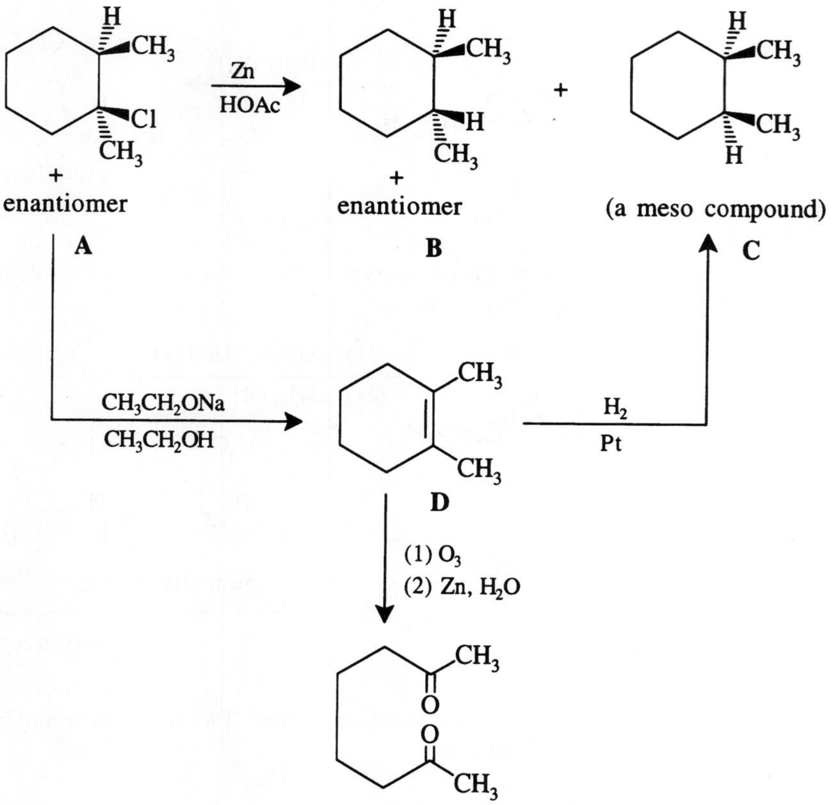

17 (a)

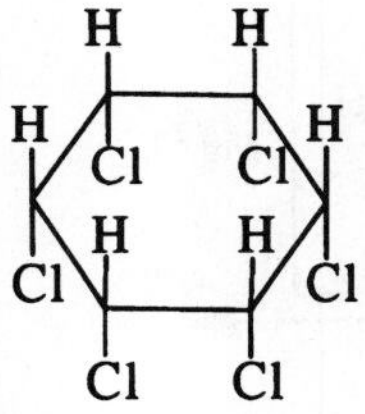

1 all *cis* meso	**2** 1-*trans* meso	**3** 1,4-*trans* meso	**4** 1,3-*trans* meso

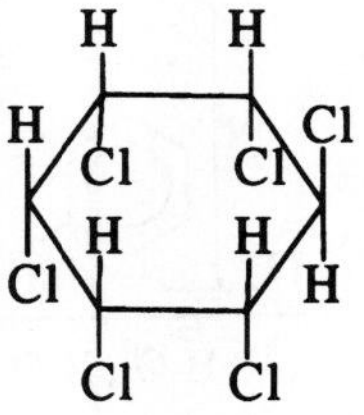

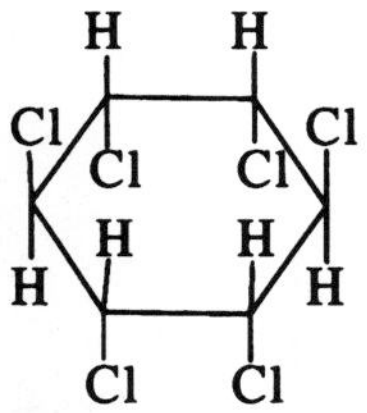

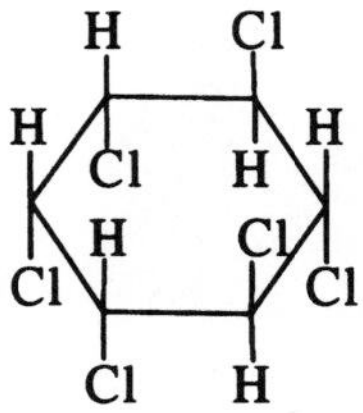

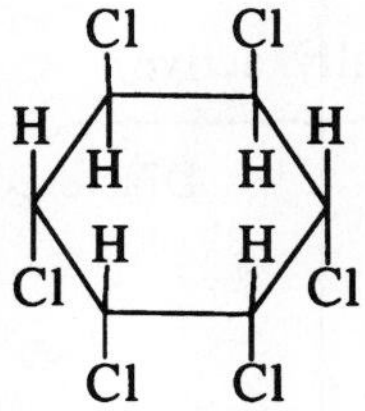

5 1,2-*trans* meso	**6** 1,2,3-*trans* meso	**7**	**8**

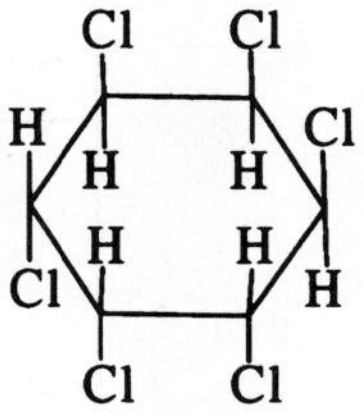

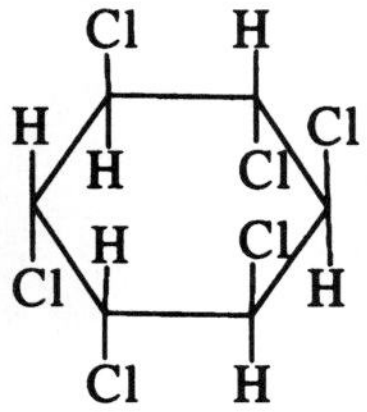

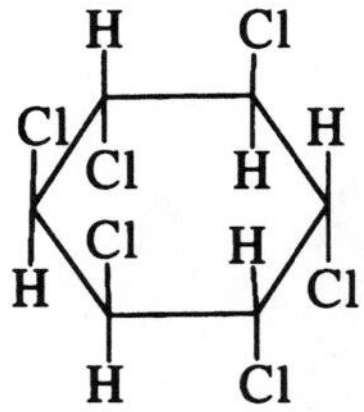

1,2,4-*trans*
Enantiomers

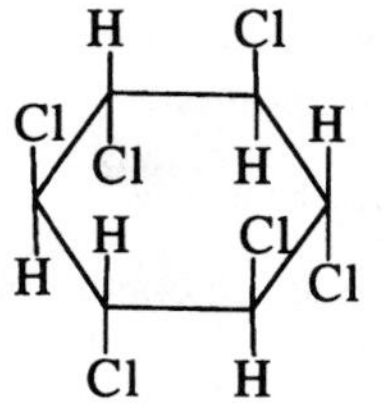

9   1,3,5-*trans*
meso

(b)   Isomer **9** is slow to react in an E2 reaction because in its more stable conformation (see following structure) all the chlorine atoms are equatorial and an anti periplanar transition state cannot be achieved. All other isomers **1-8** can have a −Cl axial and thus achieve an anti periplanar transition state.

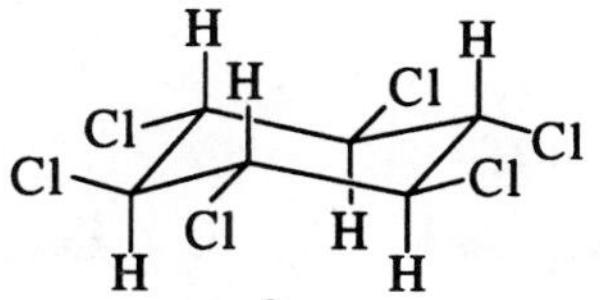

9

18   (a) $CH_3CHCH_2CH_3$ $\xrightarrow{F_2}$

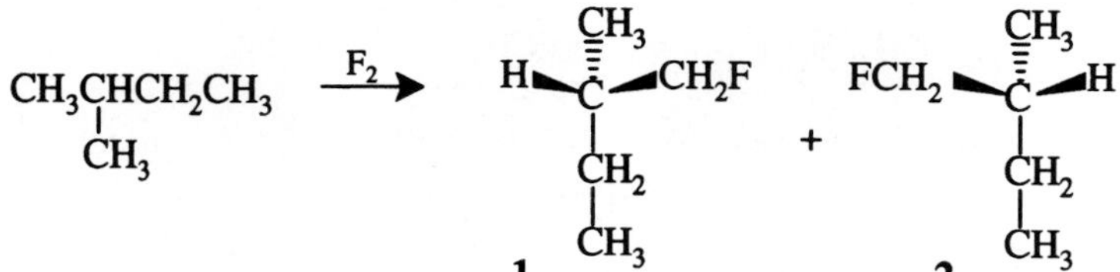

Enantiomers
(obtained in one fraction as an
optically inactive racemic form)

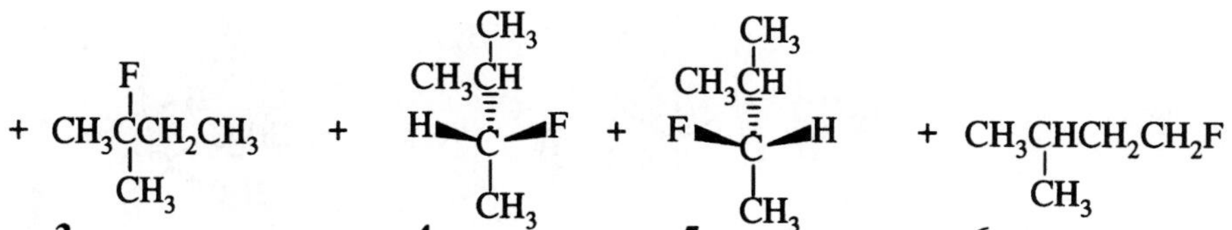

3
(achiral and, therefore,
optically inactive)

4   **5**
Enantiomers
(obtained in one fraction as an
optically inactive racemic form)

6
(achiral and, therefore,
optically inactive)

(b)   Four fractions. The enantiomeric pairs would not be separated by fractional distillation because enantiomers have the same boiling points.

(c)   All of the fractions would be optically inactive.

(d)   The fraction containing **1** and **2** and the fraction containing **4** and **5**.

19

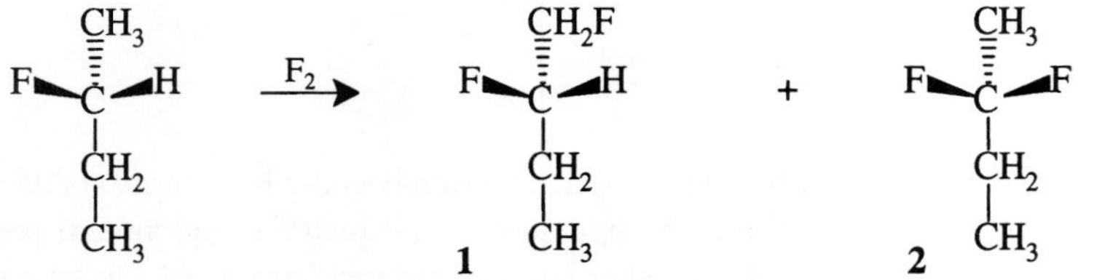

(R)-2-Fluorobutane   (optically active)   (achiral and, therefore, optically inactive)

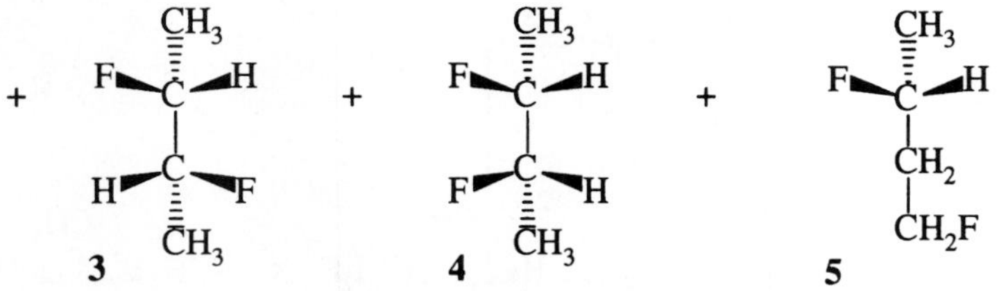

(optically active)   meso compound   (optically active)
                          (optically inactive)

(b)   Five. Compounds **3** and **4** are diastereomers. All others are constitutional isomers of each other.

(c)   See above.

20

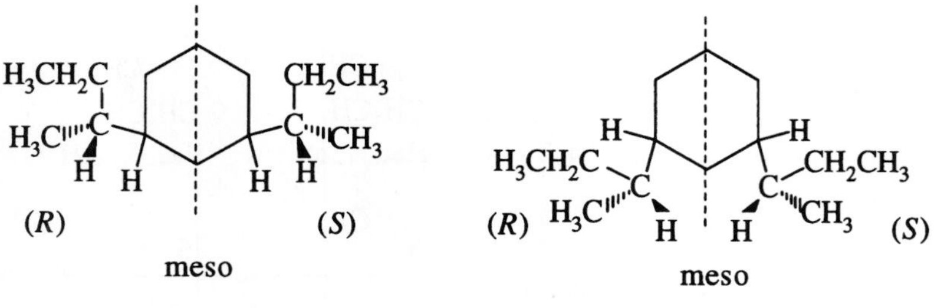

Each of the two structures just given has a plane of symmetry (indicated by the dashed line), and, therefore, each is a meso compound. The two structures are not superposable one on the other; therefore, they represent molecules of different compounds and are diastereomers.

21   Only a proton or deuteron anti to the bromine can be eliminated; that is, the two groups undergoing elimination (H and Br or D and Br) must lie in an anti periplanar arrangement. The two conformations of *erythro*-2-bromobutane-3-*d* in which a proton or deuteron is anti periplanar to the bromine are **I** and **II**.

Conformation **I** can undergo loss of HBr to yield *cis*-2-butene-2-*d*. Conformation **II** can undergo loss of DBr to yield *trans*-2-butene.

13

SPECTROSCOPIC METHODS OF STRUCTURE DETERMINATION

SOLUTIONS TO PROBLEMS

13.1 The formula, C_6H_8, tells us that **A** and **B** have six hydrogen atoms less than an alkane. This unsaturation may be due to three double bonds, one triple bond and one double bond, or combinations of two double bonds and a ring, or one triple bond and a ring. Since both **A** and **B** react with 2 mol of H_2 to yield cyclohexane, they are either cyclohexyne or cyclohexadienes. The absorption maximum of 256 nm for **A** tells us that it is conjugated. Compound **B**, with no absorption maximum beyond 200 nm, possesses isolated double bonds. We can rule out cyclohexyne because of ring strain caused by the requirement of linearity of the $-C \equiv C-$ system. Therefore, **A** is 1,3-cyclohexadiene; **B** is 1,4-cyclohexadiene.

13.2 All three compounds have an unbranched five-carbon chain, because the product of hydrogenation is unbranched pentane. The formula, C_5H_6, suggests that they have one double bond and one triple bond. Compounds **D, E,** and **F** must differ, therefore, in the way the multiple bonds are distributed in the chain. Compounds **E** and **F** have a terminal $-C \equiv CH$ [reaction with $Ag(NH_3)_2^+ OH^-$]. The absorption maximum near 230 nm for **D** and **E** suggests that in these compounds, the multiple bonds are conjugated. The structures are

$$CH_3-C \equiv C-CH=CH_2 \qquad HC \equiv C-CH=CH-CH_3 \qquad HC \equiv C-CH_2-CH=CH_2$$
$$\textbf{D} \qquad\qquad\qquad \textbf{E} \qquad\qquad\qquad\qquad \textbf{F}$$

13.3 We see below that if we replace a methyl hydrogen by some group Z, then rotation of the methyl group or turning the whole molecule end-for-end gives structures that represent the same compound. This means that all of the methyl hydrogens are equivalent.

replace H with Z

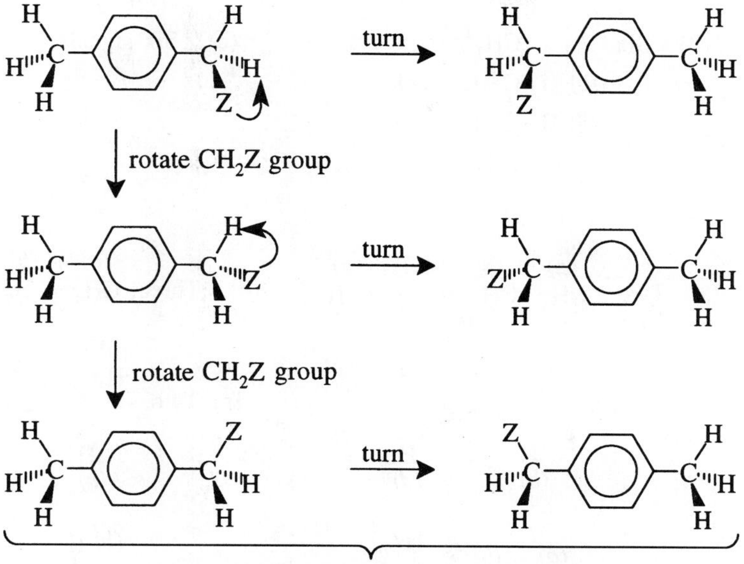

All of these are the same compound

Replacing a ring hydrogen by Z gives another compound, but replacing each ring hydrogen in turn gives the same compound. This shows that all four ring hydrogens are equivalent and that 1,4-dimethylbenzene has only two sets of chemical shift equivalent protons.

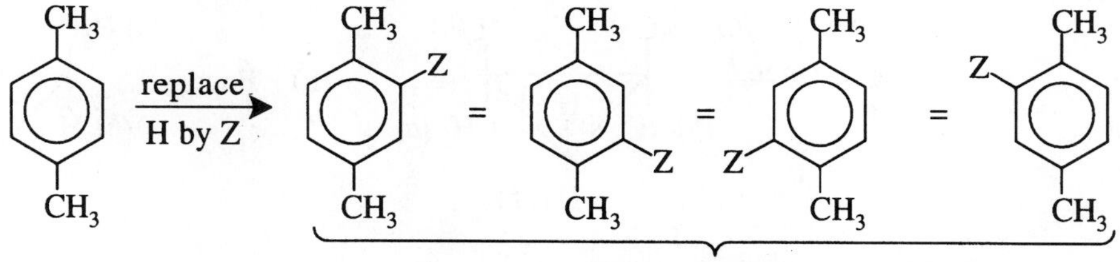

All of these are the same compound

13.4   (a)   One   (d)   One

(b)   Two   (e)   Two

(c)   Two   (f)   Two

13.5   (a)

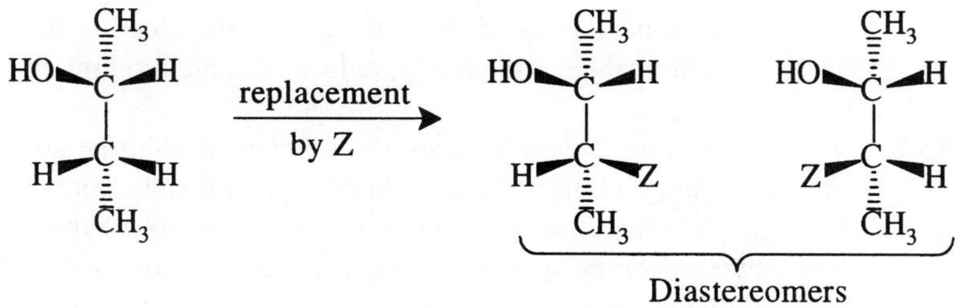

Diastereomers

(b) Six

$$\begin{array}{c} (a) \\ CH_3 \\ (b)\ H-\overset{|}{\underset{|}{C}}-OH\ (c) \\ (d)\ H-\overset{|}{\underset{|}{C}}-H\quad (e) \\ CH_3 \\ (f) \end{array}$$

(c) Six signals

13.6 (a) Two, $\overset{(a)}{CH_3}-\overset{(b)}{CH_2}-\overset{(b)}{CH_2}-\overset{(a)}{CH_3}$

(b) Three, $\overset{(a)}{CH_3}-\overset{(b)}{CH_2}-\overset{(c)}{O}-H$

(c) Four,

$$\begin{array}{ccc} (a) & & (c) \\ H_3C & & H \\ & C{=}C & \\ H & & H \\ (b) & & (d) \end{array}$$

(d) Two,

$$\begin{array}{ccc} (a) & & (b) \\ H_3C & & H \\ & C{=}C & \\ H & & CH_3 \\ (b) & & (a) \end{array}$$

(e) Four, $\overset{(a)}{CH_3}-\overset{(b)}{CHBr}-\overset{H\,(c)}{\underset{H\,(d)}{C}}-Br$

(f) Two,

(g) Three,

(h) Four,

(i) Six,

$$\begin{array}{cccc} (a) & (b) & (c) & (e) \\ CH_3-CH_2-CH_2 & & H \\ & & C{=}C \\ (d)\ H & & H\ (f) \end{array}$$

(j) Five, $\begin{array}{c} (a) \\ H \\ Cl-\overset{|}{\underset{|}{C}}-\overset{(c)}{CH}-CH_3\,(e) \\ (b)\ H \quad OH\,(d) \end{array}$

13.7 The ^{1}H NMR spectrum of $CHBr_2CHCl_2$ consists of two doublets. The doublet from the proton of the $-CHCl_2$ group should occur at lower magnetic field strength because the greater electronegativity of chlorine reduces the electron density in the vicinity of the $-CHCl_2$ proton, and consequently, reduces its shielding relative to $-CHBr_2$.

13.8 The determining factors here are the number of chlorine atoms attached to the carbon atoms bearing protons and the deshielding that results from chlorine's electronegativity. In 1,1,2-trichloroethane the proton that gives rise to the triplet is on a carbon atom that bears two chlorines, and the signal from this proton is downfield. In 1,1,2,3,3-penta-chloropropane, the proton that gives rise to the triplet is on a carbon atom that bears only one chlorine; the signal from this proton is upfield.

13.9 The signal from the three equivalent protons designated *(a)* should be split into a doublet by the proton *(b)*. This doublet, because of the electronegativity of the attached chlorines, should occur downfield.

$$\overset{(a)\qquad(b)}{(Cl_2CH)_{\overline{3}}-CH}$$

The signal for the proton designated *(b)* should be split into a quartet by the three equivalent protons *(a)*. The quartet should occur upfield.

13.10

A C$_3$H$_7$I is $\overset{(a)\ \ (b)\ \ (a)}{CH_3-CH-CH_3}$
$\ \ \ \ \ \ \ \ \ \ \ \ \ \ \ |$
$\ \ \ \ \ \ \ \ \ \ \ \ \ \ \ I$

   (a)   doublet δ 1.9
   (b)   septet δ 4.35

B C$_2$H$_4$Cl$_2$ is $\overset{(a)\ \ (b)}{CH_3-CH-Cl}$
$\ \ \ \ \ \ \ \ \ \ \ \ \ \ \ \ \ \ \ |$
$\ \ \ \ \ \ \ \ \ \ \ \ \ \ \ \ \ \ Cl$

   (a)   doublet δ 2.08
   (b)   quartet δ 5.9

C C$_3$H$_6$Cl$_2$ is $\overset{(a)\ \ \ \ (b)\ \ \ \ (a)}{Cl-CH_2-CH_2-CH_2-Cl}$
   (a)   triplet δ 3.7
   (b)   quintet δ 2.2

13.11 *(a)* $J_{ab} = 2 J_{bc}$

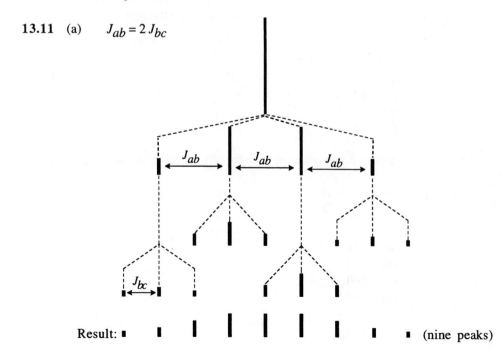

Result: ▪   ▮   ▮   ▮   ▮   ▮   ▮   ▮   ▪   (nine peaks)

(b) $J_{ab} = J_{bc}$

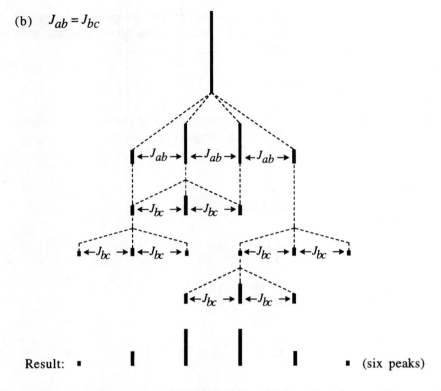

Result: ▪ | | | | ▪   (six peaks)

13.12 A single unsplit signal, because the proton is rapidly shifted from axial to equatorial positions.

13.13 **A** is 1-bromo-3-methylbutane. The following are the signal assignments:

$$\begin{array}{cccc} (d) & (c) & (b) & (a) \\ \end{array}$$
$$\text{BrCH}_2\text{CH}_2\text{CH(CH}_3)_2$$

(a) $\delta\,23$
(b) $\delta\,27$
(c) $\delta\,32$
(d) $\delta\,42$

13.14 **B** is 2-bromo-2-methylbutane. The following are the signal assignments:

$$\begin{array}{cccc} (a) & (c) & \overset{\text{Br}}{\underset{|}{}} & (b) \\ \end{array}$$
$$\text{CH}_3\text{CH}_2\overset{|}{\text{C}}(\,\text{CH}_3)_2$$
$$\qquad\quad (d)$$

(a) $\delta\,11$
(b) $\delta\,33$
(c) $\delta\,40$
(d) $\delta\,68$

C is 1-bromo-pentane. The following are the signal assignments.

$$\overset{(e)\ (d)\ (c)\ (b)\ (a)}{BrCH_2CH_2CH_2CH_2CH_3}$$

(a) $\delta\ 14$
(b) $\delta\ 23$
(c) $\delta\ 30$
(d) $\delta\ 33$
(e) $\delta\ 34$

13.14 (a)

$$\overset{(a)\quad CH_3\ (a)}{CH_3-\underset{\underset{CH_3\ (a)}{|}}{\overset{|}{C}}-OH\ (b)}$$

(a) Singlet, $\delta\ 1.28$ (9H)
(b) Singlet, $\delta\ 1.35$ (1H)

(b)

$$\overset{(a)\ \ (b)\ \ (a)}{CH_3-\underset{\underset{Br}{|}}{CH}-CH_3}$$

(a) Doublet, $\delta\ 1.71$ (6H)
(b) Septet, $\delta\ 4.32$ (1H)

(c)

$$\overset{(b)\ \ \ \ \overset{O}{\overset{||}{}}\ \ (c)\ \ (a)}{CH_3-C-CH_2-CH_3}$$

(a) Triplet, $\delta\ 1.05$ (3H) $C=O$, 1720 cm^{-1}
(b) Singlet, $\delta\ 2.13$ (3H)
(c) Quartet, $\delta\ 2.47$ (2H)

(d)

$$\overset{(b)\ \ (a)}{-CH_2-OH}$$

(c)

(a) Singlet, $\delta\ 2.43$ (1H) $O-H$, 3200-3550 cm^{-1}
(b) Singlet, $\delta\ 4.58$ (2H)
(c) Multiplet, $\delta\ 7.28$ (5H)

(e)

$$\overset{(b)\ \ (c)}{CH_3-CH-CH_2Cl}$$
$$\underset{(a)}{\underset{CH_3}{}}$$

(a) Doublet, $\delta\ 1.04$ (6H)
(b) Multiplet, $\delta\ 1.95$ (1H)
(c) Doublet, $\delta\ 3.35$ (2H)

(f)

$$\overset{(b)\ \ \overset{O}{\overset{||}{}}\ \ (a)}{C_6H_5-CH-C-CH_3}$$
$$\underset{(c)}{\underset{C_6H_5}{}}$$

(a) Singlet, $\delta\ 2.20$ (3H) $C=O$, near 1720 cm^{-1}
(b) Singlet, $\delta\ 5.08$ (1H)
(c) Multiplet, $\delta\ 7.25$ (10H)

(g)

$$\overset{(a)\ \ (b)\ \ (c)\ \ (d)}{CH_3-CH_2-\underset{\underset{Br}{|}}{CH}CO_2H}$$

(a) Triplet, $\delta\ 1.08$ (3H) $C=O$ (acid) 1715 cm^{-1}
(b) Multiplet, $\delta\ 2.07$ (2H) $O-H$, 2500-3000 cm^{-1}
(c) Triplet, $\delta\ 4.23$ (1H)
(d) Singlet, $\delta\ 10.97$ (1H)

(h)

(b) (a)
—CH₂—CH₃

(c)

(a) Triplet, δ 1.25 (3H)
(b) Quartet, δ 2.68 (2H)
(c) Multiplet, δ 7.23 (5H)

(i)

(a) (b)   (c)   (d)
CH₃–CH₂–O–CH₂CO₂H

(a) Triplet, δ 1.27 (3H)   C=O (acid) 1715 cm⁻¹
(b) Quartet, δ 3.66 (2H)   O–H, 2500-3000 cm⁻¹
(c) Singlet, δ 4.13 (2H)
(d) Singlet, δ 10.95 (1H)

(j)

(a) (b) (a)
CH₃–CH–CH₃
        |
        NO₂

(a) Doublet, δ 1.55 (6H)
(b) Septet, δ 4.67 (1H)

(k)

(a)   (b) (b)   (a)
CH₃O–CH₂CH₂–OCH₃

(a) Singlet, δ 3.25 (6H)
(b) Singlet, δ 3.45 (4H)

(l)

(b)   O   (c)
      ||
CH₃–C–CH–CH₃
           |
           CH₃ (a)

(a) Doublet, δ 1.10 (6H)   C=O, near 1720 cm⁻¹
(b) Singlet, δ 2.10 (3H)
(c) Septet, δ 2.50 (1H)

(m)

(b)   (a)
—CH–CH₃
   |
   Br

(c)

(a) Doublet, δ 2.00 (3H)
(b) Quartet, δ 5.15 (1H)
(c) Multiplet, δ 7.35 (5H)

13.15 Compound **E** is phenylacetylene, $C_6H_5C \equiv CH$. We can make the following assignments in the IR spectrum:

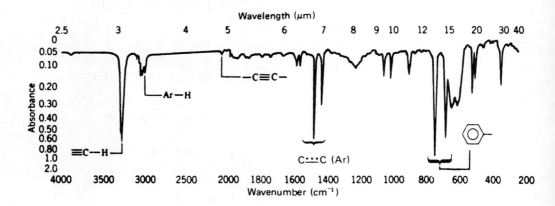

The IR spectrum of compound **E** (Problem 13.15). (Spectrum courtesy of Sadtler Research Laboratories Inc., Philadelphia.)

13.16 Compound **G** is 2-bromobutane. Assignments are as follows:

(a) (b) (d) (c)
$CH_3–CH_2–CH–CH_3$
　　　　　|
　　　　　Br

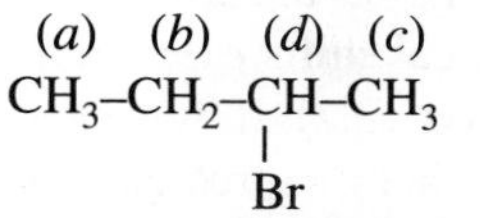

    (a)　triplet δ 1.05
    (b)　multiplet δ 1.82
    (c)　doublet δ 1.7
    (d)　multiplet δ 4.1

Compound **H** is 2,3-dibromopropene. Assignments are as follows:

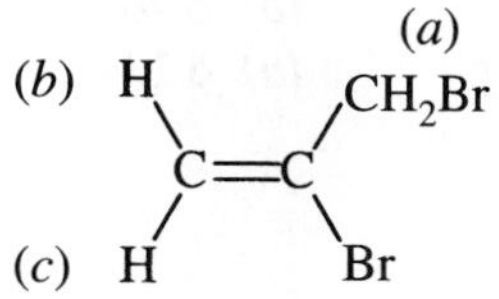

    (a)　δ 4.2
    (b)　δ 6.05
    (c)　δ 5.64

13.17 Compound **J** is *cis*-1,2-dichloroethene.

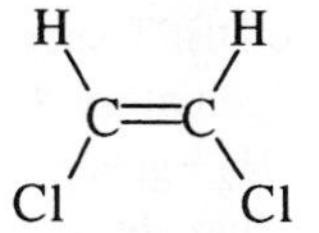

We can make the following IR assignments:

    3125 cm^{-1}, alkene C—H stretching
    1625 cm^{-1}, C=C stretching
    695 cm^{-1}, out-of-plane bending of cis double bond

13.18 (a) Compound **K** is,

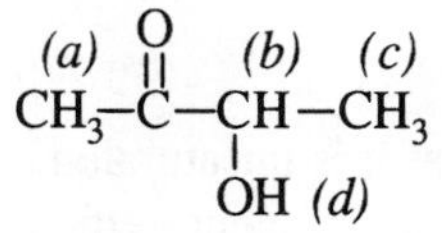

(a) O   (b) (c)
　　 ‖
$CH_3–C–CH–CH_3$
　　　　　 |
　　　　　 OH *(d)*

    (a) Singlet, δ 2.15　　C=O, 1720 cm^{-1}
    (b) Quartet, δ 4.25
    (c) Doublet, δ 1.35
    (d) Singlet, δ 3.75

(b) When the compound is dissolved in D_2O, the —OH proton (d) is replaced by a deuteron, and thus the ^{1}H NMR absorption peak disappears.

$$CH_3\overset{\text{O}}{\overset{\|}{C}}CHCH_3 + D_2O \rightleftharpoons CH_3\overset{\text{O}}{\overset{\|}{C}}CHCH_3 + DHO$$
$$\quad\quad\; | \quad\quad\quad\quad\quad\quad\quad\quad\quad | $$
$$\quad\quad OH \quad\quad\quad\quad\quad\quad\quad\quad OD$$

13.19   Run the spectrum with the spectrometer operating at a different magnetic field strength (i.e., at 30 or at 100 MHz). If the peaks are two singlets the distance between them—*when measured in hertz*—will change because chemical shifts *expressed in hertz* are proportional to the strength of the applied field (Section 13.7). If, however, the two peaks represent a doublet, then the distance that separates them, expressed in hertz, will not change because this distance represents the magnitude of the coupling constant and coupling constants are independent of the applied magnetic field (Section 13.9).

13.20   Compound **O** is 1,4-cyclohexadiene and **P** is cyclohexane.

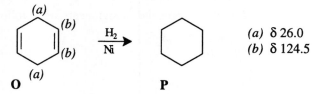

(a) δ 26.0
(b) δ 124.5

13.21   The molecular formula of **Q** (C_7H_8) indicates an index of hydrogen deficiency (Section 7.8) of four. The hydrogenation experiment suggests that **Q** contains two double bonds (or one triple bond). Compound **Q**, therefore, must contain two rings.

Bicyclo [2.2.1]hepta-2,5-diene. The following reasoning shows one way to arrive at this conclusion: There is only one signal (δ 143) in the region for a doubly bonded carbon. This fact indicates that the doubly bonded carbon atoms are all equivalent. That the signal at δ 143 is a CH group in the DEPT spectrum indicates that each of the doubly bonded carbon atoms bears one hydrogen atom. Information from the DEPT spectrum tells us that the signal at δ 75 is a $-CH_2-$group and the signal at δ 50 is a $\geq$C$-$H group. The molecular formula tells us that the compound must contain two $\geq$C$-$H groups, and since only one signal occurs in the ^{13}C spectrum, these $\geq$C$-$H groups must be equivalent. Putting this all together, we get the following:

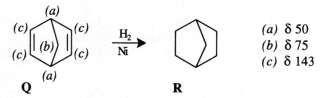

(a) δ 50
(b) δ 75
(c) δ 143

13.22   That S decolorizes bromine indicates that it is unsaturated. The molecular formula of **S** allows us to calculate an index of hydrogen deficiency equal to 1. Therefore, we can conclude that **S** has one double bond.

The ^{13}C spectrum shows the doubly bonded carbon atoms at δ 130 and δ 135. In the DEPT spectrum, one of these signals (δ 130) is a carbon that bears no hydrogen atoms; the other (δ 135) is a carbon that bears one hydrogen atom. We can now arrive at the following partial structure.

The three most upfield signals (δ 19, δ 28, and δ 31) all arise from methyl groups. The signal at δ32 is a carbon atom with no hydrogen atoms. Putting these facts together allows us to arrive at the following structure.

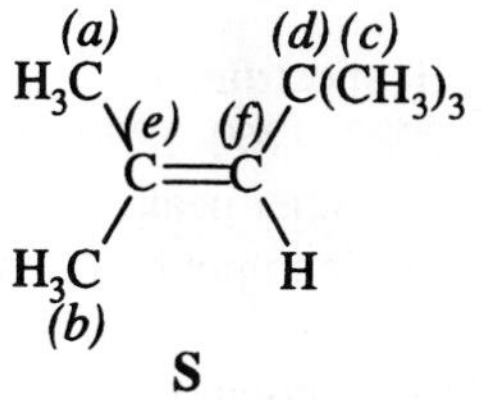

(a) δ 19   (d) δ 32
(b) δ 28   (e) δ 130
(c) δ 31   (f) δ 135

Although the structure just given is the actual compound, other reasonable structures that one might be led to are

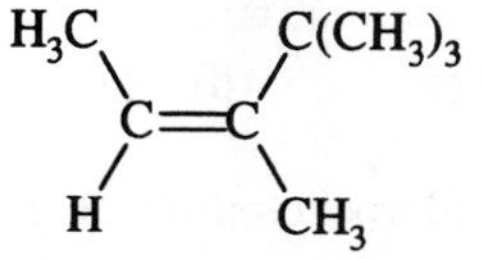

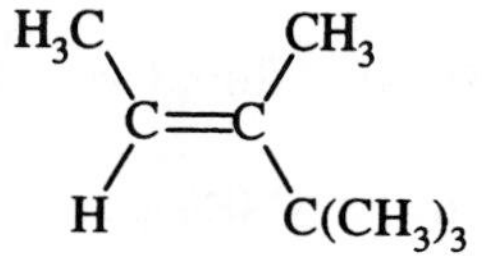

13.23 The IR absorption band at 1745 cm^{-1} indicates the presence of a $>$C=O group in a five-member ring, and the signal at δ218.2 can be assigned to the carbon of the carbonyl group.

There are only two other signals in the ^{13}C spectrum; the DEPT spectra suggest two equivalent sets of two –CH$_2$– groups each. Putting these facts together, we arrive at cyclopentanone as the structure for **T**.

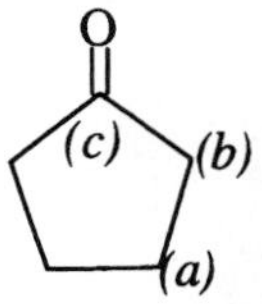

(a) δ 23.5
(b) δ 38.0
(c) δ 218.2

QUIZ

13.1   Propose a structure that is consistent with each set of following data.

(a) C_4H_9Br ^{1}H NMR spectrum
                    Singlet δ 1.7

(b) $C_4H_7Br_3$ ^{1}H NMR spectrum
                    Singlet δ 1.95 (3H)
                    Singlet δ 3.9 (4H)

(c) C_8H_{16} ^{1}H NMR spectrum   IR spectrum
                    Singlet δ 1.0 (9H)   3040, 2950, 1640 cm^{-1}
                    Singlet δ 1.75 (3H)   and other peaks.
                    Singlet δ 1.9 (2H)
                    Singlet δ 4.6 (1H)
                    Singlet δ 4.8 (1H)

(d) $C_9H_{10}O$ ^1H NMR spectrum   IR spectrum
                    Singlet δ 2.0 (3H)   3100, 3000, 1720,
                    Singlet δ 3.75 (2H)   740, 700 cm^{-1}
                    Singlet δ 7.2 (5H)   and other peaks.

(e) $C_5H_7NO_2$ ^1H NMR spectrum   IR spectrum
                    Triplet δ 1.2 (3H)   2980, 2260, 1750 cm^{-1}
                    Singlet δ 3.5 (2H)   and other peaks.
                    Quartet δ 4.2 (2H)   This compound has a nitro group.

13.2   How many ^1H NMR signals would the following compound give?

$$CH_3CHCH_2Cl$$
$$\overset{|}{CH_3}$$

(a) One   (b) Two   (c) Three   (d) Four   (e) Five

13.3   How many ^1H NMR signals would 1,1-dichlorocyclopropane give?

(a) One   (b) Two   (c) Three   (d) Four   (e) Five

F

SPECIAL TOPIC
Mass Spectrometry

SOLUTIONS TO PROBLEMS

F.1 The compound is methane, CH_4. The molecular ion is at m/z 16. (This peak also happens to be the base peak.)

$$H-\overset{\overset{\displaystyle H}{|}}{\underset{\underset{\displaystyle H}{|}}{C}}-H \quad + \quad e^- \quad \longrightarrow \quad H-\overset{\overset{\displaystyle H}{|}}{\underset{\underset{\displaystyle H}{|}}{C}}{\colon}H \quad + \quad 2\,e^-$$

$$m/z\ 16$$
$$\mathbf{M\ {\colon}}$$

The peaks at m/z 15, 14, 13, and 12 are caused by successive losses of hydrogen atoms.

$$H-\overset{\overset{\displaystyle H}{|}}{\underset{\underset{\displaystyle H}{|}}{C}}{\cdot}\,H \quad \longrightarrow \quad H-\overset{\overset{\displaystyle H}{|}}{\underset{\underset{\displaystyle H}{|}}{C}}{}^+ \quad + \quad H{\cdot}$$
$$m/z\ 15$$

$$H-\overset{\overset{\displaystyle H}{|}}{\underset{\underset{\displaystyle H}{|}}{C}}{}^+ \quad \longrightarrow \quad H-\overset{\overset{\displaystyle H}{|}}{\underset{\underset{\displaystyle \bullet}{}}{C}}{}^+ \quad + \quad H{\cdot}$$
$$m/z\ 14$$

$$H-\overset{\overset{\displaystyle H}{|}}{\underset{\underset{\displaystyle \bullet}{}}{C}}{}^+ \quad \longrightarrow \quad H-C{\colon}{}^+ \quad + \quad H{\cdot}$$
$$m/z\ 13$$

$$H-C{\colon}{}^+ \quad \longrightarrow \quad {\bullet}C{\colon}{}^+ \quad + \quad H{\cdot}$$
$$m/z\ 12$$

The small peak at m/z 17 ($\mathbf{M^+ + 1}$) comes mainly from methane molecules that contain ^{13}C.

$$\underset{\underset{\text{H}}{|}}{\overset{\overset{\text{H}}{|}}{\text{H}-\overset{13}{\text{C}}-\text{H}}} \ + \ e^- \ \longrightarrow \ \underset{\underset{\text{H}}{|}}{\overset{\overset{\text{H}}{|}}{\text{H}-\overset{13}{\text{C}}\overset{\bullet}{\text{+}}\text{H}}} \ + \ 2\,e^-$$

m/z 17

(**M** ⁺+ 1)

F.2   The compound is water.

$$\text{H}-\overset{\cdot\cdot}{\underset{\cdot\cdot}{\text{O}}}-\text{H} \ + \ e^- \ \longrightarrow \ \text{H}-\overset{\bullet+}{\underset{\cdot\cdot}{\text{O}}}-\text{H} \ + \ 2\,e^-$$

m/z 18

(**M** ⁺)

$$\text{H}-\overset{\bullet+}{\underset{\cdot\cdot}{\text{O}}}-\text{H} \ \longrightarrow \ \text{H}-\overset{\cdot\cdot}{\underset{\cdot\cdot}{\text{O}}}^+ \ + \ \text{H}\bullet$$

m/z 17

$$\text{H}-\overset{\cdot\cdot}{\underset{\cdot\cdot}{\text{O}}}^+ \ \longrightarrow \ \bullet\overset{\cdot\cdot}{\underset{\cdot\cdot}{\text{O}}}^+ \ + \ \text{H}\bullet$$

m/z 16

The peaks at *m/z* 19 and *m/z* 20 are due (primarily) to naturally occurring oxygen isotopes.

$$\text{H}-\overset{17\,\cdot\cdot}{\underset{\cdot\cdot}{\text{O}}}-\text{H} \ + \ e^- \ \longrightarrow \ \text{H}-\overset{17\,\bullet+}{\underset{\cdot\cdot}{\text{O}}}-\text{H} \ + \ 2\,e^-$$

m/z 19

(**M** ⁺+ 1)

$$\text{H}-\overset{18\,\cdot\cdot}{\underset{\cdot\cdot}{\text{O}}}-\text{H} \ + \ e^- \ \longrightarrow \ \text{H}-\overset{18\,\bullet+}{\underset{\cdot\cdot}{\text{O}}}-\text{H} \ + \ 2\,e^-$$

m/z 20

(**M** ⁺+ 2)

F.3   The compound is methyl fluoride, CH_3F.

$$CH_3-F \ + \ e^- \ \longrightarrow \ [CH_3-F]^{\bullet} \ + \ 2\,e^-$$

m/z 34

(**M** ⁺)

$$[CH_3F]^{\bullet} \ \longrightarrow \ [CH_2F]^+ \ + \ H\bullet$$

m/z 33

$$[CH_2F]^+ \ \longrightarrow \ [CHF]^{\bullet} \ + \ H\bullet$$

m/z 32

$$[CHF]^{\bullet} \ \longrightarrow \ [CF]^+ \ + \ H\bullet$$

m/z 31

$$[CH_3F] \overset{+}{\cdot} \longrightarrow [F]^+ \quad + \quad CH_3^\bullet$$
$$m/z \; 19$$

$$[CH_3F] \overset{+}{\cdot} \longrightarrow [CH_3]^+ \quad + \quad F\bullet$$
$$m/z \; 15$$

$$[CH_3]^+ \longrightarrow [CH_2] \overset{+}{\cdot} \quad + \quad H\bullet$$
$$m/z \; 14$$

F.4   (a)

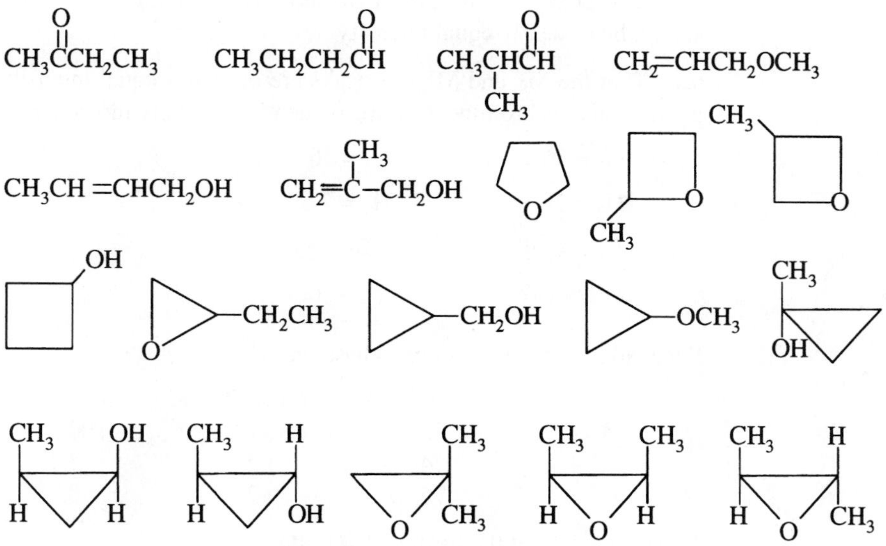

(b)   Only the first three. (The peak at 1730 cm⁻¹ is due to a C=O group.)

F.5   First, we recalculate the intensities of the peaks so as to base them on the M⁺ peak:

m/z		INTENSITY % of M⁺
86 M⁺	10.0/10.0 × 100 =	100
87	0.56/10.0 × 100 =	5.6
88	0.04/10.0 × 100 =	0.4

1. Since M⁺ is even, the compound must contain an even number of nitrogen atoms (i.e., 0, 2, 4, etc.).

2. The value of the M⁺ + 1 peak gives the number of carbon atoms.

     Number of carbon atoms = 5.6/1.1 ≃ 5

The compound must contain no nitrogen atoms because $C_5N_2 = (5 \times 12) + (2 \times 14) = 88$, and the molecular weight of the compound (from the M⁺ peak) is only 86.

3. The very low value of the M⁺ + 2 peak (0.4%) tells us that the compound does not contain S, Cl, or Br.

4. If the compound were composed only of C and H, it would have to be C_5H_{26}:

     $H = 86 - (5 \times 12) = 26$

But C_5H_{26} is impossible.

However, a formula with one oxygen gives a reasonable number of hydrogen atoms,

$$H = 86 - (5 \times 12) - 16 = 10$$

and thus our compound has the formula $C_5H_{10}O$.

F.6 (a) The $M^{\ddagger} + 2$ peak due to $CH_3-^{37}Cl$ (at m/z 52) should be almost one-third (32.5%) as large as the $M^{\ddagger}$ peak at m/z 50.

(b) The peaks due to $CH_3-^{79}Br$ and $CH_3-^{81}Br$ (at m/z 94 and m/z 96, respectively) should be of nearly equal intensity.

(c) That the $M^{\ddagger}$ and $M^{\ddagger} + 2$ peaks are of nearly equal intensity tells us that the compound contains bromine. C_3H_7Br is therefore a likely molecular formula.

$$
\begin{array}{ll}
C_3 = 36 & C_3 = 36 \\
H_7 = 7 & H_7 = 7 \\
{}^{79}Br = \underline{79} & {}^{81}Br = \underline{81} \\
m/z = 122 & m/z = 124
\end{array}
$$

F.7 Recalculating the intensities to base on $M^{\ddagger}$

PEAK	m/z	% of BASE PEAK	% of $M^{\ddagger}$
$M^{\ddagger}$	73	86.1	100
$M^{\ddagger} + 1$	74	3.2	3.72
$M^{\ddagger} + 2$	75	0.2	0.23

These data best fit the formula C_3H_7NO.

F.8 (a) First recalculating the intensities so as to base them on the $M^{\ddagger}$ peak:

m/z		INTENSITY % of $M^{\ddagger}$
78 $M^{\ddagger}$	$24/24 \times 100 =$	100
79	$0.8/24 \times 100 =$	3.3
80	$8/24 \times 100 =$	33

1. Since $M^{\ddagger}$ is even, the compound contains an even number of nitrogen atoms.

2. Number of carbon atoms $= (M^{\ddagger} + 1)/1.1 = 3.3/1.1 = 3$.

3. The intensity of the $M^{\ddagger} + 2$ peak (33%) tells us that the compound contains one chlorine atom.

4. We use the molecular weight (from the $M^{\ddagger}$ peak) to calculate the number of hydrogen atoms.

$$H = 78 - (3 \times 12) - 35 = 7$$

Thus, the formula for the compound is C_3H_7Cl.

(b) $CH_3\underset{\underset{Cl}{|}}{C}HCH_3$

F.9   (a)   A *tert*-butyl cation, $(CH_3)_3C^+$.

(b) $$\left[CH_3-\underset{\underset{\displaystyle CH_3}{|}}{\overset{\overset{\displaystyle CH_3}{|}}{C}}-CH_3 \right]^{\ddagger} \longrightarrow CH_3-\underset{\underset{\displaystyle CH_3}{|}}{\overset{\overset{\displaystyle CH_3}{|}}{C^+}} + CH_3^{\bullet}$$

m/z 57

F.10   A peak at M^+ - 15 involves the loss of a methyl radical and the formation of a 1° or 2° carbocation.

$$[CH_3CH_2\overset{\overset{\displaystyle CH_3}{|}}{C}HCH_2CH_3]^{\ddagger} \longrightarrow CH_3CH_2\overset{+}{C}HCH_2CH_3 + CH_3^{\bullet}$$

$(M\ddagger)$ $(M\ddagger - 15)$

or

$$[CH_3CH_2\overset{\overset{\displaystyle CH_3}{|}}{C}HCH_2CH_3]^{\ddagger} \longrightarrow CH_3CH_2\overset{\overset{\displaystyle CH_3}{|}}{C}HCH_2{}^+ + CH_3^{\bullet}$$

$(M\ddagger)$ $(M\ddagger - 15)$

A peak at M^+ - 29 arises from the loss of an ethyl radical and the formation of a 2° carbocation.

$$[CH_3CH_2\overset{\overset{\displaystyle CH_3}{|}}{C}HCH_2CH_3]^{\ddagger} \longrightarrow CH_3CH_2\overset{\overset{\displaystyle CH_3}{|}}{C}H^+ + CH_3CH_2^{\bullet}$$

$(M\ddagger)$ $(M\ddagger - 29)$

Since a 2° carbocation is more stable, the peak at M^+ - 29 is more intense.

F.11   Both peaks arise from allylic fragmentations:

$$^+CH_2-\overset{\bullet}{C}H-CH_2-\underset{\underset{\displaystyle CH_3}{|}}{C}HCH_2CH_3 \longrightarrow \overset{\bullet}{C}H_2-CH{=}CH_2 + {}^+\overset{\overset{\displaystyle CH_3}{|}}{C}HCH_2CH_3$$

Allyl radical m/z 57

$$CH_2{}^+{-}\overset{\bullet}{C}H{-}CH_2\ {:}\underset{\underset{\displaystyle CH_3}{|}}{C}HCH_2CH_3 \longrightarrow {}^+CH_2{-}CH{=}CH_2 + {}^{\bullet}\overset{\overset{\displaystyle CH_3}{|}}{C}HCH_2CH_3$$

m/z 41

Allyl cation

F.12   (a)   Alcohols undergo rapid cleavage of a carbon-carbon bond next to oxygen because this leads to a resonance-stabilized cation.

1° alcohol $R : CH_2 \overset{\cdot\cdot}{OH^+}$ $\xrightarrow{-R\cdot}$ $CH_2 = \overset{+}{\underset{\cdot\cdot}{O}}H$ $\longleftrightarrow$ $\overset{+}{C}H_2 - \overset{\cdot\cdot}{\underset{\cdot\cdot}{O}}H$

2° alcohol $R - \overset{R}{\overset{|}{C}}H \overset{\cdot\cdot}{OH^+}$ $\xrightarrow{-R\cdot}$ $RCH = \overset{+}{\underset{\cdot\cdot}{O}}H$ $\longleftrightarrow$ $R\overset{+}{C}H - \overset{\cdot\cdot}{\underset{\cdot\cdot}{O}}H$

3° alcohol $R - \overset{R}{\underset{\underset{R}{|}}{\overset{|}{C}}} \overset{\cdot\cdot}{OH^+}$ $\xrightarrow{-R\cdot}$ $R\overset{}{\underset{\underset{R}{|}}{C}} = \overset{+}{\underset{\cdot\cdot}{O}}H$ $\longleftrightarrow$ $R\overset{+}{\underset{\underset{R}{|}}{C}} - \overset{\cdot\cdot}{\underset{\cdot\cdot}{O}}H$

The cation obtained from a tertiary alcohol is the most stable (because of the electron-releasing R groups).

(b)   Primary alcohols give a peak at m/z 31 due to $CH_2 = \overset{+}{O}H$.

(c)   Secondary alcohols give peaks at m/z 45, 59, 73, and so forth, because ions like the following are produced.

$CH_3CH = \overset{+}{O}H$ $CH_3CH_2CH = \overset{+}{O}H$ $CH_3CH_2CH_2CH = \overset{+}{O}H$

m/z 45 m/z 59 m/z 73

(d)   Tertiary alcohols give peaks at m/z 59, 73, 87, and so forth, because ions like the following are produced.

$CH_3\overset{}{\underset{\underset{CH_3}{|}}{C}} = \overset{+}{O}H$ $CH_3CH_2\overset{}{\underset{\underset{CH_3}{|}}{C}} = \overset{+}{O}H$ $CH_3CH_2CH_2\overset{}{\underset{\underset{CH_3}{|}}{C}} = \overset{+}{O}H$

m/z 59 m/z 73 m/z 87

F.13   The spectrum given in Fig. F.12 is that of butyl isopropyl ether. The main clues are the peaks at m/z 101 and m/z 73 due to the following fragmentations.

$$\left[\underset{}{\overset{\overset{CH_3}{|}}{CH_3 - CH - OCH_2CH_2CH_2CH_3}} \right]^{\ddot{+}} \xrightarrow{-CH_3\cdot} CH_3CH = \overset{+}{O}CH_2CH_2CH_2CH_3$$
$$m/z\ 101$$

$$\left[\underset{}{\overset{\overset{CH_3}{|}}{CH_3 - CH - OCH_2CH_2CH_2CH_3}} \right]^{\ddot{+}} \xrightarrow{-CH_3CH_2CH_2\cdot} \underset{}{\overset{\overset{CH_3}{|}}{CH_3CH\overset{+}{O} = CH_2}}$$
$$m/z\ 73$$

Butyl propyl ether (Fig. F.13) has no peak at m/z 101 but has a peak at m/z 87 instead.

$$[CH_3CH_2CH_2 - O - CH_2CH_2CH_2CH_3]^{\ddot{+}} \xrightarrow{-CH_3CH_2\cdot} CH_2 = \overset{+}{O}CH_2CH_2CH_2CH_3$$
$$m/z\ 87$$

Butyl propyl ether also has a peak at m/z 73.

$$[CH_3CH_2CH_2\!-\!O\!-\!CH_2CH_2CH_2CH_3]\ \ddagger\ \xrightarrow{-CH_3CH_2CH_2\bullet}\ CH_3CH_2CH_2\overset{+}{O}=CH_2$$
$$m/z\ 73$$

[Although the observation does not help us decide, it is interesting to notice that both spectra have intense peaks at m/z 43 and m/z 57 corresponding to propyl (or isopropyl) and butyl cations formed by carbon-oxygen bond cleavage.]

F.14   The compound is butanal. The peak at m/z 44 arises from a McLafferty rearrangement.

$$m/z\ 72 \qquad\qquad\qquad m/z\ 44$$
$$(\mathbf{M\ddagger}) \qquad\qquad\qquad (\mathbf{M\ddagger-28})$$

The peak at m/z 29 arises from a fragmentation producing an acylium ion.

$$H\!-\!C\!\equiv\!\overset{+}{O}\ +\ CH_3CH_2CH_2\bullet$$
$$m/z\ 29$$

F.15   The ion, $CH_2=\overset{+}{N}H_2$, produced by the following fragmentation.

$$m/z\ 30$$

F.16   Compound **A** is *tert*-butylamine. Our first clue is the molecular ion at m/z 73 (an odd-numbered mass unit) indicating the presence of an odd number of nitrogen atoms. The base peak at m/z 58 is our second important clue. It arises from the following fragmentation.

$$m/z\ 58$$

The ^{1}H NMR spectrum confirms the structure

$$\overset{(a)\qquad (b)}{(CH_3)_3C\!-\!NH_2}$$

(a) Singlet δ 1.2 (9H)
(b) Singlet δ 1.3 (2H)

F.17 The compound is 2-methyl-2-butanol. Although the molecular ion is not discernible, we are given that it is at m/z 88. This information gives us the molecular weight of **B** and rules out the possibility of a structure with an odd number of nitrogen atoms.

The IR absorption (3200–3550 cm^{-1}) suggests the presence of an –OH group.

Two important peaks in the mass spectrum are the intense peaks at m/z 59 and m/z 73. These peaks correspond to fragmentation reactions that produce resonance-stabilized oxonium ions and strongly suggest that we have a tertiary alcohol [see Problem F.12, part (d)].

The peak at m/z 70 corresponds to the loss of a molecule of water from the molecular ion, and the peak at m/z 55 probably arises from a subsequent allylic cleavage.

The ^{1}H NMR spectrum of **B** confirms that it is 2-methyl-2-butanol.

(a) Triplet, δ 0.9 (3H)
(b) Quartet, δ 1.6 (2H)
(c) and (d) Overlapping singlets, δ 1.1 (7H)

14 AROMATIC COMPOUNDS

SOLUTIONS TO PROBLEMS

14.1 Compounds (a) and (b) would yield only one monosubstitution product.

14.2 Resonance structures are defined as being structures that differ *only* in the positions of the electrons. In the two 1,3,5-cyclohexatrienes shown, the carbon atoms are in different positions; therefore, they cannot be resonance structures.

14.3 Inscribing a square in a circle with one corner at the bottom gives the following results:

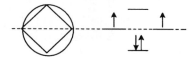

We see, therefore, that cyclobutadiene would be a diradical and would not be aromatic.

14.4 (a)

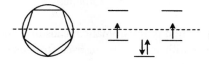

The cyclopentadienyl anion (above) should be aromatic because it has a closed bonding shell of delocalized π electrons.
(b) and (c) The cyclopentadienyl cation (below) would be a diradical. We would not expect it to be aromatic.

(d) No, 4 is not a Hückel number.

14.5   (a) The cycloheptatrienyl cation (below) would be aromatic because it would have a closed bonding shell of delocalized π electrons.

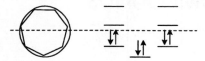

(b) No, the cycloheptatrienyl anion (below) would be a diradical.

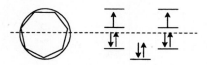

(c) No, 8 is not a Hückel number.

14.6   If the 1,3,5-cycloheptatrienyl anion *were* aromatic, we would expect it to be unusually stable. This would mean that 1,3,5-cycloheptatriene should be unusually acidic. The fact that 1,3,5-cycloheptatriene is not unusually acidic (it is less acidic than 1,3,5-heptatriene) confirms the prediction made in the previous problem, that the 1,3,5-cycloheptatrienyl anion should not be aromatic.

14.7   (a)

Br

Br

(b)

Br

$\xrightarrow[\text{--HBr}]{\text{heat}}$ + Br$^-$

Br   Tropylium bromide

These results suggest that the bonding in tropylium bromide is ionic; that is, it consists of a positive tropylium ion and a negative bromide ion.

14.8   It suggests that the cyclopropenyl cation should be aromatic.

14.9   The fact that the cyclopentadienyl cation is antiaromatic means that the following hypothetical transformation would occur with an increase in π-electron energy.

HC$_+$ $\xrightarrow[\text{energy increases}]{\pi\text{-electron}}$ +   + H$_2$

14.10 The cyclopropenyl cation (below).

14.11 The high field signal arises from the six methyl protons of 15,16-dimethyldihydropyrene, which by virtue of their location are strongly shielded by the magnetic field created by the aromatic ring current (see Figure 14.7).

14.12 Major contributors to the hybrid must be ones that involve separated charges. Contributors like the following would have separated charges, and would have aromatic five- and seven-membered rings.

14.13 (a)   (b)

14.14 Because of their symmetries, *p*-dibromobenzene would give two ^{13}C signals, *o*-dibromobenzene would give three, and *m*-dibromobenzene would give four.

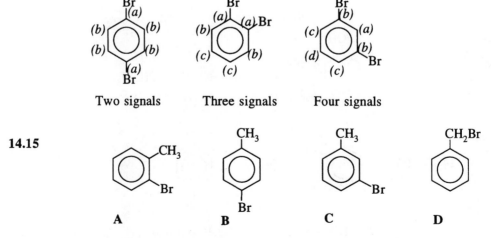

Two signals   Three signals   Four signals

14.15

A   B   C   D

A. Strong absorption at 740 cm^{-1} is characteristic of ortho substitution.

B. A very strong absorption peak at 800 cm^{-1} is characteristic of para substitution.

C. Strong absorption peaks at 680 and 760 cm^{-1} are characteristic of meta substitution.

D. Strong absorption peaks at 693 and 765 cm^{-1} are characteristic of a monosubstituted benzene ring.

14.16 (a) 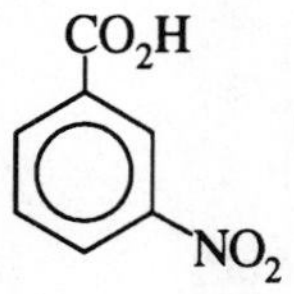 3-nitrobenzoic acid structure (CO$_2$H, NO$_2$)

(b) 4-bromotoluene structure (Br, CH$_3$)

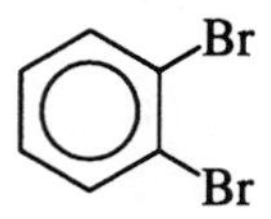

(c) 1,2-dibromobenzene structure (Br, Br)

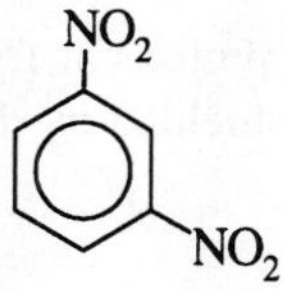

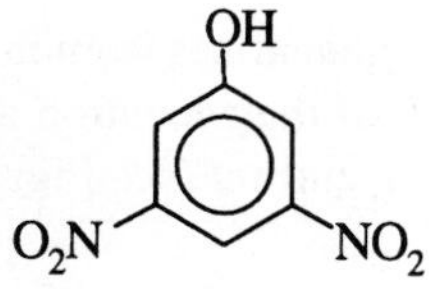

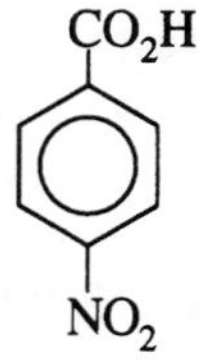

(d) 1,3-dinitrobenzene structure (NO$_2$, NO$_2$)

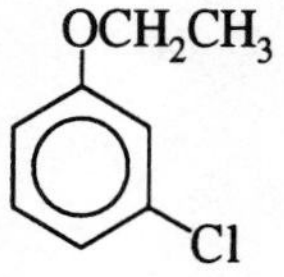

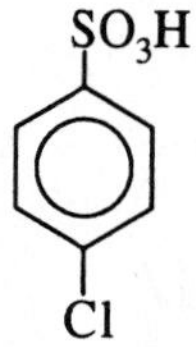

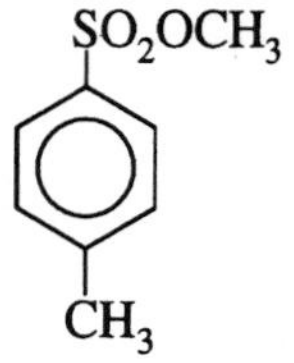

(e) 3,5-dinitrophenol structure (OH, O$_2$N, NO$_2$)

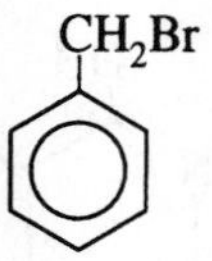

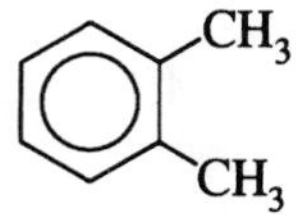

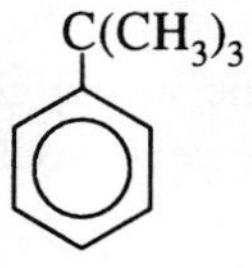

(f) 4-nitrobenzoic acid structure (CO$_2$H, NO$_2$)

(g) 3-chlorophenetole structure (OCH$_2$CH$_3$, Cl)

(h) 4-chlorobenzenesulfonic acid structure (SO$_3$H, Cl)

(i) methyl 4-methylbenzenesulfonate structure (SO$_2$OCH$_3$, CH$_3$)

(j) benzyl bromide structure (CH$_2$Br)

(k) 4-nitroaniline structure (NH$_2$, NO$_2$)

(l) o-xylene structure (CH$_3$, CH$_3$)

(m) tert-butylbenzene structure (C(CH$_3$)$_3$)

(n) 4-methylphenol structure (OH, CH$_3$)

(o) 4-bromoacetophenone structure (C=O, CH$_3$, Br)

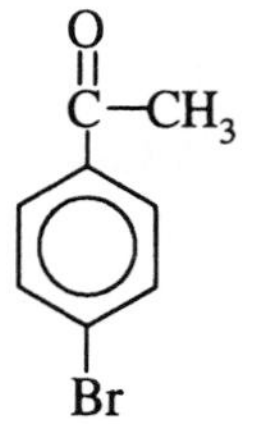

(p) 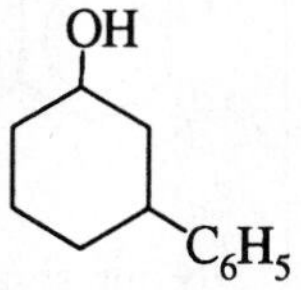 cyclohexane structure (OH, C$_6$H$_5$)

(q) 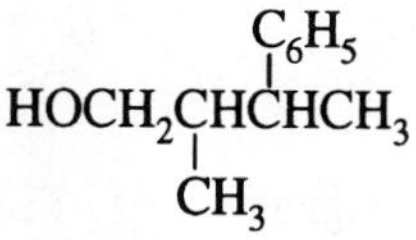 HOCH$_2$CHCHCH$_3$ with C$_6$H$_5$ and CH$_3$ substituents

(r) 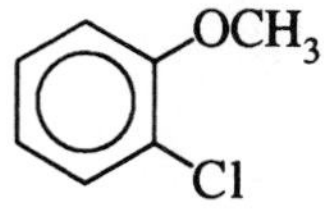 2-chloroanisole structure (OCH$_3$, Cl)

14.17 (a) 1,2,3-tribromobenzene structure (Br, Br, Br)

1,2,4-tribromobenzene structure (Br, Br, Br)

1,3,5-tribromobenzene structure (Br, Br, Br)

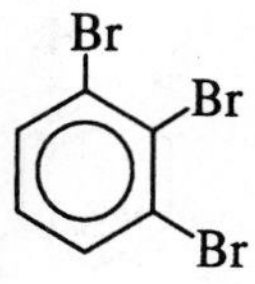

1,2,3-Tribromo-
benzene

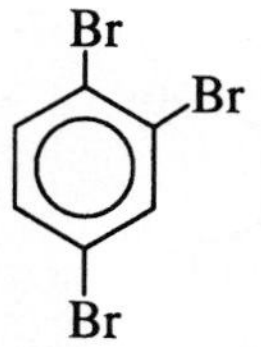

1,2,4-Tribromo-
benzene

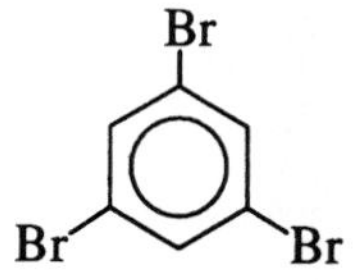

1,3,5-Tribromo-
benzene

(b)

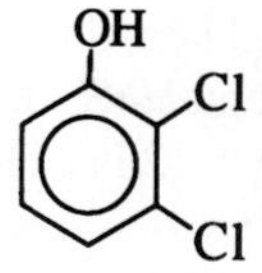

2,3-Dichloro-
phenol

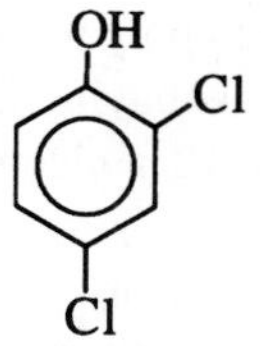

2,4-Dichloro-
phenol

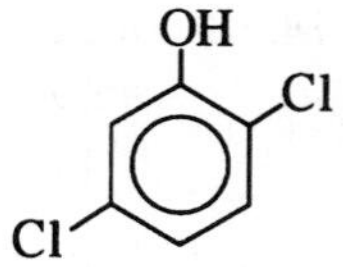

2,5-Dichloro-
phenol

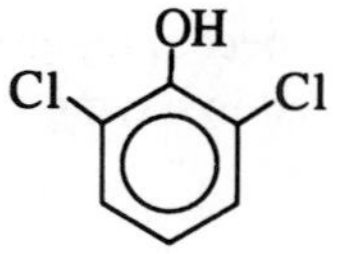

2,6-Dichloro-
phenol

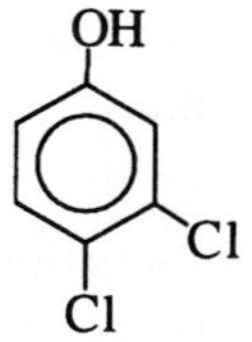

3,4-Dichloro-
phenol

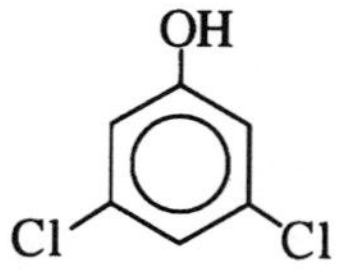

3,5-Dichloro-
phenol

(c)

4-Nitroaniline
(*p*-nitroaniline)

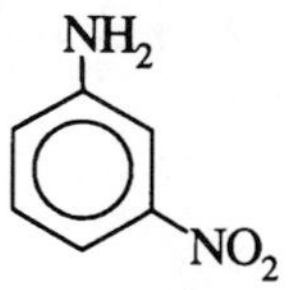

3-Nitroaniline
(*m*-nitroaniline)

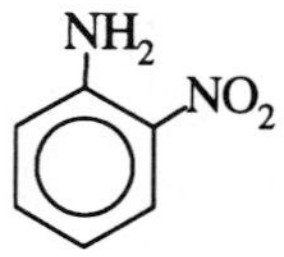

2-Nitroaniline
(*o*-nitroaniline)

(d)

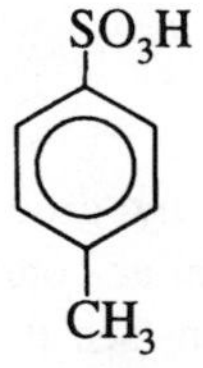

4-Methylbenzene-
sulfonic acid
(*p*-toluenesulfonic
acid)

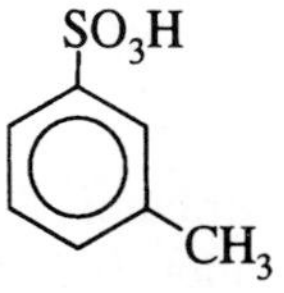

3-Methylbenzene-
sulfonic acid
(*m*-toluenesulfonic
acid)

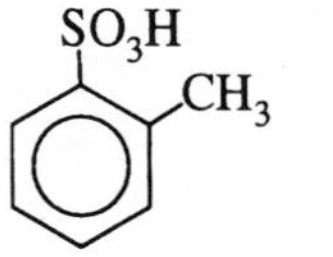

2-Methylbenzene-
sulfonic acid
(*o*-toluenesulfonic
acid)

(e)

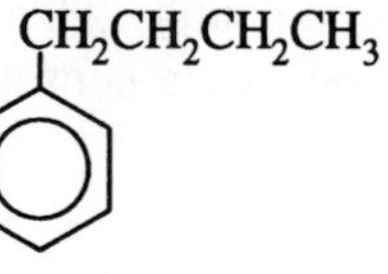

Butylbenzene

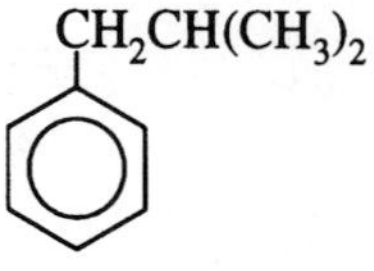

Isobutylbenzene

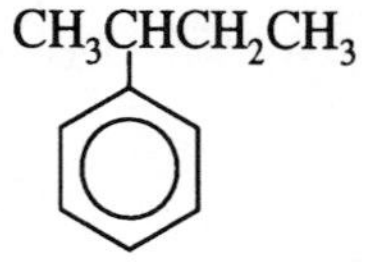

sec-Butylbenzene

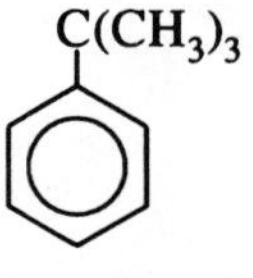

tert-Butylbenzene

14.18 Hückel's rule should apply to both pentalene and heptalene. Pentalene's antiaromaticity can be attributed to its having 8 π electrons. Heptalene's lack of aromaticity can be attributed to its having 12 π electrons. Neither 8 nor 12 is a Hückel number.

14.19 (a) The extra two electrons go into the two partly filled (nonbonding) molecular orbitals (Fig. 14.6), causing them to become filled. The dianion, therefore, is not a diradical. Moreover, the cyclooctatetraene dianion has 10 π electrons (a Hückel number), and this apparently gives it the stability of an aromatic compound. (The highest occupied molecular orbitals may become slightly lower in energy and become bonding molecular orbitals.) The stability gained by becoming aromatic is apparently large enough to overcome the extra strain involved in having the ring of the dianion become planar.

(b) The strong base (butyllithium) removes two protons from the compound on the left. This acid-base reaction leads to the formation of the 10 π electron pentalene dianion, an aromatic dianion.

Pentalene dianion

14.20 The bridging –CH_2–group causes the 10 π electron ring system (below) to become planar. This allows the ring to become aromatic.

14.21 (a) Resonance contributors that involve the carbonyl group of **I** resemble the *aromatic* cycloheptatrienyl cation and thus stabilize **I**. Similar contributors to the hybrid of **II** resemble the *antiaromatic* cyclopentadienyl cation (see Problem 14.9) and thus destabilize **II**.

Contributors like **IA** are exceptionally stable because they resemble an aromatic compound. They therefore make large stabilizing contributions to the hybrid.

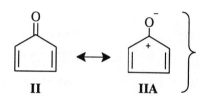

Contributors like **IIA** are exceptionally unstable because they resemble an anti-aromatic compound. Any contribution they make to the hybrid is destabilizing.

II **IIA**

(b)

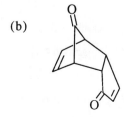

14.22 Ionization of 5-chloro-1,3-cyclopentadiene would produce a cyclopentadienyl cation, and the cyclopentadienyl cation (see Problem 14.9) would be highly unstable because it would be antiaromatic.

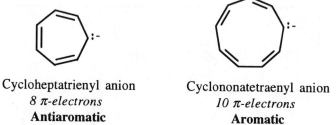

Antiaromatic ion
(highly unstable)

14.23 (a) The cycloheptatrienyl anion has 8 π electrons and does not obey Hückel's rule; the cyclononatetraenyl anion with 10 π electrons obeys Hückel's rule.

Cycloheptatrienyl anion
8 π-electrons
Antiaromatic

Cyclononatetraenyl anion
10 π-electrons
Aromatic

(b) By adding 2 π electrons, [16]annulene becomes an 18 π electron system and therefore obeys Hückel's rule.

14.24 As noted in Problem 12.25, furan can serve as the diene component of Diels-Alder reactions, readily losing all aromatic character in the process. Benzene, on the other hand, is so unreactive in a Diels-Alder reaction that it can be used as a nonreactive solvent for Diels-Alder reactions.

14.25

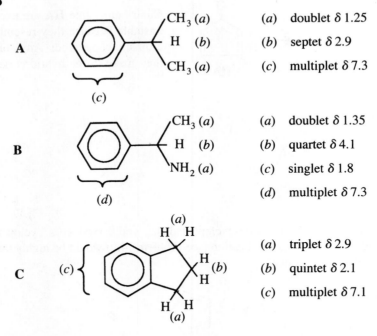

A

CH$_3$ (a)
H (b)
CH$_3$ (a)
(c)

(a) doublet δ 1.25
(b) septet δ 2.9
(c) multiplet δ 7.3

B

CH$_3$ (a)
H (b)
NH$_2$ (a)
(d)

(a) doublet δ 1.35
(b) quartet δ 4.1
(c) singlet δ 1.8
(d) multiplet δ 7.3

C

(c)
(a)
H H
H (b)
H
H H
(a)

(a) triplet δ 2.9
(b) quintet δ 2.1
(c) multiplet δ 7.1

14.26 A ^{1}H NMR signal this far upfield indicates that cyclooctatetraene is a cyclic polyene and is not aromatic.

14.27 Compound F is *p*-isopropyltoluene. ^{1}H NMR assignments are shown in the following spectrum.

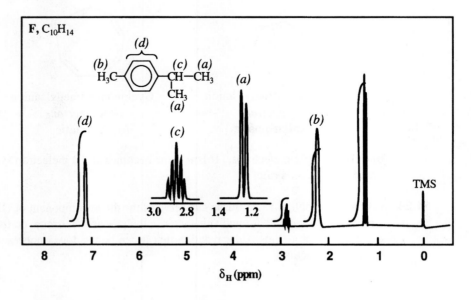

We can make the following ^{1}H NMR assignments:

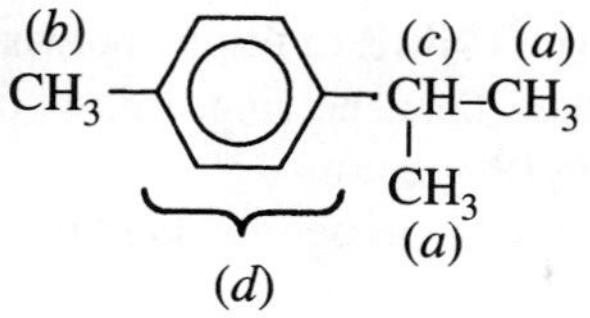

(a) doublet δ 1.25   (c) septet δ 2.85
(b) singlet δ 2.3   (d) multiplet δ 7.1

14.28 Compound **L** is allylbenzene,

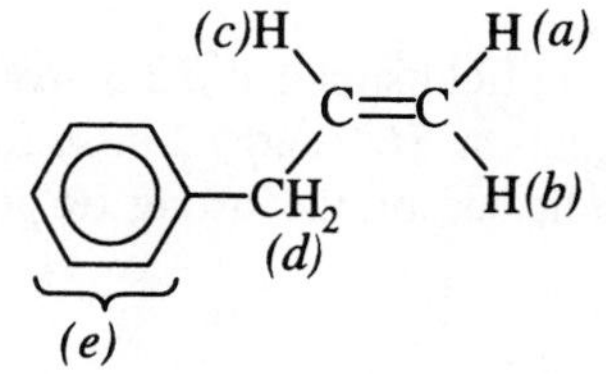

(d) Doublet, δ 3.1 (2H)
(a) or (b) Multiplet, δ 4.8
(a) or (b) Multiplet, δ 5.1
(c) Multiplet, δ 5.8
(e) Multiplet, δ 7.1 (5H)

The following IR assignments can be made.

3035 cm^{-1}, C–H stretching of benzene ring
3020 cm^{-1}, C–H stretching of –CH=CH$_2$ group
2925 cm^{-1} and 2853 cm^{-1}, C–H stretching of –CH$_2$ –group
1640 cm^{-1}, C=C stretching
990 cm^{-1} and 915 cm^{-1}, C–H bendings of –CH=CH$_2$ group
740 cm^{-1} and 695 cm^{-1}, C–H bendings of –C$_6$H$_5$ group

The UV absorbance maximum at 255 nm is indicative of a benzene ring that is not conjugated with a double bond.

14.29 Compound **M** is *m*-ethyltoluene. We can make the following assignments in the spectra.

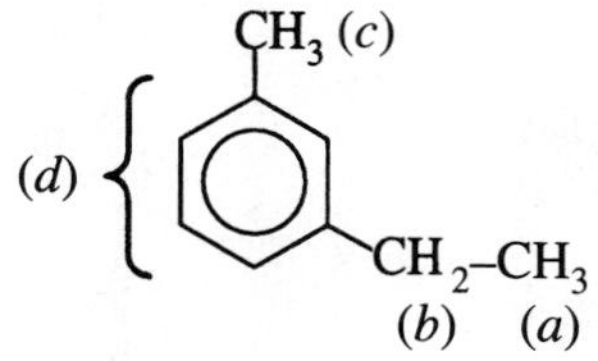

(a) triplet δ 1.4
(b) quartet δ 2.6
(c) Singlet δ 2.4
(d) multiplet δ 7.05

Meta substitution is indicated by the very strong peaks at 690 and 780 cm^{-1} in the IR spectrum.

14.30 Compound **N** is $C_6H_5CH=CHOCH_3$. The absence of absorption peaks due to O–H or C=O stretching in the IR spectrum of **N** suggests that the oxygen atom is present as part of an ether linkage. The (5H) ^{1}H NMR multiplet between δ 7.1–7.6 strongly suggests the presence of a monosubstituted benzene ring; this is confirmed by the strong peaks at ~690 and ~770 cm^{-1} in the IR spectrum.

We can make the following assignments in the ^{1}H NMR spectrum:

$$(a) \quad (b) \quad (c) \quad (d)$$
$$C_6H_5–CH=CH–OCH_3$$

  (a)   Multiplet δ 7.1–7.6
  (b)   Doublet δ 6.1
  (c)   Doublet δ 5.2
  (d)   Singlet δ 3.7

14.31 Compound **X** is *meta*-xylene. The upfield signal at δ 2.3 arises from the two equivalent methyl groups. The downfield signals at δ 6.9 and 7.1 arise from the protons of the benzene ring. Meta substitution is indicated by the strong IR peak at 680 cm^{-1} and very strong IR peak at 760 cm^{-1}.

14.32 The broad IR peak at 3400 cm^{-1} indicates a hydroxy group, and the two bands at 720 and 770 cm^{-1} suggest a monosubstituted benzene ring. The presence of these groups is also indicated by the peaks at δ 4.4 and δ 7.2 in the ^{1}H NMR spectrum. The ^{1}H NMR spectrum also shows a triplet at δ 0.85 indicating a –CH$_3$ group coupled with an adjacent –CH$_2$–group. There is a complex multiplet at δ 1.7 and there is also a triplet at δ 4.5 (1H). Putting these pieces together in the only way possible gives us the following structure for **Y**.

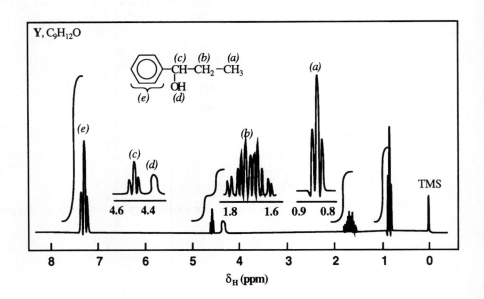

14.33   (a)   Four unsplit signals,

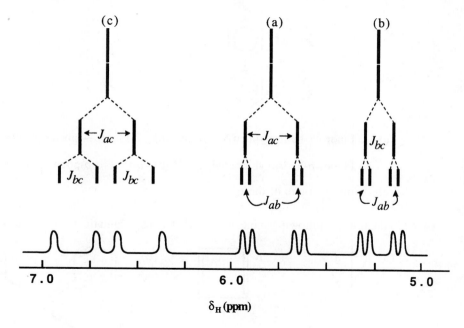

(b)   Absorptions arising from: =C–H, CH_3, and C=O groups.

14.34   The vinylic protons of *p*-chlorostyrene should give a spectrum approximately like the following:

δ_H (ppm)

QUIZ

14.1   Which of the following reactions is inconsistent with the assertion that benzene is aromatic?

(a) $Br_2/CCl_4/25°C \longrightarrow$ no reaction

(b) $H_2/Pt/25°C \longrightarrow$ no reaction

(c) $Br_2/FeBr_3 \longrightarrow C_6H_5Br + HBr$

(d) $KMnO_4/H_2O/25°C \longrightarrow$ no reaction

(e)   None of the above

14.2 Which is the correct name of the compound shown?

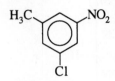

(a) 3-Chloro-5-nitrotoluene

(b) *m*-Chloro-*m*-nitrotoluene

(c) 1-Chloro-3-nitro-5-toluene

(d) *m*-Chloromethylnitrobenzene

(e) More than one of these

14.3 Which is the correct name of the compound shown?

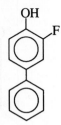

(a) 2-Fluoro-1-hydroxyphenylbenzene

(b) 2-Fluoro-4-phenylphenol

(c) *m*-Fluoro-*p*-hydroxybiphenyl

(d) *o*-Fluoro-*p*-phenylphenol

(e) More than one of these

14.4 Which of the following molecules or ions is not aromatic according to Huckel's rule?

(a)   (b)   (c) 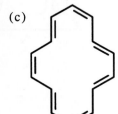   (d)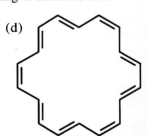

(e) All are aromatic.

14.5   Give the structure of a compound with the formula C_7H_7Cl that is capable of undergoing both S_N1 and S_N2 reactions.

14.6   Write the name of an aromatic compound that is isomeric with naphthalene.

15

REACTIONS OF AROMATIC COMPOUNDS

SOLUTIONS TO PROBLEMS

15.1

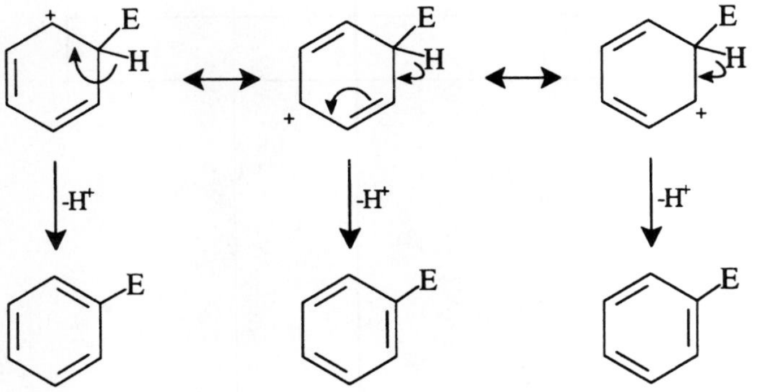

15.2 The rate is dependent on the concentration of NO_2^+ ion formed from protonated nitric acid.

$$H-\overset{+}{\underset{H}{O}}-NO_2 \ + \ HA \ \longrightarrow \ NO_2^+ \ + \ H_3O^+ \ + \ A^-$$

(where HA = HNO_3 or $HOSO_3H$)

Because H_2SO_4 ($HOSO_3H$) is a stronger acid, a mixture of it and HNO_3 will contain a higher concentration of protonated nitric acid than will nitric acid alone.

That is, the reaction,

$$H-O-NO_2 \ + \ HOSO_3H \ \rightleftharpoons \ H-\overset{+}{\underset{H}{O}}-NO_2 \ + \ HSO_4^-$$

Protonated
nitric acid

produces more protonated nitric acid than the reaction,

$$H-O-NO_2 \ + \ H-O-NO_2 \ \rightleftharpoons \ H-\overset{+}{\underset{H}{O}}-NO_2 \ + \ NO_3^-$$

15.3   *Step 1* $CH_3CH=CH_2$ + H–F $\rightleftharpoons$ $CH_3\overset{+}{C}HCH_3$ + F^-

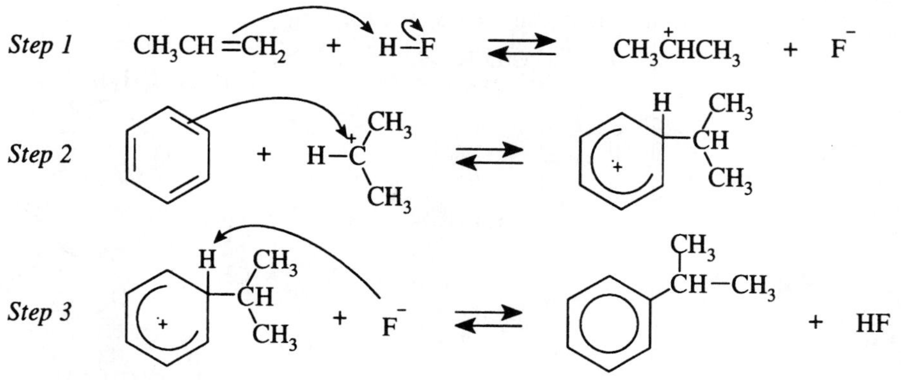

Step 2   + $H-\overset{+}{\underset{CH_3}{\overset{CH_3}{C}}}$ $\rightleftharpoons$

Step 3   + F^- $\rightleftharpoons$ + HF

15.4 $CH_3\overset{:\ddot{O}}{\overset{\|}{C}}-\ddot{\underset{..}{O}}-\overset{:\ddot{O}}{\overset{\|}{C}}CH_3$ + $AlCl_3$ $\rightleftharpoons$ $CH_3\overset{:\ddot{O}}{\overset{\|}{C}}-\overset{..}{\underset{..}{O}}-\overset{+\ddot{O}AlCl_3^-}{\overset{\|}{C}}CH_3$ $\longrightarrow$

$CH_3\overset{+}{C}\equiv\overset{..}{O}:$ + $CH_3\overset{:\ddot{O}}{\overset{\|}{C}}-\ddot{\underset{..}{O}}AlCl_3^-$

$\updownarrow$

$CH_3\overset{+}{C}=\ddot{O}:$

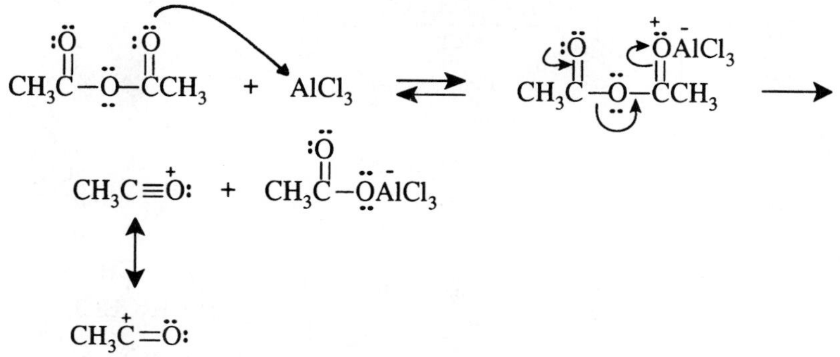

15.5   The carbocation formed by the action of $AlCl_3$ on neopentyl chloride is primary. This carbocation rearranges to the more stable tertiary carbocation before it can react with the benzene ring:

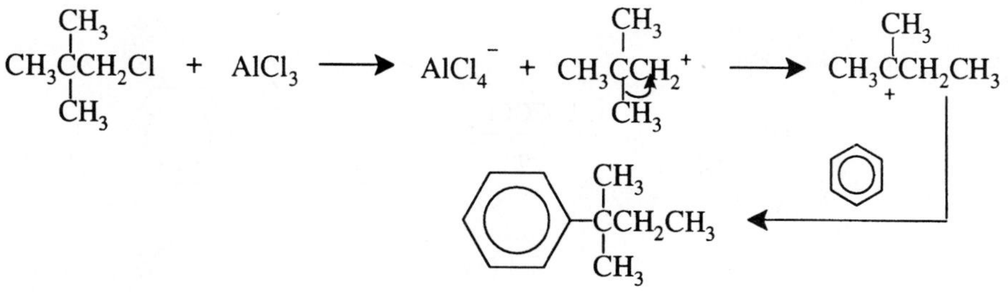

$CH_3\overset{CH_3}{\underset{CH_3}{C}}CH_2Cl$ + $AlCl_3$ $\longrightarrow$ $AlCl_4^-$ + $CH_3\overset{CH_3}{\underset{CH_3}{C}}\overset{+}{C}H_2$ $\longrightarrow$ $CH_3\overset{CH_3}{\underset{+}{C}}CH_2CH_3$

+

15.6 $CH_3CH_2CH_2-OH$ + BF_3 $\rightleftharpoons$ $CH_3CH_2CH_2^+$ + $HOBF_3$

The propyl cation can rearrange into an isopropyl cation:

$CH_3CH_2CH_2^+$ $\xrightarrow{\text{hydride shift}}$ $CH_3\overset{+}{C}HCH_3$

Both cations can then attack the ring.

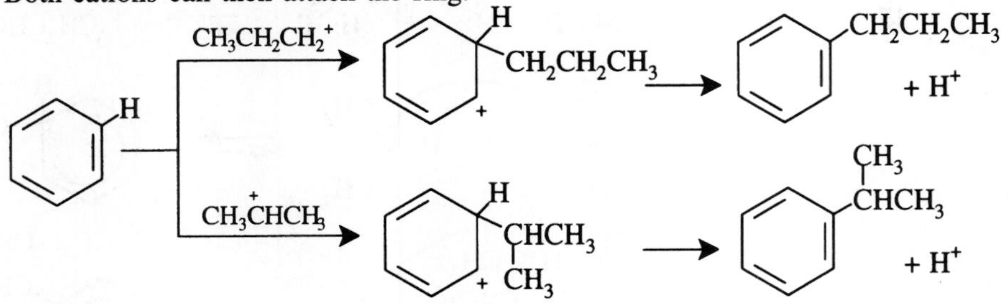

15.7 (a)

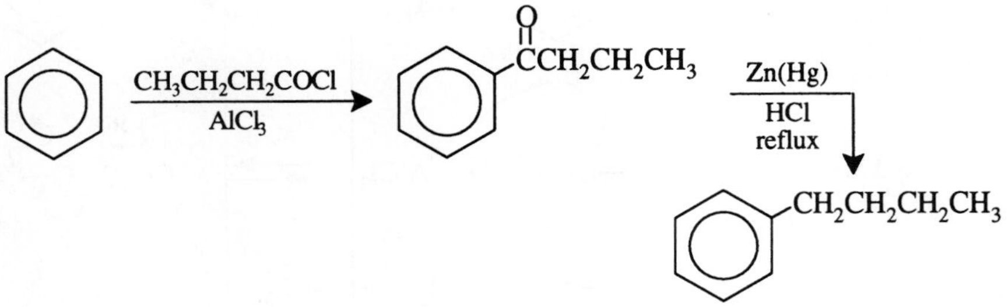

(b)

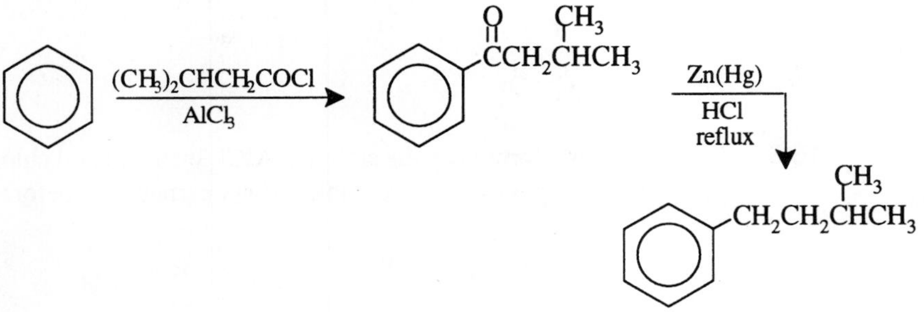

(c)

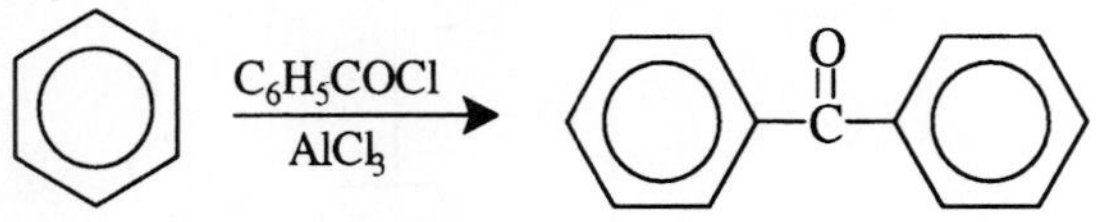

(d)

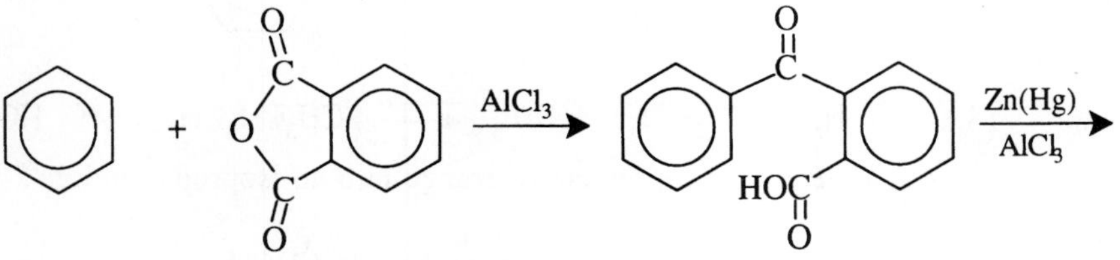

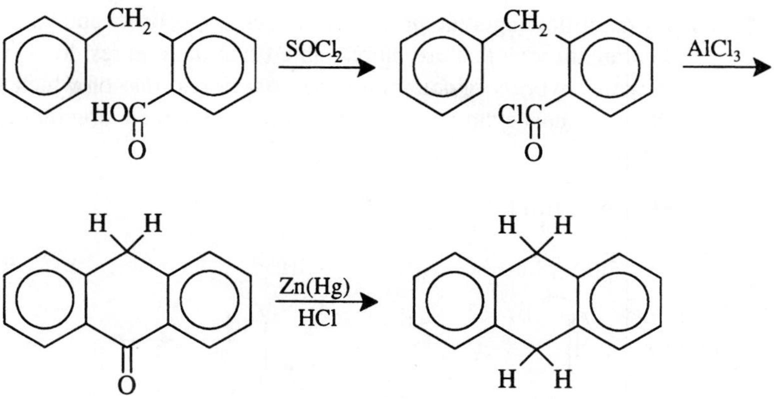

15.8 If the methyl group had no directive effect on the incoming electrophile, we would expect to obtain the products in purely statistical amounts. Since there are two ortho hydrogen atoms, two meta hydrogen atoms, and one para hydrogen, we would expect to get 40% ortho (2/5), 40% meta (2/5), and 20% para (1/5). Thus, we would expect that only 60% of the mixture of mononitrotoluenes would have the nitro group in the ortho or para position. And we would expect to obtain 40% of *m*-nitrotoluene. In actuality, we get 96% of combined *o*- and *p*-nitrotoluene and only 4% *m*-nitrotoluene. This shows the ortho-para directive effect of the methyl group.

15.9 (a)

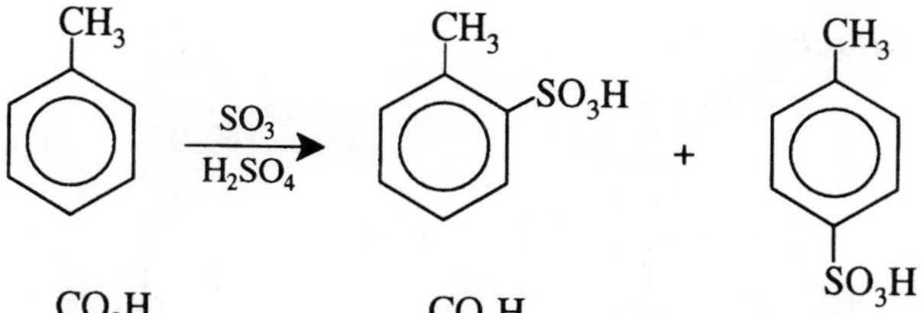

(b)

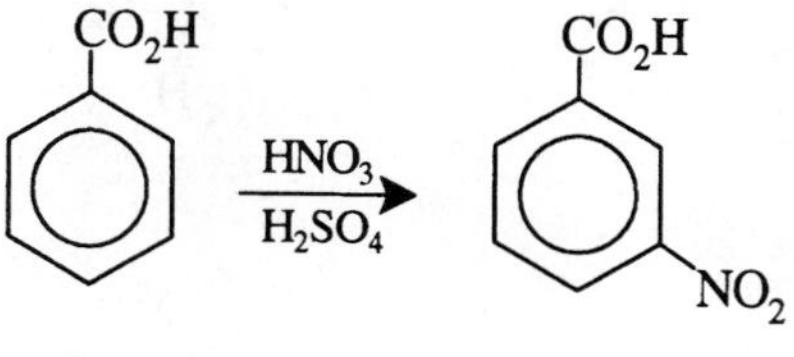

(c)

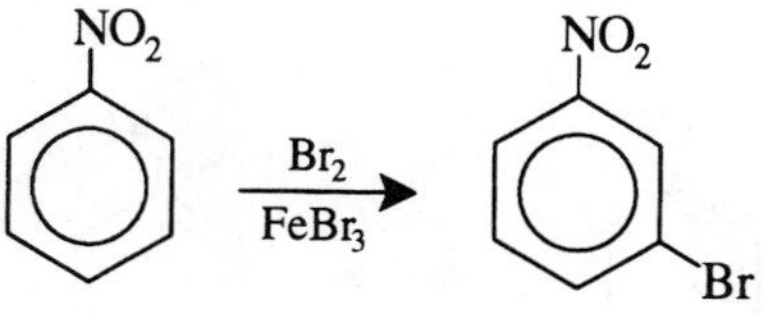

(d)

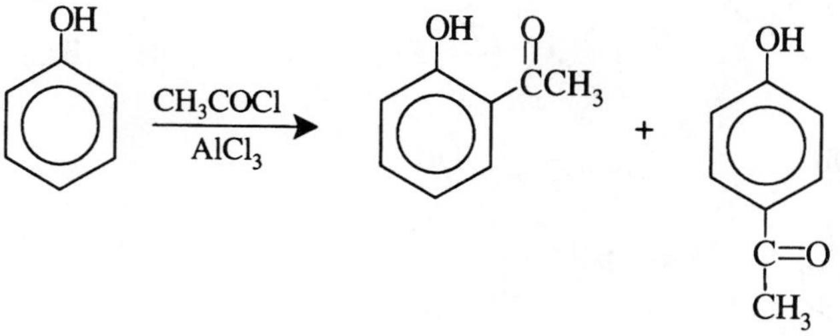

15.10 As the following structures show, attack at the ortho and para positions of phenol leads to arenium ions that are more stable (than the one resulting from meta attack) because they are hybrids of four resonance structures, one of which is relatively stable. Only three resonance structures are possible for the meta arenium ion, and none is relatively stable.

Ortho *attack*

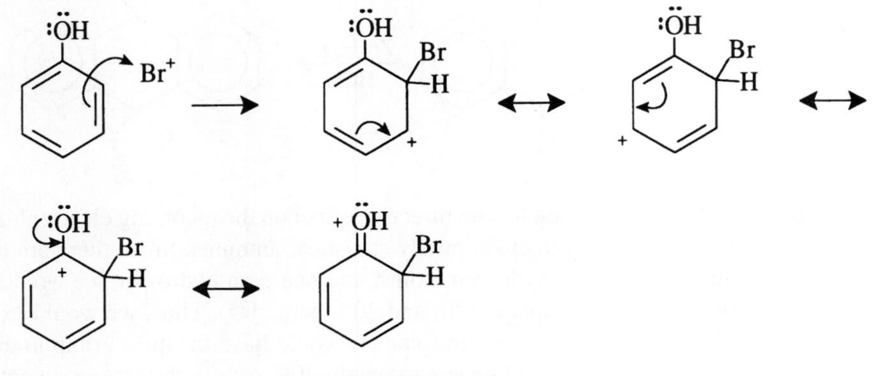

Relatively stable

Meta *attack*

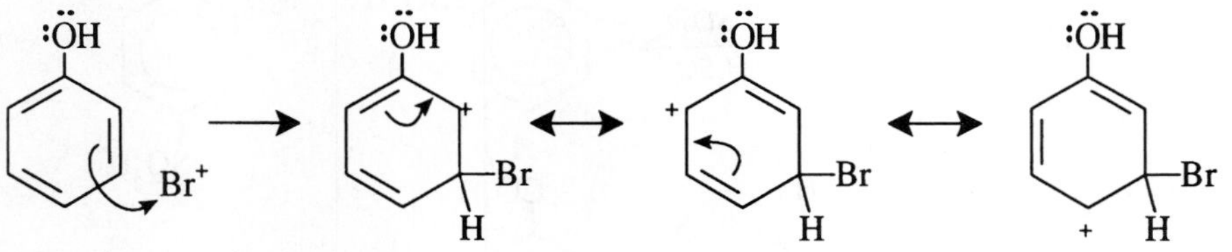

Para *attack*

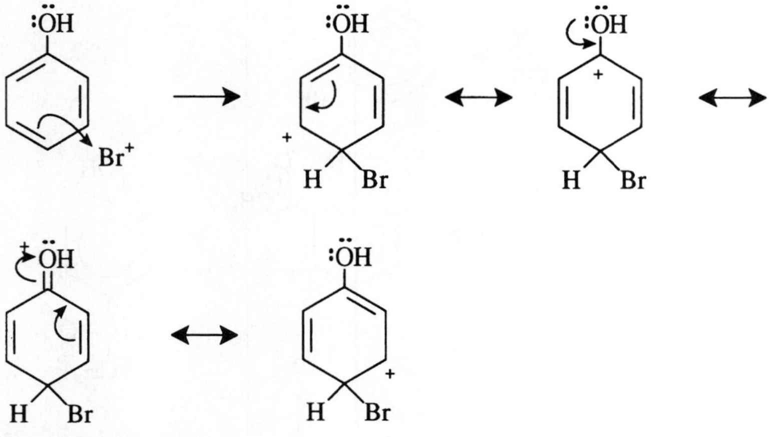

Relatively stable

15.11 (a) The atom (an oxygen atom) attached to the benzene ring has an unshared electron pair that it can donate to the arenium ions formed from ortho and para attack stabilizing them. (The arenium ions are analogous to the previous answer with a $-COCH_3$ group replacing the $-H$).

(b) Structures such as the following compete with the benzene ring for the oxygen electrons, making them less available to the benzene ring.

This effect makes the benzene ring of phenyl acetate less electron rich and, therefore, less reactive.

(c) Because the acetamido group has an unshared electron pair on the nitrogen atom that it can donate to the benzene ring it is an ortho-para director.

(d) Structures such as the following compete with the benzene ring for the nitrogen electrons, making them less available to the benzene ring.

15.12 The electron-withdrawing inductive effect of the chlorine of chloroethene makes its double bond less electron rich than that of ethene. This causes the rate of reaction of chloroethene with an electrophile (i.e., a proton) to be slower than the corresponding reaction of ethene.

When chloroethene adds a proton, the orientation is governed by a resonance effect. In theory, two carbocations can form:

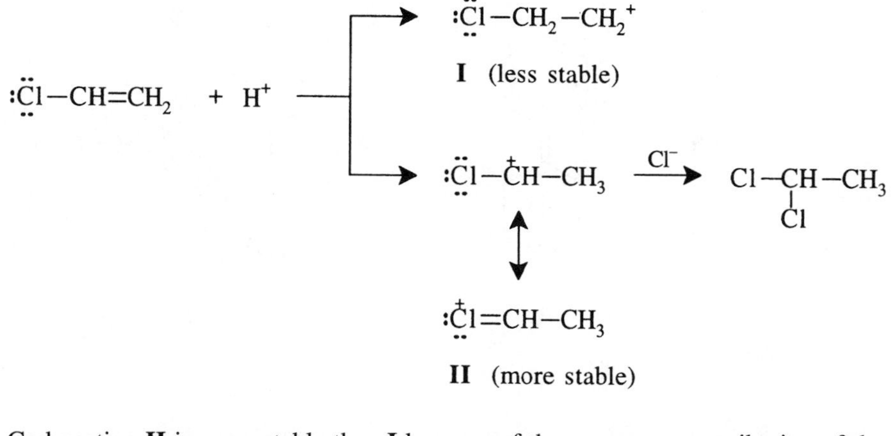

Carbocation **II** is more stable than **I** because of the resonance contribution of the extra structure just shown in which the chlorine atom donates an electron pair (see Section 15.11D).

15.13   *Ortho attack*

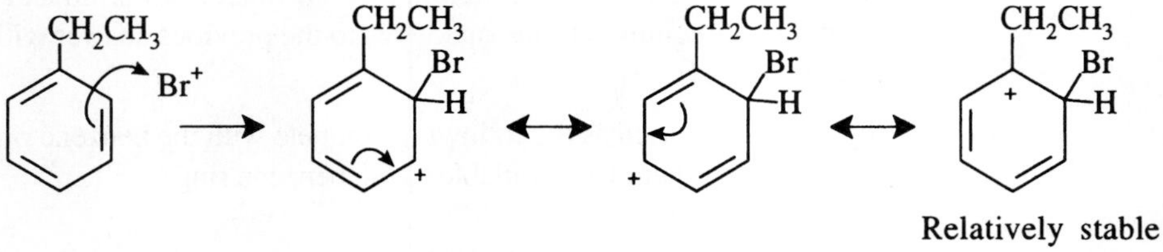

Relatively stable

Para attack

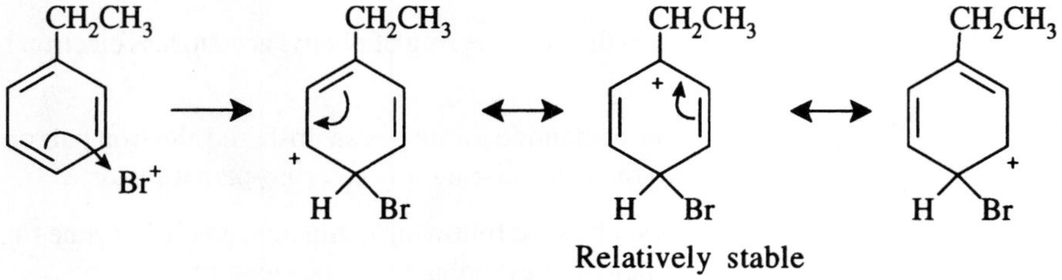

Relatively stable

15.14   The phenyl group, as the following resonance structures show, can act as an electron-releasing group and can stabilize the arenium ions formed from ortho and para attack.

Ortho attack

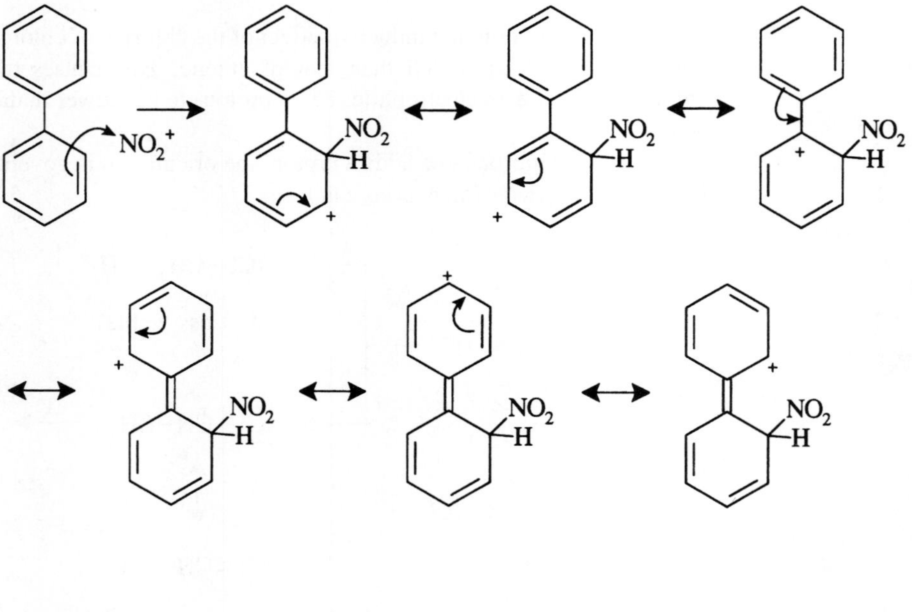

Para attack

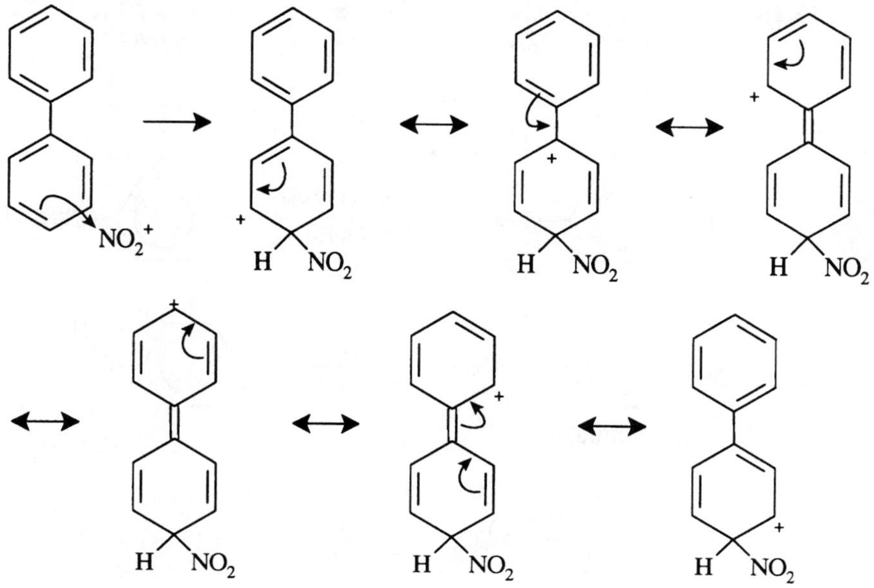

15.15

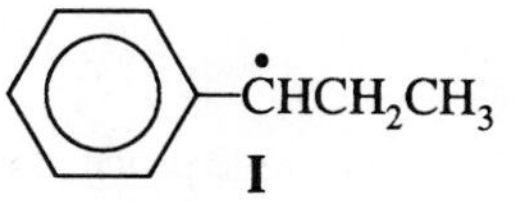

I   leads to 1-chloro-1-phenylpropane

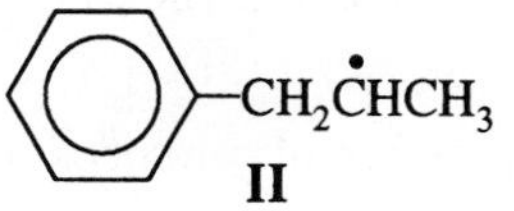

II   leads to 2-chloro-1-phenylpropane

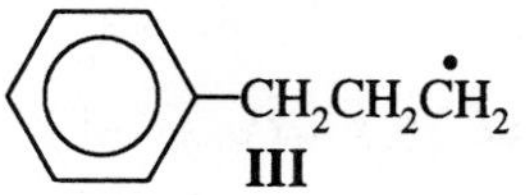

III   leads to 1-chloro-3-phenylpropane

The major product is 1-chloro-1-phenylpropane because **I** is the most stable radical. It is a benzylic radical and therefore is stabilized by resonance.

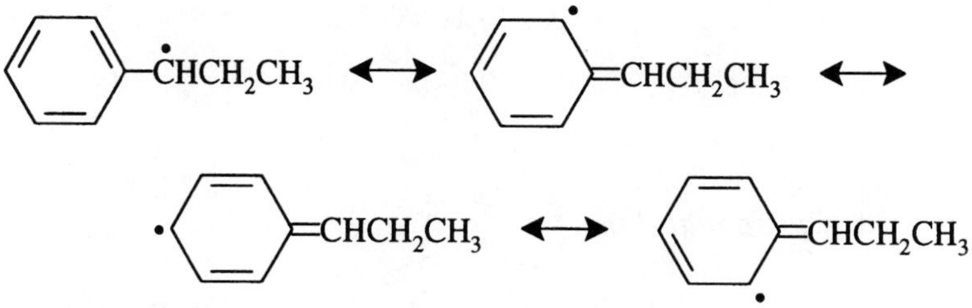

15.16 (a)

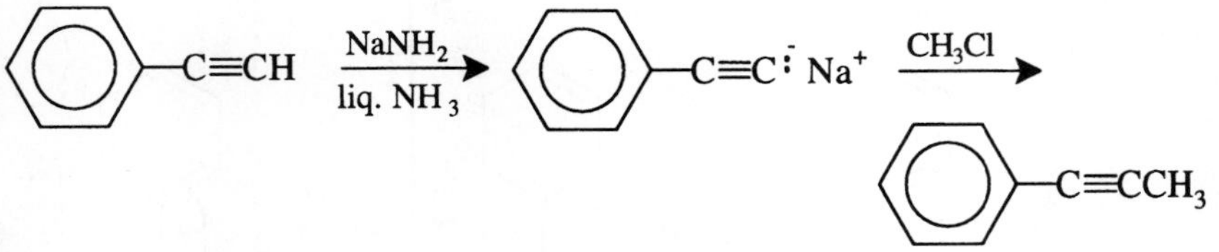

(b)

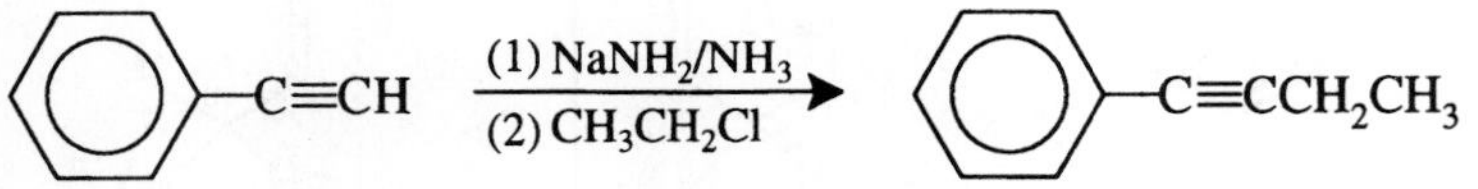

(c)

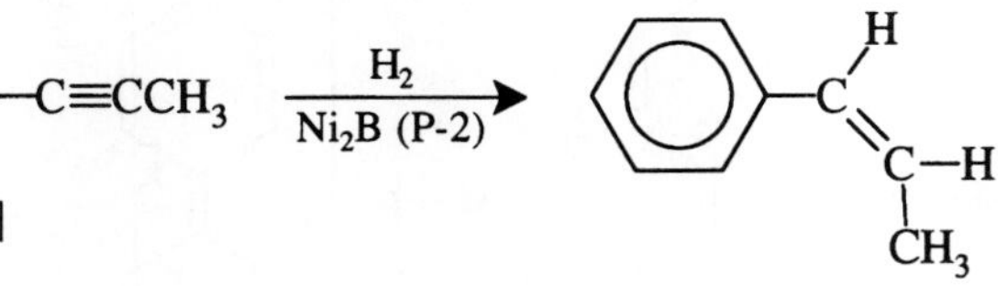

[from(a)]

(d)

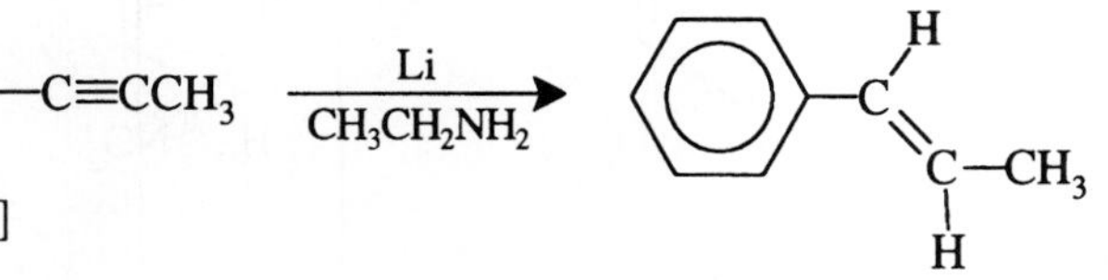

[from(a)]

15.17 The addition of hydrogen bromide to 1-phenylpropene proceeds through a benzylic radical in the presence of peroxides, and through a benzylic cation in their absence (cf., a and b as follows).

(a) Hydrogen bromide addition in the presence of peroxides.

Chain Initiation

Step 1 $R\!-\!O\!-\!O\!-\!R \longrightarrow 2\ R\!-\!O\cdot$

Step 2 $RO\cdot\ +\ H\!-\!Br \longrightarrow R\!-\!O\!-\!H\ +\ Br\cdot$

Step 3 $Br\cdot\ +\ C_6H_5CH\!=\!CHCH_3 \longrightarrow C_6H_5\overset{\cdot}{C}H\!-\!CHCH_3$
$\qquad\qquad\qquad\qquad\qquad\qquad\qquad\qquad\qquad\overset{|}{Br}$

A benzylic radical

Chain Propagation

Step 4 $C_6H_5\overset{\cdot}{C}HCHCH_3\ +\ H\!-\!Br \longrightarrow C_6H_5CH_2CHCH_3\ +\ Br\cdot$
$\qquad\qquad\qquad\overset{|}{Br}\qquad\qquad\qquad\qquad\qquad\qquad\overset{|}{Br}$

2-Bromo-1-phenylpropane

The mechanism for the addition of hydrogen bromide to 1-phenylpropene in the presence of peroxides is a chain mechanism analogous to the one we discussed when we described anti-Markovnikov addition in Section 9.9. The step that determines the orientation of the reaction is the first chain-propagating step. Bromine attacks the second carbon atom of the chain because by doing so the reaction produces a more stable benzylic radical. Had the bromine atom attacked the double bond in the opposite way, a less stable secondary radical would have been formed.

$$C_6H_5CH = CHCH_3 \quad + \quad Br \cdot \quad \xrightarrow{X} \quad C_6H_5CH - \overset{\displaystyle \cdot}{C}HCH_3$$
$$\underset{Br}{|}$$

A secondary radical

(b)   Hydrogen bromide addition in the absence of peroxides.

$$C_6H_5CH = CHCH_3 \quad + \quad HBr \quad \longrightarrow \quad C_6H_5\overset{+}{C}HCH_2CH_3 \quad + \quad Br^-$$

A benzylic cation

$$\downarrow$$

$$C_6H_5CHCH_2CH_3$$
$$\underset{Br}{|}$$

1-Bromo-1-phenylpropane

In the absence of peroxides, hydrogen bromide adds through an ionic mechanism. The step that determines the orientation in the ionic mechanism is the first, where the proton attacks the double bond to give the more stable benzylic cation. Had the proton attacked the double bond in the opposite way, a less stable secondary cation would have been formed.

$$C_6H_5CH = CHCH_3 \quad + \quad HBr \quad \xrightarrow{X} \quad C_6H_5CH - \overset{+}{C}HCH_3 \quad + \quad Br^-$$
$$\underset{H}{|}$$

A secondary cation

15.18 (a) ⬡—CHCH$_2$CH$_3$ because the more stable carbocation intermediate is the
                  |
                  Cl
benzylic carbocation, ⬡—$\overset{+}{C}$HCH$_2$CH$_3$, which then reacts with a chloride ion.

(b) ⬡—CHCH$_2$CH$_3$ because the more stable intermediate is a benzylic cation,
          |
          OH   which then reacts with H$_2$O.

15.19 (a) The first method would fail because introducing the chlorine substituent first would introduce an ortho-para directing group. Consequently, the subsequent Friedel-Crafts reaction would not then take place at the desired meta position.

The second method would fail for essentially the same reasons. Introducing the ethyl group first would introduce an ortho-para director, and subsequent ring chlorination would not take place at the desired meta position.

(b) If we introduce an acetyl group first, which we later convert to an ethyl group, we install a meta director. This allows us to put the chlorine atom in the desired position. Conversion of the acetyl group to an ethyl group is then carried out using the Clemmensen reduction.

15.20 (a) (b) (c)

15.21 (a) In concentrated base and ethanol (a relatively nonpolar solvent), the S_N2 reaction is favored. Thus, the rate depends on the concentration of both the alkyl halide and $NaOC_2H_5$. Since no carbocation is formed, the only product is

$$CH_3CH=CHCH_2OCH_2CH_3$$

(b) When the concentration of $C_2H_5O^-$ ion is small or zero, the reaction occurs through the S_N1 mechanism. The carbocation that is produced in the first step of the S_N1 mechanism is resonance hybrid.

This ion reacts with the nucleophile ($C_2H_5O^-$ or C_2H_5OH) to produce two isomeric ethers

$$CH_3CH=CHCH_2\text{-}OCH_2CH_3 \quad \text{and} \quad CH_3\overset{OCH_2CH_3}{\underset{|}{C}}HCH=CH_2$$

(c)   In the presence of water, the first step of the S_N1 reaction occurs. The reverse of this reaction produces two compounds because the positive charge on the carbocation is distributed over carbon atoms one and three

$$\left[\begin{array}{c} CH_3CH=CH\overset{+}{C}H_2 \\ \updownarrow \\ CH_3\overset{+}{C}H-CH=CH_2 \end{array}\right] + Cl^- \longrightarrow \begin{array}{c} CH_3CH=CHCH_2Cl \\ + \\ \underset{Cl}{\overset{}{C}l} \\ CH_3\overset{|}{C}HCH=CH_2 \end{array}$$

15.22 (a)   The carbocation that is produced in the S_N1 reaction is exceptionally stable because one resonance contributor is not only allylic but also tertiary.

$$\underset{CH_3C=CHCH_2Cl}{\overset{CH_3}{|}} \underset{\overrightarrow{\longleftarrow}}{\overset{S_N1}{}} \left[\underset{CH_3C=CH\overset{+}{C}H_2}{\overset{CH_3}{|}} \longleftrightarrow \underset{CH_3-\underset{+}{C}-CH=CH_2}{\overset{CH_3}{|}}\right]$$

A 3° allylic carbocation

(b) $\underset{CH_3C=CHCH_2OH}{\overset{CH_3}{|}}$ + $\underset{\underset{OH}{|}}{\overset{CH_3}{\underset{}{\overset{|}{CH_3CCH=CH_2}}}}$

15.23 Compounds that undergo reactions by an S_N1 path must be capable of forming relatively stable carbocations. Primary halides of the type $ROCH_2X$ form carbocations that are stabilized by resonance:

$$R-\overset{..}{\underset{..}{O}}-\overset{+}{C}H_2 \quad \longleftrightarrow \quad R-\overset{+}{\underset{..}{O}}=CH_2$$

15.24 The relative rates are in the order of the relative stabilities of the carbocations:

$$C_6H_5\overset{+}{C}H_2 \quad < \quad C_6H_5\overset{+}{C}HCH_3 \quad < \quad (C_6H_5)_2\overset{+}{C}H \quad < \quad (C_6H_5)_3\overset{+}{C}$$

The solvolysis reaction involves a carbocation intermediate.

15.25

15.26 (a)

+

(b)

+

(c)

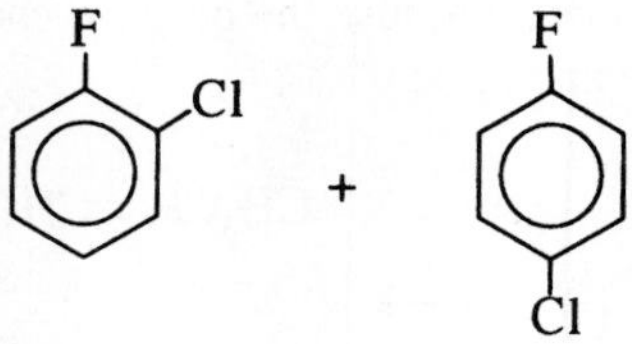

+

(d)

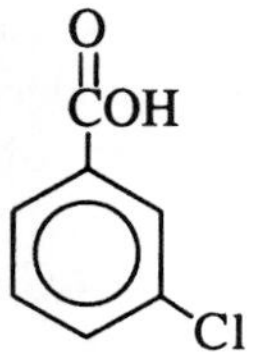

(e)

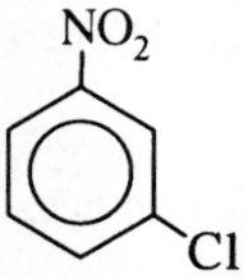

(f)

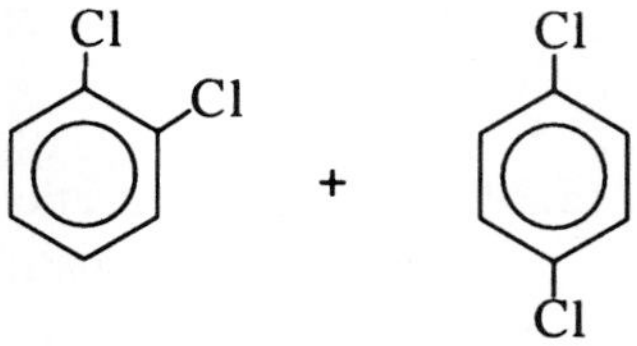

+

(g)

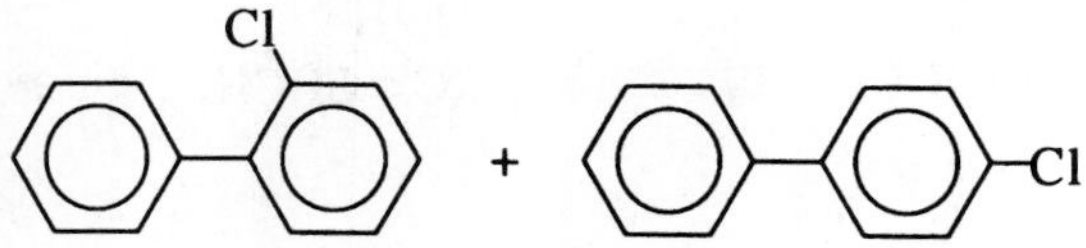

+

(h) 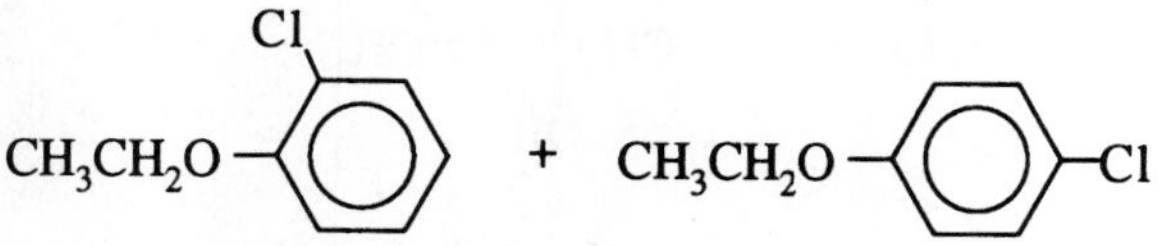 CH$_3$CH$_2$O—

+ CH$_3$CH$_2$O—

—Cl

15.27 (a)

+

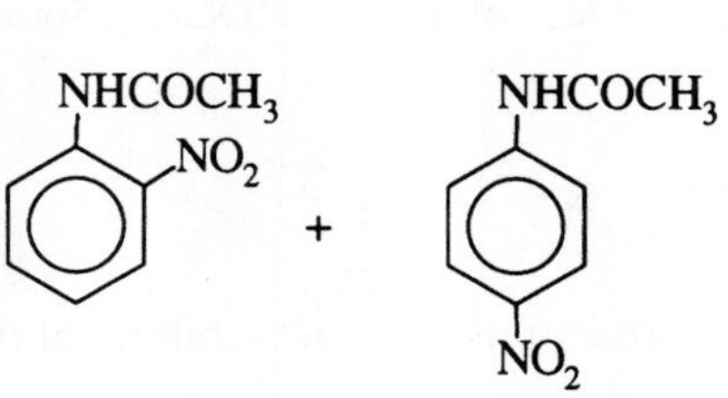

(mainly)

(b)

+

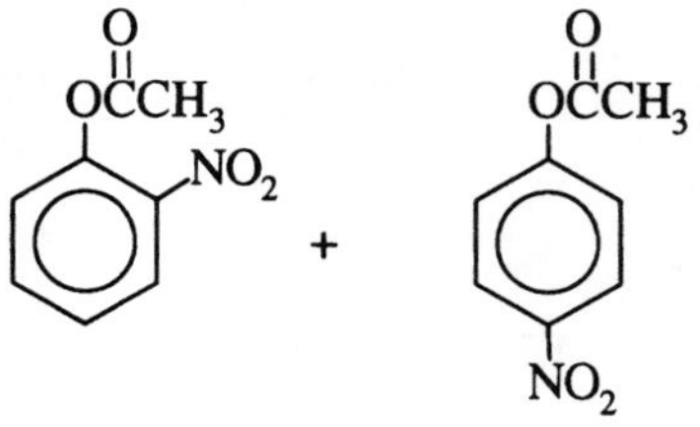

(c)

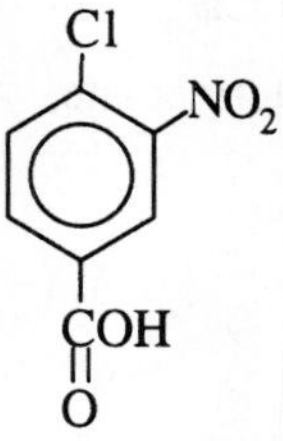

(d)

+

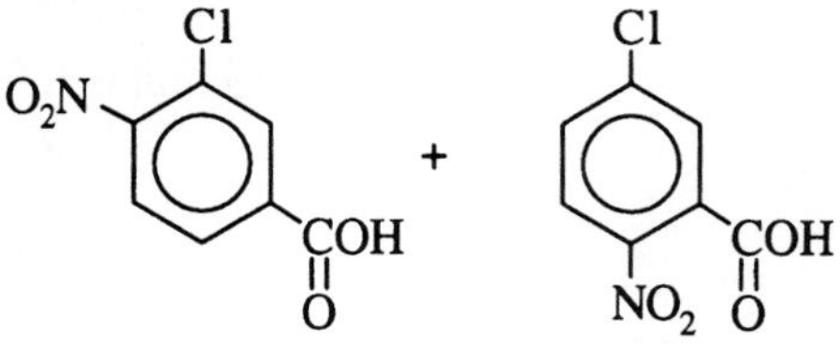

(e)

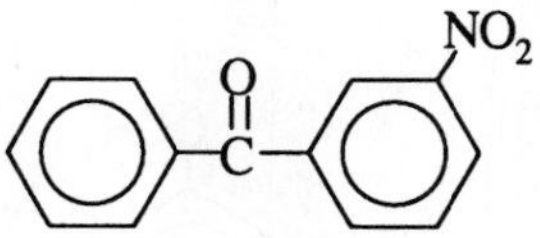

15.28 (a)

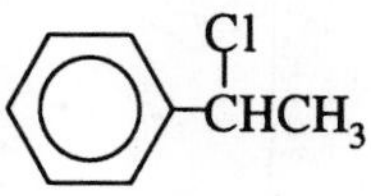

(b)

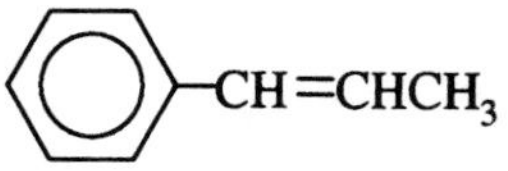

(c)

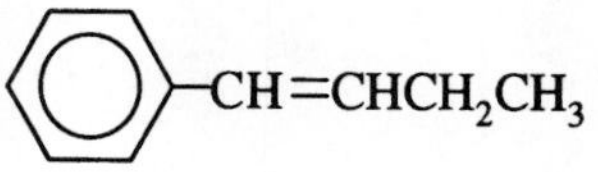

(d)

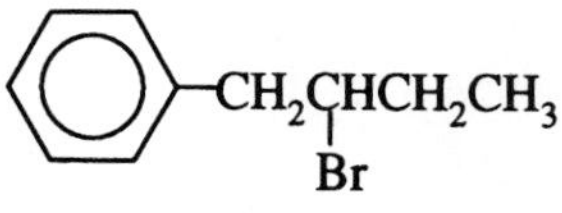

(e) 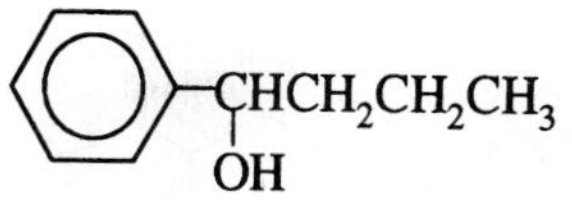 (f)

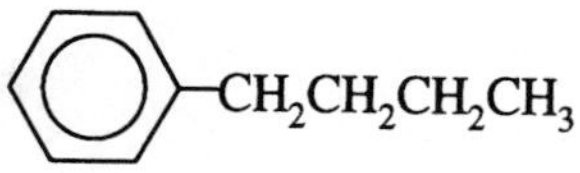

(g)

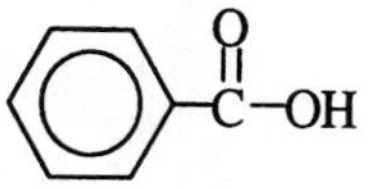

15.29 (a)

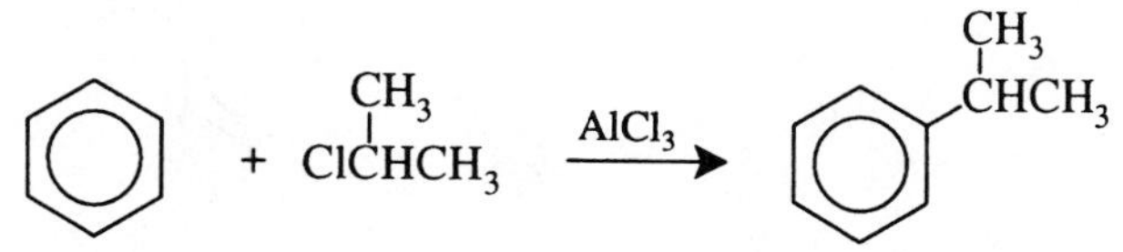

(b)

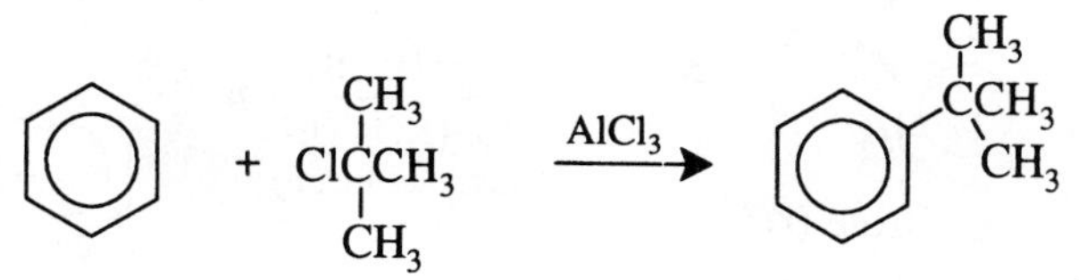

(c)

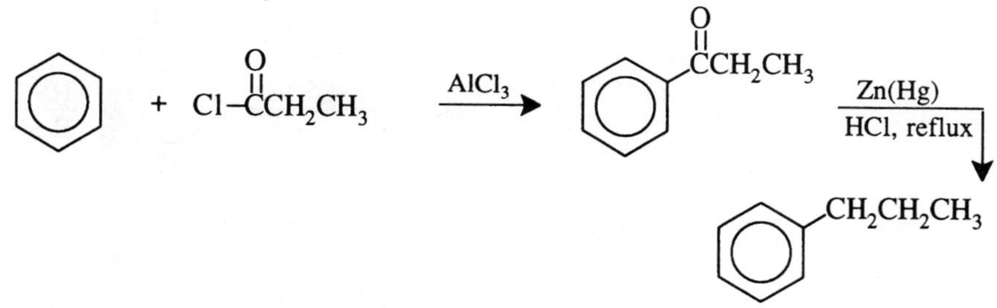

(*Note:* The use of Cl-CH$_2$CH$_2$CH$_3$ in a Friedel-Crafts synthesis gives mainly the rearranged product, isopropylbenzene.)

(d)

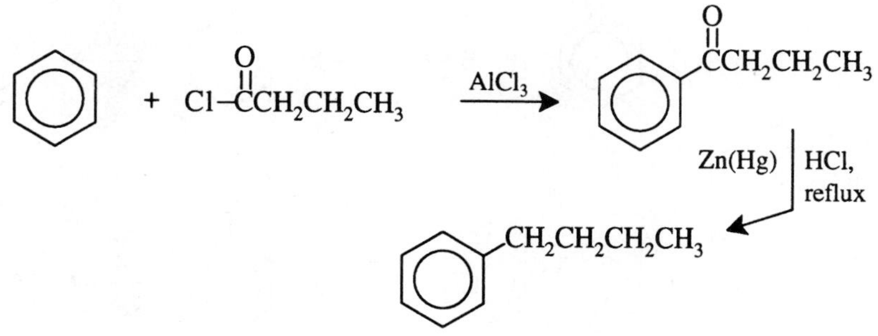

(e)

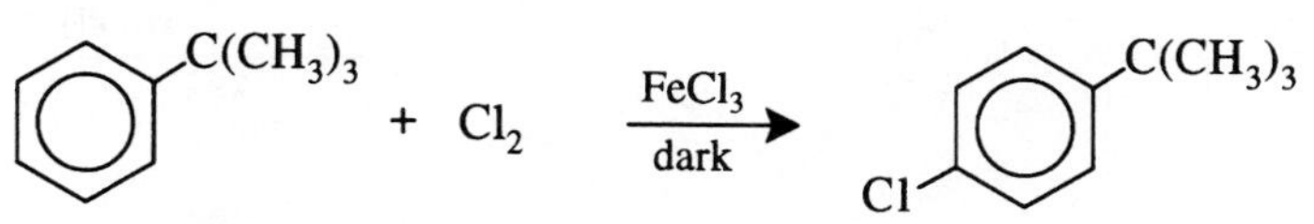

[from (b)]

(f)

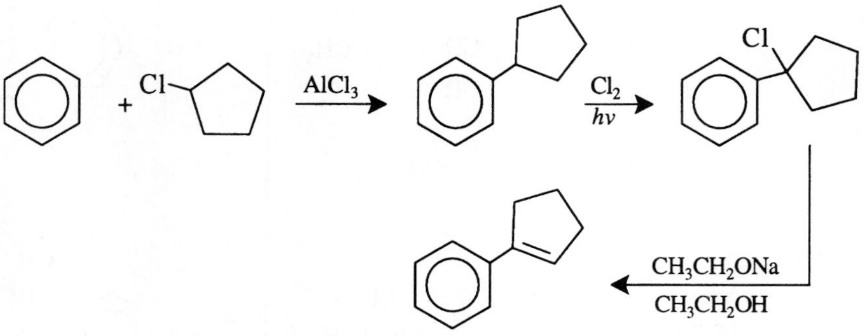

(g) [from (f)]   (1) THF: BH₃ / (2) H₂O₂, OH⁻ / (syn addition)

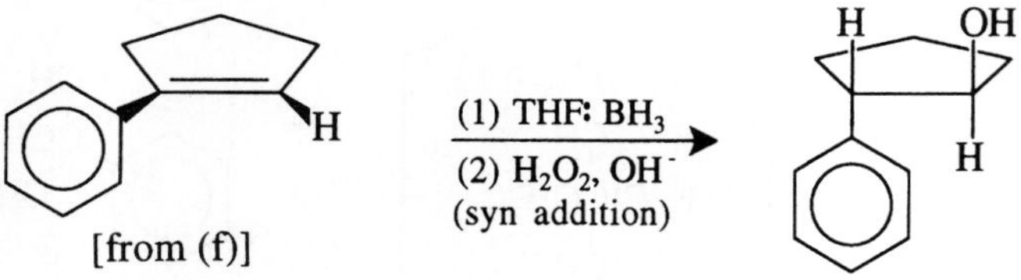

(h)   + HNO₃   H₂SO₄ →   →HNO₃/H₂SO₄

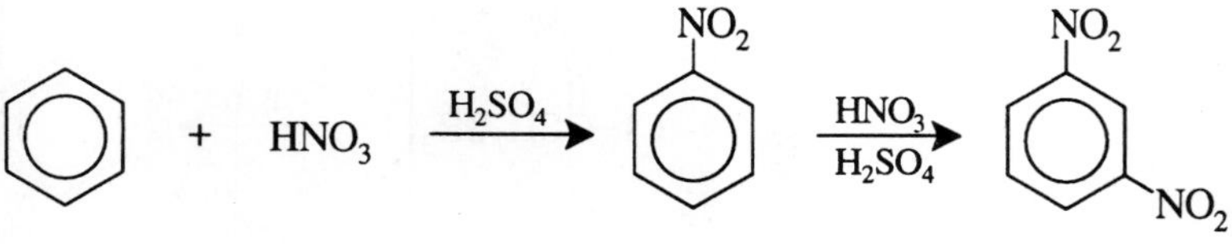

(i)   + Br₂   FeBr₃ →

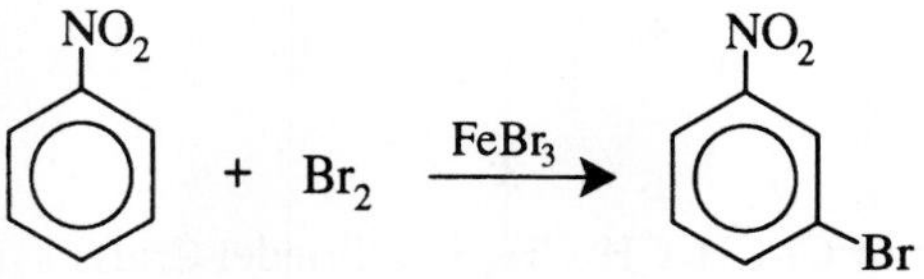

(j)   + Br₂   FeBr₃ →   HNO₃/H₂SO₄ →   + ortho

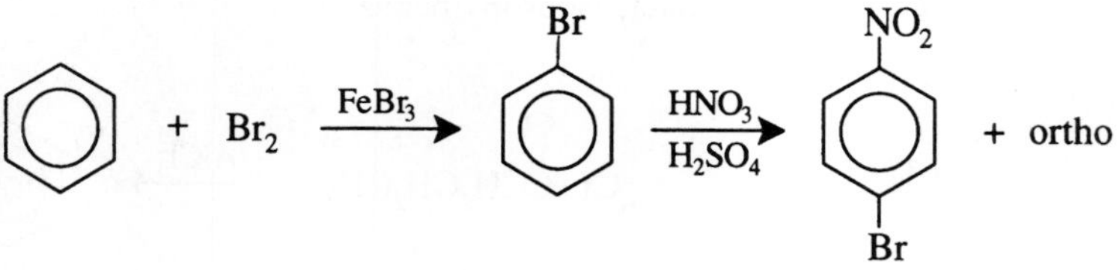

(k)   + Cl₂   FeCl₃ →   SO₃/H₂SO₄ →   (+) (separate)   H₂O/H⁺, heat

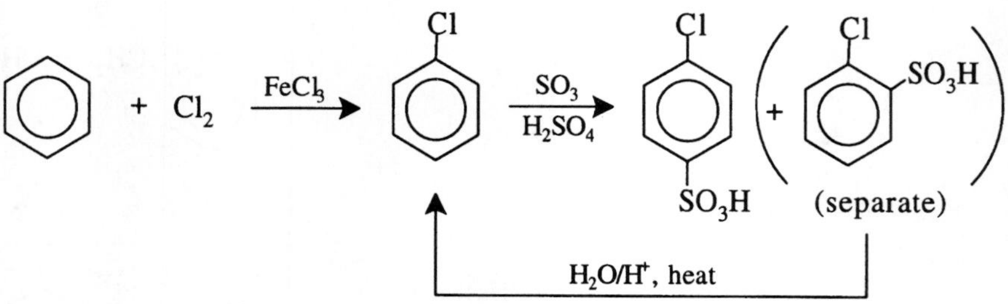

(1)

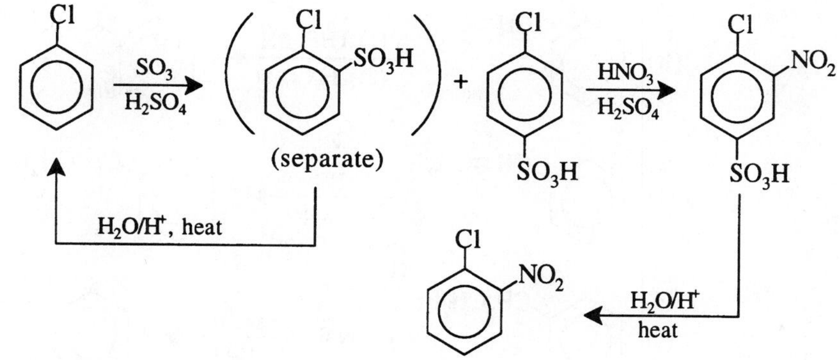

(m)

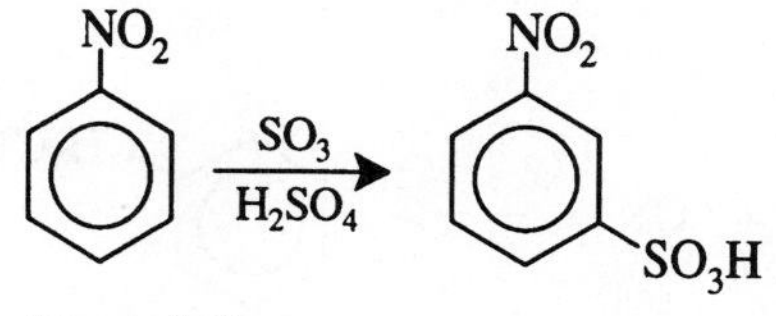

[from (h)]

15.30 (a)

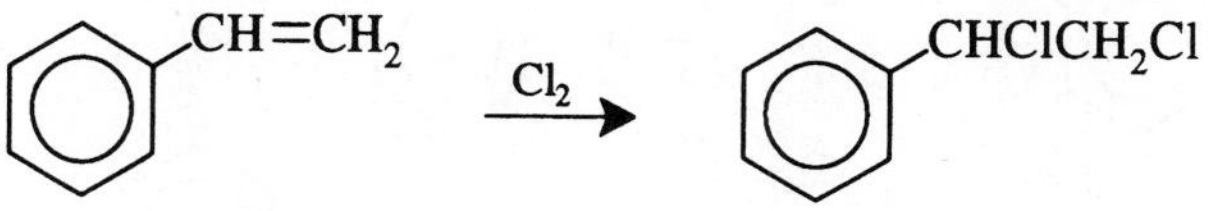

$\xrightarrow{Cl_2}$

(b)

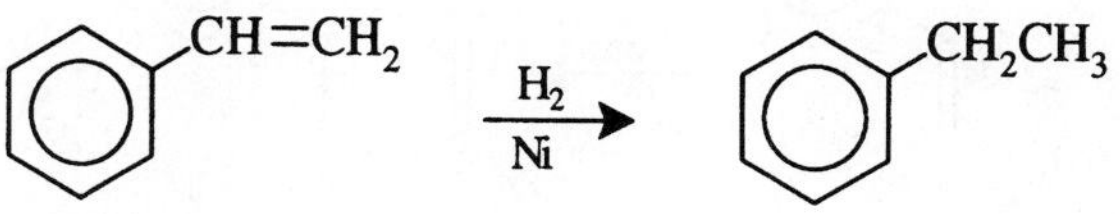

$\xrightarrow{H_2 \atop Ni}$

(c)

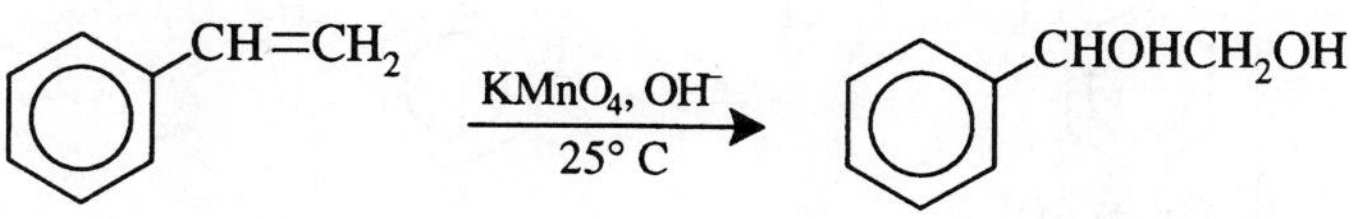

$\xrightarrow[25° C]{KMnO_4, OH^-}$

(d)

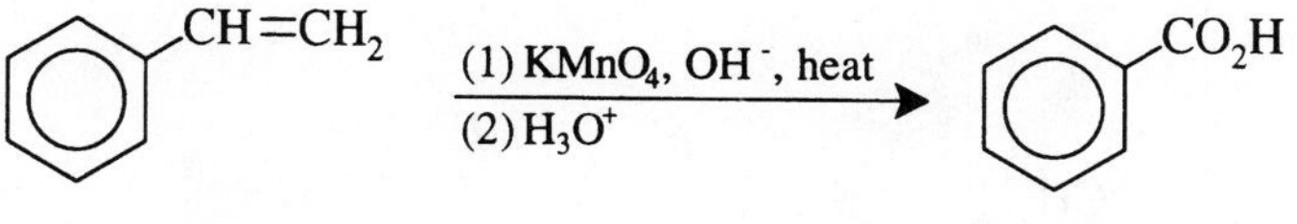

$\xrightarrow[(2) H_3O^+]{(1) KMnO_4, OH^-, heat}$

(e)

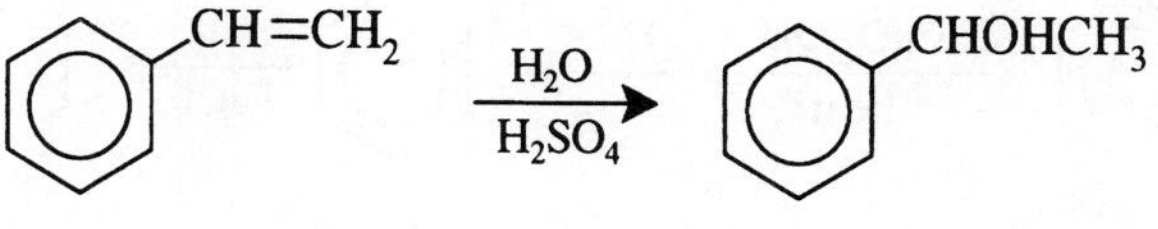

$\xrightarrow[H_2SO_4]{H_2O}$

(f)

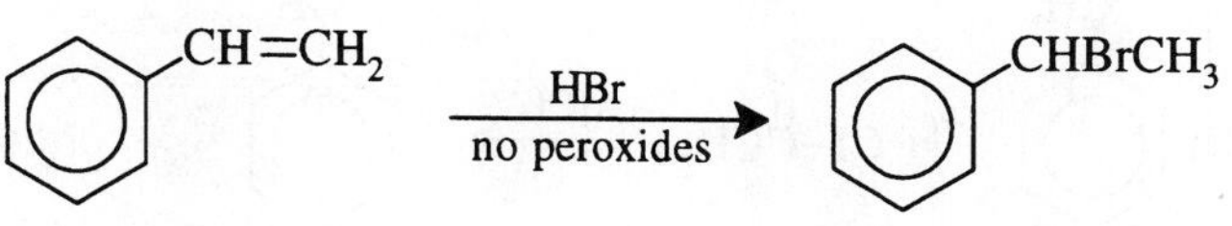

$\xrightarrow[\text{no peroxides}]{HBr}$

(g)

$\xrightarrow[(2) H_2O_2, OH^-]{(1) THF: BH_3}$

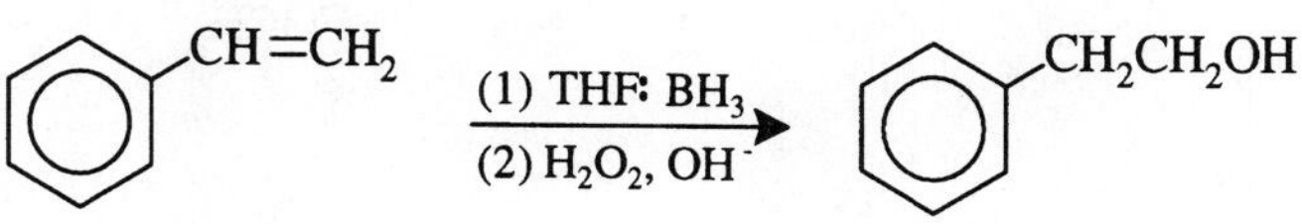

(h)

$\underset{C_6H_5}{\bigcirc}$—CH=CH$_2$ $\xrightarrow[\text{(2) CH}_3\text{CO}_2\text{D}]{\text{(1) THF: BH}_3}$ $\bigcirc$—CH$_2$CH$_2$D

(i)

$\bigcirc$—CH=CH$_2$ $\xrightarrow[\text{peroxides}]{\text{HBr}}$ $\bigcirc$—CH$_2$CH$_2$Br

(j)

$\bigcirc$—CH$_2$CH$_2$Br + NaI $\xrightarrow[\text{H}_2\text{O}]{\text{acetone}}$ $\bigcirc$—CH$_2$CH$_2$I

[from (i)]

(k)

$\bigcirc$—CH$_2$CH$_2$Br + CN$^-$ $\longrightarrow$ $\bigcirc$—CH$_2$CH$_2$CN

[from (i)]

(l)

$\bigcirc$—CH=CH$_2$ $\xrightarrow[\text{Ni}]{\text{D}_2}$ $\bigcirc$—CHDCH$_2$D

(m)

$\bigcirc$ (butadiene) + $\bigcirc$—CH=CH$_2$ $\xrightarrow{\text{heat}}$ $\bigcirc$ (cyclohexene-phenyl) $\xrightarrow[\text{Ni}]{\text{H}_2}$ $\bigcirc$ (cyclohexyl-phenyl)

(n)

$\bigcirc$—CH$_2$CH$_2$OH $\xrightarrow{\text{Na}}$ $\bigcirc$—CH$_2$CH$_2$ONa $\xrightarrow{\text{CH}_3\text{I}}$

[from (g)]

$\bigcirc$—CH$_2$CH$_2$OCH$_3$

15.31 (a)

$\underset{\text{CH}_3}{\bigcirc}$ $\xrightarrow[\text{heat}]{\text{KMnO}_4, \text{OH}^-}$ $\xrightarrow{\text{H}_3\text{O}^+}$ $\underset{\text{CO}_2\text{H}}{\bigcirc}$ $\xrightarrow[\text{FeCl}_3]{\text{Cl}_2}$ $\underset{\text{Cl}}{\overset{\text{CO}_2\text{H}}{\bigcirc}}$

(b)

$\underset{\text{CH}_3}{\bigcirc}$ + Cl–$\overset{\overset{\text{O}}{\|}}{\text{C}}CH_3$ $\xrightarrow{\text{AlCl}_3}$ $\underset{\text{COCH}_3}{\overset{\text{CH}_3}{\bigcirc}}$ + ortho

(c)

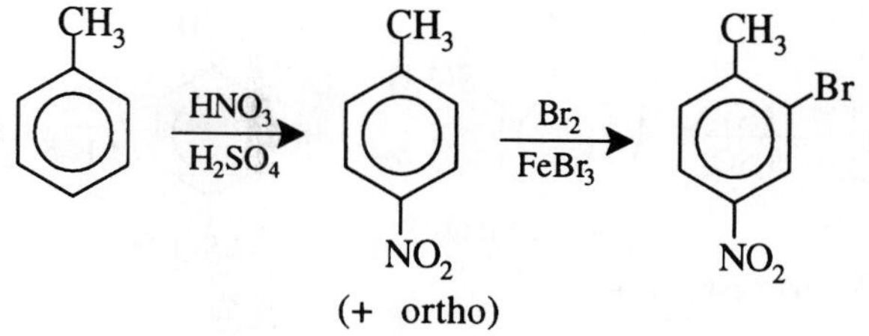

$$\xrightarrow[\text{H}_2\text{SO}_4]{\text{HNO}_3}$$

(+ ortho)

$$\xrightarrow[\text{FeBr}_3]{\text{Br}_2}$$

(d)

$$\xrightarrow[\text{FeBr}_3]{\text{Br}_2}$$

(+ ortho)

$$\xrightarrow[\text{heat}]{\text{KMnO}_4, \text{OH}^-}$$ $$\xrightarrow{\text{H}_3\text{O}^+}$$

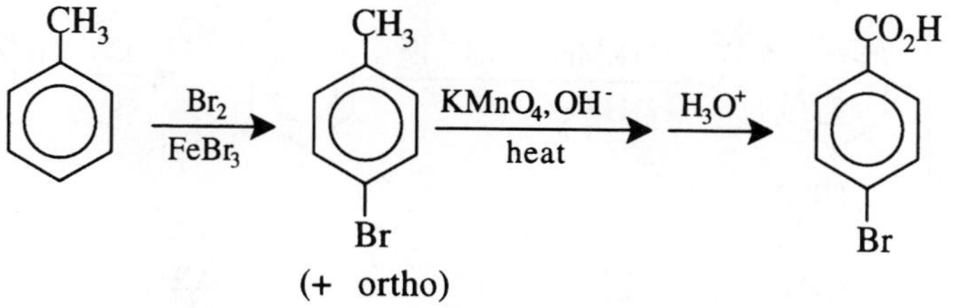

(e)

$$\xrightarrow[\text{light}]{\text{Cl}_2 \text{ (excess)}}$$

$$\xrightarrow[\text{FeCl}_3]{\text{Cl}_2}$$

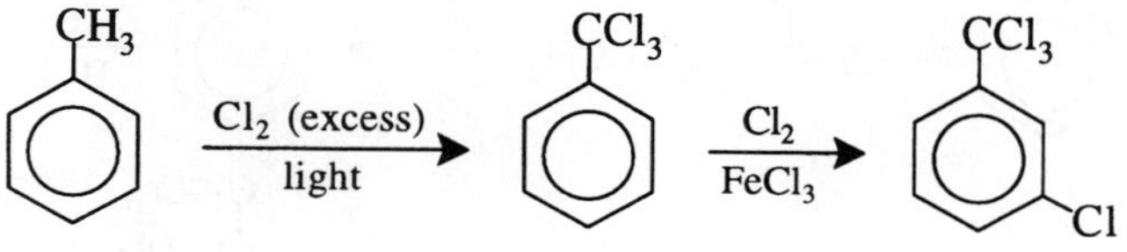

(f)   + Cl–CHCH$_3$ with CH$_3$

$$\xrightarrow{\text{AlCl}_3}$$ + ortho

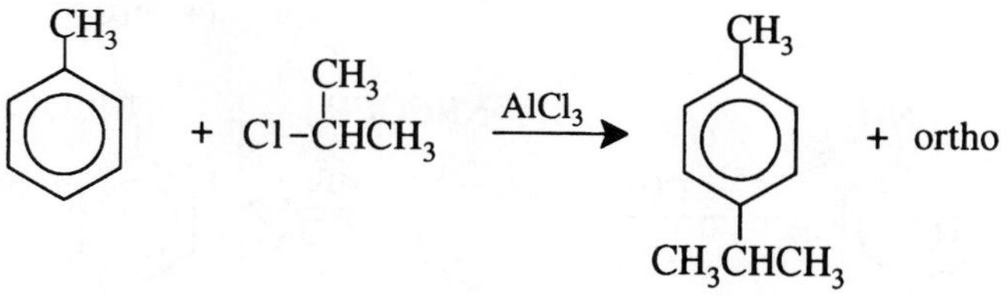

(g)   + Cl cyclohexane

$$\xrightarrow{\text{AlCl}_3}$$ CH$_3$— + ortho

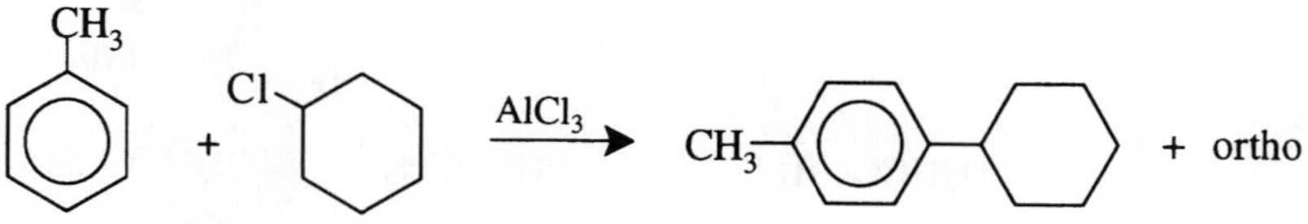

(h)

$$\xrightarrow[\text{H}_2\text{SO}_4]{\text{HNO}_3 \text{ (excess)}}$$

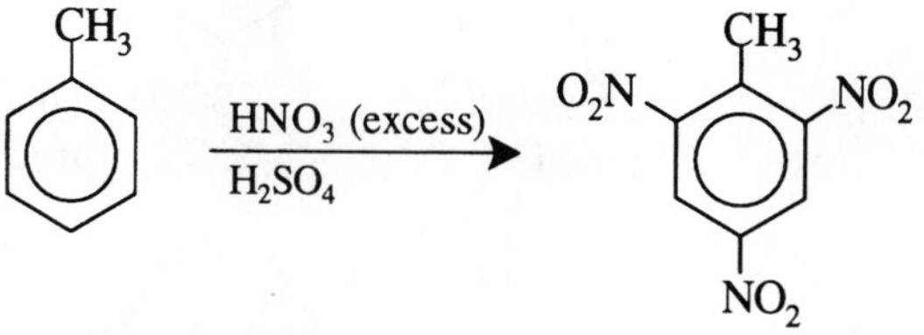

(i)

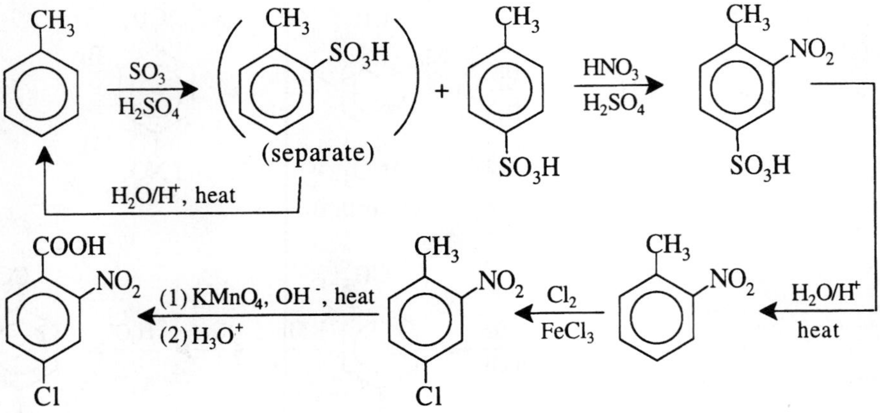

(j)

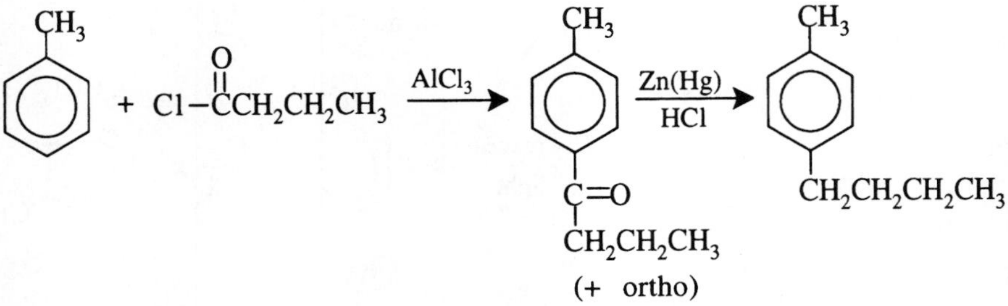

15.32 (a)

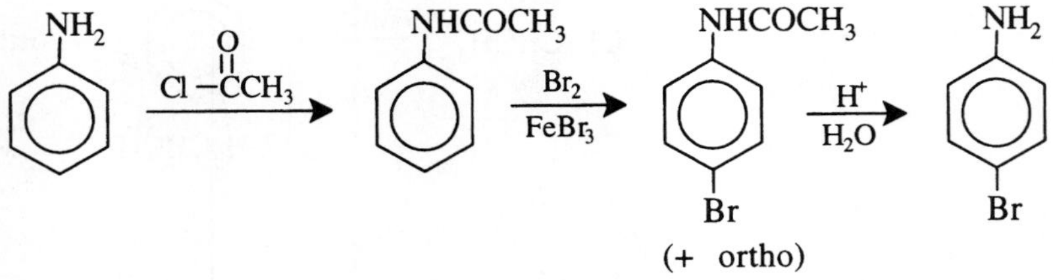

(+ ortho)

(b) 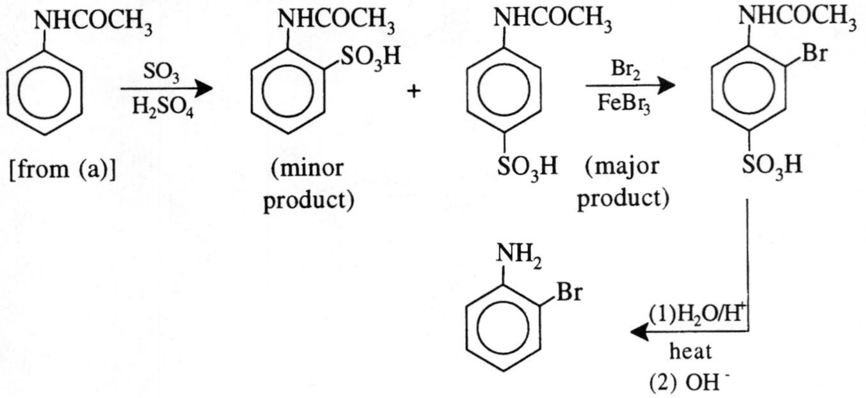 [from (a)]   (minor product)   (major product)

(c)

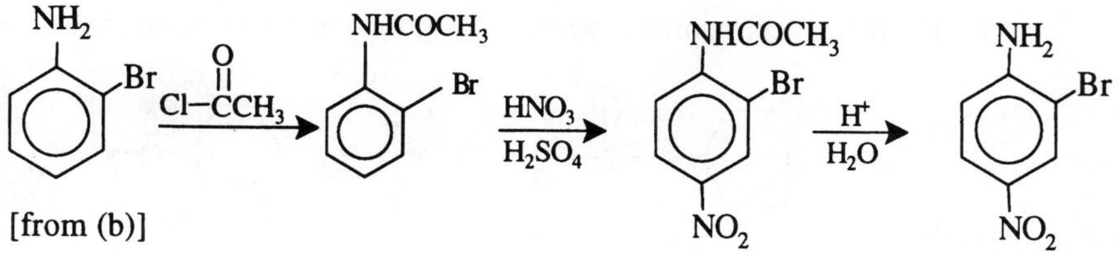

[from (b)]

(d)

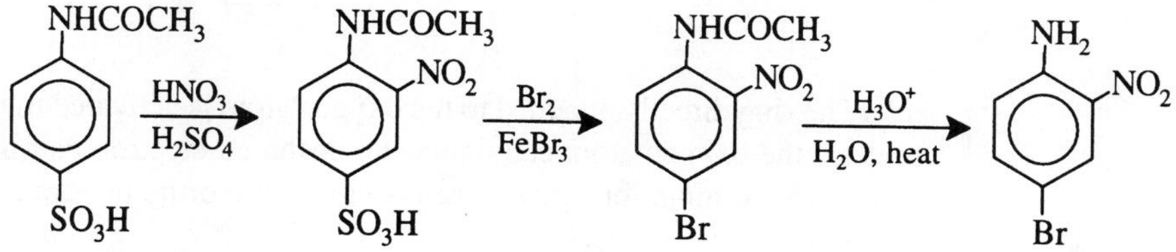

[from (b)]

(e)

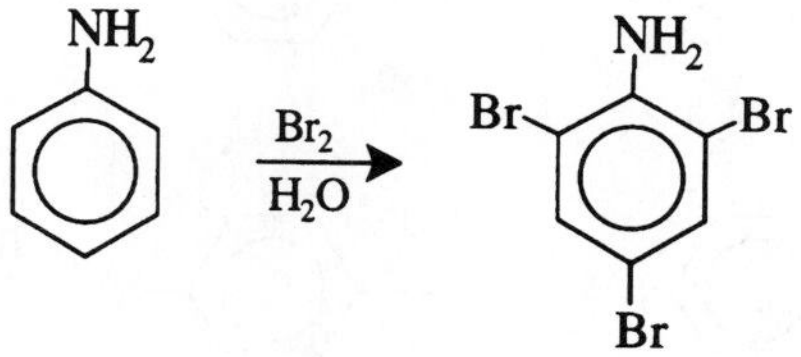

15.33 (a) Step (2) will fail because a Friedel-Crafts reaction will not take place on a ring that bears an $-NO_2$ group (or any meta director).

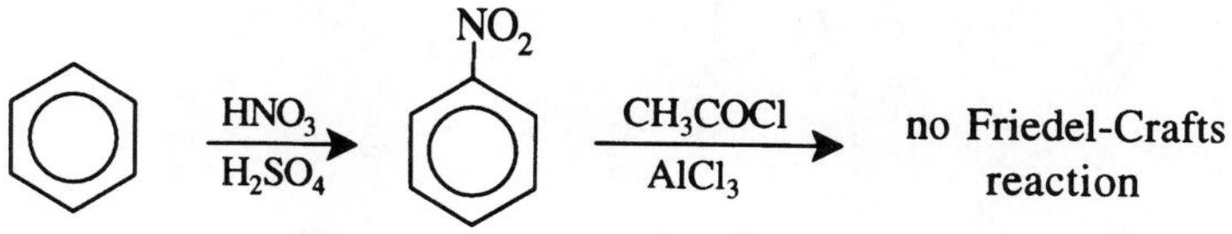

(b) The last step will first brominate the double bond.

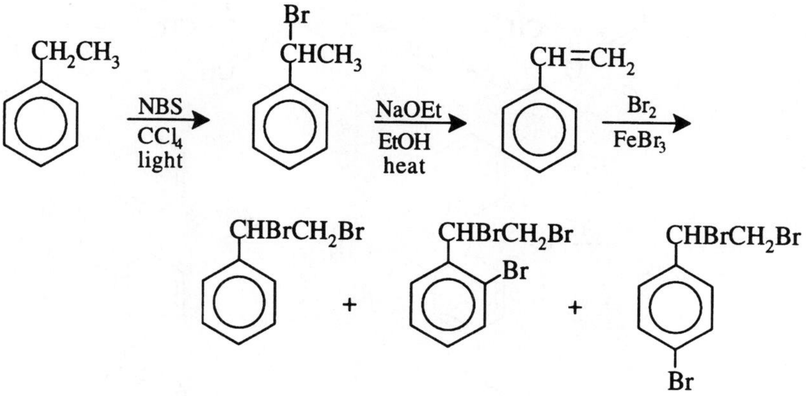

15.34 (a) Electrophilic aromatic substitution will take place as follows:

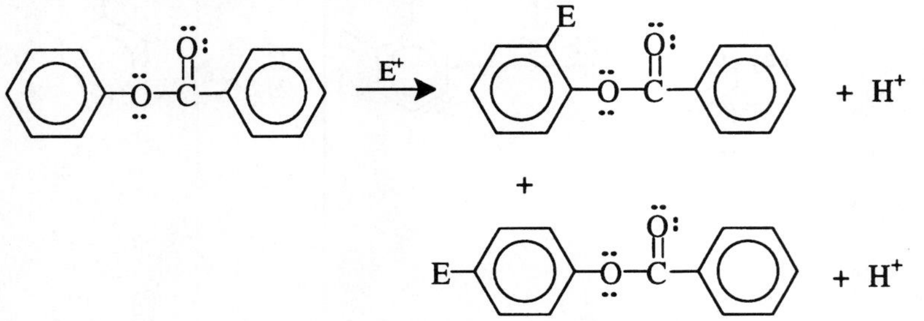

(b) The ring directly attached to the oxygen atom is activated toward electrophilic attack because the oxygen atom can donate an unshared electron pair to it and stabilize the intermediate arenium ion when attack occurs at the ortho or para position.

15.35 (a)

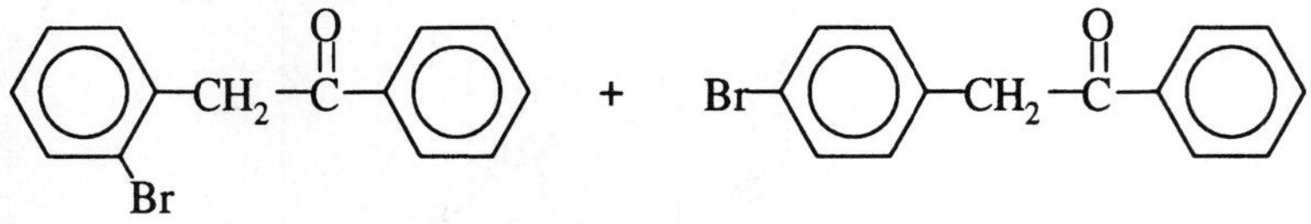

(b)

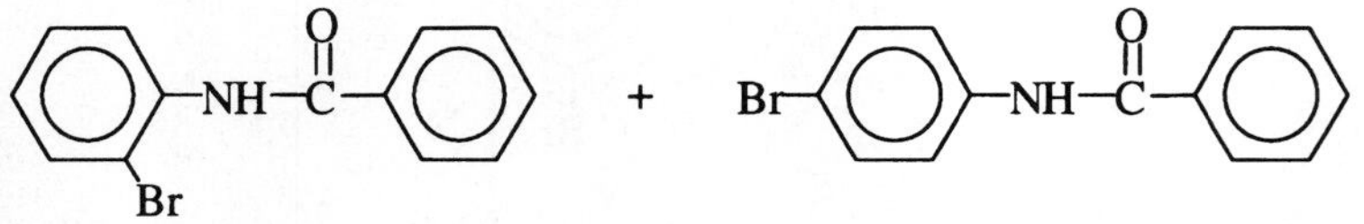

(c)

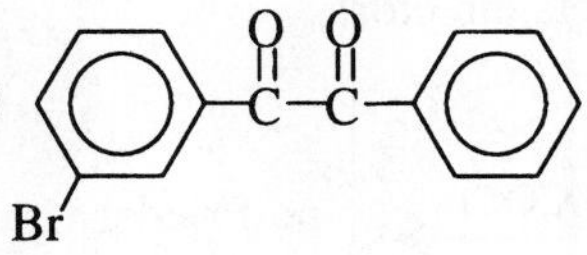

15.36

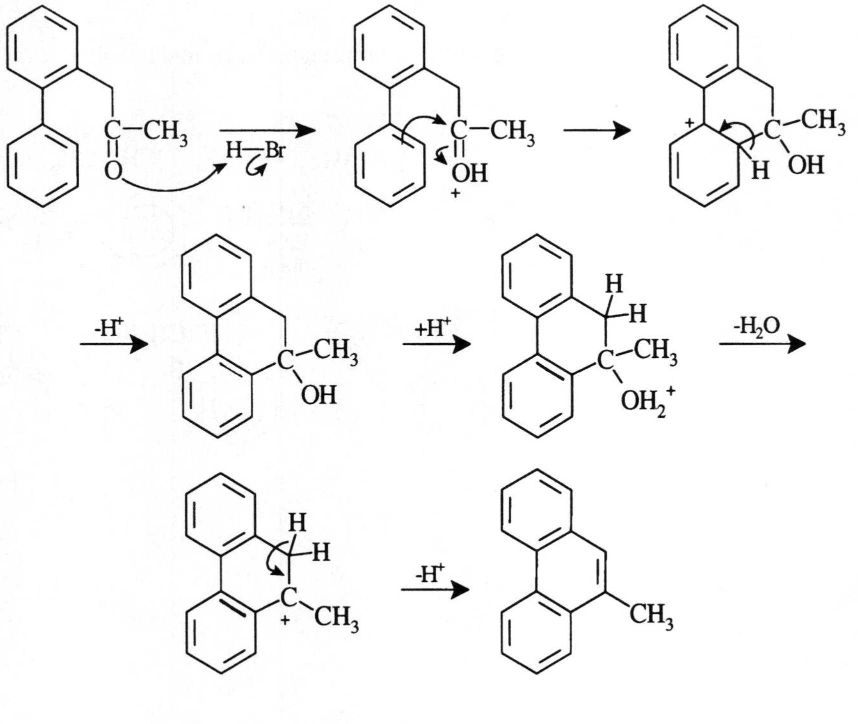

15.37 This problem serves as another illustration of the use of a sulfonic acid group as a blocking group in a synthetic sequence. Here we are able to bring about nitration between two meta substituents.

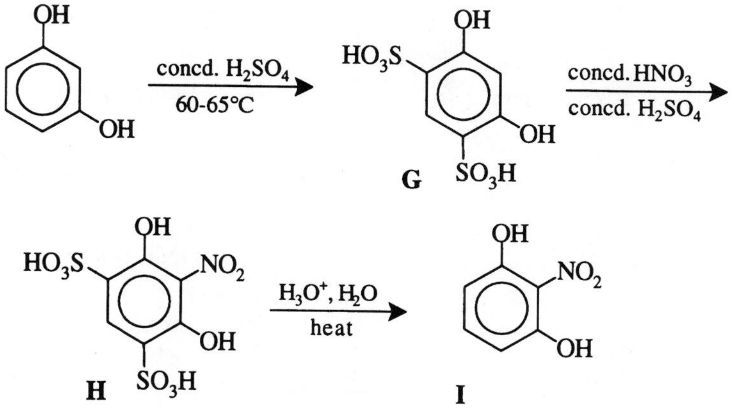

15.38

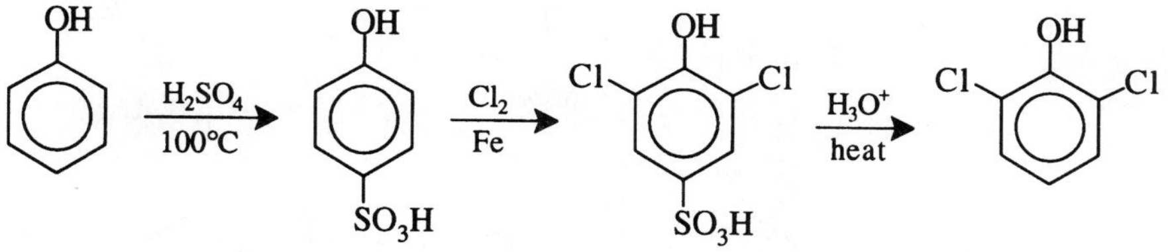

15.39 (a) (1) $C_6H_5CH=CH-CH=CH_2$ $\xrightarrow{H^+}$ $C_6H_5CH=CH-\overset{+}{C}H-CH_3$

$\updownarrow$

$C_6H_5\overset{+}{C}H-CH=CH-CH_3$

$C_6H_5\overset{\delta^+}{CH} \text{---} CH \text{---} \overset{\delta^+}{CH}-CH_3$

(2) $C_6H_5\overset{\delta^+}{CH} \text{---} CH \text{---} \overset{\delta^+}{CH}-CH_3$ $\xrightarrow{X^-}$ $C_6H_5CH=CH-\underset{X}{CH}-CH_3$

(b) 1,2 Addition.

(c) Yes. The carbocation given in (a) is a hybrid of *secondary allylic and benzylic* contributors and is therefore more stable than any other possibility; for example,

$C_6H_5CH=CH-CH=CH_2$ $\xrightarrow{H^+}$ $C_6H_5CH_2-\overset{+}{C}H-CH=CH_2$

$\updownarrow$

$C_6H_5CH_2-CH=CH-\overset{+}{C}H_2$

A hybrid of allylic contributors only

(d) Since the reaction produces only *the more stable isomer*—that is, the one in which the double bond is conjugated with the benzene ring—the reaction is likely to be under equilibrium control:

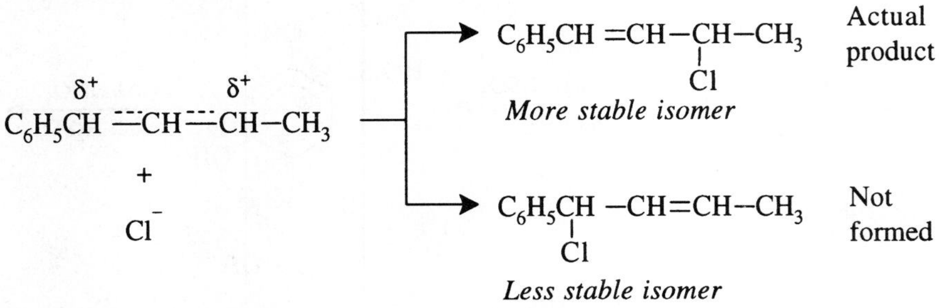
δ+ δ+
$C_6H_5CH \overset{...}{=} CH \overset{...}{=} CH{-}CH_3$
+
Cl^-
$C_6H_5CH{=}CH{-}CH(Cl){-}CH_3$
More stable isomer
Actual product
$C_6H_5CH(Cl){-}CH{=}CH{-}CH_3$
Less stable isomer
Not formed

15.40

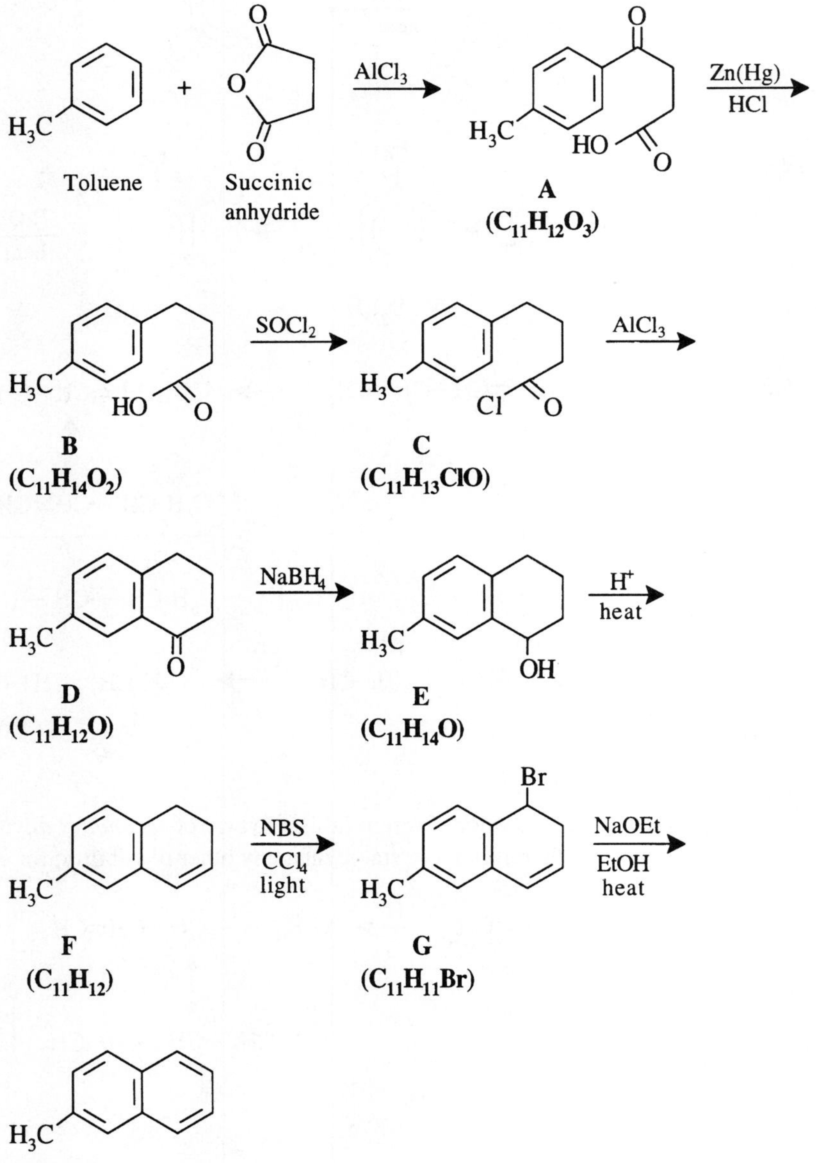
Toluene
Succinic anhydride
$AlCl_3$
A
$(C_{11}H_{12}O_3)$
Zn(Hg)
HCl
B
$(C_{11}H_{14}O_2)$
$SOCl_2$
C
$(C_{11}H_{13}ClO)$
$AlCl_3$
D
$(C_{11}H_{12}O)$
$NaBH_4$
E
$(C_{11}H_{14}O)$
H^+
heat
F
$(C_{11}H_{12})$
NBS
CCl_4
light
G
$(C_{11}H_{11}Br)$
NaOEt
EtOH
heat

2-Methylnaphthalene

15.41

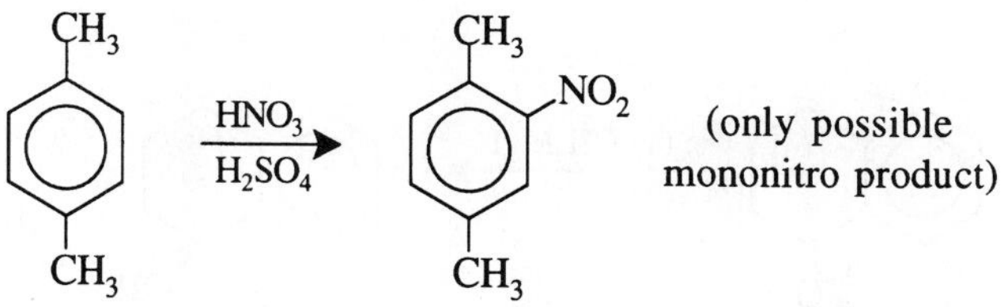

HNO₃/H₂SO₄ → (only possible mononitro product)

15.42 (a)

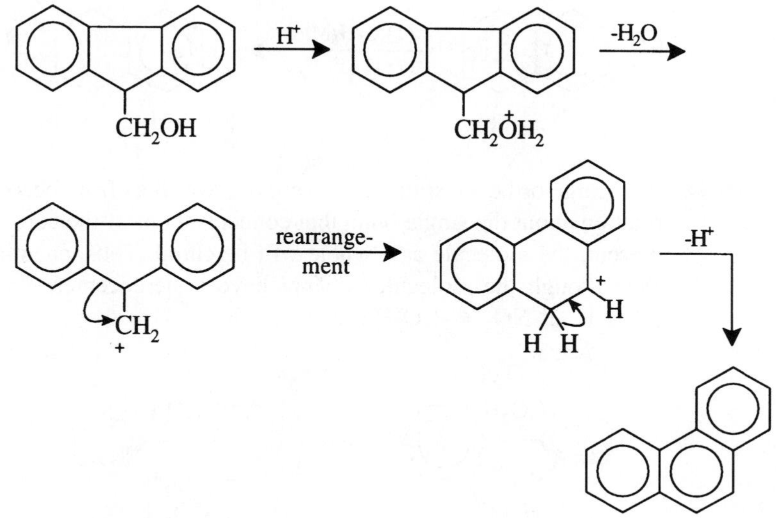

$\xrightarrow{\text{H}^+}$

$\xrightarrow{-\text{H}_2\text{O}}$

$\xrightarrow[\text{ment}]{\text{rearrange-}}$

$\xrightarrow{-\text{H}^+}$

(b)

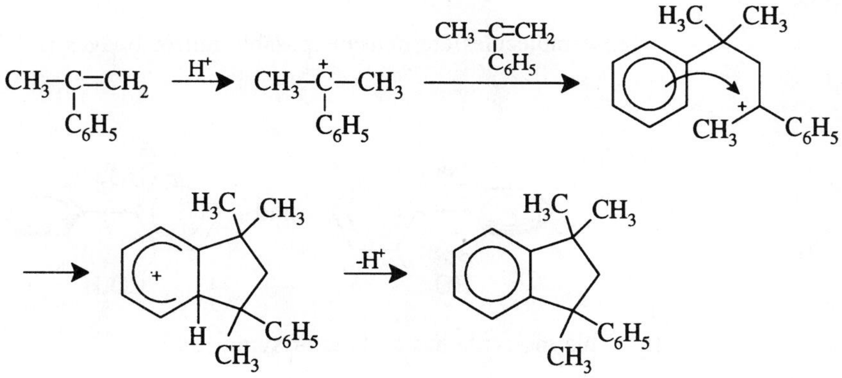

$$\text{CH}_3-\overset{\text{C}_6\text{H}_5}{\underset{}{\text{C}}}=\text{CH}_2 \xrightarrow{\text{H}^+} \text{CH}_3-\overset{+}{\underset{\text{C}_6\text{H}_5}{\text{C}}}-\text{CH}_3 \xrightarrow[\text{C}_6\text{H}_5]{\text{CH}_3-\text{C}=\text{CH}_2}$$

$\xrightarrow{}$ $\xrightarrow{-\text{H}^+}$

15.43 (a)

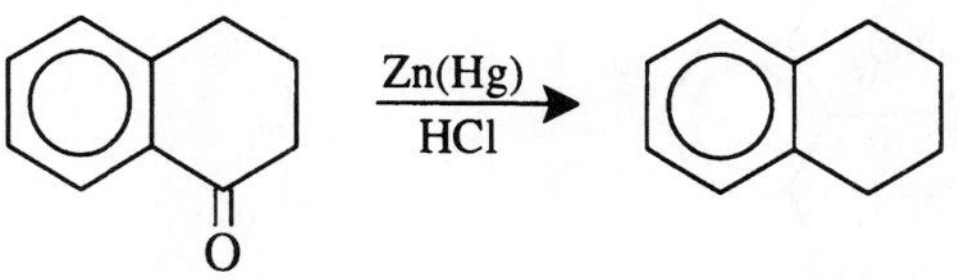

$\xrightarrow[\text{HCl}]{\text{Zn(Hg)}}$

(b)

$\xrightarrow[\text{OH}^-]{\text{NaBH}_4}$

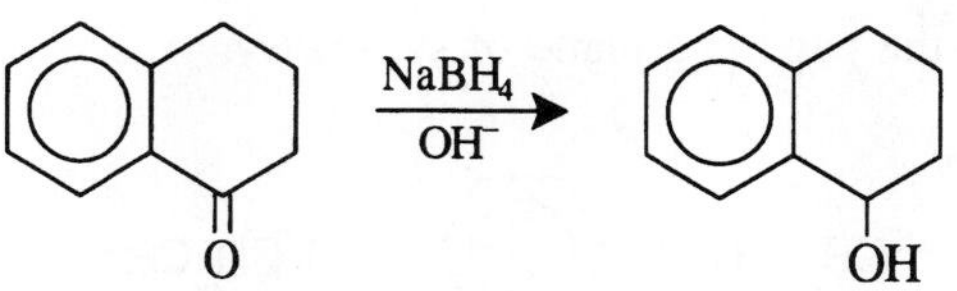

(c)

(d)

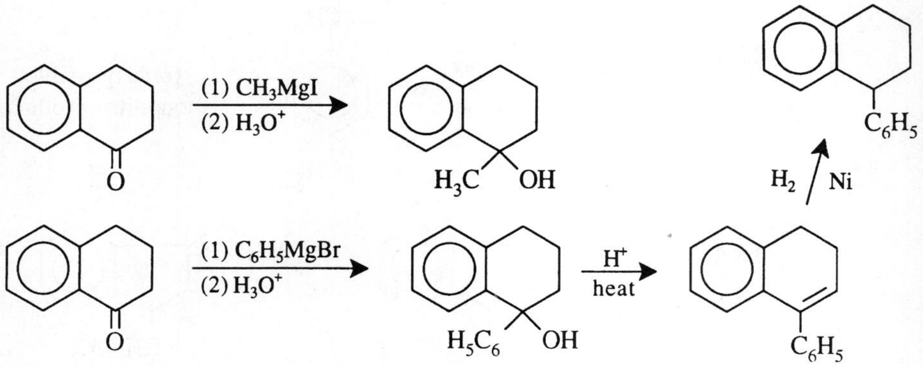

15.44 (a) Large ortho substituents prevent the two rings from becoming coplanar and prevent rotation about the single bond that connects them. If the correct substitution patterns are present, the molecule as a whole will be chiral. Thus, enantiomeric forms are possible even though the molecules do not have a stereocenter. The compound with 2-NO_2, 6-CO_2H, 2'-NO_2, 6'-CO_2H is an example,

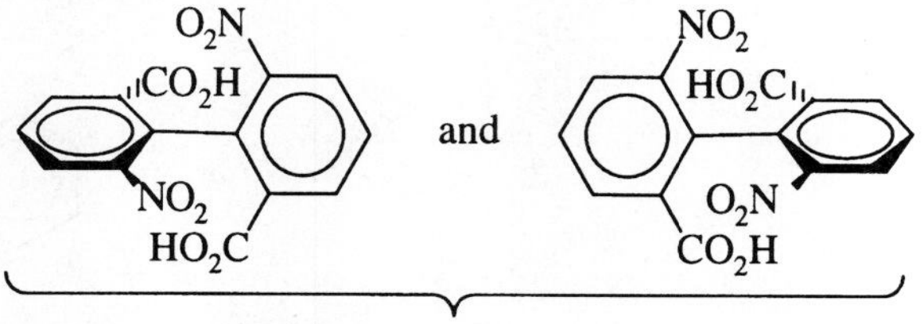

and

These molecules are nonsuperposable mirror images and, thus, are enantiomers.

(b) Yes

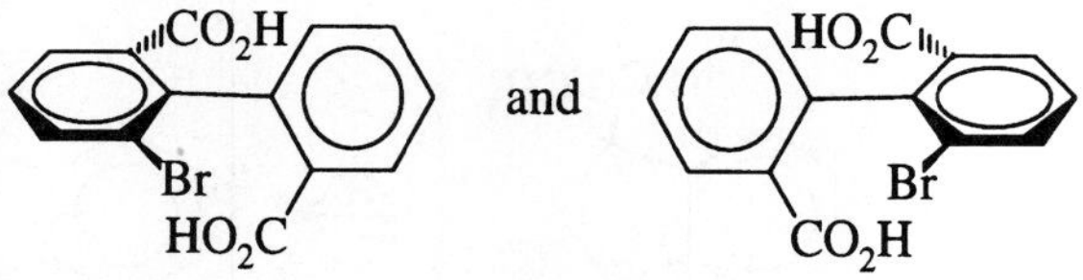

and

(c) This molecule has a plane of symmetry.

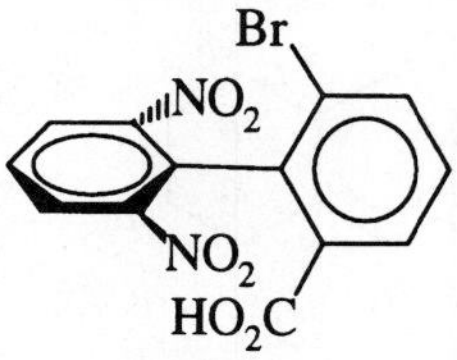

The plane of the page is a plane of symmetry.

15.45 (a)

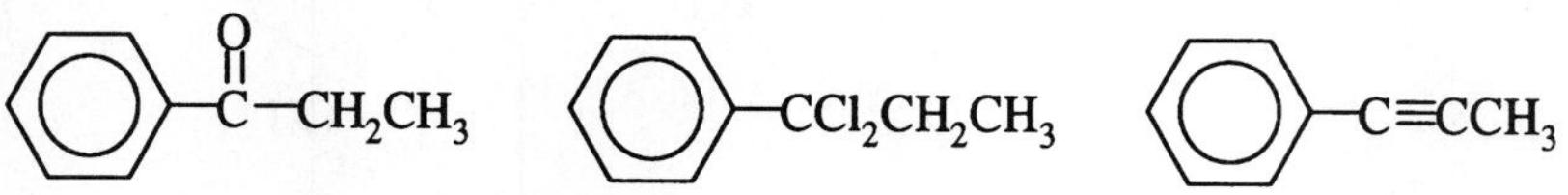

A   B   C

(b)

(c)

(d)

15.46

15.47 (a and b) The *tert*-butyl group is easily introduced by any of the variations of the Friedel-Crafts alkylation reaction, and, because of the stability of the *tert*-butyl cation, it is easily removed under acidic conditions.

(c) In contrast to the -SO$_3$H group often used as a blocking group, –C(CH$_3$)$_3$ activates the ring to further electrophilic substitution.

15.48 At the lower temperature, the reaction is kinetically controlled, and the usual o/p directive effects of the –CH$_3$ group are observed. At higher temperatures, the reaction is thermodynamically controlled. At reaction times long enough for equilibrium to be reached, the most stable isomer, *m*-toluenesulfonic acid, is the principal product.

15.49 The evidence indicates that the mechanistic step in which the C–H bond is broken is *not* rate determining. (In the case cited, it makes no difference kinetically if a C–H or C–D bond is broken in electrophilic aromatic substitution.) This evidence is consistent with the two-step mechanism given in Section 15.2. The step in which the aromatic compound reacts with the electrophile (NO$_2$+) is the slow rate-determining step. Proton (or deuteron) loss from the arenium ion to return to an aromatic system is a rapid step and has no effect on the overall rate.

15.50 (a) $C_6H_5CH_2Br \xrightarrow[\substack{DMF \\ (-NaBr)}]{NaCN} C_6H_5CH_2CN$

(b) $C_6H_5CH_2Br \xrightarrow[\substack{CH_3OH \\ (-NaBr)}]{CH_3ONa} C_6H_5CH_2OCH_3$

(c) $C_6H_5CH_2Br \xrightarrow[\substack{CH_3CO_2H \\ (-NaBr)}]{CH_3CO_2Na} C_6H_5CH_2O_2CCH_3$

(d) $C_6H_5CH_2Br \xrightarrow[\substack{acetone \\ (-NaBr)}]{NaI} C_6H_5CH_2I$

(e) $CH_2{=}CHCH_2Br \xrightarrow[\substack{acetone \\ (-NaBr)}]{NaN_3} CH_2{=}CHCH_2N_3$

(f) $CH_2{=}CHCH_2Br \xrightarrow[\substack{(CH_3)_2CHCH_2OH \\ (-NaBr)}]{NaOCH_2CH(CH_3)_2} CH_2{=}CHCH_2OCH_2CH(CH_3)_2$

15.51 $A = $ (benzene ring) $B = $ (bromobenzene ring, Br) $C = $ (toluene ring, CH₃)

15.52 $(C_6H_5)_3C{-}O{-}H \underset{-H^+}{\overset{+H^+}{\rightleftharpoons}} (C_6H_5)_3C{-}\overset{+}{O}H_2 \underset{+H_2O}{\overset{-H_2O}{\rightleftharpoons}} (C_6H_5)_3C^+ \underset{-EtOH}{\overset{+EtOH}{\rightleftharpoons}}$
Very stable carbocation

$(C_6H_5)_3C{-}\underset{\underset{H}{|}}{\overset{+}{O}}{-}Et \underset{+H^+}{\overset{-H^+}{\rightleftharpoons}} (C_6H_5)_3C{-}O{-}Et$

15.53 (a) CH₃CH₂CH=CHCH₂Br would be the most reactive in an S$_N$2 reaction because it is a 1° allylic halide. There would, therefore, be less steric hindrance to the attacking nucleophile.

(b) CH₂=CHCBr(CH₃)₂ would be the most reactive in an S$_N$1 reaction because it is a 3° allylic halide. The carbocation formed in the rate-determining step, being both 3° and allylic, would be the most stable.

15.54 The final product is *o*-nitroaniline. (The reactions are given in Section 15.14.) The presence of six signals in the ^{13}C NMR spectrum confirms that the substitution in the final product is *ortho* and not *para*. A final product with para substitution (i.e., *p*-nitroaniline) would have given only four signals in the ^{13}C NMR spectrum.

QUIZ

15.1 Which of the following compounds would be most reactive toward ring bromination?

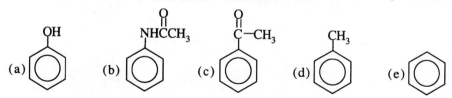

(a)   (b)   (c)   (d)   (e)

15.2 Which of the following is *not* a meta directing substituent when present on a benzene ring?

(a) $-C_6H_5$ (b) $-NO_2$ (c) $-N(CH_3)_3^+$ (d) $-C\equiv N$ (e) $-CO_2H$

15.3 The major product(s), **C**, of the reaction,

$\xrightarrow[\text{FeBr}_3]{\text{Br}_2}$ **C**, would be

(a)   (b)   (c)

(d)   Equal amounts of (a) and (b)   (e)   Equal amounts of (a) and (c)

15.4 Complete the following syntheses.

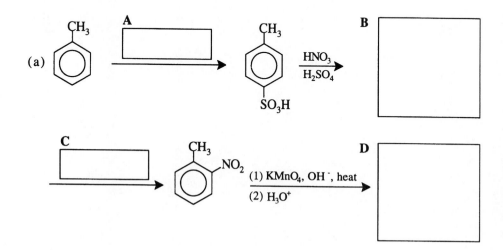

(b)

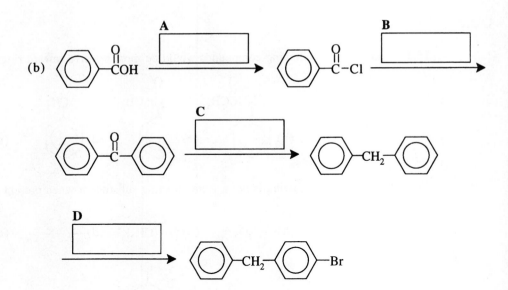

16

ALDEHYDES AND KETONES I:
NUCLEOPHILIC ADDITIONS TO THE
CARBONYL GROUP

SOLUTIONS TO PROBLEMS

16.1 (a)

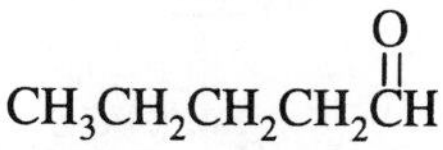

$$CH_3CH_2CH_2CH_2\overset{\overset{\displaystyle O}{\|}}{C}H$$

Pentanal

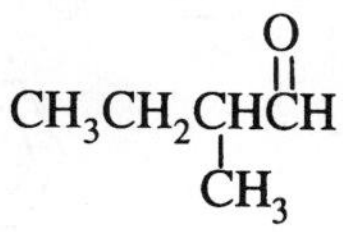

$$CH_3CH_2\overset{\overset{\displaystyle O}{\|}}{C}H\underset{\underset{\displaystyle CH_3}{|}}{}$$

2-Methylbutanal

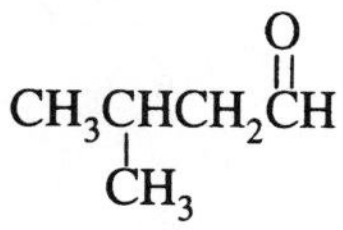

$$CH_3\overset{}{C}HCH_2\overset{\overset{\displaystyle O}{\|}}{C}H\underset{\underset{\displaystyle CH_3}{|}}{}$$

3-Methylbutanal

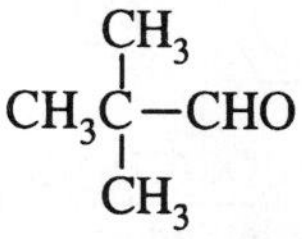

$$CH_3\overset{\overset{\displaystyle CH_3}{|}}{\underset{\underset{\displaystyle CH_3}{|}}{C}}-CHO$$

2,2-Dimethylpropanal

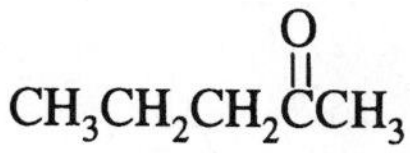

$$CH_3CH_2CH_2\overset{\overset{\displaystyle O}{\|}}{C}CH_3$$

2-Pentanone

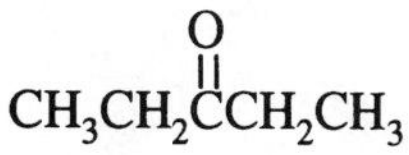

$$CH_3CH_2\overset{\overset{\displaystyle O}{\|}}{C}CH_2CH_3$$

3-Pentanone

$$CH_3\overset{}{C}H\overset{\overset{\displaystyle O}{\|}}{C}CH_3\underset{\underset{\displaystyle CH_3}{|}}{}$$

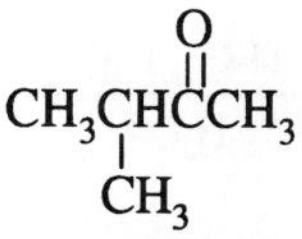

3-Methyl-2-butanone

(b)

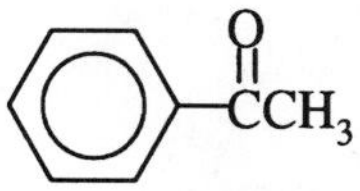

Acetophenone or
methyl phenyl ketone

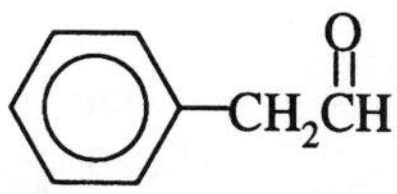

Phenylethanal or
phenylacetaldehyde

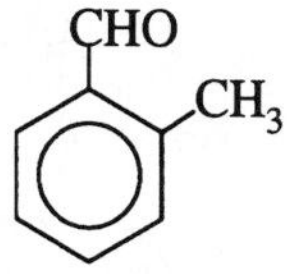

2-Methylbenzaldehyde
(*o*-tolualdehyde)

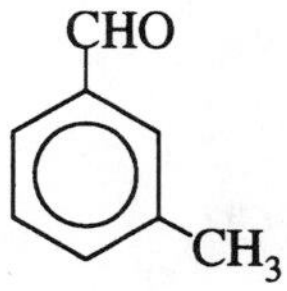

3-Methylbenzaldehyde
(*m*-tolualdehyde)

4-Methylbenzaldehyde
(*p*-tolualdehyde)

16.2 (a) 1-Pentanol, because its molecules form hydrogen bonds to each other.

(b) 2-Pentanol, because its molecules form hydrogen bonds to each other.

(c) Pentanal, because its molecules are more polar.

(d) 2-Phenylethanol, because its molecules form hydrogen bonds to each other.

(e) Benzyl alcohol because its molecules form hydrogen bonds to each other.

16.3 (a) $CH_3CH_2CH_2OH \xrightarrow[CH_2Cl_2]{PCC} CH_3CH_2\overset{\overset{\displaystyle O}{\|}}{C}H$

(b) $CH_3CH_2CO_2H \xrightarrow{SOCl_2} CH_3CH_2\overset{\overset{\displaystyle O}{\|}}{C}Cl \xrightarrow[\text{diethyl ether}]{LiAlH[OC(CH_3)_3]_3} CH_3CH_2\overset{\overset{\displaystyle O}{\|}}{C}H$

16.4 (a)

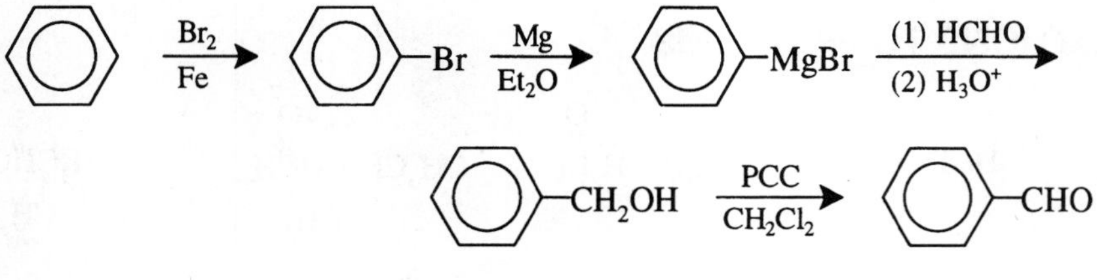

(b)

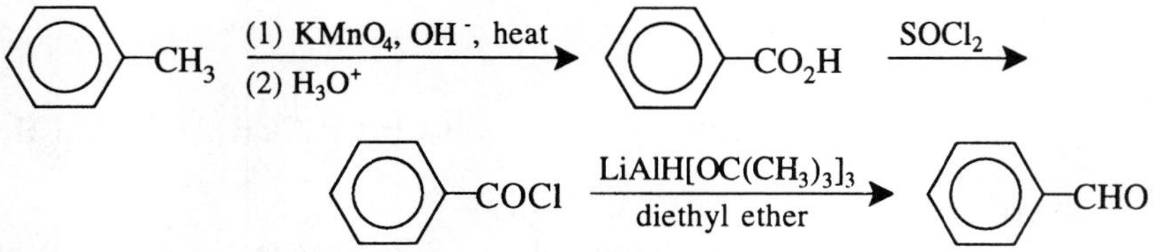

(c) $CH_3CH_2Br \xrightarrow{HC\equiv CNa} CH_3CH_2C\equiv CH \xrightarrow[H_2O]{H_3O^+,\ Hg^{2+}} CH_3CH_2\overset{\overset{\displaystyle O}{\|}}{C}CH_3$

(d) $CH_3C\equiv CCH_3 \xrightarrow[H_2O]{H_3O^+,\ Hg^{2+}} CH_3CH_2\overset{\overset{\displaystyle O}{\|}}{C}CH_3$

(e)

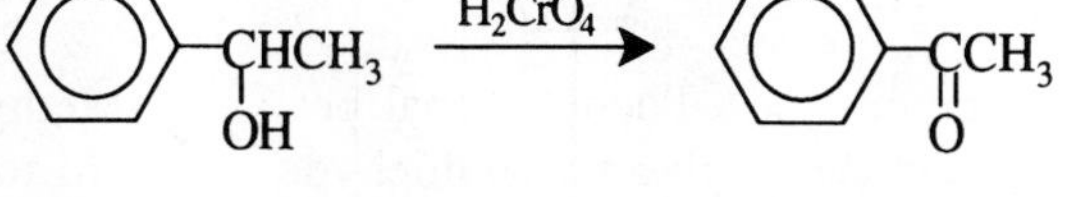

(f)

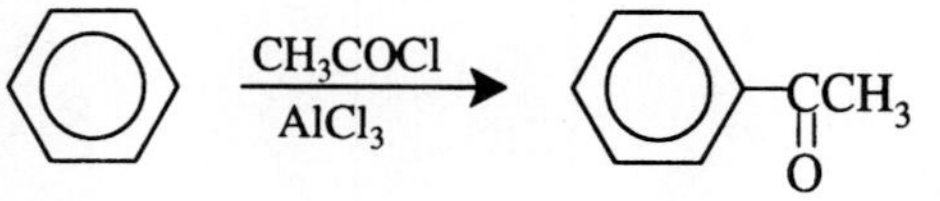

(g)

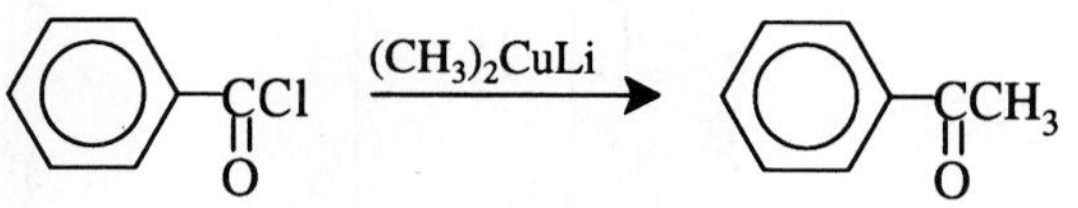

(h) C$_6$H$_5$—COH $\xrightarrow{\text{SOCl}_2}$ C$_6$H$_5$—CCl $\xrightarrow{\text{(CH}_3)_2\text{CuLi}}$ C$_6$H$_5$—CCH$_3$

(i) C$_6$H$_5$—CH$_2$Br $\xrightarrow{\text{CN}^-}$ C$_6$H$_5$—CH$_2$CN $\xrightarrow[\text{diethyl ether}]{\text{CH}_3\text{CH}_2\text{MgBr}}$

C$_6$H$_5$—CH$_2$CCH$_2$CH$_3$ $\xleftarrow{\text{H}_3\text{O}^+}$ C$_6$H$_5$—CH$_2$CCH$_2$CH$_3$ (NMgBr)

(j) C$_6$H$_5$CH$_2$CN $\xrightarrow[\text{(2) H}_3\text{O}^+]{\text{(1) } i\text{-Bu}_2\text{AlH}}$ C$_6$H$_5$CH$_2$CH (=O)

(k) CH$_3$(CH$_2$)$_4$CO$_2$CH$_3$ $\xrightarrow[\text{(2) H}_3\text{O}^+]{\text{(1)} i\text{-Bu}_2\text{AlH}}$ CH$_3$(CH$_2$)$_4$CH (=O)

16.5   (a)   The nucleophile is the negatively charged carbon of the Grignard reagent *acting as a carbanion.*

(b)   The magnesium portion of the Grignard reagent acts as a Lewis acid and accepts an electron pair of the carbonyl oxygen. This acid-base interaction makes the carbonyl carbon even more positive and, therefore, even more susceptible to nucleophilic attack.

(c)   The product that forms initially (above) is a magnesium derivative of an alcohol.

(d)   On addition of water, the organic product that forms is an alcohol.

16.6   The nucleophile is a hydride ion.

16.7

16.8 *Acid-Catalyzed Reaction*

$$CH_3-\overset{O}{\overset{||}{C}}-CH_3 \underset{-H^+}{\overset{+H^+}{\rightleftarrows}} CH_3-\overset{+}{\overset{OH}{\overset{||}{C}}}-CH_3 \underset{-H_2{}^{18}O}{\overset{+H_2{}^{18}O}{\rightleftarrows}} CH_3-\underset{H_2{}^{18}O^+}{\overset{OH}{\underset{|}{\overset{|}{C}}}}-CH_3 \underset{+H^+}{\overset{-H^+}{\rightleftarrows}}$$

$$CH_3-\underset{{}^{18}OH}{\overset{OH}{\underset{|}{\overset{|}{C}}}}-CH_3 \underset{-H^+}{\overset{+H^+}{\rightleftarrows}} CH_3-\underset{{}^{18}OH}{\overset{+OH_2}{\underset{|}{\overset{|}{C}}}}-CH_3 \underset{+H_2O}{\overset{-H_2O}{\rightleftarrows}} CH_3-\underset{{}^{18}OH^+}{\overset{C}{||}}-CH_3 \overset{-H^+}{\underset{+H^+}{\longleftarrow}}$$

$$CH_3-\underset{{}^{18}O}{\overset{C}{||}}-CH_3$$

Base-Catalyzed Reaction

$$OH^- \; + \; H_2{}^{18}O \; \rightleftarrows \; H_2O \; + \; {}^{18}OH^-$$

$$CH_3-\overset{O}{\overset{||}{C}}-CH_3 \; + \; {}^{18}OH^- \; \rightleftarrows \; CH_3-\underset{{}^{18}OH}{\overset{O^-}{\underset{|}{\overset{|}{C}}}}-CH_3 \underset{OH^-}{\overset{H_2O}{\rightleftarrows}} CH_3-\underset{{}^{18}OH}{\overset{OH}{\underset{|}{\overset{|}{C}}}}-CH_3 \underset{H_2O}{\overset{OH^-}{\rightleftarrows}}$$

$$CH_3-\underset{{}^{18}O^-}{\overset{OH}{\underset{|}{\overset{|}{C}}}}-CH_3 \underset{+OH^-}{\overset{-OH^-}{\rightleftarrows}} CH_3-\underset{{}^{18}O}{\overset{C}{||}}-CH_3$$

16.9

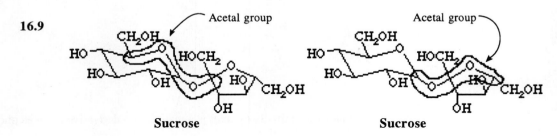

Sucrose   Sucrose

16.10

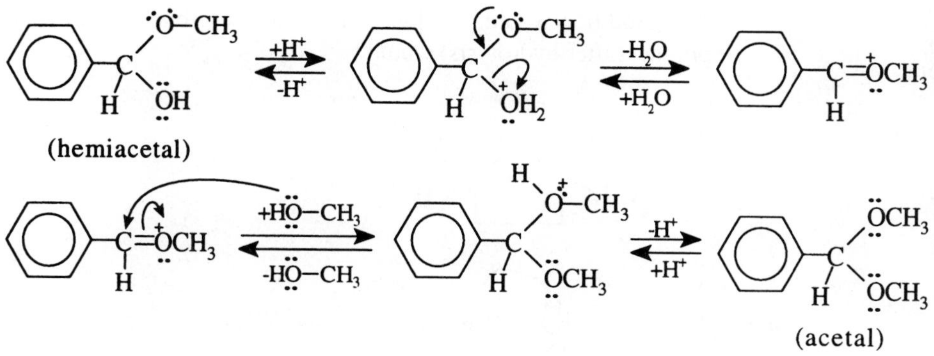

(hemiacetal)

(acetal)

16.11

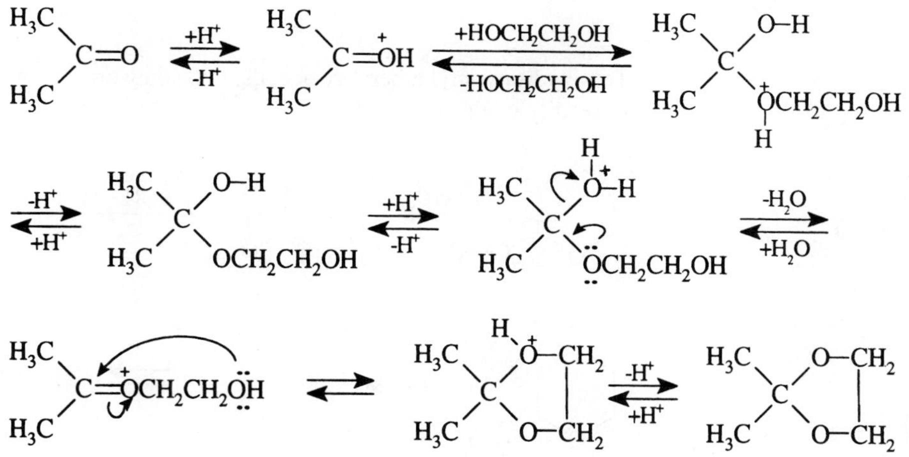

16.12   HO⟋⬡⟍CH₂OH

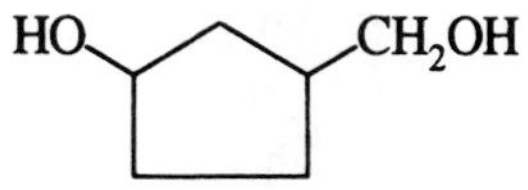

16.13   (a)

A

C

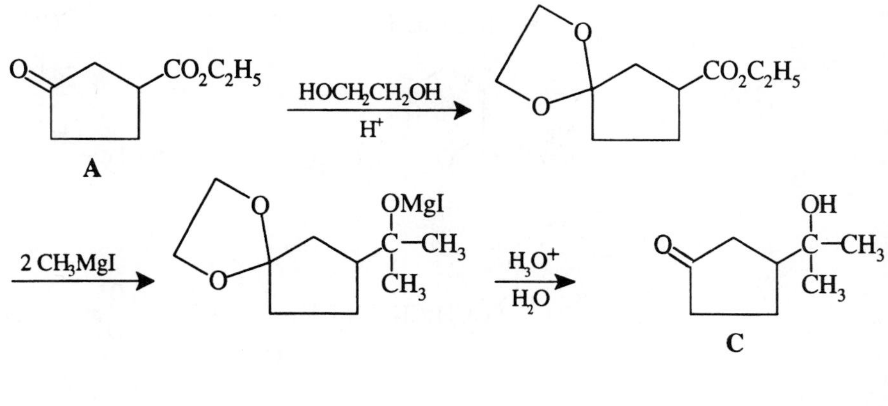

(b)   Addition would take place at the ketone group as well as at the ester group. The product (after hydrolysis) would be.

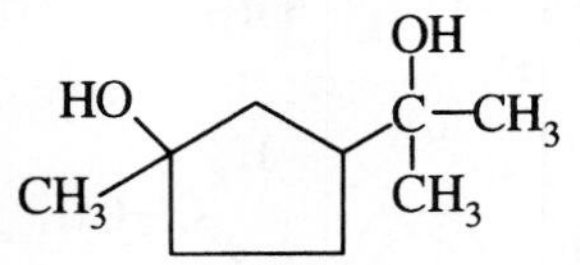

16.14   (a)

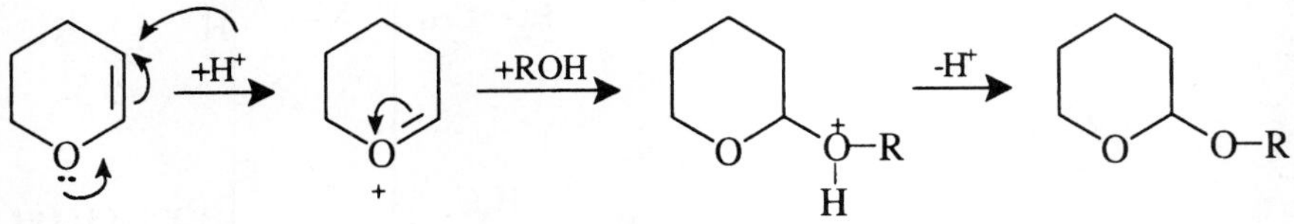

(b)   Tetrahydropyranyl ethers are acetals; thus, they are stable in aqueous base and hydrolyze readily in aqueous acid.

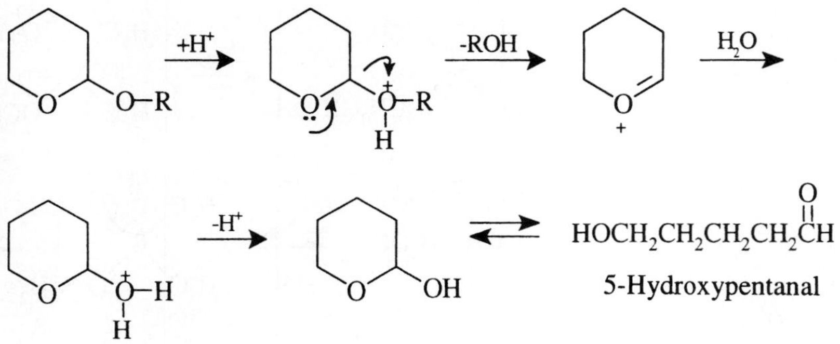

5-Hydroxypentanal

(c) $HOCH_2CH_2CH_2CH_2Cl$

$\xrightarrow[Et_2O]{Mg}$

$\xrightarrow[H_2O]{H_3O^+}$ $HOCH_2CH_2CH_2CH_2\overset{\underset{|}{CH_3}}{\underset{|}{CH_3}}{C}OH$

$(+\ HOCH_2CH_2CH_2CH_2\overset{O}{\overset{||}{C}}H)$

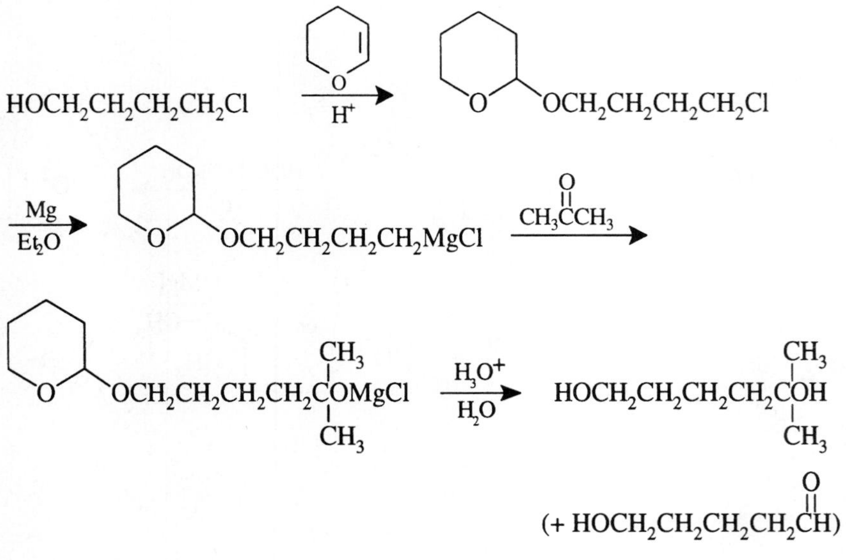

16.15 (a)

$$\text{(cyclohexanone)} + \text{HSCH}_2\text{CH}_2\text{SH} \xrightarrow{\text{BF}_3} \text{(1,4-dithiaspiro product)}$$

$$\xrightarrow[\text{H}_2]{\text{Raney Ni}} \text{(cyclohexane)} + \text{CH}_3\text{CH}_3 + \text{NiS}$$

(b)

$$\text{C}_6\text{H}_5\overset{\text{O}}{\overset{\|}{\text{C}}}\text{H} + \text{HSCH}_2\text{CH}_2\text{SH} \xrightarrow{\text{BF}_3} \text{(dithiolane product)}$$

$$\xrightarrow[\text{H}_2]{\text{Raney Ni}} \text{C}_6\text{H}_5-\text{CH}_3 + \text{CH}_3\text{CH}_3 + \text{NiS}$$

16.16 (a)

$$\text{CH}_3\overset{\text{O}}{\overset{\|}{\text{C}}}\text{H} \xrightarrow{\text{HCN}} \text{CH}_3\overset{\text{OH}}{\overset{|}{\text{C}}}\text{HCN} \xrightarrow[\text{reflux}]{\text{HCl, H}_2\text{O}} \text{CH}_3\overset{\text{OH}}{\overset{|}{\text{C}}}\text{HCO}_2\text{H}$$
$$\text{Lactic acid}$$

(b)   A racemic form

16.17 (a)

$$\text{CH}_3\text{I} \xrightarrow[\text{(2) RLi}]{\text{(1) (C}_6\text{H}_5)_3\text{P}} \overset{..}{:}\text{CH}_2-\overset{+}{\text{P}}(\text{C}_6\text{H}_5)_3 \xrightarrow{\text{C}_6\text{H}_5\overset{\text{O}}{\overset{\|}{\text{C}}}\text{CH}_3} \text{C}_6\text{H}_5\overset{|}{\underset{\text{CH}_3}{\text{C}}}=\text{CH}_2$$

(b)

$$\text{CH}_3\text{CH}_2\text{Br} \xrightarrow[\text{(2) RLi}]{\text{(1) (C}_6\text{H}_5)_3\text{P}} \text{CH}_3\overset{..}{\text{C}}\text{H}-\overset{+}{\text{P}}(\text{C}_6\text{H}_5)_3 \xrightarrow{\text{C}_6\text{H}_5\overset{\text{O}}{\overset{\|}{\text{C}}}\text{CH}_3} \text{C}_6\text{H}_5\overset{|}{\underset{\text{CH}_3}{\text{C}}}=\text{CHCH}_3$$

(c)

$$\overset{..}{:}\text{CH}_2-\overset{+}{\text{P}}(\text{C}_6\text{H}_5)_3 \xrightarrow{\text{CH}_3\overset{\text{O}}{\overset{\|}{\text{C}}}\text{CH}_3} \overset{\text{H}_3\text{C}}{\underset{\text{H}_3\text{C}}{>}}\text{C}=\text{CH}_2$$

[from part (a)]

(d)

$$\overset{..}{:}\text{CH}_2-\overset{+}{\text{P}}(\text{C}_6\text{H}_5)_3 \xrightarrow{\text{(cyclopentanone)}} \text{(methylenecyclopentane)}$$

[from part (a)]

(e)

$$\text{CH}_3\text{CH}_2\text{CH}_2\text{Br} \xrightarrow[\text{(2) RLi}]{\text{(1) (C}_6\text{H}_5)_3\text{P}} \text{CH}_3\text{CH}_2\overset{..}{\text{C}}\text{H}-\overset{+}{\text{P}}(\text{C}_6\text{H}_5)_3 \xrightarrow{\text{CH}_3\overset{\text{O}}{\overset{\|}{\text{C}}}\text{CH}_2\text{CH}_3}$$

$$\text{CH}_3\text{CH}_2\text{CH}=\overset{\overset{\text{CH}_3}{|}}{\text{C}}\text{CH}_2\text{CH}_3$$

(f) $CH_2\!=\!CHCH_2Br$ $\xrightarrow[\text{(2) RLi}]{\text{(1)}(C_6H_5)_3P}$ $CH_2\!=\!CH\overset{..}{C}H-\overset{+}{P}(C_6H_5)_3$ $\xrightarrow{C_6H_5\overset{O}{\overset{\|}{C}}H}$

$$C_6H_5CH=CHCH=CH_2$$

(g) $C_6H_5CH_2Br$ $\xrightarrow[\text{(2) RLi}]{\text{(1)}(C_6H_5)_3P}$ $C_6H_5\overset{..}{C}H-\overset{+}{P}(C_6H_5)_3$ $\xrightarrow{C_6H_5\overset{O}{\overset{\|}{C}}H}$

$$C_6H_5CH=CHC_6H_5$$

16.18 $(C_6H_5)_3P:$ + $C_6H_5\overset{O}{\overset{\diagup\diagdown}{CH-CHCH_3}}$ $\longrightarrow$ $C_6H_5CH-\overset{O^-}{\underset{(C_6H_5)_3\overset{+}{P}}{CHCH_3}}$ $\longrightarrow$

$$\underset{(C_6H_5)_3P-O}{C_6H_5CH-CHCH_3} \longrightarrow C_6H_5CH=CHCH_3 \ + \ (C_6H_5)_3P=O$$

16.19 (a) $(CH_3)_2C=O$ + $BrCH_2CO_2CH_2CH_3$ $\xrightarrow[\text{benzene}]{Zn}$ $(CH_3)_2\overset{OZnBr}{\underset{}{C}}CH_2CO_2CH_2CH_3$

$$\xrightarrow{H_3O^+} (CH_3)_2\overset{OH}{\underset{}{C}}CH_2CO_2CH_2CH_3$$

(b) $=O$ + $Br\overset{}{\underset{CH_3}{C}}HCO_2CH_2CH_3$ $\xrightarrow[\text{(2) }H_3O^+]{\text{(1) Zn, benzene}}$

(c) $CH_3CH_2\overset{O}{\overset{\|}{C}}H$ + $BrCH_2CO_2CH_2CH_3$ $\xrightarrow[\text{(2)}H_3O^+,\ \text{heat}]{\text{(1) Zn, benzene}}$

$$CH_3CH_2CH_2CH_2CO_2CH_2CH_3 \xleftarrow[Pt]{H_2} CH_3CH_2CH=CHCO_2CH_2CH_3$$

16.20

16.21   The product is a lactone, formed as follows:

(a lactone)

16.22 $CH_3\overset{O}{\overset{\|}{C}}-O-\underset{\underset{CH_3}{|}}{CHCH_3}$ The isopropyl group has a greater migratory aptitude than the methyl group. The mechanism is as follows:

16.23   (a)   HCHO   Methanal

(b) CH_3CHO Ethanal

(c) $C_6H_5CH_2CHO$ Phenylethanal

(d) CH_3COCH_3 Propanone

(e) $CH_3COCH_2CH_3$ Butanone

(f) $CH_3COC_6H_5$ 1-Phenylethanone or methyl phenyl ketone

(g) $C_6H_5COC_6H_5$ Diphenylmethanone or diphenyl ketone

(h)
2-Hydroxybenzaldehyde or o-hydroxybenzaldehyde

(i)
4-Hydroxy-3-methoxybenzaldehyde

(j) $CH_3CH_2COCH_2CH_3$ 3-Pentanone

(k) $CH_3CH_2COCH(CH_3)_2$ 2-Methyl-3-pentanone

(l) $(CH_3)_2CHCOCH(CH_3)_2$ 2,4-Dimethyl-3-pentanone

(m) $CH_3(CH_2)_3CO(CH_2)_3CH_3$ 5-Nonanone

(n) $CH_3(CH_2)_2CO(CH_2)_2CH_3$ 4-Heptanone

(o) $C_6H_5CH=CHCHO$ 3-Phenyl-2-propenal

16.24   (a) $CH_3CH_2CH_2OH$ (i) $CH_3CH_2CHOHCH_2CO_2C_2H_5$

(b) $CH_3CH_2CHOHC_6H_5$ (j) $CH_3CH_2CO_2^- \, NH_4^+ + Ag\downarrow$

(c) $CH_3CH_2CH_2OH$ (k) $CH_3CH_2CH=NOH$

(d) $CH_3CH_2\overset{\overset{\displaystyle O}{\|}}{C}-O^-$ (l) $CH_3CH_2CH=NNHCONH_2$

(e) $CH_3CH_2CH=CH_2$ (m) $CH_3CH_2CH=NNHC_6H_5$

(f) $CH_3CH_2CH_2OH$ (n) $CH_3CH_2CO_2H$

(g) $CH_3CH_2CH{\begin{smallmatrix}O-CH_2\\ \,\\O-CH_2\end{smallmatrix}}$ (o) $CH_3CH_2CH{\begin{smallmatrix}S-CH_2\\ \,\\S-CH_2\end{smallmatrix}}$

(h) $CH_3CH_2CH=CHCH_3$ (p) $CH_3CH_2CH_3 + CH_3CH_3 + NiS$

16.25   (a) $CH_3CHOHCH_3$ (j)   No reaction

(b) $C_6H_5\underset{\underset{\displaystyle CH_3}{|}}{C}OHCH_3$ (k) $CH_3\underset{\underset{\displaystyle CH_3}{|}}{C}=NOH$

(c) $CH_3CHOHCH_3$ (l) $CH_3\underset{\underset{\displaystyle CH_3}{|}}{C}=NNHCONH_2$

(d)   No reaction   (m) $CH_3\underset{\underset{\displaystyle CH_3}{|}}{C}=NNHC_6H_5$

(e) $CH_3\overset{\overset{\displaystyle CH_3}{|}}{C}=CH_2$ (n)   No reaction

(f) $CH_3CHOHCH_3$ (o) $\begin{smallmatrix}H_3C\\ \,\\H_3C\end{smallmatrix}{>}C{\begin{smallmatrix}S-CH_2\\ \,\\S-CH_2\end{smallmatrix}}$

(g) $\begin{smallmatrix}H_3C\\ \,\\H_3C\end{smallmatrix}{>}C{\begin{smallmatrix}O-CH_2\\ \,\\O-CH_2\end{smallmatrix}}$ (p) $CH_3CH_2CH_3 + CH_3CH_3 + NiS$

(h) $CH_3CH=C(CH_3)_2$

(i) $CH_3\underset{\underset{\displaystyle CH_3}{|}}{\overset{\overset{\displaystyle OH}{|}}{C}}CH_2CO_2C_2H_5$

16.26 (a)

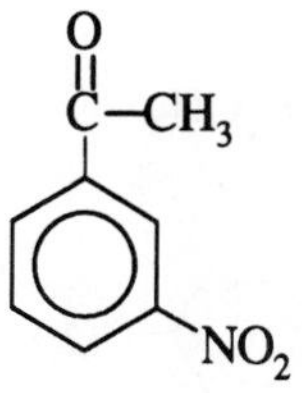

(b)

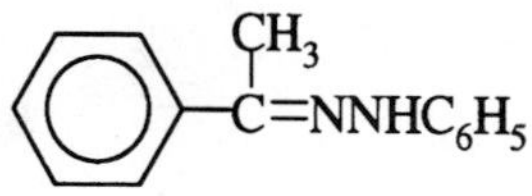

(c)

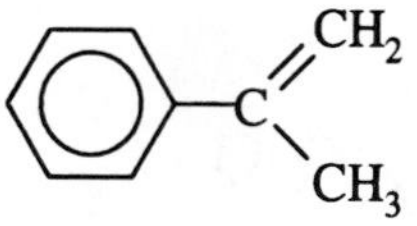

(d)

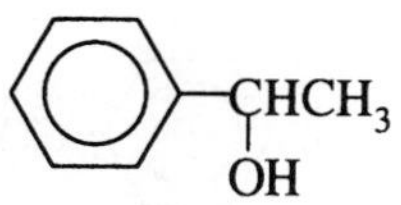

(e)

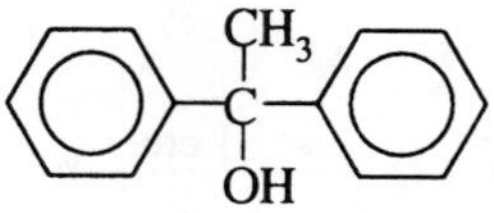

16.27 (a)

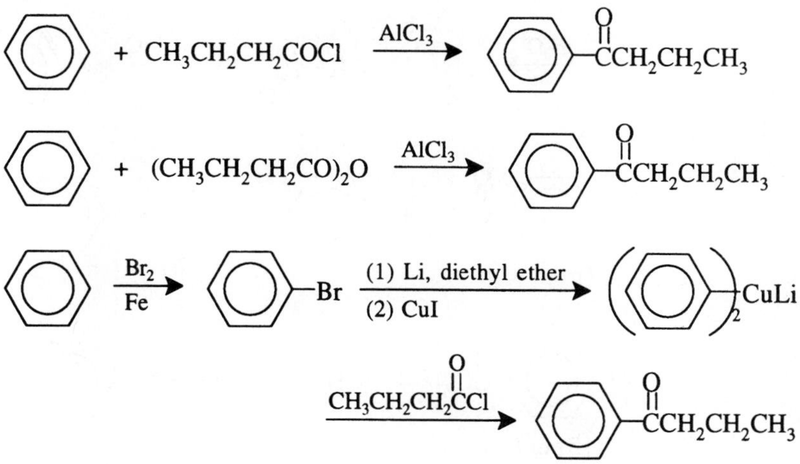

(b)

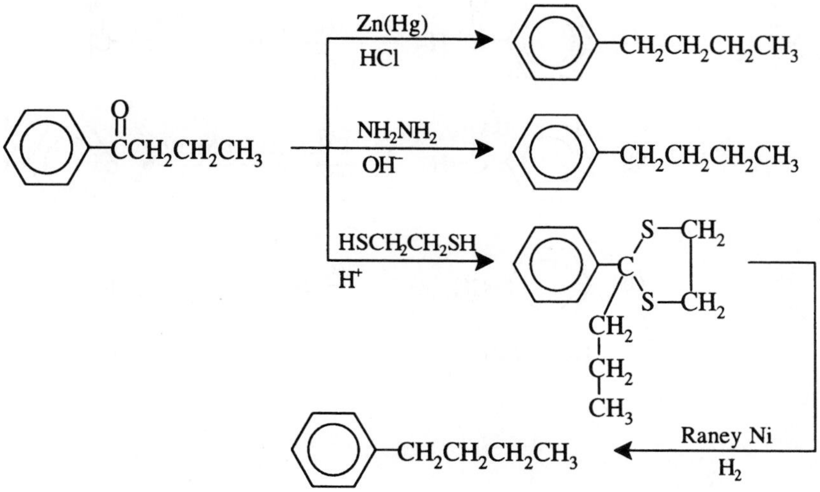

16.28 (a) C_6H_5-CHO $\xrightarrow{\text{NaBH}_4}$ C_6H_5-CH$_2$OH

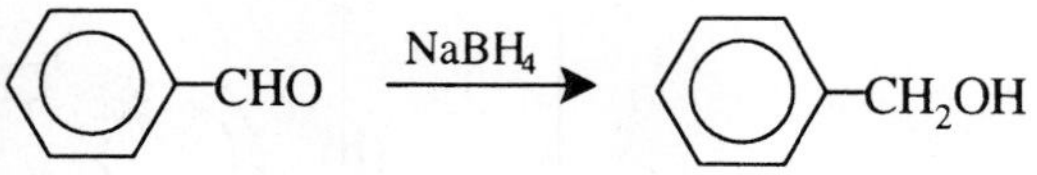

(b) C_6H_5-CHO $\xrightarrow[\text{NH}_3]{\text{Ag(NH}_3)_2^+}$ $\xrightarrow{\text{H}_3\text{O}^+}$ C_6H_5-CO$_2$H

(c) C_6H_5-CO$_2$H $\xrightarrow{\text{SOCl}_2}$ C_6H_5-COCl

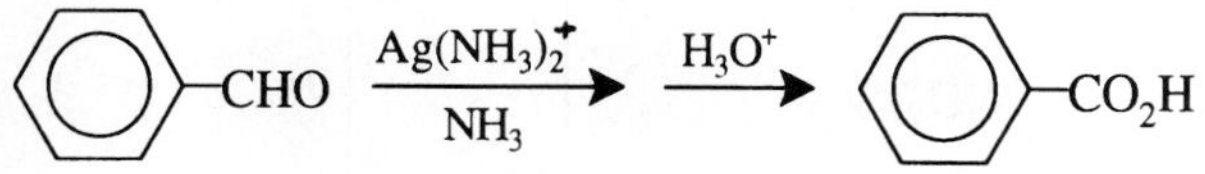

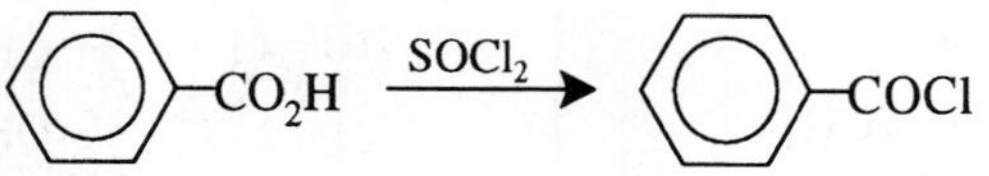

[from (b)]

(d) C_6H_5-CHO + C_6H_5-MgBr $\xrightarrow{\text{diethyl ether}}$ C_6H_5-CH($\overset{\text{OMgBr}}{|}$)-$C_6H_5$

$\xrightarrow{\text{H}_3\text{O}^+}$ C_6H_5-CH($\overset{\text{OH}}{|}$)-C_6H_5 $\xrightarrow{\text{H}_2\text{CrO}_4}$ C_6H_5-C(=O)-C_6H_5

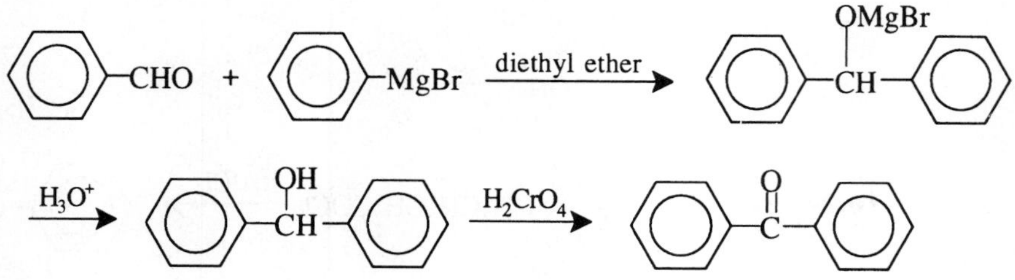

or

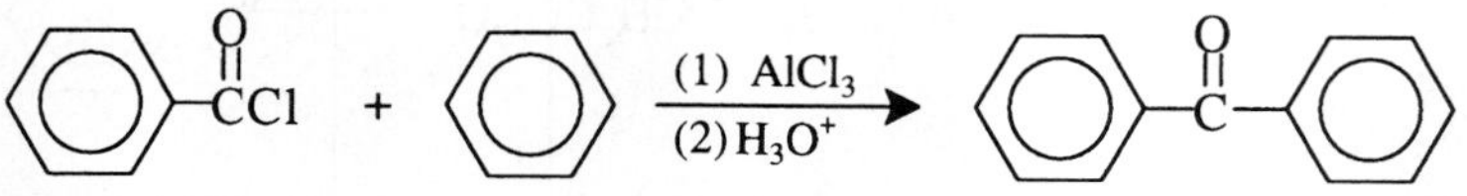

C_6H_5-CCl(=O) + C_6H_6 $\xrightarrow[(2)\,\text{H}_3\text{O}^+]{(1)\,\text{AlCl}_3}$ C_6H_5-C(=O)-C_6H_5

[from (c)]

(e) C_6H_5-CCl(=O) + (CH$_3$)$_2$CuLi $\longrightarrow$ C_6H_5-CCH$_3$(=O) + CH$_3$Cu + LiCl

[from (c)]

(f) C_6H_5-C(=O)-H $\xrightarrow[(2)\,\text{H}_3\text{O}^+]{(1)\,\text{CH}_3\text{MgI}}$ C_6H_5-CHCH$_3$($\overset{\text{OH}}{|}$)

(g) C_6H_5-C(=O)-H $\xrightarrow[(2)\,\text{H}_3\text{O}^+]{(1)\,(\text{CH}_3)_2\text{CHCH}_2\text{MgBr}}$ C_6H_5-CHCH$_2$CHCH$_3$ ($\overset{\text{OH}}{|}$ ... $\overset{\text{CH}_3}{|}$)

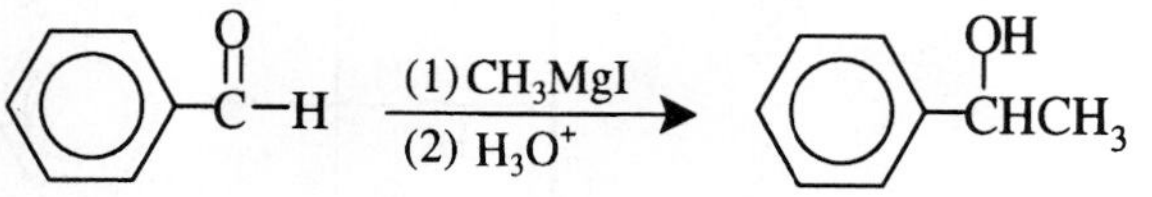

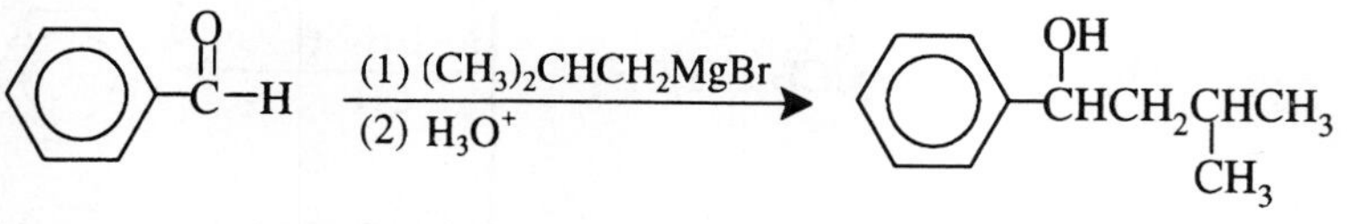

(h) C_6H_5-CH$_2$OH $\xrightarrow{\text{PBr}_3}$ C_6H_5-CH$_2$Br

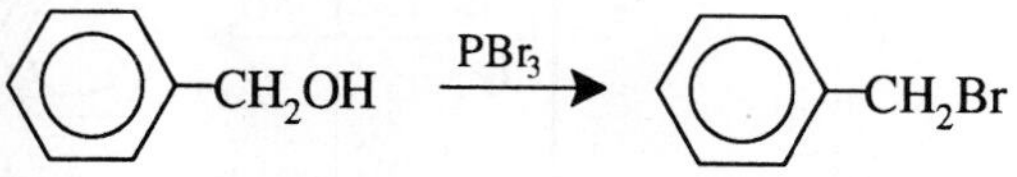

[from (a)]

(i) 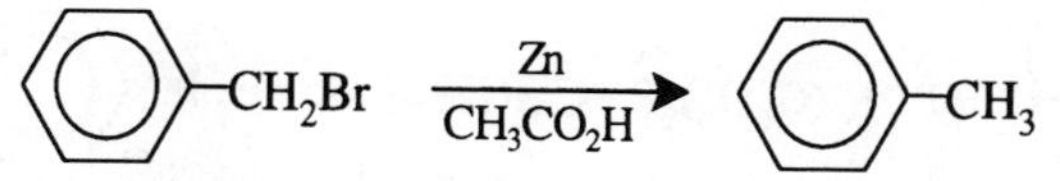 C_6H_5—CH_2Br $\xrightarrow[CH_3CO_2H]{Zn}$ C_6H_5—CH_3

[from (h)]

or

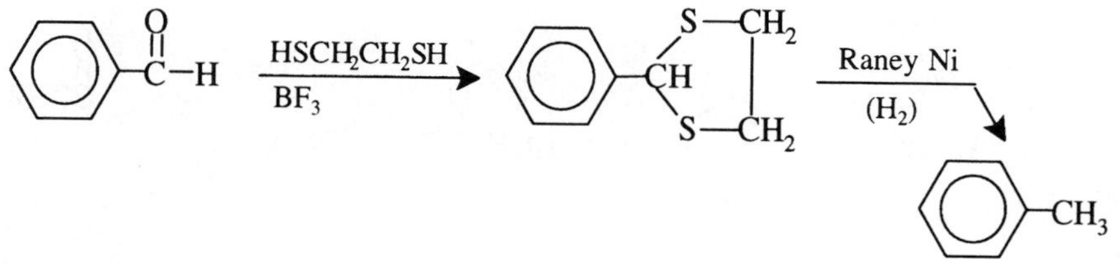

C_6H_5—C(=O)—H $\xrightarrow[BF_3]{HSCH_2CH_2SH}$ [dithiolane] $\xrightarrow[(H_2)]{Raney\ Ni}$ C_6H_5—CH_3

(j) 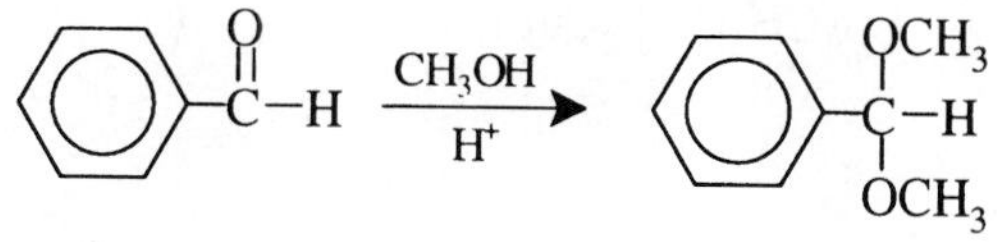 C_6H_5—C(=O)—H $\xrightarrow[H^+]{CH_3OH}$ C_6H_5—CH(OCH$_3$)$_2$

(k) 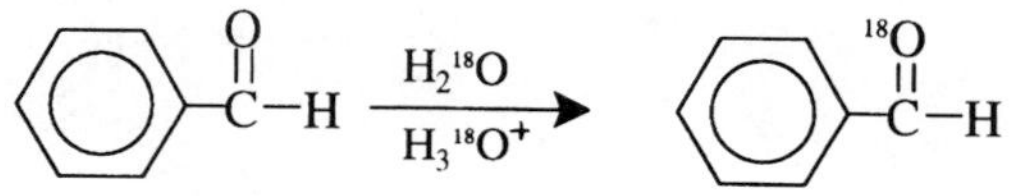 C_6H_5—C(=O)—H $\xrightarrow[H_3{}^{18}O^+]{H_2{}^{18}O}$ C_6H_5—C(^{18}O)—H   (See problem 16.8 for the mechanism)

(l) 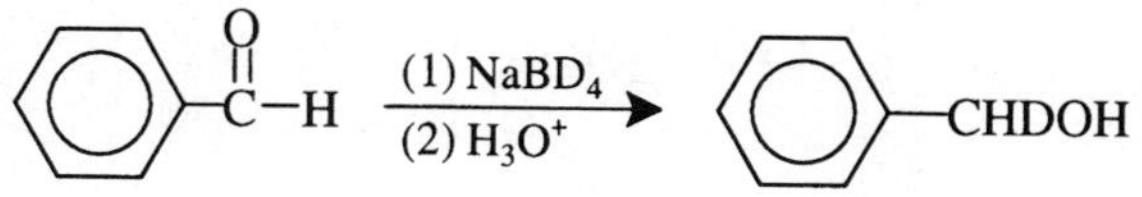 C_6H_5—C(=O)—H $\xrightarrow[(2)\ H_3O^+]{(1)\ NaBD_4}$ C_6H_5—CHDOH

(m) 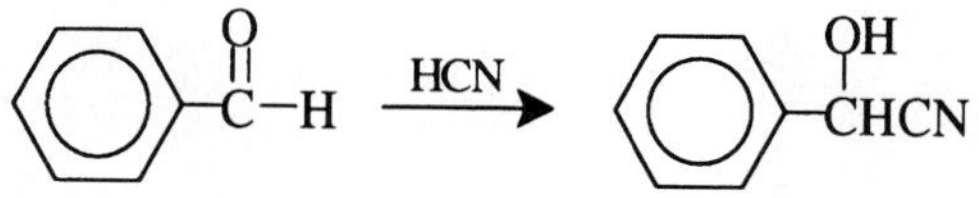 C_6H_5—C(=O)—H $\xrightarrow{HCN}$ C_6H_5—CHCN with OH   (a cyanohydrin)

(n) 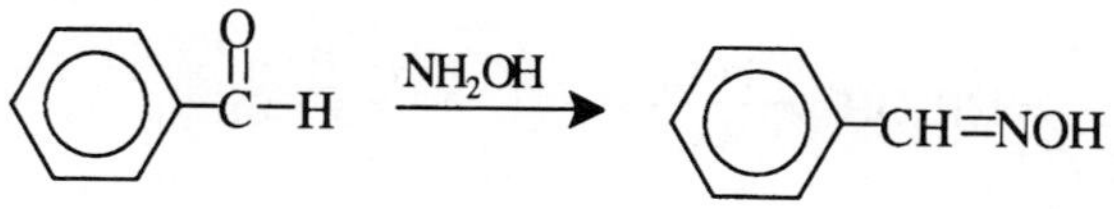 C_6H_5—C(=O)—H $\xrightarrow{NH_2OH}$ C_6H_5—CH=NOH   (an oxime)

(o) 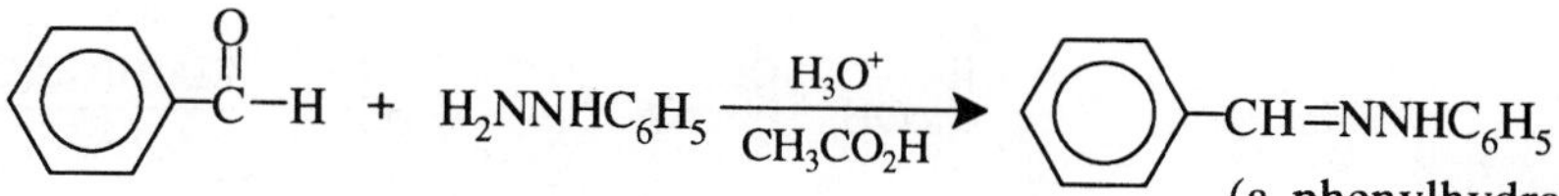 C_6H_5—C(=O)—H + $H_2NNHC_6H_5$ $\xrightarrow[CH_3CO_2H]{H_3O^+}$ C_6H_5—CH=NNHC$_6$H$_5$ (a phenylhydrazone)

(p) 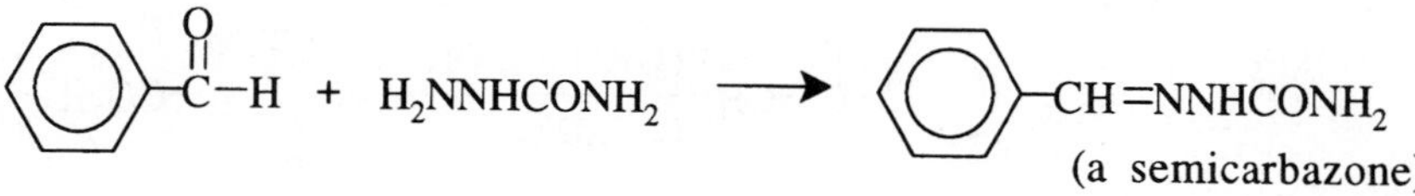 C_6H_5—C(=O)—H + $H_2NNHCONH_2$ $\longrightarrow$ C_6H_5—CH=NNHCONH$_2$ (a semicarbazone)

(q) $C_6H_5\overset{O}{\overset{\|}{C}}-H$ + $(C_6H_5)_3\overset{+}{P}-\overset{\cdot\cdot}{C}HCH=CH_2$ $\longrightarrow$ $C_6H_5-CH=CHCH=CH_2$

(a Wittig reagent)

16.29 (a) benzene + $CH_3CH_2\overset{O}{\overset{\|}{C}}Cl$ $\xrightarrow{AlCl_3}$ $\overset{O}{\overset{\|}{C}CH_2CH_3}$

(b) $\overset{O}{\overset{\|}{C}-Cl}$ + $(CH_3CH_2)_2CuLi$ $\longrightarrow$ $\overset{O}{\overset{\|}{C}CH_2CH_3}$

(c) $C\equiv N$ + CH_3CH_2Li $\xrightarrow{(2)\ H_3O^+}$ $\overset{O}{\overset{\|}{C}CH_2CH_3}$

(d) CHO + CH_3CH_2MgBr $\xrightarrow{(2)\ H_3O^+}$ $\overset{OH}{\overset{|}{C}HCH_2CH_3}$ $\xrightarrow{H_2CrO_4}$ $\overset{O}{\overset{\|}{C}CH_2CH_3}$

16.30 (a) $-CH_2OH$ $\xrightarrow[CH_2Cl_2]{PCC}$ $\overset{O}{\overset{\|}{C}-H}$

(b) $\overset{O}{\overset{\|}{C}-OH}$ $\xrightarrow{SOCl_2}$ $\overset{O}{\overset{\|}{C}-Cl}$ $\xrightarrow{LiAlH[OC(CH_3)_3]_3}$ $\overset{O}{\overset{\|}{C}-H}$

(c) $-C\equiv CH$ $\xrightarrow[(2)\ H_3O^+]{(1)\ KMnO_4,\ OH^-}$ $-CO_2H$ $\xrightarrow[(b)]{[as\ in}$ $\overset{O}{\overset{\|}{C}-H}$

(d) $\text{C}_6\text{H}_5{-}\text{CH}{=}\text{CH}_2$ $\xrightarrow[\text{(2) H}_3\text{O}^+]{\text{(1) KMnO}_4,\ \text{OH}^-}$ $\text{C}_6\text{H}_5{-}\text{CO}_2\text{H}$ $\xrightarrow{\text{[as in (b)]}}$ benzaldehyde

(e) benzoic acid methyl ester $\xrightarrow[\text{(2) H}_2\text{O}]{\text{(1)}\,(i\text{-Bu})_2\text{AlH, hexane, -78°C}}$ benzaldehyde

(f) $\text{C}_6\text{H}_5{-}\text{C}{\equiv}\text{N}$ $\xrightarrow[\text{(2)H}_2\text{O}]{\text{(1)}\,(i\text{-Bu})_2\text{AlH, hexane, -78°C}}$ benzaldehyde

16.31 cyclohexanol $\xrightarrow[\text{acetone}]{\text{H}_2\text{CrO}_4}$ cyclohexanone **A** $\xrightarrow[\text{(2) H}_3\text{O}^+]{\text{(1) CH}_3\text{MgI}}$ 1-methylcyclohexanol **B** $\xrightarrow[\text{heat}]{\text{H}^+}$

1-methylcyclohexene **C** $\xrightarrow[\text{(2) Zn, H}_2\text{O}]{\text{(1) O}_3}$ $\text{O}{=}\text{CHCH}_2\text{CH}_2\text{CH}_2\text{CH}_2\overset{\text{O}}{\overset{\|}{\text{C}}}\text{CH}_3$ **D** $\xrightarrow[\text{(2) H}^+]{\text{(1) Ag}_2\text{O, OH}^-}$

$\text{HO}{-}\overset{\text{O}}{\overset{\|}{\text{C}}}\text{CH}_2\text{CH}_2\text{CH}_2\text{CH}_2\overset{\text{O}}{\overset{\|}{\text{C}}}\text{CH}_3$
E

16.32 $\text{CH}_3\overset{\text{O}}{\overset{\|}{\text{CH}}}$ $\xrightarrow[\text{Zn}]{\text{BrCH}_2\text{CO}_2\text{Et}}$ $\xrightarrow{\text{H}_3\text{O}^+}$ $\text{CH}_3\overset{\text{OH}}{\underset{}{\text{CHCH}}}\text{CH}_2\text{CO}_2\text{Et}$ **K** $\xrightarrow[\text{heat}]{\text{H}^+}$

$\text{CH}_3\text{CH}{=}\text{CHCO}_2\text{Et}$ $\xrightarrow[\text{Pt}]{\text{H}_2}$ $\text{CH}_3\text{CH}_2\text{CH}_2\text{CO}_2\text{Et}$ $\xrightarrow[\text{(2) H}_3\text{O}^+]{\text{(1) DIBAL-H}}$
L   **M**

$\text{CH}_3\text{CH}_2\text{CH}_2\overset{\text{O}}{\overset{\|}{\text{CH}}}$

16.33

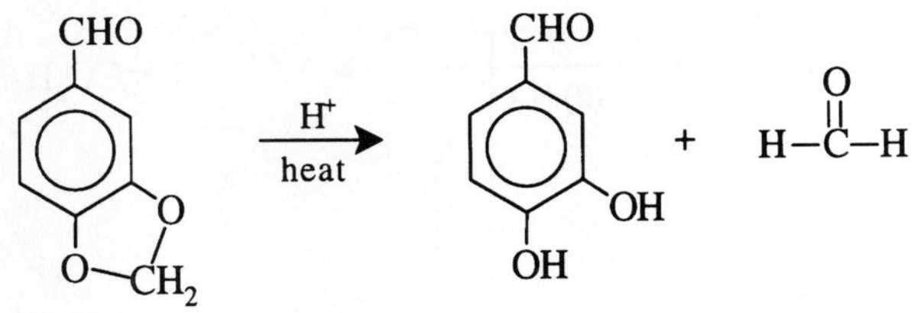

The compound $C_7H_6O_3$ is 3,4-dihydroxybenzaldehyde. The reaction involves hydrolysis of the acetal of formaldehyde.

16.34 (a)

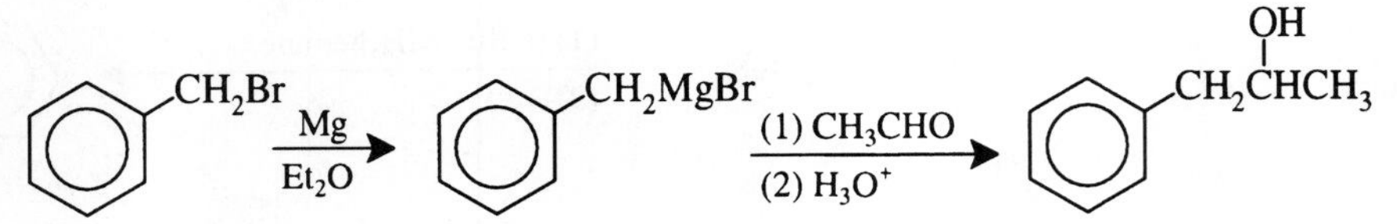

(b)

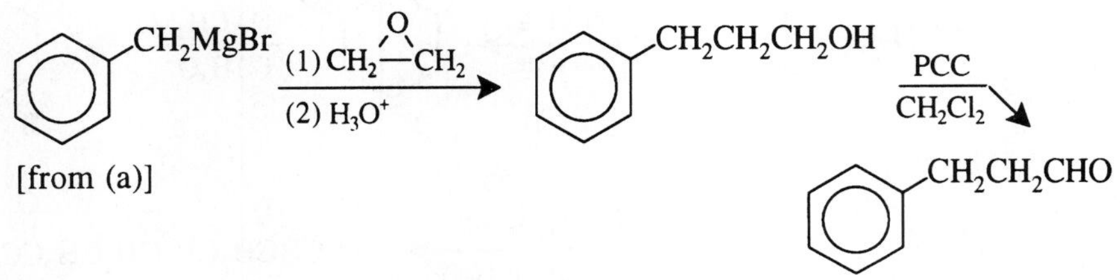

[from (a)]

(c)

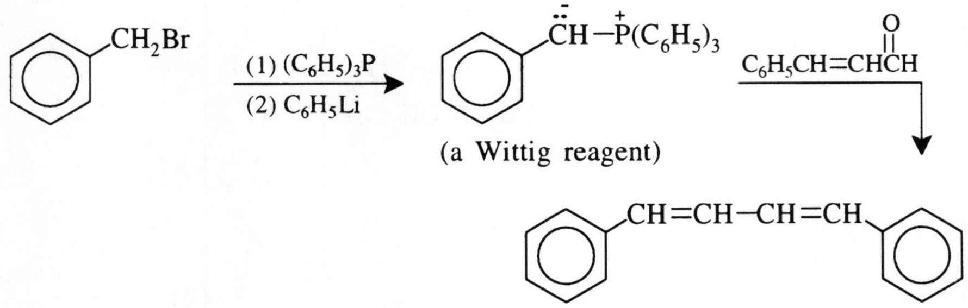

(a Wittig reagent)

(d)

[from (a)]

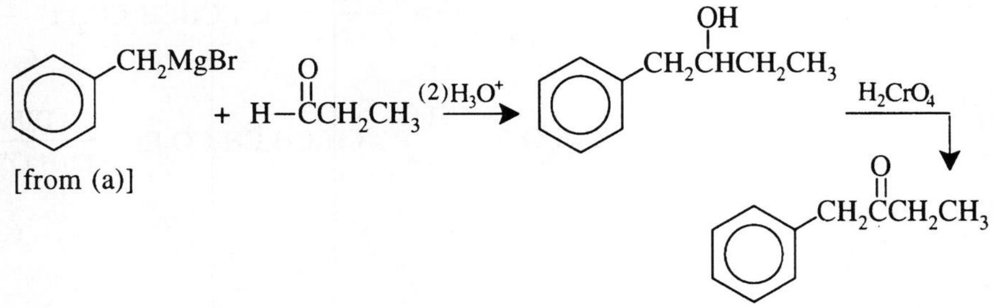

16.35 $BrCH_2CH_2CH_2\overset{\overset{O}{\|}}{C}-H$ $\xrightarrow{HOCH_2CH_2OH,H^+}$ $BrCH_2CH_2CH_2\overset{\overset{O}{|}}{\underset{O}{\overset{|}{C}}H}$ $\xrightarrow[Et_2O]{Mg}$

A

$BrMgCH_2CH_2CH_2\overset{\overset{O}{|}}{\underset{O}{\overset{|}{C}}H}$ $\xrightarrow[(2)H_3O^+,H_2O]{(1)CH_3CHO}$ $CH_3\overset{\overset{OH}{|}}{C}HCH_2CH_2CH_2\overset{\overset{O}{\|}}{C}-H$

B   **C**

$\rightleftharpoons$ (a hemiacetal) $\xrightarrow[(2)\ H^+]{(1)CH_3OH}$ **D** (an acetal)

16.36   (a) $(CH_3)_2SO_4$, NaOH, or CH_3I, NaOH

(b) PCC/CH_2Cl_2 (c)   Zn, $Br\overset{\overset{CH_3}{|}}{C}HCO_2Et$, then H_3O^+

(d) $LiAlH_4$, then H_2O Intermediate is $CH_3O-\langle\bigcirc\rangle-CHO$

16.37 $CH_2{=}CHCH_2OH$ $\xrightarrow[CH_2Cl_2]{PCC}$ $CH_2{=}CH\overset{\overset{O}{\|}}{C}H$ $\xrightarrow[H^+]{CH_3OH}$ $CH_2{=}CH-\overset{\overset{OCH_3}{|}}{\underset{OCH_3}{\overset{|}{C}}H}$

A   **B**

$\xrightarrow[\text{cold, dilute}]{KMnO_4,\ OH^-}$ $\overset{\ }{\underset{OH}{\overset{|}{C}}H_2}-\overset{\ }{\underset{OH}{\overset{|}{C}}H}-\overset{\overset{OCH_3}{|}}{\underset{OCH_3}{\overset{|}{C}}H}$ $\xrightarrow[H_2O]{H_3O^+}$ $\overset{\ }{\underset{OH}{\overset{|}{C}}H_2}-\overset{\ }{\underset{OH}{\overset{|}{C}}H}-\overset{\overset{O}{\|}}{C}H$

C   Glyceraldehyde

The product would be racemic as no chiral reagents were used.

16.38 $CH_3\overset{\overset{O}{\|}}{C}-C\overset{\cdots H}{\underset{CH_2CH_3}{|}}$ $\xrightarrow{NaBH_4}$

(*R*)-3-Phenyl-2-pentanone

(*R*) (*R*)   +   (*S*) (*R*)

Diastereomers

16.39 $BrCH_2(CH_2)_7CH_2Br$ $\xrightarrow[\text{(2) RLi}]{\text{(1) 2 }(C_6H_5)_3P}$ $(C_6H_5)_3\overset{+}{P}-\overset{\cdot\cdot}{C}H(CH_2)_7\overset{\cdot\cdot}{C}H-\overset{+}{P}(C_6H_5)_3$

A

$\xrightarrow{2\ CH_3(CH_2)_{11}\overset{\displaystyle O}{\overset{\|}{C}}CH_3}$ $CH_3(CH_2)_{11}\overset{CH_3}{\overset{|}{C}}=CH(CH_2)_7CH=\overset{CH_3}{\overset{|}{C}}(CH_2)_{11}CH_3$ $\xrightarrow[\text{Pt}]{H_2}$

B

$CH_3(CH_2)_{11}\overset{CH_3}{\overset{|}{C}}H(CH_2)_9\overset{CH_3}{\overset{|}{C}}H(CH_2)_{11}CH_3$

C

16.40 (a) $Ag(NH_3)_2{}^+OH^-$ (positive test with benzaldehyde)

(b) $Ag(NH_3)_2{}^+OH^-$ (positive test with hexanal)

(c) Concentrated H_2SO_4 (2-hexanone is soluble)

(d) CrO_3 in H_2SO_4 (positive test with 2-hexanol)

(e) Br_2 in CCl_4 (decolorization with $C_6H_5CH=CHCOC_6H_5$)

(f) $Ag(NH_3)_2{}^+OH^-$ (positive test with pentanal)

(g) Br_2 in CCl_4 (immediate decolorization occurs with enol form)

(h) $Ag(NH_3)_2{}^+OH^-$ (positive test with cyclic hemiacetal)

16.41 Compound W is

multiplet δ 7.3

singlet δ 3.4

IR peak near 1715 cm^{-1}

Compound X is

multiplet δ 7.5

triplet δ 2.5

triplet δ 3.1

(1)KMnO$_4$, OH$^-$,heat
(2) H$_3$O$^+$

Phthalic acid

16.42 Each ^{1}H NMR spectrum (Figs. 16.4 and 16.5) has a five-hydrogen peak near δ 7.2, suggesting the **Y** and **Z** each has a C_6H_5-group. The IR spectrum of each compound shows a strong peak near 1710 cm^{-1}. This absorption indicates that each compound has a C=O group not adjacent to the phenyl group. We have, therefore, the following pieces,

and

If we subtract the atoms of these pieces from the molecular formula,

$$C_{10}H_{12}O$$
$$-C_7H_5O \quad (C_6H_5 + C=O)$$

We are left with, C_3H_7

In the ^1H NMR spectrum of **Y**, we see an ethyl group [triplet, δ 1.0 (3H) and quartet, δ 2.45 (2H)] and an unsplit $-CH_2-$ group [singlet, δ 3.7 (2H)]. This means that **Y** must be

1-Phenyl-2-butanone

In the ^1H NMR spectrum of **Z,** we see an unsplit $-CH_3$ group [singlet, δ 2.1 (3H)] and two triplets at δ 2.7 and 2.9. This means **Z** must be

4-Phenyl-2-butanone

16.43 That compound **A** forms a phenylhydrazone, gives a negative Tollens' test, and gives an IR band near 1710 cm^{-1} indicates that **A** is a ketone. The ^{13}C spectrum of **A** contains only four signals indicating that A has a high degree of symmetry. The information from the DEPT ^{13}C NMR spectra enables us to conclude that **A** is diisobutyl ketone:

$$\underset{(d)}{\underset{\displaystyle (CH_3)_2\overset{(a)}{C}H\overset{(b)}{C}H_2\overset{(c)}{\overset{\displaystyle O}{\overset{\|}{C}}}CH_2CH(CH_3)_2}{}}$$

Assignments:

        (a) δ 22.6

        (b) δ 24.4

        (c) δ 52.3

        (d) δ 210.0

16.44 That the ^{13}C spectrum of **B** contains only three signals indicates that **B** has a highly symmetrical structure. The information from DEPT spectra indicates the presence of equivalent methyl groups (CH_3 at δ 18.8), equivalent $-\overset{|}{\underset{|}{C}}-$ groups (at δ 70.4), and equivalent $\overset{\diagup}{C}=O$ groups (at δ 215.0). These features allow only one possible structure for **B**:

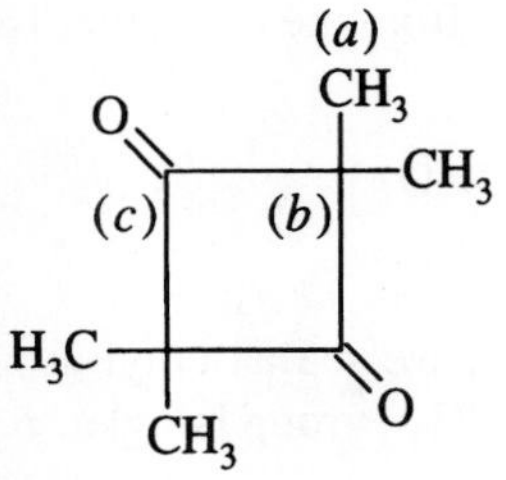

Assignments:

(a)   δ 18.8

(b)   δ 70.4

(c)   δ 215.0

16.45 The two nitrogen atoms of semicarbazide that are adjacent to the C=O group bear partial positive charges because of resonance contributions made by the second and third structures below.

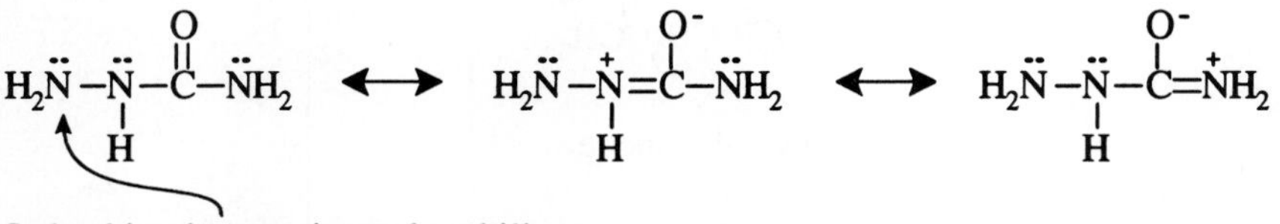

Only this nitrogen is nucleophilic.

16.46 Hydrolysis of the acetal linkage of multistriatin produces the ketodiol below.

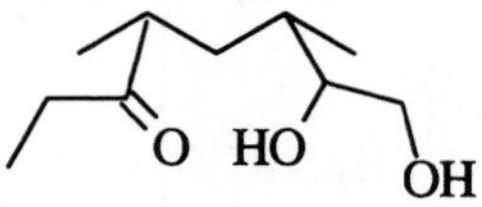

16.47

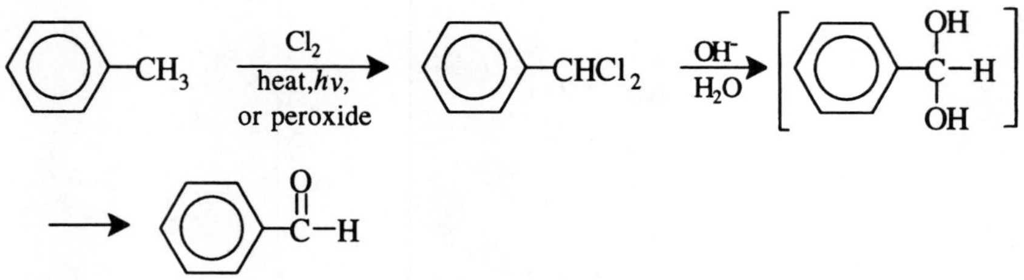

The *gem*-diol formed in the alkaline hydrolysis step readily loses water to form the aldehyde.

16.48

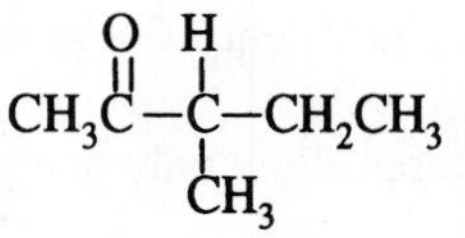

$$CH_3\overset{\text{O}}{\overset{\|}{C}}-\overset{\text{H}}{\underset{\underset{CH_3}{|}}{C}}-CH_2CH_3 \quad \text{or its enantiomer}$$

16.49 The general formula for an oxime is

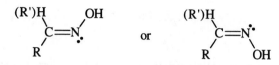

Both carbon and nitrogen are *sp²* hybridized; the electron pair on nitrogen occupies one *sp²* orbital. Aldoximes and ketoximes can exist in either of these two stereoisomeric forms:

$$
\begin{array}{ccc}
\underset{R}{\overset{(R')H}{C}}\!=\!\!\overset{\cdot\cdot}{N}\!-\!OH & \text{or} & \underset{R}{\overset{(R')H}{C}}\!=\!\overset{\cdot\cdot}{N} \\
& & \quad\quad OH
\end{array}
$$

This type of stereoisomerism is also observed in the case of other compounds that possess the C=N group, for example, phenylhydrazones and semicarbazones.

QUIZ

16.1 Which Wittig reagent could be used to synthesize $C_6H_5CH=CHCH_2CH_3$? (Assume any other needed reagents are available.)

(a) $C_6H_5\overset{\cdot\cdot}{\underset{}{C}}H\overset{+}{P}(C_6H_5)_3$

(b) $C_6H_5CH=CH\overset{\cdot\cdot}{C}H\overset{+}{P}(C_6H_5)_3$

(c) $CH_3CH_2\overset{\cdot\cdot}{C}H\overset{+}{P}(C_6H_5)_3$

(d) More than one of these

(e) None of these

16.2 Which compound is an acetal?

(a) $C_6H_5\underset{\underset{OH}{|}}{C}HOCH_3$

(b)

(c)

(d) More than one of these

(e) None of these

16.3 Which reaction sequence could be used to convert $C_6H_{13}C\equiv CH$ to $C_6H_{13}\overset{\overset{\displaystyle O}{\|}}{C}CH_3$?

(a) O_3, then Zn, H_2O, then $AlCl_3$, then CH_3CO_2H

(b) H_2SO_4, $HgSO_4$, H_2O, heat

(c) HCl, then CH_3CO_2H

(d) O_3, then Zn, H_2O, then H_2SO_4, $HgSO_4$, H_2O, heat

(e) CH_3CO_2H, then H_2O_2, OH^-/H_2O

16.4 Complete the following syntheses. If more than one step is required for a transformation, list them as (1), (2), (3), and so on.

(a)

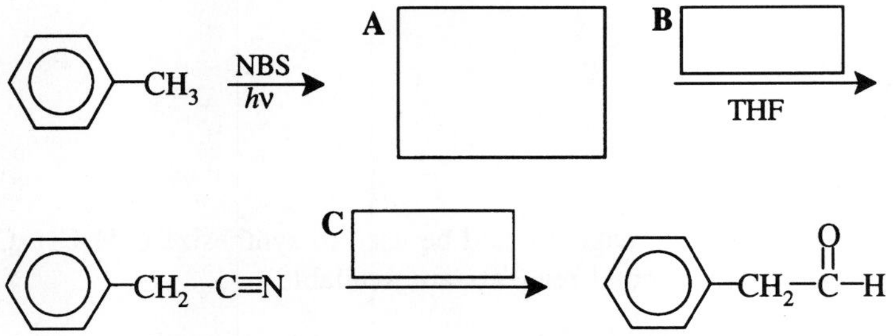

(b)

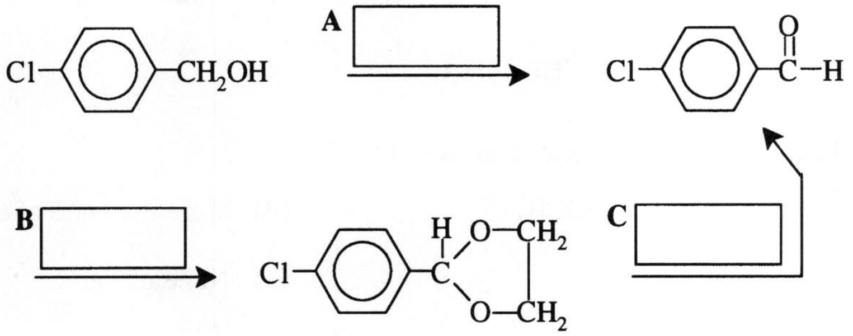

(c)

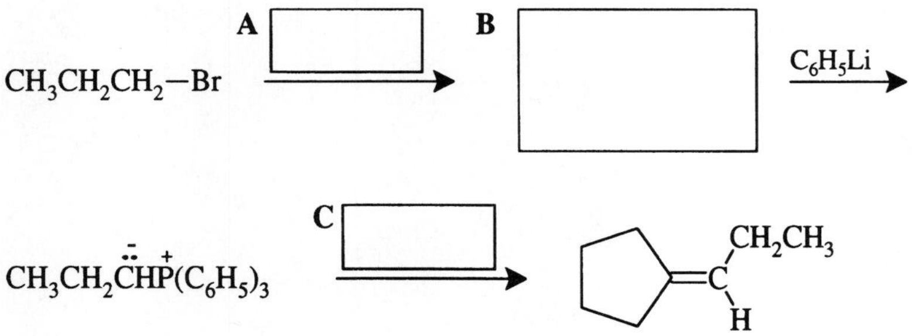

(d)

$C_6H_5-\overset{\overset{\displaystyle O}{\|}}{C}-Cl$ $\xrightarrow{\text{A} \boxed{}}$ $C_6H_5-\overset{\overset{\displaystyle O}{\|}}{C}-CH_3$ $\xrightarrow{\text{B} \boxed{}}$

$C_6H_5-\overset{\overset{\displaystyle OH}{|}}{\underset{\underset{\displaystyle N}{\|}}{\underset{\displaystyle C}{C}}}-CH_3$ $\xrightarrow{\text{C} \boxed{}}$ $C_6H_5-\overset{\overset{\displaystyle OH}{|}}{\underset{\underset{\displaystyle CH_2NH_2}{|}}{C}}-CH_3$

17 ALDEHYDES AND KETONES II. ALDOL REACTIONS

SOLUTIONS TO PROBLEMS

17.1

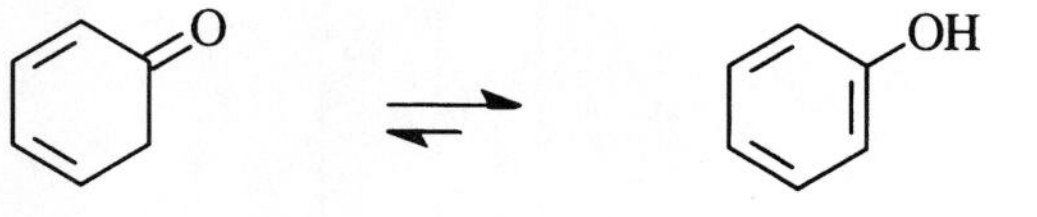

2,4-Cyclohexadien-1-one
(keto form)

Phenol
(enol form)

The enol form is aromatic, and it is therefore stabilized by the resonance energy of the benzene ring.

17.2 No.

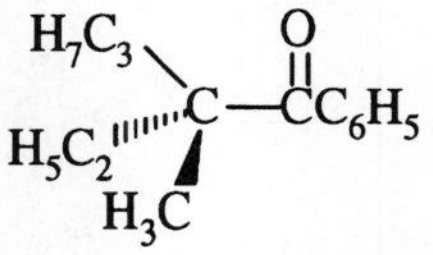

does not have a hydrogen attached to its α-carbon atom (which is a stereocenter) and thus enol formation involving the stereocenter is not possible. With

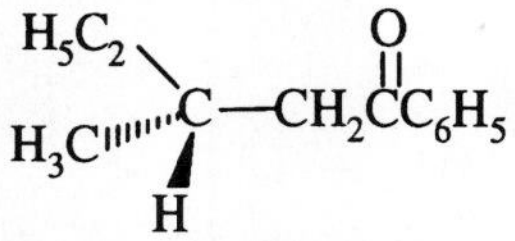

the stereocenter is a β carbon and thus enol formation does not affect it.

17.3 In OD⁻/D₂O:

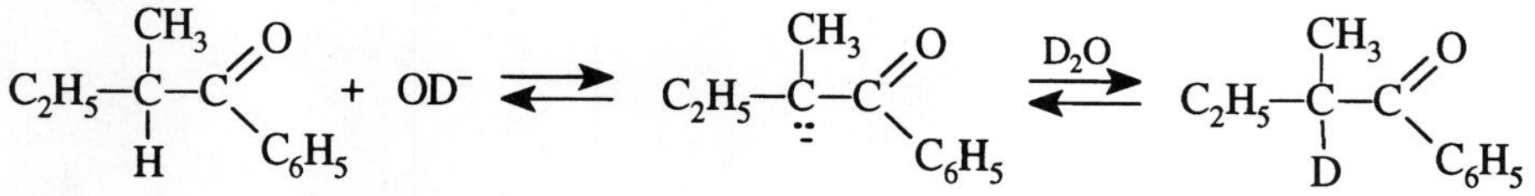

In D₃O⁺/D₂O:

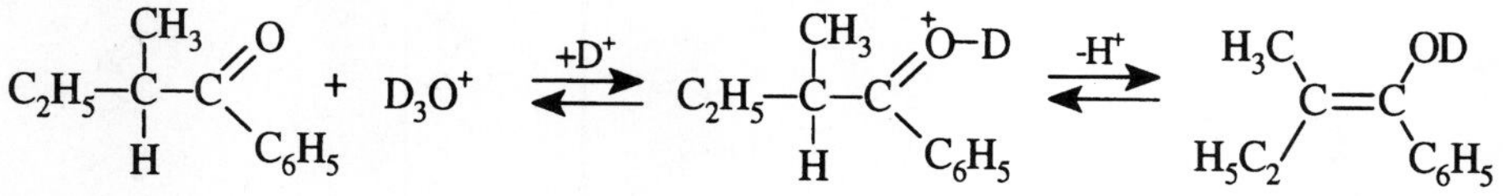

$$\xrightleftharpoons{+D^+} \quad C_2H_5-\overset{\overset{\displaystyle CH_3}{|}}{\underset{\underset{\displaystyle D}{|}}{C}}-C\overset{\displaystyle \overset{+}{O}-D}{\underset{\displaystyle C_6H_5}{}} \quad \xrightleftharpoons{-D^+} \quad C_2H_5-\overset{\overset{\displaystyle CH_3}{|}}{\underset{\underset{\displaystyle D}{|}}{C}}-C\overset{\displaystyle O}{\underset{\displaystyle C_6H_5}{}}$$

17.4 The reaction is said to be "base promoted" because base is consumed as the reaction takes place. A catalyst is, by definition, not consumed.

17.5 (a) The slow step in base-catalyzed racemization is the same as that in base-promoted halogenation—*the formation of an enolate ion.* (Formation of an enolate ion from *sec*-butyl phenyl ketone leads to racemization because the enolate ion is achiral. When it accepts a proton, it yields a racemic form.) The slow step in acid-catalyzed racemization is also the same as that in acid-catalyzed halogenation—*the formation of an enol.* (The enol, like the enolate ion, is achiral and tautomerizes to yield a racemic form of the ketone.)

(b) According to the mechanism given, the slow step for acid-catalyzed iodination (formation of the enol) is the same as that for acid-catalyzed bromination. Thus, we would expect both reactions to occur at the same rate.

(c) Again, the slow step for both reactions (formation of the enolate ion) is the same, and consequently, both reactions take place at the same rate.

17.6 (a) Acetone, $CH_3\overset{\overset{\displaystyle O}{||}}{C}CH_3$

(b) Acetophenone, $C_6H_5\overset{\overset{\displaystyle O}{||}}{C}CH_3$

(d) 2-Pentanone, $CH_3CH_2CH_2\overset{\overset{\displaystyle O}{||}}{C}CH_3$

(f) 1-Phenylethanol, $C_6H_5\overset{\overset{\displaystyle OH}{|}}{C}HCH_3$

(h) 2-Butanol, $CH_3CH_2\overset{\overset{\displaystyle OH}{|}}{C}HCH_3$

(i) Methyl 2-naphthyl ketone,

$$\text{(naphthalene ring)}\;\overset{\overset{\displaystyle O}{||}}{C}CH_3$$

17.7 (a) $\overset{\beta}{C}H_3\overset{\alpha}{C}H_2\overset{\overset{\displaystyle O}{||}}{C}H \;+\; OH^- \;\rightleftharpoons\; CH_3\overset{..}{C}H\overset{\overset{\displaystyle O}{||}}{C}H \;+\; H_2O$

$$CH_3CH_2\overset{O}{\overset{\|}{C}}H \; + \; :\overset{O}{\underset{\overset{|}{CH_3}}{\overset{\|}{CHCH}}} \; \rightleftarrows \; CH_3CH_2\overset{O^-}{\underset{\overset{|}{CH_3}}{CH}}\overset{O}{\overset{\|}{CHCH}}$$

$$CH_3CH_2\overset{O^-}{\underset{\overset{|}{CH_3}}{CH}}\overset{O}{\overset{\|}{CHCH}} \; + \; H_2O \; \rightleftarrows \; CH_3CH_2\overset{OH}{\underset{\overset{|}{CH_3}}{CH}}\overset{O}{\overset{\|}{CHCH}} \; + \; OH^-$$

(b) For $CH_3CH_2\overset{OH}{\underset{|}{C}}HCH_2CH_2\overset{O}{\overset{\|}{C}}H$ to form, a hydroxide ion would have to remove a β proton in the first step. This does not happen because the anion that would be produced, that is, $:CH_2CH_2CHO$, cannot be stabilized by resonance.

(c) $CH_3CH_2CH{=}\overset{O}{\underset{\overset{|}{CH_3}}{\overset{\|}{CCH}}}$

17.8 CH_3CHO $\xrightarrow[5°C]{10\% \; NaOH}$ $CH_3\overset{OH}{\underset{|}{C}}HCH_2CHO$ $\xrightarrow{heat}$ $CH_3CH{=}CHCHO$ $\xrightarrow[Ni]{H_2}$
(aldol)

$$CH_3CH_2CH_2CH_2OH$$

17.9 (a) $2\,CH_3CH_2CH_2CHO$ $\xrightarrow[H_2O]{OH^-}$ $CH_3CH_2CH_2\overset{OH}{\underset{\overset{|}{CH_2}}{CH}}\overset{}{\underset{\overset{|}{CH_3}}{CH}}CHO$

(b) Product of (a) $\xrightarrow[-H_2O]{H^+}$ $CH_3CH_2CH_2CH{=}\overset{}{\underset{\overset{|}{CH_2}}{\underset{\overset{|}{CH_3}}{C}}}CHO$ $\xrightarrow[(2)\,H_2O]{(1)\,LiAlH_4,\,Et_2O}$

$$CH_3CH_2CH_2CH{=}\overset{}{\underset{\overset{|}{CH_2}}{\underset{\overset{|}{CH_3}}{C}}}CH_2OH$$

(c) Product of (b) $\xrightarrow[Pt]{H_2}$ $CH_3CH_2CH_2CH_2\overset{}{\underset{\overset{|}{CH_2}}{\underset{\overset{|}{CH_3}}{C}}}HCH_2OH$

(d) Product of (a) $\xrightarrow{NaBH_4}$ $CH_3CH_2CH_2\overset{OH}{\underset{\overset{|}{CH_2}}{\underset{\overset{|}{CH_3}}{C}}}HCHCH_2OH$

17.10

$$C_{11}H_{14}O$$

$$C_{14}H_{18}O \qquad \qquad \text{Lily aldehyde } (C_{14}H_{20}O)$$

17.11

17.12 Three successive aldol additions occur.

First
Aldol
Addition

$$CH_3\overset{O}{\overset{\|}{C}}H \;+\; OH^- \;\rightleftharpoons\; {}^-\!:\!CH_2\overset{O}{\overset{\|}{C}}H \;+\; H_2O$$

$$H\overset{O}{\overset{\|}{C}}H \;+\; {}^-\!:\!CH_2\overset{O}{\overset{\|}{C}}H \;\rightleftharpoons\; {}^-\!OCH_2CH_2\overset{O}{\overset{\|}{C}}H$$

$${}^-\!OCH_2CH_2\overset{O}{\overset{\|}{C}}H \;+\; H_2O \;\rightleftharpoons\; HOCH_2CH_2\overset{O}{\overset{\|}{C}}H \;+\; OH^-$$

Second
Aldol
Addition

$$HOCH_2CH_2\overset{O}{\overset{\|}{C}}H \;+\; OH^- \;\rightleftharpoons\; HOCH_2\overset{..}{C}H\overset{O}{\overset{\|}{C}}H \;+\; H_2O$$

$$H\overset{O}{\overset{\|}{C}}H \;+\; HOCH_2\overset{..}{C}H\overset{O}{\overset{\|}{C}}H \;\rightleftharpoons\; HOCH_2\overset{CH_2O^-}{\overset{|}{C}}HCHO$$

$$HOCH_2\overset{CH_2O^-}{\overset{|}{C}}HCHO \;+\; H_2O \;\rightleftharpoons\; HOCH_2\overset{CH_2OH}{\overset{|}{C}}HCHO \;+\; OH^-$$

Third
Aldol
Addition

$$HOCH_2\overset{CH_2OH}{\overset{|}{C}}HCHO \;+\; OH^- \;\rightleftharpoons\; HOCH_2\overset{CH_2OH}{\overset{|}{\underset{..}{C}}}\!-CHO \;+\; H_2O$$

$$H\overset{O}{\overset{\|}{C}}H \;+\; HOCH_2\overset{CH_2OH}{\overset{|}{\underset{..}{C}}}\!-CHO \;\rightleftharpoons\; HOCH_2\!-\!\overset{CH_2OH}{\underset{CH_2O^-}{\overset{|}{\underset{|}{C}}}}\!-CHO$$

$$HOCH_2\!-\!\overset{CH_2OH}{\underset{CH_2O^-}{\overset{|}{\underset{|}{C}}}}\!-CHO \;+\; H_2O \;\rightleftharpoons\; HOCH_2\!-\!\overset{CH_2OH}{\underset{CH_2OH}{\overset{|}{\underset{|}{C}}}}\!-CHO \;+\; OH^-$$

17.13 (a) $CH_3CO_2H + BF_3 \rightleftharpoons CH_3CO_2BF_3^- + H^+$

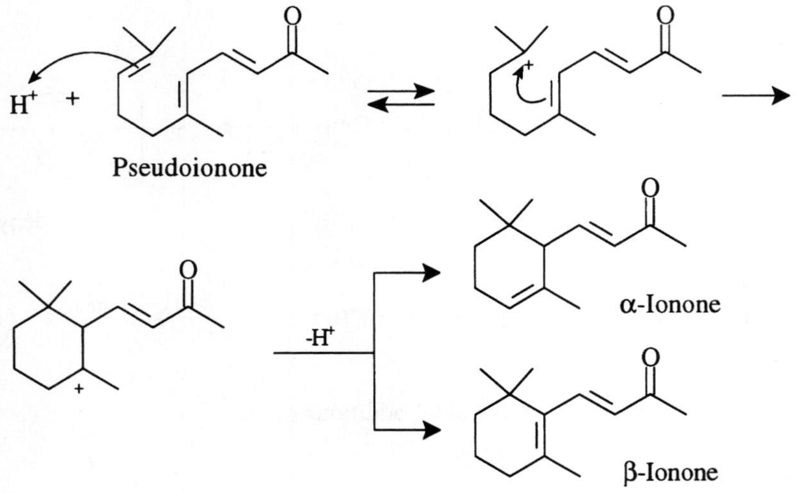

Pseudoionone

α-Ionone

β-Ionone

(b) In β-ionone both double bonds and the carbonyl group are conjugated; thus it is more stable.

(c) β-Ionone, because it is a fully conjugated unsaturated system.

17.14 (a)

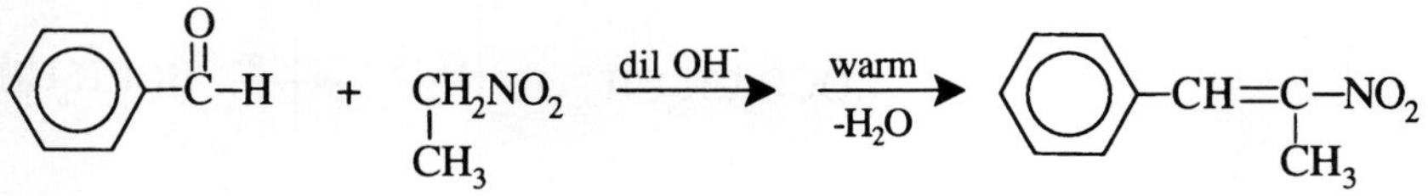

(b) $H\overset{O}{\overset{\|}{C}}H + CH_3NO_2 \xrightarrow{\text{dil OH}^-} HOCH_2CH_2NO_2$

(c)

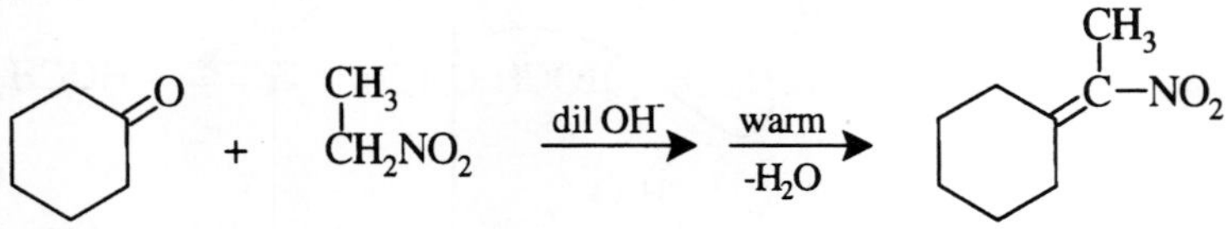

17.15 (a) $^-\!:CH_2\!\!-\!\!C\!\equiv\!\!N\!: \longleftrightarrow CH_2\!\!=\!\!C\!\!=\!\!\ddot{N}\!:^-$

(b) $CH_3\!\!-\!\!C\!\equiv\!\!N\!: \overset{\text{EtO}^-}{\rightleftharpoons} \left[^-\!:CH_2\!\!-\!\!C\!\equiv\!\!N\!: \longleftrightarrow CH_2\!\!=\!\!C\!\!=\!\!\ddot{N}\!:^- \right] + \text{EtOH}$

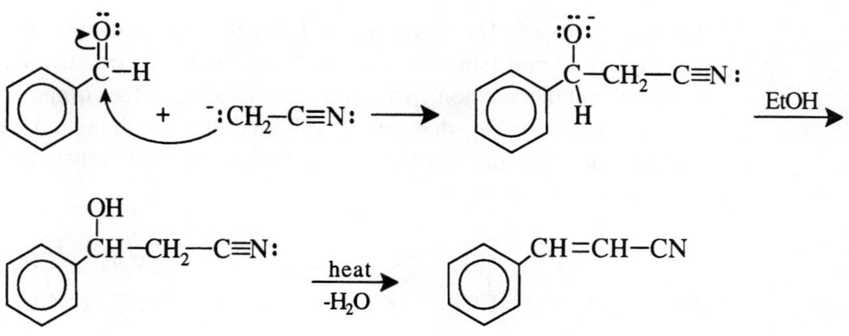

17.16

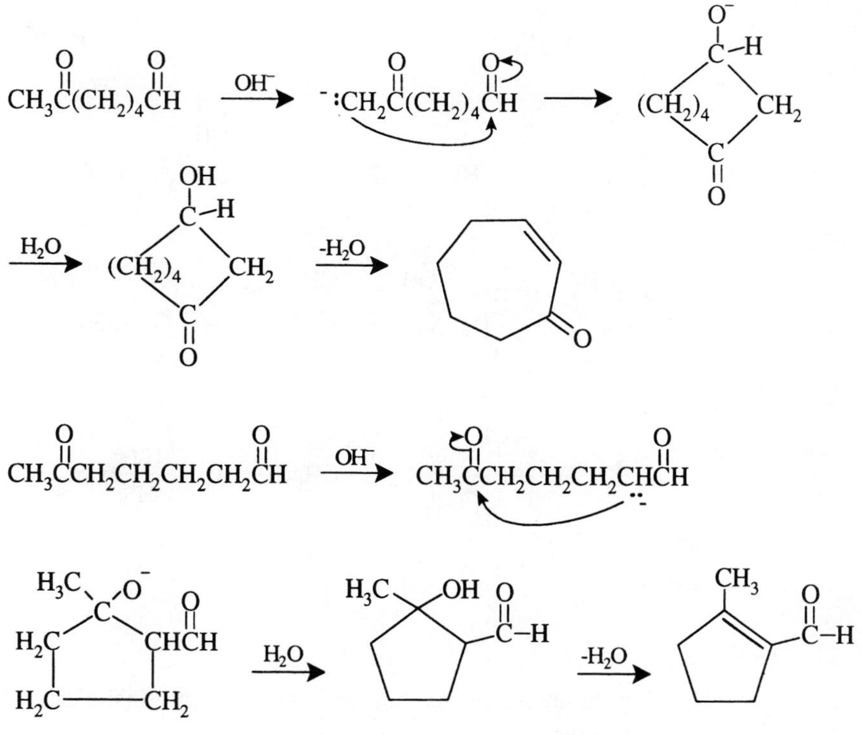

17.17 (a)

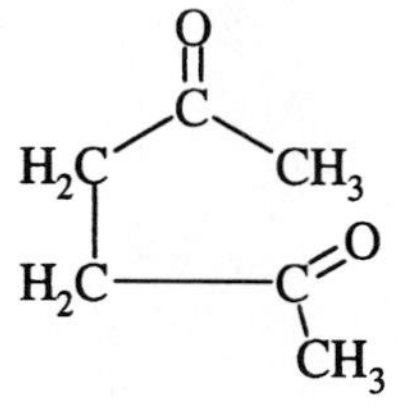

(b)

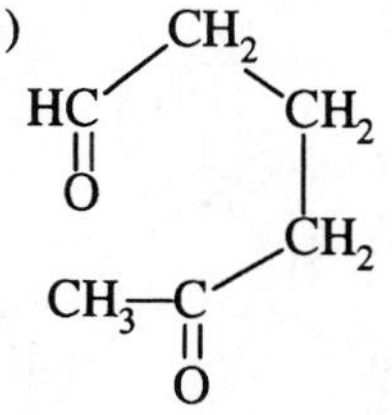

(c)

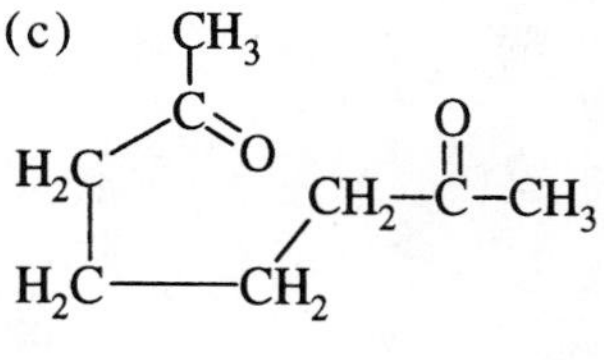

Notice that starting compounds are drawn so as to indicate which atoms are involved in the cyclization reaction.

17.18 It is necessary for conditions to favor the intramolecular reaction rather than the intermolecular one. One way to create these conditions is to use very dilute solutions when we carry out the reaction. When the concentration of the compound to be cyclized is very low (i.e., when we use what we call a "high dilution technique"), the probability is greater that one end of a molecule will react with the other end of that same molecule rather than with a different molecule.

17.19

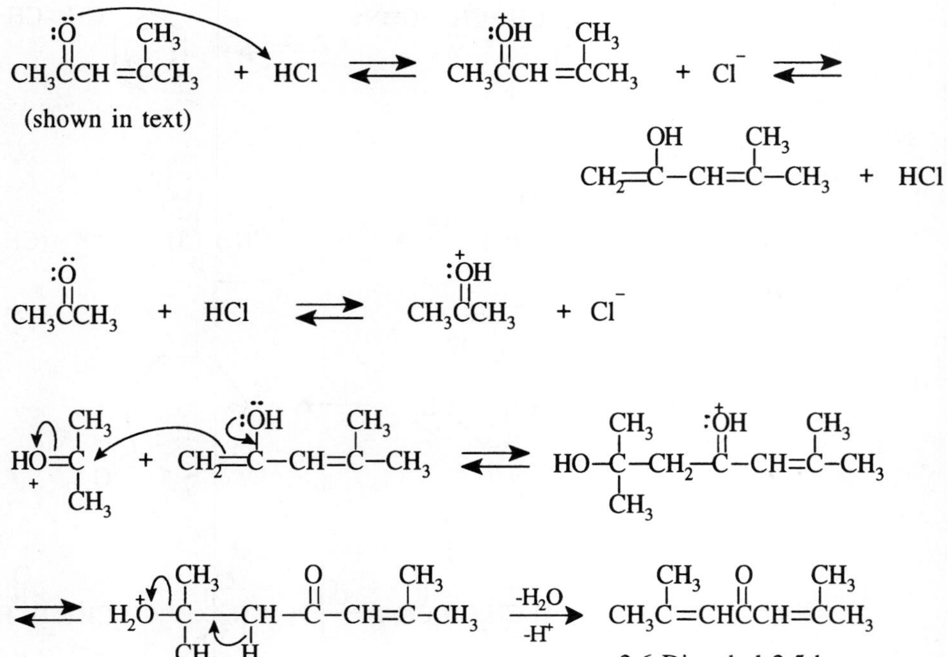

(shown in text)

17.20

Drawing the molecules as they will appear in the final product helps to visualize the necessary steps:

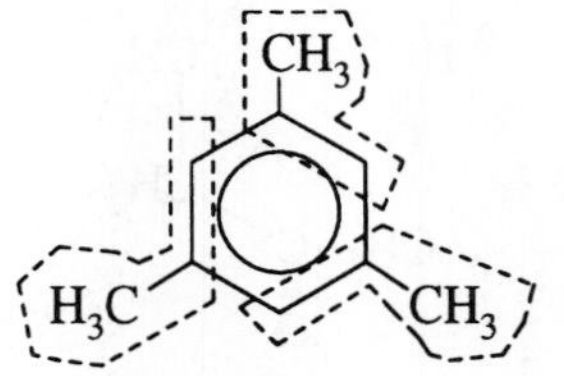

Mesitylene

The two molecules that lead to mesitylene are shown as follows:

This molecule (4-methyl-3-penten-2-one) is formed by an acid-catalyzed condensation between two molecules of acetone as shown in the text.

The mechanism is,

17.21 (a)

(b)   2-Methyl-1,3-cyclohexanedione is more acidic because its enolate ion is stabilized by an additional resonance structure.

17.22   (a) $C_6H_5\overset{O}{\overset{\|}{C}}CH_3 \underset{+H^+}{\overset{-H^+}{\rightleftharpoons}} C_6H_5\overset{O}{\overset{\|}{C}}CH_2{:}^-$

$C_6H_5\overset{O}{\overset{\|}{C}}CH_2{:}^- + C_6H_5CH{=}CH\overset{O}{\overset{\|}{C}}C_6H_5 \rightleftharpoons C_6H_5CH{-}CH{=}\overset{-O}{C}C_6H_5$

$\overset{|}{C}H_2$

$\overset{|}{C}{=}O$

$\overset{|}{C}_6H_5$

$\underset{+H^+}{\overset{-H^+}{\updownarrow}}$

$C_6H_5CHCH_2\overset{O}{\overset{\|}{C}}C_6H_5$

$\overset{|}{C}H_2$

$\overset{|}{C}{=}O$

$\overset{|}{C}_6H_5$

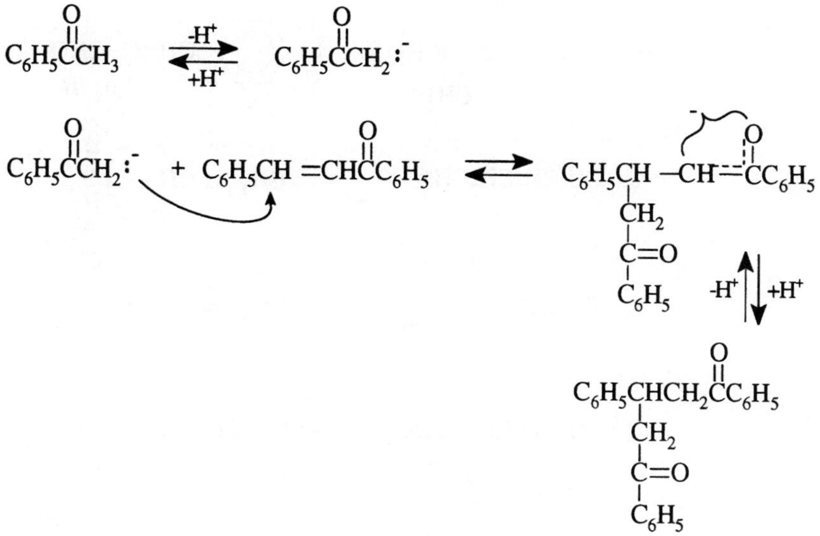

(b)

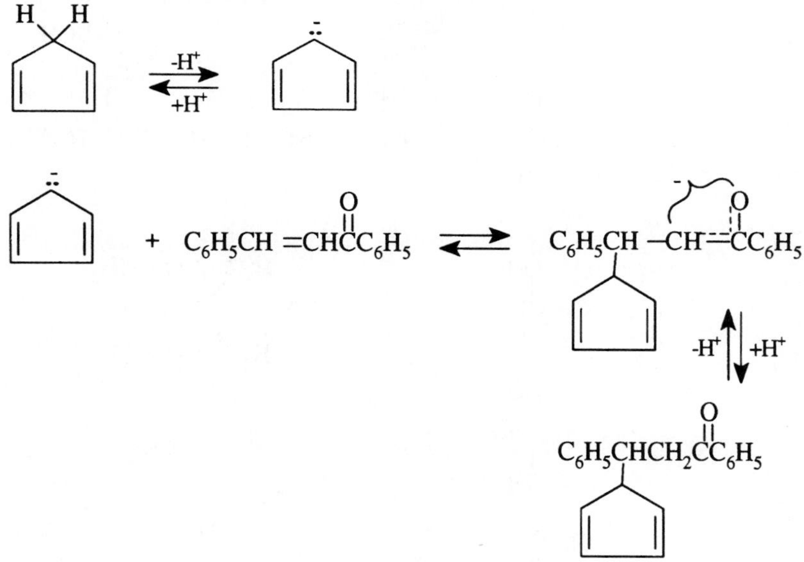

$+ C_6H_5CH{=}CH\overset{O}{\overset{\|}{C}}C_6H_5 \rightleftharpoons C_6H_5CH{-}CH{=}\overset{-O}{C}C_6H_5$

$\underset{+H^+}{\overset{-H^+}{\updownarrow}}$

$C_6H_5CHCH_2\overset{O}{\overset{\|}{C}}C_6H_5$

17.23 $H_2\ddot{N}{-}\ddot{N}H_2 + CH_2{=}CH{-}\overset{O}{\overset{\|}{C}}H \xrightarrow[\text{addition}]{\text{conjugate}}$

$\rightleftharpoons$ $\xrightarrow{-H_2O}$

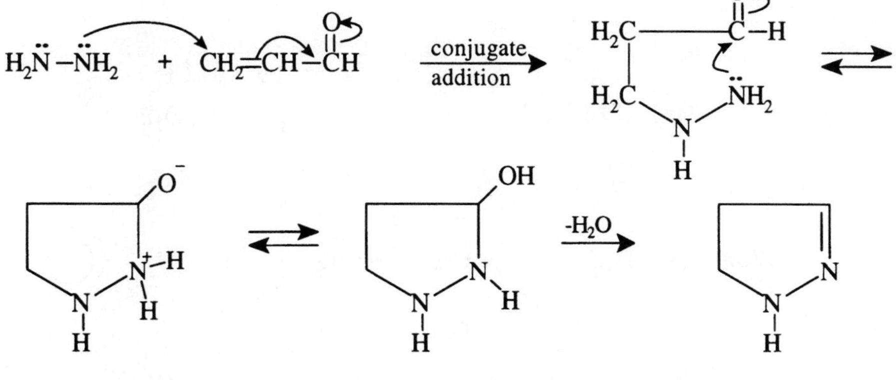

17.24 (a) $CH_3CH_2\overset{\displaystyle OH}{\underset{\displaystyle CH_3}{\overset{|}{\underset{|}{C}}HCHCHO}}$

(b) $\langle\!\langle\bigcirc\!\rangle\!\rangle\!-CH\!=\!\overset{\displaystyle}{\underset{\displaystyle CH_3}{\overset{}{\underset{|}{C}}}}-CHO$

(c) $CH_3CH_2\overset{\displaystyle OH}{\overset{|}{C}HCN}$

(d) $CH_3CH_2CH_2OH$

(e) $CH_3CH_2\overset{\displaystyle O-CH_2}{\underset{\displaystyle O-CH_2}{\overset{}{C}H{\Big\langle}}}$

(f) $CH_3CH_2\overset{\displaystyle O}{\overset{\|}{C}}-OH$

(g) $CH_3CH_2\overset{\displaystyle OH}{\overset{|}{C}HCH_3}$

(h) $CH_3CH_2\overset{\displaystyle O}{\overset{\|}{C}}-OH$

(i) $CH_3CH_2CH=NOH$

(j) $CH_3CH_2CH=CHC_6H_5$

(k) $CH_3CH_2\overset{\displaystyle OH}{\overset{|}{C}H}-C_6H_5$

(l) $CH_3CH_2\overset{\displaystyle OH}{\overset{|}{C}HC\equiv CH}$

(m) $CH_3CH_2CH_3$

(n) $CH_3CH_2\overset{\displaystyle OH}{\overset{|}{C}H}-\underset{\displaystyle CH_2CH_3}{\overset{|}{C}HCO_2Et}$

17.25 (a) $CH_3\overset{\displaystyle CH_3}{\underset{\displaystyle OH}{\overset{|}{\underset{|}{C}}}}-CH_2\overset{\displaystyle O}{\overset{\|}{C}}CH_3$

(b) $C_6H_5CH=CH\overset{\displaystyle O}{\overset{\|}{C}}CH_3$

(c) $CH_3\overset{\displaystyle OH}{\underset{\displaystyle CN}{\overset{|}{\underset{|}{C}}}}CH_3$

(d) $CH_3\overset{\displaystyle OH}{\overset{|}{C}HCH_3}$

(e) $\underset{\displaystyle H_3C}{\overset{\displaystyle H_3C}{{>}}}C{\overset{\displaystyle O-CH_2}{\underset{\displaystyle O-CH_2}{{\Big\langle}}}}$

(f) No reaction

(g) $CH_3\overset{\displaystyle OH}{\underset{\displaystyle CH_3}{\overset{|}{\underset{|}{C}}}}CH_3$

(h) No reaction

(i) $CH_3\overset{\displaystyle NOH}{\overset{\|}{C}}CH_3$

(j) $\underset{\displaystyle H_3C}{\overset{\displaystyle H_3C}{{>}}}C=CHC_6H_5$

(k) $CH_3\overset{\displaystyle OH}{\underset{\displaystyle CH_3}{\overset{|}{\underset{|}{C}}}}C_6H_5$

(l) $CH_3\overset{\displaystyle OH}{\underset{\displaystyle CH_3}{\overset{|}{\underset{|}{C}}}}C\equiv CH$

(m) $CH_3CH_2CH_3$

(n) $CH_3\overset{\displaystyle HO}{\overset{|}{\underset{\displaystyle CH_3}{\underset{|}{C}}}}-\overset{\displaystyle CH_2CH_3}{\overset{|}{C}HCO_2Et}$

17.26 (a) $H_3C-\langle\!\langle\bigcirc\!\rangle\!\rangle\!-CH=CHCHO$

(b) $H_3C-\langle\!\langle\bigcirc\!\rangle\!\rangle\!-\overset{\displaystyle OH}{\overset{|}{C}HC\equiv CCH_3}$

(c) $H_3C-\langle\!\langle\bigcirc\!\rangle\!\rangle\!-\overset{\displaystyle OH}{\overset{|}{C}HCH_2CH_3}$

(d) $H_3C-\langle\!\langle\bigcirc\!\rangle\!\rangle\!-\overset{\displaystyle O}{\overset{\|}{C}}-OH$

(e)

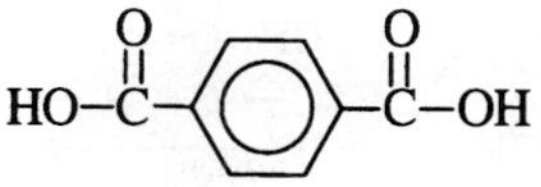

(f)

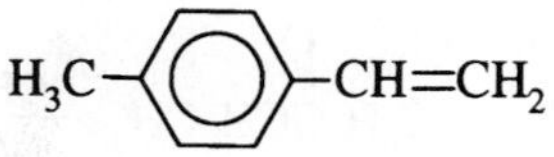

(g)

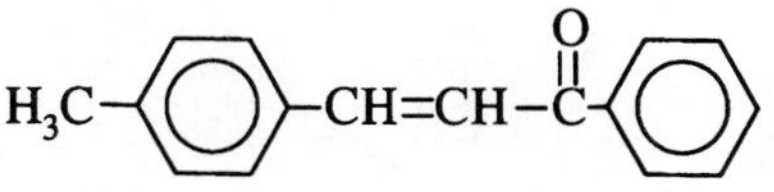

(h)

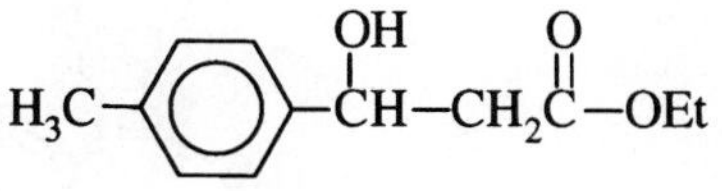

17.27 (a)

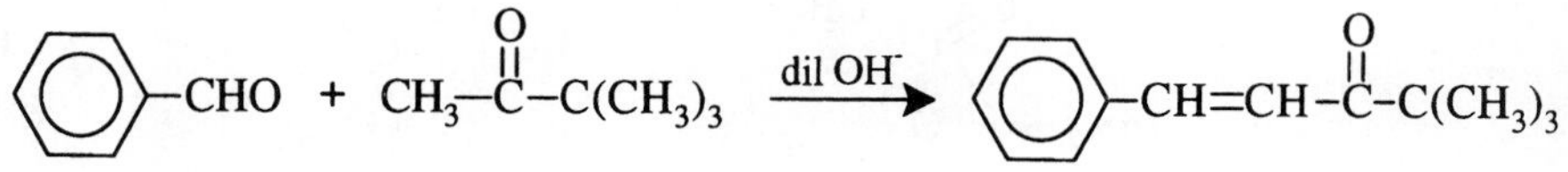

(b)

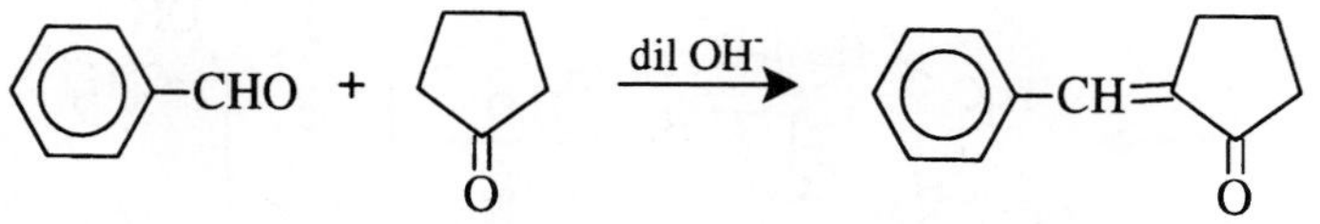

(c)

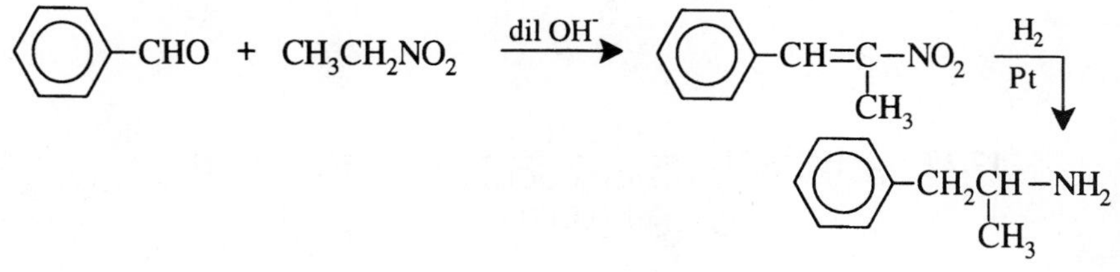

(d)

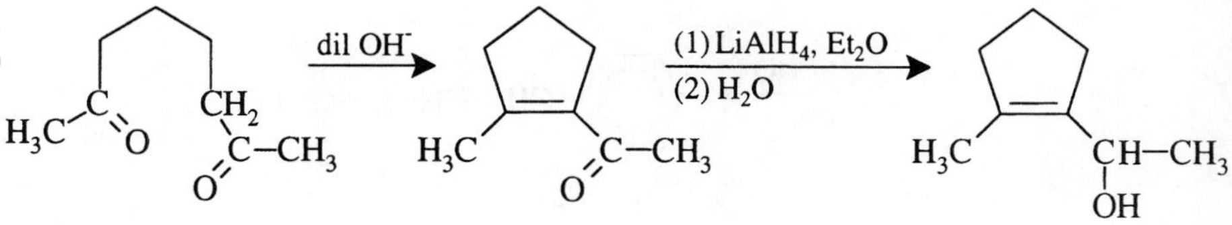

(e) 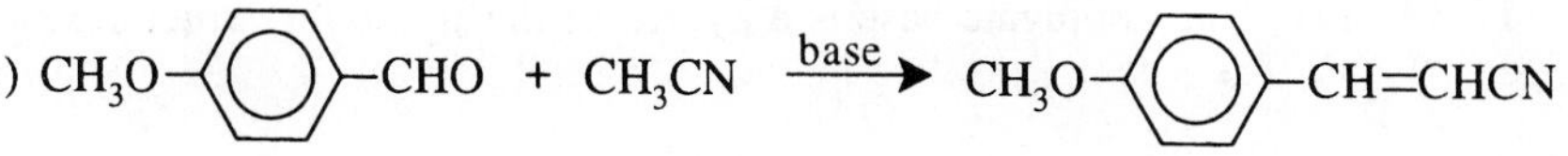 CH_3O—⟨⟩—CHO + CH_3CN $\xrightarrow{base}$ CH_3O—⟨⟩—$CH=CHCN$

(f) 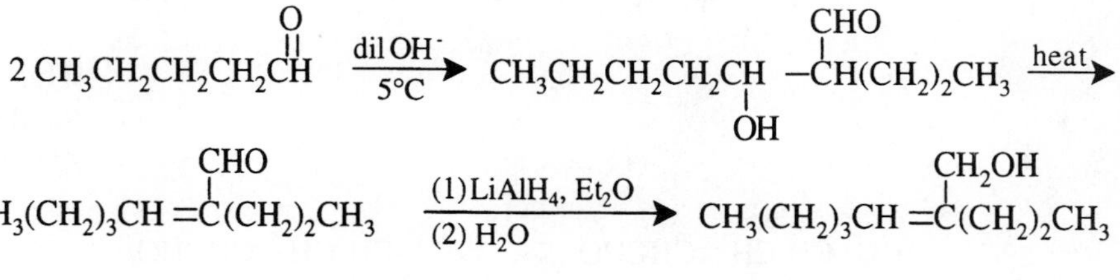 $2\ CH_3CH_2CH_2CH_2\overset{O}{\overset{\|}{C}}H$ $\xrightarrow[5°C]{dil\ OH^-}$ $CH_3CH_2CH_2CH_2CH\underset{OH}{\overset{CHO}{\underset{|}{-}}}CH(CH_2)_2CH_3$ $\xrightarrow{heat}$

$CH_3(CH_2)_3CH\overset{CHO}{=}C(CH_2)_2CH_3$ $\xrightarrow[(2)\ H_2O]{(1)LiAlH_4,\ Et_2O}$ $CH_3(CH_2)_3CH\overset{CH_2OH}{=}C(CH_2)_2CH_3$

(g)

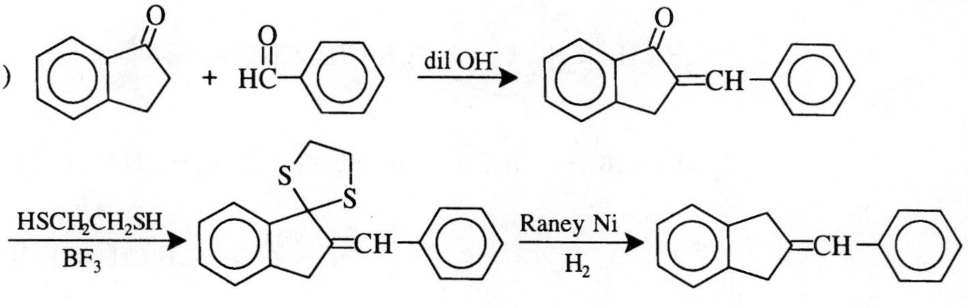

17.28 $C_6H_5\overset{O}{\overset{\|}{C}}CH_2CH_3$ $\xrightarrow{OH^-}$ $C_6H_5\overset{O}{\overset{\|}{C}}-\overset{..}{C}HCH_3$ $\longrightarrow$

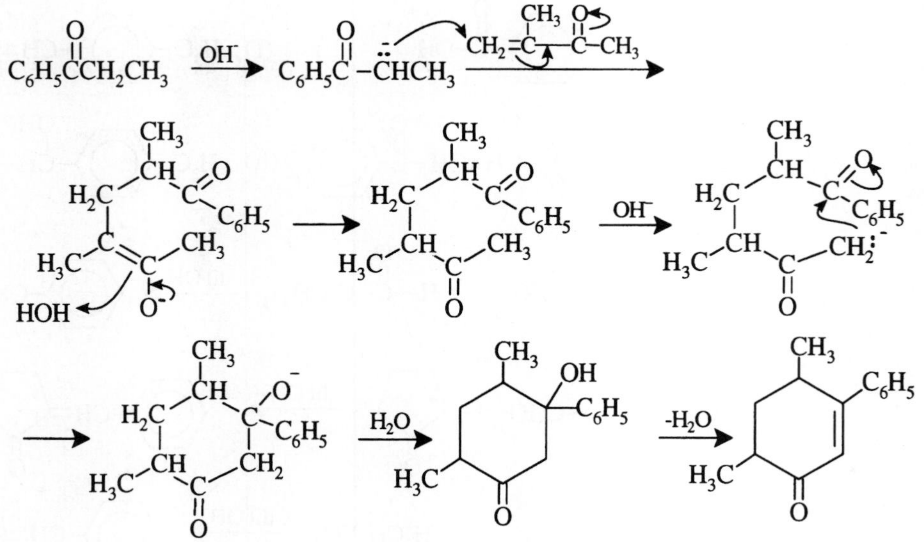

$\longrightarrow$

17.29 $HC\equiv CH$ $\xrightarrow[\substack{(2)\,CH_3COCH_3 \\ (3)\,NH_4Cl/H_2O}]{(1)\,NaNH_2}$ $HC\equiv C-\overset{CH_3}{\underset{CH_3}{\overset{|}{\underset{|}{C}}}}-OH$ $\xrightarrow[H_2O]{Hg^{2+},\,H_3O^+}$ $CH_3\overset{O}{\overset{\|}{C}}-\overset{CH_3}{\underset{CH_3}{\overset{|}{\underset{|}{C}}}}-OH$

A   **B**

$\xrightarrow[OH^-]{C_6H_5CHO}$

C

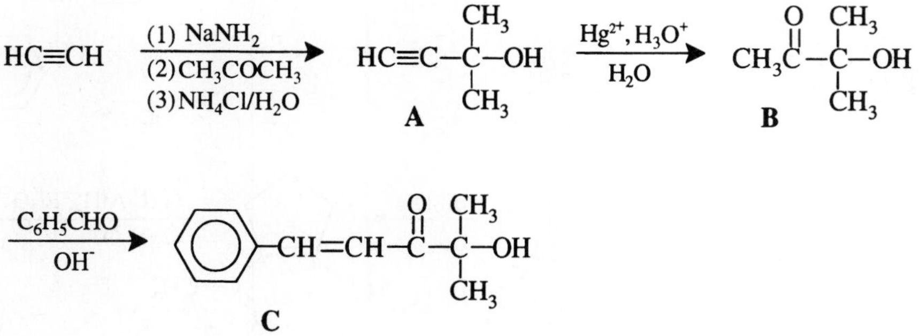

17.30   (a)   The conjugate base is a hybrid of the following structures:

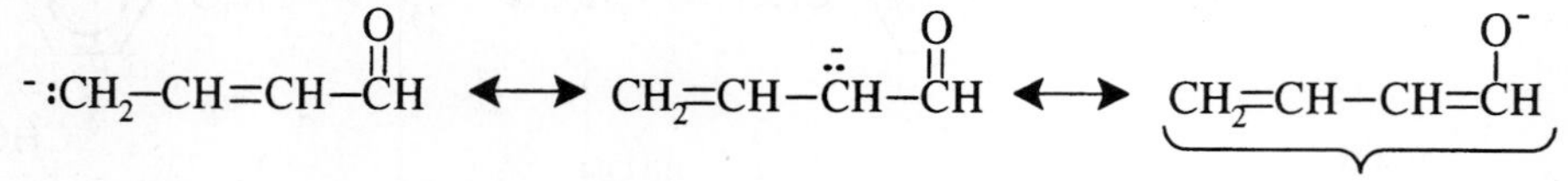

$^-\!:CH_2-CH=CH-\overset{O}{\overset{\|}{C}}H$ $\longleftrightarrow$ $CH_2=CH-\overset{..}{C}H-\overset{O}{\overset{\|}{C}}H$ $\longleftrightarrow$ $\underbrace{CH_2=CH-CH=\overset{O^-}{\underset{|}{C}}H}$

This structure is especially
stable because the negative
charge is on the oxygen atom.

(b) $CH_3CH=CHCHO$ $\underset{+H^+}{\overset{-H^+}{\rightleftarrows}}$ $^-\!:CH_2CH=CHCHO$

$C_6H_5CH=CH\overset{O}{\overset{\|}{C}}H$ $+$ $^-\!:CH_2CH=CHCHO$ $\rightleftarrows$

$C_6H_5CH=CH\overset{O^-}{\underset{|}{C}}H-CH_2CH=CHCHO$ $\underset{-H^+}{\overset{+H^+}{\rightleftarrows}}$ $C_6H_5CH=CH\overset{OH}{\underset{|}{C}}H-CH_2CH=CHCHO$

$\xrightarrow{-H_2O}$ $C_6H_5CH=CHCH=CHCH=CHCHO$

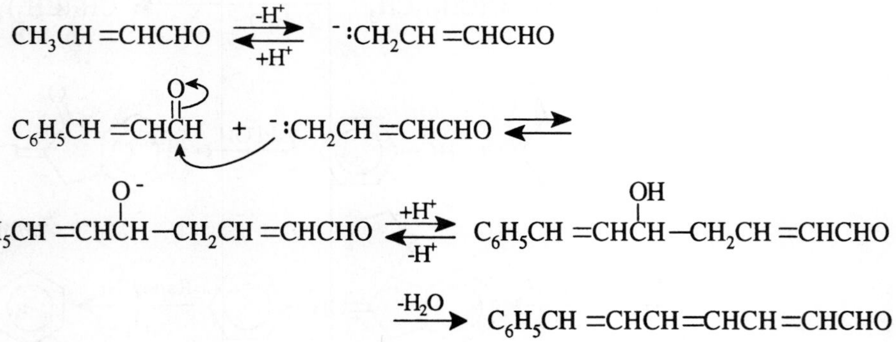

17.31 (a)

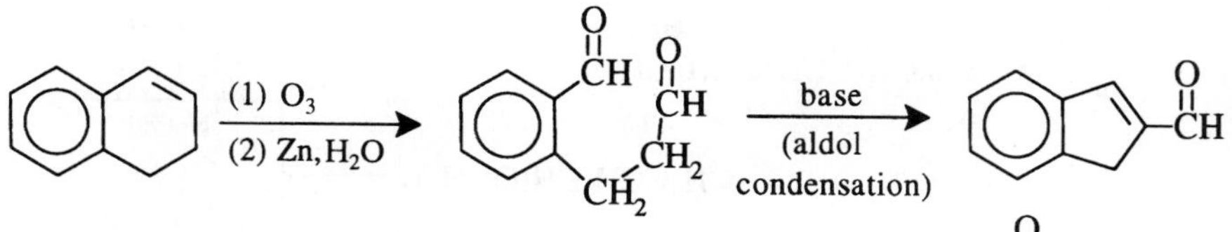

(b)

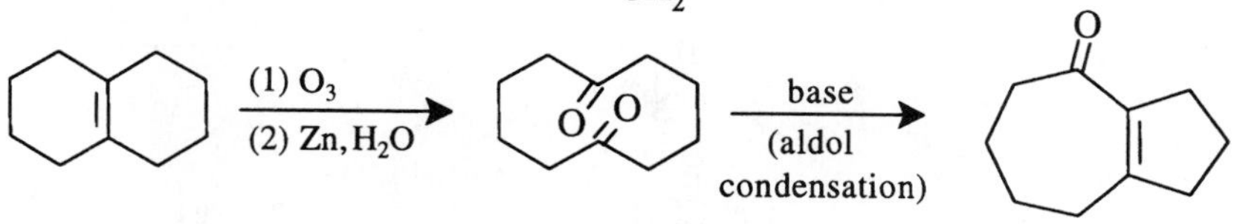

(c)

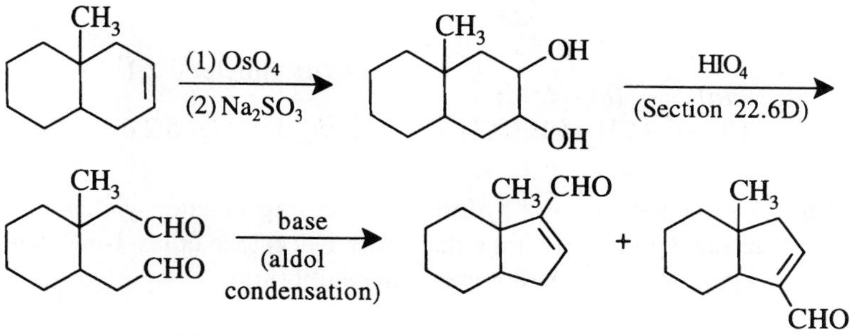

(d)

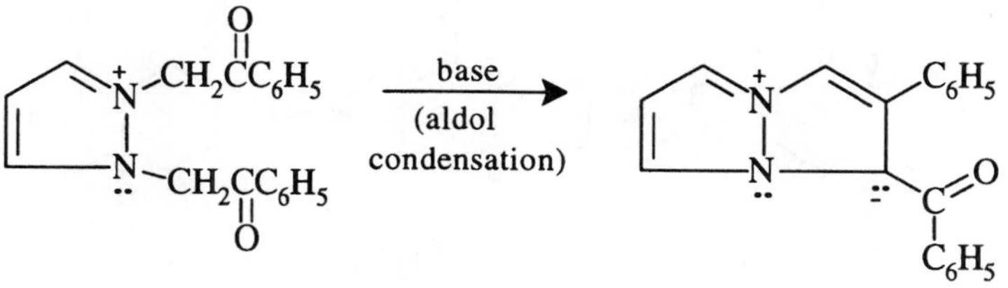

17.32 (a) In simple addition, the carbonyl peak (1665–1780-cm^{-1} region) does not appear in the product; in conjugate addition it does.

(b) As the reaction takes place, the long-wavelength absorption arising from the conjugated system should disappear. One could follow the rate of the reaction by following the rate at which this absorption peak disappears.

17.33 (a) Compound **U** is ethyl phenyl ketone: (b) Compound **V** is benzyl methyl ketone:

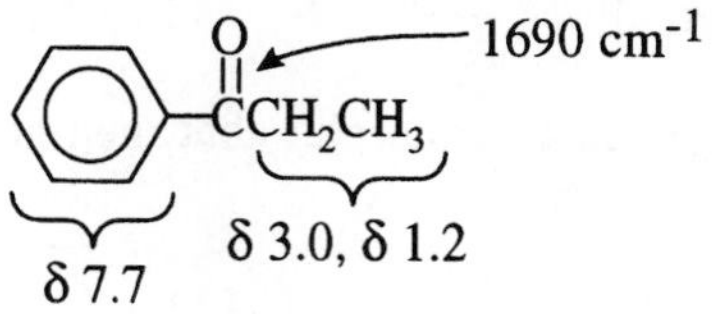

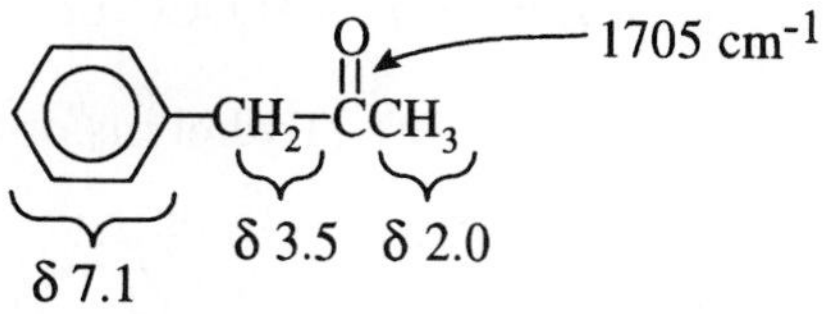

17.34 A is $CH_3\overset{O}{\overset{\|}{C}}CH_2CH(OCH_3)_2$

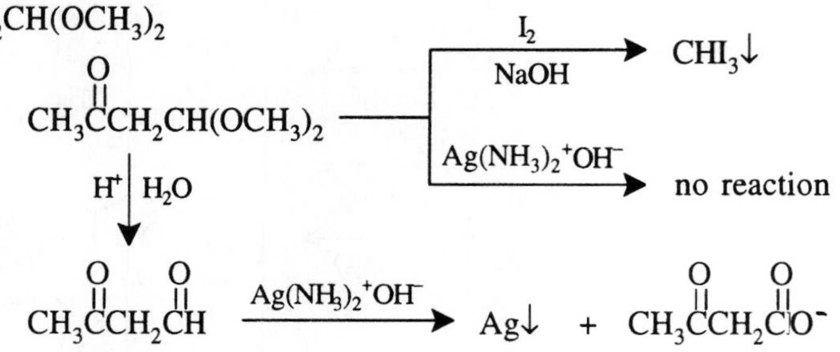

$$CH_3\overset{O}{\overset{\|}{C}}CH_2CH(OCH_3)_2 \quad \xrightarrow[\text{NaOH}]{I_2} \quad CHI_3\downarrow$$

$$\xrightarrow{Ag(NH_3)_2{}^+OH^-} \quad \text{no reaction}$$

$H^+ \Big| H_2O$

$$CH_3\overset{O}{\overset{\|}{C}}CH_2\overset{O}{\overset{\|}{C}}H \xrightarrow{Ag(NH_3)_2{}^+OH^-} Ag\downarrow \; + \; CH_3\overset{O}{\overset{\|}{C}}CH_2\overset{O}{\overset{\|}{C}}O^-$$

(a) $\overset{O}{\overset{\|}{}}$ (b) (c) (d)

$CH_3-C-CH_2-CH(OCH_3)_2$

(a) Singlet δ 2.1   (c) Triplet δ 4.7

(b) Doublet δ 2.6   (d) Singlet δ 3.2

17.35 Abstraction of an α hydrogen at the ring junction yields an enolate ion that can then accept a proton to form either *trans*-1-decalone or *cis*-1-decalone. Since *trans*-1-decalone is more stable, it predominates at equilibrium.

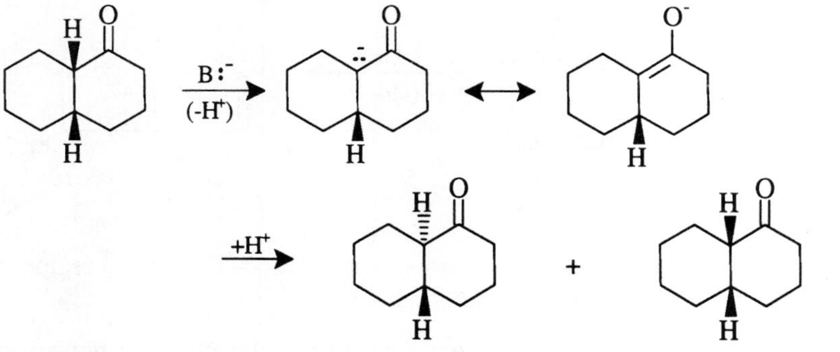

(95%)   (5%)

trans-1-Decalone   *cis*-1-Decalone

(more stable)   (less stable)

17.36 (a) $CH_3OCH_2Br \; + \; (C_6H_5)_3P \xrightarrow{\text{(2) RLi}} CH_3OCH=P(C_6H_5)_3$

(b) Hydrolysis of the ether yields a hemiacetal that then goes on to form an aldehyde.

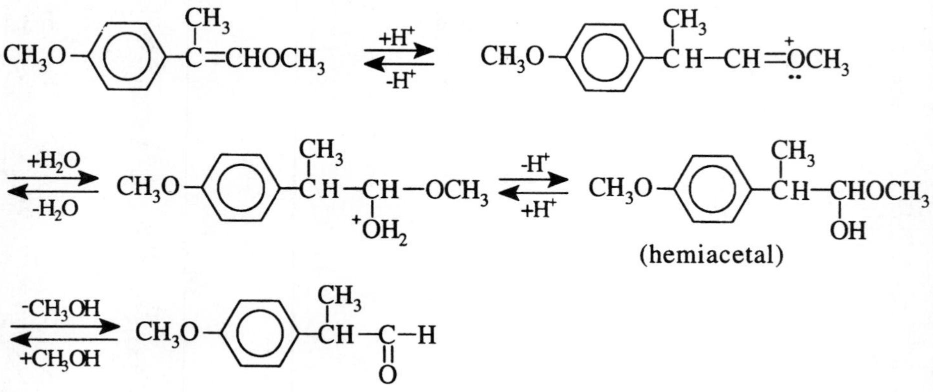

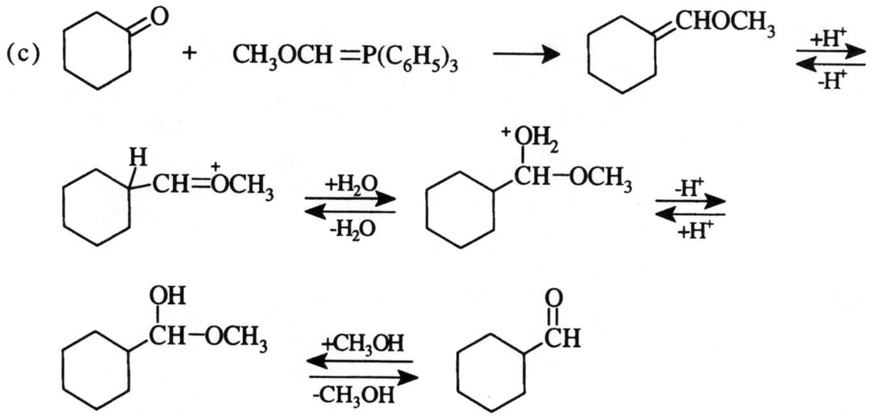

(c)

17.37 (a)   The hydrogen atom that is added to the aldehyde carbon atom in the reduction must come from the other aldehyde rather than from the solvent. It must be transferred as a hydride ion and directly from molecule to molecule, since if it were ever a free species it would react immediately with the solvent. A possible mechanism is the following:

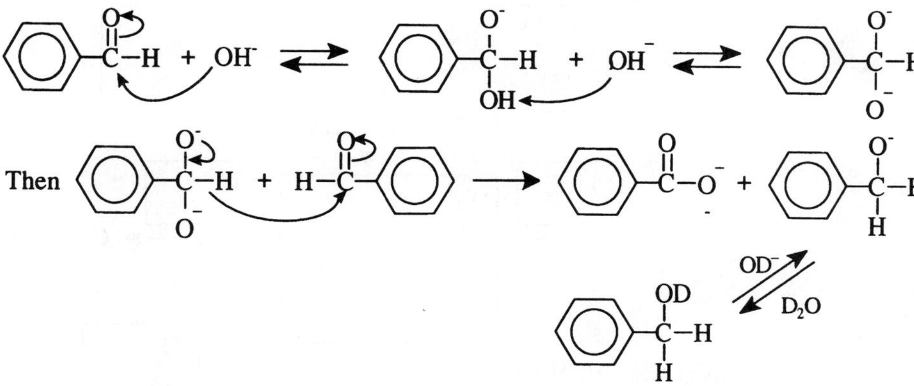

(b)   Although an aldol reaction occurs initially, the aldol reaction is reversible. The Cannizzaro reaction, though slower, is irreversible. Eventually, all the product is in the form of the alcohol and the carboxylate ion.

17.38 This difference in behavior indicates that, for acetaldehyde, the capture of a proton from the solvent (the reverse of the reaction by which the enolate ion is formed) occurs much more slowly than the attack by the enolate ion on another molecule.

When acetone is used, the equilibrium for the formation of the enolate ion is unfavorable, but more importantly, enolate attack on another acetone molecule is disfavored due to steric hindrance. Here proton capture (actually deuteron capture) competes very well with the aldol reaction.

17.39

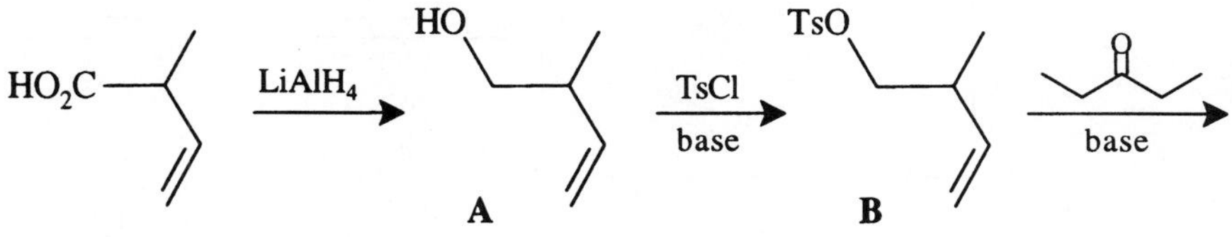

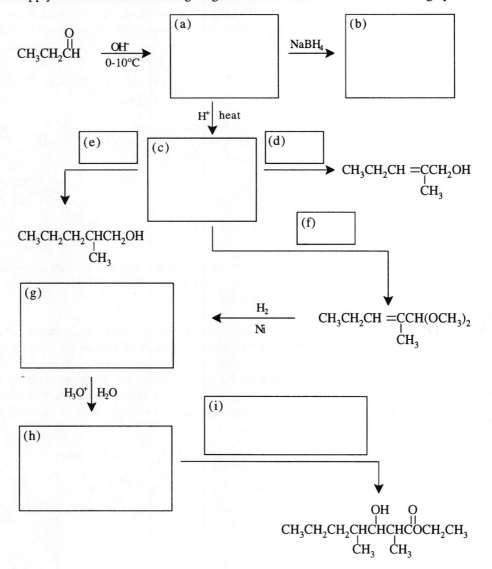

17.40 X is:

$$\langle\text{Ph}\rangle\text{—CH}=\text{CH—}\overset{\overset{\text{O}}{\|}}{\text{C}}\text{—CH}=\text{CH—}\langle\text{Ph}\rangle$$

QUIZ

17.1 Supply formulas for the missing reagents and intermediates in the following synthesis.

17.2 Supply formulas for the missing reagents and intermediates in the following synthesis.

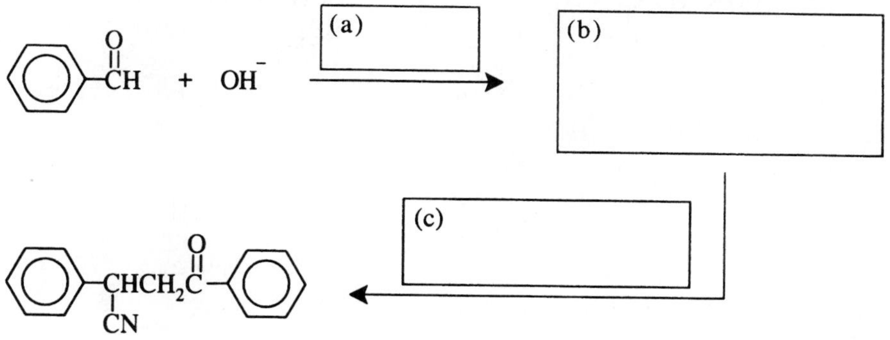

17.3 Supply formulas for the missing reagents and intermediates in the following synthesis.

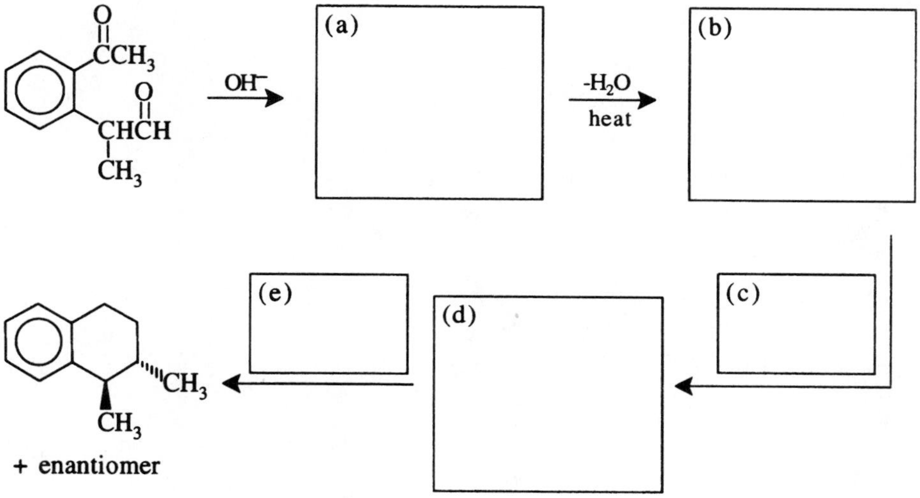

+ enantiomer

17.4 Which would be formed in the following reaction?

$$CH_3\overset{O}{\overset{\|}{C}}H \quad + \quad CH_3CH_2\overset{O}{\overset{\|}{C}}H \quad \xrightarrow[25°C]{OH^-} \quad ?$$

(a) $CH_3\overset{OH}{\underset{|}{C}}HCH_2\overset{O}{\overset{\|}{C}}H$

(b) $CH_3CH_2\overset{OH}{\underset{|}{C}}H\overset{O}{\overset{\|}{C}}H$
　　　　　　　　　$\underset{CH_3}{}$

(c) $CH_3\overset{OH}{\underset{|}{C}}HCH\overset{O}{\overset{\|}{C}}H$
　　　　　　　　$\underset{CH_3}{}$

(d) $CH_3CH_2\overset{OH}{\underset{|}{C}}HCH_2\overset{O}{\overset{\|}{C}}H$

(e) All of these will be formed.

17.5 What would be the major product of the following reaction?

$$C_6H_5\overset{O}{\overset{\|}{C}}\overset{}{\underset{|}{C}}HCH_3 \quad + \quad Br_2 \quad + \quad OH^- \quad \longrightarrow$$
　　　　　$\underset{CH_3}{}$

(a) $C_6H_5\overset{\overset{\displaystyle O}{\|}}{C}\underset{\underset{\displaystyle CH_3}{|}}{C}BrCH_3$

(b) $C_6H_5\overset{\overset{\displaystyle O}{\|}}{C}\underset{\underset{\displaystyle CH_3}{|}}{C}HCH_2Br$

(c) $C_6H_5\overset{\overset{\displaystyle O}{\|}}{C}\underset{\underset{\displaystyle CH_3}{|}}{C}HCH_2OH$

(d) $C_6H_5\underset{\overset{\displaystyle |}{\underset{\displaystyle CH_3}{}}}{C}\overset{\overset{\displaystyle CH_3}{|}}{Br_2}CHCH_3$

(e)   None of these

SOLUTIONS TO PROBLEMS

H.1

Enolate from
cyclohexanone

$C_6H_5CH_2Br$

O-Alkylated product

C-Alkylated product

H.2

(a) [cyclopentanone with CH₃] $\xrightarrow{\text{LDA}}$ [lithium enolate] Kinetic enolate $\xrightarrow[\text{(2) H}_2\text{O}]{\text{(1) CH}_3\text{CH}}$ [product]

(b) $CH_3CH_2\overset{\text{O}}{\overset{\|}{C}}CH_3$ $\xrightarrow{\text{LDA}}$ $CH_3CH_2\overset{\text{O}^-\text{Li}^+}{\overset{|}{C}}=CH_2$ $\xrightarrow[\text{(2) H}_2\text{O}]{\text{(1) C}_6\text{H}_5\text{CH}}$ $CH_3CH_2\overset{\text{O}}{\overset{\|}{C}}CH_2\overset{\text{OH}}{\overset{|}{C}}HC_6H_5$

Kinetic enolate

(c) $CH_3\overset{\text{CH}_3}{\underset{|}{C}}H\overset{\text{O}}{\overset{\|}{C}}CH_3$ $\xrightarrow{\text{LDA}}$ $CH_3\overset{\text{CH}_3}{\underset{|}{C}}H\overset{\text{O}^-\text{Li}^+}{\overset{|}{C}}=CH_2$ $\xrightarrow[\text{(2) H}_2\text{O}]{\text{(1) CH}_3\text{CH}_2\text{CH}}$ $CH_3\overset{\text{CH}_3}{\underset{|}{C}}H\overset{\text{O}}{\overset{\|}{C}}CH_2\overset{\text{OH}}{\overset{|}{C}}HCH_2CH_3$

Kinetic enolate

(d) $CH_3CH=CH\overset{\text{O}}{\overset{\|}{C}}CH_3$ $\xrightarrow{\text{LDA}}$ $CH_3CH=CH\overset{\text{O}^-\text{Li}^+}{\overset{|}{C}}=CH_2$ $\xrightarrow[\text{(2) H}_2\text{O}]{\text{(1) CH}_3\text{CH}}$ $CH_3CH=CH\overset{\text{O}}{\overset{\|}{C}}CH_2\overset{\text{OH}}{\overset{|}{C}}HCH_3$

H.3   (a)

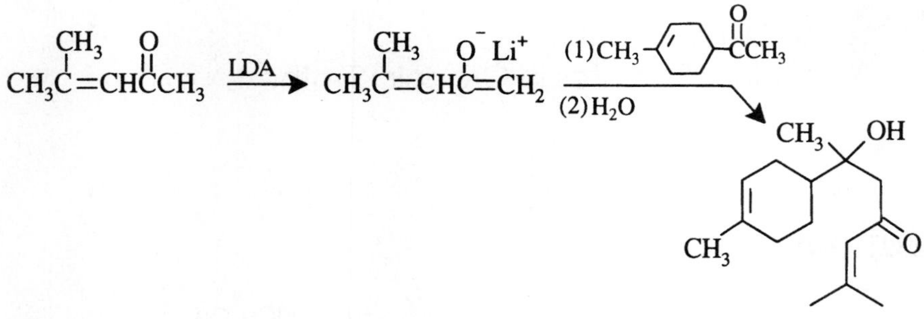

α-Bisabolanone

(b)

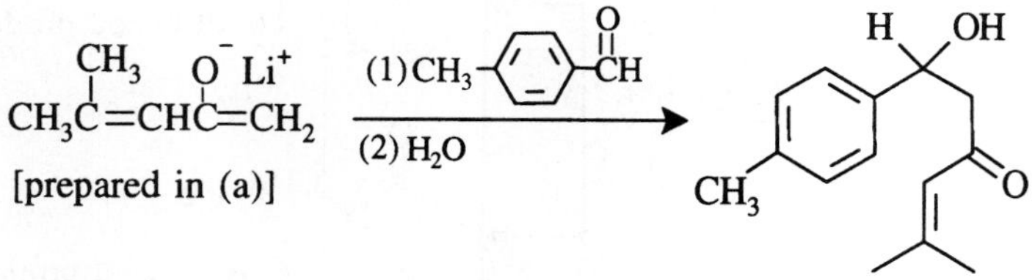

[prepared in (a)]

Ocimenone

H.4

Step 1

Step 2

Step 3

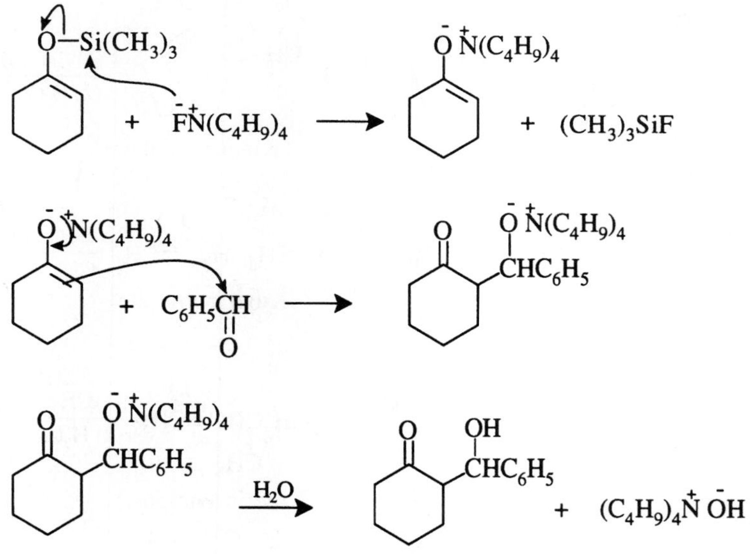

H.5

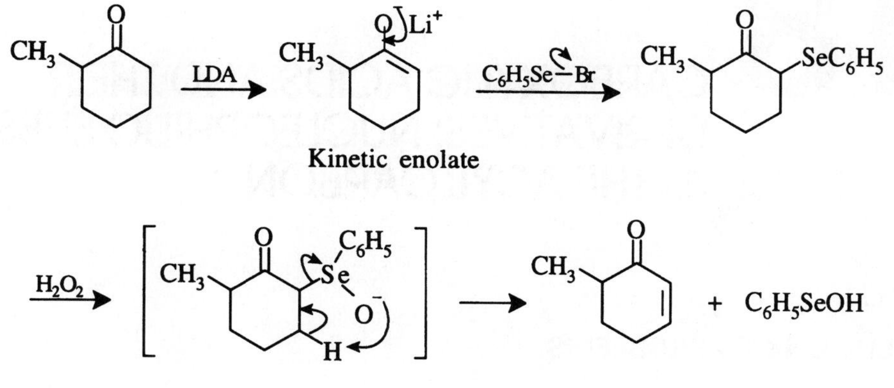
CH3
O
LDA
CH3
O
Li+
Kinetic enolate
C6H5Se—Br
CH3
O
SeC6H5
H2O2
CH3
O
C6H5
Se
O−
H
CH3
O
+ C6H5SeOH

18

CARBOXYLIC ACIDS AND THEIR DERIVATIVES: NUCLEOPHILIC SUBSTITUTION AT THE ACYL CARBON

SOLUTIONS TO PROBLEMS

18.1 (a) 2-Methylbutanoic acid

(b) 3-Pentenoic acid

(c) Sodium 4-bromobutanoate

(d) 5-Phenylpentanoic acid

(e) 3-Methyl-3-pentenoic acid

18.2 Acetic acid, in the absence of solvating molecules, exists as a dimer owing to the formation of two intermolecular hydrogen bonds:

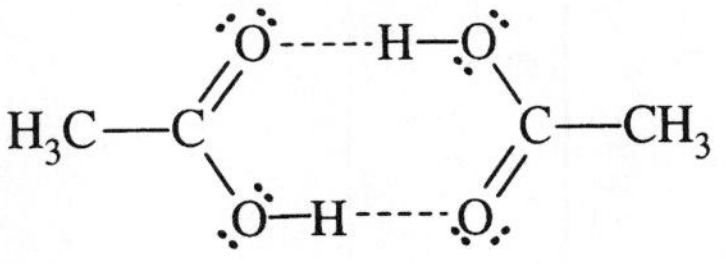

At temperatures much above the boiling point, the dimer dissociates into the individual molecules.

18.3 (a) CH_2FCO_2H (F– is more electronegative than H–)

(b) CH_2FCO_2H (F– is more electronegative than Cl–)

(c) CH_2ClCO_2H (Cl– is more electronegative than Br–)

(d) $CH_3CHFCH_2CO_2H$ (F– is closer to –CO_2H)

(e) $CH_3CH_2CHFCO_2H$ (F– is closer to –CO_2H)

(f) 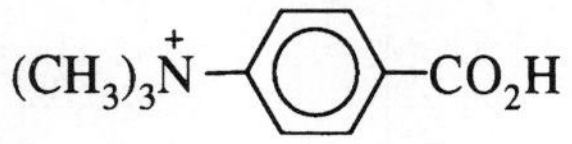 $(CH_3)_3\overset{+}{N}$—⬡—CO_2H [$(CH_3)_3\overset{+}{N}$– is more electronegative than H–]

(g) 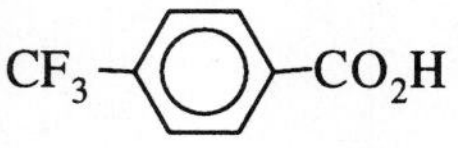 CF_3—⬡—CO_2H (CF_3– is more electronegative than CH_3–)

18.4 (a) The carboxyl group is an electron-withdrawing group; thus, in a dicarboxylic acid such as those in Table 18.3, one carboxyl group increases the acidity of the other.

(b) As the distance between the carboxyl groups increases the acid-strengthening, inductive effect decreases.

18.5 (a) $CH_3CH_2\overset{\displaystyle O}{\overset{\|}{C}}-OCH_3$

(b) $O_2N-\langle\bigcirc\rangle-\overset{\displaystyle O}{\overset{\|}{C}}-OCH_2CH_3$

(c) $CH_3O-\overset{\displaystyle O}{\overset{\|}{C}}CH_2\overset{\displaystyle O}{\overset{\|}{C}}-OCH_3$

(d) $\langle\bigcirc\rangle-\overset{\displaystyle O}{\overset{\|}{C}}-N(CH_3)_2$

(e) $CH_3CH_2CH_2CH_2C\equiv N$

(f) $\begin{array}{c}\langle\bigcirc\rangle\overset{\displaystyle O}{\overset{\|}{C}}-OCH_3\\ \underset{\displaystyle O}{\underset{\|}{C}}-OCH_3\end{array}$

(g) $\begin{array}{c}H-C\overset{\displaystyle O}{\overset{\|}{-C}}-OCH_2CH_2CH_3\\ \|\\ H-C\underset{\displaystyle O}{\underset{\|}{-C}}-OCH_2CH_2CH_3\end{array}$

(h) $H-\overset{\displaystyle O}{\overset{\|}{C}}-N(CH_3)_2$

(i) $CH_3\underset{\displaystyle Br}{\underset{|}{CH}}\overset{\displaystyle O}{\overset{\|}{C}}-Br$

(j) $\begin{array}{c}H_2C\overset{\displaystyle O}{\overset{\|}{-C}}-OCH_2CH_3\\ |\\ H_2C\underset{\displaystyle O}{\underset{\|}{-C}}-OCH_2CH_3\end{array}$

18.6 (a) $\langle\bigcirc\rangle-CH_2CH_3 \xrightarrow[\text{(2) H}_3O^+]{\text{(1)KMnO}_4,\ OH^-,\ \text{heat}} \langle\bigcirc\rangle-\overset{\displaystyle O}{\overset{\|}{C}}OH + CO_2$

(b) $\langle\bigcirc\rangle-Br \xrightarrow[\text{Et}_2O]{\text{Mg}} \langle\bigcirc\rangle-MgBr \xrightarrow{CO_2} \langle\bigcirc\rangle-\overset{\displaystyle O}{\overset{\|}{C}}OMgBr \xrightarrow{H_3O^+}$

$\langle\bigcirc\rangle-\overset{\displaystyle O}{\overset{\|}{C}}OH$

(c) $\langle\bigcirc\rangle-\overset{\displaystyle O}{\overset{\|}{C}}-CH_3 \xrightarrow[\text{(2) H}_3O^+]{\text{(1)Cl}_2/\text{NaOH}} \langle\bigcirc\rangle-\overset{\displaystyle O}{\overset{\|}{C}}OH + CHCl_3$

(d) $\langle\bigcirc\rangle-CH=CH_2 \xrightarrow[\text{(2) H}_3O^+]{\text{(1)KMnO}_4,\ OH^-,\ \text{heat}} \langle\bigcirc\rangle-\overset{\displaystyle O}{\overset{\|}{C}}OH$

(e) $\langle\bigcirc\rangle-CH_2OH \xrightarrow[\text{(2) H}_3O^+]{\text{(1)KMnO}_4,\ OH^-,\ \text{heat}} \langle\bigcirc\rangle-\overset{\displaystyle O}{\overset{\|}{C}}OH$

(f) $\langle\bigcirc\rangle-\overset{\displaystyle O}{\overset{\|}{CH}} \xrightarrow[\text{(2) H}_3O^+]{\text{(1)KMnO}_4,\ OH^-} \langle\bigcirc\rangle-\overset{\displaystyle O}{\overset{\|}{C}}OH$

18.7 These syntheses are easy to see if we work backward.

(a) $C_6H_5CH_2CO_2H$ $\xleftarrow[\text{(2) H}^+]{\text{(1) CO}_2}$ $C_6H_5CH_2MgBr$ $\xleftarrow[\text{Et}_2\text{O}]{\text{Mg}}$ $C_6H_5CH_2Br$

(b) $CH_3CH_2CH_2\overset{\overset{\displaystyle CH_3}{|}}{\underset{\underset{\displaystyle CH_3}{|}}{C}}CO_2H$ $\xleftarrow[\text{(2) H}^+]{\text{(1) CO}_2}$ $CH_3CH_2CH_2\overset{\overset{\displaystyle CH_3}{|}}{\underset{\underset{\displaystyle CH_3}{|}}{C}}MgBr$ $\xleftarrow[\text{Et}_2\text{O}]{\text{Mg}}$ $CH_3CH_2CH_2\overset{\overset{\displaystyle CH_3}{|}}{\underset{\underset{\displaystyle CH_3}{|}}{C}}Br$

(c) $CH_2{=}CHCH_2CO_2H$ $\xleftarrow[\text{(2) H}^+]{\text{(1) CO}_2}$ $CH_2{=}CHCH_2MgBr$ $\xleftarrow[\text{Et}_2\text{O}]{\text{Mg}}$ $CH_2{=}CHCH_2Br$

(d) H_3C—⬡—CO_2H $\xleftarrow[\text{(2) H}^+]{\text{(1) CO}_2}$ H_3C—⬡—$MgBr$ $\xleftarrow[\text{Et}_2\text{O}]{\text{Mg}}$ H_3C—⬡—Br

(e) $CH_3CH_2CH_2CH_2CH_2CO_2H$ $\xleftarrow[\text{(2) H}^+]{\text{(1) CO}_2}$ $CH_3CH_2CH_2CH_2CH_2MgBr$ $\xleftarrow[\text{Mg}]{\text{Et}_2\text{O}}$ $CH_3CH_2CH_2CH_2CH_2Br$

18.8 (a) $C_6H_5CH_2CO_2H$ $\xleftarrow[\text{(2) H}^+,\ \text{H}_2\text{O, heat}]{\text{(1) CN}^-}$ $C_6H_5CH_2Br$

$CH_2{=}CHCH_2CO_2H$ $\xleftarrow[\text{(2) H}^+,\ \text{H}_2\text{O, heat}]{\text{(1) CN}^-}$ $CH_2{=}CHCH_2Br$

$CH_3CH_2CH_2CH_2CH_2CO_2H$ $\xleftarrow[\text{(2) H}^+,\ \text{H}_2\text{O, heat}]{\text{(1) CN}^-}$ $CH_3CH_2CH_2CH_2CH_2Br$

(b) A nitrile synthesis. Preparation of a Grignard reagent from $HOCH_2CH_2CH_2CH_2Br$ would not be possible because of the presence of the acidic hydroxyl group.

18.9 Since maleic acid is a cis dicarboxylic acid, dehydration occurs readily:

Maleic acid   Maleic anhydride

Being a trans dicarboxylic acid, fumaric acid must undergo isomerization to maleic acid first. This isomerization requires a higher temperature.

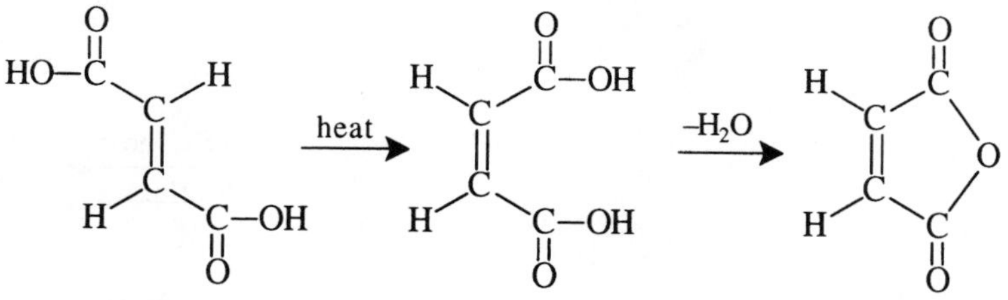

Fumaric acid

18.10 The labeled oxygen atom should appear in the carboxyl group of the acid. (Follow the reverse steps of the mechanism in Section 18.7A of the text using $H_2{}^{18}O$.)

18.11

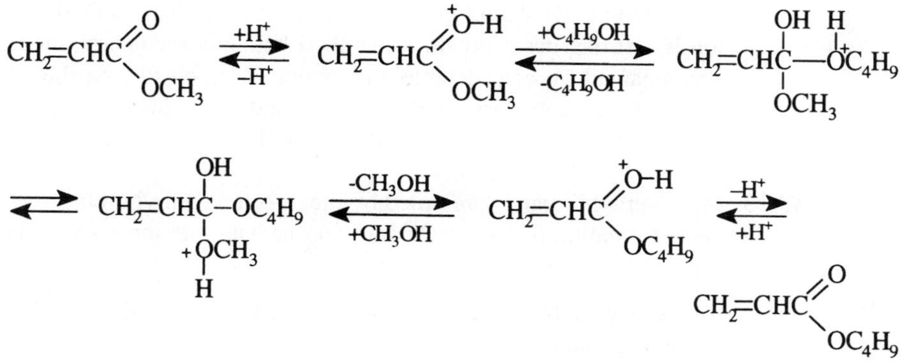

18.12 (a) (1)

(2)

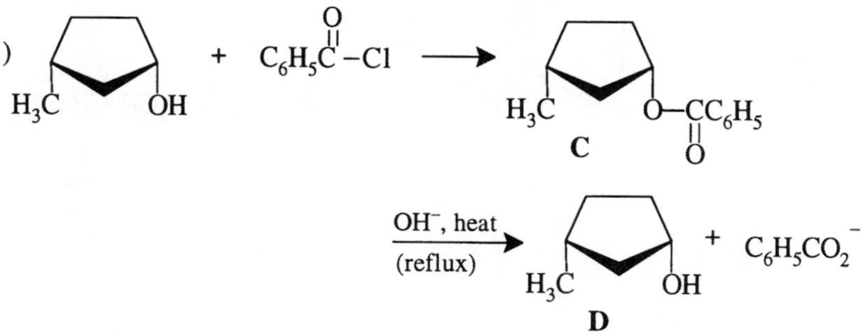

(3) $\underset{C_6H_{13}}{\overset{H}{\underset{CH_3}{\rvert}}} C-Br + CH_3\overset{O}{\overset{\|}{C}}O^- Na^+ \xrightarrow{\text{(inversion)}} CH_3\overset{O}{\overset{\|}{C}}O-\underset{\overset{}{CH_3}}{\overset{H}{C}}\text{''''}C_6H_{13}$

E

$\xrightarrow[\text{(reflux)}]{OH^-, \text{heat}} HO-\underset{\overset{}{CH_3}}{\overset{H}{C}}\text{''''}C_6H_{13}$

F

(4) $\underset{C_6H_{13}}{\overset{H}{\underset{CH_3}{\rvert}}} C-Br \xrightarrow[\text{(inversion)}]{OH^-, \text{heat}} HO-\underset{\overset{}{CH_3}}{\overset{H}{C}}\text{''''}C_6H_{13}$

F

(b)   Method (3) should give a higher yield of **F** than method (4). Since the hydroxide ion is a strong base and since the alkyl halide is secondary, method (4) is likely to be accompanied by considerable elimination. Method (3), on the other hand, employs a weaker base, acetate ion, in the S_N2 step and is less likely to be complicated by elimination. Hydrolysis of the ester **E** that results should also proceed in high yield.

18.13 (a)   Steric hindrance presented by the di-ortho methyl groups of methyl mesitoate prevents formation of the tetrahedral intermediate that must accompany attack at the acyl carbon.

(b)   Carry out hydrolysis with labeled $^{18}OH^-$ in labeled $H_2^{18}O$. The label should appear in the methanol.

18.14 (a) $C_6H_5\overset{O}{\overset{\|}{C}}N(CH_2CH_3)_2$

$\xrightarrow[H_2O]{OH^-} C_6H_5CO_2^- + (CH_3CH_2)_2NH$

$\xrightarrow[H_2O]{H^+} C_6H_5CO_2H + (CH_3CH_2)_2\overset{+}{N}H_2$

(b)

$\xrightarrow[H_2O]{OH^-} {}^-O\overset{O}{\overset{\|}{C}}CH_2CH_2CH_2CH_2NH_2$

$\xrightarrow[H_2O]{H^+} HO\overset{O}{\overset{\|}{C}}CH_2CH_2CH_2CH_2\overset{+}{N}H_3$

(c) $HO_2CCH-NHCCHNH_2$ with O on the second carbonyl, CH_3 and $CH_2C_6H_5$ substituents

$\xrightarrow[H_2O]{OH^-}$ $^-O_2CCHNH_2$ (CH_3) + $^-O_2CCHNH_2$ ($CH_2C_6H_5$)

$\xrightarrow[H_2O]{H^+}$ $HO_2CCHNH_3^+$ (CH_3) + $HO_2CCHNH_3^+$ ($CH_2C_6H_5$)

18.15 (a) $(CH_3)_3CCO_2H$ $\xrightarrow{SOCl_2}$ $(CH_3)_3CCOCl$ $\xrightarrow{NH_3}$ $(CH_3)_3CCONH_2$

$\xrightarrow[\text{heat}]{P_4O_{10}}$ $(CH_3)_3CC\equiv N$

(b)   An elimination reaction would take place because CN^- is a strong base.

CN^- + $H-CH_2-C(CH_3)_2-Br$ $\longrightarrow$ HCN + $CH_2=C(CH_3)_2$ + Br^-

18.16 (a)   benzyl CH_2OH + $O=C=N$–phenyl $\longrightarrow$ benzyl $CH_2O-C(=O)-N(H)$–phenyl

(b) $Cl-C(=O)-Cl$ + $4\,CH_3NH_2$ $\longrightarrow$ $CH_3N(H)-C(=O)-N(H)CH_3$ + $2\,CH_3NH_3^+$ + $2\,Cl^-$

(c)   benzyl $CH_2O-C(=O)-Cl$ + $H_3^+NCH_2CO_2^-$ $\xrightarrow{OH^-}$ benzyl $CH_2O-C(=O)-N(H)CH_2CO_2H$ + Cl^-

(d)   benzyl $CH_2OC(=O)NHCH_2CO_2H$ $\xrightarrow[Pd]{H_2}$ $H_3^+NCH_2CO_2^-$ + CO_2 + toluene (CH_3–phenyl)

(e) 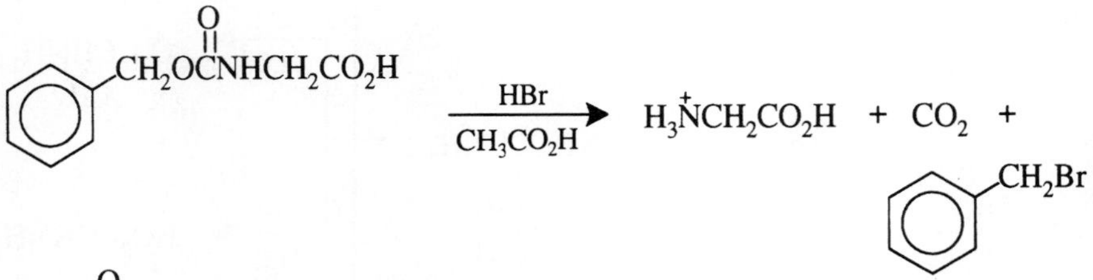
$$\text{C}_6\text{H}_5\text{CH}_2\text{O}\overset{\text{O}}{\underset{\|}{\text{C}}}\text{NHCH}_2\text{CO}_2\text{H} \xrightarrow[\text{CH}_3\text{CO}_2\text{H}]{\text{HBr}} \overset{+}{\text{H}_3\text{N}}\text{CH}_2\text{CO}_2\text{H} + \text{CO}_2 +$$

CH₂Br (benzyl bromide)

(f) $\text{H}_2\text{N}-\overset{\text{O}}{\underset{\|}{\text{C}}}-\text{NH}_2 \xrightarrow[\text{heat}]{\text{OH}^-, \text{H}_2\text{O}} 2\,\text{NH}_3 + \text{CO}_3^{2-}$

18.17   (a)   By decarboxylation of a β-keto acid:

$$\text{CH}_3(\text{CH}_2)_3\overset{\text{O}}{\underset{\|}{\text{C}}}\text{CH}_2\overset{\text{O}}{\underset{\|}{\text{C}}}\text{OH} \xrightarrow{100\text{-}150°\text{C}} \text{CH}_3(\text{CH}_2)_3\overset{\text{O}}{\underset{\|}{\text{C}}}\text{CH}_3 + \text{CO}_2$$

(b)   By decarboxylation of a substituted malonic acid:

$$\text{CH}_3\text{CH}_2\overset{\overset{\displaystyle\text{CO}_2\text{H}}{|}}{\underset{\underset{\displaystyle\text{CH}_3}{|}}{\text{C}}}-\text{CO}_2\text{H} \xrightarrow{100\text{-}150°\text{C}} \text{CH}_3\text{CH}_2\overset{}{\underset{\underset{\displaystyle\text{CH}_3}{|}}{\text{CH}}}\text{CO}_2\text{H} + \text{CO}_2$$

(c)   By decarboxylation of a β-keto acid:

$$\xrightarrow{100\text{-}150°\text{C}}$$ + CO₂

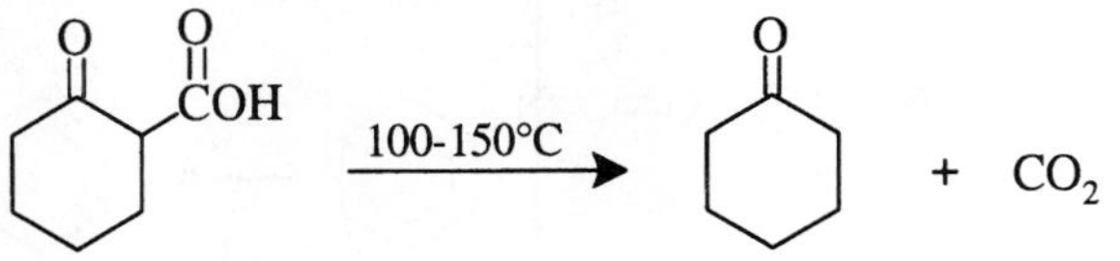

(d)   By decarboxylation of a substituted malonic acid:

$$\text{CH}_3\text{CH}_2\text{CH}_2\overset{\overset{\displaystyle\text{CO}_2\text{H}}{|}}{\underset{\underset{\displaystyle\text{CO}_2\text{H}}{|}}{\text{CH}}} \xrightarrow{100\text{-}150°\text{C}} \text{CH}_3\text{CH}_2\text{CH}_2\text{CH}_2\text{CO}_2\text{H} + \text{CO}_2$$

18.18   (a)   The oxygen-oxygen bond of the diacyl peroxide has a low homolytic bond dissociation energy ($DH° \simeq 35$ kcal mol⁻¹). This allows the following reaction to occur at a moderate temperature.

$$\text{R}-\overset{\text{O}}{\underset{\|}{\text{C}}}-\text{O}-\text{O}-\overset{\text{O}}{\underset{\|}{\text{C}}}-\text{R} \longrightarrow 2\,\text{R}-\overset{\text{O}}{\underset{\|}{\text{C}}}-\text{O}\cdot \qquad \Delta H° \simeq 35 \text{ kcal mol}^{-1}$$

(b)   By decarboxylation of the carboxylate radical produced in part (a).

$$\text{R}-\overset{\text{O}}{\underset{\|}{\text{C}}}-\text{O}\cdot \longrightarrow \text{R}\cdot + \text{CO}_2$$

(c) *Chain Initiation*

Step 1 $R-\overset{\overset{O}{\|}}{C}-O-O-\overset{\overset{O}{\|}}{C}-R$ $\xrightarrow{\text{heat}}$ $2\,R-\overset{\overset{O}{\|}}{C}-O\bullet$

Step 2 $R-\overset{\overset{O}{\|}}{C}-O\bullet$ $\longrightarrow$ $R\bullet$ + CO_2

Chain Propagation

Step 3 $R\bullet$ + $CH_2\!=\!CH_2$ $\longrightarrow$ $RCH_2CH_2\bullet$

Step 4 $RCH_2CH_2\bullet$ + $CH_2\!=\!CH_2$ $\longrightarrow$ $RCH_2CH_2CH_2CH_2\bullet$

Step 3, 4, 3, 4, and so on.

18.19 (a) $CH_3(CH_2)_4CO_2H$

(b) $CH_3(CH_2)_4CONH_2$

(c) $CH_3(CH_2)_4CONHC_2H_5$

(d) $CH_3(CH_2)_4CON(C_2H_5)_2$

(e) $CH_3CH_2CH\!=\!CHCH_2CO_2H$

(f) $CH_3CH\!=\!CHCH_2\underset{\underset{CH_3}{|}}{C}HCO_2H$

(g) $HO_2CCH_2CH_2CH_2CH_2CO_2H$

(h) 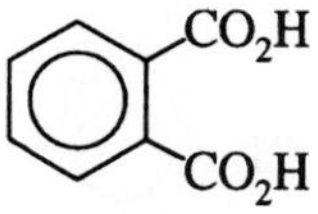 benzene ring with two adjacent CO_2H groups

(i) 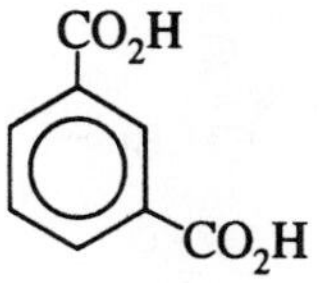 benzene ring with meta CO_2H groups

(j)  benzene ring with para CO_2H groups

(k) $C_2H_5O_2C-CO_2C_2H_5$

(l) $C_2H_5O_2C(CH_2)_4CO_2C_2H_5$

(m) $CH_3CH_2CO_2CH_2CH(CH_3)_2$

(n) 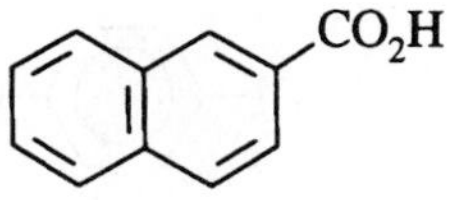 naphthalene with CO_2H group

(o) 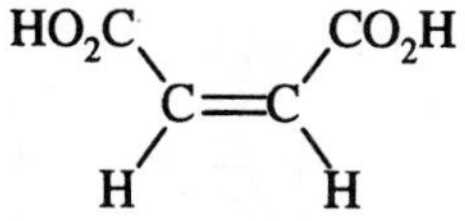 $\underset{H}{\overset{HO_2C}{\diagdown}}C\!=\!C\underset{H}{\overset{CO_2H}{\diagup}}$

(p) $HO_2CCHOHCH_2CO_2H$

(q) 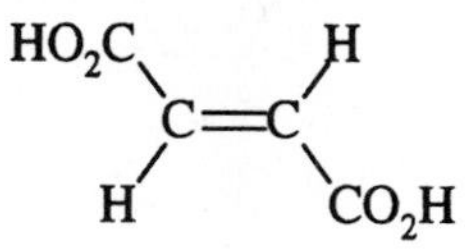 $\underset{H}{\overset{HO_2C}{\diagdown}}C\!=\!C\underset{CO_2H}{\overset{H}{\diagup}}$

(r) $HO_2CCH_2CH_2CO_2H$

(s)

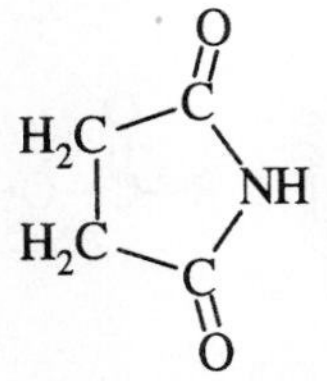

(t) $HO_2CCH_2CO_2H$

(u) $C_2H_5O_2CCH_2CO_2C_2H_5$

18.20   (a)   Benzoic acid

(b)   Benzoyl chloride

(c)   Benzamide

(d)   Benzoic anhydride

(e)   Benzyl benzoate

(f)   Phenyl benzoate

(g)   Isopropyl acetate

(h)   *N,N*-Dimethylacetamide

(i)   Acetonitrile

18.21   (a)

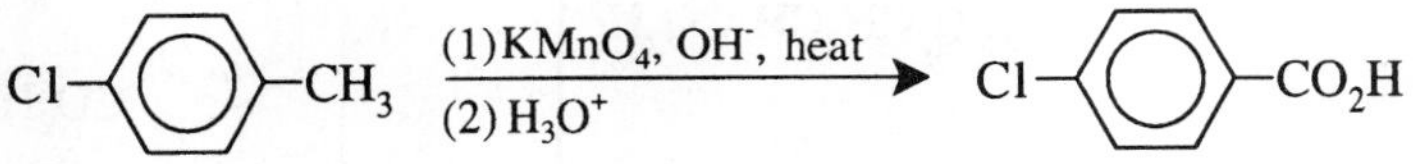

(b)

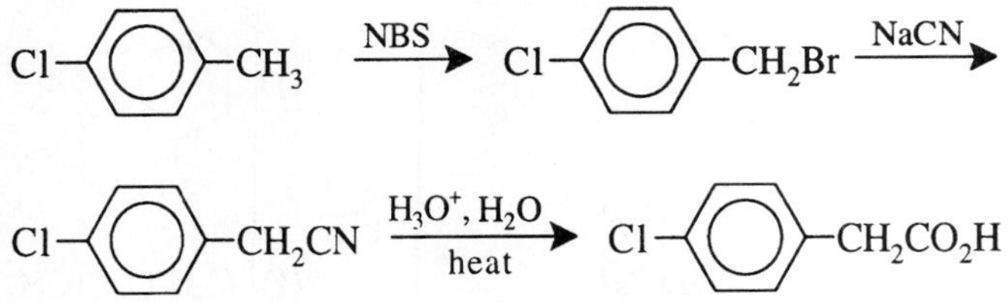

(c)

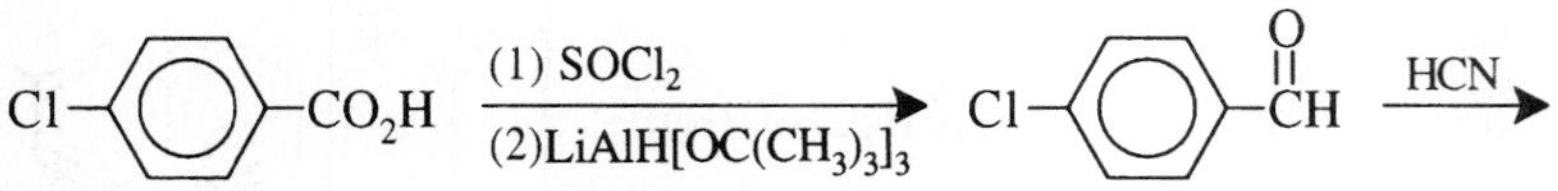

[from (a)]

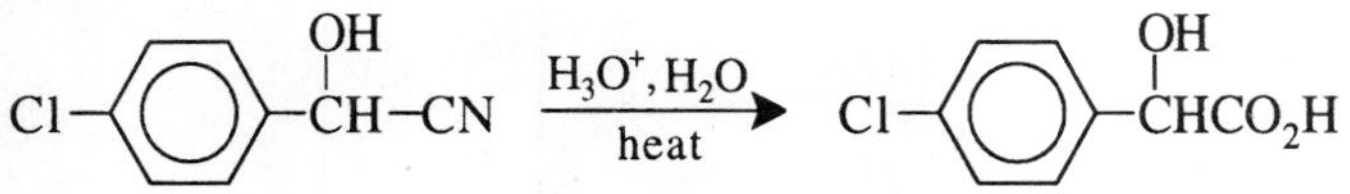

(d) $Cl-\langle\bigcirc\rangle-CH_2CO_2H \xrightarrow[\text{(2)LiAlH[OC(CH}_3)_3]_3]{\text{(1) SOCl}_2} Cl-\langle\bigcirc\rangle-CH_2\overset{\overset{\displaystyle O}{\|}}{C}H \xrightarrow{\text{HCN}}$

[from (b)]

$Cl-\langle\bigcirc\rangle-CH_2\overset{\overset{\displaystyle OH}{|}}{C}HCN \xrightarrow[\text{heat}]{H_3O^+, H_2O} \left[Cl-\langle\bigcirc\rangle-CH_2\overset{\overset{\displaystyle OH}{|}}{C}HCO_2H \right] \longrightarrow$

$Cl-\langle\bigcirc\rangle-CH=CHCO_2H$

18.22 (a) $H_3C-CO_2H \xrightarrow[P]{Br_2} CH_2Br-CO_2H \xrightarrow{NaOH} CH_2Br-CO_2Na \xrightarrow{NaCN}$

$\underset{\overset{\displaystyle |}{CN}}{CH_2}-CO_2Na \xrightarrow[\text{heat}]{H_3O^+, H_2O} HO_2C-CH_2-CO_2H$

(b) $HO-CH_2-CH_2-CH_2-CH_2-OH \xrightarrow[\text{(2) H}_3O^+]{\text{(1) KMnO}_4, \text{ OH}^-, \text{ heat}} HO-\overset{\overset{\displaystyle O}{\|}}{C}-CH_2-CH_2-\overset{\overset{\displaystyle O}{\|}}{C}-OH$

(c) $\langle\bigcirc\rangle{-}OH \xrightarrow[\text{(2) H}_3O^+]{\text{(1) KMnO}_4, \text{ OH}^-, \text{ heat}}$ (dicarboxylic acid structure)

18.23 (a) $CH_3CH_2CH_2CH_2CH_2OH \xrightarrow[\text{(2) H}_3O^+]{\text{(1) KMnO}_4, \text{ OH}^-, \text{ heat}} CH_3CH_2CH_2CH_2CO_2H$

(b) $CH_3CH_2CH_2CH_2Br \xrightarrow[\text{(2) CO}_2]{\text{(1) Mg, Et}_2O} CH_3CH_2CH_2CH_2CO_2MgBr \xrightarrow{H_3O^+}$

$CH_3CH_2CH_2CH_2CO_2H$

$CH_3CH_2CH_2CH_2Br \xrightarrow{CN^-} CH_3CH_2CH_2CH_2CN \xrightarrow[\text{heat}]{H_3O^+, H_2O}$

$CH_3CH_2CH_2CH_2CO_2H$

(c) $CH_3(CH_2)_3CH=CH(CH_2)_3CH_3 \xrightarrow[\text{(2) H}_3O^+]{\text{(1) KMnO}_4, \text{ OH}^-, \text{ heat}} 2\ CH_3(CH_2)_3CO_2H$

(d) $CH_3CH_2CH_2CH_2CHO \xrightarrow[\text{(2) H}_3O^+]{\text{(1) Ag(NH}_3)_2^+ \text{ OH}^-} CH_3CH_2CH_2CH_2CO_2H$

18.24 (a) $CH_3CO_2H + HCl$ (b) $CH_3CO_2H + AgCl$

(c) $CH_3CO_2CH_2(CH_2)_2CH_3$ (d) CH_3CONH_2

(e) 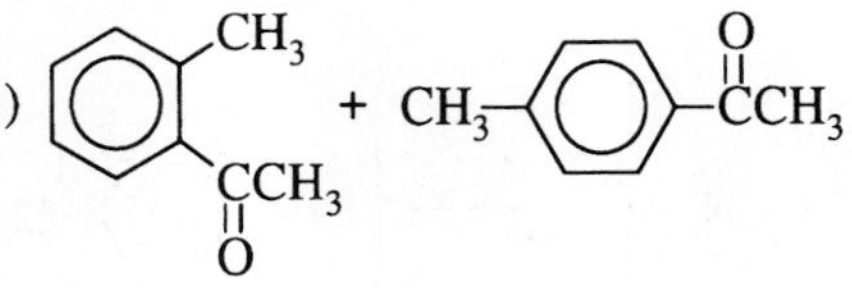   (f) CH_3CHO

(g) CH_3COCH_3 (h) CH_3CO_2Na

(i) $CH_3CONHCH_3$ (j) $CH_3CONHC_6H_5$

(k) $CH_3CON(CH_3)_2$ (l) $CH_3CO_2CH_2CH_3$

(m) $(CH_3CO)_2O$ (n) $(CH_3CO)_2O$

(o) $CH_3CO_2C_6H_5$ (p) $BrCH_2COCl$

18.25 (a) $CH_3CONH_2 + CH_3CO_2NH_4$

(b) $2\ CH_3CO_2H$

(c) $CH_3CO_2CH_2CH_2CH_3 + CH_3CO_2H$

(d) $C_6H_5COCH_3 + CH_3CO_2H$

(e) $CH_3CONHCH_2CH_3 + CH_3CO_2^-\ CH_3CH_2NH_3^+$

(f) $CH_3CON(CH_2CH_3)_2 + CH_3CO_2^-\ (CH_3CH_2)_2NH_2^+$

18.26 (a)
$$\begin{array}{l} CONH_2 \\ | \\ CH_2 \\ | \\ CH_2 \\ | \\ CO_2^-\ NH_4^+ \end{array}$$

(b)
$$\begin{array}{l} CO_2H \\ | \\ CH_2 \\ | \\ CH_2 \\ | \\ CO_2H \end{array}$$

(c)
$$\begin{array}{l} CO_2CH_2CH_2CH_3 \\ | \\ CH_2 \\ | \\ CH_2 \\ | \\ CO_2H \end{array}$$

(d)

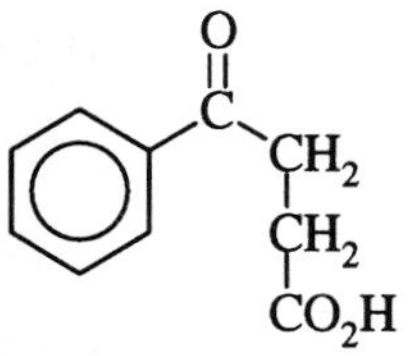

(e)
$$\begin{array}{l} CONHCH_2CH_3 \\ | \\ CH_2 \\ | \\ CH_2 \\ | \\ CO_2^-\ CH_3CH_2NH_3^+ \end{array}$$

(f)
$$\begin{array}{l} CON(CH_2CH_3)_2 \\ | \\ CH_2 \\ | \\ CH_2 \\ | \\ CO_2^-\ (CH_3CH_2)_2NH_2^+ \end{array}$$

See text, p. 670.

18.27

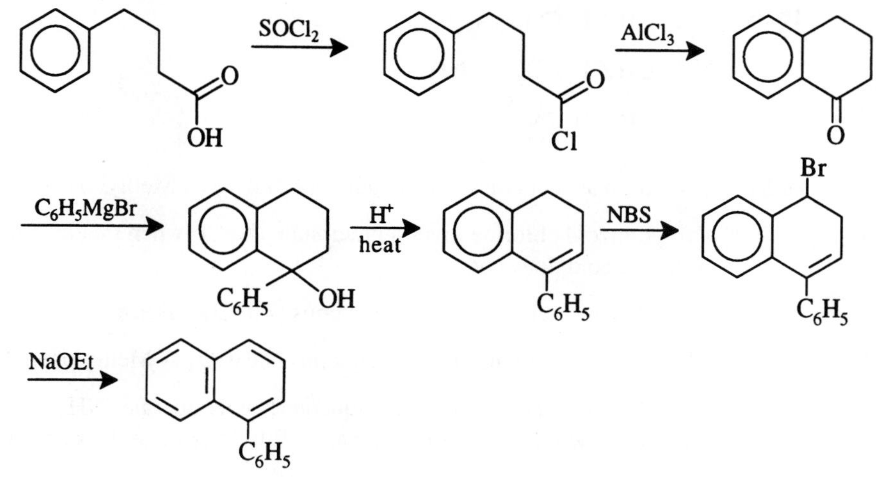

18.28 (a)

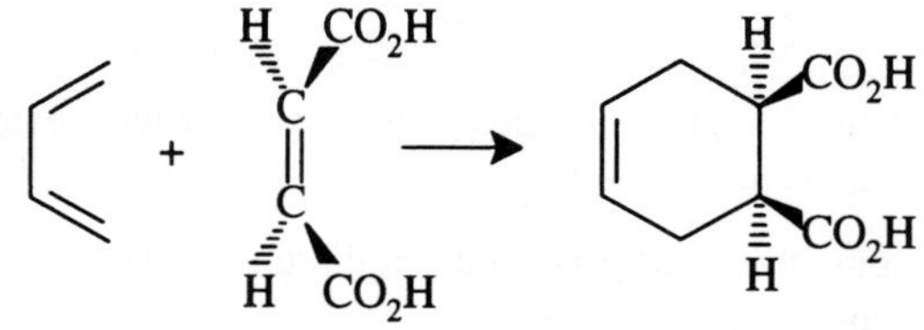

(b)

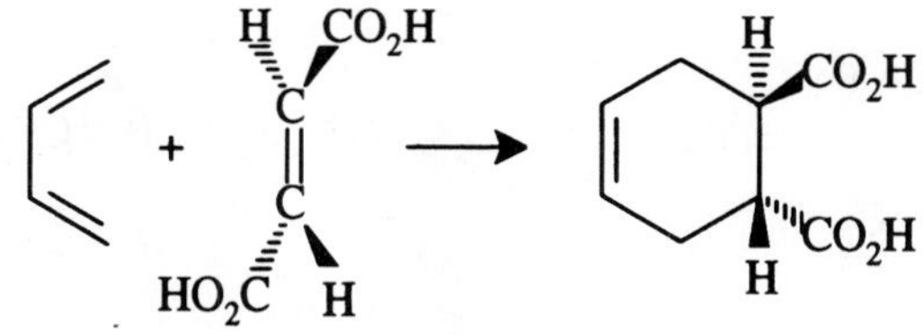

(c)

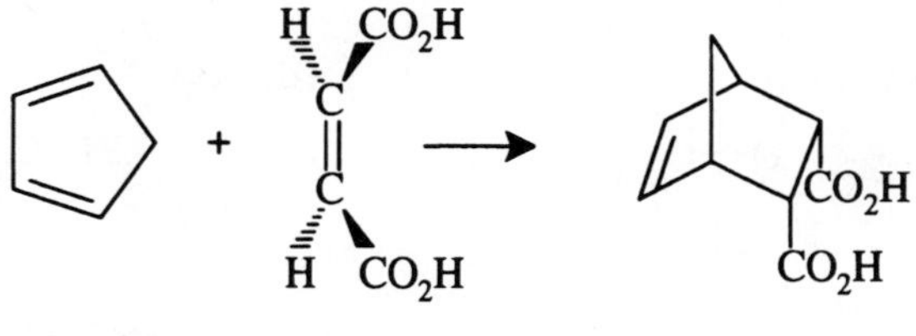

18.29 (a) $CH_3CH_2CO_2H + CH_3CH_2OH$

(b) $CH_3CH_2CO_2^- + CH_3CH_2OH$

(c) $CH_3CH_2CO_2(CH_2)_7CH_3 + CH_3CH_2OH$

(d) $CH_3CH_2CONHCH_3 + CH_3CH_2OH$

(e) $CH_3CH_2CH_2OH + CH_3CH_2OH$

(f) $CH_3CH_2\overset{\displaystyle C_6H_5}{\underset{\displaystyle OH}{C}}-C_6H_5 + CH_3CH_2OH$

18.30 (a) $CH_3CH_2CO_2H + NH_4^+$

(b) $CH_3CH_2CO_2^- + NH_3$

(c) CH_3CH_2CN

18.31   (a)   Benzoic acid dissolves in aqueous $NaHCO_3$. Methyl benzoate does not.

(b)   Benzoyl chloride gives a precipitate (AgCl) when treated with alcoholic $AgNO_3$. Benzoic acid does not.

(c)   Benzoic acid dissolves in aqueous $NaHCO_3$. Benzamide does not.

(d)   Benzoic acid dissolves in aqueous $NaHCO_3$. 4-Methylphenol (*p*-cresol) does not.

(e)   Refluxing benzamide with aqueous NaOH liberates NH_3, which can be detected in the vapors with moist red litmus paper. Ethyl benzoate does not liberate NH_3.

(f)   Cinnamic acid, because it has a double bond, decolorizes Br_2 in CCl_4. Benzoic acid does not.

(g)   Benzoyl chloride gives a precipitate (AgCl) when treated with alcoholic $AgNO_3$. Ethyl benzoate does not.

(h)   2-Chlorobutanoic acid gives a precipitate (AgCl) when treated with alcoholic silver nitrate. Butanoic acid does not.

18.32   (a)

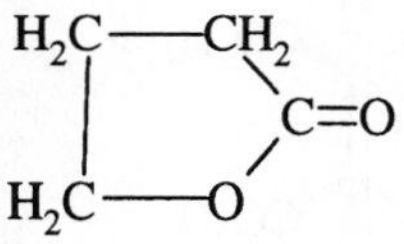

(b) $CH_3CH = CHCO_2H$

(c)

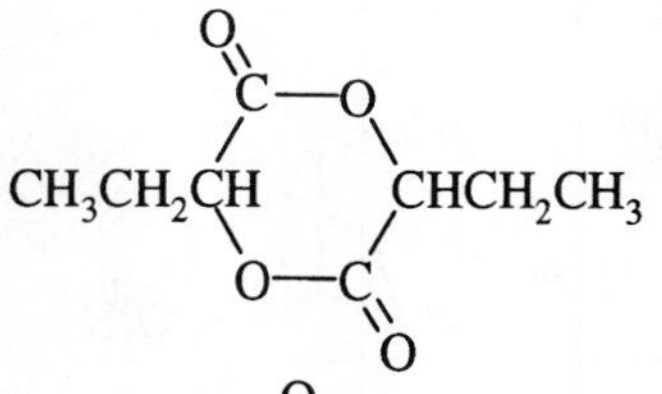

(d)

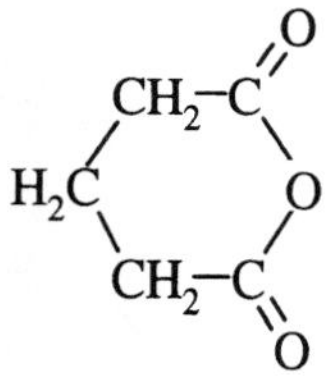

(e)

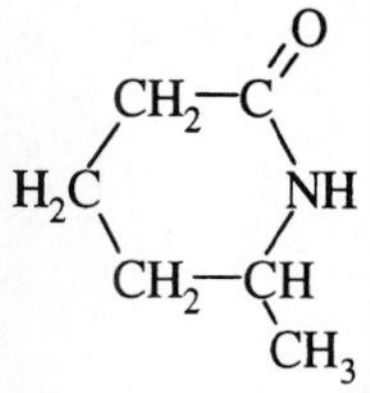

(f)

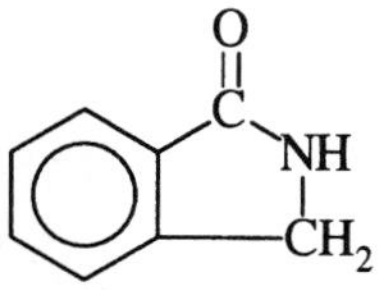

18.33 (a)

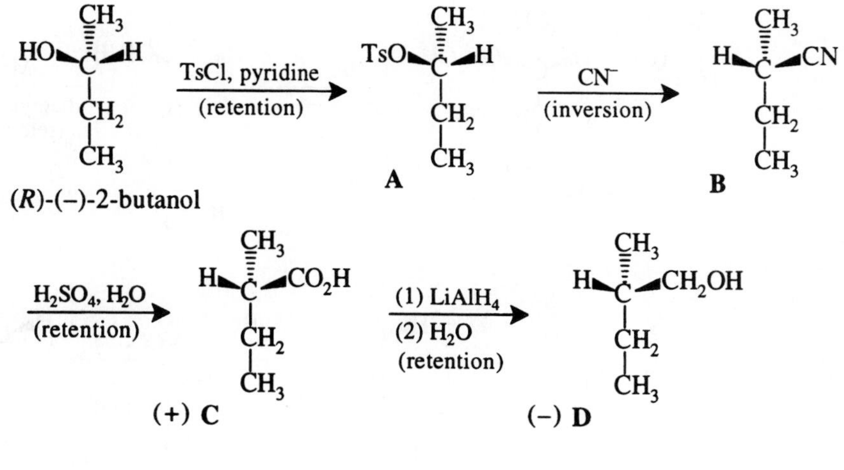

(R)-(−)-2-butanol

$\xrightarrow[\text{(retention)}]{\text{TsCl, pyridine}}$ **A**

$\xrightarrow[\text{(inversion)}]{CN^-}$ **B**

$\xrightarrow[\text{(retention)}]{H_2SO_4, H_2O}$ (+) **C**

$\xrightarrow[\substack{\text{(2) }H_2O \\ \text{(retention)}}]{\text{(1) LiAlH}_4}$ (−) **D**

(b)

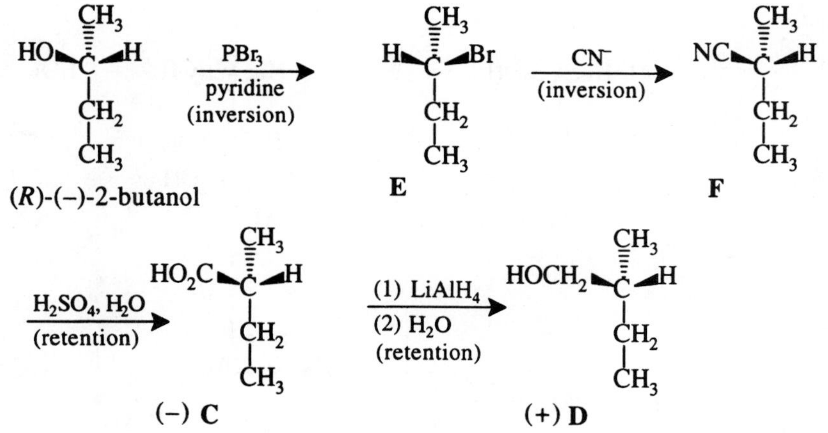

(R)-(−)-2-butanol

$\xrightarrow[\substack{\text{pyridine} \\ \text{(inversion)}}]{\text{PBr}_3}$ **E**

$\xrightarrow[\text{(inversion)}]{CN^-}$ **F**

$\xrightarrow[\text{(retention)}]{H_2SO_4, H_2O}$ (−) **C**

$\xrightarrow[\substack{\text{(2) }H_2O \\ \text{(retention)}}]{\text{(1) LiAlH}_4}$ (+) **D**

(c)

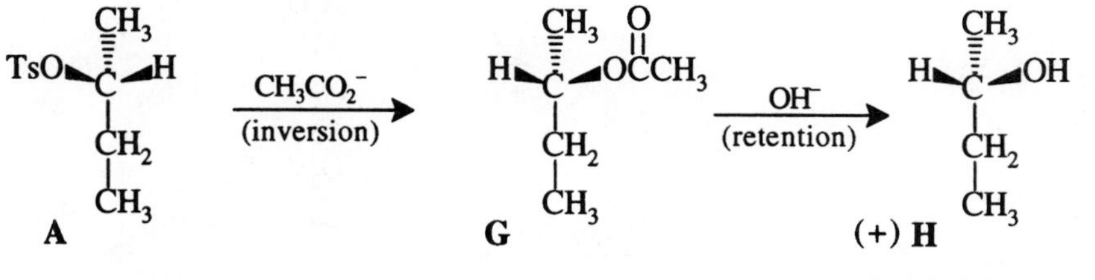

A

$\xrightarrow[\text{(inversion)}]{CH_3CO_2^-}$ **G**

$\xrightarrow[\text{(retention)}]{OH^-}$ (+) **H**

(S)-(+)-2-butanol

(d)

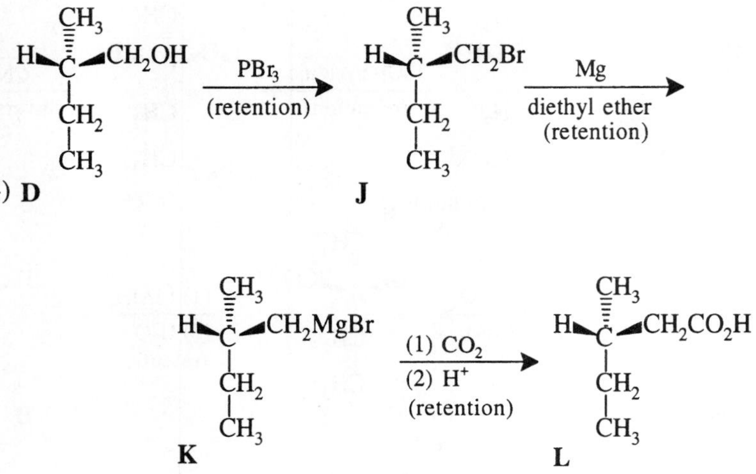

$$\underset{(-)\ \textbf{D}}{\overset{\underset{\displaystyle CH_3}{\overset{\displaystyle CH_3}{\big|}}}{H\cdots C\cdots CH_2OH}} \xrightarrow[\text{(retention)}]{PBr_3} \underset{\textbf{J}}{\overset{\underset{\displaystyle CH_3}{\overset{\displaystyle CH_3}{\big|}}}{H\cdots C\cdots CH_2Br}} \xrightarrow[\substack{\text{diethyl ether}\\\text{(retention)}}]{Mg}$$

$$\underset{\textbf{K}}{\overset{\underset{\displaystyle CH_3}{\overset{\displaystyle CH_3}{\big|}}}{H\cdots C\cdots CH_2MgBr}} \xrightarrow[\substack{(2)\ H^+\\\text{(retention)}}]{(1)\ CO_2} \underset{\textbf{L}}{\overset{\underset{\displaystyle CH_3}{\overset{\displaystyle CH_3}{\big|}}}{H\cdots C\cdots CH_2CO_2H}}$$

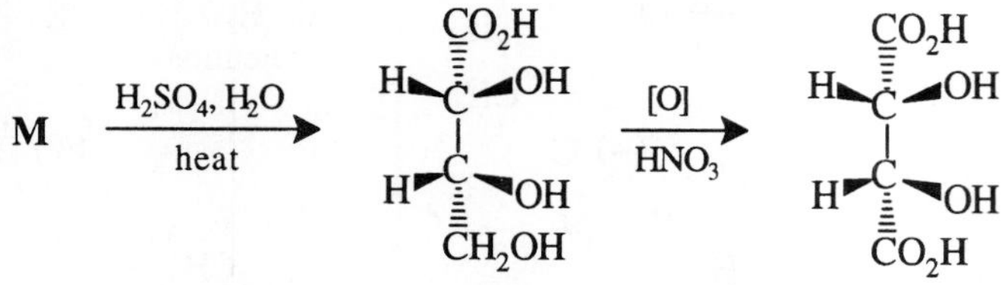

(e)

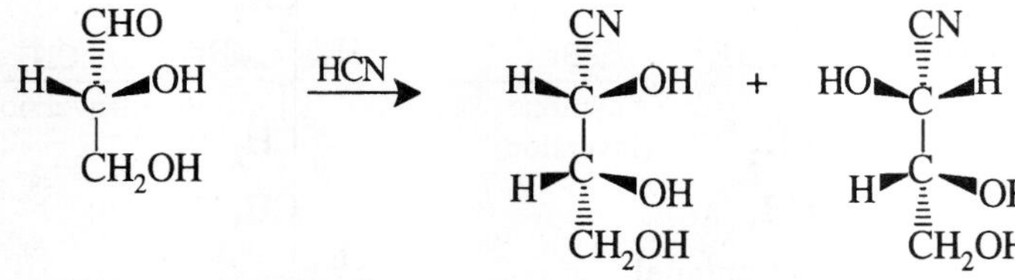

$$\underset{(R)\text{-}(+)\text{-Glyceraldehyde}}{\overset{\underset{\displaystyle CH_2OH}{\overset{\displaystyle CHO}{\big|}}}{H\cdots C\cdots OH}} \xrightarrow{HCN} \underset{\textbf{M}}{\text{structure}} + \underset{\textbf{N}}{\text{structure}}$$

(f)   **M** $\xrightarrow[\text{heat}]{H_2SO_4,\ H_2O}$ **P** $\xrightarrow[\text{HNO}_3]{[O]}$ *meso*-Tartaric acid

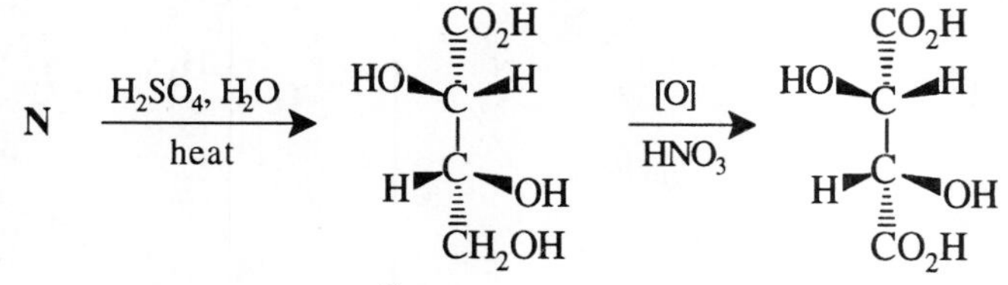

(g)   **N** $\xrightarrow[\text{heat}]{H_2SO_4,\ H_2O}$ **Q** $\xrightarrow[\text{HNO}_3]{[O]}$ (–)-Tartaric acid

18.34 (a) CH$_3$CHCHO + HCH $\xrightarrow[\text{H}_2\text{O}]{\text{K}_2\text{CO}_3}$ CH$_3$CCHO $\xrightarrow{\text{HCN}}$ CH$_3$C——CHCN

CH$_3$ (on CH$_3$CHCHO); O on HCH; CH$_3$ and CH$_2$OH on A; CH$_3$, OH, CH$_2$OH on (±)-B

A   **(±)-B**

$\xrightarrow[\text{heat}]{\text{H}_3\text{O}^+}$ $\left[\text{CH}_3\text{C——CHCO}_2\text{H} \right]$ $\xrightarrow{-\text{H}_2\text{O}}$ H$_3$C—C——CH $\xrightarrow{\text{H}_3\overset{+}{\text{N}}\text{CH}_2\text{CH}_2\overset{O}{\overset{\|}{\text{C}}}\text{O}^-}$

CH$_3$ OH and CH$_2$OH on (±)-C; CH$_3$, OH, H$_2$C, O, C=O on (±)-D

(±)-C   **(±)-D**

(±)-pantothenic
acid

(±)-pantetheine $\longleftarrow$ H$_2$NCH$_2$CH$_2\overset{O}{\overset{\|}{\text{C}}}$NHCH$_2CH_2$SH

(b) (CH$_3$)$_2$C—C—C—NHCH$_2$CH$_2$C—NHCH$_2$CH$_2$SH

HOH$_2$C, OH, O (over first C); H (below); O (over last C)

(c) $\xrightarrow[\text{heat}]{\text{OH}^-, \text{H}_2\text{O}}$ (CH$_3$)$_2$C—C—CO$_2^-$ + H$_2$NCH$_2$CH$_2$CO$_2^-$ + H$_2$NCH$_2$CH$_2$S$^-$

HOH$_2$C, OH (over C); H (below)

18.35 CH$_3$CH$_2$O-⟨○⟩-NH-C-CH$_3$ $\xrightarrow[\text{reflux}]{\text{OH}^-, \text{H}_2\text{O}}$ CH$_3$CH$_2$O-⟨○⟩-NH$_2$ +

O (over C)

Phenacetin   Phenetidine   CH$_3$CO$_2^-$

An interpretation of the ^{1}H NMR spectral data for phenacetin is as follows:

(a) (c)   (b)
CH$_3$—CH$_2$—O-⟨○⟩-NH-C-CH$_3$
                    (e)
            (d)
O (over C)

(a) triplet δ 1.4

(b) singlet δ 2.1

(c) quartet δ 3.95

(d) multiplet δ 6.8-7.4

(e) broad singlet δ 9.0

18.36 (a) $\underset{\underset{(a)\ \ (c)}{}}{CH_3CH_2-O}-\overset{\overset{O}{\|}}{C}-\underset{\underset{(b)\ \ (b)}{}}{CH_2CH_2}-\overset{\overset{O}{\|}}{C}-\underset{\underset{(c)\ \ (a)}{}}{O-CH_2CH_3}$

Interpretation:

(a) Triplet δ 1.2 (6H) $2-\overset{\overset{O}{\|}}{C}-O-$, 1740 cm^{-1}
(b) Singlet δ 2.5 (4H)
(c) Quartet δ 4.1 (4H)

(b)

$\overset{\overset{O}{\|}}{C}-O-\underset{(c)}{CH_2}-\underset{(b)}{\overset{\overset{CH_3\ (a)}{|}}{CH}}-\underset{(a)}{CH_3}$

(d)

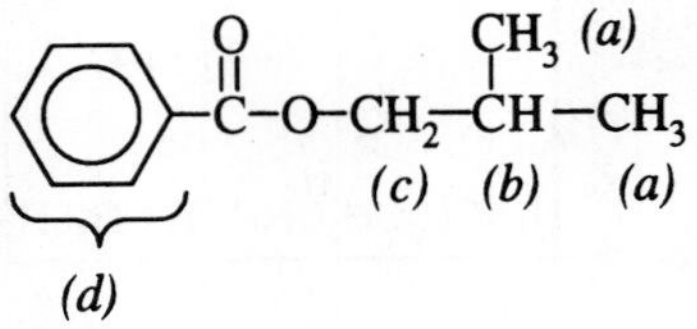

Interpretation:

(a) Doublet δ 1.0 (6H) $-\overset{\overset{O}{\|}}{C}-O-$, 1720 cm^{-1} (ester)
(b) Multiplet δ 2.1 (1H)
(c) Doublet δ 4.1 (2H)
(d) Multiplet δ 7.8 (5H)

(c)

$-\underset{(b)}{CH_2}-\overset{\overset{O}{\|}}{C}-O-\underset{(c)\ \ (a)}{CH_2CH_3}$

(d)

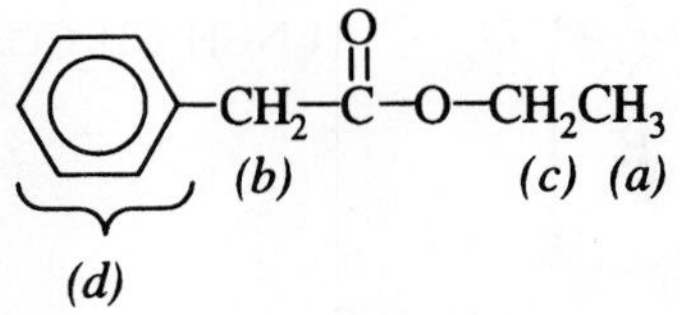

Interpretation:

(a) Triplet δ 1.2 (3H) $-\overset{\overset{O}{\|}}{C}-O-$, 1740 cm^{-1} (ester)
(b) Singlet   δ 3.5 (2H)
(c) Quartet δ 4.1 (2H)
(d) Multiplet δ 7.3 (5H)

(d) $\underset{(a)}{Cl-\overset{\overset{Cl}{|}}{CH}}-\underset{(b)}{CO_2H}$

Interpretation: $-OH$, 2500-2700 cm^{-1}

(a) Singlet δ 6.0
(b) Singlet δ 11.7 $-\overset{\overset{O}{\|}}{C}-O-$, 1705 cm^{-1} (acid)

(e) $\underset{\overset{(b)}{}\quad\underset{(c)}{}\;\underset{(a)}{}}{Cl-CH_2-\overset{\overset{\displaystyle O}{\|}}{C}-O-CH_2CH_3}$

Interpretation:

(a) Triplet δ 1.3 $-\overset{\overset{\displaystyle O}{\|}}{C}-O-$, 1745 cm^{-1} (ester)
(b) Singlet δ 4.0
(c) Quartet δ 4.2

18.37

$\underset{CH_3}{\overset{CO_2H}{\bigcirc}}$ + SOCl$_2$ $\longrightarrow$ $\underset{CH_3}{\overset{COCl}{\bigcirc}}$ $\xrightarrow{(C_2H_5)_2NH}$ $\underset{CH_3}{\overset{CON(CH_2CH_3)_2}{\bigcirc}}$

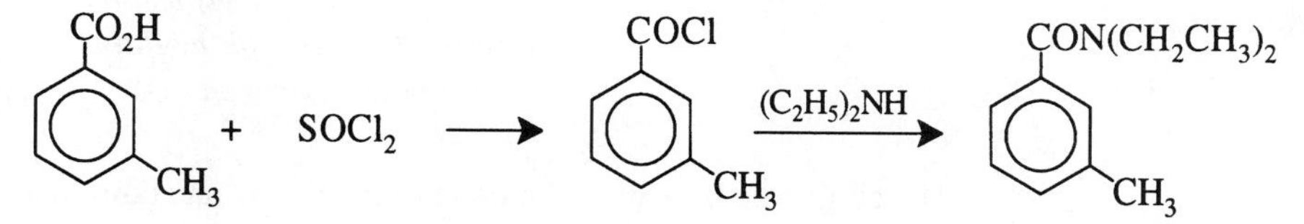

18.38 Alkyl groups are electron releasing; they help disperse the positive charge of an alkyl-ammonium salt and thereby help to stabilize it.

$$R\ddot{N}H_2 + H_3O^+ \longrightarrow R{\rightarrow}NH_3^+ + H_2O$$

*Stabilized by
electron-releasing
alkyl group*

Consequently, alkylamines are somewhat stronger bases than ammonia.

Amides, on the other hand, have acyl groups, $R-\overset{\overset{\displaystyle O}{\|}}{C}-$, attached to nitrogen, and acyl groups are electron withdrawing. They are especially electron withdrawing because of resonance contributions of the kind shown here,

$$R-\overset{\overset{\displaystyle :\ddot{O}}{\|}}{C}-\ddot{N}H_2 \quad\longleftrightarrow\quad R-\overset{\overset{\displaystyle :\ddot{O}:^-}{|}}{C}=\overset{+}{N}H_2$$

This kind of resonance also *stabilizes* the amide. The tendency of the acyl group to be electron withdrawing, however, *destabilizes* the conjugate acid of an amide, and reactions such as the following do not take place to an appreciable extent.

$$R-\overset{\overset{\displaystyle O}{\|}}{C}-\ddot{N}H_2 + H_3O^+ \;\overset{\longleftarrow}{\longrightarrow}\; R-\overset{\overset{\displaystyle O}{\|}}{C}-NH_3^+ + H_2O$$

*Stabilized
by
resonance*   *Destabilized by
electron-withdrawing
acyl group*

18.39 (a)   The conjugate base of an amide is stabilized by resonance.

$$R-\overset{\displaystyle :\overset{..}{O}:}{\underset{}{C}}-\overset{..}{N}H_2 \ + \ :B^- \ \rightleftharpoons \ R-\overset{\displaystyle :\overset{..}{O}:}{\underset{}{C}}-\overset{..}{N}H^- \ + \ BH$$

$$R-\overset{\displaystyle :\overset{..}{O}:^-}{\underset{}{C}}=\overset{..}{N}H$$

*This structure is especially
stable because the negative
charge is on oxygen.*

(b)   The conjugate base of an imide is stabilized by an additional resonance structure,

$$R-\overset{\displaystyle :\overset{..}{O}:}{\underset{}{C}}-\overset{..}{N}H-\overset{\displaystyle :\overset{..}{O}:}{\underset{}{C}}-R \ + \ OH^- \ \rightleftharpoons \ R-\overset{\displaystyle :\overset{..}{O}:}{\underset{}{C}}-\overset{..}{N}:-\overset{\displaystyle :\overset{..}{O}:}{\underset{}{C}}-R \ + \ H_2O$$

An imide

$$R-\overset{\displaystyle :\overset{..}{O}:^-}{\underset{}{C}}=N-\overset{\displaystyle :\overset{..}{O}:}{\underset{}{C}}-R$$

$$R-\overset{\displaystyle :\overset{..}{O}:}{\underset{}{C}}-N=\overset{\displaystyle :\overset{..}{O}:^-}{\underset{}{C}}-R$$

18.40   That compound **X** does not dissolve in aqueous sodium bicarbonate indicates that **X** is not a carboxylic acid. That **X** has an IR absorption peak at 1740 cm^{-1} indicates the presence of a carbonyl group, probably that of an ester (Table 18.5). That the molecular formula of **X** ($C_7H_{12}O_4$) contains four oxygen atoms suggests that **X** is a diester.

The ^{13}C spectrum shows only four signals indicating a high degree of symmetry for **X**. The single signal at δ 166.7 is that of an ester carbonyl carbon, indicating that both ester groups of **X** are equivalent.

Putting these observations together with the information gathered from DEPT ^{13}C spectra and the molecular formula leads us to the conclusion that **X** is diethyl malonate. The assignments are

$$\underset{(a)\ \ (c)\ \ \ (d)(b)\ (d)\ \ (c)\ \ (a)}{CH_3CH_2O\overset{\displaystyle O}{\overset{\|}{C}}CH_2\overset{\displaystyle O}{\overset{\|}{C}}OCH_2CH_3}$$

(a) δ 14.2
(b) δ 41.6
(c) δ 61.3
(d) δ 166.7

18.41 (a) *Chain Initiation*

Step 1 RO—OR $\longrightarrow$ 2 RO·

Step 2 $CH_3\overset{\overset{O}{\|}}{C}SH$ + RO· $\longrightarrow$ $CH_3\overset{\overset{O}{\|}}{C}S·$ + ROH

Chain Propagation

Step 3 $CH_3\overset{\overset{O}{\|}}{C}S·$ + $CH_2{=}CHR$ $\longrightarrow$ $CH_3\overset{\overset{O}{\|}}{C}SCH_2\overset{·}{C}HR$

Step 4 $CH_3\overset{\overset{O}{\|}}{C}SCH_2\overset{·}{C}HR$ + $CH_3\overset{\overset{O}{\|}}{C}SH$ $\longrightarrow$ $CH_3\overset{\overset{O}{\|}}{C}SCH_2CH_2R$ + $CH_3\overset{\overset{O}{\|}}{C}S·$

(b) $\overset{\overset{\displaystyle CH_3}{\displaystyle |}}{CH_3C}{=}CHCH_3$ + $CH_3\overset{\overset{O}{\|}}{C}SH$ $\xrightarrow{ROOR}$ $CH_3\underset{\underset{\underset{\displaystyle O}{\displaystyle \|}}{\displaystyle SCCH_3}}{\overset{\overset{\displaystyle CH_3}{\displaystyle |}}{C}HCHCH_3}$ $\xrightarrow[\text{(2) } H_3O^+]{\text{(1) } OH^-, \text{ heat}}$

$CH_3\overset{\overset{O}{\|}}{C}OH$ + $CH_3\underset{\underset{\displaystyle SH}{\displaystyle |}}{\overset{\overset{\displaystyle CH_3}{\displaystyle |}}{C}HCHCH_3}$

18.42 *cis*-4-Hydroxycyclohexanecarboxylic acid can assume a boat conformation that permits lactone formation.

Neither of the chair conformations nor the boat form of *trans*-4-hydroxycyclohexanecarboxylic acid places the –OH group and the –CO₂H group close enough together to permit lactonization.

18.43

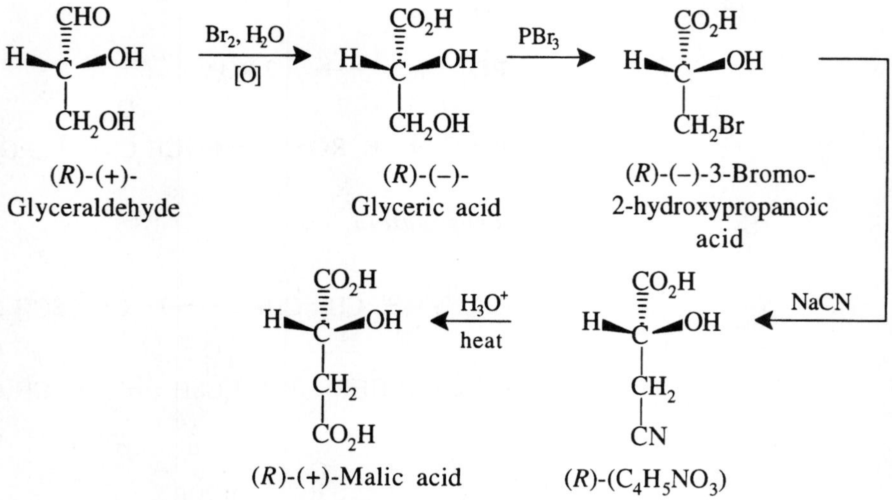
Br2, H2O
[O]
PBr3
NaCN
H3O+
heat
(R)-(+)-
Glyceraldehyde
(R)-(−)-
Glyceric acid
(R)-(−)-3-Bromo-
2-hydroxypropanoic
acid
(R)-(+)-Malic acid
(R)-(C4H5NO3)

18.44

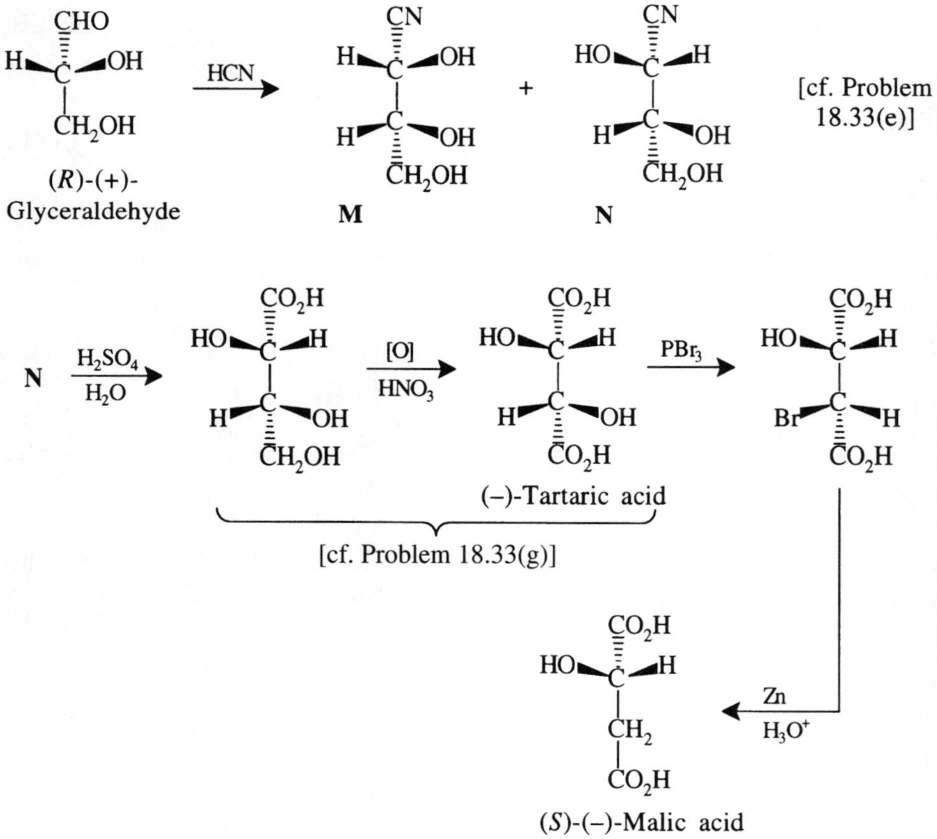
HCN
(R)-(+)-
Glyceraldehyde
M
N
[cf. Problem
18.33(e)]
H2SO4
H2O
[O]
HNO3
PBr3
(−)-Tartaric acid
[cf. Problem 18.33(g)]
Zn
H3O+
(S)-(−)-Malic acid

(b) Replacement of either alcoholic –OH by a reaction that proceeds with inversion produces the same stereoisomer.

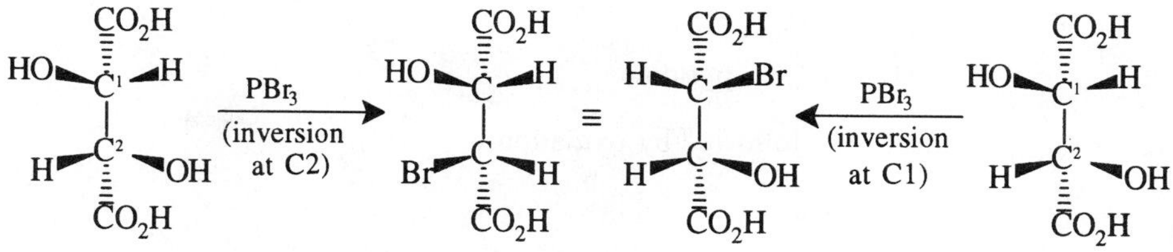

(Structural diagrams showing Fischer projections of malic/bromo-succinic acid derivatives with PBr₃ inversion at C2 and C1)

(c)   Two. The stereoisomer given in (b) and the one given next, below.

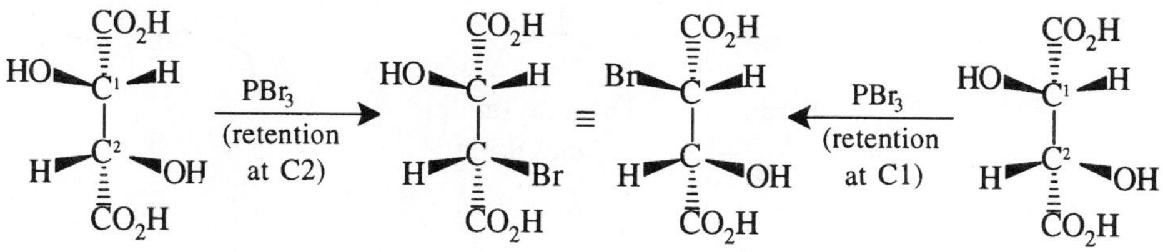

(Structural diagrams showing Fischer projections with PBr₃ retention at C2 and C1)

(d)   It would have made no difference because treating either isomer (or both together) with zinc and acid produces (−) malic acid.

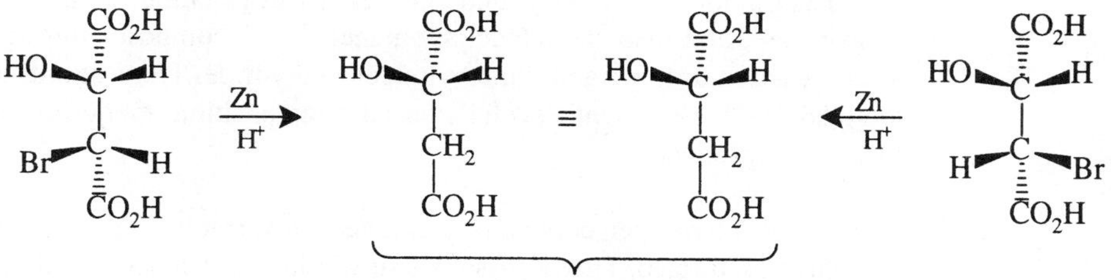

(Structural diagrams showing Zn/H⁺ reactions producing (−)-Malic acid)

(−)-Malic acid

18.45 (a) $CH_3O_2C-C{\equiv}C-CO_2CH_3$. This is a Diels-Alder reaction.

(b) H_2, Pd. The disubstituted double bond is less hindered than the tetrasubstituted double bond and hence is more reactive.

(c) $CH_2{=}CH-CH{=}CH_2$. Another Diels-Alder rection.

(d) $LiAlH_4$

(e) $CH_3-\overset{\overset{O}{\|}}{\underset{\underset{O}{\|}}{S}}-Cl$ and pyridine

(f) $CH_3CH_2S^-$

(g) OsO_4

(h)   Raney Ni

(i)   Base. This is an aldol condensation.

(j) C_6H_5Li (or C_6H_5MgBr) followed by H_3O^+

(k) H_3O^+. This is an acid-catalyzed rearrangement of an allylic alcohol.

(l) CH₃CCl, pyridine

(m) O₃, followed by oxidation

(n) Heat

18.46 (a)

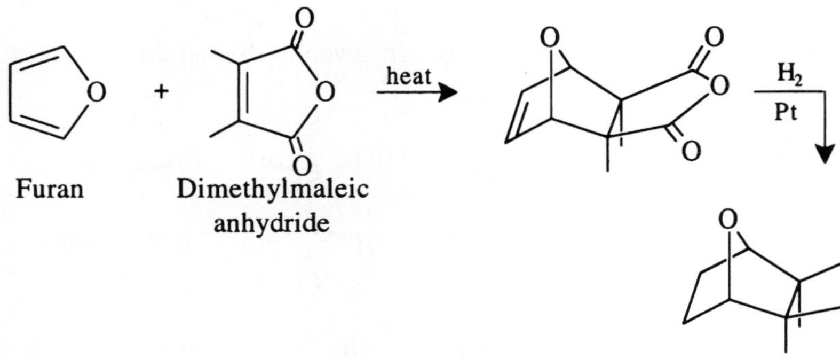

Furan   Dimethylmaleic
             anhydride

Cantharidin

(b) Cantharidin apparently undergoes dehydrogenation to the Diels-Alder adduct shown here, and then the adduct spontaneously decomposes through a reverse Diels-Alder reaction to furan and dimethylmaleic anhydride. These results suggest that the attempted Diels-Alder synthesis fails because the position of equilibrium favors reactants rather than products.

18.47 The very low hydrogen content of the molecular formula of **Y** (C₈H₄O₃) indicates that **Y** is highly unsaturated. That **Y** dissolves slowly in warm aqueous NaHCO₃ suggests that **Y** is a carboxylic acid anhydride that hydrolyses and dissolves because it forms a carboxylate salt:

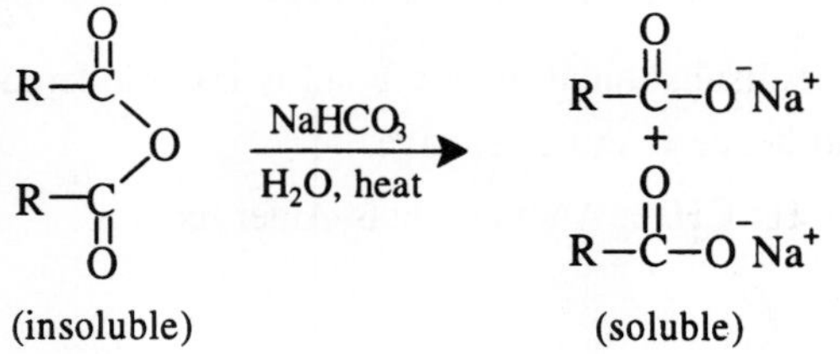

(insoluble)   (soluble)

The infrared absorption peaks at 1779 and 1854 cm⁻¹ are consistent with those of an aromatic carboxylic anhydride (Table 18.5).

That only four signals appear in the ¹³C spectrum of **Y** indicates a high degree of symmetry for **Y**. Three of the signals occur in the aromatic region (δ 120–δ 140) and one signal is downfield (δ 163.1)

These signals and the information from the DEPT ¹³C NMR spectra lead us to conclude that **Y** is phthalic anhydride. The assignments are

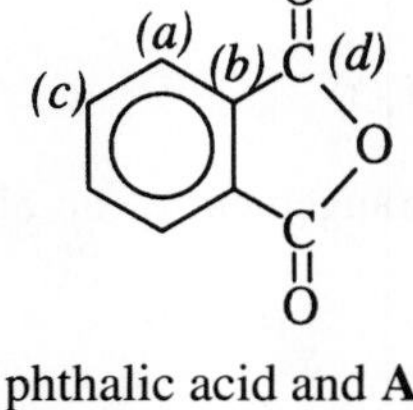

(a) δ 124.3
(b) δ 131.1
(c) δ 136.1
(d) δ 163.1

Z is phthalic acid and **AA** is ethyl hydrogen phthalate.

18.48   (a)   Ethyl acetate   (b)   Acetic anhydride   (c)   *N*-Ethylacetamide.

18.49   In the first instance, nucleophilic attack by the amine occurs preferentially at the less hindered carbon of the formyl group. (Recall that aldehydes are more reactive than ketones toward nucleophiles for the same reason.) In the second case, $CF_3CO_2^-$ is a better leaving group than $CH_3CO_2^-$ since the former is the conjugate base of the stronger acid.

18.50

18.51 $C_6H_6 + (CH_3)_2CHCCl$ (O) $\xrightarrow{AlCl_3}$ [phenyl]$-CCH(CH_3)_2$ (O) $\xrightarrow[\text{Wolff-Kishner}]{\text{Clemmensen or}}$

18.52   A = [phenyl]$-CCH(CH_3)_2$ (O)   C = $(CH_3)_2CHCH_2-$[phenyl]$-CCH_3$ (O)

B = [phenyl]$-CH_2CH(CH_3)_2$ D = $(CH_3)_2CHCH_2-$[phenyl]$-C(OH)CN$ with CH_3

In the last step, HI/red P accomplishes both the reduction of $-OH$ to $-H$ and the hydrolysis of the nitrile function.

18.53   (a)   The signal at δ 193.8 is consistent with the carbonyl carbon of an aldehyde and shows that the PCC reaction produced cinnamaldehyde.

(b)   The signal at δ 164.5 is consistent with the carbonyl carbon of a carboxylic acid, and suggests that the oxidation with $K_2Cr_2O_7$ in sulfuric acid produced cinnamic acid.

QUIZ

18.1 Which of the following would be the strongest acid?

(a) Benzoic acid   (b) 4-Nitrobenzoic acid   (c) 4-Methylbenzoic acid

(d) 4-Methoxybenzoic acid   (e) 4-Ethylbenzoic acid

18.2 Which of the following would yield (S)-2-butanol?

(a) (R)-2-Bromobutane + $CH_3CO_2^-$ Na^+ $\longrightarrow$ product $\xrightarrow[\text{heat}]{OH^-, H_2O}$

(b) (R)-2-Bromobutane $\xrightarrow[\text{heat}]{OH^-, H_2O}$

(c) (S)-2-Butyl acetate $\xrightarrow[\text{heat}]{OH^-, H_2O}$

(d) All of the above

(e) None of the above

18.3 Which reagent would serve as the basis for a simple chemical test to distinguish between hexanoic acid and hexanamide?

(a) Cold dilute NaOH   (b) Cold dilute $NaHCO_3$

(c) Cold concd. H_2SO_4 (d) More than one of these

(e) None of these

18.4 Give an acceptable name for:

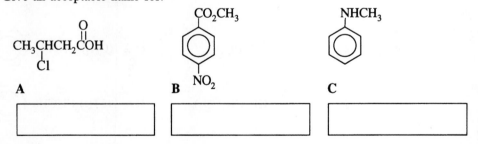

A   **B**   **C**

18.5 Complete the following syntheses.

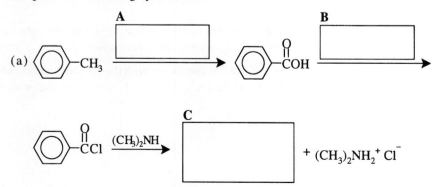

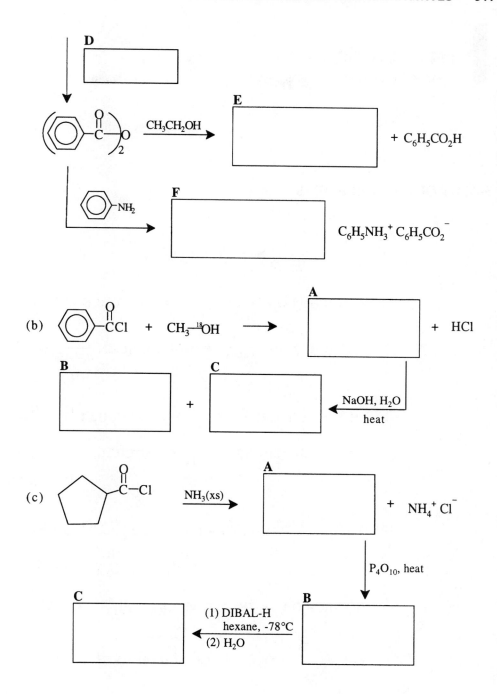

(b)

(c)

SPECIAL TOPIC
Step-Growth Polymers

SOLUTIONS TO PROBLEMS

I.1 (a) cyclohexanone $\xrightarrow[\text{H}_2\text{Cr}_2\text{O}_7]{\text{[O]}}$ $HO_2C(CH_2)_4CO_2H$

(b) $HO_2C(CH_2)_4CO_2H + 2\ NH_3 \longrightarrow NH_4O_2C(CH_2)_4CO_2NH_4 \xrightarrow[\text{-2H}_2\text{O}]{\text{heat}}$

$H_2N\overset{\text{O}}{\overset{\|}{C}}(CH_2)_4\overset{\text{O}}{\overset{\|}{C}}NH_2 \xrightarrow[\text{catalyst}]{350°C} N\equiv C(CH_2)_4C\equiv N \xrightarrow[\text{catalyst}]{4\ H_2}$

$H_2NCH_2(CH_2)_4CH_2NH_2$

(c) $CH_2{=}CH{-}CH{=}CH_2 \xrightarrow{Cl_2} ClCH_2CH{=}CHCH_2Cl \xrightarrow{2\ NaCN}$

$N\equiv CCH_2CH{=}CHCH_2C\equiv N \xrightarrow[\text{Ni}]{H_2} N\equiv C(CH_2)_4C\equiv N \xrightarrow[\text{catalyst}]{4\ H_2}$

$H_2NCH_2(CH_2)_4CH_2NH_2$

(d) tetrahydrofuran $\xrightarrow{2\ HCl} ClCH_2CH_2CH_2CH_2Cl \xrightarrow{2\ NaCN} N\equiv C(CH_2)_4C\equiv N$

$\xrightarrow[\text{catalyst}]{4\ H_2} H_2NCH_2(CH_2)_4CH_2NH_2$

I.2 (a) $HOCH_2CH_2OH + :B^- \rightleftharpoons HOCH_2CH_2O^- + HB$

$RO\overset{\text{O}}{\overset{\|}{C}}{-}\!\!\bigcirc\!\!{-}\overset{\text{O}}{\overset{\|}{C}}OCH_3 + {}^-OCH_2CH_2OH \rightleftharpoons$

$RO\overset{\text{O}}{\overset{\|}{C}}{-}\!\!\bigcirc\!\!{-}\underset{\underset{\text{OCH}_3}{|}}{\overset{\overset{\text{O}^-}{|}}{C}}{-}OCH_2CH_2OH \rightleftharpoons RO\overset{\text{O}}{\overset{\|}{C}}{-}\!\!\bigcirc\!\!{-}\overset{\text{O}}{\overset{\|}{C}}OCH_2CH_2OH$

$+\ CH_3O^-$

$CH_3O^- + HB \rightleftharpoons CH_3OH + :B^-$

$[R = CH_3\text{- or } HOCH_2CH_2\text{-}]$

(b)

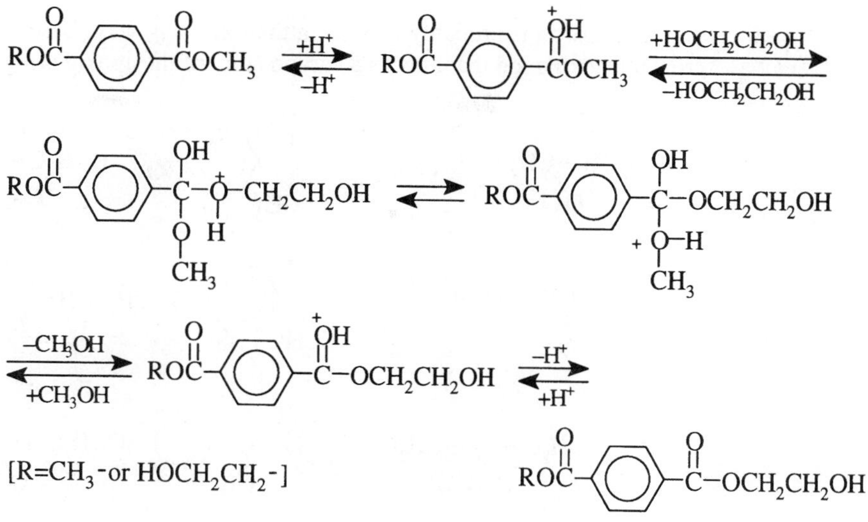

[R=CH$_3$-or HOCH$_2$CH$_2$-]

I.3 (a)

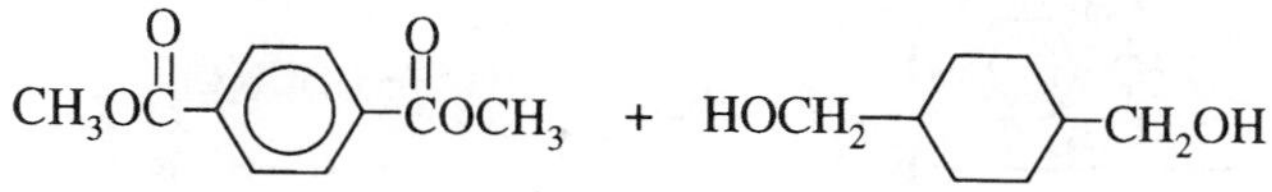

(b) By high-pressure catalytic hydrogenation

I.4

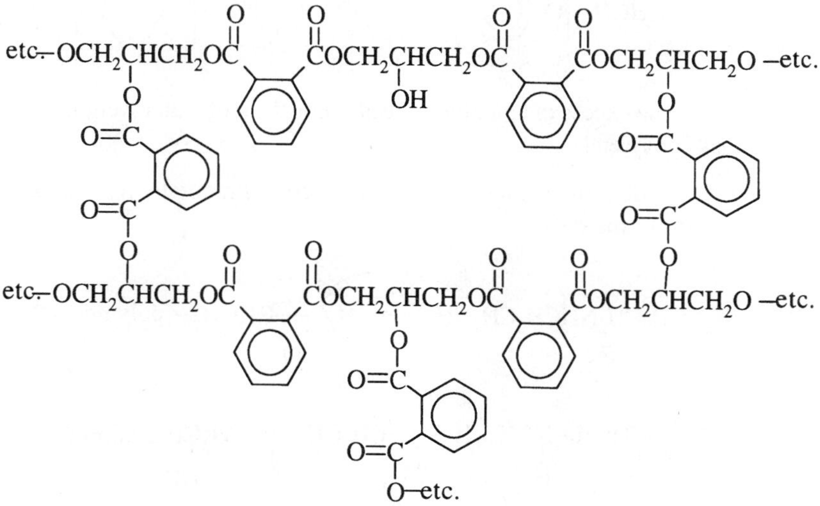

I.5

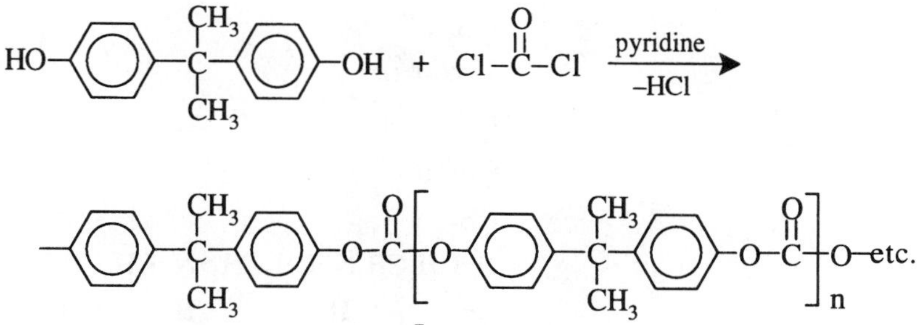

Lexan

I.6   (a)   The resin is probably formed in the following way. Base converts the bisphenol A to a phenoxide ion that attacks a carbon atom of the epoxide ring of epichlorohydrin:

$$ClCH_2CH\overset{\displaystyle O}{-}CH_2 \;+\; {}^-O\!\!-\!\!\!\bigcirc\!\!-\!\!\underset{CH_3}{\overset{CH_3}{C}}\!\!-\!\!\bigcirc\!\!-\!\!O^- \;+\; H_2C\overset{\displaystyle O}{-}CHCH_2Cl \longrightarrow$$

$$Cl\overset{\frown}{-}CH_2\!-\!CH\!-\!CH_2\!-\!O\!-\!\!\bigcirc\!\!-\!\!\underset{CH_3}{\overset{CH_3}{C}}\!\!-\!\!\bigcirc\!\!-\!\!O\!-\!CH_2\!-\!CH\!-\!CH_2\!-\!Cl$$

$$\xrightarrow{-2Cl^-} H_2C\overset{\displaystyle O}{-}CHCH_2O\!-\!\!\bigcirc\!\!-\!\!\underset{CH_3}{\overset{CH_3}{C}}\!\!-\!\!\bigcirc\!\!-\!\!OCH_2CH\overset{\displaystyle O}{-}CH_2$$

$${}^-O\!\!-\!\!\bigcirc\!\!-\!\!\underset{CH_3}{\overset{CH_3}{C}}\!\!-\!\!\bigcirc\!\!-\!\!O^- \qquad \text{then} \qquad H_2C\overset{\displaystyle O}{-}CHCH_2Cl \longrightarrow$$

$$H_2C\overset{\displaystyle O}{-}CHCH_2\!\!\left[\!O\!-\!\!\bigcirc\!\!-\!\!\underset{CH_3}{\overset{CH_3}{C}}\!\!-\!\!\bigcirc\!\!-\!\!OCH_2\underset{OH}{CHCH_2}\!\right]_n\!\!O\!-\!\!\bigcirc\!\!-\!\!\underset{CH_3}{\overset{CH_3}{C}}\!\!-\!\!\bigcirc\!\!-\!\!OCH_2CH\overset{\displaystyle O}{-}CH_2$$

(b)   The excess of epichlorohydrin limits the molecular weight and ensures that the resin has epoxy ends.

(c)   Adding the hardener brings about cross linking by reacting at the terminal epoxide groups of the resin:

$$H_2NCH_2CH_2NHCH_2CH_2\overset{..}{N}H_2 \;+\; H_2C\overset{\displaystyle O}{-}CHCH_2\!-\![\text{polymer}]\!-\!CH_2CH\overset{\displaystyle O}{-}CH_2 \longrightarrow$$

$$-CH_2\!-\!\underset{OH}{CHCH_2}\!-\!\underset{H}{NCH_2CH_2}\!-\!\underset{\underset{\text{[polymer]}}{\overset{\displaystyle CH_2}{\underset{\displaystyle CHOH}{\overset{\displaystyle CH_2}{\mid}}}}}{N}\!-\!CH_2CH_2\!-\!\underset{H}{N}\!-\!CH_2\underset{OH}{CHCH_2}[\text{polymer}]CH_2\underset{OH}{CHCH_2}\!-\!\text{etc.}$$

$$\underset{H}{N}\!-\!CH_2CH_2\underset{H}{N}\!-\!CH_2CH_2\!-\!\underset{H}{N}\!-\!CH_2\underset{OH}{CHCH_2}[\text{polymer}]CH_2\underset{OH}{CHCH_2} \qquad \text{etc.}$$

I.7

(a)

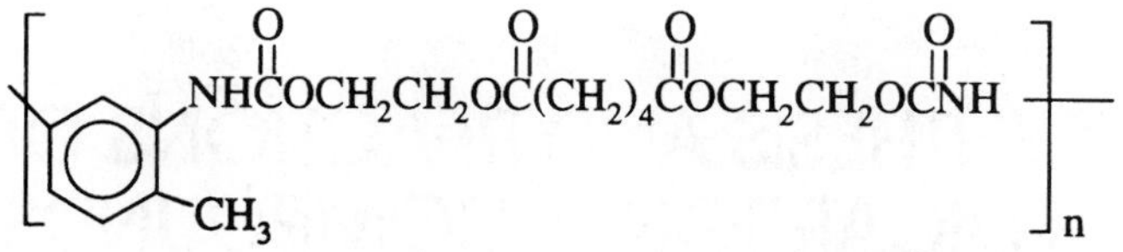

$$\left[\underset{CH_3}{\overset{}{\bigcirc}} NHCOCH_2CH_2OC(CH_2)_4COCH_2CH_2OCNH \right]_n$$

(b) To ensure that the polyester chain has –CH₂OH end groups.

I.8 Because the para position is occupied by a methyl group, cross-linking does not occur and the resulting polymer remains thermoplastic (See Section I.4.)

I.9

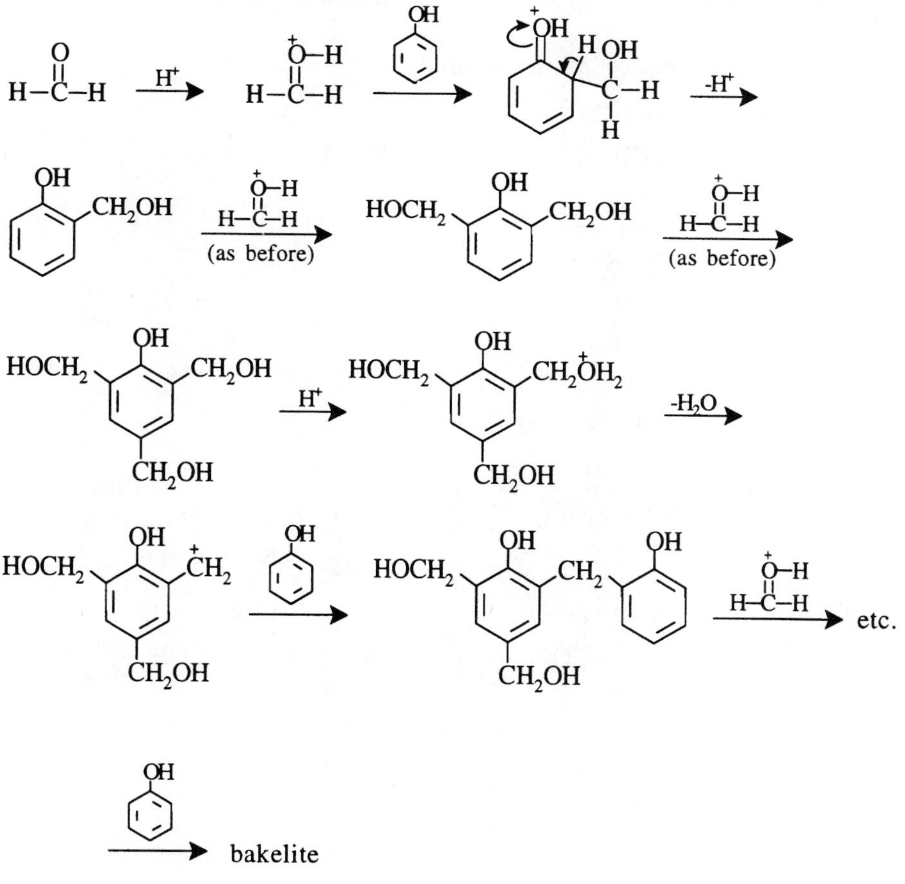

$$H-\overset{O}{\overset{\|}{C}}-H \xrightarrow{\;H^+\;} H-\overset{\overset{+}{O}-H}{\overset{\|}{C}}-H \longrightarrow \;\cdots\; \xrightarrow{\;-H^+\;}$$

(reaction sequence of phenol with formaldehyde under acid catalysis)

... $\overset{OH}{\bigcirc}CH_2OH \xrightarrow[\text{(as before)}]{H-\overset{\overset{+}{O}-H}{\overset{\|}{C}}-H} HOCH_2\overset{OH}{\bigcirc}CH_2OH \xrightarrow[\text{(as before)}]{H-\overset{\overset{+}{O}-H}{\overset{\|}{C}}-H}$

... $HOCH_2\overset{OH}{\bigcirc}\underset{CH_2OH}{CH_2OH} \xrightarrow{\;H^+\;} HOCH_2\overset{OH}{\bigcirc}\underset{CH_2OH}{CH_2\overset{+}{O}H_2} \xrightarrow{\;-H_2O\;}$

... $HOCH_2\overset{OH}{\bigcirc}\underset{CH_2OH}{\overset{+}{C}H_2} + \overset{OH}{\bigcirc} \longrightarrow HOCH_2\overset{OH}{\bigcirc}\underset{CH_2OH}{CH_2}\overset{OH}{\bigcirc} \xrightarrow{\;H-\overset{\overset{+}{O}-H}{\overset{\|}{C}}-H\;} \text{etc.}$

$\overset{OH}{\bigcirc} \longrightarrow$ bakelite

19

SYNTHESIS AND REACTIONS OF β-DICARBONYL COMPOUNDS: MORE CHEMISTRY OF ENOLATE IONS

SUMMARY OF ACETOACETIC ESTER AND MALONIC ESTER SYNTHESES

A. Acetoacetic Ester Synthesis

$$CH_3\overset{O}{\overset{\|}{C}}CH_2\overset{O}{\overset{\|}{C}}OEt \xrightarrow[\text{(2) RX}]{\text{(1) NaOEt}} CH_3\overset{O}{\overset{\|}{C}}\overset{O}{\overset{\|}{C}}HCOEt \xrightarrow[\text{(2) R'X}]{\text{(1)}(CH_3)_3COK}$$

$$CH_3\overset{O}{\overset{\|}{C}}-\overset{R'}{\underset{R}{C}}-\overset{O}{\overset{\|}{C}}OEt \xrightarrow[\text{(2) H}_3\text{O}^+]{\text{(1)OH}^-, \text{H}_2\text{O}} CH_3\overset{O}{\overset{\|}{C}}-\overset{R'}{\underset{R}{C}}-\overset{O}{\overset{\|}{C}}OH \xrightarrow[\text{-CO}_2]{\text{heat}} CH_3\overset{O}{\overset{\|}{C}}\overset{}{\underset{R}{C}}HR'$$

B. Malonic Ester Synthesis

$$EtO\overset{O}{\overset{\|}{C}}CH_2\overset{O}{\overset{\|}{C}}OEt \xrightarrow[\text{(2) RX}]{\text{(1) NaOEt}} EtO\overset{O}{\overset{\|}{C}}\overset{O}{\overset{\|}{C}}HCOEt \xrightarrow[\text{(2) R'X}]{\text{(1)}(CH_3)_3COK}$$

$$EtO\overset{O}{\overset{\|}{C}}-\overset{R'}{\underset{R}{C}}-\overset{O}{\overset{\|}{C}}OEt \xrightarrow[\text{(2) H}_3\text{O}^+]{\text{(1) OH,H}_2\text{O}} HO\overset{O}{\overset{\|}{C}}-\overset{R'}{\underset{R}{C}}-\overset{O}{\overset{\|}{C}}OH \xrightarrow[\text{-CO}_2]{\text{heat}} HO\overset{O}{\overset{\|}{C}}\overset{}{\underset{R}{C}}HR'$$

SOLUTIONS TO PROBLEMS

19.1 (a) *Step 1* $CH_3\overset{}{\underset{H}{C}}H-\overset{O}{\overset{\|}{C}}OC_2H_5 + {}^-OC_2H_5 \rightleftarrows CH_3\overset{..}{C}H-\overset{O}{\overset{\|}{C}}OC_2H_5 + C_2H_5OH$

$$\updownarrow$$

$$CH_3CH=\overset{O^-}{\overset{}{C}}OC_2H_5$$

Step 2   CH₃CH₂C〈O / OC₂H₅   +   :CHCOC₂H₅ / CH₃   ⇌   CH₃CH₂C(O⁻)−CH−COC₂H₅ / C₂H₅O⁺CH₃

C₂H₅O⁻   +   CH₃CH₂C−CH−COC₂H₅ / CH₃

Step 3   CH₃CH₂C−C−COC₂H₅ (O H O / CH₃)   +   ⁻OC₂H₅   ⇌   CH₃CH₂C=C−COC₂H₅ (O O / CH₃) + C₂H₅OH

(b)   CH₃CH₂C−CH−COC₂H₅ (O O / CH₃)   +   CH₃CH₂C=C−COC₂H₅ (OH O / CH₃)

19.2   (a)

(b)   To undergo a Dieckmann condensation, diethyl glutarate would have to form a highly strained four-membered ring.

19.3   CH₃COC₂H₅ (O)   +   C₂H₅O⁻   ⇌   :CH₂COC₂H₅ (O)   +   C₂H₅OH

C₆H₅COC₂H₅ (O)   +   :CH₂COC₂H₅ (O)   ⇌   C₆H₅C(O⁻)−CH₂COC₂H₅ / OC₂H₅

C₆H₅C−CH₂COC₂H₅ (O O)   +   C₂H₅O⁻   ⇌   C₆H₅C=CH−COC₂H₅ (O O)   +   C₂H₅OH

→ (H⁺)   C₆H₅C−CH₂COC₂H₅ (O O)

C₆H₅CH₂COC₂H₅ (O)   +   C₂H₅O⁻   ⇌   C₆H₅CHCOC₂H₅ (⁻ O)   +   C₂H₅OH

$$C_6H_5\overset{\cdot\cdot}{C}H\overset{O}{\overset{\|}{C}}OC_2H_5 \;+\; C_2H_2O\overset{O}{\overset{\|}{C}}OC_2H_5 \;\rightleftarrows\;$$

$$\begin{array}{c} C_2H_5O-\overset{O^-}{\underset{\|}{C}}-OC_2H_5 \\ C_6H_5\overset{}{C}H \\ \overset{}{\underset{\|}{C}}OC_2H_5 \\ \overset{}{O} \end{array} \;\longleftrightarrow\;$$

$$\begin{array}{c} \overset{O}{\overset{\|}{C}}-OC_2H_5 \\ C_6H_5\overset{}{C}H \\ \overset{}{\underset{\|}{C}}OC_2H_5 \\ \overset{}{O} \end{array} \;+\; C_2H_5O^- \;\rightleftarrows\; \begin{array}{c} \overset{O}{\overset{\|}{C}}-OC_2H_5 \\ C_6H_5\overset{}{C}:^- \\ \overset{}{\underset{\|}{C}}OC_2H_5 \\ \overset{}{O} \end{array} \;+\; C_2H_5OH$$

Resonance
stabilized

$$\overset{H^+}{\longrightarrow}\; \begin{array}{c} \overset{O}{\overset{\|}{C}}-OC_2H_5 \\ C_6H_5\overset{}{C}H \\ \overset{}{\underset{\|}{C}}OC_2H_5 \\ \overset{}{O} \end{array}$$

19.4 (a) $CH_3CH_2\overset{O}{\overset{\|}{C}}OC_2H_5 \;+\; C_2H_5O\overset{O}{\overset{\|}{C}}-\overset{O}{\overset{\|}{C}}OC_2H_5 \;\xrightarrow[\;(2)\,H^+\;]{(1)\,NaOCH_2CH_3}\; \begin{array}{c} CH_3\overset{O}{\overset{}{C}}H\overset{O}{\overset{\|}{C}}OC_2H_5 \\ \overset{}{\underset{\|}{C}}-\overset{}{\underset{\|}{C}}OC_2H_5 \\ \overset{}{O}\;\;\overset{}{O} \end{array}$

(b) $CH_3\overset{O}{\overset{\|}{C}}OC_2H_5 \;+\; H\overset{O}{\overset{\|}{C}}OC_2H_5 \;\xrightarrow[\;(2)\,H^+\;]{(1)\,NaOCH_2CH_3}\; H\overset{O}{\overset{\|}{C}}CH_2\overset{O}{\overset{\|}{C}}OC_2H_5$

19.5 (a) $+\; H\overset{O}{\overset{\|}{C}}OC_2H_5 \;\xrightarrow[\;(2)\,H^+\;]{(1)\,NaOCH_2CH_3}$

(b) $CH_3CH_2\overset{O}{\overset{\|}{C}}CH_2CH_2CH_2\overset{O}{\overset{\|}{C}}OC_2H_5 \;\xrightarrow[\;(2)\,H^+\;]{(1)\,NaOCH_2CH_3}$

(c) $C_2H_5O_2CCH_2\underset{\underset{CH_3}{|}}{\overset{\overset{CH_3}{|}}{C}}CH_2CO_2C_2H_5 \;+\; C_2H_5O\overset{O}{\overset{\|}{C}}-\overset{O}{\overset{\|}{C}}OC_2H_5 \;\xrightarrow[\;(2)\,H^+\;]{(1)\,NaOCH_2CH_3}$

$\xrightarrow[\;(2)\,H^+\;]{(1)\,NaOCH_2CH_3}$

19.6 $CH_3CCH_2CH_2CH_2CH_2COC_2H_5$ + $C_2H_5O^-$ $\underset{-C_2H_5OH}{\longleftrightarrow}$ [cyclic β-keto ester structures shown]

[reaction scheme with cyclopentane ring intermediates]

$\rightleftarrows$ [enolate] + C_2H_5OH $\xrightarrow{H^+}$ [2-acetylcyclopentanone]

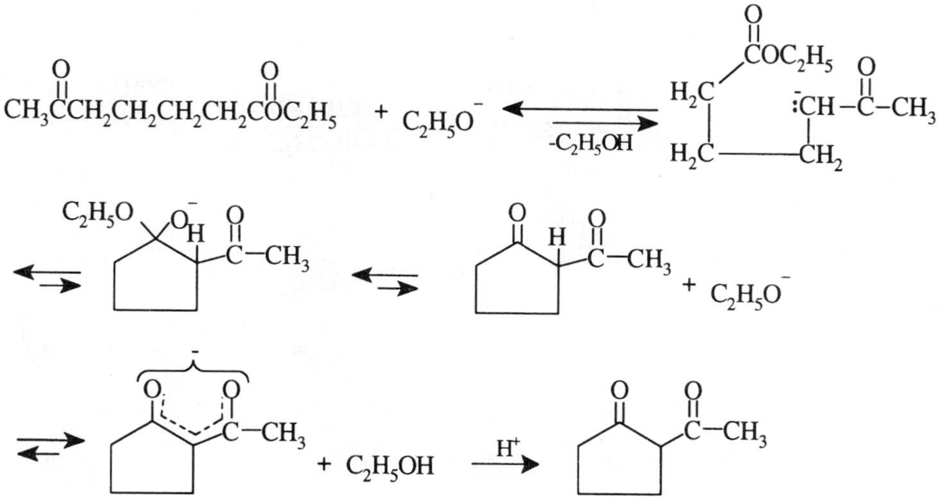

19.7 The partially negative oxygen atom of sodioacetoacetic ester acts as the nucleophile.

$$CH_3\overset{O}{\overset{\|}{C}}-\overset{\cdot\cdot}{\overset{-}{C}H}-\overset{O}{\overset{\|}{C}}-OC_2H_5 \quad\longleftrightarrow\quad CH_3\overset{O^-}{\overset{|}{C}}=CH-\overset{O}{\overset{\|}{C}}-OC_2H_5$$

19.8 Again, working backward,

(a) 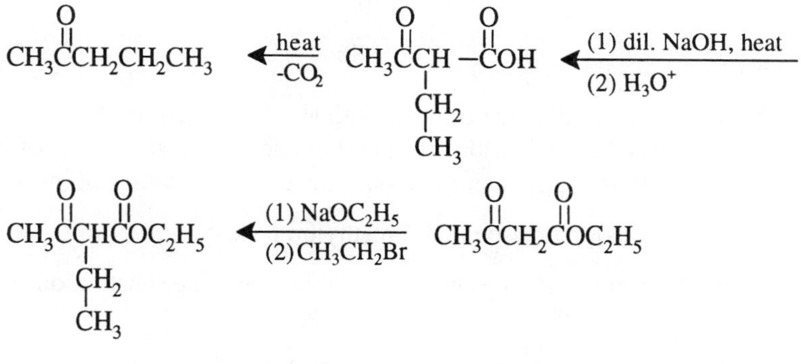 $CH_3\overset{O}{\overset{\|}{C}}CH_2CH_2CH_3$ $\underset{-CO_2}{\overset{heat}{\longleftarrow}}$ $CH_3\overset{O}{\overset{\|}{C}}CH-\overset{O}{\overset{\|}{C}}OH$ $\underset{(2)\,H_3O^+}{\overset{(1)\,dil.\,NaOH,\,heat}{\longleftarrow}}$
   with $\overset{|}{C}H_2$ / CH_3 branch

$CH_3\overset{O}{\overset{\|}{C}}CH\overset{O}{\overset{\|}{C}}OC_2H_5$ $\underset{(2)\,CH_3CH_2Br}{\overset{(1)\,NaOC_2H_5}{\longleftarrow}}$ $CH_3\overset{O}{\overset{\|}{C}}CH_2\overset{O}{\overset{\|}{C}}OC_2H_5$
   with $\overset{|}{C}H_2$ / CH_3 branch

(b) 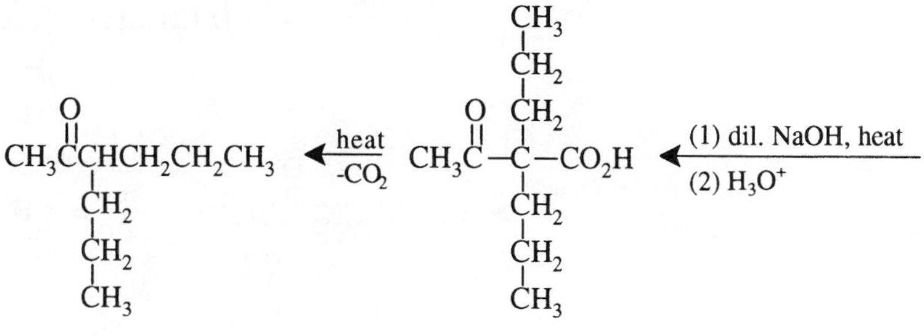 $CH_3\overset{O}{\overset{\|}{C}}CHCH_2CH_2CH_3$ $\underset{-CO_2}{\overset{heat}{\longleftarrow}}$ $CH_3\overset{O}{\overset{\|}{C}}-\overset{|}{C}-CO_2H$ $\underset{(2)\,H_3O^+}{\overset{(1)\,dil.\,NaOH,\,heat}{\longleftarrow}}$
   with CH_2/CH_2/CH_3 branches on both

$$CH_3\overset{O}{\overset{\|}{C}}-\overset{\overset{\overset{CH_3}{|}}{\overset{CH_2}{|}}{\overset{CH_2}{|}}}{\underset{\underset{\underset{CH_3}{|}}{\overset{CH_2}{|}}}{C}}-CO_2C_2H_5 \quad \xleftarrow[\text{(2) } CH_3CH_2CH_2Br]{\text{(1) } (CH_3)_3COK} \quad CH_3\overset{O}{\overset{\|}{C}}-\underset{\underset{\underset{CH_3}{|}}{\overset{CH_2}{|}}}{CH}\overset{O}{\overset{\|}{C}}OC_2H_5$$

$$\xleftarrow[\text{(2) } CH_3CH_2CH_2Br]{\text{(1) } NaOC_2H_5} \quad CH_3\overset{O}{\overset{\|}{C}}CH_2\overset{O}{\overset{\|}{C}}OC_2H_5$$

(c) $CH_3\overset{O}{\overset{\|}{C}}CH_2CH_2C_6H_5 \quad \xleftarrow[-CO_2]{\text{heat}} \quad CH_3\overset{O}{\overset{\|}{C}}\underset{\underset{\underset{C_6H_5}{|}}{\overset{CH_2}{|}}}{CH}\overset{O}{\overset{\|}{C}}OH \quad \xleftarrow[\text{(2) } H_3O^+]{\text{(1) } OH^-, \text{ heat}}$

$$CH_3\overset{O}{\overset{\|}{C}}\underset{\underset{\underset{C_6H_5}{|}}{\overset{CH_2}{|}}}{CH}\overset{O}{\overset{\|}{C}}OC_2H_5 \quad \xleftarrow[\text{(2) } C_6H_5CH_2Br]{\text{(1) } NaOC_2H_5} \quad CH_3\overset{O}{\overset{\|}{C}}CH_2\overset{O}{\overset{\|}{C}}OC_2H_5$$

19.9 (a) Reactivity is the same as with any S_N2 reaction. With primary halides substitution is highly favored, with secondary halides elimination competes with substitution, and with tertiary halides elimination is the exclusive course of reaction.

(b) Acetoacetic ester and 2-methylpropene

(c) Bromobenzene is unreactive to nucleophilic substitution.

19.10 $CH_3CH_2CH_2\overset{O}{\overset{\|}{C}}OC_2H_5 \quad \xrightarrow[\text{(2) } H^+]{\text{(1) } NaOC_2H_5} \quad CH_3CH_2CH_2\overset{O}{\overset{\|}{C}}\underset{\underset{\underset{CH_3}{|}}{\overset{CH_2}{|}}}{C}H\overset{O}{\overset{\|}{C}}OC_2H_5$

$\xrightarrow[\text{(2) } H_3O^+]{\text{(1) NaOH , } H_2O, \text{ heat}} \quad CH_3CH_2CH_2\overset{O}{\overset{\|}{C}}\underset{\underset{\underset{CH_3}{|}}{\overset{CH_2}{|}}}{C}H\overset{O}{\overset{\|}{C}}OH \quad \xrightarrow[-CO_2]{\text{heat}} \quad CH_3CH_2CH_2\overset{O}{\overset{\|}{C}}CH_2CH_2CH_3$

19.11 The carboxyl group that is lost more readily is the one that is β to the keto group (cf. Section 18.11 of the text).

19.12 $CH_3CCH_2CH_2CC_6H_5$ $\xleftarrow[\text{-CO}_2]{\text{heat}}$ $CH_3CCHCOH$ $\xleftarrow[\text{(2) H}_3O^+]{\text{(1) OH}^-, \text{H}_2O, \text{heat}}$

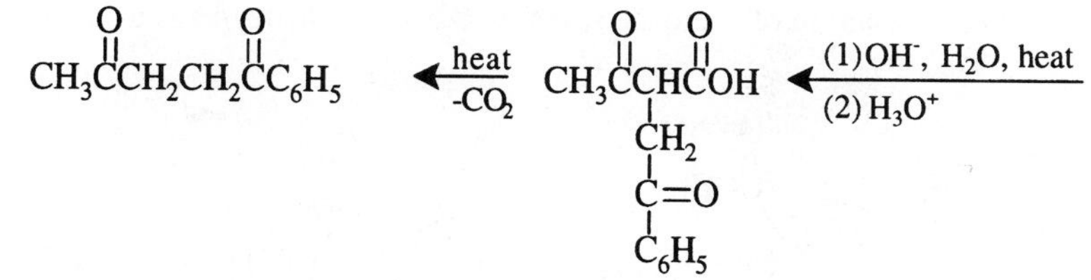

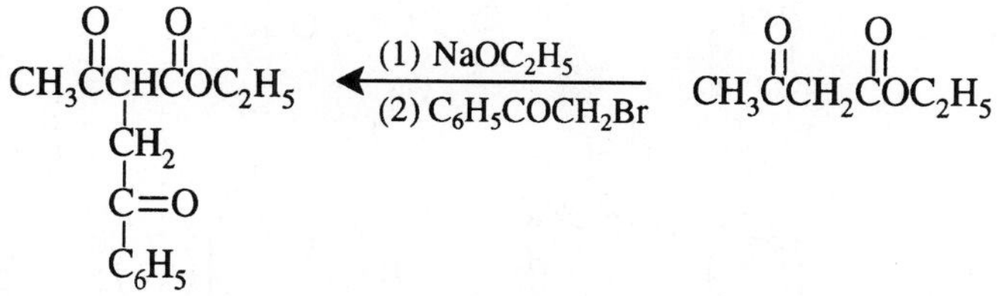

with pendant group:
CH_2
$C=O$
C_6H_5

$CH_3CCHCOC_2H_5$ $\xleftarrow[\text{(2) C}_6H_5COCH_2Br]{\text{(1) NaOC}_2H_5}$ $CH_3CCH_2COC_2H_5$

with pendant group:
CH_2
$C=O$
C_6H_5

19.13 $CH_3CCH_2CC_6H_5$ $\xleftarrow[\text{-CO}_2]{\text{heat}}$ $CH_3CCHCOH$ $\xleftarrow[\text{(2) H}_3O^+]{\text{(1) OH}^-, \text{H}_2O, \text{heat}}$

with pendant group:
$C=O$
C_6H_5

$CH_3CCHCOC_2H_5$ $\xleftarrow[\text{(2) C}_6H_5COCl}]{\text{(1) NaH}}$ $CH_3CCH_2COC_2H_5$

with pendant group:
$C=O$
C_6H_5

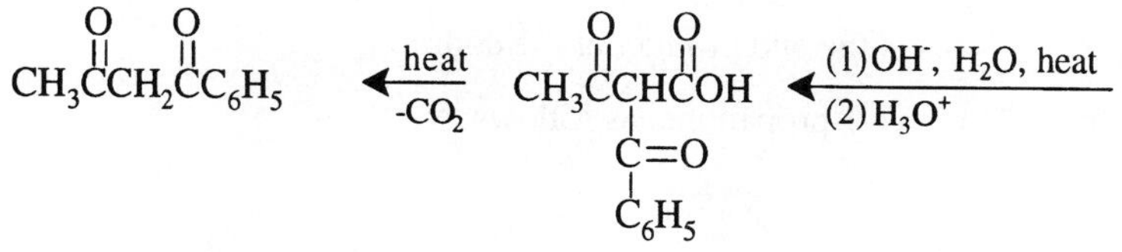

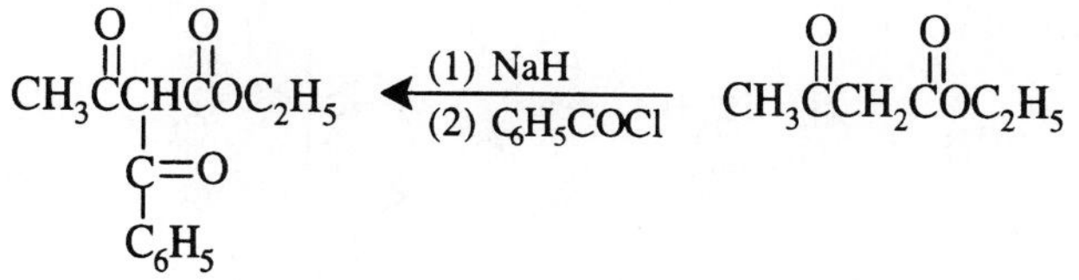

19.14   (a)   One molar equivalent of NaNH$_2$ converts acetoacetic ester to its anion,

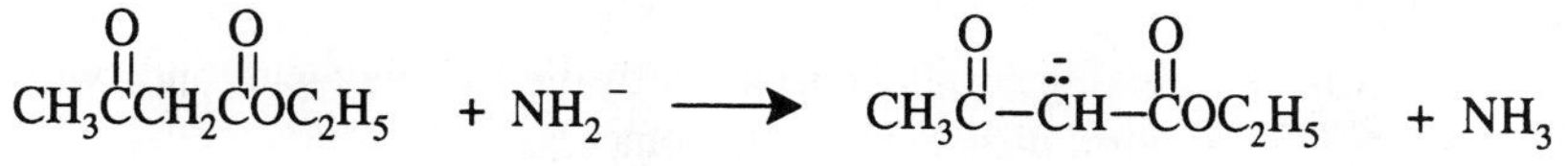

$CH_3CCH_2COC_2H_5$ $+$ NH_2^- $\longrightarrow$ $CH_3C-\overset{\cdot\cdot}{C}H-COC_2H_5$ $+$ NH_3

and one molar equivalent of NaNH$_2$ converts bromobenzene to benzyne (cf. Section 21.11B):

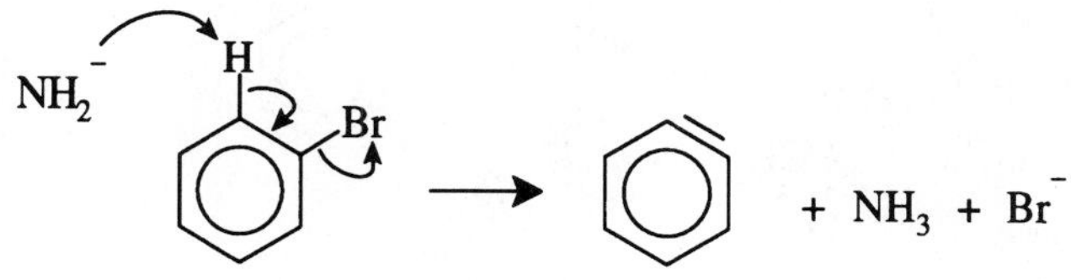

NH_2^- ... $\longrightarrow$ ... $+$ NH_3 $+$ Br^-

Then the anion of acetoacetic ester adds to the benzyne as it forms in the mixture.

This is the end product of the addition.

(b)   1-phenyl-2-propanone, as follows:

$$\text{(1) OH}^-, \text{H}_2\text{O, heat} \quad \text{(2) H}_3\text{O}^+$$

$$\xrightarrow[\text{-CO}_2]{\text{heat}}$$

(c)   By treating bromobenzene with diethyl malonate and two molar equivalents of NaNH₂ to form diethyl phenylmalonate.

$$+ \quad \xrightarrow{2 \text{ NaNH}_2}$$

[The mechanism for this reaction is analogous to that given in part (a).]

Then hydrolysis and decarboxylation will convert diethyl phenylmalonate to phenylacetic acid.

$$\xrightarrow[\text{(2) H}_3\text{O}^+]{\text{(1) OH}^-, \text{H}_2\text{O, heat}} \quad \xrightarrow{\text{heat}} \quad + \text{ CO}_2$$

19.15 Here we alkylate the dianion,

$$CH_3\overset{O}{\overset{||}{C}}-CH_2-\overset{O}{\overset{||}{C}}OC_2H_5 \xrightarrow[\text{liq. NH}_3]{\text{2 KNH}_2} \;:CH_2\overset{O}{\overset{||}{C}}-\overset{..}{C}H-\overset{O}{\overset{||}{C}}OC_2H_5 \xrightarrow[(2)\,NH_4Cl]{(1)\,C_6H_5CH_2Cl}$$

$$C_6H_5CH_2CH_2\overset{O}{\overset{||}{C}}CH_2\overset{O}{\overset{||}{C}}OC_2H_5$$

19.16 Working backward,

(a) $CH_3CH_2CH_2CH_2CO_2H$ $\xleftarrow[-CO_2]{\text{heat}}$ $CH_3CH_2CH_2\underset{\overset{|}{CO_2H}}{\overset{\overset{CO_2H}{|}}{CH}}$ $\xleftarrow[(2)\,H_3O^+]{(1)\,OH^-,\,H_2O,\,\text{heat}}$

$CH_3CH_2CH_2\underset{\overset{|}{CO_2C_2H_5}}{\overset{\overset{CO_2C_2H_5}{|}}{CH}}$ $\xleftarrow[(2)\,CH_3CH_2CH_2Br]{(1)\,NaOC_2H_5}$ $\underset{\overset{|}{CO_2C_2H_5}}{\overset{\overset{CO_2C_2H_5}{|}}{CH_2}}$

(b) $CH_3CH_2CH_2\underset{\overset{|}{CH_3}}{CHCO_2H}$ $\xleftarrow[-CO_2]{\text{heat}}$ $CH_3CH_2CH_2\underset{\overset{|}{CO_2H}}{\overset{\overset{CO_2H}{|}}{C}}CH_3$ $\xleftarrow[(2)\,H_3O^+]{(1)\,OH^-,\,H_2O,\,\text{heat}}$

$CH_3CH_2CH_2\underset{\overset{|}{CO_2C_2H_5}}{\overset{\overset{CO_2C_2H_5}{|}}{C}}CH_3$ $\xleftarrow[(CH_3)_3COK]{CH_3I}$ $CH_3CH_2CH_2\underset{\overset{|}{CO_2C_2H_5}}{\overset{\overset{CO_2C_2H_5}{|}}{CH}}$

$\xleftarrow[(2)\,CH_3CH_2CH_2Br]{(1)\,NaOC_2H_5}$ $\underset{\overset{|}{CO_2C_2H_5}}{\overset{\overset{CO_2C_2H_5}{|}}{CH_2}}$

(c) $CH_3\underset{\overset{|}{CH_3}}{CH}CH_2CH_2CO_2H$ $\xleftarrow[-CO_2]{\text{heat}}$ $CH_3\underset{\overset{|}{CH_3}}{CH}CH_2\underset{\overset{|}{CO_2H}}{\overset{\overset{CO_2H}{|}}{CH}}$ $\xleftarrow[(2)\,H_3O^+]{(1)\,OH^-,\,H_2O,\,\text{heat}}$

$CH_3\underset{\overset{|}{CH_3}}{CH}CH_2\underset{\overset{|}{CO_2C_2H_5}}{\overset{\overset{CO_2C_2H_5}{|}}{CH}}$ $\xleftarrow[(2)\,CH_3CHCH_2Br\;\;(CH_3)]{(1)\,NaOC_2H_5}$ $\underset{\overset{|}{CO_2C_2H_5}}{\overset{\overset{CO_2C_2H_5}{|}}{CH_2}}$

19.17 $2\;H_3C-CH_2-CH_2-Br \;+\; H_2C\overset{\displaystyle CN}{\underset{\displaystyle \underset{\overset{||}{O}}{COEt}}{}} \xrightarrow[-2HBr]{\text{base}}$

$$\xrightarrow[\text{-CO}_2, \text{-EtOH}]{\text{OH}^-, \text{ heat}}$$ —CN $$\xrightarrow[\text{- NH}_3]{\text{OH}^-, \text{ heat}}$$ —CO$_2$H

Valproic acid

19.18 (a) Formaldehyde, $\overset{\overset{\text{O}}{\|}}{\text{H}-\text{C}-\text{H}}$

(b)

$$\xrightarrow[\text{- C}_4\text{H}_{10}]{\text{C}_4\text{H}_9\text{Li}}$$

$$\xrightarrow[\text{-LiBr}]{\text{C}_6\text{H}_5\text{CH}_2\text{Br}}$$

$$\xrightarrow[\text{-HSCH}_2\text{CH}_2\text{CH}_2\text{SH}]{\text{HgCl}_2, \text{CH}_3\text{OH}, \text{H}_2\text{O}}$$ $\text{C}_6\text{H}_5\text{CH}_2\overset{\overset{\text{O}}{\|}}{\text{CH}}$

(c) $\text{C}_6\text{H}_5\overset{\overset{\text{O}}{\|}}{\text{CH}}$ + $\text{HSCH}_2\text{CH}_2\text{CH}_2\text{SH}$ $\xrightarrow{\text{H}^+}$

$$\xrightarrow[(2)\text{CH}_3\text{I}]{(1)\text{C}_4\text{H}_9\text{Li}}$$

$$\xrightarrow{\text{HgCl}_2, \text{CH}_3\text{OH}, \text{H}_2\text{O}}$$ $\text{C}_6\text{H}_5\overset{\overset{\text{O}}{\|}}{\text{C}}\text{CH}_3$ + $\text{HSCH}_2\text{CH}_2\text{CH}_2\text{SH}$

19.19 By treating the thioketal with Raney nickel.

$$\xrightarrow[(2)\text{ R'CH}_2\text{Br}]{(1)\text{C}_4\text{H}_9\text{Li}}$$

$$\xrightarrow[\text{H}_2]{\text{Raney Ni}}$$ $\text{RCH}_2\text{CH}_2\text{R'}$
+
$\text{CH}_3\text{CH}_2\text{CH}_3$

19.20 (a)

$$\xrightarrow[\text{H}^+]{2\,\text{HSCH}_2\text{CH}_2\text{CH}_2\text{SH}}$$

A

$$\xrightarrow[(2)\text{BrCH}_2-\bigcirc-\text{CH}_2\text{Br}]{(1)\,2\,\text{C}_4\text{H}_9\text{Li}}$$

$$\xrightarrow{\text{hydrolysis}}$$

B

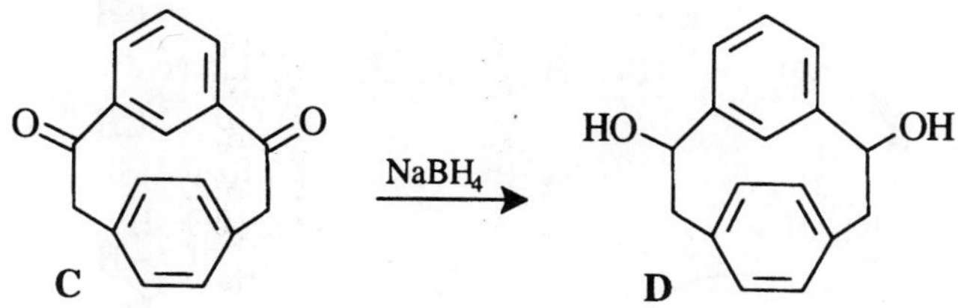

C →(NaBH₄)→ D

(b)

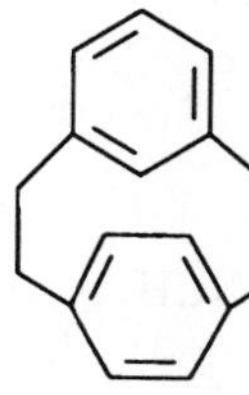

19.21

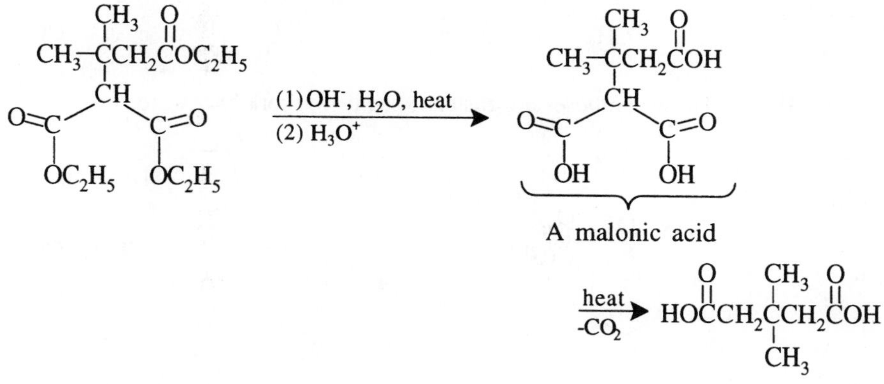

$$CH_3\text{--}\underset{\underset{\displaystyle C=O}{|}}{\overset{\underset{\displaystyle CH}{|}}{C}}CH_2COC_2H_5 \quad \xrightarrow[\text{(2) }H_3O^+]{\text{(1) OH}^-,\ H_2O,\ \text{heat}} \quad$$

A malonic acid

$$\xrightarrow[-CO_2]{\text{heat}} \quad HOCCH_2CCH_2COH$$
$$CH_3$$

19.22 (a)

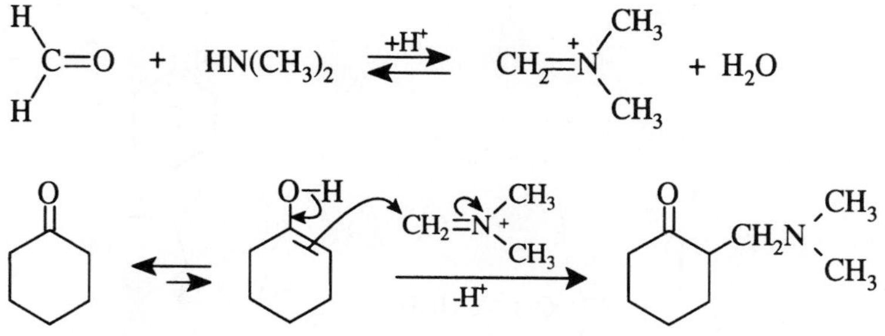

$$\overset{H}{\underset{H}{}}C=O \ + \ HN(CH_3)_2 \ \underset{}{\overset{+H^+}{\rightleftharpoons}} \ CH_2{=}\overset{+}{N}\overset{CH_3}{\underset{CH_3}{}} \ + \ H_2O$$

(b)

$$\overset{H}{\underset{H}{}}C=O \ + \ H{-}N\big\rangle \ \underset{}{\overset{+H^+}{\rightleftharpoons}} \ CH_2{=}\overset{+}{N}\big\rangle \ + \ H_2O$$

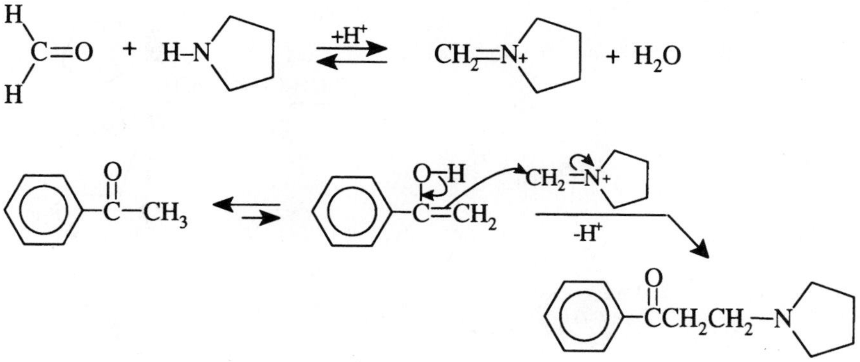

(c)

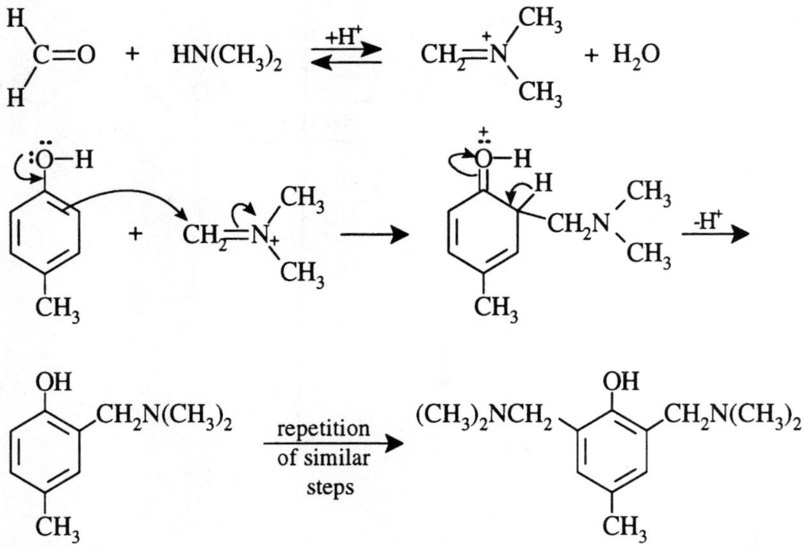

$$\underset{H}{\overset{H}{>}}C=O \ + \ HN(CH_3)_2 \ \underset{\Longleftarrow}{\overset{+H^+}{\Longrightarrow}} \ CH_2\overset{+}{=}N\overset{CH_3}{\underset{CH_3}{<}} \ + \ H_2O$$

19.23 These syntheses are easier to see if we work backward.

(a)

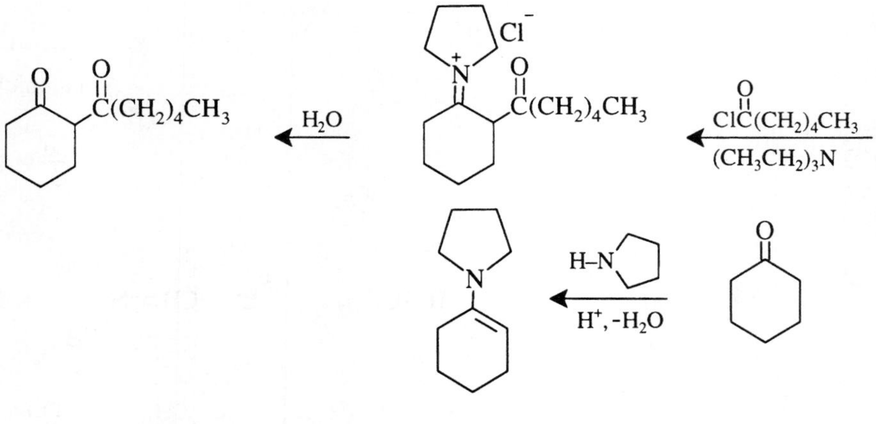

(b)

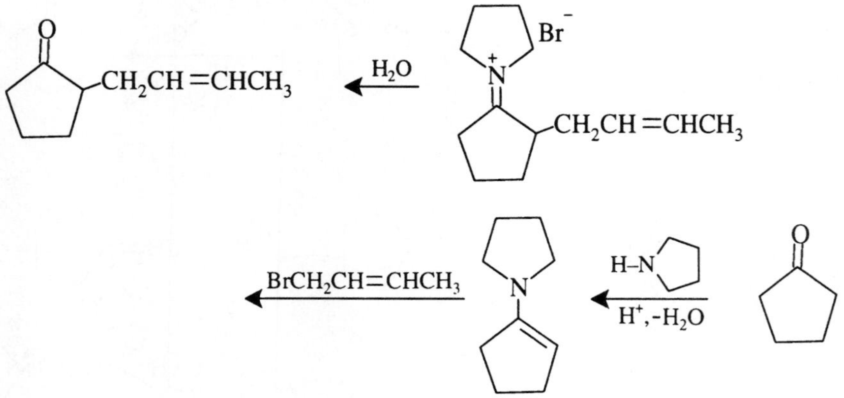

(c) 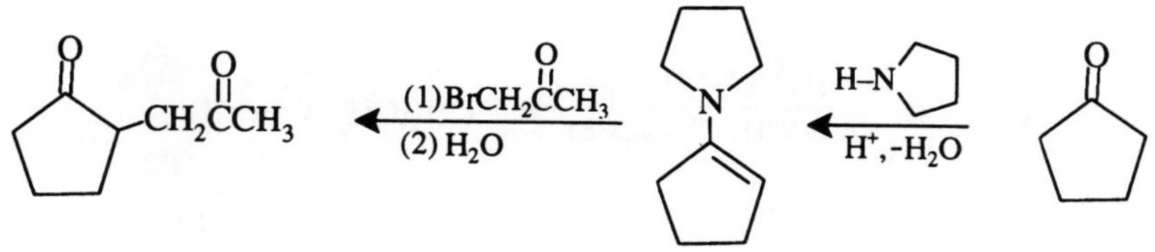

(d) 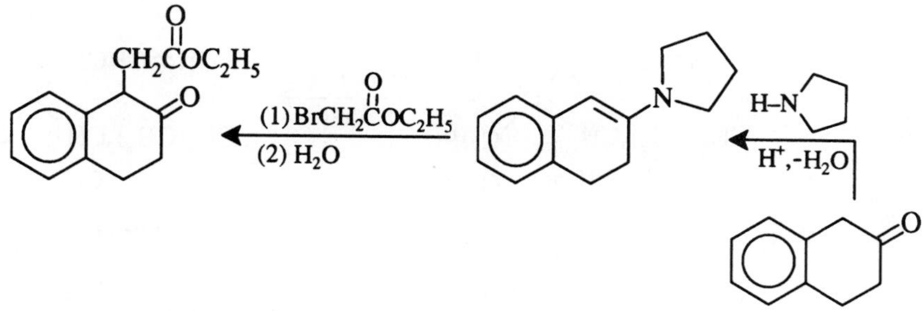

19.24 (a) 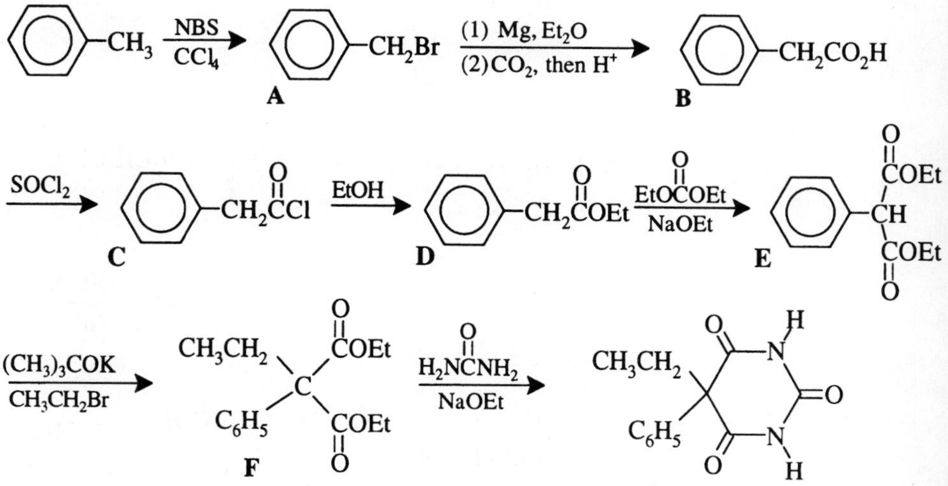

Phenobarbital

(b) See Section 19.3.

19.25 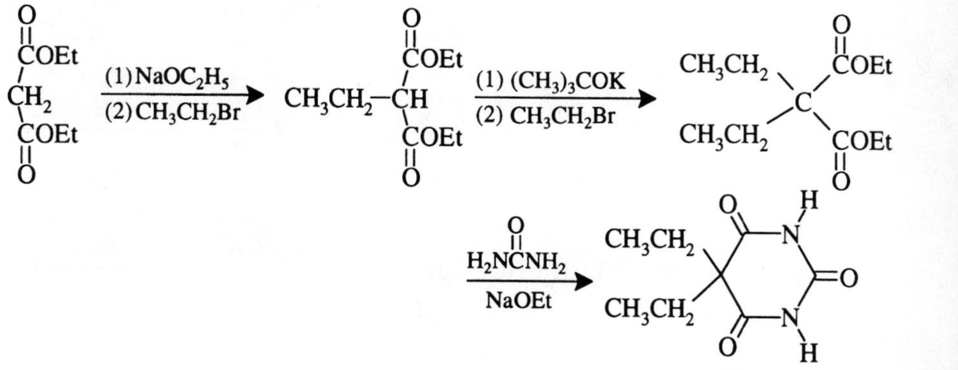

Veronal

Seconal

19.26 (a) $CH_3CH_2CH_2\overset{O}{\overset{\|}{C}}-\underset{\underset{CH_3}{|}}{\overset{CH_2}{\underset{|}{CH}}}-\overset{O}{\overset{\|}{C}}OC_2H_5$ $\xleftarrow[\text{(2) H}^+]{\text{(1) NaOEt}}$ $CH_3CH_2CH_2\overset{O}{\overset{\|}{C}}OC_2H_5$

(b) $CH_3CH_2CH_2\overset{O}{\overset{\|}{C}}CH_2CH_2CH_3$ $\xleftarrow[-CO_2]{\text{heat}}$ $CH_3CH_2CH_2\overset{O}{\overset{\|}{C}}-\underset{\underset{CH_3}{|}}{\underset{CH_2}{\overset{|}{CH}}}-\overset{O}{\overset{\|}{C}}OH$

$\xleftarrow[\text{(2) H}_3O^+]{\text{(1) OH}^-,\ H_2O,\ \text{heat}}$ product of (a)

(c) $\underset{\underset{CH_3}{|}}{C_6H_5\overset{CH_3}{\overset{|}{C}}HCO_2H}$ $\xleftarrow[-CO_2]{\text{heat}}$ $\underset{H_5C_6}{\overset{H_3C}{}}C\underset{CO_2H}{\overset{CO_2H}{}}$ $\xleftarrow[\text{(2) H}_3O^+]{\text{(1) OH}^-,\ H_2O,\ \text{heat}}$

$\underset{H_5C_6}{\overset{H_3C}{}}C\underset{CO_2C_2H_5}{\overset{CO_2C_2H_5}{}}$ $\xleftarrow[\text{(2) CH}_3I]{\text{(1) NaOC}_2H_5}$ $C_6H_5-\underset{CO_2C_2H_5}{\overset{CO_2C_2H_5}{\overset{|}{CH}}}$

$\xleftarrow[\text{(2) H}_3O^+]{\text{(1)}C_2H_5O\overset{O}{\overset{\|}{C}}OC_2H_5,\ NaOC_2H_5}$ $C_6H_5CH_2\overset{O}{\overset{\|}{C}}OC_2H_5$

(d) $CH_3CH_2\underset{\underset{O\ \ O}{\overset{\|\ \ \|}{C-COC_2H_5}}}{\overset{O}{\overset{\|}{CHCOC_2H_5}}}$ $\xleftarrow[\text{(2) H}_3O^+]{\text{(1)}C_2H_5O\overset{O\ \ O}{\overset{\|\ \ \|}{C-C}}OC_2H_5,\ NaOC_2H_5}$ $CH_3CH_2CH_2\overset{O}{\overset{\|}{C}}OC_2H_5$

(e) $CH_3CH_2CH_2\overset{O}{\underset{}{C}}-\overset{O}{\underset{}{C}}OC_2H_5$ $\xleftarrow[C_2H_5OH]{H^+}$ $CH_3CH_2CH_2\overset{O}{\underset{}{C}}-\overset{O}{\underset{}{C}}OH$

$\xleftarrow[-CO_2]{heat}$ $CH_3CH_2\underset{\underset{O\ \ O}{\underset{\|\ \ \|}{C-COH}}}{\overset{O}{\underset{}{CHCOH}}}$ $\xleftarrow[\text{(2) } H_3O^+]{\text{(1) } OH^-, H_2O, \text{ heat}}$ product of (d)

(f) $C_6H_5\underset{\underset{O}{\underset{\|}{CH}}}{\overset{O}{\underset{}{CHCOC_2H_5}}}$ $\xleftarrow[\text{(2) } H_3O^+]{\text{(1) } H\overset{O}{\underset{}{C}}OC_2H_5, NaOC_2H_5}$ $C_6H_5CH_2\overset{O}{\underset{}{C}}OC_2H_5$

(g) [cyclopentanone derivative with $\overset{O}{\underset{}{C}}CH_3$] $\xleftarrow{H_2O}$ [pyrrolidine enamine with $\overset{O}{\underset{}{C}}CH_3$] $\xleftarrow[(R_3N)]{CH_3\overset{O}{\underset{}{C}}Cl}$ [enamine of cyclopentanone] $\xleftarrow[H^+, -H_2O]{H-N\text{(pyrrolidine)}}$ [cyclopentanone]

(h) [cyclopentanone with CH_3 and $\underset{\underset{O}{\|}}{C-CH_3}$] $\xleftarrow[(CH_3)_3COK]{CH_3I}$ product of (g)

(i) [cyclohexanone with CH_2CH_3] $\xleftarrow[-CO_2]{heat}$ [cyclohexanone with CH_2CH_3 and CO_2H] $\xleftarrow[\text{(2) } H_3O^+]{\text{(1) } OH^-, H_2O, \text{ heat}}$

[cyclohexanone with CH_2CH_3 and $CO_2C_2H_5$] $\xleftarrow[NaOC_2H_5]{CH_3CH_2Br}$ [cyclohexanone with $\overset{O}{\underset{}{C}}OC_2H_5$]

19.27 (a) $CH_3\underset{\underset{CH_3}{\|}}{\overset{O}{\underset{}{C}}-\underset{\underset{CH_3}{}}{C}-CH_3}$ $\xleftarrow[H^+]{Zn}$ $CH_3\overset{O}{\underset{}{C}}-\underset{\underset{CH_3}{}}{\overset{\overset{CH_3}{}}{C}}-CH_2Br$ $\xleftarrow{PBr_3}$ $CH_3\overset{O}{\underset{}{C}}-\underset{\underset{CH_3}{}}{\overset{\overset{CH_3}{}}{C}}-CH_2OH$

$\xleftarrow[\text{(2) } H_3O^+]{\text{(1) } LiAlH_4}$ $CH_3-\underset{}{\overset{\overset{O\diagup\diagdown O}{}}{C}}-\underset{\underset{CH_3}{}}{\overset{\overset{CH_3}{}}{C}}-CO_2C_2H_5$ $\xleftarrow[H^+]{HOCH_2CH_2OH}$ $CH_3\overset{O}{\underset{}{C}}-\underset{\underset{CH_3}{}}{\overset{\overset{CH_3}{}}{C}}-CO_2C_2H_5$

$\xleftarrow[KOC(CH_3)_3]{CH_3I}$ $CH_3\overset{O}{\underset{}{C}}-\underset{\underset{CH_3}{|}}{CH}-CO_2C_2H_5$ $\xleftarrow[NaOEt]{CH_3I}$ $CH_3\overset{O}{\underset{}{C}}CH_2\overset{O}{\underset{}{C}}OC_2H_5$

(b) $CH_3\overset{O}{\overset{\|}{C}}CH_2CH_2CH_2CH_3$ $\xleftarrow[-CO_2]{heat}$ $CH_3\overset{O}{\overset{\|}{C}}-\overset{|}{C}H-\overset{O}{\overset{\|}{C}}OH$ (with $-CH_2-CH_2-CH_3$ chain) $\xleftarrow[(2)\ H_3O^+]{(1)\ OH^-,\ H_2O,\ heat}$

$CH_3\overset{O}{\overset{\|}{C}}-\overset{|}{C}H-\overset{O}{\overset{\|}{C}}OC_2H_5$ (with $-CH_2-CH_2-CH_3$ chain) $\xleftarrow[CH_3CH_2CH_2Br]{NaOC_2H_5}$ $CH_3\overset{O}{\overset{\|}{C}}CH_2\overset{O}{\overset{\|}{C}}OC_2H_5$

(c) $CH_3\overset{O}{\overset{\|}{C}}CH_2CH_2\overset{O}{\overset{\|}{C}}CH_3$ $\xleftarrow[-CO_2]{heat}$ $CH_3\overset{O}{\overset{\|}{C}}-\overset{|}{C}H-\overset{O}{\overset{\|}{C}}OH$ (with $-CH_2-\overset{O}{\overset{\|}{C}}-CH_3$ chain) $\xleftarrow[(2)\ H_3O^+]{(1)\ OH^-,\ H_2O,\ heat}$

$CH_3\overset{O}{\overset{\|}{C}}-\overset{|}{C}H-\overset{O}{\overset{\|}{C}}OC_2H_5$ (with $-CH_2-\overset{O}{\overset{\|}{C}}-CH_3$ chain) $\xleftarrow[(2)CH_3COCH_2Br]{(1)\ NaOC_2H_5}$ $CH_3\overset{O}{\overset{\|}{C}}CH_2\overset{O}{\overset{\|}{C}}OC_2H_5$

(d) $CH_3\overset{OH}{\overset{|}{C}}HCH_2CH_2CO_2H$ $\xleftarrow{NaBH_4}$ $CH_3\overset{O}{\overset{\|}{C}}CH_2CH_2\overset{O}{\overset{\|}{C}}OH$ $\xleftarrow[-CO_2]{heat}$

$CH_3\overset{O}{\overset{\|}{C}}-\overset{|}{C}H-\overset{O}{\overset{\|}{C}}OH$ (with $-CH_2-CO_2H$ chain) $\xleftarrow[(2)\ H_3O^+]{(1)\ OH^-,\ H_2O,\ heat}$ $CH_3\overset{O}{\overset{\|}{C}}-\overset{|}{C}H-\overset{O}{\overset{\|}{C}}OC_2H_5$ (with $-CH_2-\overset{O}{\overset{\|}{C}}OC_2H_5$ chain)

$\xleftarrow[(2)\ BrCH_2CO_2C_2H_5]{(1)\ NaOC_2H_5}$ $CH_3\overset{O}{\overset{\|}{C}}CH_2\overset{O}{\overset{\|}{C}}OC_2H_5$

(e) $CH_3\overset{OH}{\overset{|}{C}}H\overset{|}{C}HCH_2OH$ (with $-C_2H_5$ chain) $\xleftarrow[(2)\ H_3O^+]{(1)\ LiAlH_4}$ $CH_3\overset{O}{\overset{\|}{C}}\overset{|}{C}H\overset{O}{\overset{\|}{C}}OC_2H_5$ (with $-C_2H_5$ chain) $\xleftarrow[(2)CH_3CH_2Br]{(1)\ NaOC_2H_5}$

$CH_3\overset{O}{\overset{\|}{C}}CH_2\overset{O}{\overset{\|}{C}}OC_2H_5$

(f) $CH_3\overset{OH}{\overset{|}{C}}HCH_2\overset{OH}{\overset{|}{C}}HC_6H_5$ $\xleftarrow{NaBH_4}$ $CH_3\overset{O}{\overset{\|}{C}}CH_2\overset{O}{\overset{\|}{C}}C_6H_5$ $\xleftarrow{}$ compare Problem 19.13

19.28 (a)

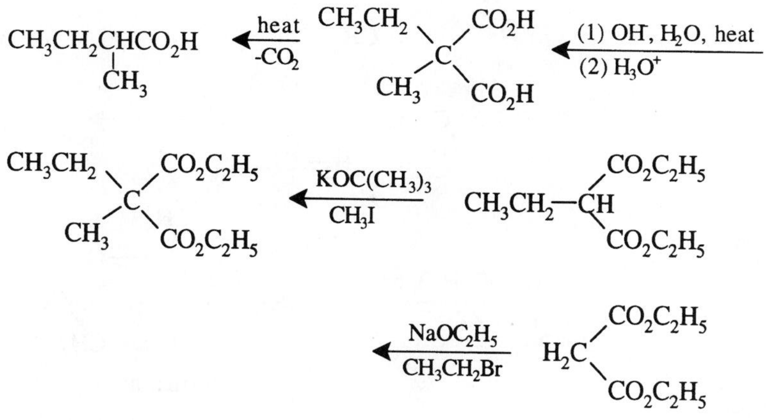

$$CH_3CH_2\underset{\underset{CH_3}{|}}{C}HCO_2H \xleftarrow[-CO_2]{\text{heat}} CH_3CH_2\underset{CH_3}{\overset{CO_2H}{\underset{|}{C}}}\overset{CO_2H}{\underset{CO_2H}{}} \xleftarrow[\text{(2) } H_3O^+]{\text{(1) OH}^-, H_2O, \text{heat}}$$

$$CH_3CH_2\underset{CH_3}{\overset{CO_2C_2H_5}{\underset{|}{C}}}CO_2C_2H_5 \xleftarrow[CH_3I]{KOC(CH_3)_3} CH_3CH_2-\underset{}{C}H\overset{CO_2C_2H_5}{\underset{CO_2C_2H_5}{}}$$

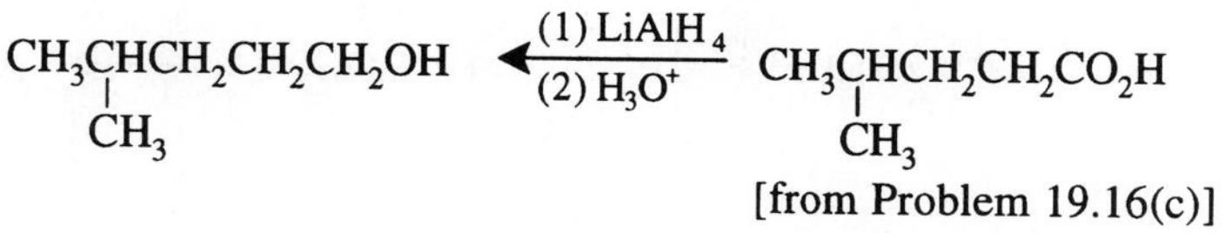

$$\xleftarrow[CH_3CH_2Br]{NaOC_2H_5} H_2C\overset{CO_2C_2H_5}{\underset{CO_2C_2H_5}{}}$$

(b) $CH_3\underset{\underset{CH_3}{|}}{C}HCH_2CH_2CH_2OH \xleftarrow[\text{(2) } H_3O^+]{\text{(1) LiAlH}_4} CH_3\underset{\underset{CH_3}{|}}{C}HCH_2CH_2CO_2H$

[from Problem 19.16(c)]

(c) $CH_3CH_2\underset{\underset{CH_2OH}{|}}{C}HCH_2OH \xleftarrow[\text{(2) } H_3O^+]{\text{(1) LiAlH}_4} CH_3CH_2-\underset{}{C}H\overset{CO_2C_2H_5}{\underset{CO_2C_2H_5}{}}$

[from (a) above]

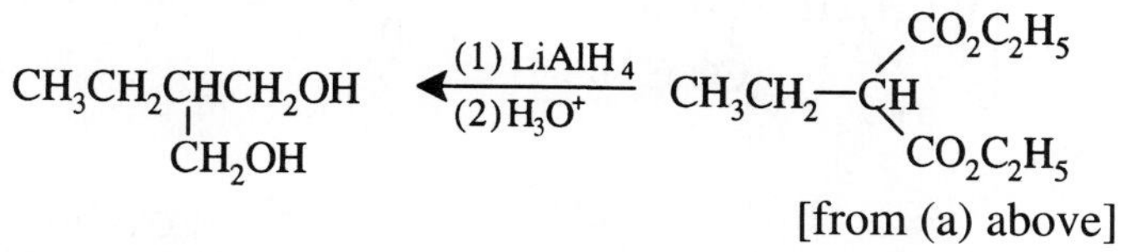

(d) $HOCH_2CH_2CH_2CH_2OH \xleftarrow[\text{(2) } H_3O^+]{\text{(1) LiAlH}_4} HO_2CCH_2CH_2CO_2H \xleftarrow[-CO_2]{\text{heat}}$

$$HO_2C\overset{}{\underset{HO_2C}{}}CHCH_2CO_2H \xleftarrow[\text{heat}]{HCl} C_2H_5O_2C\overset{}{\underset{C_2H_5O_2C}{}}CHCH_2CO_2C_2H_5$$

$$\xleftarrow{} C_2H_5O_2C\overset{}{\underset{C_2H_5O_2C}{}}CH_2 + NaOC_2H_5 + BrCH_2CO_2C_2H_5$$

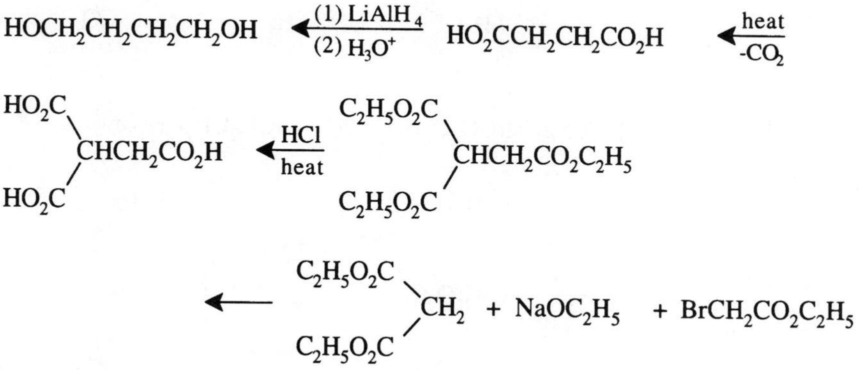

19.29 The following reaction took place:

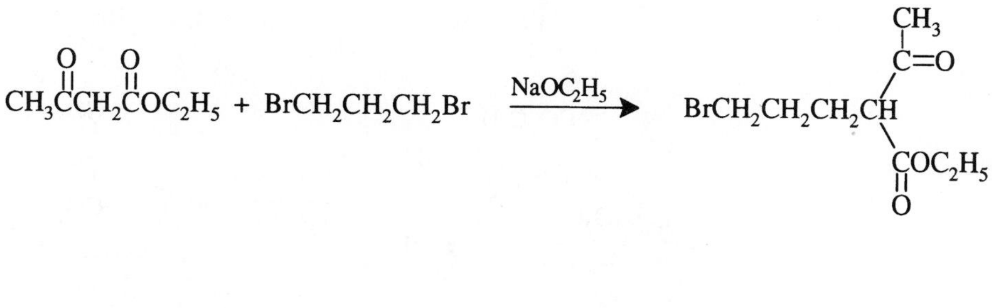

$$CH_3\overset{O}{\overset{\|}{C}}CH_2\overset{O}{\overset{\|}{C}}OC_2H_5 + BrCH_2CH_2CH_2Br \xrightarrow{NaOC_2H_5} BrCH_2CH_2CH_2\underset{}{C}H\overset{\overset{CH_3}{\underset{|}{C=O}}}{\underset{\underset{O}{\overset{\|}{C}}OC_2H_5}{}}$$

$$\xrightarrow[\text{-H}^+]{\text{NaOC}_2\text{H}_5}$$

Perkin's ester

$$\xrightarrow[\text{(2) H}_3\text{O}^+]{\text{(1) OH}^-, \text{H}_2\text{O, heat}}$$

Perkin's acid

19.30 (a) $\text{BrCH}_2\text{CH}_2\text{Br} + \underset{\text{CO}_2\text{C}_2\text{H}_5}{\overset{\text{CO}_2\text{C}_2\text{H}_5}{\text{CH}_2}} + \text{NaOC}_2\text{H}_5 \longrightarrow$

$$\left[\text{BrCH}_2\text{CH}_2-\underset{\text{CO}_2\text{C}_2\text{H}_5}{\overset{\text{CO}_2\text{C}_2\text{H}_5}{\text{CH}}} \right] \xrightarrow[\text{-H}^+]{(\text{CH}_3)_3\text{COK}} \left[\text{BrCH}_2\text{CH}_2-\underset{\text{CO}_2\text{C}_2\text{H}_5}{\overset{\text{CO}_2\text{C}_2\text{H}_5}{\text{C}:^-}} \right]$$

$$\longrightarrow \underset{\text{H}_2\text{C}}{\overset{\text{H}_2\text{C}}{\text{C}}} \underset{\text{CO}_2\text{C}_2\text{H}_5}{\overset{\text{CO}_2\text{C}_2\text{H}_5}{\text{C}}} \xrightarrow[\substack{(2) \text{ H}_3\text{O}^+ \\ (3) \text{ heat, -CO}_2}]{(1) \text{ OH}^-, \text{H}_2\text{O, heat}} \triangle\text{-CO}_2\text{H}$$

(b) $2\ \text{NaCH(CO}_2\text{C}_2\text{H}_5)_2 + \text{BrCH}_2\text{CH}_2\text{CH}_2\text{Br} \longrightarrow$

$$\underset{\text{C}_2\text{H}_2\text{O}_2\text{C}}{\overset{\text{C}_2\text{H}_5\text{O}_2\text{C}}{}} \text{H}-\text{CCH}_2\text{CH}_2\text{CH}_2\text{C}-\text{H} \overset{\text{CO}_2\text{C}_2\text{H}_5}{\underset{\text{CO}_2\text{C}_2\text{H}_5}{}}$$

A

$$\xrightarrow[\text{Br}_2]{\text{NaOC}_2\text{H}_5} \left[\underset{\text{C}_2\text{H}_5\text{O}_2\text{C}}{\overset{\text{C}_2\text{H}_5\text{O}_2\text{C}}{}} \text{H}-\text{CCH}_2\text{CH}_2\text{CH}_2\text{C}-\text{Br} \overset{\text{CO}_2\text{C}_2\text{H}_5}{\underset{\text{CO}_2\text{C}_2\text{H}_5}{}} \right] \xrightarrow{\text{NaOC}_2\text{H}_5}$$

B $\xrightarrow[\text{(2) H}_3\text{O}^+]{\text{(1) OH}^-, \text{H}_2\text{O}}$ **C** $\xrightarrow[\text{-2 CO}_2]{\text{heat}}$

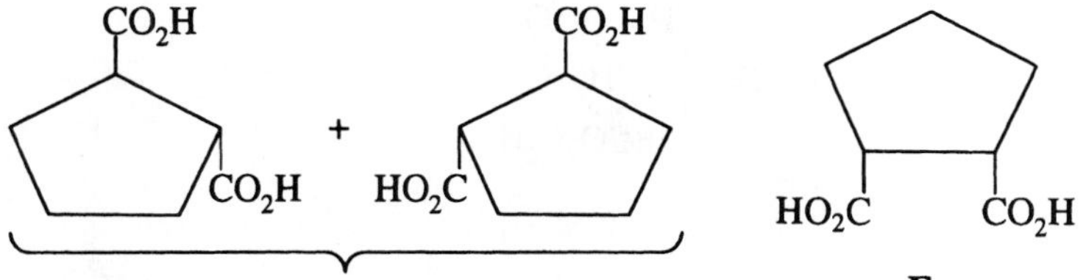

D
Racemic form

E
Meso compound

(c) $BrCH_2CH_2CH_2CH_2Br$ $\xrightarrow{NaCH(CO_2C_2H_5)_2}$ $BrCH_2CH_2CH_2CH_2CH(CO_2C_2H_5)(CO_2C_2H_5)$

$\xrightarrow{KOC(CH_3)_3}$ (cyclopentane with $CO_2C_2H_5$, $CO_2C_2H_5$) $\xrightarrow[\text{(2) }H_3O^+]{\text{(1) }OH^-, H_2O}$ (cyclopentane with $-CO_2H$)
(3) heat, $-CO_2$

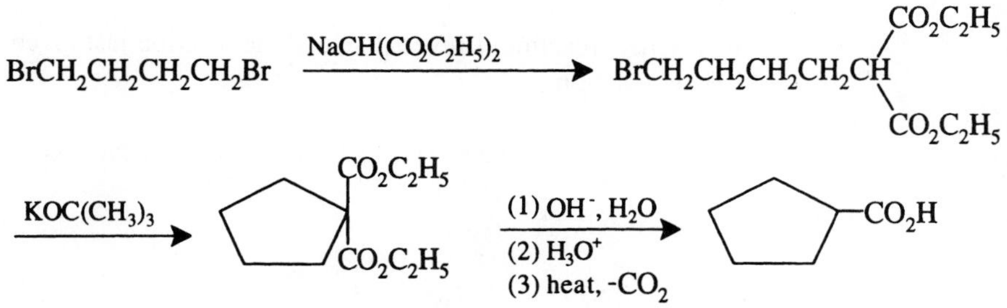

19.31 (a) $CH_2(CO_2C_2H_5)_2 + {}^-OC_2H_5 \rightleftharpoons {}^:CH(CO_2C_2H_5)_2 + C_2H_5OH$

$C_6H_5CH=CH-\overset{O}{\underset{}{C}}OC_2H_5 + {}^:CH(CO_2C_2H_5)_2 \rightleftharpoons C_6H_5CHCH=\overset{O}{\underset{}{C}}OC_2H_5$
$CH(CO_2C_2H_5)_2$

$\xrightarrow{+H^+} C_6H_5CHCH_2-\overset{O}{\underset{}{C}}OC_2H_5$
$CH(CO_2C_2H_5)_2$

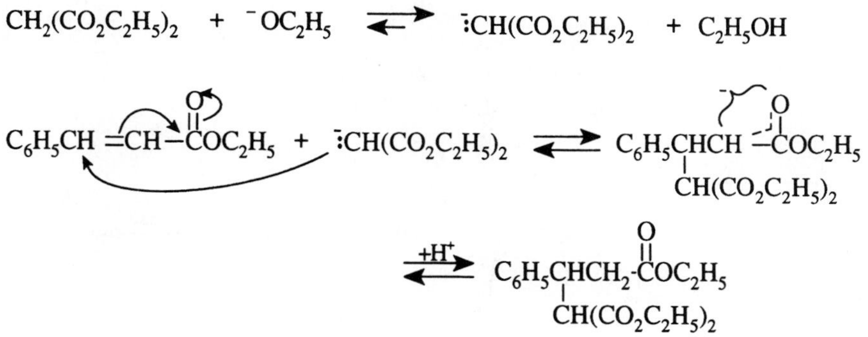

(b) $CH_3\ddot{N}H_2 + CH_2=CH-\overset{O}{\underset{}{C}}OCH_3 \rightleftharpoons CH_3-\overset{H}{\underset{H}{N^+}}-CH_2-CH=\overset{O}{\underset{}{C}}OCH_3 \rightleftharpoons$

$CH_3\underset{H}{\overset{}{N}}-CH_2-CH_2-\overset{O}{\underset{}{C}}OCH_3 \xrightleftharpoons[]{CH_2=CH-\overset{O}{C}OCH_3} CH_3N(CH_2CH_2CO_2CH_3)_2$

$\xrightarrow[\text{base}]{}$ (structure) $\xrightarrow[\text{(several steps)}]{\text{Dieckmann condensation}}$ (ring structure with CO_2CH_3, CH_3-N, $=O$)

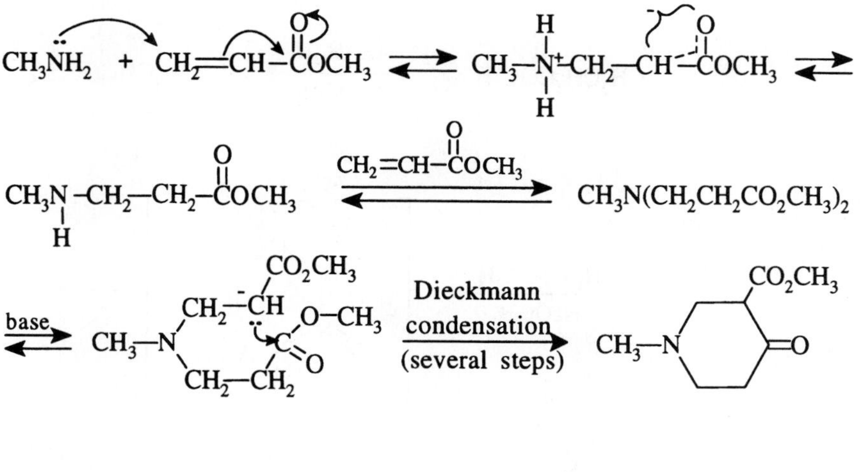

(c) $\underset{\underset{\text{CH(CO}_2\text{C}_2\text{H}_5)_2}{|}}{\overset{\overset{\text{CH}_3}{|}}{\text{CH}_3\text{-C-CH}_2\text{-COC}_2\text{H}_5}}$ + $^-$OC$_2$H$_5$ $\rightleftharpoons$ $\underset{\underset{\text{CH(CO}_2\text{C}_2\text{H}_5)_2}{|}}{\overset{\overset{\text{CH}_3}{|}}{\text{CH}_3\text{-C-CH=COC}_2\text{H}_5}}$

$+$ C$_2$H$_5$OH

$\underset{\underset{\text{CH(CO}_2\text{C}_2\text{H}_5)_2}{|}}{\overset{\overset{\text{CH}_3}{|}}{\text{CH}_3\text{-C-CH=COC}_2\text{H}_5}}$ $\rightleftharpoons$ $\overset{\overset{\text{CH}_3}{|}}{\text{CH}_3\text{-C=CH-COC}_2\text{H}_5}$ $+$ $^:$CH(CO$_2$C$_2$H$_5$)$_2$

The Michael reaction is reversible, and the reaction just given is an example of a reverse Michael reaction.

19.32 Two reactions take place. The first is a normal Knoevenagel condensation,

$$\underset{\underset{\text{R'}}{|}}{\text{R-C=O}} + \text{CH}_2(\text{COCH}_3)_2 \xrightarrow[-\text{H}_2\text{O}]{\text{base}} \underset{\underset{\text{R'}}{|}}{\text{R-C=C}}\overset{\overset{\overset{\text{O}}{||}}{\text{CCH}_3}}{\underset{\underset{\underset{\text{O}}{||}}{\text{CCH}_3}}{}}$$

Then the α, β-unsaturated diketone reacts with a second mole of the active methylene compound in a Michael addition.

$$\underset{\underset{\text{R'}}{|}}{\text{R-C=C}}\overset{\overset{\overset{\text{O}}{||}}{\text{CCH}_3}}{\underset{\underset{\underset{\text{O}}{||}}{\text{CCH}_3}}{}} + \text{CH}_2(\text{COCH}_3)_2 \xrightarrow{\text{base}} \underset{\underset{\text{R'}}{}}{\overset{\text{R}}{}}\text{C}\overset{\text{CH(COCH}_3)_2}{\underset{\text{CH(COCH}_3)_2}{}}$$

19.33 $\overset{\overset{\text{O}}{||}}{\text{CH}_3\text{CH}_2\text{COC}_2\text{H}_5}$ $\xrightarrow[\text{NaOC}_2\text{H}_5]{\overset{\overset{\text{O}}{||}}{\text{HCOC}_2\text{H}_5}}$ $\underset{\underset{\underset{\text{H}}{|}}{\overset{\overset{}{}}{\text{C=O}}}}{\overset{\overset{\text{O}}{||}}{\text{CH}_3\text{CHCOC}_2\text{H}_5}}$

$\xrightarrow[\text{NaOC}_2\text{H}_5]{\overset{\overset{\text{O}}{||}}{\text{H}_2\text{N-C-NH}_2}}$ $\left[\begin{array}{c} \overset{\text{C}_2\text{H}_5\text{O}}{\underset{\text{HN}}{}}\text{C}\overset{\text{CH}_3}{\underset{\text{CH}}{\text{C}}} \\ \text{O=C}\underset{\underset{\text{H}}{\text{N}}}{} \end{array} \right]$ $\longrightarrow$ thymine

19.34

$$CH_3\overset{\overset{O}{\|}}{C}(CH_2)_5CHO \xrightarrow[\text{pyridine}]{CH_2(CO_2H)_2} CH_3\overset{\overset{O}{\|}}{C}(CH_2)_5CH=CHCO_2H$$

$$\textbf{C} \qquad\qquad\qquad\qquad \text{Queen substance}$$

$$\xrightarrow[Pd]{H_2} CH_3\overset{\overset{O}{\|}}{C}(CH_2)_7CO_2H \xrightarrow[(2)\,H_3O^+]{(1)\,I_2/NaOH} HO_2C(CH_2)_7CO_2H$$

$$\textbf{D} \qquad\qquad\qquad\qquad\qquad \textbf{E}$$

19.35

$$CH_2\!\!=\!\!\underset{\underset{CH_3}{|}}{C}\!\!-\!\!CH=CH_2 + HBr \longrightarrow CH_3\underset{\underset{\textbf{F}\;CH_3}{}}{C}\!\!=\!\!CHCH_2Br \xrightarrow{\underset{\overset{}{CO_2C_2H_5}}{Na^+\;\,{}^-\!:\!\overset{\overset{\overset{O}{\|}}{CCH_3}}{CH}}}$$

$$CH_3\underset{\underset{CH_3}{|}}{C}\!\!=\!\!CHCH_2\underset{\underset{CO_2C_2H_5}{|}}{CH}\overset{\overset{O}{\|}}{C}CH_3 \xrightarrow[(2)\,H_3O^+,\,(3)\,\text{heat}]{(1)\,\text{dil. NaOH}} CH_3\underset{\underset{CH_3}{|}}{C}\!\!=\!\!CHCH_2CH_2\overset{\overset{O}{\|}}{C}CH_3$$

$$\textbf{G} \qquad\qquad\qquad\qquad\qquad\qquad \textbf{H}$$

$$\xrightarrow[(2)\,H_3O^+]{(1)\,LiC\!\equiv\!CH} CH_3\underset{\underset{CH_3}{|}}{C}\!\!=\!\!CHCH_2CH_2\underset{\underset{CH_3}{|}}{\overset{\overset{OH}{|}}{C}}C\!\equiv\!CH \xrightarrow[\substack{\text{Lindlar's}\\\text{catalyst}}]{H_2} \text{linalool}$$

$$\textbf{I}$$

19.36

$$(C_{10}H_{17}BrO_4)$$

$$\xrightarrow{NaOC_2H_5}$$

$$(C_{10}H_{16}O_4)$$

$$\xrightarrow[(2)\,H_2O]{(1)\,LiAlH_4}$$

$$(C_6H_{12}O_2) \qquad \xrightarrow{HBr}$$

$$(C_6H_{10}Br_2)$$

$$\xrightarrow[2\,NaOC_2H_5]{CH_2(CO_2C_2H_5)_2}$$

$$(C_{13}H_{20}O_4)$$

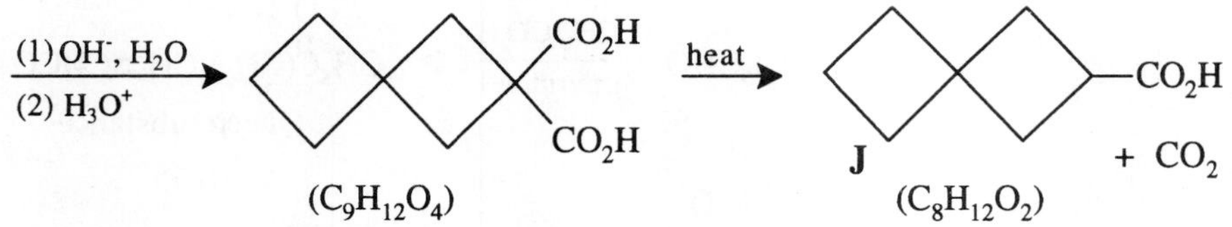

$$\xrightarrow[\text{(2) H}_3\text{O}^+]{\text{(1) OH}^-, \text{H}_2\text{O}} \quad \text{(structure with } \text{CO}_2\text{H, CO}_2\text{H)} \quad \xrightarrow{\text{heat}} \quad \text{(structure)} \text{—CO}_2\text{H}$$

$$(\text{C}_9\text{H}_{12}\text{O}_4) \qquad\qquad\qquad \textbf{J} \quad + \text{CO}_2$$
$$(\text{C}_8\text{H}_{12}\text{O}_2)$$

19.37 (a) $\text{ClCH}_2\text{CO}_2\text{C}_2\text{H}_5 \; + \; \text{C}_2\text{H}_5\text{O}^- \; \rightleftharpoons \; \text{Cl—}\ddot{\overset{..}{\text{C}}}\text{HCO}_2\text{C}_2\text{H}_5 \; + \; \text{C}_2\text{H}_5\text{OH}$

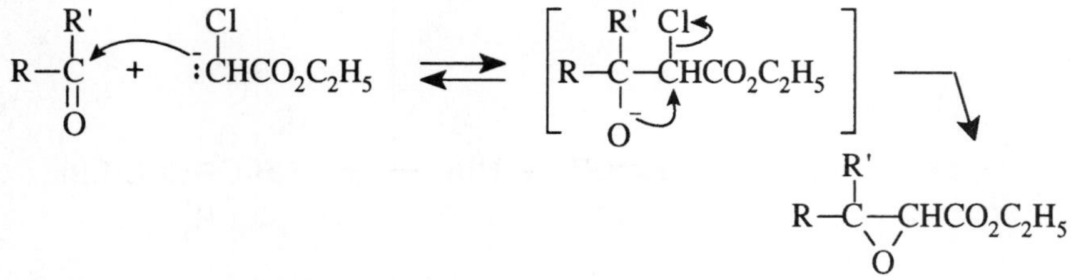

(b) Decarboxylation of the epoxy acid gives an enol anion which, on protonation, gives an aldehyde.

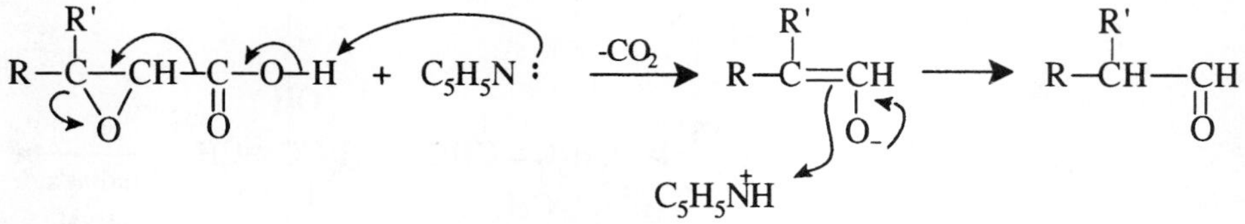

(c)

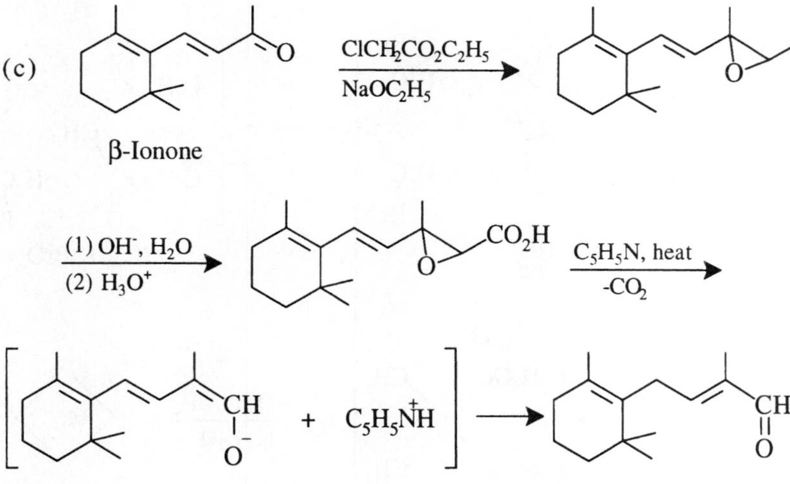

β-Ionone

19.38 (a)

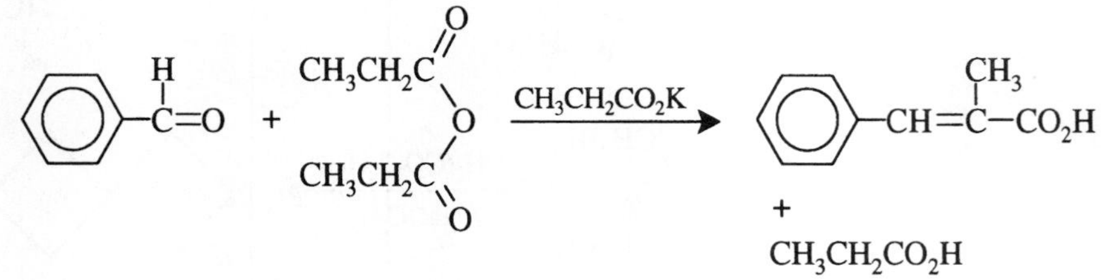

(b)

$$Cl-\langle\bigcirc\rangle-\overset{H}{\underset{}{C}}{=}O \quad + \quad \begin{matrix}CH_3C\overset{O}{\diagdown}\\ \quad\quad O\\ CH_3C\diagup\\ \quad\quad O\end{matrix} \quad\xrightarrow{CH_3CO_2K}\quad Cl-\langle\bigcirc\rangle-CH{=}CHCO_2H$$

$$+$$

$$CH_3CO_2H$$

19.39 (a) $CH_2{=}\overset{CH_3}{\underset{|}{C}}{-}CO_2CH_3$ (b) $KMnO_4$, OH^-, then H_3O^+

(c) CH_3OH, H^+ (d) CH_3ONa, then H^+

(e) and (f)

and

(g) OH^-, H_2O, then H_3O^+ (h) heat ($-CO_2$)

(i) CH_3OH, H^+ (j) $BrCH_2CO_2CH_3$, Zn, then H_3O^+

(k)

(l) H_2, Pt

(m) CH_3ONa, then H^+ (n) $2\ NaNH_2 + 2\ CH_3I$

19.40

$$\langle\bigcirc\rangle-\overset{CH_3}{\underset{\underset{O}{\|}}{\overset{|}{C}}}{-}CH_2 \quad + \quad HCHO \quad + \quad HN(CH_3)_2 \quad\xrightarrow[(-H_2O)]{\text{Mannich reaction}}$$

$$\langle\bigcirc\rangle-\overset{CH_3}{\underset{\underset{O}{\|}}{\overset{|}{C}}}{-}CH{-}CH_2{-}N(CH_3)_2 \quad\xrightarrow[(2)\ H_3O^+]{(1)\ C_6H_5CH_2MgBr}\quad \langle\bigcirc\rangle-CH_2{-}\overset{\overset{\langle\bigcirc\rangle}{|}}{\underset{\underset{OH}{|}}{C}}{-}\overset{CH_3}{\underset{}{CH}}{-}CH_2{-}N(CH_3)_2$$

$$\xrightarrow{(CH_3CH_2CO)_2O}\quad \langle\bigcirc\rangle-CH_2{-}\overset{\overset{\langle\bigcirc\rangle}{|}}{\underset{\underset{\underset{O}{\|}}{\overset{|}{O}-C-CH_2-CH_3}}{C}}{-}\overset{CH_3}{\underset{}{CH}}{-}CH_2{-}N(CH_3)_2$$

Darvon

19.41 In a polar solvent, such as water, the keto form is stabilized by solvation. When the interaction with the solvent becomes minimal, the enol form achieves stability by internal hydrogen bonding.

19.42 Intramolecular cyclization (which would give a product of formula $C_6H_8O_3$) is not favored because of ring strain. The formula of the product actually obtained suggests a 1:1 intermolecular reaction:

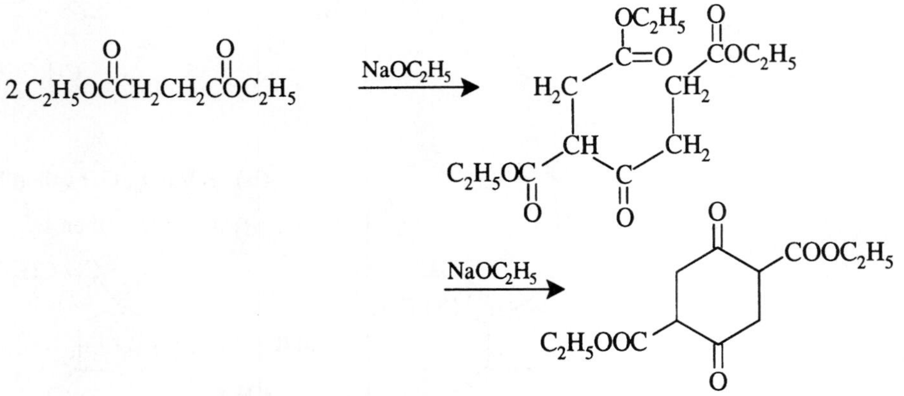

19.43 A gamma hydrogen is abstracted by base (as is an alpha hydrogen in the usual Claisen) to give a resonance-stabilized species:

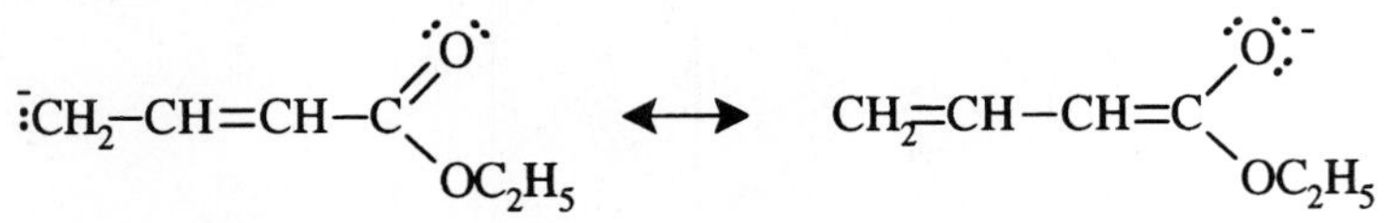

Ethyl crotonate differs from ethyl acetate by $-CH=CH-$, a vinyl group. The transmission of the stabilizing effect of the $-COOC_2H_5$ group is an example of the **principle of vinylogy.**

19.44 The synthesis actually involves two sequential Claisen condensations, with ethyl acetate serving as the source of the carbanionic species.

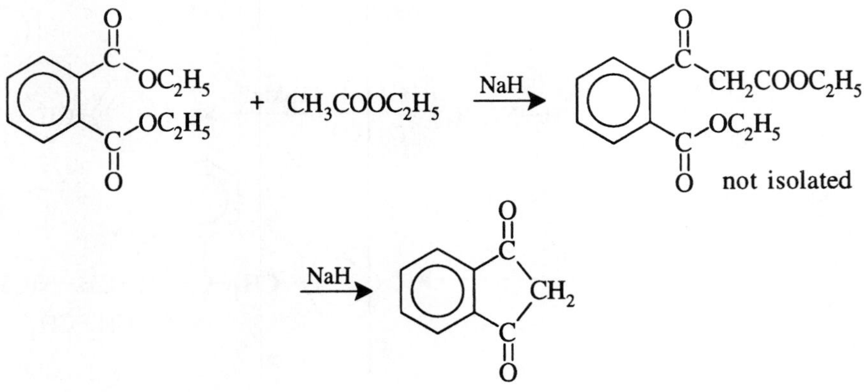

QUIZ

19.1 Which hydrogen atoms in the following ester are most acidic?

$$\overset{a}{CH_3}-\overset{b}{CH_2}-\overset{O}{\overset{\|}{C}}-\overset{c}{CH_2}-\overset{O}{\overset{\|}{C}}-O\overset{d}{CH_2}-\overset{e}{CH_3}$$

(a) *a*   (b) *b*   (c) *c*   (d) *d*   (e) *e*

19.2 What would be the product of the following reaction?

$$CH_3CH_2\overset{O}{\overset{\|}{C}}OEt \quad \xrightarrow[(2)\,H^+]{(1)\,NaOC_2H_5} \quad ?$$

(a) $CH_3CH_2\overset{O}{\overset{\|}{C}}CH_2CH_2\overset{O}{\overset{\|}{C}}OEt$

(b) $CH_3CH_2\overset{O}{\overset{\|}{C}}CH_2\overset{O}{\overset{\|}{C}}CH_3$

(c) $CH_3CH_2\overset{O}{\overset{\|}{C}}CH_2\overset{O}{\overset{\|}{C}}OEt$

(d) $CH_3\overset{O}{\overset{\|}{C}}CH_2CH_2\overset{O}{\overset{\|}{C}}OEt$

(e) $CH_3CH_2\overset{O}{\overset{\|}{C}}\underset{\underset{CH_3}{|}}{CH}\overset{O}{\overset{\|}{C}}OEt$

19.3 What starting materials could be used in a crossed Claisen condensation to prepare the following compound?

$$EtO-\overset{O}{\overset{\|}{C}}-\overset{O}{\overset{\|}{C}}-\underset{\underset{CH_3}{|}}{CH}-\overset{O}{\overset{\|}{C}}OEt$$

(a) CH_3CO_2Et and $EtO-\overset{O}{\overset{\|}{C}}-\overset{O}{\overset{\|}{C}}-CH_2CH_3$

(b) $CH_3CH_2CO_2Et$ and EtO_2C-CO_2Et

(c) $CH_3CH_2CO_2Et$ and HCO_2Et

(d) EtO_2CCHCO_2Et and HCO_2Et
      $\underset{\underset{CH_3}{|}}{}$

(e) More than one of the above

19.4 Supply the missing reagents, intermediates, and products.

(a) $CH_3CH_2CH_2\overset{O}{\overset{\|}{C}}OEt$ + $EtO\overset{O}{\overset{\|}{C}}OEt$ $\xrightarrow[(2)\,H^+\,(-EtOH)]{(1)\,NaOC_2H_5}$

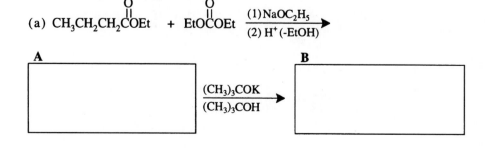

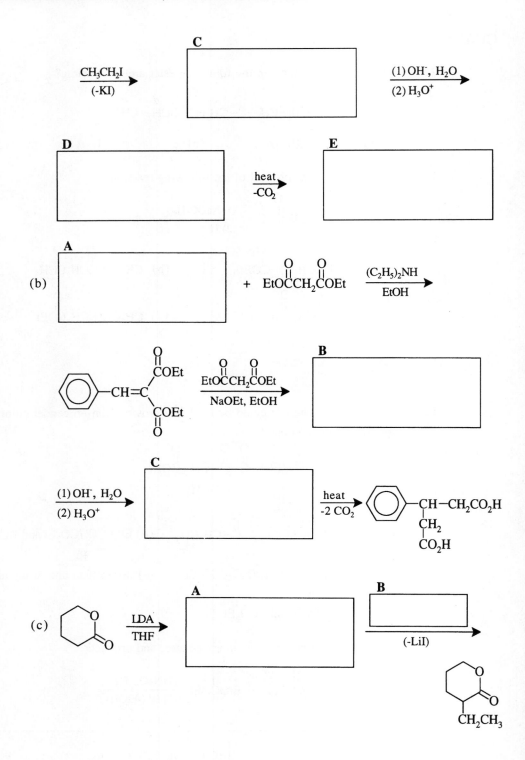

(d)

20 AMINES

PREPARATION AND REACTIONS OF AMINES

A. Preparation

1. Preparation via nucleophilic substitution reactions.

2. Preparation through reduction of nitro compounds.

3. Preparation via reductive amination.

$$\underset{R'}{\overset{R}{>}}C=O \; + \; NH_3 \;\xrightarrow[Ni]{H_2}\; R-\underset{R'}{\overset{|}{C}}HNH_2$$

4. Preparation of amines through reduction of amides, oximes, and nitriles.

$$RCH_2Br \xrightarrow{CN^-} RCH_2CN \xrightarrow[Ni]{H_2} RCH_2CH_2NH_2$$

$$\underset{R'}{\overset{R}{>}}C=O \xrightarrow{NH_2OH} \underset{R'}{\overset{R}{>}}C=NOH \xrightarrow{Na/ethanol} R-\underset{R'}{\overset{|}{C}}HNH_2$$

$$R-NH_2 \; + \; R'\overset{O}{\overset{||}{C}}Cl \longrightarrow R-NH\overset{O}{\overset{||}{C}}R' \xrightarrow[(2)\,H_2O]{(1)\,LiAlH_4} RNHCH_2R'$$

5. Preparation through the Hofmann rearrangement of amides.

$$RCOH \xrightarrow{SOCl_2} RCCl \xrightarrow{NH_3} RCNH_2 \xrightarrow[\text{(NaOBr)}]{Br_2/NaOH} RNH_2 + CO_2^{2-}$$

B. Reaction of Amines

1. As a base or a nucleophile

As a base

$$\underset{/}{\overset{\backslash}{-}}N: + H-\underset{H}{\overset{|}{O}}-H \rightleftharpoons -\overset{|}{\underset{|}{N^+}}-H + H_2O$$

As a nucleophile in alkylation

$$\underset{/}{\overset{\backslash}{-}}N: + RCH_2{\overset{\frown}{-}}X \longrightarrow -\overset{|}{\underset{|}{N^+}}-CH_2R + X^-$$

As a nucleophile in acylation

$$\underset{H}{\overset{\backslash}{-}}N: + R-\overset{O}{\overset{\|}{C}}-Cl \xrightarrow{\text{(-HCl)}} -\overset{|}{\underset{\cdot\cdot}{N}}-\overset{O}{\overset{\|}{C}}-R$$

2. With nitrous acid

$$R-NH_2 \xrightarrow[HX]{HONO} R-N_2^+X^- \xrightarrow{-N_2} R^+ \longrightarrow \begin{array}{l}\text{alkenes,}\\\text{alcohols,}\\\text{and so on}\end{array}$$

1° aliphatic   (unstable)

$$Ar-NH_2 \xrightarrow[HX\ (0-5°C)]{HONO} Ar-N_2^+X^-$$

1° aromatic

- $\xrightarrow{CuCl} ArCl + N_2$
- $\xrightarrow{CuBr} ArBr + N_2$
- $\xrightarrow{CuCN} ArCN + N_2$
- $\xrightarrow{KI} ArI + N_2$
- $\xrightarrow{HBF_4} ArN_2^+ BF_4^- \xrightarrow{\Delta} \begin{cases}ArF +\\N_2 +\\BF_3\end{cases}$
- $\xrightarrow[Cu^{2+},\ H_2O]{Cu_2O} ArOH + N_2$
- $\xrightarrow{H_3PO_2} ArH + N_2$
- $\xrightarrow{\text{⬡-OH}} Ar-N=N-\text{⬡}-OH$
- $\xrightarrow{\text{⬡-NR}_2} Ar-N=N-\text{⬡}-NR_2$

$$R_2NH \xrightarrow{\text{HONO}} R_2N-N=O$$

2° aliphatic

$$ArNHR \xrightarrow{\text{HONO}} \overset{\overset{\displaystyle N=O}{\displaystyle |}}{ArN-R}$$

2° aromatic

$$R_3N \xrightarrow[\text{HX}]{\text{NaNO}_2} R_3NH^+ X^- + R_3\overset{+}{N}-N=O \ X^-$$

3° aliphatic

$$R_2N-\hexagon \xrightarrow{\text{HONO}} R_2N-\hexagon-N=O$$

3° aromatic

3. With sulfonyl chlorides

$$R-NH_2 + ArSO_2Cl \xrightarrow[\text{(-HCl)}]{} RNHSO_2Ar \underset{H^+}{\overset{OH^-}{\rightleftarrows}} [RNSO_2Ar]^- + H_2O$$

1° amine

$$R_2NH + ArSO_2Cl \xrightarrow[\text{(-HCl)}]{} R_2NSO_2Ar$$

2° amine

4. The Hofmann elimination

$$HO^- + \ \overset{\overset{\displaystyle H}{\displaystyle |}}{-\underset{\underset{\overset{\displaystyle |}{\displaystyle +}}{N(CH_3)_3}}{\overset{\displaystyle |}{C}}-\overset{\displaystyle |}{\underset{\displaystyle |}{C}}-} \ \xrightarrow{\text{heat}} \ \overset{\diagdown}{\diagup}C=C\overset{\diagup}{\diagdown} + (CH_3)_3N + H_2O$$

SOLUTIONS TO PROBLEMS

20.1 Dissolve both compounds in diethyl ether and extract with aqueous HCl. This procedure gives an ether layer that contains cyclohexane and an aqueous layer that contains hexylaminium chloride. Cyclohexane may then be recovered from the ether layer by distillation. Hexylamine may be recovered from the aqueous layer by adding aqueous NaOH (to convert hexylaminium chloride to the hexylamine) and then by ether extraction and distillation.

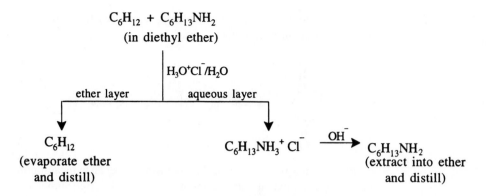

20.2 We begin by dissolving the mixture in a water-immiscible organic solvent such as CH_2Cl_2 or diethyl ether. Then, extractions with aqueous acids and bases allow us to separate the components. [We separate 4-methylphenol (*p*-cresol) from benzoic acid by taking advantage of benzoic acid's solubility in the more weakly basic aqueous $NaHCO_3$, whereas *p*-cresol requires the more strongly basic, aqueous NaOH.]

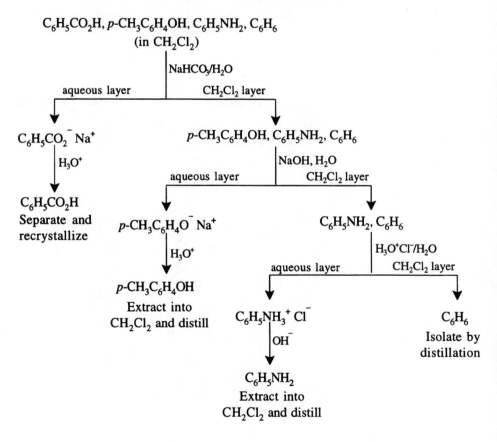

20.3   (a)   Neglecting Kekulé forms of the ring, we can write the following resonance structures for the phthalimide anion.

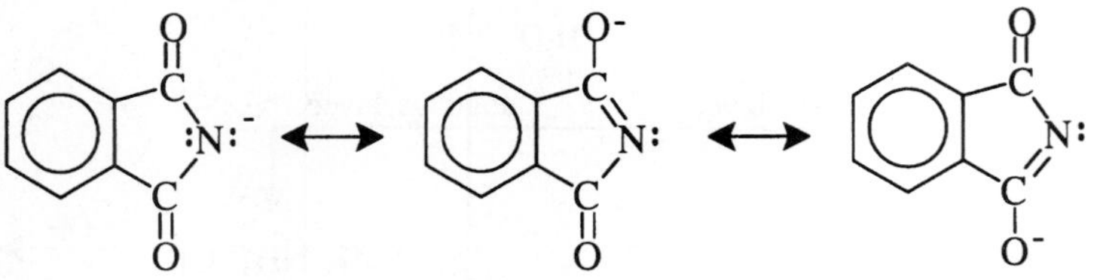

(b)   Phthalimide is more acidic than benzamide because its anion is stabilized by resonance to a greater extent than the anion of benzamide. (Benzamide has only one carbonyl group attached to the nitrogen atom, and thus fewer resonance contributors are possible.)

(c)

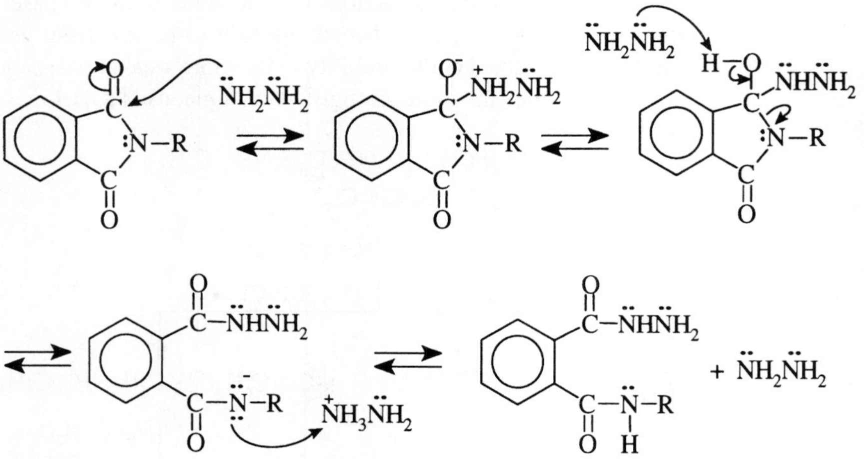

Then,

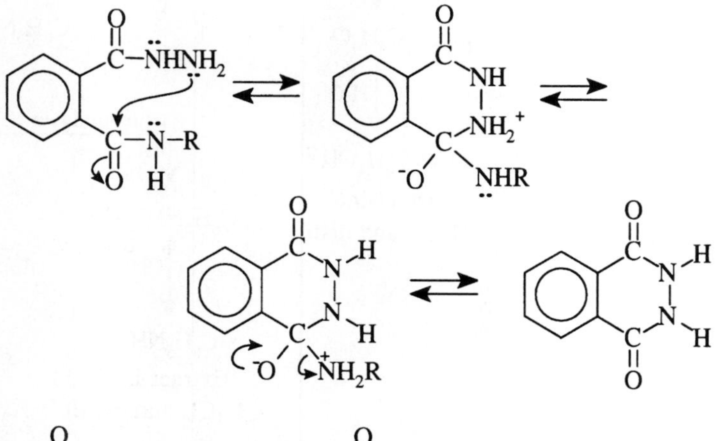

20.4

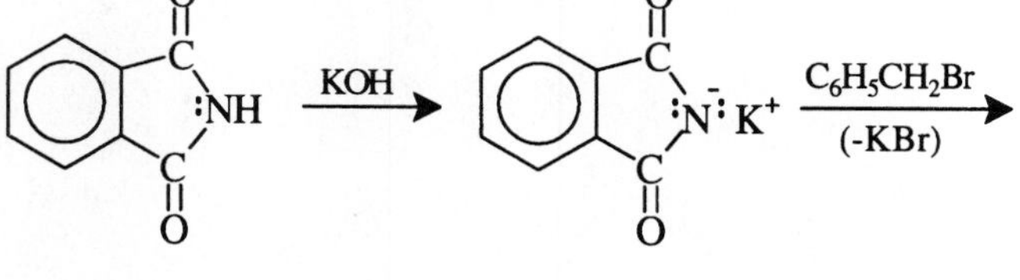

20.5 (a) $CH_3(CH_2)_3CHO + NH_3 \xrightarrow[Ni]{H_2} CH_3(CH_2)_3CH_2NH_2$

(b) $C_6H_5CH_2\overset{O}{\underset{\|}{C}}CH_3 + NH_3 \xrightarrow[Ni]{H_2} C_6H_5CH_2\overset{}{\underset{NH_2}{C}}HCH_3$

(c) $CH_3(CH_2)_4CHO + C_6H_5NH_2 \xrightarrow[CH_3OH]{LiBH_3CN} CH_3(CH_2)_4CH_2NHC_6H_5$

(d) $C_6H_5CHO + (CH_3)_2NH \xrightarrow[CH_3OH]{LiBH_3CN} C_6H_5CH_2N(CH_3)_2$

20.6 The reaction of a secondary halide with ammonia would inevitably be accompanied by considerable elimination, thus decreasing the yield.

20.7 (a) $C_6H_5CO_2H \xrightarrow{SOCl_2} C_6H_5COCl \xrightarrow{CH_3CH_2NH_2}$

$C_6H_5CONHCH_2CH_3 \xrightarrow{LiAlH_4} C_6H_5CH_2NHCH_2CH_3$

(b) $CH_3CH_2CH_2CH_2CH_2Br \xrightarrow{NaCN} CH_3CH_2CH_2CH_2CH_2CN$

$\xrightarrow{LiAlH_4} CH_3CH_2CH_2CH_2CH_2CH_2NH_2$

(c) $CH_3CH_2CO_2H \xrightarrow{SOCl_2} CH_3CH_2COCl \xrightarrow{(CH_3CH_2CH_2)_2NH}$

$CH_3CH_2CON(CH_2CH_2CH_3)_2 \xrightarrow{LiAlH_4} N(CH_2CH_2CH_3)_3$

(d) $CH_3\overset{O}{\underset{\|}{C}}CH_2CH_3 \xrightarrow{NH_2OH} CH_3\overset{NOH}{\underset{\|}{C}}CH_2CH_3 \xrightarrow{Na/ethanol} CH_3\overset{NH_2}{\underset{|}{C}}HCH_2CH_3$

20.8 (a) CH_3O—⟨⟩ $\xrightarrow[H_2SO_4]{HNO_3}$ CH_3O—⟨⟩—NO_2 $\xrightarrow[HCl]{Fe}$ CH_3O—⟨⟩—NH_2

(b) CH_3O—⟨⟩ $\xrightarrow[AlCl_3]{CH_3COCl}$ CH_3O—⟨⟩—$\overset{\overset{O}{\|}}{C}CH_3$ $\xrightarrow[Ni]{NH_3, H_2}$

CH_3O—⟨⟩—$\underset{NH_2}{\overset{}{C}HCH_3}$

(c) ⟨⟩—CH_3 $\xrightarrow[hv]{Cl_2}$ ⟨⟩—CH_2Cl $\xrightarrow{(CH_3)_3N}$ ⟨⟩—$CH_2\overset{+}{N}(CH_3)_3Cl^-$
    (excess)

(d) O_2N—⟨⟩—CH_3 $\xrightarrow[(2)\,H^+]{(1)\,KMnO_4,\,OH^-}$ O_2N—⟨⟩—CO_2H $\xrightarrow{SOCl_2}$

O_2N—⟨⟩—$\overset{\overset{O}{\|}}{C}$-Cl $\xrightarrow{NH_3}$ O_2N—⟨⟩—$\overset{\overset{O}{\|}}{C}NH_2$ $\xrightarrow[OH^-]{Br_2}$ O_2N—⟨⟩—NH_2

(e) ⟨⟩—CH_3 + NBS $\xrightarrow{ROOR}$ ⟨⟩—CH_2Br $\xrightarrow{KCN}$

⟨⟩—CH_2CN $\xrightarrow[(2)\,H_2O]{(1)\,LiAlH_4,\,Et_2O}$ ⟨⟩—$CH_2CH_2NH_2$

20.9 An amine acting as a base.

$$CH_3CH_2\overset{..}{N}H_2 \;+\; H_3O^+ \;\rightleftharpoons\; CH_3CH_2NH_3^+ \;+\; H_2O$$

An amine acting as a nucleophile in an alkylation reaction.

$$(CH_3CH_2)_3N: \;+\; CH_3{-}I \;\longrightarrow\; (CH_3CH_2)_3\overset{+}{N}CH_3\; I^-$$

An amine acting as a nucleophile in an acylation reaction.

$$(CH_3)_2\overset{..}{N}H \;+\; CH_3\overset{\overset{O}{\|}}{C}{\diagdown}{}^{O}_{Cl} \;\longrightarrow\; (CH_3)_2N\overset{\overset{O}{\|}}{C}CH_3 \;+\; (CH_3)_2\overset{+}{N}H_2\; Cl^-$$
(excess)

An amino group acting as an activating group and as an ortho-para director in electrophilic aromatic substitution.

20.10 (a, b) $^-O-N=O + H_3O^+ \rightleftharpoons HO-N=O + H_2O$

$$HO-N=O + H_3O^+ \rightleftharpoons \overset{+}{\underset{H}{HO}}-N=O + H_2O$$

$$\overset{+}{\underset{H}{HO}}-N=O \rightleftharpoons H_2O + \overset{+}{N}=O$$

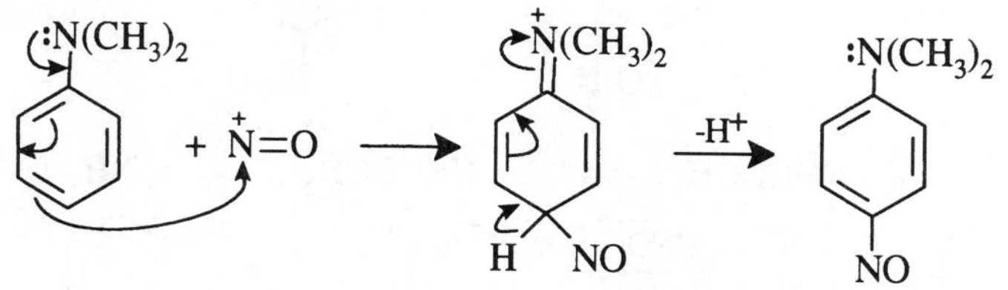

(c) The $\overset{+}{N}O$ ion is a weak electrophile. For it to react with an aromatic ring, the ring must have a powerful activating group such as $-OH$ or $-NR_2$.

20.11 (a)

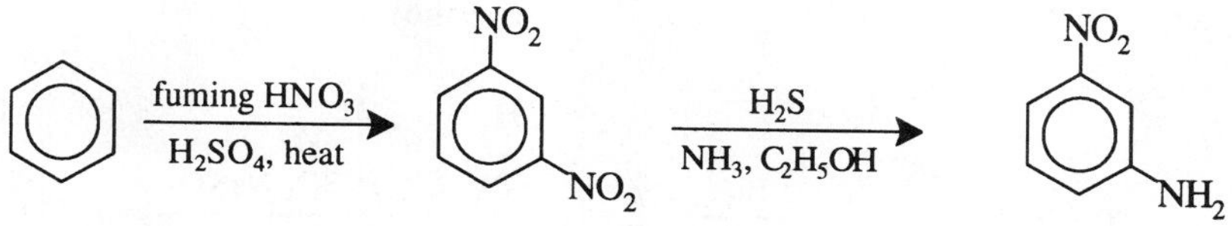

(b)

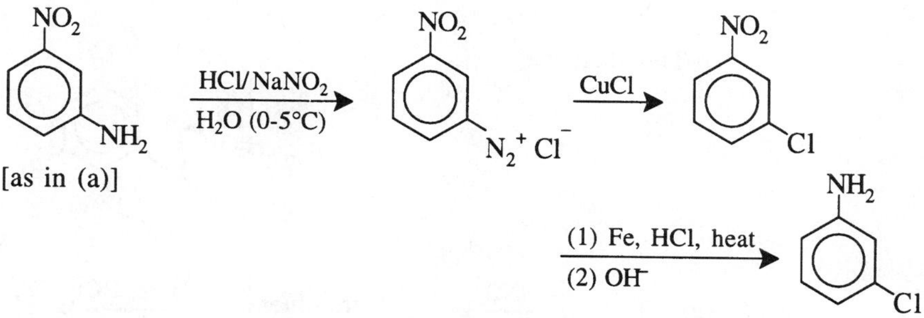

[as in (a)]

(c)

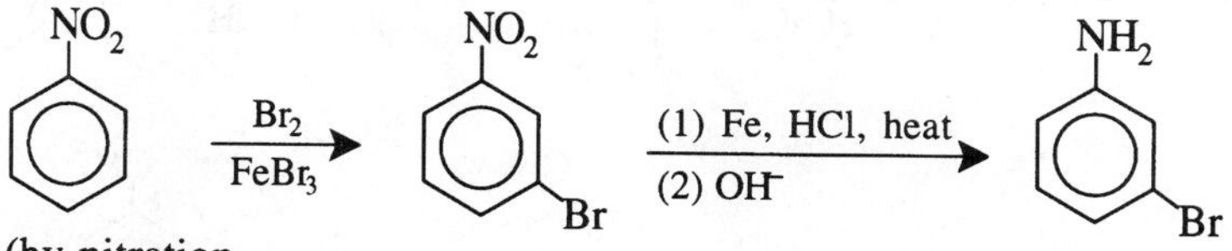

(by nitration of benzene)

(d)

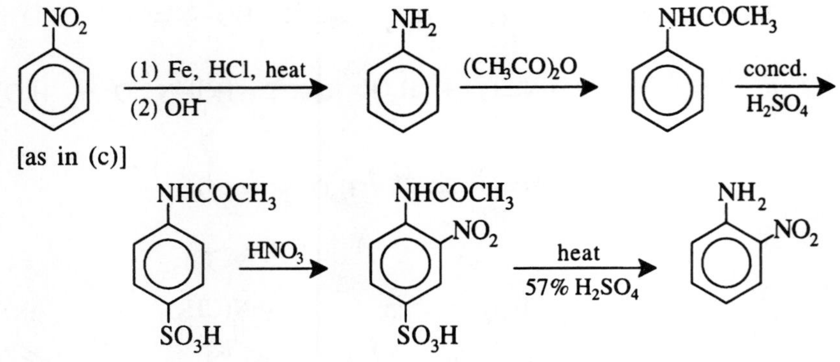

[as in (c)]

(e)

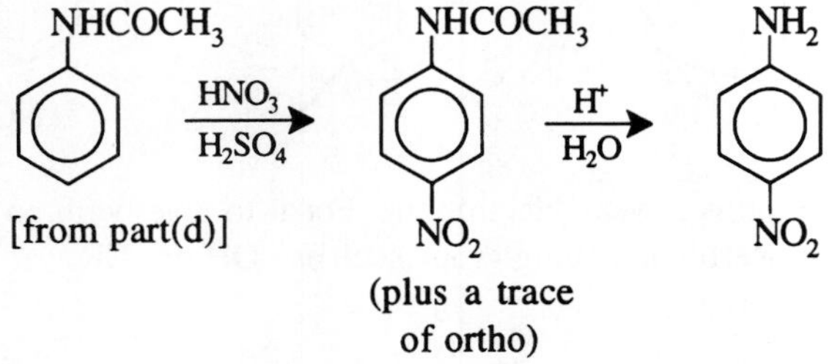

[from part(d)]

(plus a trace
of ortho)

20.12

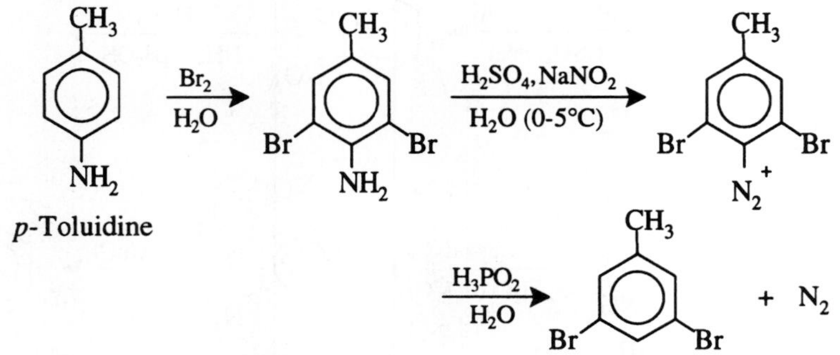

p-Toluidine

20.13 (a)   Toluene $\xrightarrow[\text{H}_2\text{SO}_4]{\text{HNO}_3}$ *p*-Nitrotoluene $\xrightarrow[\text{(2) OH}^-]{\text{(1) Fe, HCl}}$ $\xrightarrow{\text{(CH}_3\text{CO)}_2\text{O}}$

(+ *o*-nitrotoluene)

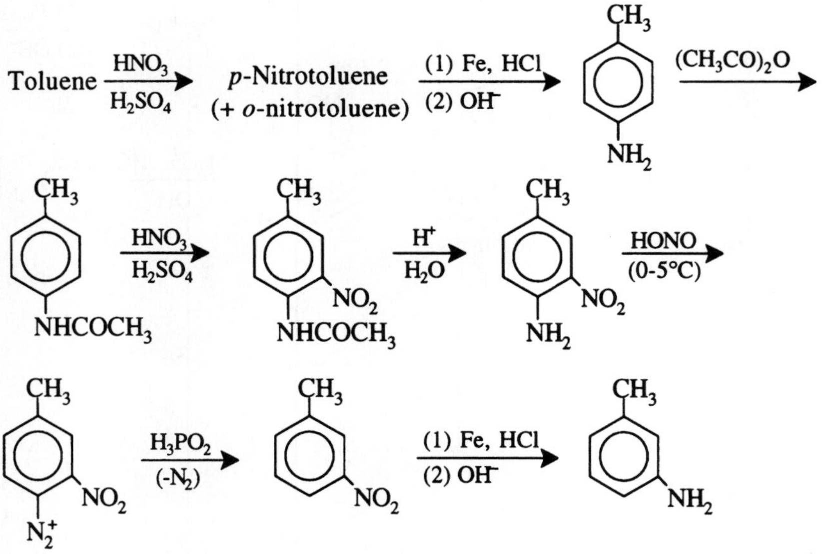

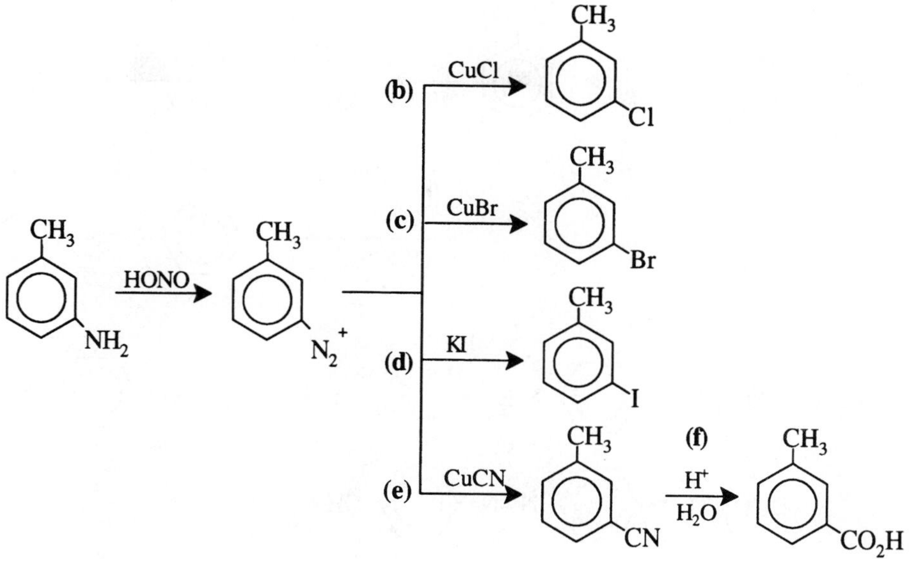
CH3
NH2
HONO
N2+
(b)
CuCl
Cl
(c)
CuBr
Br
(d)
KI
I
(e)
CuCN
CN
(f)
H+
H2O
CO2H

20.14

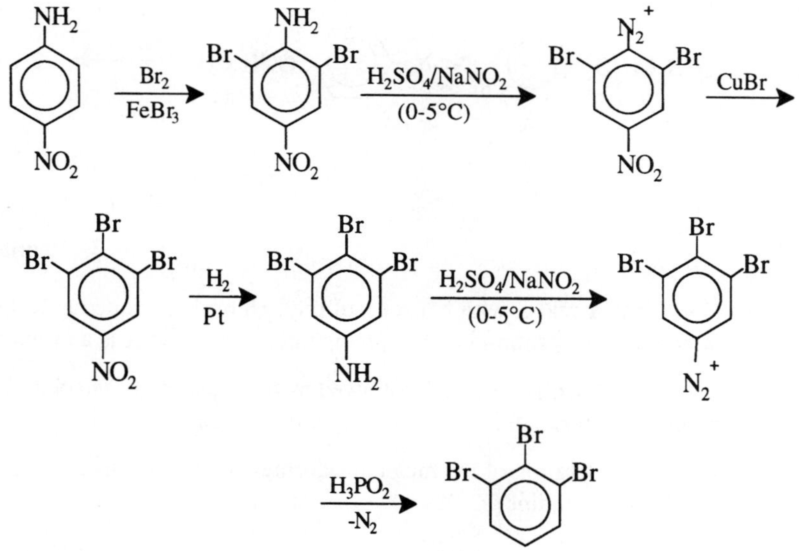
NH2
NO2
Br2
FeBr3
Br
H2SO4/NaNO2
(0-5°C)
N2+
CuBr
H2
Pt
H3PO2
-N2

20.15

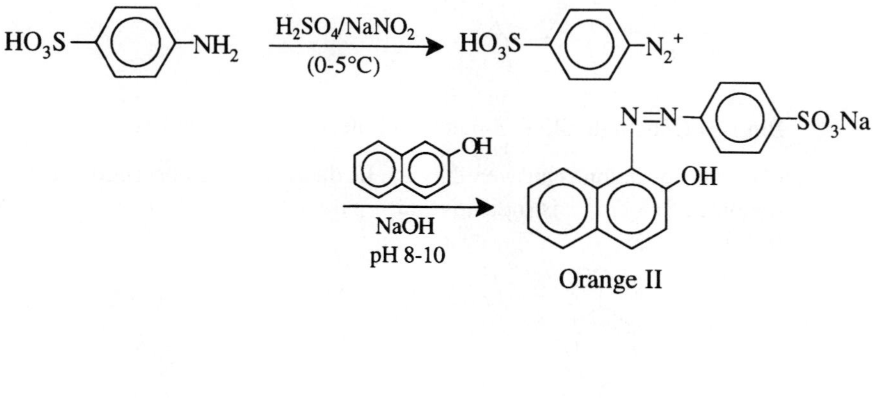
HO3S
NH2
H2SO4/NaNO2
(0-5°C)
N2+
OH
NaOH
pH 8-10
N=N
SO3Na
Orange II

20.16

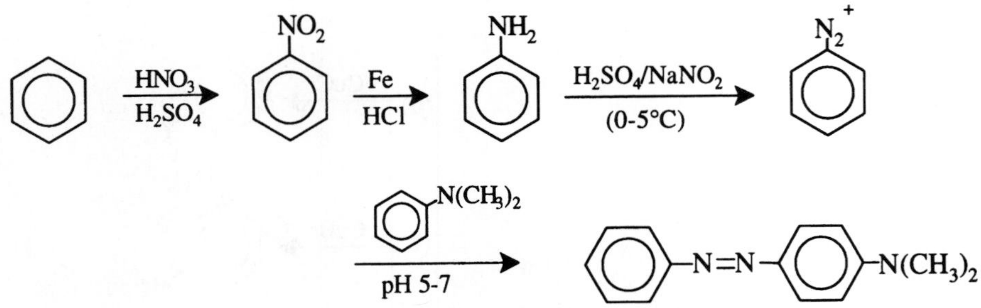

20.17

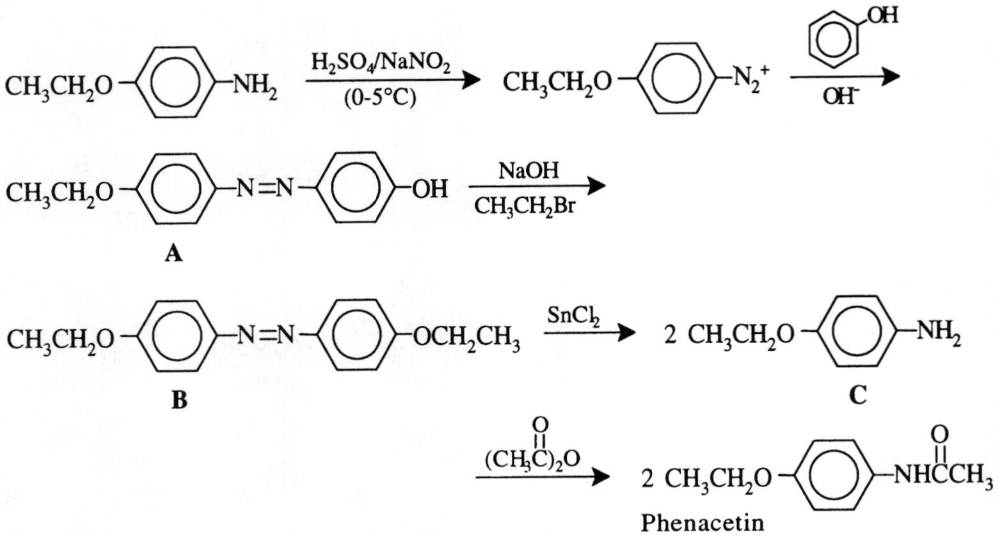

20.18 (1) That **A** reacts with benzenesulfonyl chloride in aqueous KOH to give a clear solution, which on acidification yields a precipitate, shows that **A** is a primary amine.

(2) That diazotization of **A** followed by treatment with 2-naphthol gives an intensely colored precipitate shows that **A** is a primary aromatic amine; that is, **A** is a substituted aniline.

(3) Consideration of the molecular formula of **A** leads us to conclude that **A** is a methylaniline (i.e., a toluidine).

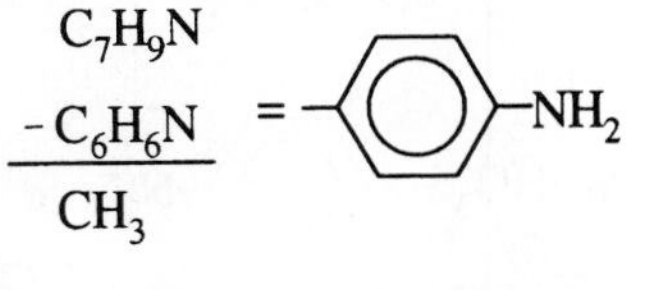

But is **A** 2-methylaniline, 3-methylaniline, or 4-methylaniline?

(4) This question is answered by the IR data. A single absorption peak in the 680–840 cm^{-1} region at 815 cm^{-1} is indicative of a para substituted benzene. Thus, **A** is 4-methylaniline (*p*-toluidine).

20.19 First convert the sulfonamide to its anion, then alkylate the anion with an alkyl halide, then remove the $-SO_2C_6H_5$ group by hydrolysis. For example,

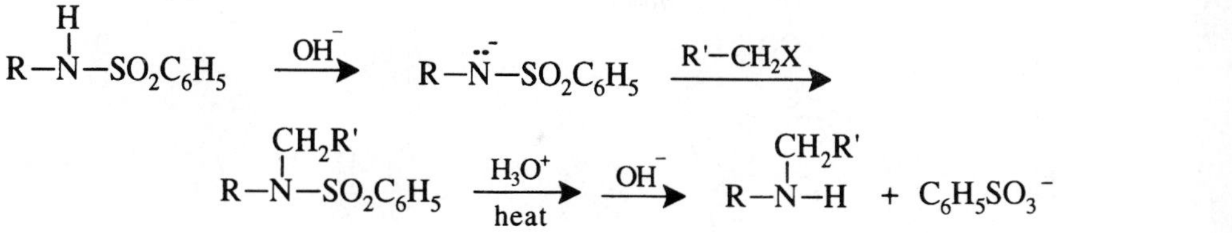

$$R-\underset{\overset{|}{H}}{N}-SO_2C_6H_5 \xrightarrow{OH^-} R-\ddot{N}^--SO_2C_6H_5 \xrightarrow{R'-CH_2X}$$

$$R-\underset{\overset{|}{CH_2R'}}{N}-SO_2C_6H_5 \xrightarrow[\text{heat}]{H_3O^+} \xrightarrow{OH^-} R-\underset{\overset{|}{CH_2R'}}{N}-H \ + \ C_6H_5SO_3^-$$

20.20 (a)

Aniline

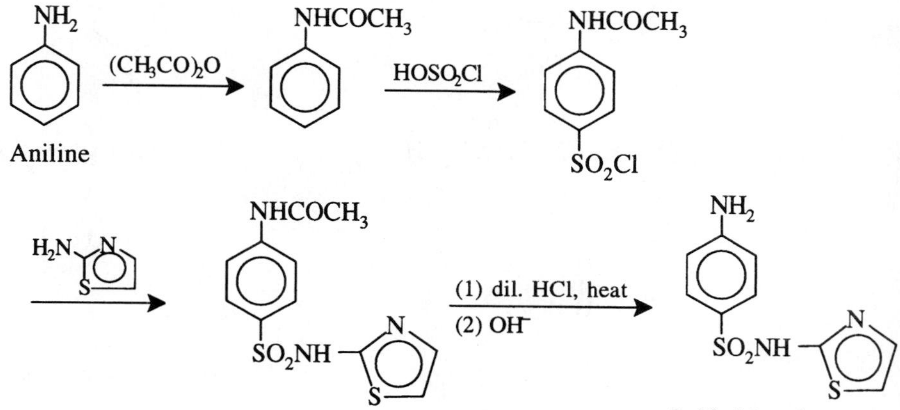

Sulfathiazole

(b)

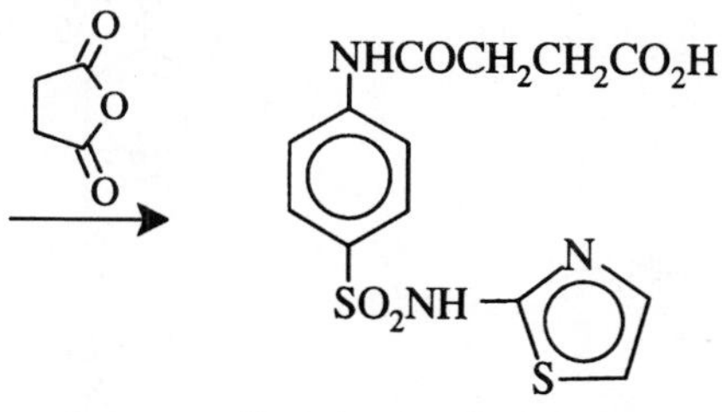

Succinylsulfathiazole

20.21 (a) $C_6H_5CH_2NHCH_3$

(b) 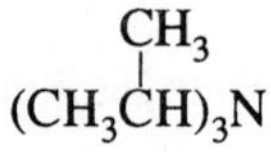 $(CH_3\underset{\overset{|}{CH}}{CH_3})_3N$ ... $\underset{CH_3}{(CH_3CH)_3N}$

(c)

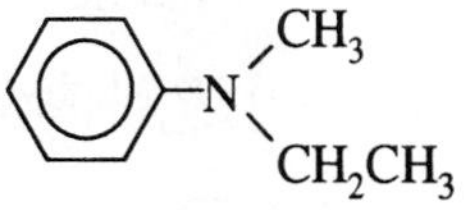

(d)

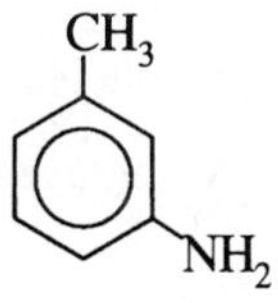

(e)

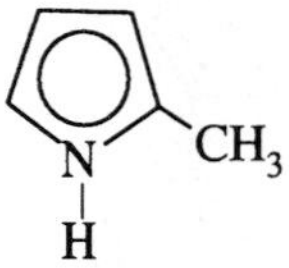

(f)

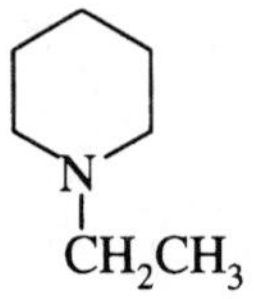

(g) 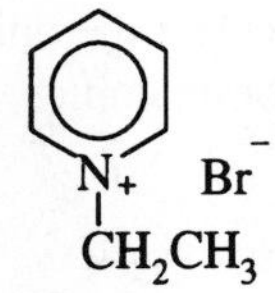 [pyridinium structure with N+ CH₂CH₃ and Br⁻]

(h) 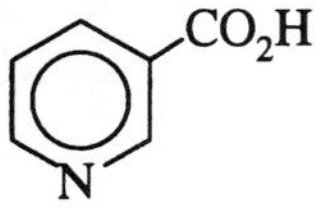 [nicotinic acid structure with CO₂H and N]

(i) 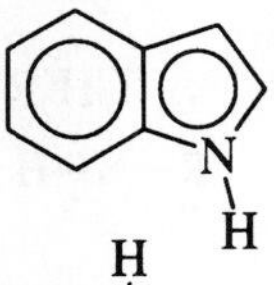 [indole structure with N, H]

(j) 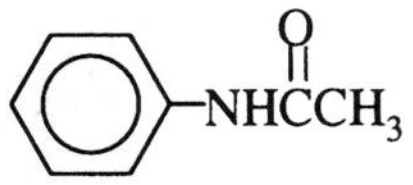 [acetanilide structure: benzene-NHCCH₃ with O]

(k) 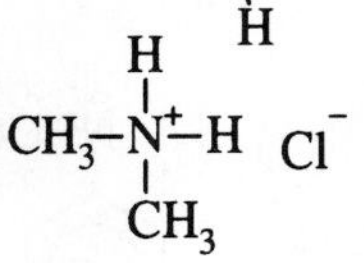 $CH_3 - \overset{\overset{H}{|}}{\underset{\underset{CH_3}{|}}{N}}^+ - H$ Cl^-

(l) 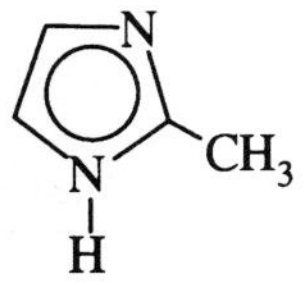 [imidazole structure with N, CH₃, N, H]

(m) $H_2NCH_2CH_2CH_2OH$

(n) $(CH_3CH_2CH_2)_4N^+$ Cl^-

(o) 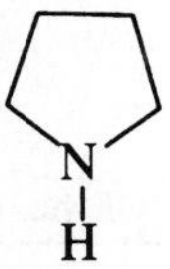 [pyrrolidine structure with N, H]

(p) 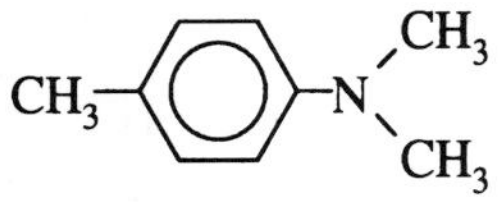 $CH_3 -$ [benzene] $- N \overset{CH_3}{\underset{CH_3}{}}$

(q) 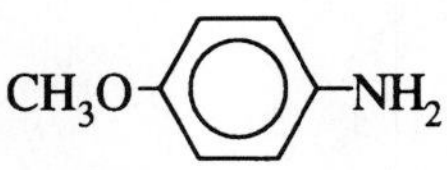 $CH_3O -$ [benzene] $- NH_2$

(r) $(CH_3)_4N^+$ OH^-

(s) 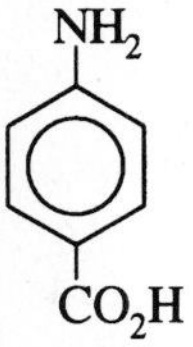 [aniline structure with NH₂ and CO₂H]

(t) 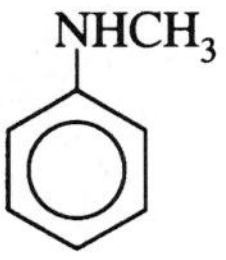 [benzene with NHCH₃]

20.22 (a) Propylamine

(b) *N*-methylaniline

(c) Isopropyltrimethylammonium iodide

(d) 2-Methylaniline (*o*-toluidine)

(e) 2-Methoxyaniline (or *o*-methoxyaniline)

(f) Pyrazole

(g) 2-Aminopyrimidine

(h) Benzylaminium chloride

(i) *N,N*-Dipropylaniline

(j) Benzenesulfonamide

(k) Methylaminium acetate

(l) 3-Amino-1-propanol

(m) Purine

(n) *N*-Methylpyrrole

20.23 (a) 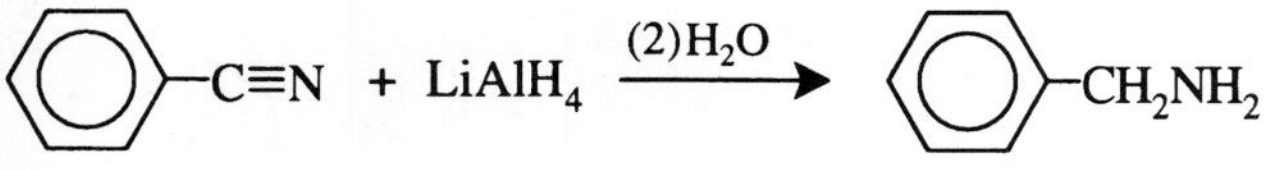 [benzene]$-C\equiv N$ $+$ $LiAlH_4$ $\xrightarrow{(2)H_2O}$ [benzene]$-CH_2NH_2$

(b) 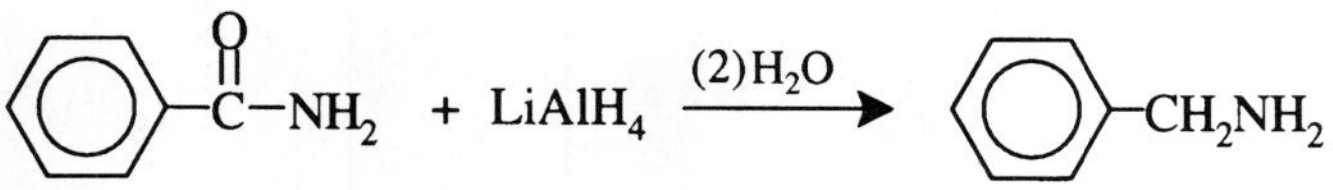 [benzene]$-\overset{\overset{O}{||}}{C}-NH_2$ $+$ $LiAlH_4$ $\xrightarrow{(2)H_2O}$ [benzene]$-CH_2NH_2$

(c) ⬡—CH$_2$Br + NH$_3$ (excess) ⟶ ⬡—CH$_2$NH$_2$

⬡—CH$_2$Br + (phthalimide)NK ⟶ ⬡—CH$_2$—N(phthalimide)

$\xrightarrow{NH_2NH_2}$ ⬡—CH$_2$NH$_2$ + (phthalhydrazide)

(d) ⬡—CH$_2$OTs + NH$_3$ (excess) ⟶ ⬡—CH$_2$NH$_2$

(e) ⬡—CHO + NH$_3$ $\xrightarrow[Ni]{H_2}$ ⬡—CH$_2$NH$_2$

(f) ⬡—CH$_2$NO$_2$ + 3 H$_2$ $\xrightarrow{Pt}$ ⬡—CH$_2$NH$_2$

(g) ⬡—CH$_2$CNH$_2$ (with C=O) $\xrightarrow[OH^-]{Br_2}$ ⬡—CH$_2$NH$_2$ + CO$_3^{2-}$

20.24 (a) ⬡ $\xrightarrow{HNO_3 / H_2SO_4}$ ⬡—NO$_2$ $\xrightarrow[(2)\ OH^-]{(1)\ Fe,\ HCl,\ heat}$ ⬡—NH$_2$

(b) ⬡—CONH$_2$ $\xrightarrow[OH^-]{Br_2}$ ⬡—NH$_2$

20.25 (a) $CH_3(CH_2)_2CH_2OH$ $\xrightarrow{PBr_3}$ $CH_3(CH_2)_2CH_2Br$ $\longrightarrow$

$NCH_2(CH_2)_2CH_3$ $\xrightarrow{NH_2NH_2}$ $CH_3(CH_2)_2CH_2NH_2$

$+$

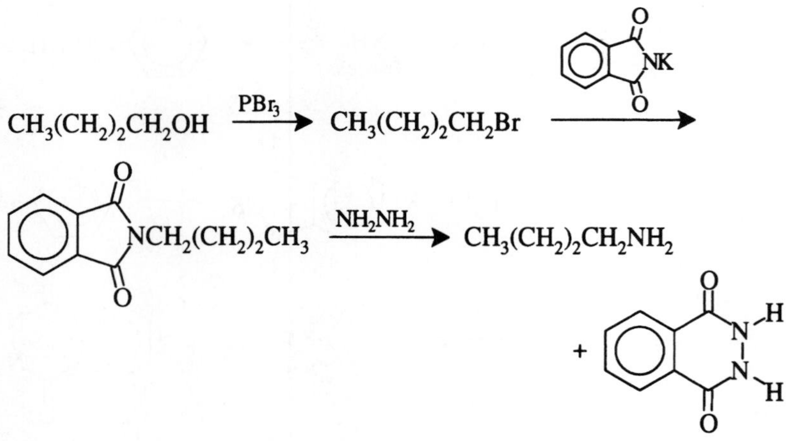

(b) $CH_3(CH_2)_2CH_2Br$ $\xrightarrow{NaCN}$ $CH_3(CH_2)_3CN$ $\xrightarrow[(2)\,H_2O]{(1)\,LiAlH_4,\,Et_2O}$ $CH_3(CH_2)_3CH_2NH_2$
[from part(a)]

(c) $CH_3(CH_2)_2CH_2OH$ $\xrightarrow[(2)\,H^+]{(1)\,KMnO_4,\,OH^-}$ $CH_3CH_2CH_2CO_2H$

$\xrightarrow[(2)\,NH_3]{(1)\,SOCl_2}$ $CH_3CH_2CH_2CONH_2$ $\xrightarrow[OH^-]{Br_2}$ $CH_3CH_2CH_2NH_2$

(d) $CH_3(CH_2)_2CH_2OH$ $\xrightarrow[CH_2Cl_2]{PCC}$ $CH_3CH_2CH_2CHO$ $\xrightarrow[H_2,\,Ni]{CH_3NH_2}$

$CH_3CH_2CH_2CH_2NHCH_3$

20.26 (a) $\xrightarrow{(CH_3CO)_2O}$

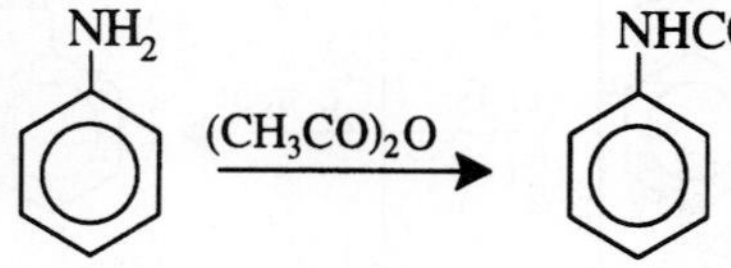

(b) $+$ $\xrightarrow{heat}$

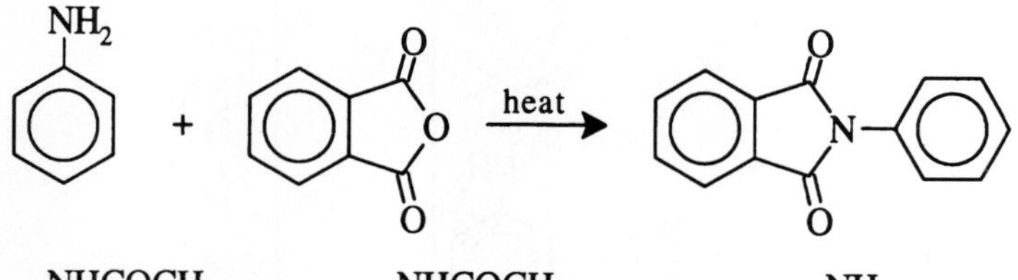

(c) $\xrightarrow[H_2SO_4]{HNO_3}$ $\xrightarrow[(2)\,OH^-]{(1)\,H^+,\,H_2O}$
[from part(a)]

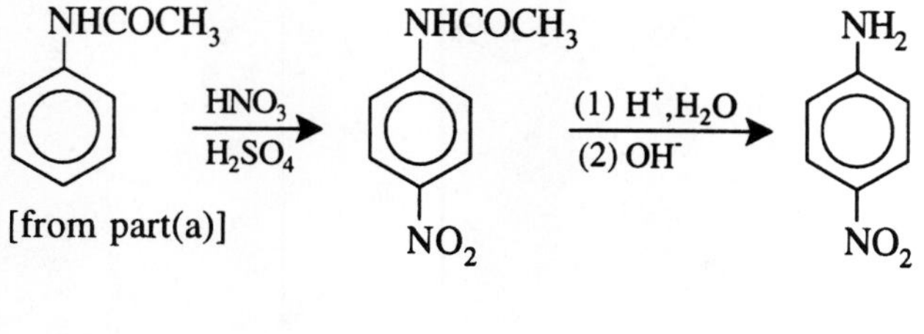

(d)

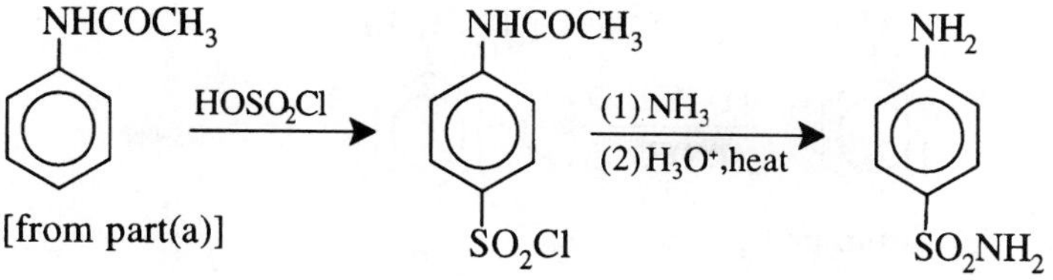

NHCOCH$_3$ [from part(a)] $\xrightarrow{\text{HOSO}_2\text{Cl}}$ NHCOCH$_3$ / SO$_2$Cl $\xrightarrow[\text{(2) H}_3\text{O}^+\text{,heat}]{\text{(1) NH}_3}$ NH$_2$ / SO$_2$NH$_2$

(e)

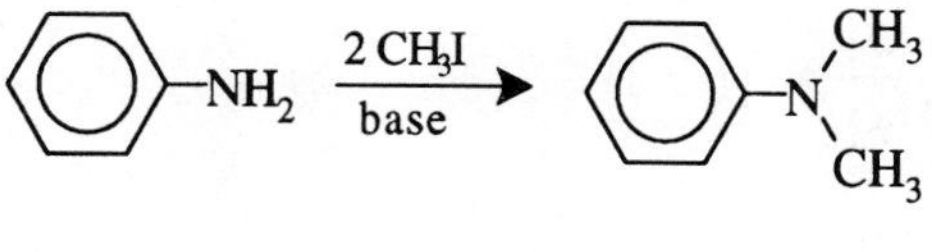

⟨⟩—NH$_2$ $\xrightarrow[\text{base}]{\text{2 CH}_3\text{I}}$ ⟨⟩—N(CH$_3$)CH$_3$

(f)

NH$_2$ $\xrightarrow[\text{(0-5°C)}]{\text{HONO}}$ N$_2^+$ X$^-$ $\xrightarrow{\text{HBF}_4}$ N$_2^+$ BF$_4^-$ $\xrightarrow{\text{heat}}$ F

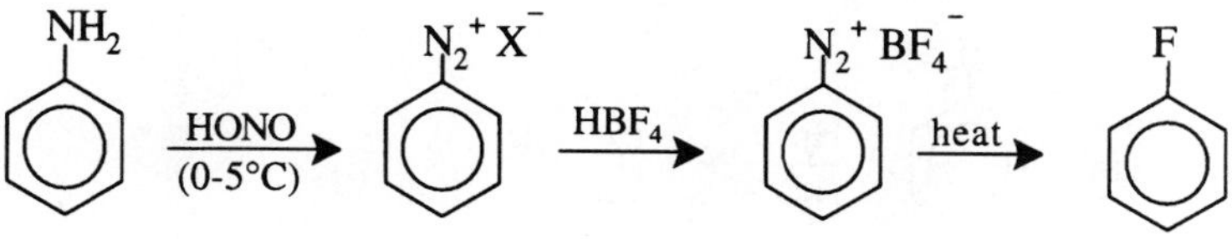

(g)

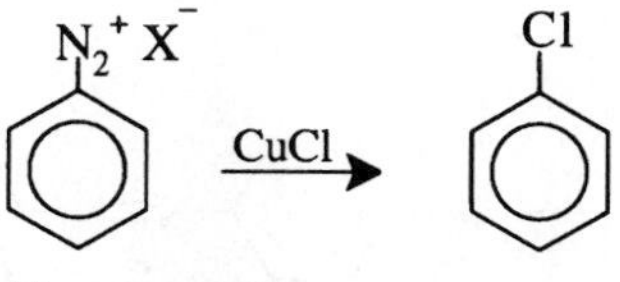

N$_2^+$ X$^-$ [from part(f)] $\xrightarrow{\text{CuCl}}$ Cl

(h)

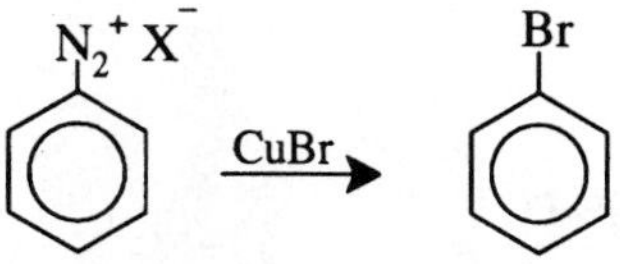

N$_2^+$ X$^-$ [from part(f)] $\xrightarrow{\text{CuBr}}$ Br

(i)

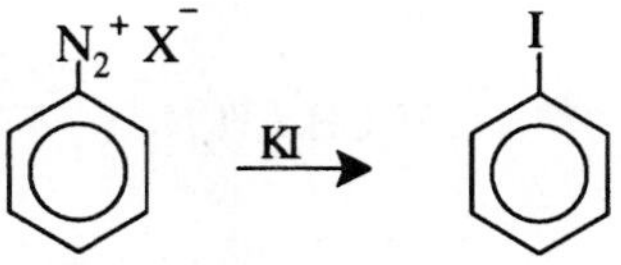

N$_2^+$ X$^-$ [from part(f)] $\xrightarrow{\text{KI}}$ I

(j)

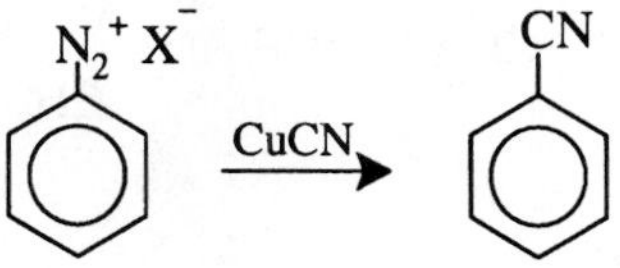

N$_2^+$ X$^-$ [from part(f)] $\xrightarrow{\text{CuCN}}$ CN

(k)

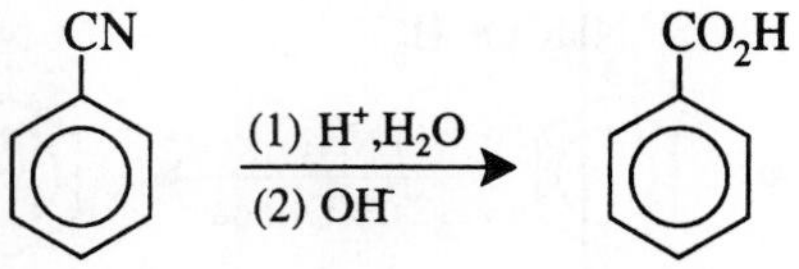

[from part(j)]

(l)

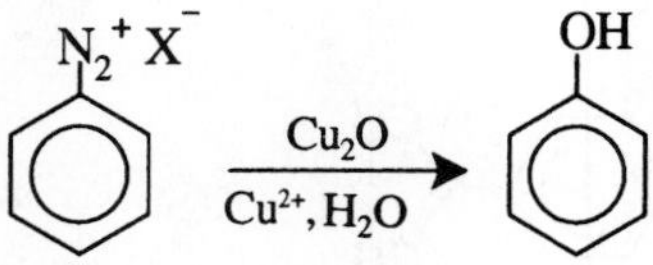

[from part(f)]

(m)

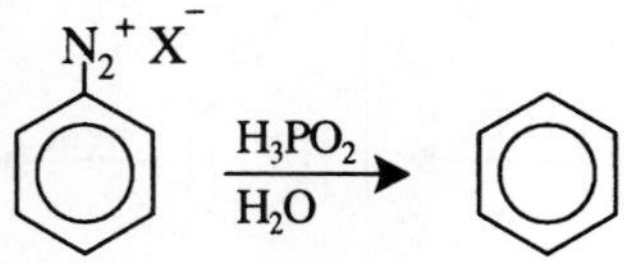

[from part(f)]

(n)

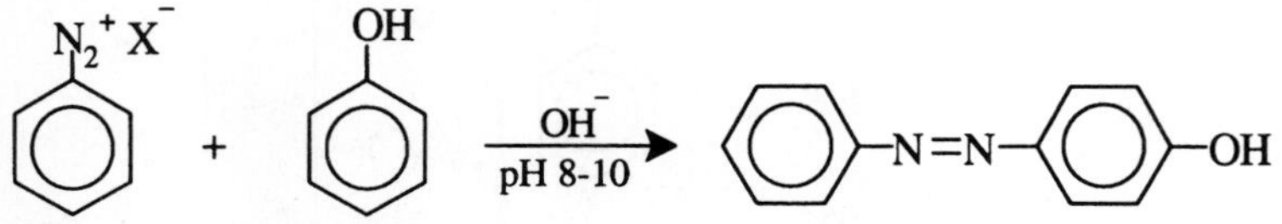

[from part(f)]   [from part(l)]

(o)

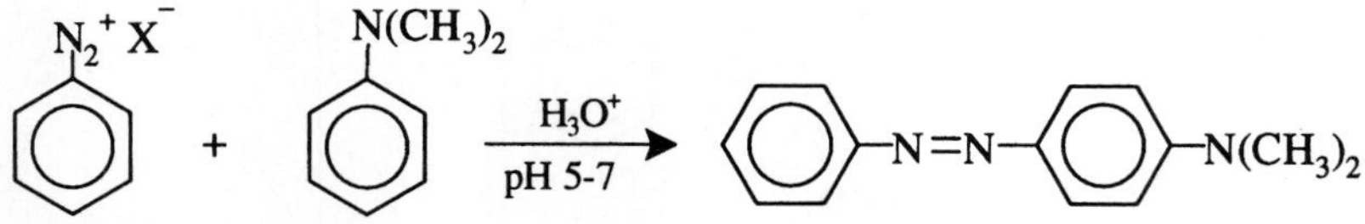

[from part(f)]   [from part(e)]

20.27 (a) $CH_3CH_2CH_2NH_2$ $\xrightarrow[\text{NaNO}_2/\text{HCl}]{\text{HONO}}$ $[CH_3CH_2CH_2N_2{}^+]$ $\xrightarrow{-N_2}$

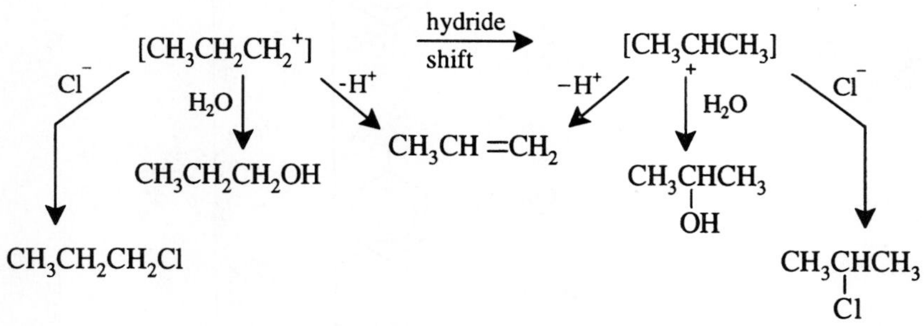

$CH_3CH_2CH_2Cl$

(b) $(CH_3CH_2CH_2)_2NH$ $\xrightarrow[\text{NaNO}_2/\text{HCl}]{\text{HONO}}$ $(CH_3CH_2CH_2)_2N-N=O$

(c)

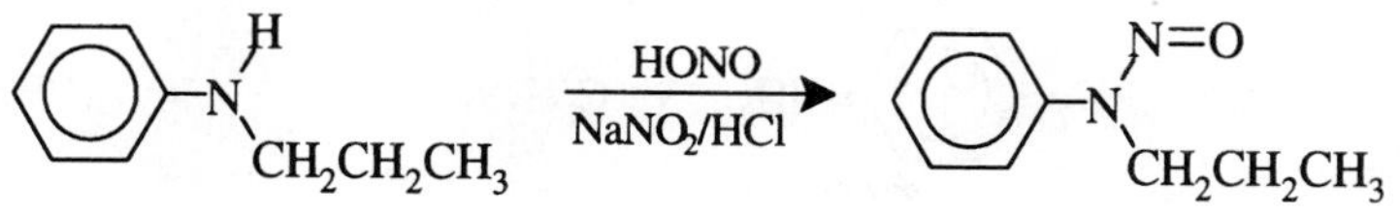

(d)

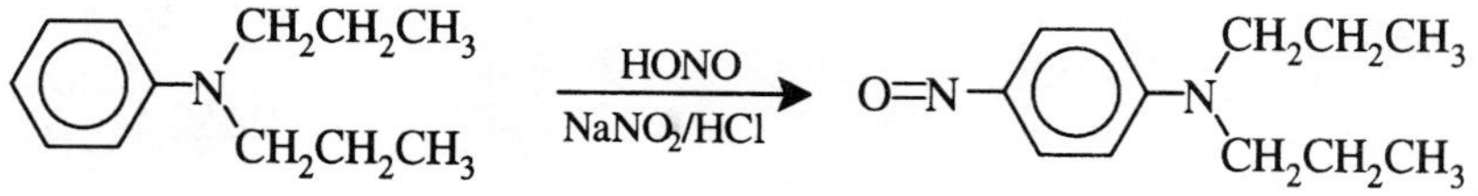

(e) 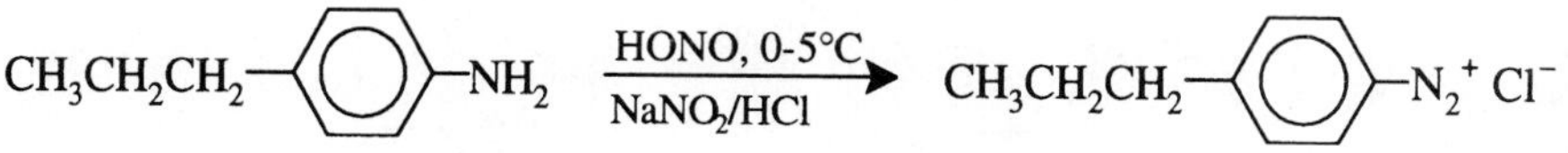 $CH_3CH_2CH_2-$⬡$-NH_2$ $\xrightarrow[\text{NaNO}_2/\text{HCl}]{\text{HONO, 0-5°C}}$ $CH_3CH_2CH_2-$⬡$-N_2^+ Cl^-$

20.28 (a) $CH_3CH_2CH_2NH_2$ + $C_6H_5SO_2Cl$ $\xrightarrow[\text{H}_2\text{O}]{\text{KOH}}$ $CH_3CH_2CH_2\overset{-}{N}SO_2C_6H_5$
$\overset{+}{K}$
Clear solution

$\xrightarrow{\text{H}_3\text{O}^+}$ $CH_3CH_2CH_2NHSO_2C_6H_5$
Precipitate

(b) $(CH_3CH_2CH_2)_2NH$ + $C_6H_5SO_2Cl$ $\xrightarrow[\text{H}_2\text{O}]{\text{KOH}}$ $(CH_3CH_2CH_2)_2NSO_2C_6H_5$
Precipitate

$\xrightarrow{\text{H}_3\text{O}^+}$ No reaction
(precipitate remains)

(c) 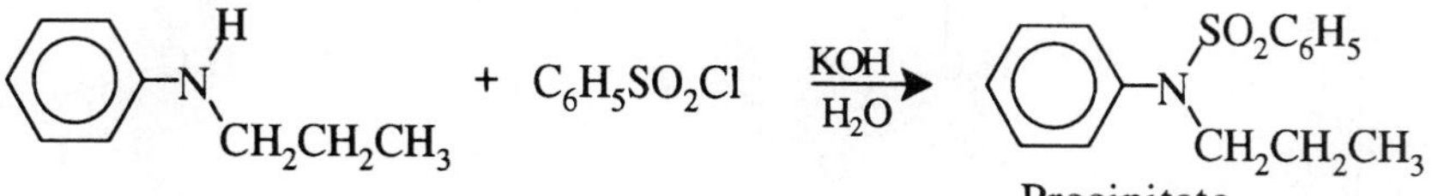 + $C_6H_5SO_2Cl$ $\xrightarrow[\text{H}_2\text{O}]{\text{KOH}}$
Precipitate

$\xrightarrow{\text{H}_3\text{O}^+}$ No reaction
(precipitate remains)

(d) 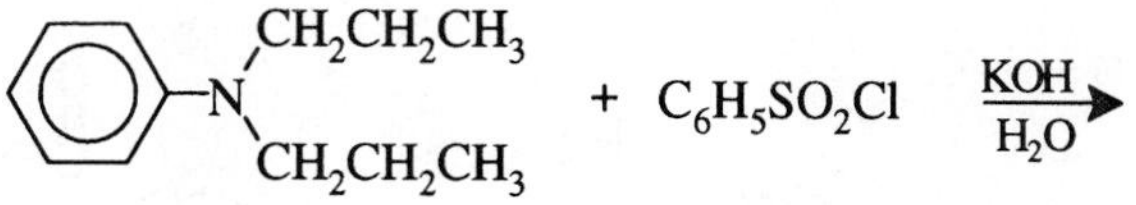 + $C_6H_5SO_2Cl$ $\xrightarrow[\text{H}_2\text{O}]{\text{KOH}}$ No reaction
(3° amine is insoluble)

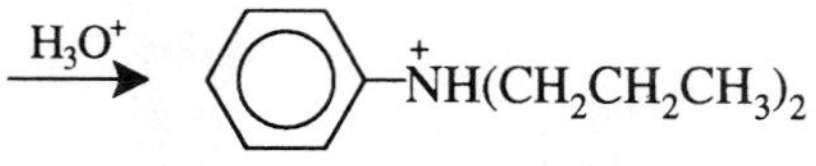

$\xrightarrow{\text{H}_3\text{O}^+}$ ⬡$-\overset{+}{N}H(CH_2CH_2CH_3)_2$

3° Amine dissolves

(e) 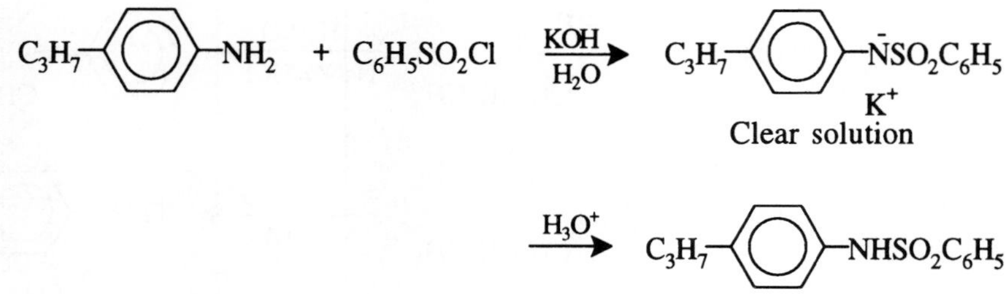 C_3H_7—⟨benzene ring⟩—NH_2 + $C_6H_5SO_2Cl$ $\xrightarrow[H_2O]{KOH}$ C_3H_7—⟨benzene ring⟩—$\overset{-}{N}SO_2C_6H_5$

K^+

Clear solution

$\xrightarrow{H_3O^+}$ C_3H_7—⟨benzene ring⟩—$NHSO_2C_6H_5$

Precipitate

20.29 (a)  ⟨piperidine⟩$N-H$ $\xrightarrow[NaNO_2/HCl]{HONO}$ ⟨piperidine⟩$N-N=O$

(b) 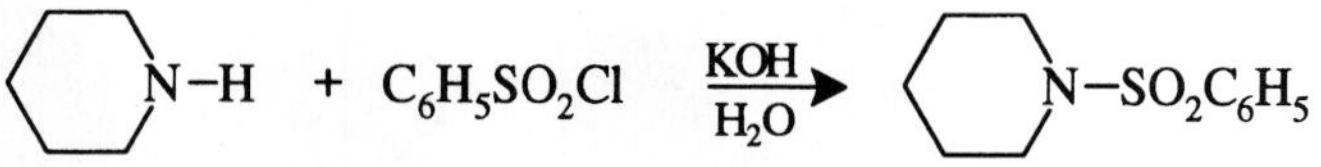 ⟨piperidine⟩$N-H$ + $C_6H_5SO_2Cl$ $\xrightarrow[H_2O]{KOH}$ ⟨piperidine⟩$N-SO_2C_6H_5$

20.30 (a) $2\ CH_3CH_2NH_2$ + C_6H_5COCl $\longrightarrow$ $CH_3CH_2NHCOC_6H_5$ +

$CH_3CH_2NH_3^+\ Cl^-$

(b) $2\ CH_3NH_2$ + $(CH_3\overset{O}{\overset{\|}{C}})_2O$ $\longrightarrow$ $CH_3NH\overset{O}{\overset{\|}{C}}CH_3$ + $CH_3\overset{+}{N}H_3\ CH_3\overset{O}{\overset{\|}{C}}O^-$

(c) 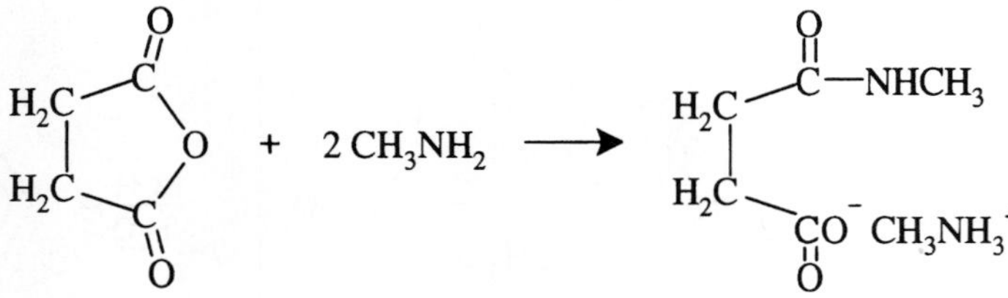 ⟨succinic anhydride⟩ + $2\ CH_3NH_2$ $\longrightarrow$ $\begin{array}{c} H_2C-\overset{O}{\overset{\|}{C}}-NHCH_3 \\ H_2C-\underset{\overset{\|}{O}}{C}O^-\ CH_3NH_3^+ \end{array}$

(d) 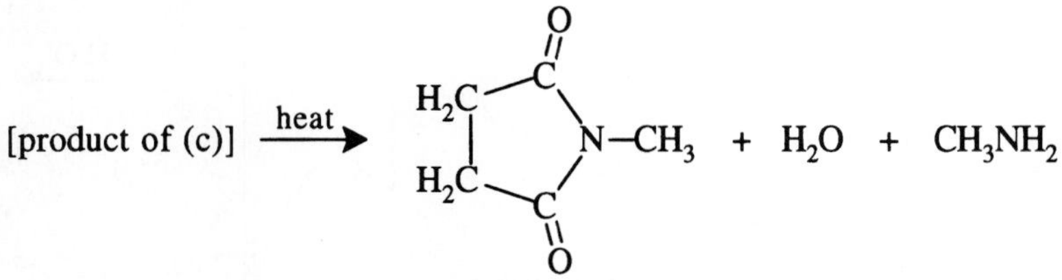 [product of (c)] $\xrightarrow{heat}$ ⟨N-methylsuccinimide⟩$N-CH_3$ + H_2O + CH_3NH_2

(e) 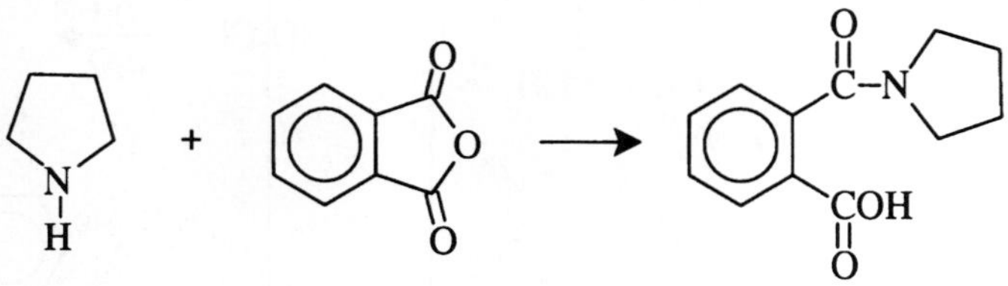 ⟨pyrrolidine⟩ + ⟨phthalic anhydride⟩ $\longrightarrow$ ⟨benzene ring⟩$\overset{O}{\overset{\|}{C}}-N$⟨pyrrolidine⟩ with $\overset{}{C}OH$ ($\overset{\|}{O}$)

(f) 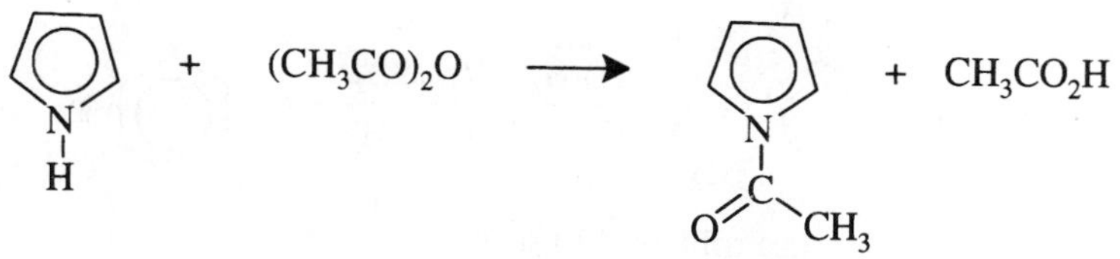 [pyrrole, N–H] + $(CH_3CO)_2O$ $\longrightarrow$ [N-acetylpyrrole] + CH_3CO_2H

(g) 2 [benzene]–NH_2 + $CH_3CH_2\overset{O}{\overset{\|}{C}}Cl$ $\longrightarrow$ [benzene]–$NH\overset{O}{\overset{\|}{C}}CH_2CH_3$ +

[benzene]–NH_3^+ Cl^-

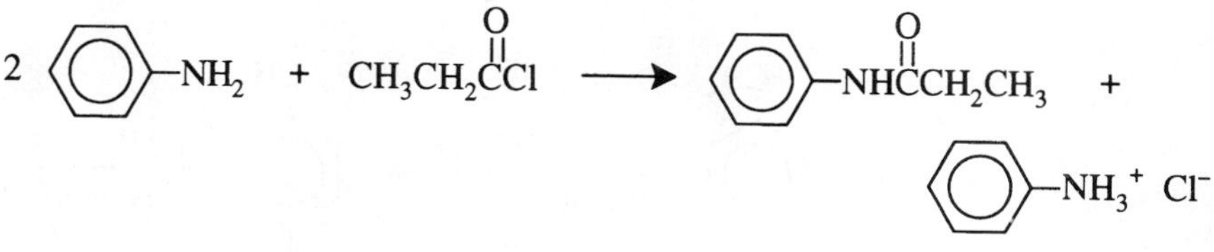

(h) $CH_3CH_2\overset{\overset{\displaystyle CH_2CH_3}{|}}{\underset{\underset{\displaystyle CH_2CH_3}{|}}{\overset{+}{N}}}{-}CH_2CH_3$ OH^- $\xrightarrow{\text{heat}}$ $CH_2{=}CH_2$ + $(CH_3CH_2)_3N$ + H_2O

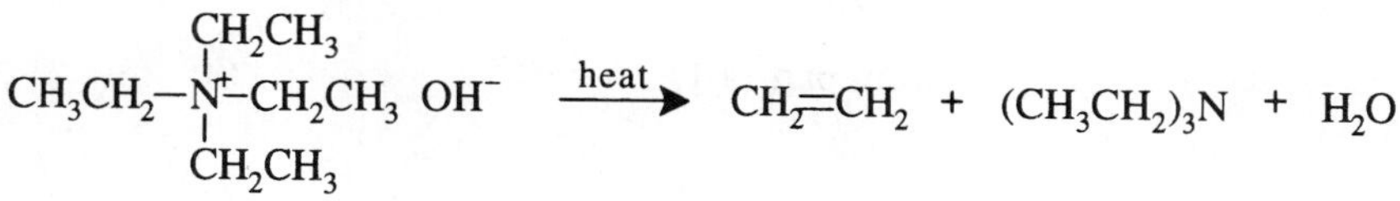

(i) 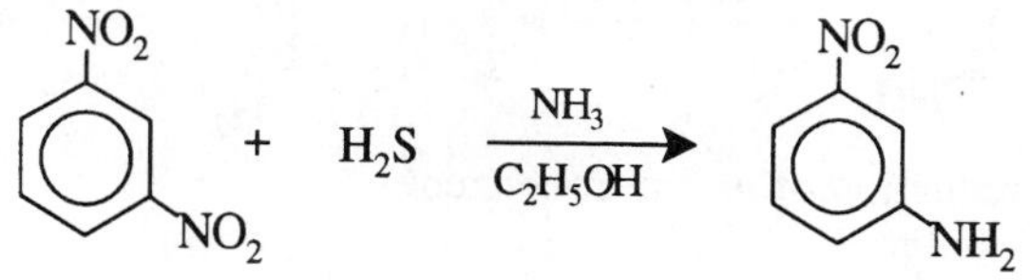 [benzene, NO_2 top, NO_2 bottom] + H_2S $\xrightarrow[C_2H_5OH]{NH_3}$ [benzene, NO_2 top, NH_2 bottom]

(j) 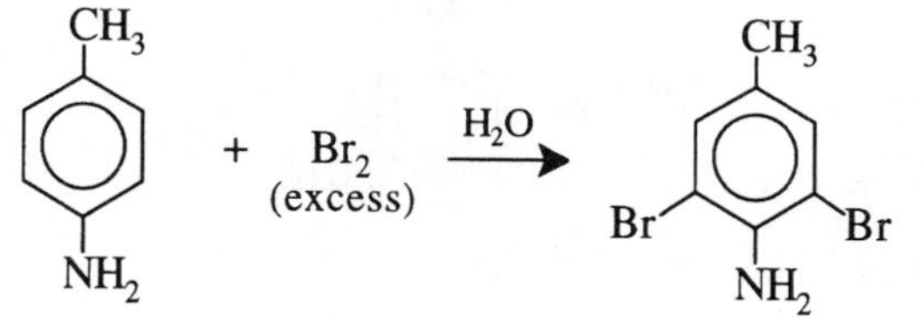 [benzene, CH_3 top, NH_2 bottom] + Br_2 (excess) $\xrightarrow{H_2O}$ [benzene, CH_3 top, Br, Br, NH_2 bottom]

20.31 (a) 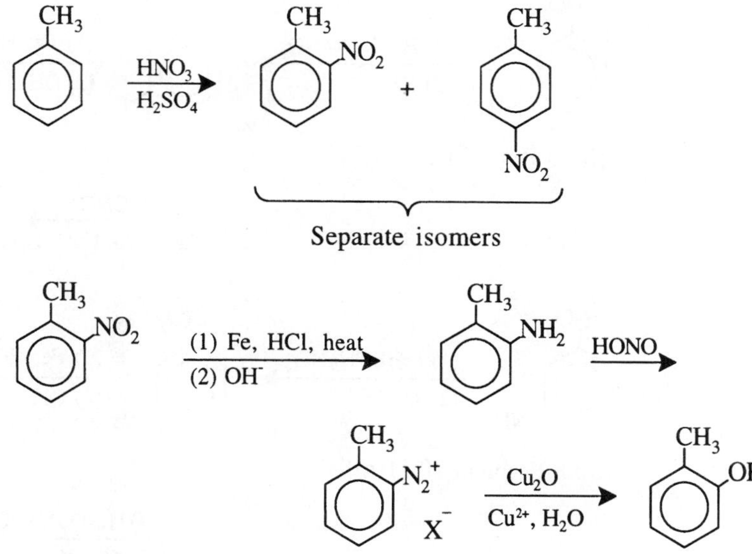 [benzene, CH_3] $\xrightarrow[H_2SO_4]{HNO_3}$ [benzene, CH_3, NO_2] + [benzene, CH_3 top, NO_2 bottom]

Separate isomers

[benzene, CH_3, NO_2] $\xrightarrow[\text{(2) } OH^-]{\text{(1) Fe, HCl, heat}}$ [benzene, CH_3, NH_2] $\xrightarrow{HONO}$

[benzene, CH_3, N_2^+ X^-] $\xrightarrow[Cu^{2+},\ H_2O]{Cu_2O}$ [benzene, CH_3, OH]

(b)

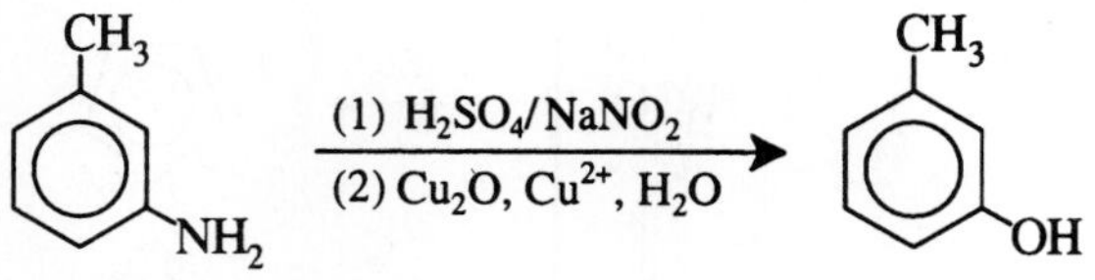

$$\xrightarrow[\text{(2) Cu}_2\text{O, Cu}^{2+}, \text{H}_2\text{O}]{\text{(1) H}_2\text{SO}_4/\text{NaNO}_2}$$

[from Problem 20.13(a)]

(c)

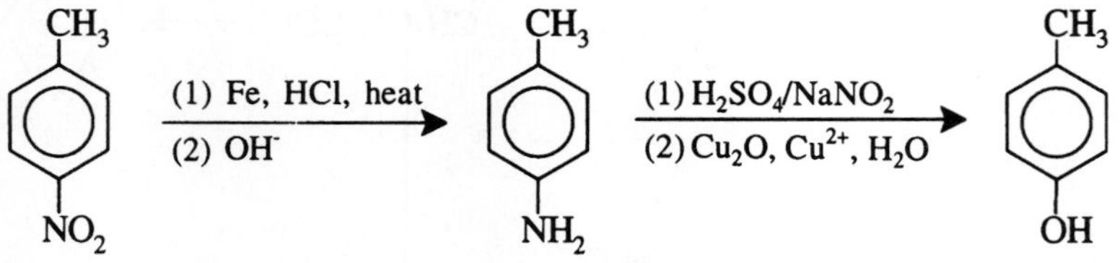

$$\xrightarrow[\text{(2) OH}^-]{\text{(1) Fe, HCl, heat}}$$

$$\xrightarrow[\text{(2) Cu}_2\text{O, Cu}^{2+}, \text{H}_2\text{O}]{\text{(1) H}_2\text{SO}_4/\text{NaNO}_2}$$

[from part (a)]

(d)

$$\xrightarrow{\text{HCl/NaNO}_2}$$

$$\xrightarrow{\text{CuCl}}$$

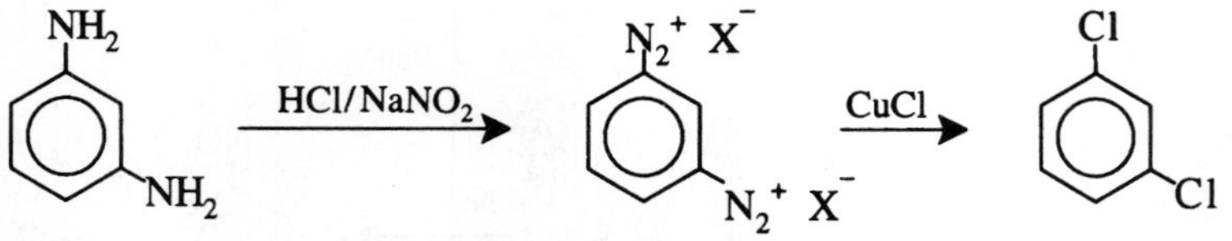

[by reduction of *m*-dinitrobenzene,
cf. Problem 20.11(a)]

(e)

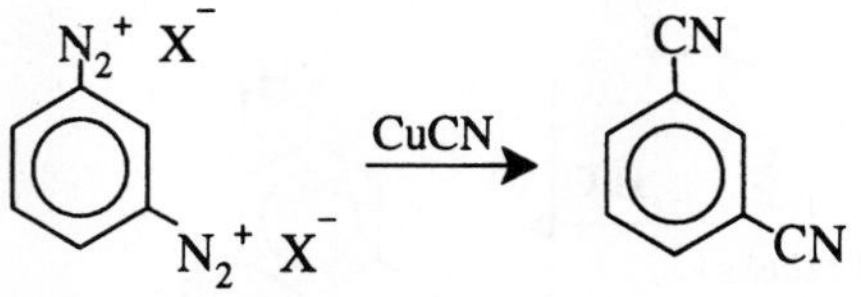

$$\xrightarrow{\text{CuCN}}$$

[from part (d)]

(f)

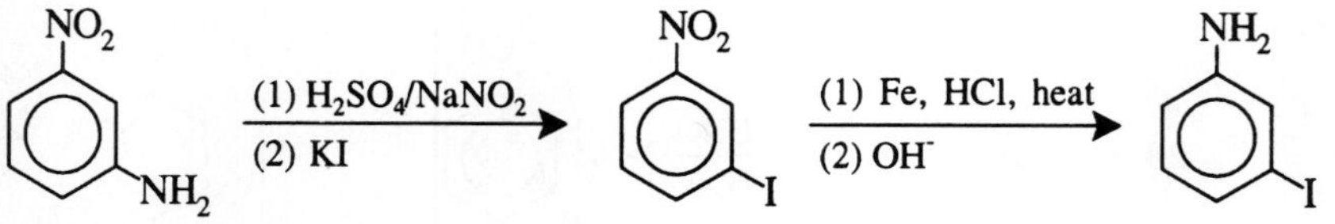

$$\xrightarrow[\text{(2) KI}]{\text{(1) H}_2\text{SO}_4/\text{NaNO}_2}$$

$$\xrightarrow[\text{(2) OH}^-]{\text{(1) Fe, HCl, heat}}$$

[from Problem 20.11(a)]

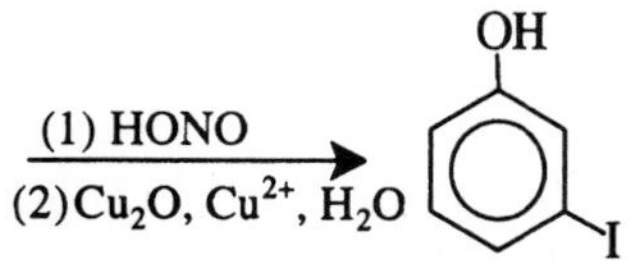

$$\xrightarrow[\text{(2) Cu}_2\text{O, Cu}^{2+}, \text{H}_2\text{O}]{\text{(1) HONO}}$$

(g)

$$\xrightarrow[\text{(2) CuBr}]{\text{(1)   HBr/NaNO}_2}$$

$$\xrightarrow[\text{(2) OH}^-]{\text{(1) Fe, HCl, heat}}$$

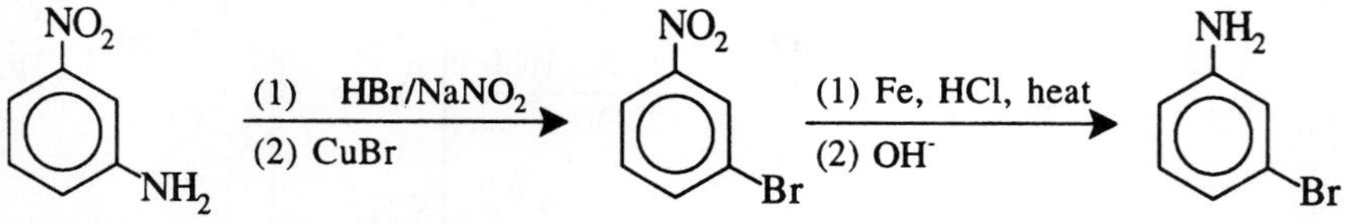

[from Problem 20.11(a)]

$$\xrightarrow[\text{(2) CuCN}]{\text{(1) H}_2\text{SO}_4/\text{NaNO}_2}$$

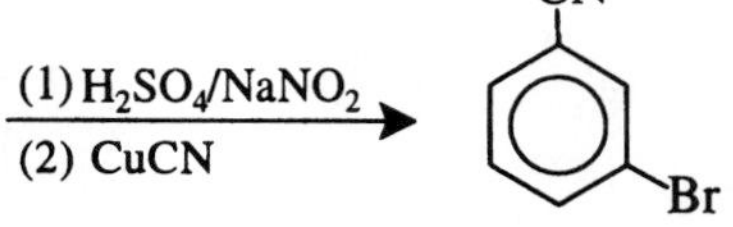

(h)

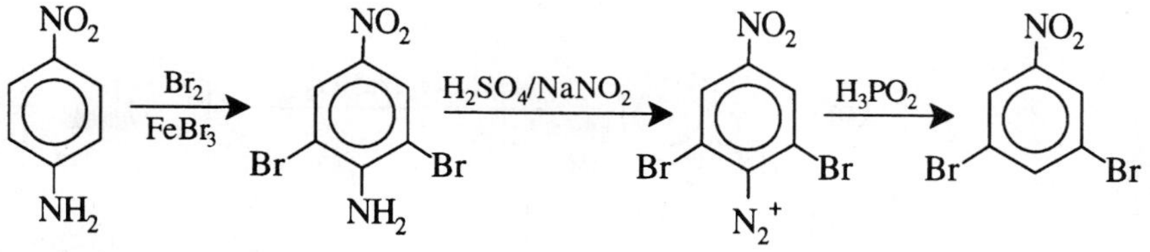

[from Problem 20.11(e)]

(i)

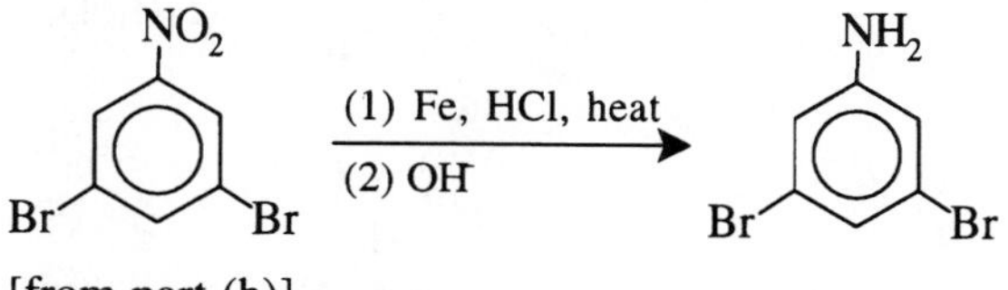

[from part (h)]

(j)

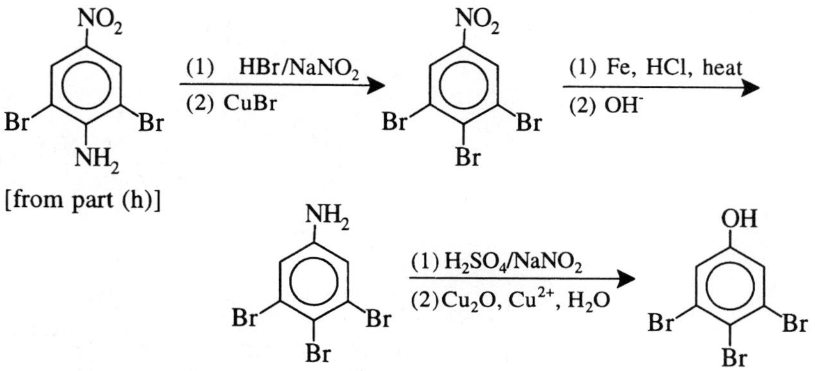

[from part (h)]

(k)

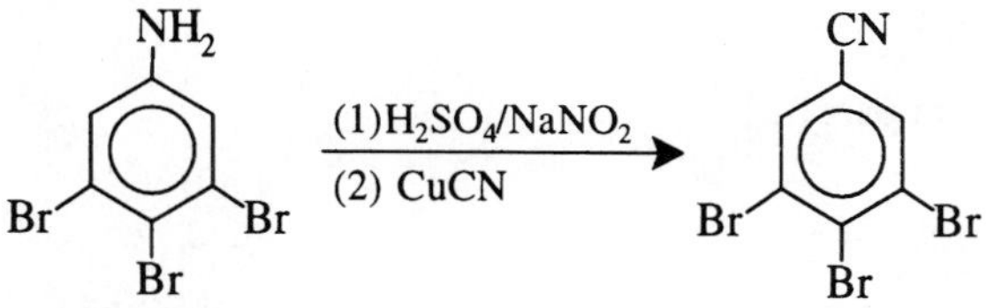

[from part (j)]

(l)

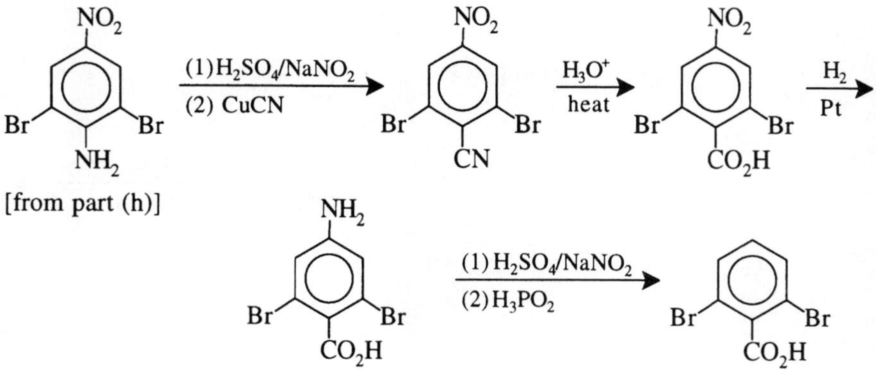

[from part (h)]

(m) 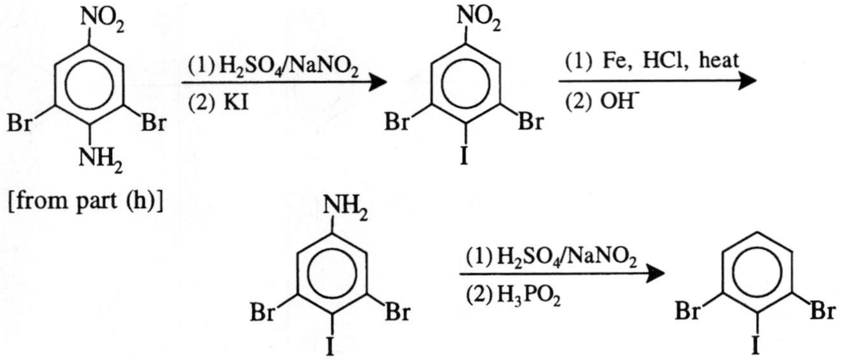 [from part (h)]

(n)

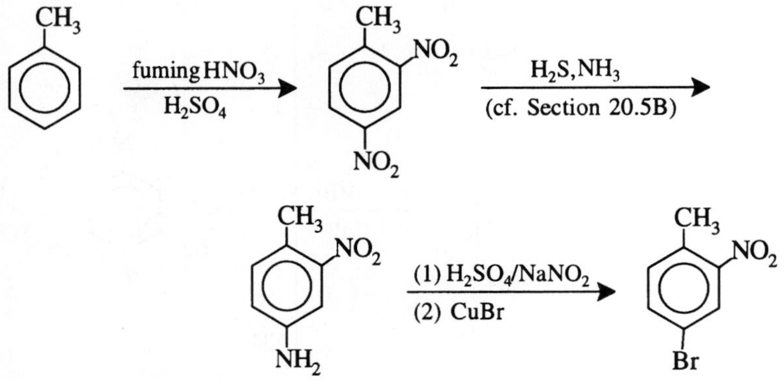

(o) 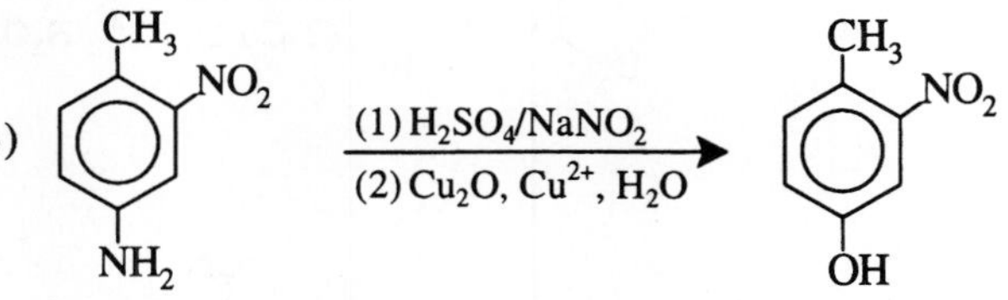 [from part (n)]

(p) 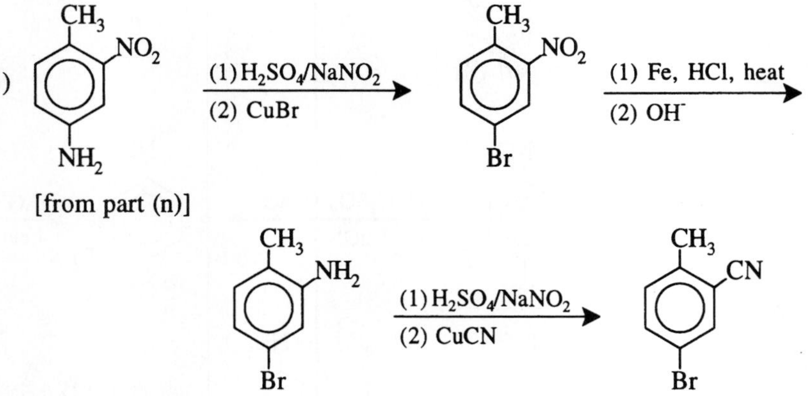 [from part (n)]

(q) H_3C—⬡—NH_2 $\xrightarrow{H_2SO_4/NaNO_2}$ H_3C—⬡—$N_2^+\ X^-$

[from part (c)]

⬡—OH, pH 8-10 ↓

H_3C—⬡—$N{=}N$—⬡—OH

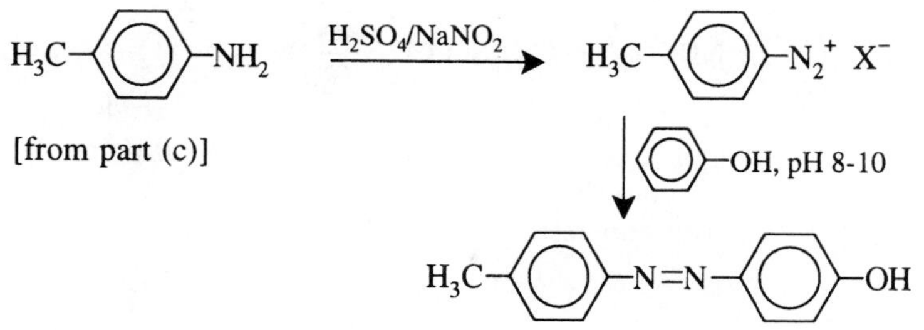

(r) H_3C—⬡—N_2^+ $\underset{X^-}{}$

[from part (q)]

H_3C—⬡—OH

$\xrightarrow[\text{pH 8-10}]{\text{[from part (c)]}}$

H_3C—⬡—$N{=}N$—⬡(HO)(CH_3)

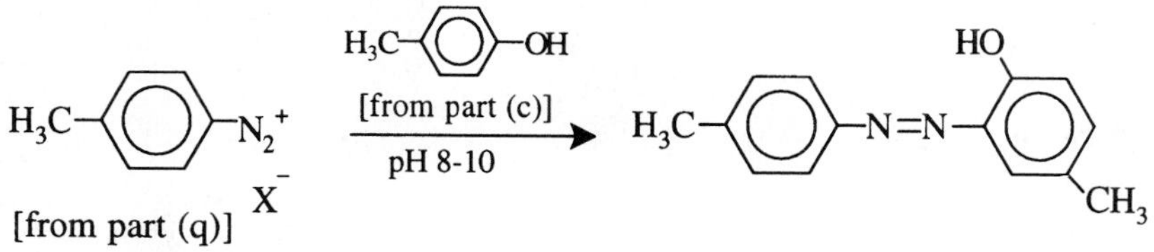

20.32 (a) Benzylamine dissolves in dilute HCl at room temperature,

$$C_6H_5CH_2NH_2 \ + \ H_3O^+ \ + \ Cl^- \ \xrightarrow{25°C} \ C_6H_5CH_2\overset{+}{N}H_3Cl^-$$

benzamide does not dissolve:

$$C_6H_5CONH_2 \ + \ H_3O^+ \ + \ Cl^- \ \xrightarrow{25°C} \ \text{No reaction}$$

(b) Allylamine reacts with (and decolorizes) bromine in carbon tetrachloride instantly,

$$CH_2{=}CHCH_2NH_2 \ + \ Br_2 \ \xrightarrow{CCl_4} \ \underset{\overset{|}{Br}\ \overset{|}{Br}}{CH_2CHCH_2NH_2}$$

propylamine does not:

$$CH_3CH_2CH_2NH_2 \ + \ Br_2 \ \xrightarrow{CCl_4} \ \begin{array}{l}\text{No reaction if the mixture}\\\text{is not heated or irradiated}\end{array}$$

(c) The Hinsberg test:

H_3C—⬡—NH_2 $+\ C_6H_5SO_2Cl$ $\xrightarrow[H_2O]{KOH}$ H_3C—⬡—$\overset{K^+}{\overset{|}{N}}SO_2C_6H_5$ $\xrightarrow{H_3O^+}$

Soluble

H_3C—⬡—$NHSO_2C_6H_5$

Precipitate

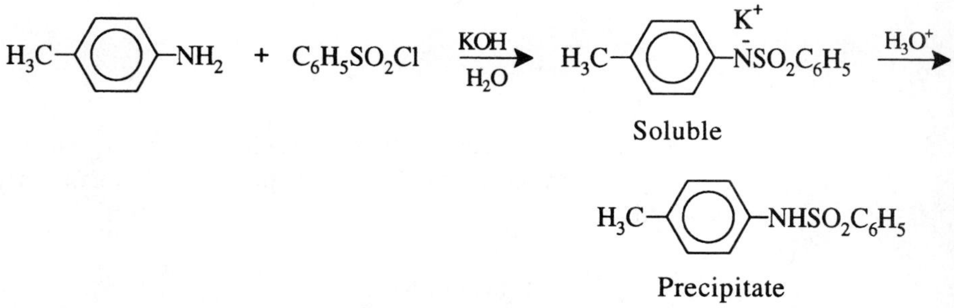

$$\text{Ar-NHCH}_3 + \text{C}_6\text{H}_5\text{SO}_2\text{Cl} \xrightarrow{\substack{\text{KOH} \\ \text{H}_2\text{O}}} \text{Ar-NSO}_2\text{C}_6\text{H}_5 \ (\text{CH}_3) \xrightarrow{\text{H}_3\text{O}^+} \begin{array}{l} \text{Precipitate} \\ \text{remains} \end{array}$$

Precipitate

(d)   The Hinsberg test:

$$\text{Ar-NH}_2 + \text{C}_6\text{H}_5\text{SO}_2\text{Cl} \xrightarrow{\substack{\text{KOH} \\ \text{H}_2\text{O}}} \text{Ar-}\overset{-}{\text{N}}\text{SO}_2\text{C}_6\text{H}_5 \ \text{K}^+ \xrightarrow{\text{H}_3\text{O}^+}$$

Soluble

$$\text{Ar-NHSO}_2\text{C}_6\text{H}_5$$

Precipitate

$$\text{R}_2\text{N-H} + \text{C}_6\text{H}_5\text{SO}_2\text{Cl} \xrightarrow{\substack{\text{KOH} \\ \text{H}_2\text{O}}} \text{R}_2\text{N-SO}_2\text{C}_6\text{H}_5 \xrightarrow{\text{H}_3\text{O}^+} \begin{array}{l} \text{Precipitate} \\ \text{remains} \end{array}$$

Precipitate

(e)   Pyridine dissolves in dilute HCl,

$$\text{C}_5\text{H}_5\text{N} + \text{H}_3\text{O}^+ + \text{Cl}^- \longrightarrow \text{C}_5\text{H}_5\overset{+}{\text{N}}\text{-H} \ \text{Cl}^-$$

benzene does not:

$$\text{C}_6\text{H}_6 + \text{H}_3\text{O}^+ + \text{Cl}^- \longrightarrow \text{No reaction}$$

(f)   Aniline reacts with nitrous acid at 0-5°C to give a stable diazonium salt that couples with 2-naphthol, yielding an intensely colored azo compound.

$$\text{Ar-NH}_2 \xrightarrow[\text{0-5°C}]{\text{H}_2\text{SO}_4/\text{NaNO}_2} \text{Ar-N}_2^+ \xrightarrow{\text{2-naphthol}} \text{Ar-N=N-C}_{10}\text{H}_6\text{OH}$$

Cyclohexylamine reacts with nitrous acid at 0-5°C to yield a highly unstable diazonium salt—one that decomposes so rapidly that the addition of 2-naphthol gives no azo compound.

$$\text{C}_6\text{H}_{11}\text{-NH}_2 \xrightarrow[\text{0-5°C}]{\text{H}_2\text{SO}_4/\text{NaNO}_2} [\text{C}_6\text{H}_{11}\text{-N}_2^+] \xrightarrow{-\text{N}_2} [\text{C}_6\text{H}_{11}^+] \longrightarrow$$

$$\text{alkenes, alcohols, and so on} \xrightarrow{\text{2-naphthol}} \text{No reaction}$$

(g)   The Hinsberg test:

$(C_2H_5)_3N$ + $C_6H_5SO_2Cl$ $\xrightarrow[\text{H}_2\text{O}]{\text{KOH}}$ No reaction $\xrightarrow{\text{H}_3\text{O}^+}$ $(C_2H_5)_3\overset{+}{N}H$
                                                                                                            Soluble

$(C_2H_5)_2NH$ + $C_6H_5SO_2Cl$ $\xrightarrow[\text{H}_2\text{O}]{\text{KOH}}$ $(C_2H_5)_2NSO_2C_6H_5$ $\xrightarrow{\text{H}_3\text{O}^+}$ Precipitate
                                                                                       Precipitate   remains

(h)   Tripropylaminium chloride reacts with aqueous NaOH to give a water insoluble tertiary amine.

$(CH_3CH_2CH_2)_3NH^+$ Cl^- $\xrightarrow[\text{H}_2\text{O}]{\text{NaOH}}$ $(CH_3CH_2CH_2)_3N$
Water soluble   Water insoluble

Tetrapropylammonium chloride does not react with aqueous NaOH (at room temperature), and the tetrapropylammonium ion remains in solution.

$(CH_3CH_2CH_2)_4N^+$ Cl^- $\xrightarrow[\text{H}_2\text{O}]{\text{NaOH}}$ $(CH_3CH_2CH_2)_4N^+$ $[Cl^-$ or $OH^-]$
Water soluble   Water soluble

(i)   Tetrapropylammonium chloride dissolves in water to give a neutral solution. Tetrapropylammonium hydroxide dissolves in water to give a strongly basic solution.

20.33   Follow the procedure outlined in the answer to Problem 20.2. Toluene will show the same solubility behavior as benzene.

20.34

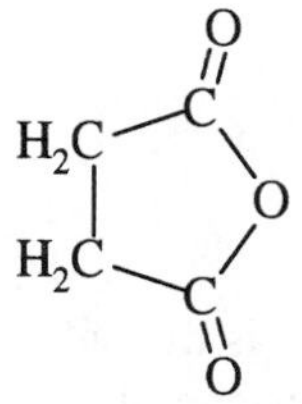

$\xrightarrow[(2)\ \text{H}^+]{}$ $H_2N\overset{O}{\overset{\|}{C}}CH_2CH_2\overset{O}{\overset{\|}{C}}OH$

$\xrightarrow[-\text{CO}_3^{2-}]{\text{Br}_2,\ \text{OH}^-}$ $H_2NCH_2CH_2CO_2^-$ $\xrightarrow{\text{H}^+}$ $H_3\overset{+}{N}CH_2CH_2CO_2^-$

20.35   (a) $HOCH_2(CH_2)_8CH_2OH$ $\xrightarrow{2\ \text{PBr}_3}$ $BrCH_2(CH_2)_8CH_2Br$ $\xrightarrow{2(\text{CH}_3)_3\text{N}}$

$(CH_3)_3\overset{+}{N}CH_2(CH_2)_8CH_2\overset{+}{N}(CH_3)_3$ 2 Br^-

(b) $HO_2CCH_2CH_2CO_2H$ + $2BrCH_2CH_2OH$ $\xrightarrow{\text{H}^+}$

$BrCH_2CH_2O_2CCH_2CH_2CO_2CH_2CH_2Br$ $\xrightarrow{2(\text{CH}_3)_3\text{N}}$

$(CH_3)_3\overset{+}{N}CH_2CH_2O_2CCH_2CH_2CO_2CH_2CH_2\overset{+}{N}(CH_3)_3$ 2 Br^-

20.36

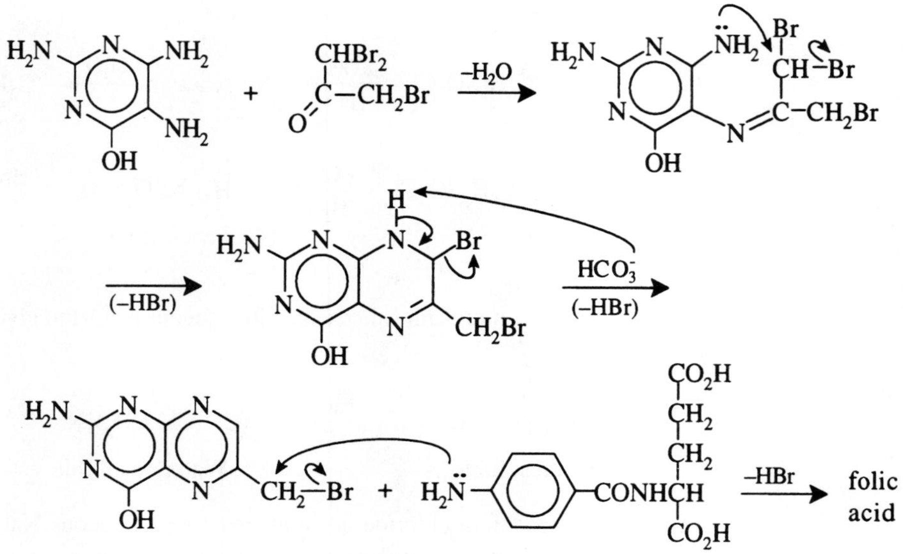

20.37 The results of the Hinsberg test indicate that compound **W** is a tertiary amine. The ^{1}H NMR provides evidence for the following:

(1) Two different C_6H_5–groups (one absorbing at δ 7.2 and one at δ 6.7).
(2) A CH_3CH_2–group (the quartet at δ 3.5 and the triplet at δ 1.2).
(3) An unsplit $–CH_2–$group (the singlet at δ 4.5).

There is only one reasonable way to pull all of this together.

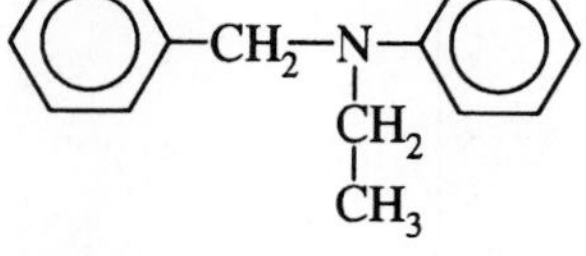

Thus **W** is *N*-benzyl-*N*-ethylaniline.

20.38 Compound **X** is benzyl bromide, $C_6H_5CH_2Br$. This is the only structure consistent with the ^{1}H NMR and IR data. (The monosubstituted benzene ring is strongly indicated by the (5H), δ 7.3 ^{1}H NMR absorption and is confirmed by the peaks at 690 and 770 cm^{-1} in the IR spectrum.)

Compound **Y**, therefore, must be phenylacetonitrile, ($C_6H_5CH_2CN$) and **Z** must be 2-phenylethylamine, $C_6H_5CH_2CH_2NH_2$.

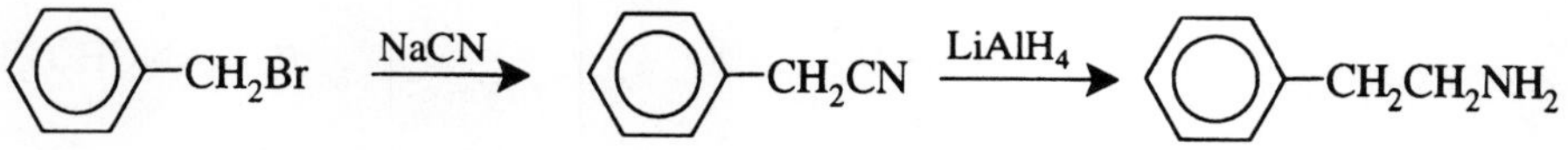

X C_7H_7Br | **Y** C_8H_7N | **Z** $C_8H_{11}N$

Interpretations of the IR and ^{1}H NMR spectra of **Z** are as follows.

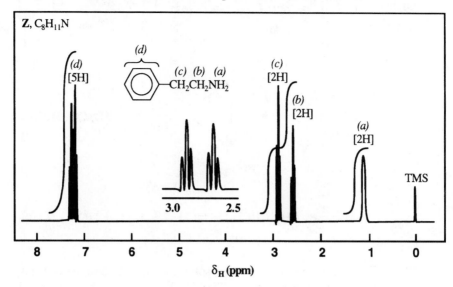

(a) singlet δ 1.0
(b) triplet δ 2.7
(c) triplet δ 2.9
(d) multiplet δ 7.25

20.39

20.40

20.41 That **A** contains nitrogen and is soluble in dilute HCl suggests that **A** is an amine. The two IR absorption bands in the 3300–3500-cm^{-1} region suggest that **A** is a primary amine. The ^{13}C spectrum shows only two signals in the upfield aliphatic region. There are four signals downfield in the aromatic region. The information from the DEPT spectra suggests an ethyl group *or two equivalent ethyl groups*. Assuming the latter, and assuming that **A** is a primary amine, we can conclude from the molecular formula that **A** is 2,6-diethylaniline. The assignments are

(a) δ 12.9
(b) δ 24.2
(c) δ 118.1
(d) δ 125.9
(e) δ 127.4
(f) δ 141.5

(An equally plausible answer would be that **A** is 3,5-diethylaniline.)

20.42 That **B** dissolves in dilute HCl suggests that **B** is an amine. That the IR spectrum of **B** lacks bands in the 3300–3500-cm^{-1} region suggests that **B** is a tertiary amine. The upfield signals in the ^{13}C spectrum and the DEPT information suggest two equivalent ethyl groups (as was also true of **A** in the preceding problem). The DEPT information for the downfield peaks (in the aromatic region) is consistent with a monosubstituted benzene ring. Putting all of these observations together with the molecular formula leads us to conclude that **B** is *N,N*-diethylaniline. The assignments are

(a) δ 12.5
(b) δ 44.2
(c) δ 112.0
(d) δ 115.5
(e) δ 128.1
(f) δ 147.8

20.43 That **C** gives a positive Tollens' test indicates the presence of an aldehyde group; the solubility of **C** in aqueous HCl suggests that **C** is also an amine. The absence of bands in the 3300–3500-cm^{-1} region of the IR spectrum of **C** suggests that **C** is a tertiary amine. The signal at δ 189.7 in the ^{13}C spectrum can be assigned to the aldehyde group. The signal at δ 39.7 is the only one in the aliphatic region and is consistent with a methyl group or with two equivalent methyl groups. The remaining signals are in the aromatic region. If we assume that **C** has a benzene ring containing a –CH group and a –N(CH$_3$)$_2$ group, then the aromatic signals and their DEPT spectra are consistent with **C** being *p*-(*N*,*N*-dimethylamino)benzaldehyde. The assignments are

(a) δ 39.7
(b) δ 110.8
(c) δ 124.9
(d) δ 131.6
(e) δ 154.1
(f) δ 189.7

20.44 (CH$_3$)$_2$NH + [epoxide] $\longrightarrow$ H$_3$C–N(CH$_3$)–CH$_2$–CH$_2$–OH $\xrightarrow[\text{base (–HCl)}]{\text{CH}_3\text{COCl}}$

H$_3$C–N(CH$_3$)–CH$_2$–CH$_2$–O–C(=O)–CH$_3$ $\xrightarrow{\text{CH}_3\text{I}}$ H$_3$C–N$^+$(CH$_3$)$_2$–CH$_2$–CH$_2$–O–C(=O)–CH$_3$ I$^-$

Acetylcholine iodide

20.45 [epoxide] $\xrightarrow{\text{NH}_3}$ HOCH$_2$CH$_2$NH$_2$ $\xrightarrow{\text{[epoxide]}}$ (HOCH$_2$CH$_2$)$_2$NH

20.46 [benzene] $\xrightarrow[\text{AlCl}_3]{\text{CH}_3\text{CH}_2\text{COCl}}$ [C$_6$H$_5$–CCH$_2$CH$_3$(=O)] $\xrightarrow[\text{Br}_2]{\text{CH}_3\text{CO}_2\text{H}}$ [C$_6$H$_5$–CCHCH$_3$(=O) with Br]

$\xrightarrow{\text{(CH}_3\text{CH}_2)_2\text{NH}}$ [C$_6$H$_5$–C(=O)–CH(CH$_3$)–N(C$_2$H$_5$)$_2$]

Diethylpropion

20.47 Carry out the Hofmann reaction using a mixture of ^{15}N labeled benzamide, $C_6H_5CO*NH_2$, and p-chlorobenzamide. If the process is intramolecular, only labeled aniline, $C_6H_5*NH_2$, and p-chloroaniline, will be produced.

If the process is one in which the migrating moiety truly separates from the remainder of the molecule, then, in addition to the two products mentioned above, there should be produced both the unlabeled aniline and labeled p-chloroaniline, p-$ClC_6H_4*NH_2$.

Note: When this experiment is actually carried out, analysis of the reaction mixture by mass spectrometry shows that the process is intramolecular.

20.48

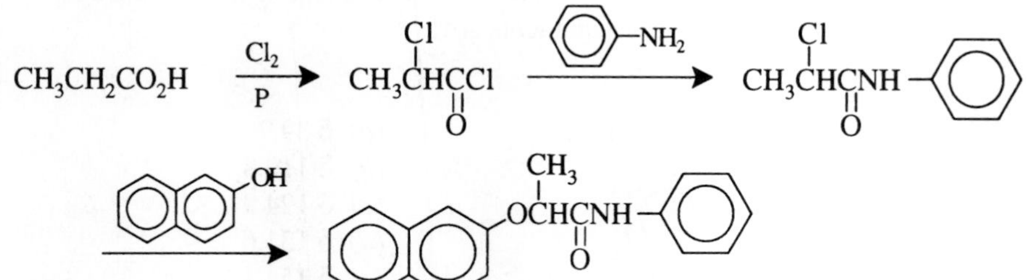

QUIZ

20.1 Which of the following would be soluble in dilute aqueous HCl?

(a) $C_6H_5NH_2$

(b) $C_6H_5CH_2NH_2$

(c) $C_6H_5\overset{\overset{O}{\|}}{C}NH_2$

(d) Two of the above

(e) All of the above

20.2 Which would yield propylamine?

(a) CH_3CH_2Br $\xrightarrow[\text{(2) LiAlH}_4]{\text{(1) NaCN}}$

(b) $CH_3CH_2\overset{\overset{O}{\|}}{C}H$ $\xrightarrow[\text{H}_2/\text{Ni}]{\text{NH}_3}$

(c) $CH_3CH_2CH_2\overset{\overset{O}{\|}}{C}NH_2$ $\xrightarrow[\text{OH}^-]{\text{Br}_2}$

(d) Two of the above

(e) All of the above

20.3 Select the reagent from the list below that could be the basis of a simple chemical test that would distinguish between each of the following:

(a) $C_6H_5\overset{\overset{\displaystyle O}{\|}}{C}NH_2$ and $C_6H_5CH_2NH_2$

(b) $C_6H_5CH_2NH_2$ and $C_6H_5NHCH_3$

(c) —NH_2 and   —NH_2

1. Cold dilute $NaHCO_3$
2. Cold dilute HCl
3. $NaNO_2$, HCl, 5°C, then 2-naphthol
4. $C_6H_5SO_2Cl$, OH⁻, then HCl
5. Cold dilute NaOH

20.4 Complete the following syntheses:

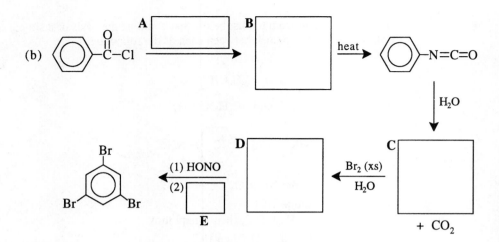

20.5   Select the stronger base from each pair:

(a)   H₃C—⟨O⟩—NH₂   or   ⟨O⟩—CH₂NH₂

      (1)   **(2)**

(b)   ⟨O⟩—C(=O)NH₂   or   ⟨O⟩—CH₂NH₂

      (1)   **(2)**

(c)   H₃C—⟨O⟩—NH₂   or   O₂N—⟨O⟩—NH₂

      (1)   **(2)**

J

SPECIAL TOPIC
Reactions and Syntheses of Heterocyclic Amines

SOLUTIONS TO PROBLEMS

J.1 (a)

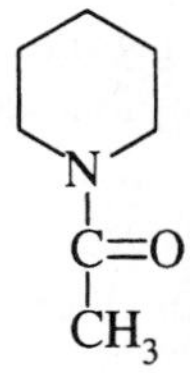

(b)

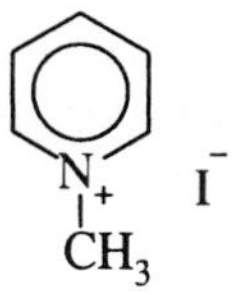

(c)

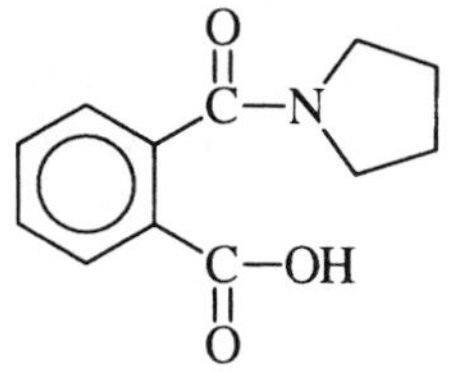

(d)

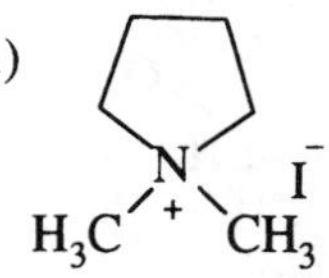

(e) $CH_2=CHCH_2CH_2\underset{\underset{CH_3}{|}}{N}-CH_3$

J.2 (a) The cyclopentadienyl anion.

(b) The pyrrole anion is a resonance hybrid of the following structures:

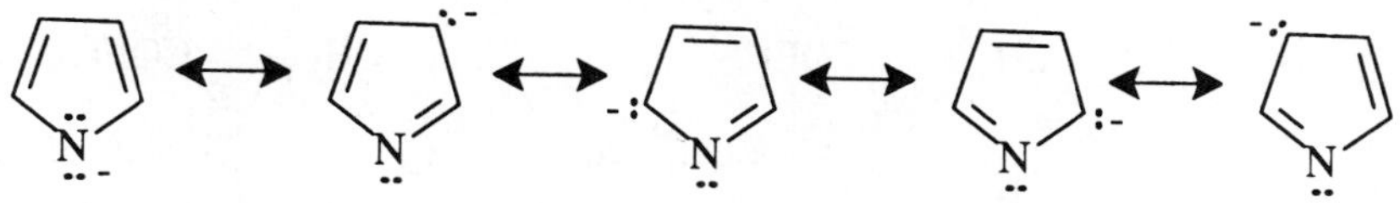

The imidazole anion is a hybrid of these:

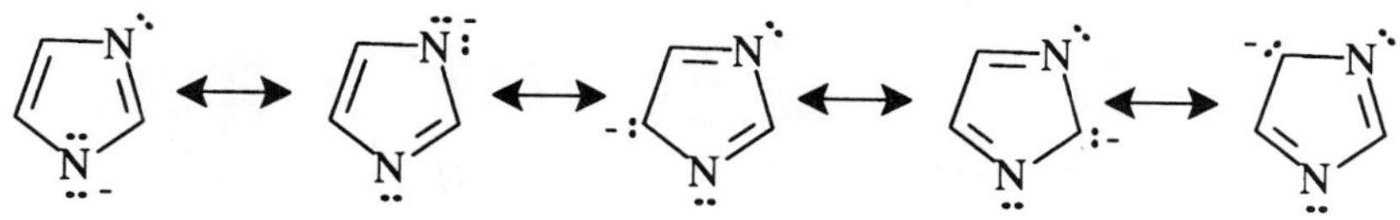

J.3 A mechanism involving a "pyridyne" intermediate would involve a net loss (of 50%) of the deuterium label.

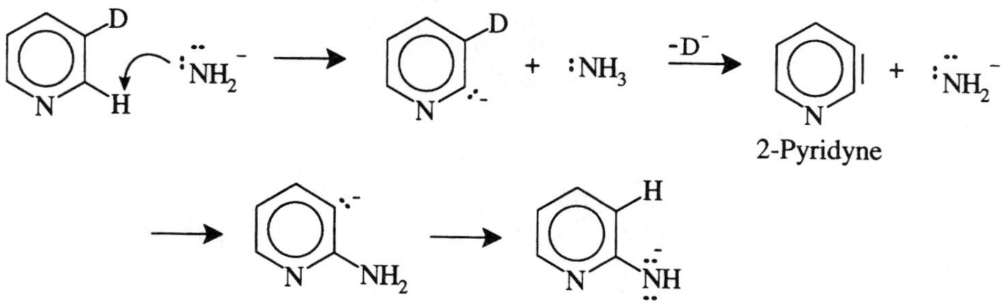

2-Pyridyne

Since in the actual experiment there was no loss of deuterium, this mechanism was disallowed.

The mechanism given in Section J.4 would not be expected to result in a loss of deuterium; thus, it is consistent with the labeling experiment.

J.4 When pyridine undergoes nucleophilic substitution, the leaving group is a hydride ion—an ion that is a strong base and, consequently, a poor leaving group. With 2-halopyridines, on the other hand, the leaving groups are halide ions—ions that are weak bases and thus good leaving groups.

J.5 If we write the reactants in the following way, we can better see how the reaction occurs.

$$CH_3CCH_2NH_3{}^+ \ Cl^- \ + \ OH^- \longrightarrow CH_3CCH_2NH_2$$

(reaction schemes with structures)

$$\xrightarrow{-H_2O}$$

$$\xrightarrow[-H^+]{OH^-}$$

$$\xrightarrow[-H^+]{+H^+}$$

$$\xrightarrow{-H_2O}$$

$$\xrightarrow{\text{tauterism}}$$ tautomerism

J.6 (a) structure $+ \ (NH_4)_2CO_3 \ \xrightarrow{100°C}$ A $+ \ 2\,H_2O \ + \ NH_4HCO_3$

(b) structure $\xrightarrow{\text{base}}$ B $+ \ 2\,H_2O$

(c) structure $+$ structure $\xrightarrow[H_2O]{H^+}$ C $+ \ 4\,CH_3OH$

(d)

(e)

(f)

SPECIAL TOPIC
Alkaloids

SOLUTIONS TO PROBLEMS

K.1 (a) The first step is similar to a crossed Claisen condensation (see Section 19.2A):

$$\text{(pyridine)}-\overset{\overset{\displaystyle O}{\|}}{C}OCH_2CH_3 \;+\; \text{(lactam)} \xrightarrow{\;C_2H_5ONa\;} \text{(product)}$$

(b) This step involves hydrolysis of an amide (lactam) and can be carried out with either acid or base. Here we use acid.

$$\xrightarrow[\;H_2O\;]{\;H_3O^+\;} \text{(pyridine)}-\overset{\overset{\displaystyle O}{\|}}{C}-\underset{\underset{\displaystyle CO_2H}{|}}{CH}CH_2CH_2NHCH_3 \xrightarrow{\;heat\;} \text{(pyridine)}-\overset{\overset{\displaystyle O}{\|}}{C}-CH_2CH_2CH_2NHCH_3$$

(c) This step is the decarboxylation of a substituted malonic acid; it requires only the application of heat and takes place during the acid hydrolysis of step (b).

(d) This is the reduction of a ketone to a secondary alcohol. A variety of reducing agents can be used, sodium borohydride, for example.

$$\xrightarrow{\;NaBH_4\;} \text{(pyridine)}-\underset{\underset{\displaystyle OH}{|}}{CH}CH_2CH_2CH_2NHCH_3$$

(e) Here we convert the secondary alcohol to an alkyl bromide with hydrogen bromide; this reagent also gives a hydrobromide salt of the aliphatic amine.

$$\xrightarrow[\;heat\;]{\;HBr\;} \text{(pyridine)}-\underset{\underset{\displaystyle Br}{|}}{CH}CH_2CH_2CH_2\overset{\overset{\displaystyle H}{|}}{\underset{\underset{\displaystyle H}{|}}{N}}{}^+\!-CH_3 \; Br^-$$

(f) Treating the salt with base produces the secondary amine; it then acts as a nucleophile and attacks the carbon atom bearing the bromine. This reaction leads to the formation of a five-membered ring and (±) nicotine.

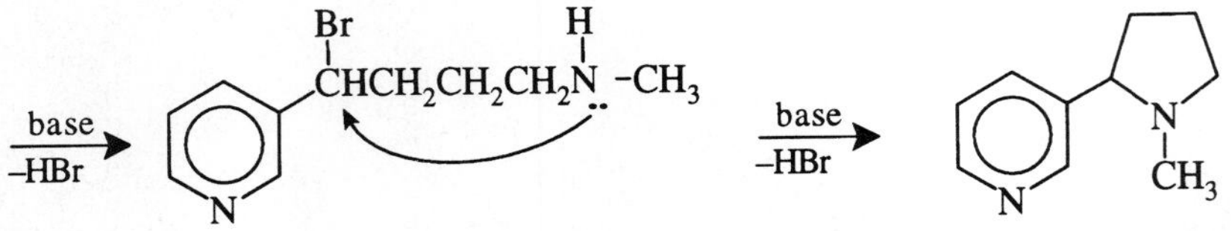

K.2 (a) The stereocenter adjacent to the ester carbonyl group is racemized by base (probably through the formation of an anion that can undergo inversion of configuration; cf. Section 17.3).

(b)

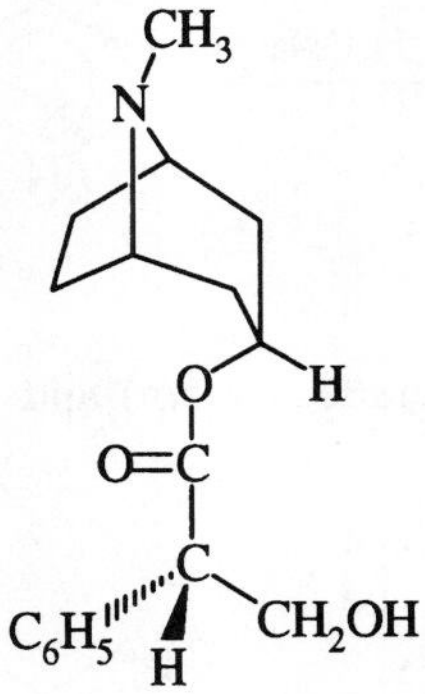

K.3 (a)

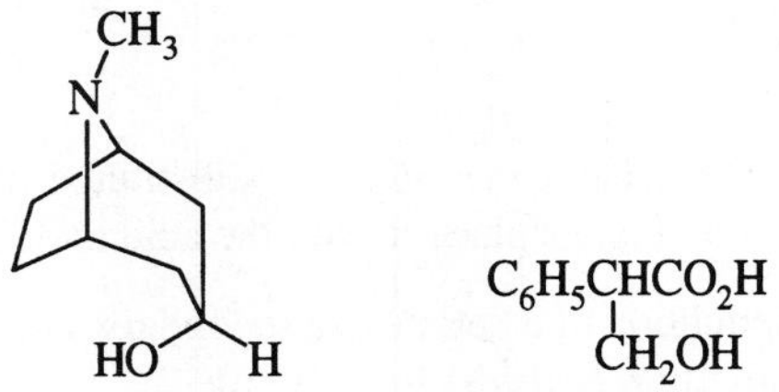

Tropine   (±) Tropic acid

(b) Tropine is a meso compound; it has a plane of symmetry that passes through the $>$CHOH group, the $>$NCH$_3$ group, and between the two $-$CH$_2$ $-$ groups of the five-membered ring.

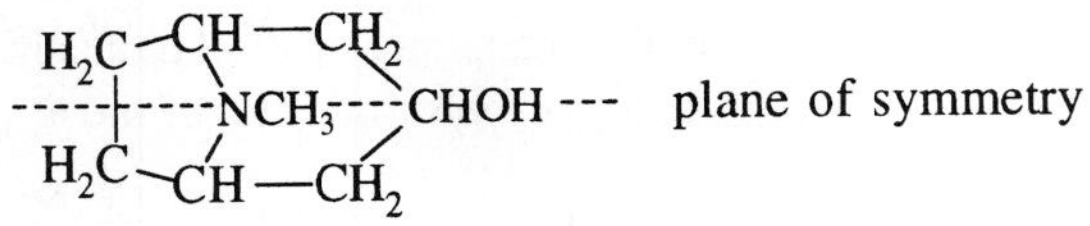

(c)

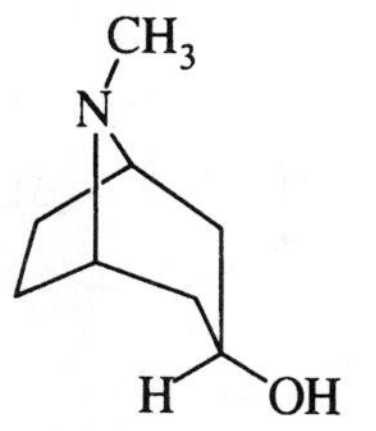

ψ-Tropine

K.4

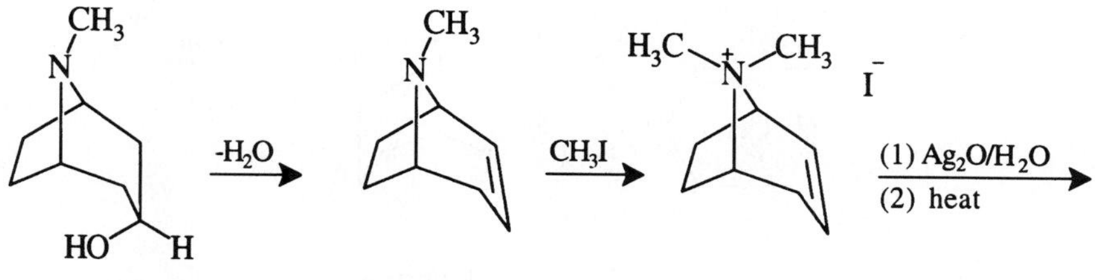

Tropine $C_8H_{13}N$ $C_9H_{16}NI$

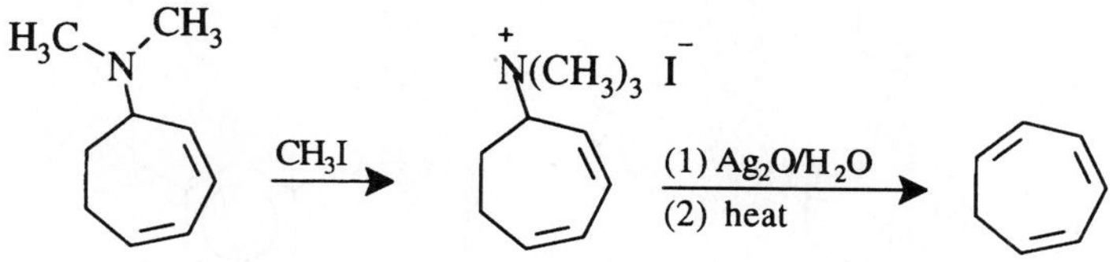

$C_9H_{15}N$ $C_{10}H_{18}NI$

K.5 One possible sequence of steps is the following:

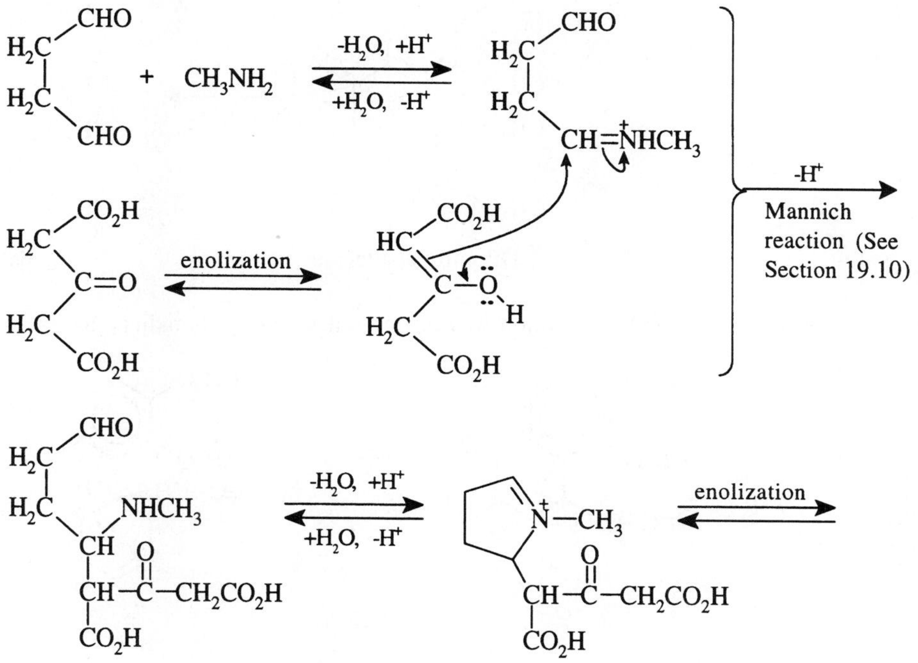

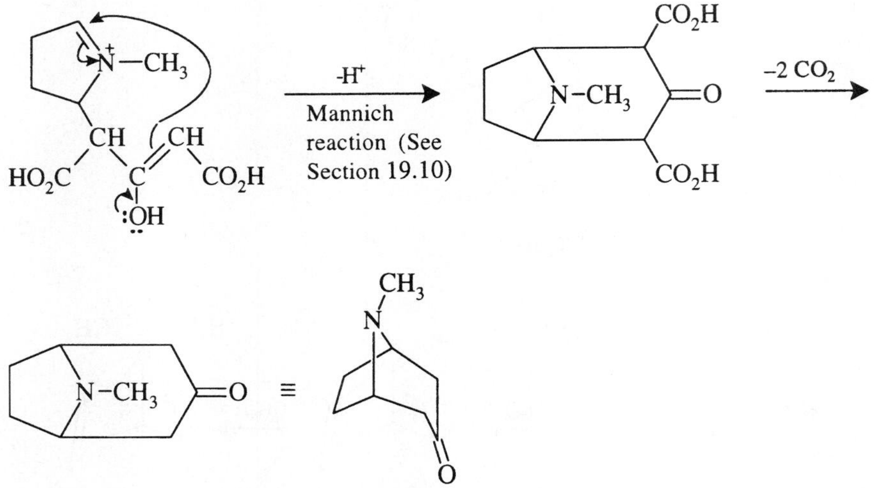

Tropinone

K.6

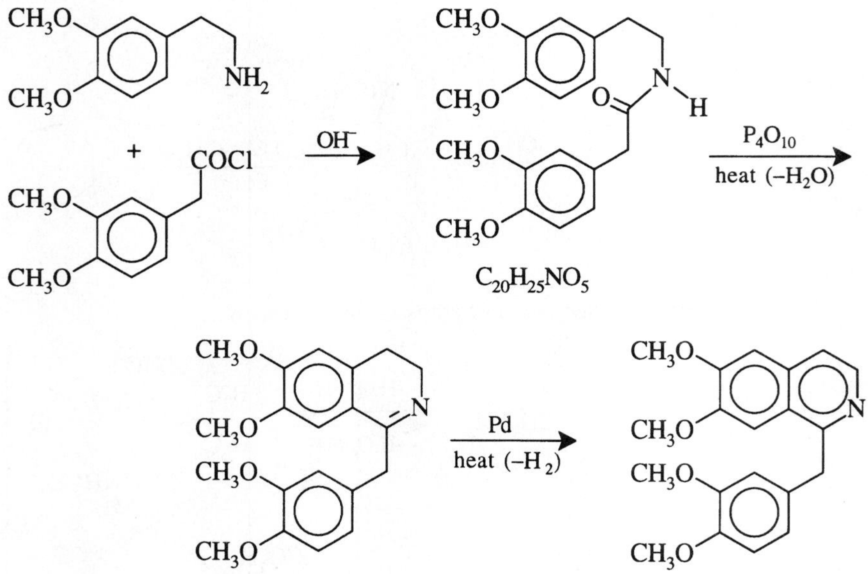

$C_{20}H_{25}NO_5$

Dihydropapaverine

Papaverine

K.7 A Diels-Alder reaction was carried out using 1,3-butadiene as the diene component.

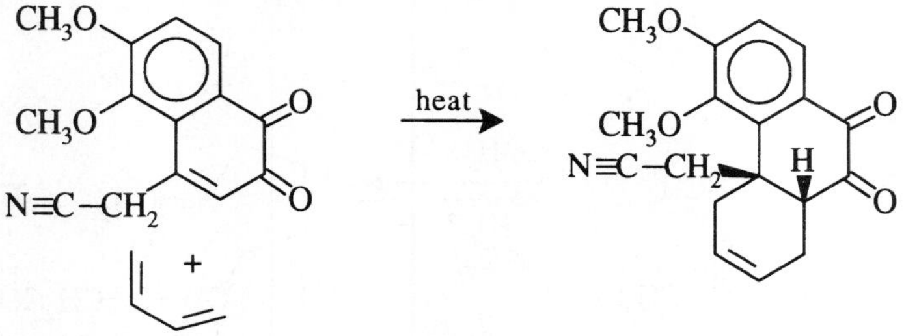

K.8   Acetic anhydride acetylates both –OH groups.

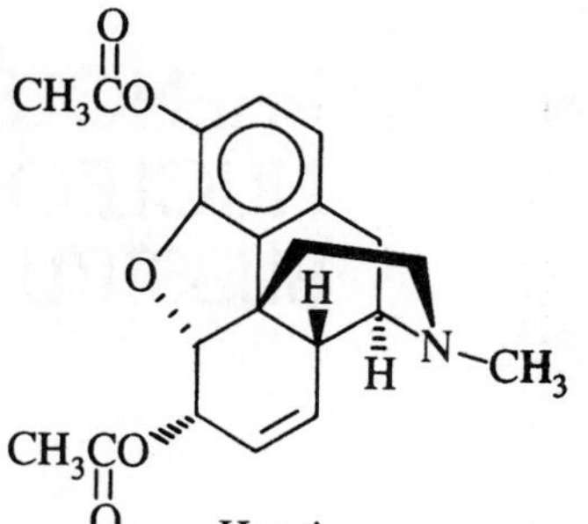

$$CH_3CO$$

Heroin

K.9   (a)   A Mannich reaction (see Section 19.10).

(b) CH_2O + $HN(CH_3)_2$ $\underset{\xrightarrow{\hspace{1cm}}}{\overset{-H_2O,\ +H^+}{\rightleftharpoons}}$ $CH_2\overset{+}{=}N(CH_3)_2$

$\xrightarrow{-H^+}$

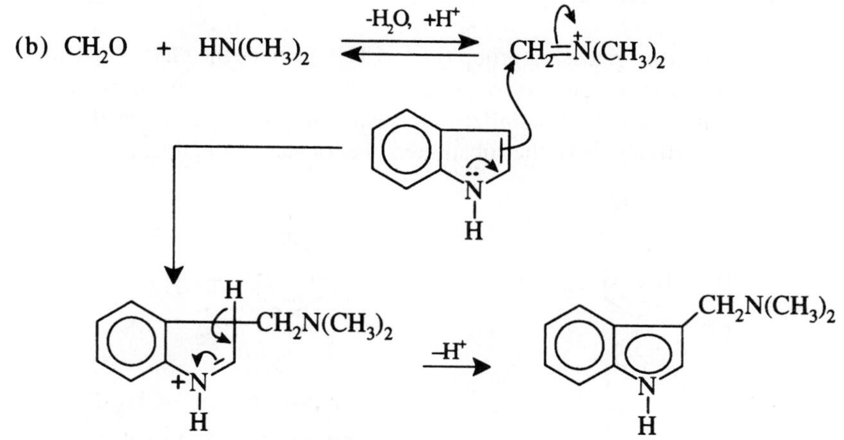

Gramine

21

PHENOLS AND ARYL HALIDES: NUCLEOPHILIC AROMATIC SUBSTITUTION

SOLUTIONS TO PROBLEMS

21.1 An electron-releasing group (i.e., –CH$_3$) changes the charge distribution in the molecule so as to make the hydroxyl oxygen less positive, causing the proton to be held more strongly; it also destabilizes the phenoxide anion by intensifying its negative charge. These effects make the substituted phenol less acidic than phenol itself.

Electron-releasing –CH$_3$ destabilizes the anion more than the acid—pK$_a$ is larger than for phenol.

21.2 An electron-withdrawing group such as chlorine changes the charge distribution in the molecule so as to make the hydroxyl oxygen more positive, causing the proton to be held less strongly; it also can stabilize the phenoxide ion by dispersing its negative charge. These effects make the substituted phenol more acidic than phenol itself.

Electron-withdrawing chlorine stabilizes the anion by dispersing the negative charge. pK$_a$ is smaller than for phenol.

Nitro groups are very powerful electron-withdrawing groups by their inductive and resonance effects. Resonance structures (**B–D**) below place a positive charge on the hydroxyl oxygen. This effect makes the hydroxyl oxygen dramatically more positive, causing the proton to be held much less strongly. These contributions explain why 2,4,6-trinitrophenol (picric acid) is so exceptionally acidic.

A   **B**

C   **D**

21.3 Dissolve the mixture in a solvent such as CH_2Cl_2 (one that is immiscible with water). Using a separatory funnel, extract this solution with an aqueous solution of sodium bicarbonate. This extraction will remove the benzoic acid from the CH_2Cl_2 solution and transfer it (as a sodium benzoate) to the aqueous bicarbonate solution. Acidification of this aqueous extract will cause benzoic acid to precipitate; it can then be separated by filtration and purified by recrystallization.

The CH_2Cl_2 solution can now be extracted with an aqueous solution of sodium hydroxide. This will remove the 4-methylphenol (as its sodium salt). Acidification of the aqueous extract will cause the formation of 4-methylphenol as a water-insoluble layer. The 4-methylphenol can then be extracted into ether, the ether removed, and the 4-methylphenol purified by distillation.

The CH_2Cl_2 solution will now contain only toluene (and CH_2Cl_2). These can be separated easily by fractional distillation.

21.4 (a) The para-sulfonated phenol, because it is the major product at the higher temperature—when the reaction is under equilibrium control.

(b) For ortho sulfonation, because it is the major reaction pathway at the lower temperature—when the reaction is under rate control.

21.5   If the mechanism involved dissociation into an allyl cation and a phenoxide ion, then re-combination would lead to two products: one in which the labeled carbon atom is bonded to the ring and one in which an unlabeled carbon atom is bonded to the ring.

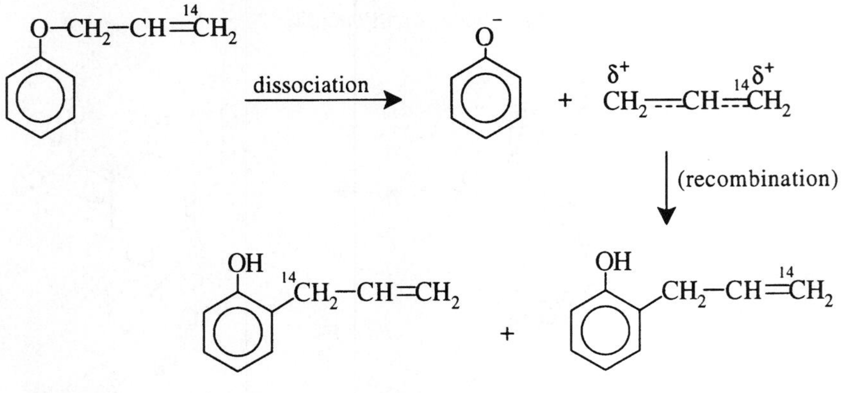

The fact that all of the product has the labeled carbon atom bonded to the ring eliminates this mechanism from consideration.

21.6

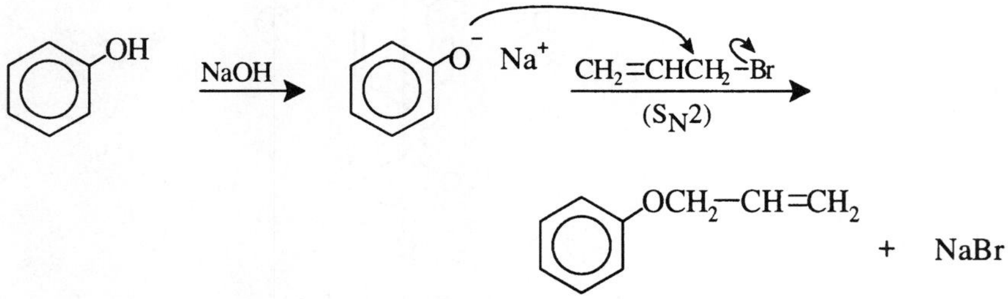

21.7 (a)

(b)

(c)

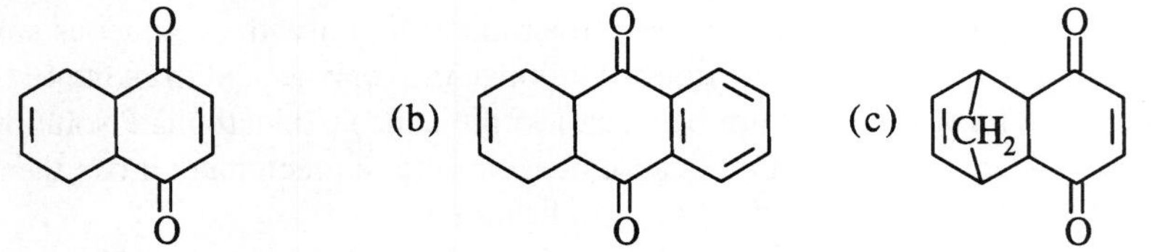

21.8

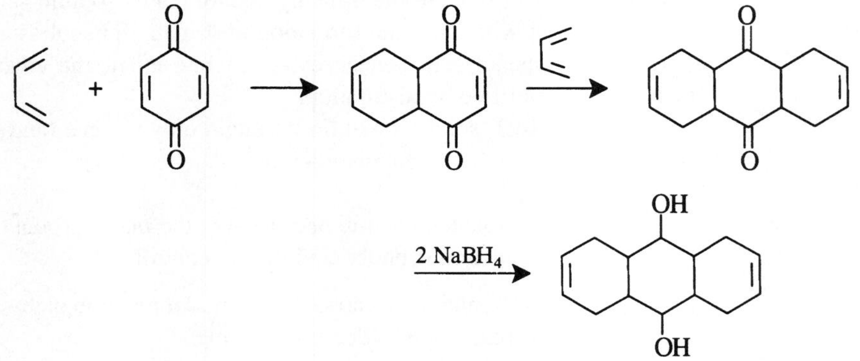

21.9 (a) 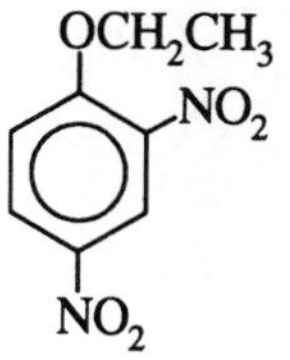 (b) 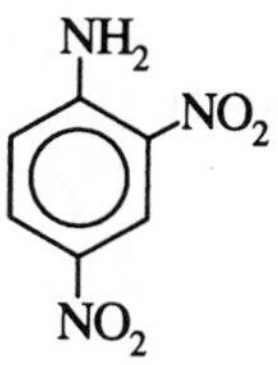 (c) 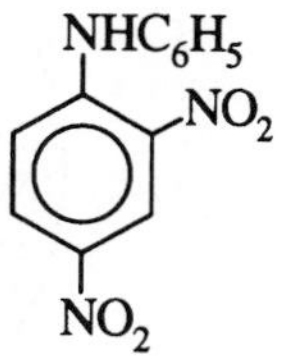 (d)

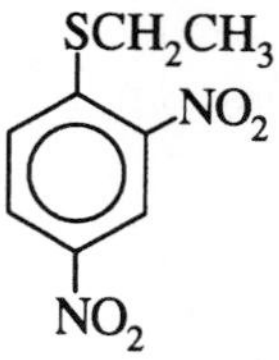

21.10 That *o*-chlorotoluene leads to the formation of two products (*o*-cresol and *m*-cresol) when submitted to the conditions used in the Dow process suggests that an elimination-addition mechanism takes place.

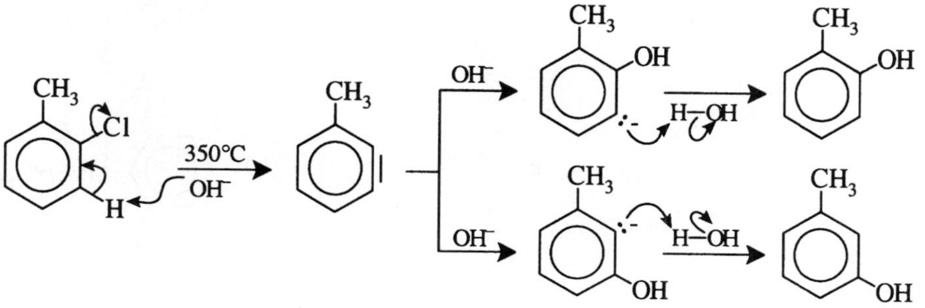

Inasmuch as chlorobenzene and *o*-chlorotoluene should have similar reactivities under these conditions, it is reasonable to assume that chlorobenzene reacts by an elimination-addition mechanism in the Dow process.

21.11 2-Bromo-1,3-dimethylbenzene, because it has no α-hydrogen, cannot undergo an elimination. Its lack of reactivity toward sodium amide in liquid ammonia suggests that those compounds (e.g., bromobenzene) that do react do so by a mechanism that begins with an elimination.

21.12 (a) 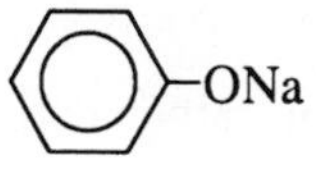 —ONa + CH_3CH_2OH (b) 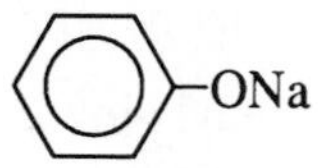 —ONa + H_2O

(c) 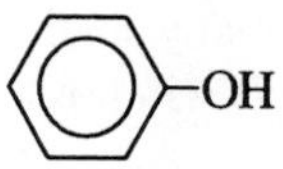 —OH + NaCl   (d)

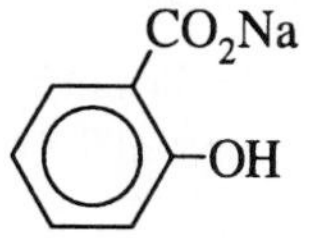

21.13 (a)

(major)

(b)

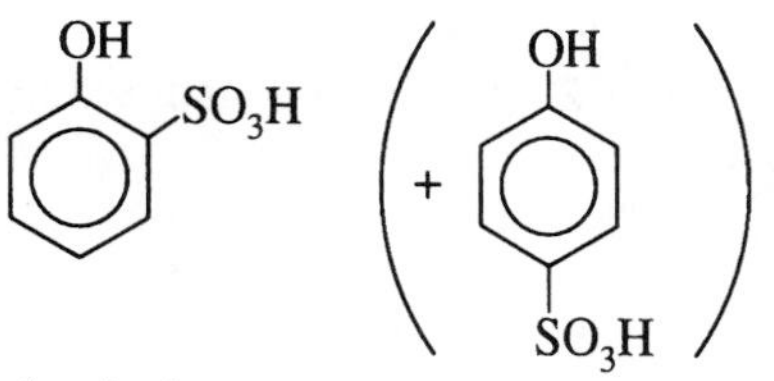

(major)

(c)

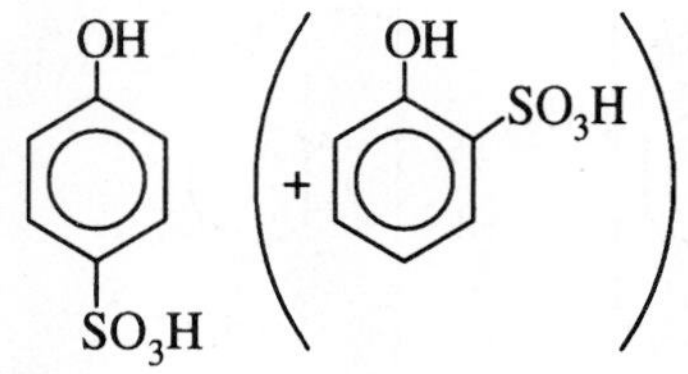

(major)

(d) 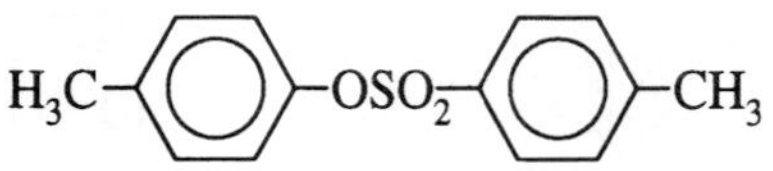 H_3C—⟨⟩—OSO_2—⟨⟩—CH_3

(e)

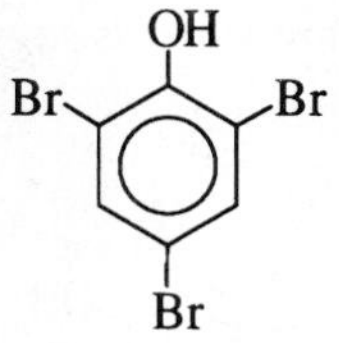

(f)

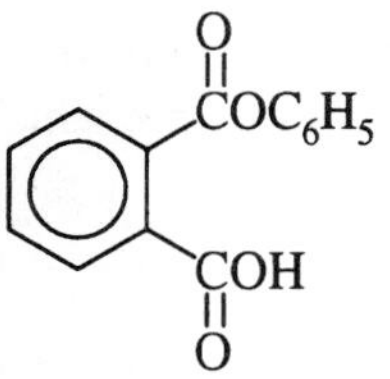

(g)

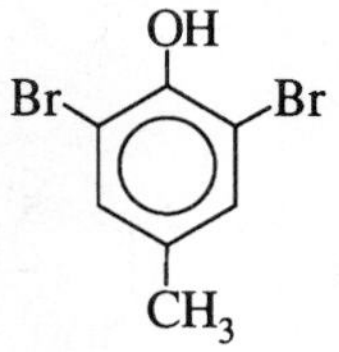

(h) 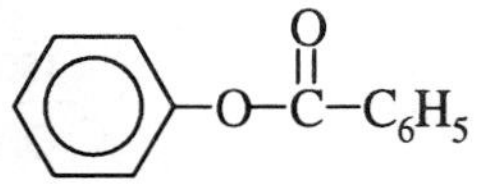   ⟨⟩—O—$\overset{\overset{\displaystyle O}{\|}}{C}$—$C_6H_5$

(i)   Same as (h)

(j) 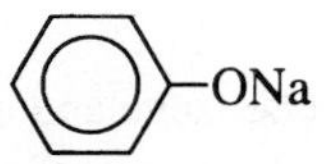   ⟨⟩—ONa

(k) 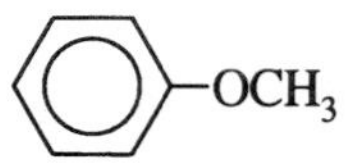   ⟨⟩—OCH_3

(l)   Same as (k)

(m) 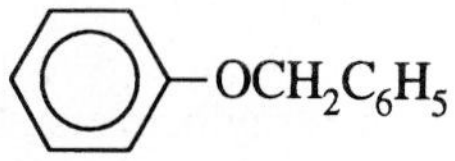   ⟨⟩—$OCH_2C_6H_5$

21.14   (a)   4-Chlorophenol will dissolve in aqueous NaOH; 4-chloro-1-methylbenzene will not.

(b)   4-Methylbenzoic acid will dissolve in aqueous $NaHCO_3$; 4-methylphenol will not.

(c)   Phenyl vinyl ether will react with bromine in carbon tetrachloride by addition (thus decolorizing the solution); ethyl phenyl ether will not.

(d)   2,4,6-Trinitrophenol, because it is so exceptionally acidic ($pK_a = 0.38$), will dissolve in aqueous $NaHCO_3$; 4-methylphenol ($pK_a = 10.17$) will not.

(e)   4-Ethylphenol will dissolve in aqueous NaOH; ethyl phenyl ether will not.

21.15   Both o- and m-toluidine are formed in the following way:

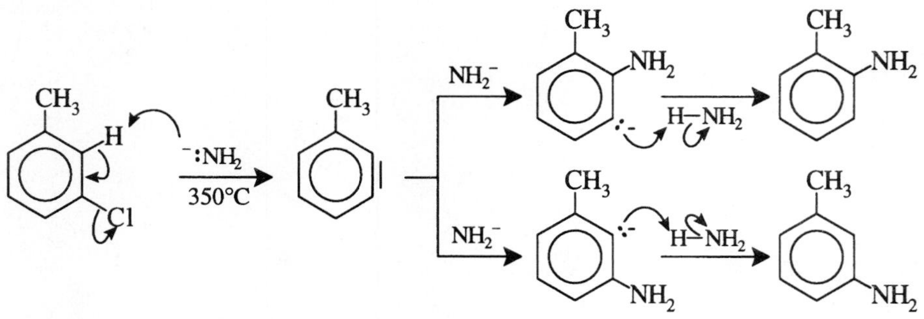

Both *m*- and *p*-toluidine are formed from another benzyne-type intermediate.

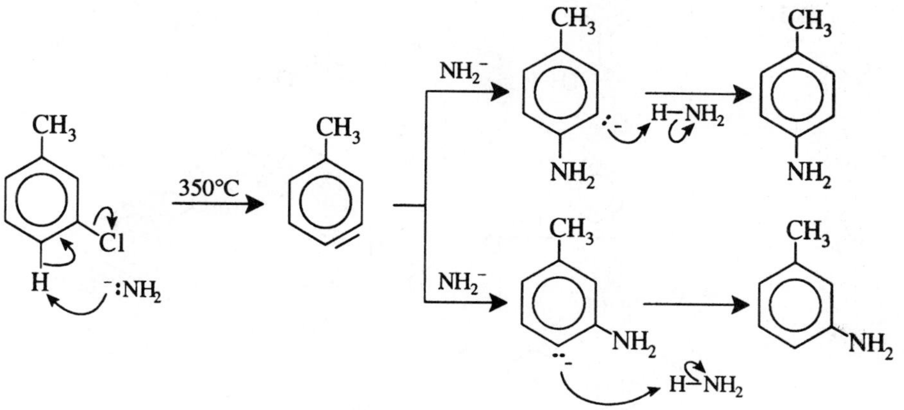

21.16 (a)   4-Fluorophenol because a fluorine substituent is more electron withdrawing than a methyl group.

(b)   4-Nitrophenol because a nitro group is more electron withdrawing than a methyl group.

(c)   4-Nitrophenol because the nitro group can exert an electron-withdrawing effect through a resonance effect.

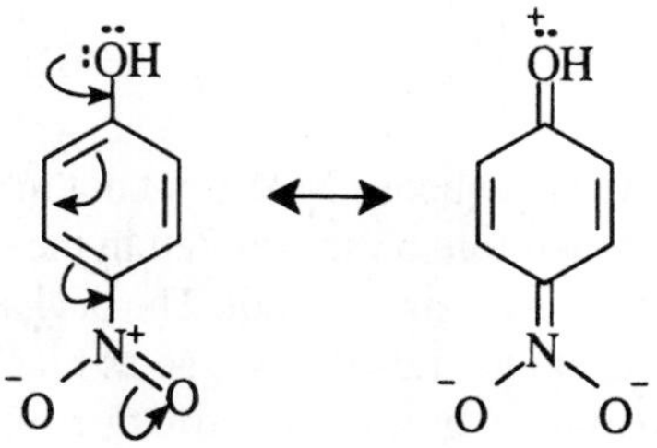

Whereas in 3-nitrophenol this resonance effect is not possible.

(d)   4-Methylphenol because it is a phenol, not an alcohol.

(e)   4-Fluorophenol because fluorine is more electronegative than bromine.

21.17 (a)

(b)   Because the phenolic radical is highly stabilized by resonance (see the following structures), it is relatively unreactive.

21.18

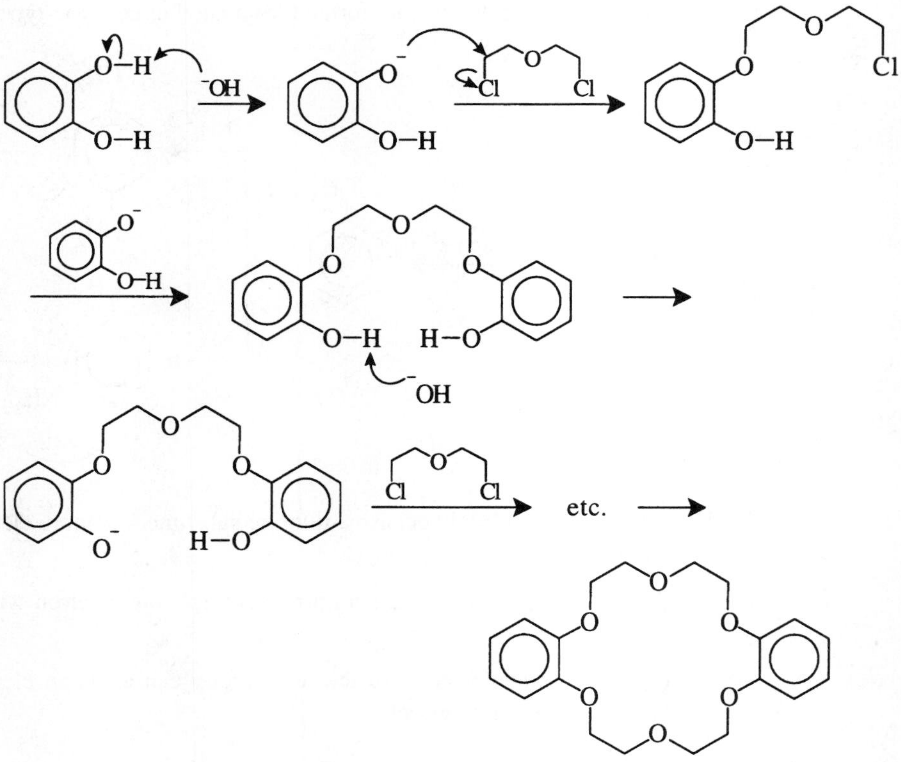

Dibenzo-18-crown-6

21.19 **X** is a phenol because it dissolves in aqueous NaOH but not in aqueous NaHCO$_3$. It gives a dibromo derivative and must therefore be substituted in the ortho or para position. The broad IR peak at 3250 cm^{-1} also suggests a phenol. The peak at 830 cm^{-1} indicates para substitution. The ^{1}H NMR singlet at δ 1.3 (9H) suggests nine equivalent methyl hydrogen atoms, which must be a *tert*-butyl group. The structure of **X** is

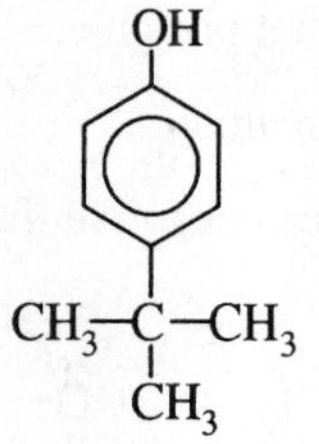

21.20   (a)

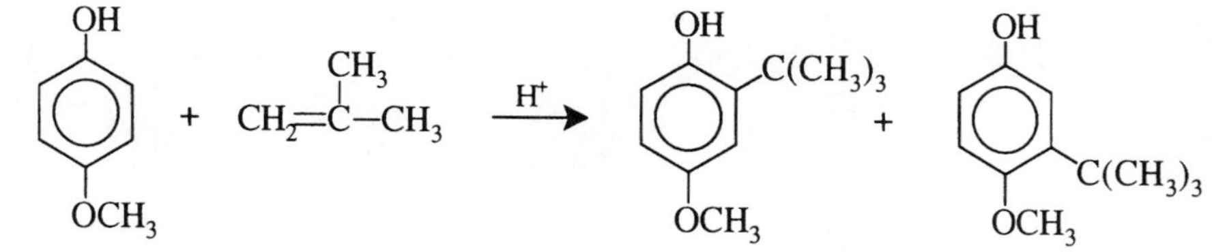

BHA

(b)

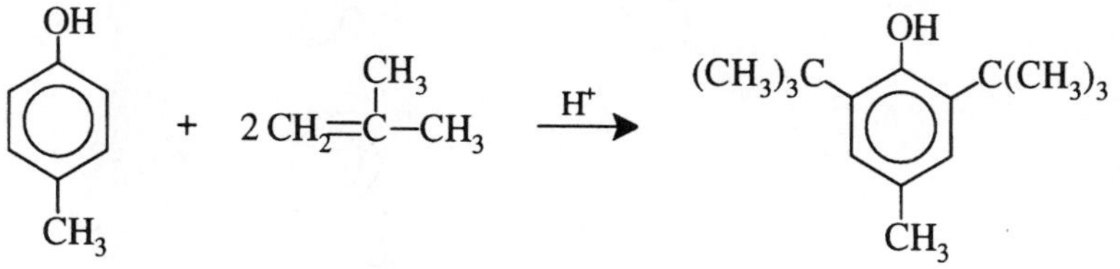

BHT

Notice that both reactions are Friedel-Crafts alkylations.

21.21

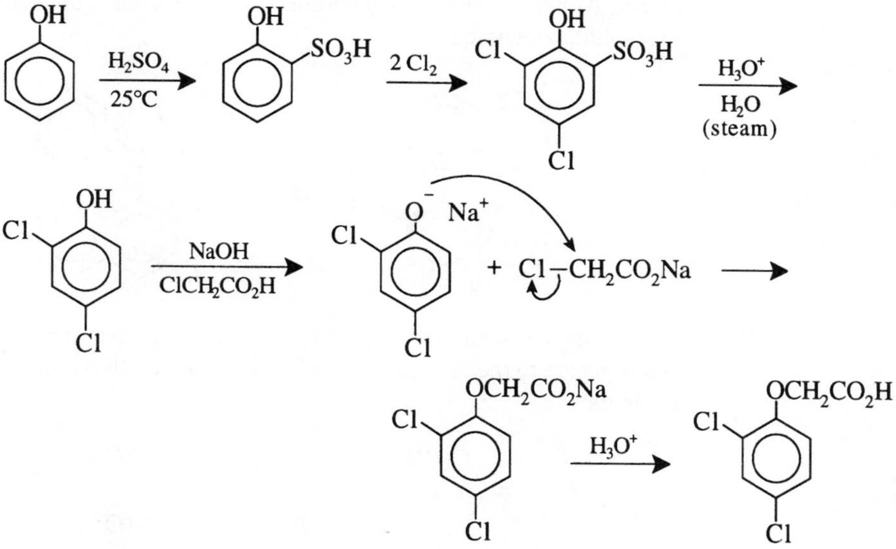

2,4-D

21.22   The broad IR peak at 3200–3600 cm^{-1} suggests a hydroxyl group. The two ^{1}H NMR peaks at δ 1.67 and δ 1.74 are not a doublet because their separation is not equal to other splittings; therefore, these peaks are singlets. Reaction with Br_2/CCl_4 suggests an alkene. If we put these bits of information together, we conclude that **Z** is 3-methyl-2-buten-1-ol.

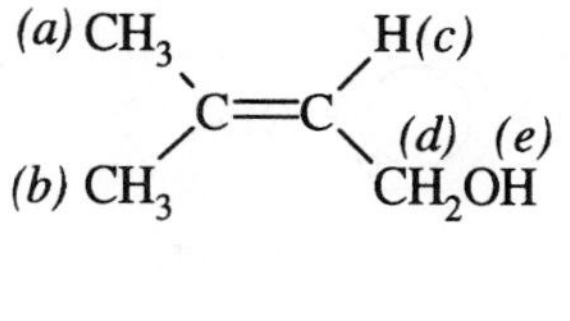

(a) and (b)   singlets δ 1.67 and δ 1.74
(c)   multiplet δ 5.40
(d)   doublet δ 4.10
(e)   singlet δ 2.3

21.23

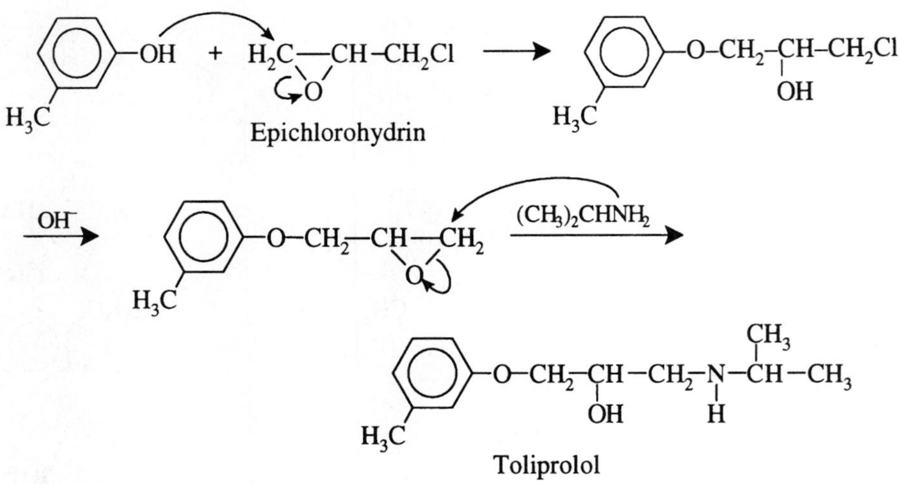

Epichlorohydrin

Toliprolol

21.24 The proximity of the two -OH groups results in the two naphthalene nuclei being non-coplanar. As a result, the two enantiomeric forms are nonequivalent and can be separated by a resolution technique.

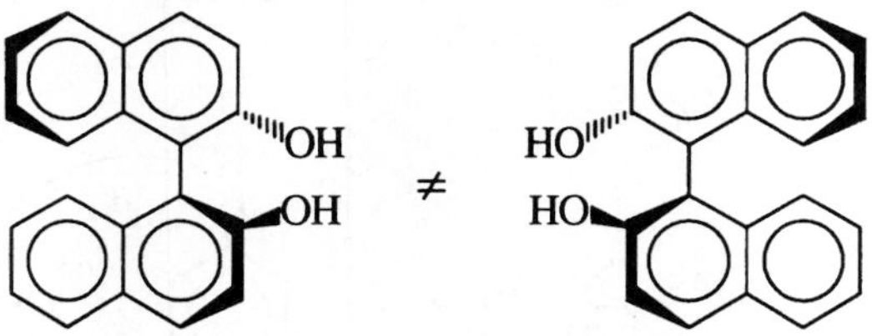

21.25 One can draw a resonance structure for the 2,6-di-*tert*-butylphenoxide ion which shows the carbon para to the oxygen as a nucleophilic center; that is, the species is an ambident nucleophile.

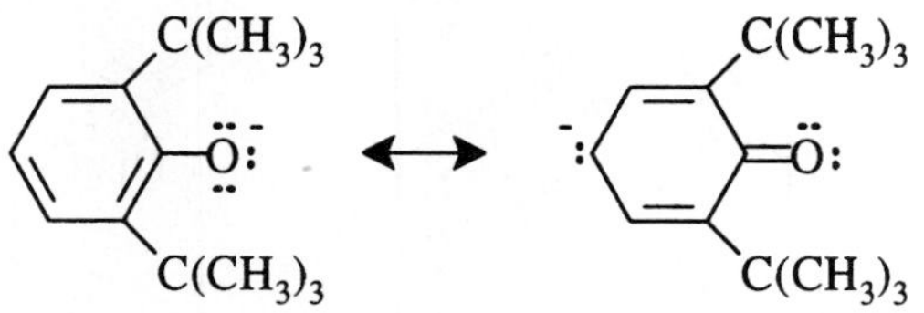

Given the steric hindrance about the oxygen, the nucleophilic character of the para carbon is dominant and an S_NAr reaction occurs at that position to produce this biphenyl derivative:

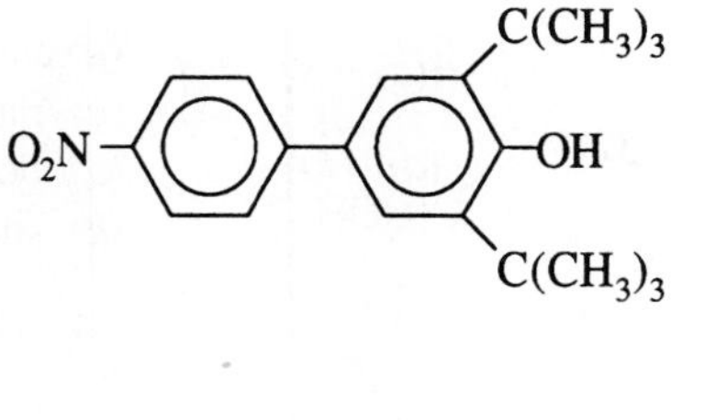

21.26 The phenoxide ion has nucleophilic character both at the oxygen and at the carbon para to it; it is ambident.

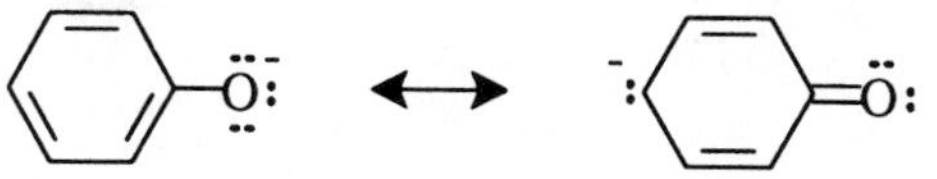

The benzyne produced by the elimination phase of the reaction can undergo addition then by attack by either nucleophilic center, reaction at oxygen producing 1 and reaction at carbon producing 2.

21.27 H^- is expected to be a very poor leaving group because of the strong basicity of the hydride ion. In this case the reaction is favored not only by the activating nitro groups but also by the oxidant ferricyanide ion, which converts H^-, as it forms, to H_2.

QUIZ

21.1 Which of the following would be the strongest acid?

(a) 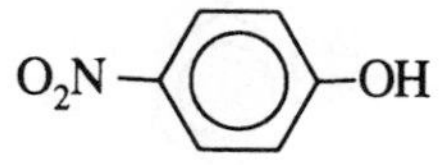 O_2N—⟨⟩—OH (b) 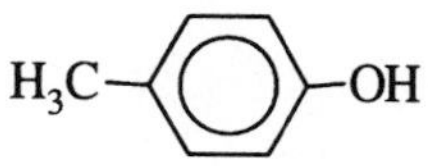 H_3C—⟨⟩—OH (c) 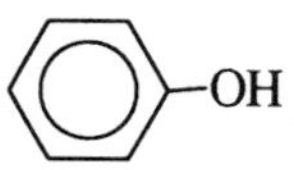 ⟨⟩—OH

(d) 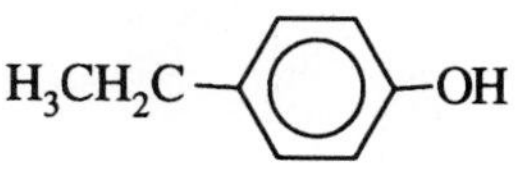 H_3CH_2C—⟨⟩—OH (e) 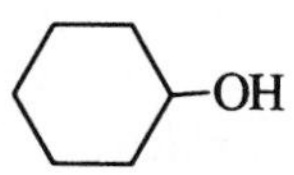 ⟨⟩—OH

21.2 What products would you expect from the following reaction?

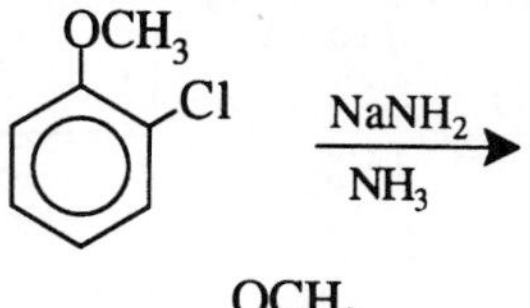

(a) 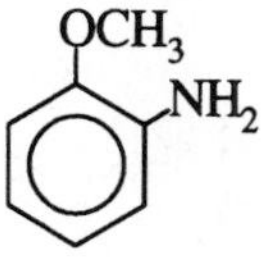 ⟨⟩ alone   (b) 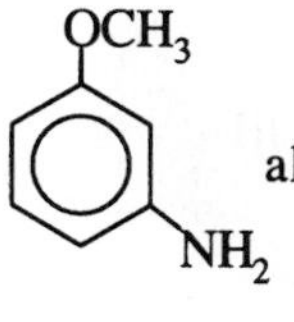 ⟨⟩ alone   (c)  ⟨⟩ alone

(d)   More than one of the above   (e)   All of the above

21.3 Which of the reagents listed here would serve as the basis for a simple chemical test to distinguish between

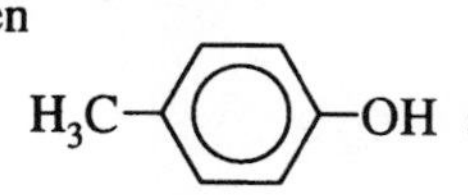

$$H_3C-⟨⟩-OH \text{ and } (CH_3)_3CCH_2OH?$$

(a) $Ag(NH_3)_2OH$ (d)   Cold concd. H_2SO_4

(b) $NaOH/H_2O$ (e)   None of the above

(c)   Dilute HCl

21.4   Indicate the correct product, if any, of the following reaction.

$$H_3C-\langle\bigcirc\rangle-OH \quad + \quad HBr \longrightarrow \quad ?$$

(a) $H_3C-\langle\bigcirc\rangle-Br$ (b) $H_3C-\langle\bigcirc\rangle-OH$ with Br   (c) $H_3C-\langle\bigcirc\rangle-OH$ with Br

(d) $H_3C-\langle\bigcirc\rangle-OH$ with Br, Br   (e)   There is no reaction.

21.5   Complete the following synthesis:

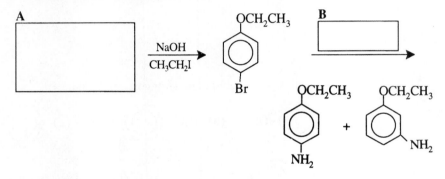

A [   ] $\xrightarrow[\text{CH}_3\text{CH}_2\text{I}]{\text{NaOH}}$ OCH$_2$CH$_3$ / Br   B [   ] $\longrightarrow$

OCH$_2$CH$_3$ / NH$_2$ +   OCH$_2$CH$_3$ / NH$_2$

21.6   Give the products:

$$CH_3O-\langle\bigcirc\rangle-Br \xrightarrow[\text{heat}]{\text{concd. HBr}}$$ [   ] + [   ]

21.7   Select the stronger acid.

(a) $\langle\bigcirc\rangle-\overset{\overset{\displaystyle O}{\|}}{C}OH$ or $H_3C-\langle\bigcirc\rangle-OH$

        (1)   **(2)**

(b) $O_2N-\langle\bigcirc\rangle-OH$ or $H_3C-\langle\bigcirc\rangle-OH$

        (1)   **(2)**

ANSWERS TO SECOND REVIEW PROBLEM SET

The problems review concepts from Chapters 13-21.

1. ⇒ Increasing acidity

(a) $CH_3\overset{O}{\overset{\|}{C}}CH_3$ < CH_3CH_2OH < $CH_3O\overset{O}{\overset{\|}{C}}CH_2\overset{O}{\overset{\|}{C}}OCH_3$ < $CH_3\overset{O}{\overset{\|}{C}}OH$

(b) 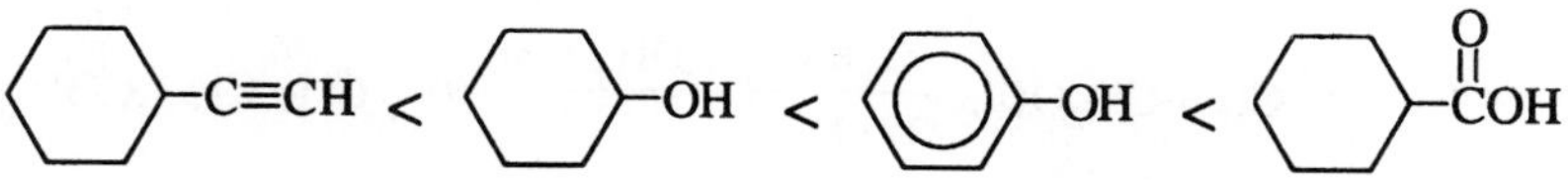 ⬡—C≡CH < ⬡—OH < ⬠—OH < ⬡—$\overset{O}{\overset{\|}{C}}OH$

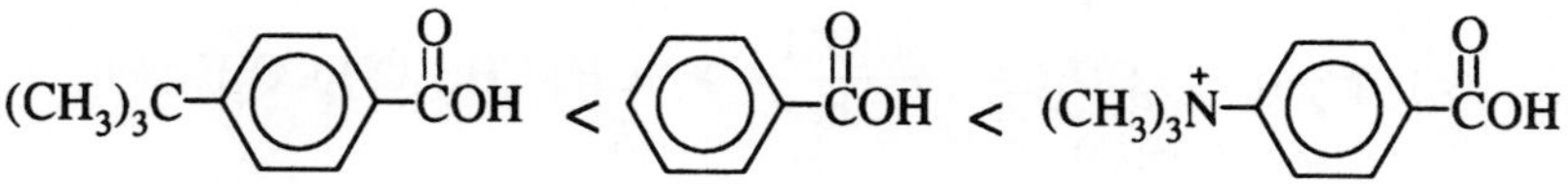

(c) $(CH_3)_3C$—⬠—$\overset{O}{\overset{\|}{C}}OH$ < ⬠—$\overset{O}{\overset{\|}{C}}OH$ < $(CH_3)_3\overset{+}{N}$—⬠—$\overset{O}{\overset{\|}{C}}OH$

(d) $CH_3CH_2\overset{O}{\overset{\|}{C}}OH$ < $CH_3CHCl\overset{O}{\overset{\|}{C}}OH$ < $CH_3CCl_2\overset{O}{\overset{\|}{C}}OH$

(e) 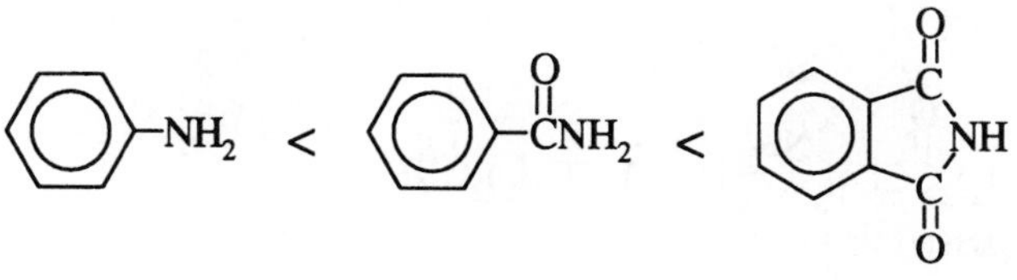 ⬠—NH_2 < ⬠—$\overset{O}{\overset{\|}{C}}NH_2$ < (phthalimide) NH

2. ⇒ Increasing basicity

(a) $CH_3\overset{O}{\overset{\|}{C}}NH_2$ < NH_3 < $CH_3CH_2NH_2$

(b) 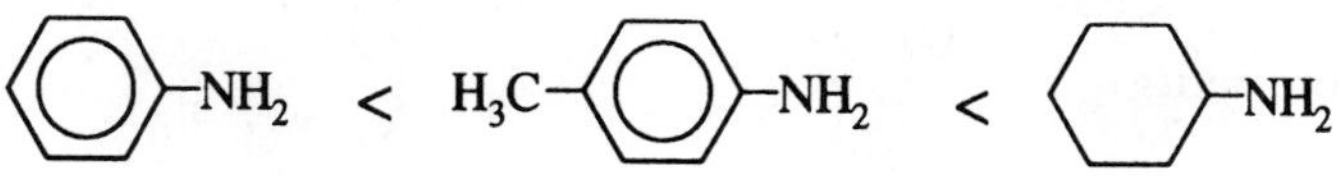 ⬠—NH_2 < H_3C—⬠—NH_2 < ⬡—NH_2

(c) 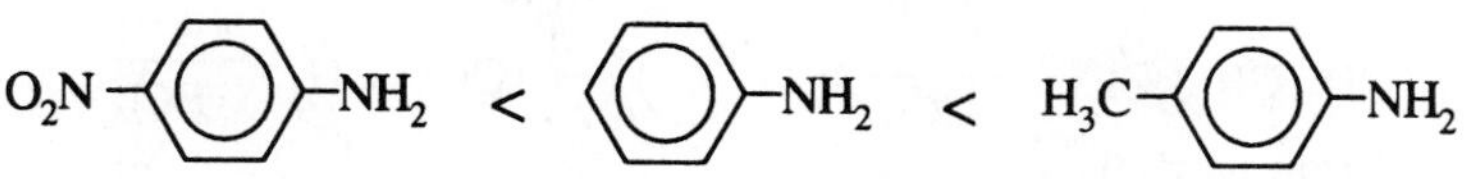 O_2N—⬠—NH_2 < ⬠—NH_2 < H_3C—⬠—NH_2

(d) $CH_3CH_2CH_3$ < CH_3OCH_3 < CH_3NHCH_3

425

3. (a) $CH_3(CH_2)_2CH_2OH \xrightarrow[\text{(or HBr)}]{PBr_3} CH_3(CH_2)_2CH_2Br$

(b) $CH_3(CH_2)_2CH_2Br$ $CH_3(CH_2)_2CH_2-N$
[from part(a)]

$\xrightarrow[\text{heat}]{H_2NNH_2} CH_3(CH_2)_2CH_2NH_2$ +

(c) $CH_3(CH_2)_2CH_2Br \xrightarrow{NaCN} CH_3(CH_2)_2CH_2CN \xrightarrow{LiAlH_4} CH_3(CH_2)_3CH_2NH_2$
[from part(a)]

(d) $CH_3(CH_2)_2CH_2OH \xrightarrow[\text{(2) H}^+]{\text{(1) KMnO}_4,\ OH^-,\ heat} CH_3(CH_2)_2\overset{O}{\overset{\|}{C}}OH$

(e) $CH_3(CH_2)_2CH_2CN \xrightarrow[\text{heat}]{H_3O^+,H_2O} CH_3(CH_2)_2CH_2CO_2H$ + NH_4^+
[from part(c)]

(f) $CH_3(CH_2)_2\overset{O}{\overset{\|}{C}}OH \xrightarrow{SOCl_2} CH_3(CH_2)_2\overset{O}{\overset{\|}{C}}Cl$
[from part(d)]

(g) $CH_3(CH_2)_2\overset{O}{\overset{\|}{C}}Cl \xrightarrow{NH_3} CH_3(CH_2)_2\overset{O}{\overset{\|}{C}}NH_2$
[from part(f)]

(h) $CH_3(CH_2)_2\overset{O}{\overset{\|}{C}}Cl \xrightarrow[\text{base}]{CH_3(CH_2)_2CH_2OH} CH_3(CH_2)_2\overset{O}{\overset{\|}{C}}OCH_2(CH_2)_2CH_3$
[from part(f)]

(i) $CH_3(CH_2)_2\overset{O}{\overset{\|}{C}}NH_2 \xrightarrow[\text{(2) H}_3O^+]{\text{(1) Br}_2,\ OH^-} CH_3CH_2CH_2NH_2$ + CO_3^{2-}
[from part(g)]

(j) $CH_3(CH_2)_2\overset{O}{\overset{\|}{C}}Cl \xrightarrow{AlCl_3}$ $\xrightarrow[\text{HCl}]{Zn(Hg)}$

(k) $CH_3(CH_2)_2\overset{O}{\overset{\|}{C}}Cl$ $\xrightarrow{CH_3(CH_2)_2\overset{O}{\overset{\|}{C}}ONa}$ $[CH_3(CH_2)_2\overset{O}{\overset{\|}{C}}]_2O$

[from part(f)]

(l) $CH_3(CH_2)_2CH_2Br$ $\xrightarrow{Na^+ \colon \overset{CO_2Et}{\underset{CO_2Et}{CH}}}$ $CH_3(CH_2)_2CH_2\overset{CO_2Et}{\underset{CO_2Et}{CH}}$ $\xrightarrow[\text{(2) } H_3O^+]{\text{(1) } OH^-, H_2O, \text{ heat}}$

[from part(a)]

$CH_3(CH_2)_2CH_2\overset{CO_2H}{\underset{CO_2H}{CH}}$ $\xrightarrow[-CO_2]{\text{heat}}$ $CH_3(CH_2)_2CH_2CH_2CO_2H$

4. (a) $H_3C-\langle\bigcirc\rangle$ $\xrightarrow[FeBr_3]{Br_2}$ $H_3C-\langle\bigcirc\rangle-Br$ $\xrightarrow[Et_2O]{Mg}$ $H_3C-\langle\bigcirc\rangle-MgBr$

(separate from
ortho isomer)

$\xrightarrow[\text{(2) } H_3O^+]{\text{(1) } H_2C\overset{O}{\overset{\triangle}{-}}CH_2}$ $H_3C-\langle\bigcirc\rangle-CH_2CH_2OH$ $\xrightarrow[CH_2Cl_2]{PCC}$ $H_3C-\langle\bigcirc\rangle-CH_2\overset{O}{\overset{\|}{C}H}$

$\xrightarrow[\text{(aldol condensation)}]{OH^-}$ $H_3C-\langle\bigcirc\rangle-CH_2CH=\overset{}{\underset{\underset{CH_3}{\langle\bigcirc\rangle}}{C}}-\overset{O}{\overset{\|}{C}}H$

(b) $\langle\bigcirc\rangle$ $\xrightarrow[HF]{CH_2=CHCH_3}$ $\langle\bigcirc\rangle-\overset{CH_3}{\underset{CH_3}{CH}}$ $\xrightarrow[CCl_4 \ (hv)]{NBS}$ $\langle\bigcirc\rangle-\overset{CH_3}{\underset{CH_3}{\overset{|}{C}}}-Br$

$\xrightarrow{\text{base}}$ $\langle\bigcirc\rangle-\overset{}{\underset{CH_3}{C}}=CH_2$ $\xrightarrow[\text{(2)} H_2O_2, OH^-]{\text{(1)} THF: BH_3}$ $\langle\bigcirc\rangle-\overset{}{\underset{CH_3}{C}HCH_2OH}$

$\xrightarrow[\text{(2) } CH_3CH_2Br]{\text{(1) NaH}}$ $\langle\bigcirc\rangle-\overset{}{\underset{CH_3}{C}HCH_2OCH_2CH_3}$

(c) $\overset{NH_2}{\underset{\langle\bigcirc\rangle}{}}$ $\xrightarrow{(CH_3\overset{O}{\overset{\|}{C}})_2O}$ $\overset{\overset{O}{\overset{\|}{NHCCH_3}}}{\underset{\langle\bigcirc\rangle}{}}$ $\xrightarrow[\text{(separate from ortho isomer)}]{Cl_2, \ FeCl_3}$ $Cl-\langle\bigcirc\rangle-\overset{O}{\overset{\|}{NHCCH_3}}$

$$Cl\text{-}\langle\bigcirc\rangle\text{-}NH_2 \xrightarrow[\text{H}_2\text{O}]{\text{OH}^-,\ \text{heat}}$$

(d) $CH_3\text{-}\langle\bigcirc\rangle \xrightarrow[\substack{(\text{separate from}\\ \text{ortho isomer})}]{\text{HNO}_3,\ \text{H}_2\text{SO}_4} CH_3\text{-}\langle\bigcirc\rangle\text{-}NO_2 \xrightarrow[(2)\ \text{H}^+]{(1)\ \text{KMnO}_4,\ \text{OH}^-,\ \text{heat}}$

$$O_2N\text{-}\langle\bigcirc\rangle\text{-}\overset{O}{\overset{\|}{C}}OH \xrightarrow{\text{SOCl}_2} O_2N\text{-}\langle\bigcirc\rangle\text{-}\overset{O}{\overset{\|}{C}}Cl \xrightarrow[\text{diethyl ether}]{\text{LiAlH}[\text{OC}(\text{CH}_3)_3]_3}$$

$$O_2N\text{-}\langle\bigcirc\rangle\text{-}\overset{O}{\overset{\|}{C}}H \xrightarrow[\text{OH}^-]{C_6H_5\overset{O}{\overset{\|}{C}}CH_3} O_2N\text{-}\langle\bigcirc\rangle\text{-}CH=CH\overset{O}{\overset{\|}{C}}\text{-}\langle\bigcirc\rangle$$

(e) $\langle\bigcirc\rangle\text{-}CH_3 \xrightarrow[\text{CCl}_4\ (h\nu)]{\text{NBS}} \langle\bigcirc\rangle\text{-}CH_2Br \xrightarrow{\text{NaC}\equiv\text{CH}} \langle\bigcirc\rangle\text{-}CH_2C\equiv CH$

$$\xrightarrow[\text{H}_2\text{O}]{\text{Hg}^{2+},\ \text{H}_3\text{O}^+} \langle\bigcirc\rangle\text{-}CH_2\overset{O}{\overset{\|}{C}}CH_3 \xrightarrow{\text{HCN}} \langle\bigcirc\rangle\text{-}CH_2\overset{OH}{\underset{CN}{\overset{|}{C}}}CH_3 \xrightarrow[\text{heat}]{\text{H}_3\text{O}^+}$$

$$\left[\langle\bigcirc\rangle\text{-}CH_2\overset{OH}{\underset{CO_2H}{\overset{|}{C}}}CH_3\right] \xrightarrow{-\text{H}_2\text{O}} \langle\bigcirc\rangle\text{-}CH=\underset{CO_2H}{\overset{|}{C}}CH_3$$

5. $H_3C\diagdown\diagup\diagdown$ + $\underset{\substack{\\ \text{Diethyl fumarate}}}{\overset{\displaystyle EtO\overset{O}{\overset{\|}{C}}\diagup H}{\underset{H\diagup\overset{\|}{C}OEt}{\diagdown\diagup}}} \xrightarrow[\text{reaction}]{\text{Diels-Alder}}$ **A** + enantiomer

2-Methyl-1,3-butadiene

$$\xrightarrow[(2)\ \text{H}_3\text{O}^+]{(1)\ \text{LiAlH}_4} \text{ } \mathbf{B} \text{ + enantiomer} \xrightarrow{\text{PBr}_3} \mathbf{C} \text{ + enantiomer} \xrightarrow[\text{H}^+]{\text{Zn}}$$

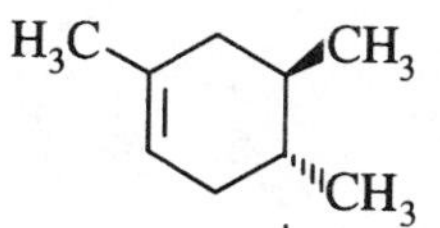

D
+
enantiomer

6. (a) **A** is $CH_2{=}CHC\underset{\underset{\displaystyle OH}{|}}{\overset{\overset{\displaystyle CH_3}{|}}{C}}C{\equiv}CH$ **C** is $BrMgOCH_2CH{=}\underset{}{\overset{\overset{\displaystyle CH_3}{|}}{C}}C{\equiv}CMgBr$

(b) **A** is an allylic alcohol and thus forms a carbocation readily. **B** is a conjugated enyne and is therefore more stable than **A**.

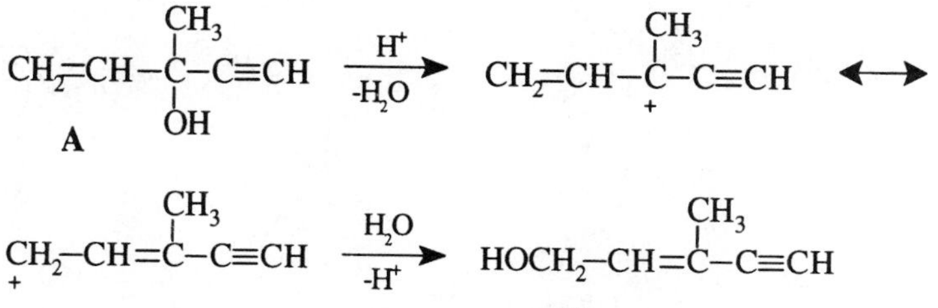

$$CH_2{=}CH{-}\underset{\underset{\displaystyle OH}{|}}{\overset{\overset{\displaystyle CH_3}{|}}{C}}{-}C{\equiv}CH \quad\underset{-H_2O}{\overset{H^+}{\longrightarrow}}\quad CH_2{=}CH{-}\underset{\underset{\displaystyle +}{}}{\overset{\overset{\displaystyle CH_3}{|}}{C}}{-}C{\equiv}CH \quad\longleftrightarrow$$

A

$$\underset{\displaystyle +}{CH_2}{-}CH{=}\overset{\overset{\displaystyle CH_3}{|}}{C}{-}C{\equiv}CH \quad\underset{-H^+}{\overset{H_2O}{\longrightarrow}}\quad HOCH_2{-}CH{=}\overset{\overset{\displaystyle CH_3}{|}}{C}{-}C{\equiv}CH$$

B

7.

$+\ BrMgOCH_2CH{=}\overset{\overset{\displaystyle CH_3}{|}}{C}C{\equiv}CMgBr \xrightarrow[\ (2)\,H^+\ ]{}$

D $\xrightarrow[Ni_2B\ (P\text{-}2)]{H_2}$

E $\xrightarrow{(CH_3\overset{\overset{\displaystyle O}{\|}}{C})_2O}$

F $\xrightarrow{-H_2O}$

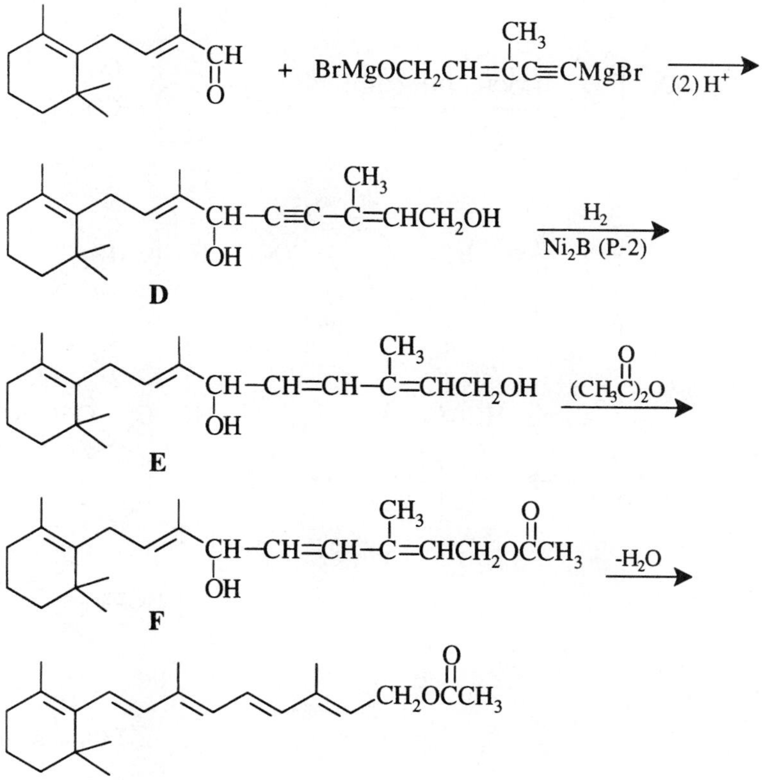

Vitamin A acetate

8. $CH_3\overset{O}{\overset{\|}{C}}CH_3$ + H^+ $\rightleftharpoons$ $HO\overset{CH_3}{\overset{|}{=}}\overset{+}{C}\overset{|}{\underset{CH_3}{}}$ $\longleftrightarrow$ $HO-\overset{CH_3}{\overset{|}{\underset{CH_3}{C+}}}$ ⬡—OH $\longrightarrow$

$HO-\overset{CH_3}{\overset{|}{\underset{CH_3}{C}}}\overset{+}{\underset{H}{⬡}}$—OH $\xrightarrow{-H^+}$ $HO-\overset{CH_3}{\overset{|}{\underset{CH_3}{C}}}$—⬡—OH $\xrightarrow{H^+}$ $H-\overset{+}{\underset{H}{O}}-\overset{CH_3}{\overset{|}{\underset{CH_3}{C}}}$—⬡—OH

$\xrightarrow{-H_2O}$ $+\overset{CH_3}{\overset{|}{\underset{CH_3}{C}}}$—⬡—OH $\xrightarrow{⬡-OH}$ $HO-⬡\overset{+}{\underset{H}{}}\overset{CH_3}{\overset{|}{\underset{CH_3}{C}}}$—⬡—OH $\xrightarrow{-H^+}$

HO—⬡—$\overset{CH_3}{\overset{|}{\underset{CH_3}{C}}}$—⬡—OH

Bisphenol A

9. O_2N—⬡—CH_3 $\xrightarrow[\text{(2) }H^+]{\text{(1) KMnO}_4\text{, OH}^-\text{, heat}}$ O_2N—⬡—$\overset{O}{\overset{\|}{C}}OH$ $\xrightarrow{SOCl_2}$

A

O_2N—⬡—$\overset{O}{\overset{\|}{C}}Cl$ $\xrightarrow{HOCH_2CH_2N(C_2H_5)_2}$ O_2N—⬡—$\overset{O}{\overset{\|}{C}}OCH_2CH_2N(C_2H_5)_2$

B **C**

$\xrightarrow[\text{cat.}]{H_2}$ H_2N—⬡—$\overset{O}{\overset{\|}{C}}OCH_2CH_2N(C_2H_5)_2$

Procaine

10. $\xrightarrow[\text{(2) }H^+]{\text{(1) HC}\equiv\text{CNa}}$ $\underset{\textbf{A}}{\overset{HC\equiv C \quad OH}{⬡}}$ $\xrightarrow{\overset{O}{\overset{\|}{Cl\overset{}{C}Cl}}}$ $\underset{\textbf{B}}{\overset{HC\equiv C \quad O-\overset{O}{\overset{\|}{C}}-Cl}{⬡}}$

$\xrightarrow{NH_3}$ $\overset{HC\equiv C \quad O-\overset{O}{\overset{\|}{C}}-NH_2}{⬡}$

Ethinamate

11. **(a)**

$$\underset{O}{\overset{\overset{O}{\parallel}}{C_6H_5CH}} \xrightarrow[\text{(2) } H_3O^+]{\text{(1) } C_6H_5MgBr} \underset{\underset{C_6H_5}{\overset{|}{\mathbf{A}}}}{C_6H_5CHOH} \xrightarrow{PBr_3} \underset{\underset{C_6H_5}{\overset{|}{\mathbf{B}}}}{C_6H_5CHBr}$$

$$\xrightarrow[\text{-HBr}]{(CH_3)_2NCH_2CH_2OH} \underset{\underset{C_6H_5}{\overset{|}{}}}{C_6H_5CHOCH_2CH_2N(CH_3)_2}$$

Diphenhydramine

(b) The last step probably takes place by an S_N1 mechanism. Diphenylmethyl bromide, **B**, ionizes readily because it forms the resonance-stabilized benzylic carbocation,

$$\underset{\overset{|}{C_6H_5}}{C_6H_5\overset{+}{C}H}$$

12. **(a)** For this synthesis we need to prepare the benzylic halide, $Br-\langle\bigcirc\rangle-\underset{\overset{|}{C_6H_5}}{CHBr}$, and then allow it to react with $(CH_3)_2NCH_2CH_2OH$ as in Problem 11.

This benzylic halide can be made as follows:

(b) For this synthesis we can prepare the requisite benzylic halide in two ways:

or

We shall then allow the benzylic halide to react with $(CH_3)_2NCH_2CH_2OH$ as in Problem 11.

13. $\underset{\overset{|}{Br}}{CH_3CHCO_2C_2H_5} \xrightarrow{CH_2(CO_2C_2H_5)_2,\ EtO^-} \begin{array}{c} C_2H_5O_2C \diagdown \qquad \diagup CH_3 \\ CH \\ | \\ CH \\ C_2H_5O_2C \diagup \qquad \diagdown CO_2C_2H_5 \\ \mathbf{A} \end{array} \xrightarrow[\text{(Michael addition)}]{CH_2=CHCN,\ EtO^-}$

$$\underset{\textbf{B}}{\overset{\text{N}}{\text{C}}}\text{—}\text{CH}_2\text{—}\text{C}(\text{CO}_2\text{C}_2\text{H}_5)_2\text{—}\text{CH}_2\text{—}\text{CH}(\text{CH}_3)\text{—}\text{CO}_2\text{C}_2\text{H}_5 \xrightarrow[\substack{\text{(converts }-\text{C}\equiv\text{N to} \\ -\text{CO}_2\text{C}_2\text{H}_5)}]{\text{C}_2\text{H}_5\text{OH, H}^+}$$

C

$$\xrightarrow[\substack{\text{(Dieckmann} \\ \text{condensation)}}]{\text{EtO}^-}$$

D

$$\xrightarrow[\text{(2) H}_3\text{O}^+]{\text{(1) OH}^-,\ \text{H}_2\text{O, heat}}$$

$$\xrightarrow[-2\text{CO}_2]{\text{(3) heat}}$$

14. $\underset{\text{O}}{\overset{\text{O}}{\text{CH}_3\text{CCH}_3}} \xrightarrow[\substack{\text{(acid-catalyzed} \\ \text{aldol condensation)}}]{\text{HCl}} \underset{\textbf{A}}{\text{CH}_3\text{C}=\text{CHCCH}_3} \xrightarrow[\text{(Michael addition)}]{\text{CH}_3\text{CCH}_2\text{COEt, base}}$

B
$$\xrightarrow[\substack{\text{(intramolecular} \\ \text{aldol condensation)}}]{\text{base}}$$

C

$$\xrightarrow[\substack{\text{(hydrolysis and decarboxylation} \\ \text{of }\beta\text{-keto ester)}}]{\text{H}^+,\ \text{H}_2\text{O, heat}}$$

D $+$ CO$_2$

15. $\text{C}_6\text{H}_5\text{CH=O} + \underset{\substack{| \\ \text{NO}_2}}{\text{CH}_2\text{CH}_2\text{OH}} \xrightarrow[\substack{\text{(aldol-type} \\ \text{condensation)}}]{\text{EtO}^-} \underset{\textbf{A}}{\text{C}_6\text{H}_5\text{CH(OH)CH(NO}_2)\text{CH}_2\text{OH}} \xrightarrow[\text{cat.}]{\text{H}_2}$

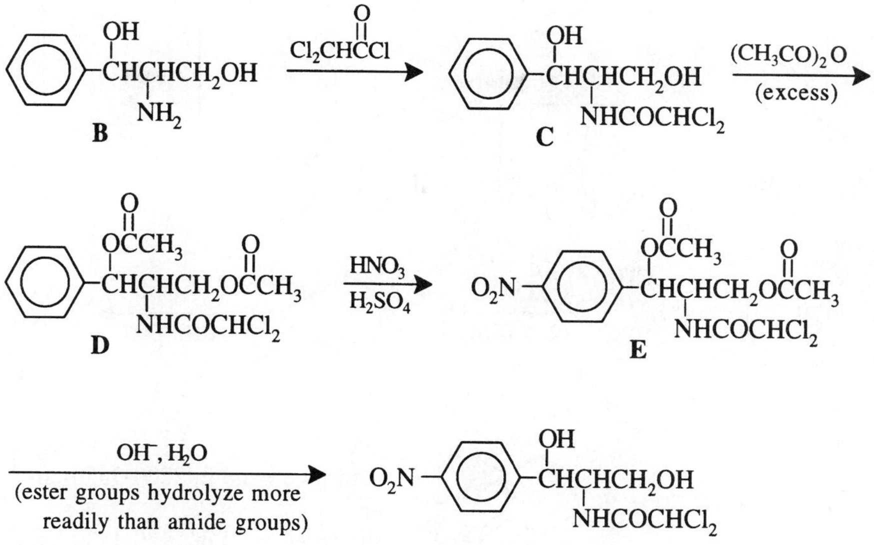

Chloramphenicol

16.

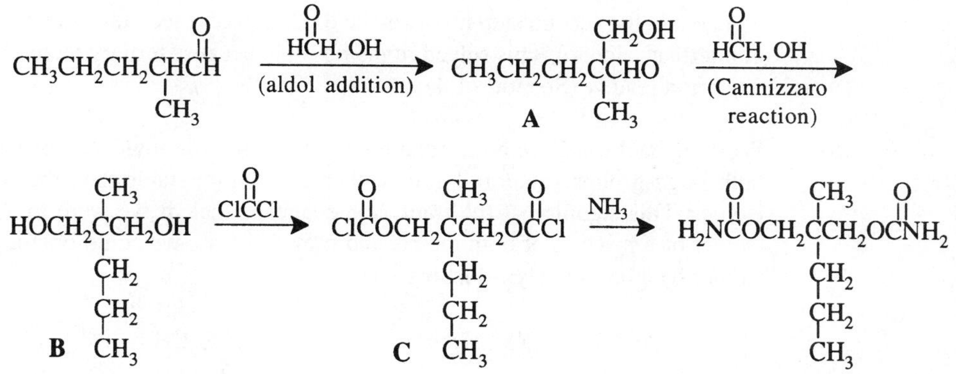

Meprobamate

17.

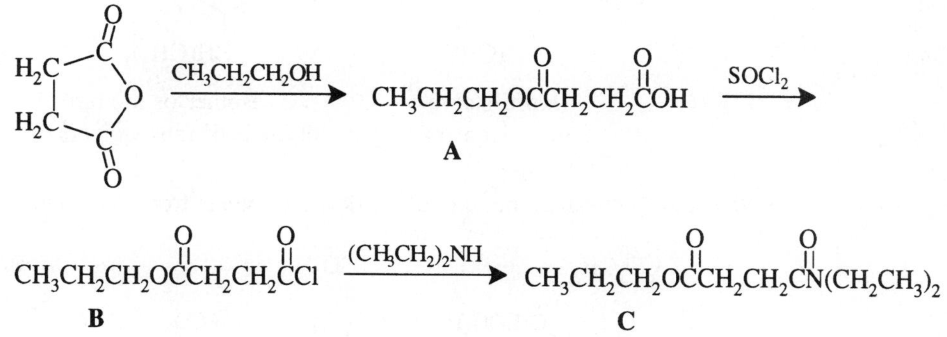

18.

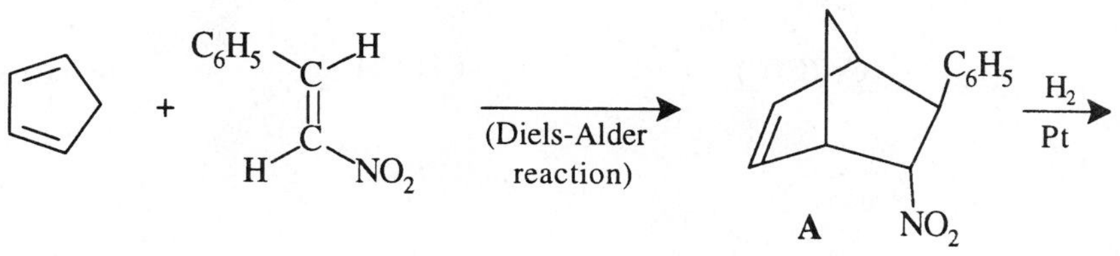

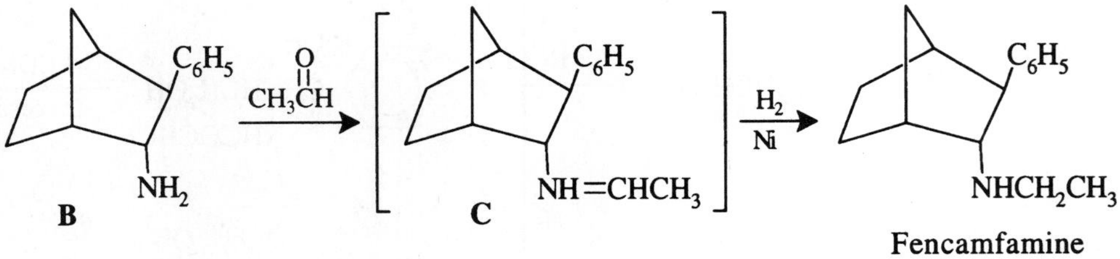

Fencamfamine

19.

A

Infrared band in 3200-3550-cm^{-1} region

Infrared band in 1650-1730-cm^{-1} region

B

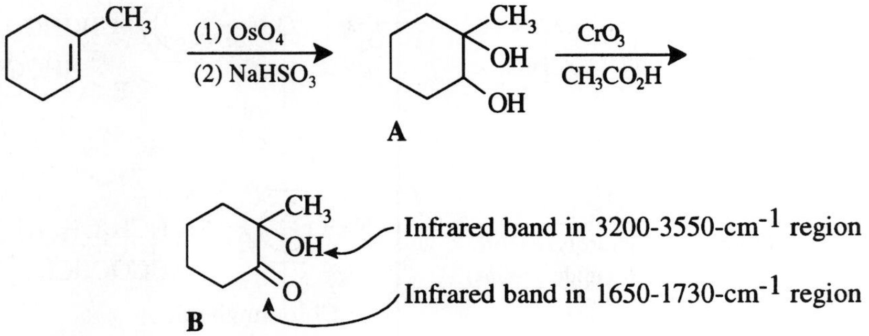

Notice that the second step involves the oxidation of a secondary alcohol in the presence of a tertiary alcohol. This selectivity is possible because tertiary alcohols do not undergo oxidation readily (Section 11.4).

20.   Working backward, we notice that methyl *trans*-4-isopropylcyclohexanecarboxylate has both large groups equatorial and is, therefore, more stable than the corresponding cis isomer. This stability of the trans isomer means that, if we were to synthesize the cis isomer or a mixture of both the cis and trans isomers, we could obtain the desired trans isomer by a base-catalyzed isomerization:

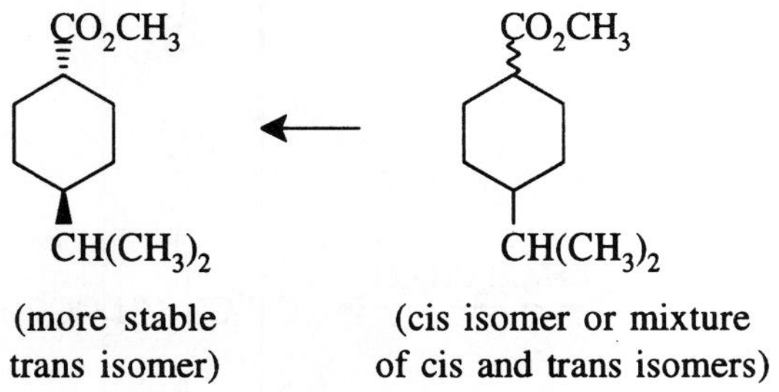

(more stable trans isomer)

(cis isomer or mixture of cis and trans isomers)

We could synthesize a mixture of the desired isomers from phenol in the following way:

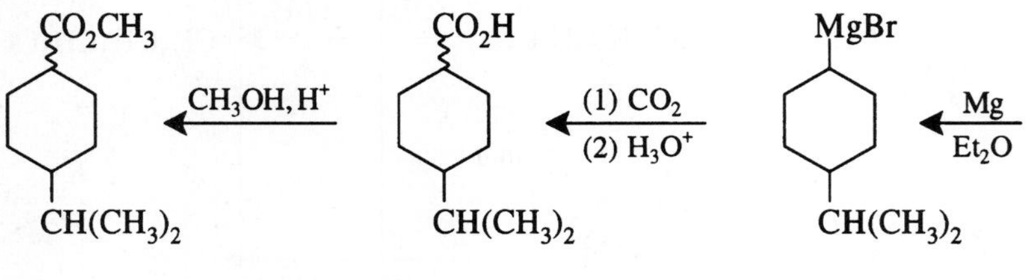

21. The positive iodoform test and the strong IR absorption of **X** indicate that it contains a $-\overset{\underset{\parallel}{O}}{C}CH_3$ group. Subtracting this from the molecular formula, $C_5H_{10}O$, leaves only C_3H_7.

$$\begin{array}{r} C_5H_{10}O \\ -C_2H_3O \\ \hline C_3H_7 \end{array}$$

This could be either a propyl group or an isopropyl group. The information from the DEPT spectra is consistent only with an isopropyl group; hence **X** is isopropyl methyl ketone. The assignments are the following:

$$\underset{(a)\quad\;(c)(d)(b)}{(CH_3)_2CH\overset{\overset{O}{\parallel}}{C}CH_3}$$

(a) δ 18.1
(b) δ 27.3
(c) δ 41.5
(d) δ 211.8

22. That **Y** gives a green opaque solution when treated with CrO_3 in aqueous H_2SO_4 indicates that **Y** is a primary or secondary alcohol. That **Y** gives a negative iodoform test indicates that **Y** does not contain the grouping $-\underset{\underset{OH}{|}}{C}HCH_3$. The ^{13}C spectrum of **Y** contains

only four signals indicating that some of the carbons in **Y** are equivalent. The information from DEPT spectra helps us conclude that **Y** is 2-ethyl-1-butanol.

$$\underset{(CH_3CH_2)_2CHCH_2OH}{\overset{(a)\;(b)\quad(c)\;(d)}{}}$$

(a) δ 11.1
(b) δ 23.0
(c) δ 43.6
(d) δ 64.6

Notice that the most downfield signal is a CH_2 group. This indicates that this carbon atom bears the $-OH$ group and that **Y** is a primary alcohol. The most upfield signals indicate the presence of the ethyl groups.

23. That **Z** decolorizes bromine in CCl_4 indicates that **Z** is an alkene. We are told that **Z** is the more stable isomer of a pair of stereoisomers. This fact suggests that **Z** is a *trans* alkene. That the ^{13}C spectrum contains only three signals, even though **Z** contains eight carbon atoms, indicates that **Z** is highly symmetric. The information from DEPT spectra indicates that the upfield signals of the alkyl groups arise from equivalent isopropyl groups. We conclude, therefore, that **Z** is *trans*-2,5-dimethyl-3-hexene.

(a) δ 22.8
(b) δ 31.0
(c) δ 134.5

SPECIAL TOPIC
Thiols, Thioethers, and Thiophenols

SOLUTIONS TO PROBLEMS

L.1 (a)

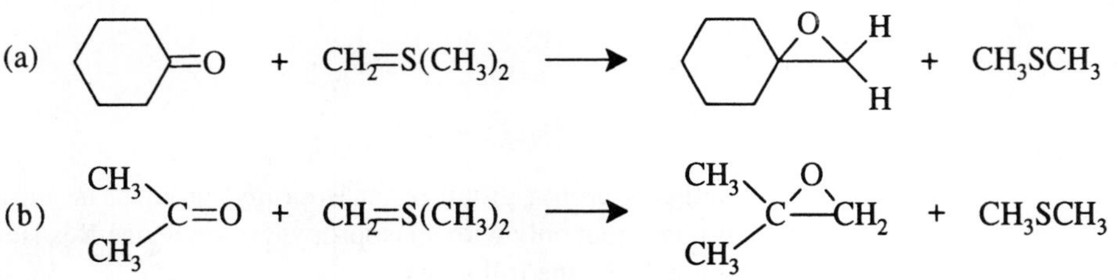

$$\text{cyclohexanone} =O \;+\; CH_2\!\!=\!\!S(CH_3)_2 \longrightarrow \text{(epoxide)} + CH_3SCH_3$$

(b)

$$\underset{CH_3}{\overset{CH_3}{\diagdown}} C=O \;+\; CH_2\!\!=\!\!S(CH_3)_2 \longrightarrow \underset{CH_3}{\overset{CH_3}{\diagdown}} C\!\!-\!\!CH_2 \;+\; CH_3SCH_3$$

L.2 (a)

$$\text{Ph}-CH_2-\overset{+}{S}=\underset{NH_2}{\overset{NH_2}{C}} \quad Br^-$$

(b) $\text{Ph}-CH_2SH$

(c) $\text{Ph}-CH_2-S-S-CH_2-\text{Ph}$

(d) $\text{Ph}-CH_2-S^-\ Na^+$

(e) $\text{Ph}-CH_2-S-CH_2-\text{Ph}$

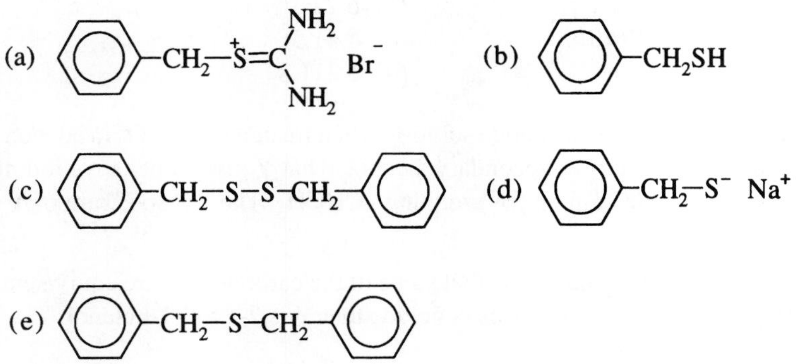

L.3
$$CH_2\!\!=\!\!CHCH_2Br \;+\; S=\underset{NH_2}{\overset{NH_2}{C}} \xrightarrow[\;(2)OH^-,\,H_2O\;]{(1)CH_3CH_2OH} CH_2\!\!=\!\!CHCH_2SH$$

$$\xrightarrow{H_2O_2} CH_2\!\!=\!\!CHCH_2-S-S-CH_2CH=CH_2$$

L.4
$$CH_2\!\!=\!\!CHCH_2OH \xrightarrow{Br_2} CH_2BrCHBrCH_2OH \xrightarrow{NaSH} \underset{SH\quad SH}{CH_2-CH-CH_2OH}$$

L.5 (a) $\underset{\textstyle\overset{\text{O}}{\|}}{}$ ClCH$_2$CH$_2$C(CH$_2$)$_4$CO$_2$C$_2$H$_5$ (This step is the Friedel-Crafts alkylation of an alkene.)

(b) SOCl$_2$

(c) 2 C$_6$H$_5$CH$_2$SH and KOH

(d) H$_3$O$^+$

(e)

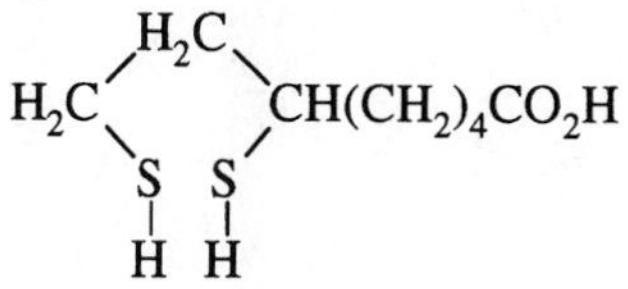

L.6

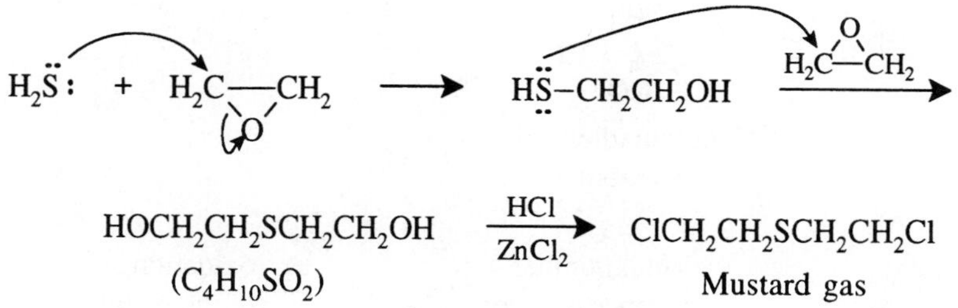

HOCH$_2$CH$_2$SCH$_2$CH$_2$OH $\xrightarrow[\text{ZnCl}_2]{\text{HCl}}$ ClCH$_2$CH$_2$SCH$_2$CH$_2$Cl
(C$_4$H$_{10}$SO$_2$)   Mustard gas

SOLUTIONS TO PROBLEMS

M.1

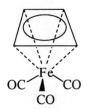

Cyclobutadiene iron
tricarbonyl

$$d^n = \begin{array}{l} \text{Total number of} \\ \text{valence electrons} \\ \text{(both } s \text{ and } d \text{ electrons)} \\ \text{of elemental iron} \end{array} - \begin{array}{l} \text{oxidation state} \\ \text{of the metal} \\ \text{in the complex} \end{array}$$

$$d^n = 8 - 0 = 8$$

$$\begin{array}{l} \text{Total number} \\ \text{of valence electrons} \\ \text{of iron in the} \\ \text{complex} \end{array} = \begin{array}{l} \text{electrons} \\ d^n + \text{donated by} \\ \text{ligands} \end{array}$$

$$= 8 + 3(CO) + \text{cyclobutadiene}$$

$$= 8 + 3(2) + 4 = 18$$

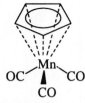

Cyclopentadienylmanganese
tricarbonyl

$$d^n = 7 - 1 = 6$$

$$\begin{array}{l} \text{Total number of} \\ \text{valence electrons} \\ \text{of Mn in complex} \end{array} = 6 + 3(CO) + Cp$$

$$= 6 + 3(2) + 6 = 18$$

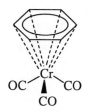

Benzene chromium
tricarbonyl

$d^n = 6 - 0 = 6$

Total number of valence electrons of Cr in complex	=	$6 + 3(CO) + $ benzene
	=	$6 + 3(2) + 6 = 18$

M.2   A syn addition of D_2 to the *trans* alkene would produce the following racemic form.

$$
\underset{\substack{\\ \text{EtO}_2\text{C}}}{\overset{\text{H}}{\underset{}{\text{C}}}} \!\!=\!\! \overset{\text{CO}_2\text{Et}}{\underset{\text{H}}{\text{C}}}
\quad + \quad D_2
\quad \xrightarrow{\text{Rh}(Ph_3P)_3Cl} \quad
\underset{\substack{\text{H} \\ \text{CO}_2\text{Et}}}{\overset{\substack{\text{CO}_2\text{Et} \\ \text{H} \quad D}}{\text{C}}}
\quad + \quad
\underset{\substack{\text{D} \\ \text{CO}_2\text{Et}}}{\overset{\substack{\text{CO}_2\text{Et} \\ \text{D} \quad \text{H}}}{\text{C}}}
$$

M.3 $(Ph_3P)_3RhCl$ + CH_3Li $\xrightarrow[\text{exchange}]{\text{ligand}}$ $(Ph_3P)_3RhCH_3$ +   LiCl

(16 electrons)   (16 electrons)

Rh^I Rh^I $\bigg\downarrow$ I   Oxidative addition

$(Ph_3P)_3Rh(CH_3)I$

(18 electrons)

Rh^{III}

$\xleftarrow[\text{elimination}]{\text{reductive}}$

toluene + $(Ph_3P)_3RhI$

(16 electrons)

Rh^I

M.4 $(Ph_3P)_2Rh(CO)Cl$ + CH_3Li $\xrightarrow{\text{(a)}}$ $(Ph_3P)_2Rh(CO)(CH_3)$ + LiCl

   1   **2**

(16 electrons)   (16 electrons)

Rh^I Rh^I

(b) $\Big\downarrow$ $C_6H_5\overset{\overset{\displaystyle O}{\|}}{C}Cl$

$(Ph_3P)_2Rh(CO)Cl$ + $C_6H_5\overset{\overset{\displaystyle O}{\|}}{C}CH_3$ $\xleftarrow{\text{(c)}}$ $(Ph_3P)_2Rh(CO)(COC_6H_5)(CH_3)Cl$

                                                  3

(16 electrons)   (18 electrons)

Rh^I Rh^{III}

(a)   Is a ligand exchange

(b)   Is an oxidative addition

(c)   Is a reductive elimination

M.5 1. $(Ph_3P)_3Rh(CO)H$ $\xrightarrow{-Ph_3P}$ $(Ph_3P)_2Rh(CO)H$ **Ligand dissociation**

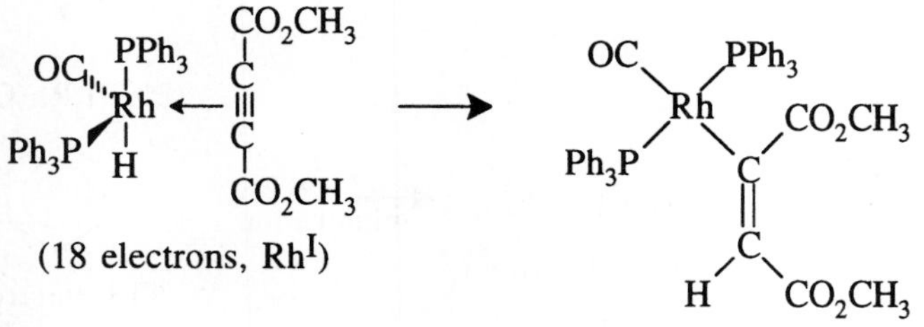

   (18 electrons, Rh^I)   (16 electrons, Rh^I)

2. $(Ph_3P)_2Rh(CO)H$ $\xrightarrow{CH_3O\overset{\overset{\displaystyle O}{\|}}{C}C\equiv C\overset{\overset{\displaystyle O}{\|}}{C}OCH_3}$

   (16 electrons, Rh^I)

Ligand association

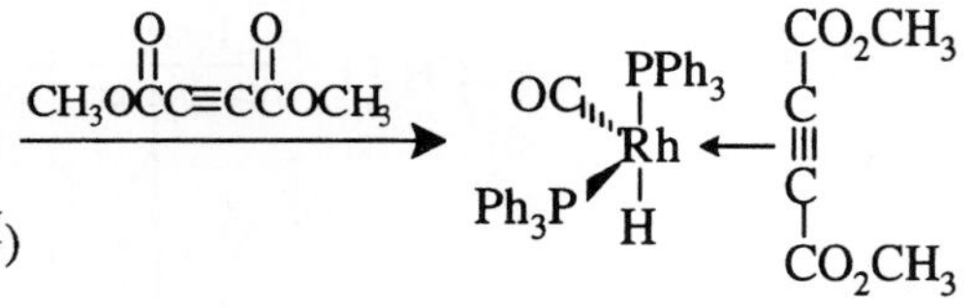

(18 electrons, Rh^I)

3.

(18 electrons, Rh^I) $\longrightarrow$

Insertion

(16 electrons, Rh^I)

4.

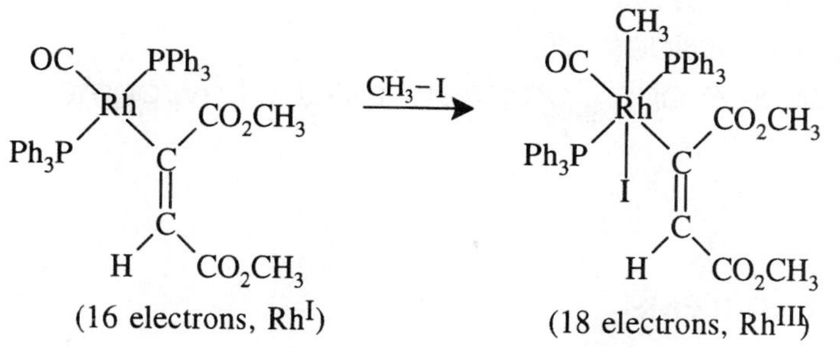

$$\underset{\text{(16 electrons, Rh}^{I})}{\underset{\text{Ph}_3\text{P}}{\overset{\text{OC}}{\text{Rh}}}}\quad\xrightarrow{\text{CH}_3-\text{I}}\quad\underset{\text{(18 electrons, Rh}^{III})}{\text{Rh complex}}$$

Oxidative addition

5.

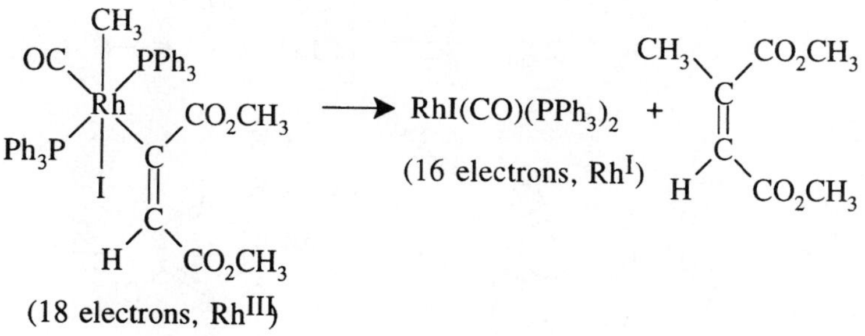

$$\underset{\text{(18 electrons, Rh}^{III})}{\text{Rh complex}}\quad\longrightarrow\quad \text{RhI(CO)(PPh}_3)_2 \;+\; \text{alkene}$$

(16 electrons, Rh^{I})

Reductive elimination

M.6 $(\text{CH}_3)_2\text{CuLi} \;+\; \text{C}_6\text{H}_5-\text{I} \;\longrightarrow\; \left[\text{CH}_3-\overset{\text{I}}{\underset{\text{CH}_3}{\text{Cu}}}-\text{C}_6\text{H}_5 \right]^{-} \text{Li}^{+}$ **Oxidative addition**

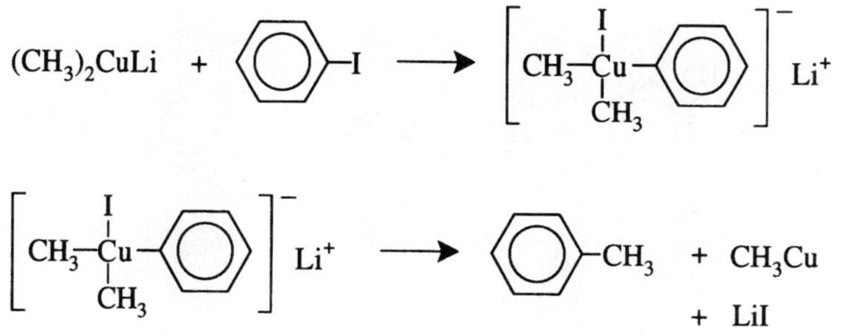

$$\left[\text{CH}_3-\overset{\text{I}}{\underset{\text{CH}_3}{\text{Cu}}}-\text{C}_6\text{H}_5 \right]^{-} \text{Li}^{+} \;\longrightarrow\; \text{C}_6\text{H}_5-\text{CH}_3 \;+\; \text{CH}_3\text{Cu} \;+\; \text{LiI}$$

Reductive elimination

N SPECIAL TOPIC
Organic Halides and Organometallic Compounds in the Environment

SOLUTIONS TO PROBLEMS

N.1

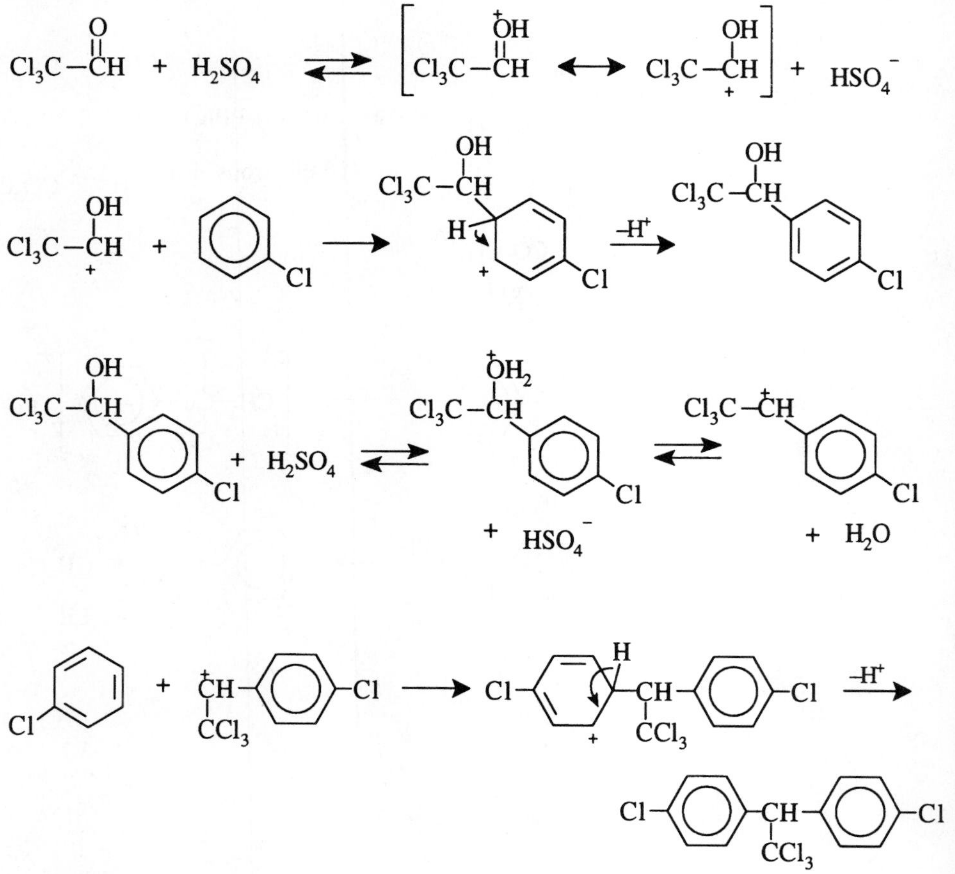

N.2 An elimination reaction.

N.3 (a)

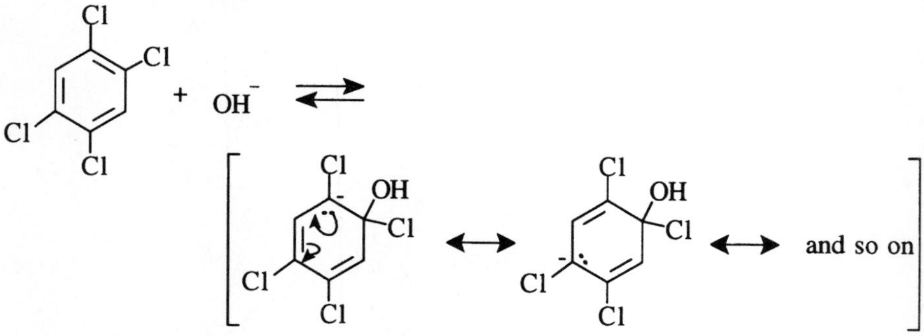

(b)

N.4 An S_N2 reaction:

O
SPECIAL TOPIC
Electrocyclic and Cycloaddition Reactions

SOLUTIONS TO PROBLEMS

O.1 According to the Woodward-Hoffmann rule for electrocyclic reactions of $4n$ π-electron systems (Section O.2A), the photochemical cyclization of *cis,trans*-2,4-hexadiene should proceed with *disrotatory motion*. Thus, it should yield *trans*-3,4-dimethylcyclobutene:

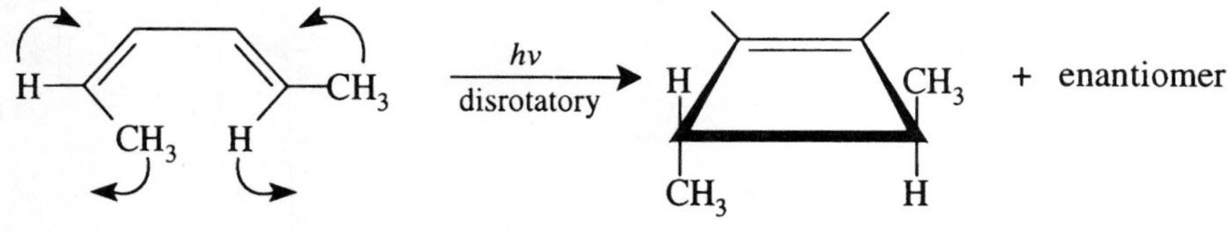

cis, trans-2,4-Hexadiene   *trans*-3,4-Dimethylcyclobutene

O.2 (a)

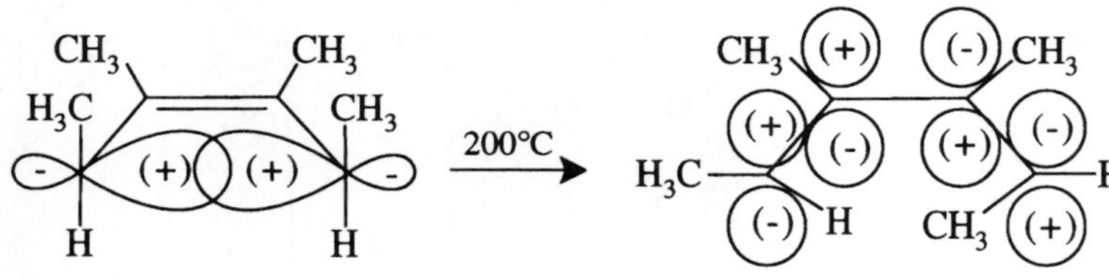

ψ_2 of a hexadiene
(Section O.2A)

(b) This is a thermal electrocyclic reaction of a $4n$ π-electron system; it should, *and does*, proceed with conrotatory motion.

O.3

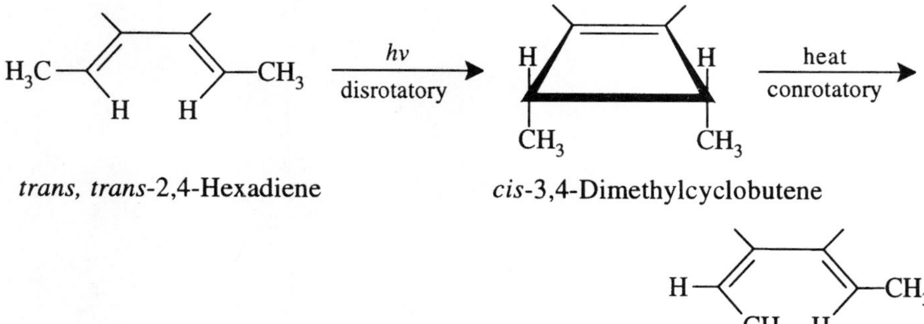

trans, trans-2,4-Hexadiene   *cis*-3,4-Dimethylcyclobutene

cis, trans-2,4-Hexadiene

Here we find that two consecutive electrocyclic reactions (the first photochemical, the second thermal), provide a stereospecific synthesis of *cis,trans*-2,4-hexadiene from *trans, trans*-2,4-hexadiene.

O.4 (a) This is a photochemical electrocyclic reaction of an eight π-electron system—a $4n$ π-electron system where $n = 2$. It should, therefore, proceed with disrotatory motion.

cis-7, 8-Dimethyl-1, 3, 5-cyclooctatriene

(b) This is a thermal electrocyclic reaction of the eight π-electron system. It should proceed with conrotatory motion.

cis-7, 8-Dimethyl-1, 3, 5-cyclooctatriene

O.5 (a) This is conrotatory motion, and since this is a $4n$ π-electron system (where $n = 1$) it should occur under the influence of heat.

(b) This is conrotatory motion, and since this is also a $4n$ π-electron system (where $n = 2$) it should occur under the influence of heat.

+ enantiomer

(c) This is disrotatory motion. This, too, is a $4n$ π-electron system (where $n = 1$); thus it should occur under the influence of light.

O.6   (a)   This is a $(4n + 2)$ π-electron system (where $n = 1$); a thermal reaction should take place with disrotatory motion:

$$\xrightarrow[\text{(disrotatory)}]{\text{heat}}$$

(b)   This is also a $(4n + 2)$ π-electron system; a photochemical reaction should take place with conrotatory motion.

$$\xrightarrow[\text{(conrotatory)}]{h\nu}$$

O.7   Here we need a conrotatory ring opening of *trans*-5,6-dimethyl-1,3-cyclohexadiene (to produce *trans,cis,trans*-2,4,6-octatriene); then we need a disrotatory cyclization to produce *cis*-5,6-dimethyl-1,3-cyclohexadiene.

$$\xrightarrow[\text{(conrotatory)}]{h\nu} \qquad \xrightarrow[\text{(disrotatory)}]{\text{heat}}$$

trans-5,6-Dimethyl-1,3-
cyclohexadiene

trans, cis, trans-2,4,6-
Octatriene

cis-5,6-Dimethyl-1,3-
cyclohexadiene

Since both reactions involve $(4n + 2)$ π-electron systems, we apply light to accomplish the first step and heat to accomplish the second. It would also be possible to use heat to produce *trans,cis,cis*-2,4,6-octatriene and then use light to produce the desired product.

O.8   The first electrocyclic reaction is a thermal, conrotatory ring opening of a $4n$ π-electron system. The second electrocyclic reaction is a thermal, disrotatory ring closure of a $(4n + 2)$ π-electron system.

O.9   (a)   There are two possible products that can result from a concerted cycloaddition. They are formed when *cis*-2-butene molecules come together in the following ways:

(b)   There are two possible products that can be obtained from *trans*-2-butene as well.

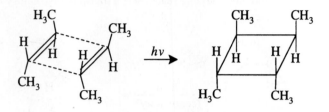

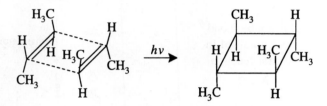

O.10   This is an intramolecular [2 + 2] cycloaddition.

O.11   (a)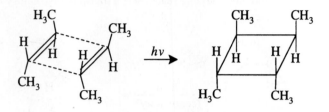

(b)

+

Enantiomer

22 CARBOHYDRATES

SUMMARY OF SOME REACTIONS OF MONOSACCHARIDES

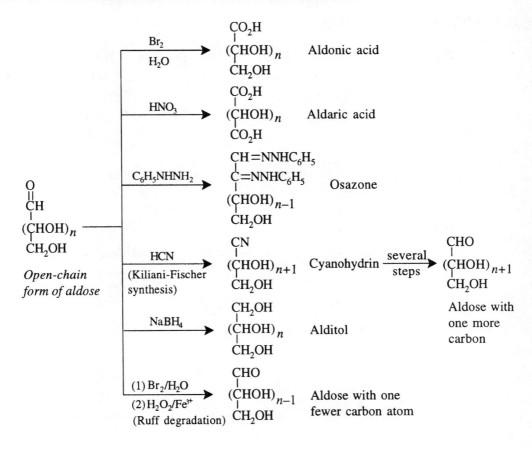

Cyclic form of D-glucose   Methyl glucoside

The reaction sequence at the top of the page:

$$\text{(pyranose with } CH_2OCH_3, CH_3O, CH_3O, OCH_3, CHOCH_3 \text{ substituents)} \xrightarrow[\text{H}_2\text{O}]{\text{H}_3\text{O}^+} \text{(pyranose with } CH_2OCH_3, CH_3O, CH_3O, CH_3, CHOH) \longleftrightarrow$$

Open-chain form:

$$\begin{array}{c} \overset{O}{\underset{||}{C}}H \\ H \!-\!\!-\! OCH_3 \\ CH_3O \!-\!\!-\! H \\ H \!-\!\!-\! OCH_3 \\ H \!-\!\!-\! OH \\ CH_2OCH_3 \end{array}$$

SOLUTIONS TO PROBLEMS

22.1 (a) Two,
$$\begin{array}{c} CHO \\ *CHOH \\ *CHOH \\ CH_2OH \end{array}$$

(b) Two,
$$\begin{array}{c} CH_2OH \\ C\!=\!O \\ *CHOH \\ *CHOH \\ CH_2OH \end{array}$$

(c) There would be four stereoisomers (two sets of enantiomers) with each general structure: $2^2 = 4$.

22.2

Row 1 (aldoses, CHO top, CH$_2$OH bottom):

$$\begin{array}{cccc}
\text{CHO} & \text{CHO} & \text{CHO} & \text{CHO} \\
H\!-\!C\!-\!OH & HO\!-\!C\!-\!H & HO\!-\!C\!-\!H & H\!-\!C\!-\!OH \\
H\!-\!C\!-\!OH & HO\!-\!C\!-\!H & H\!-\!C\!-\!OH & HO\!-\!C\!-\!H \\
CH_2OH & CH_2OH & CH_2OH & CH_2OH \\
\textbf{D} & \textbf{L} & \textbf{D} & \textbf{L}
\end{array}$$

Row 2 (ketoses, CH$_2$OH / C=O top, CH$_2$OH bottom):

$$\begin{array}{cccc}
CH_2OH & CH_2OH & CH_2OH & CH_2OH \\
C\!=\!O & C\!=\!O & C\!=\!O & C\!=\!O \\
H\!-\!C\!-\!OH & HO\!-\!C\!-\!H & HO\!-\!C\!-\!H & H\!-\!C\!-\!OH \\
H\!-\!C\!-\!OH & HO\!-\!C\!-\!H & H\!-\!C\!-\!OH & HO\!-\!C\!-\!H \\
CH_2OH & CH_2OH & CH_2OH & CH_2OH \\
\textbf{D} & \textbf{L} & \textbf{D} & \textbf{L}
\end{array}$$

22.3 (a)

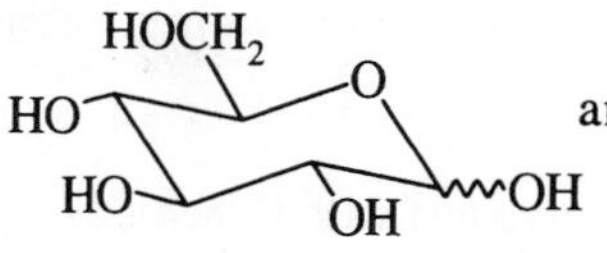

and

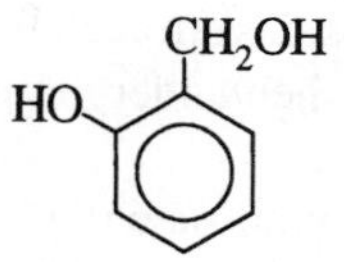

D-(+)-Glucose 2-Hydroxybenzyl alcohol

(b)

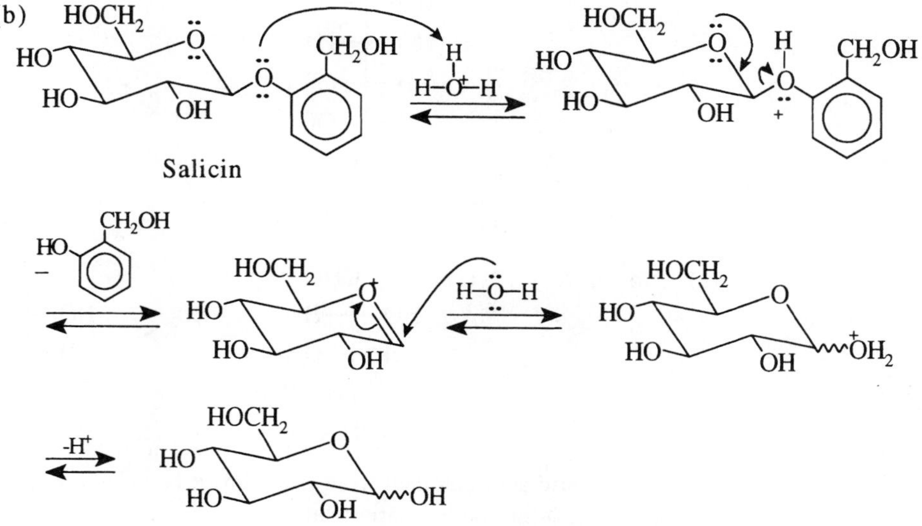

Salicin

22.4 Dissolve D-glucose in ethanol and then bubble in gaseous HCl.

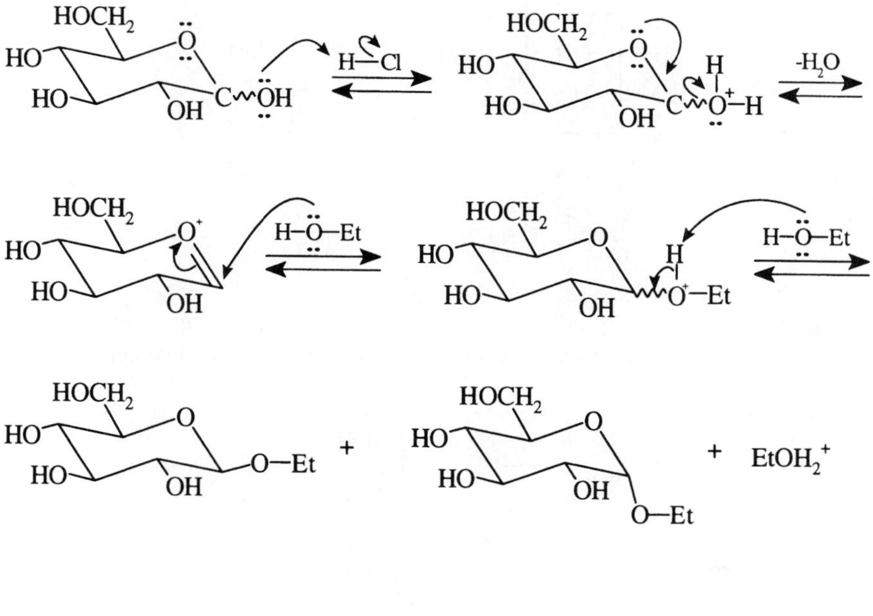

22.5 Since glycosides are acetals, they undergo hydrolysis in aqueous acid to form cyclic hemiacetals that then undergo mutarotation.

22.6 α-D-Glucopyranose will give a positive test with Benedict's or Tollens' solution because it is a cyclic hemiacetal. Methyl α-D-glucopyranoside, because it is a cyclic acetal, will not.

22.7 (a) Yes (b) (c) Yes (d)

$$
\begin{array}{c}
CO_2H \\
HO-\!\!\!\!-H \\
HO-\!\!\!\!-H \\
H-\!\!\!\!-OH \\
H-\!\!\!\!-OH \\
CO_2H
\end{array}
$$

D-Mannaric acid

$$
\begin{array}{c}
CO_2H \\
H-\!\!\!\!-OH \\
H-\!\!\!\!-OH \\
CO_2H
\end{array}
$$

(e) No (f)

$$
\begin{array}{c}
CHO \\
HO-\!\!\!\!-H \\
H-\!\!\!\!-OH \\
CH_2OH
\end{array}
\xrightarrow{HNO_3}
\begin{array}{c}
CO_2H \\
HO-\!\!\!\!-H \\
H-\!\!\!\!-OH \\
CO_2H
\end{array}
$$

D-Threose   D-Tartaric acid

(g) The aldaric acid obtained from D-erythrose is *meso*-tartaric acid; the aldaric acid obtained from D-threose is D-tartaric acid.

22.8

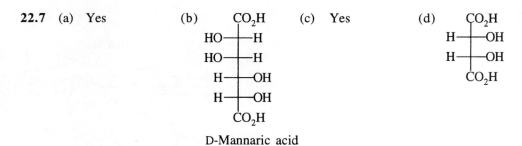

and

22.9 One way of predicting the products from a periodate oxidation is to place an –OH group on each carbon atom at the point where C–C bond cleavage has occurred:

$$
\begin{array}{c}
-\!\!\overset{|}{C}\!-OH \\
-\!\!\overset{|}{C}\!-OH
\end{array}
\xrightarrow{IO_4^-}
\begin{array}{c}
-\!\!\overset{|}{C}\!-OH \\
OH
\end{array}
$$

$$
+
$$

$$
\begin{array}{c}
OH \\
-\!\!\overset{|}{C}\!-OH
\end{array}
$$

Then if we recall (Section 16.7A) that *gem*-diols are usually unstable and lose water to produce carbonyl compounds, we get the following results:

$$-\overset{|}{\underset{\underset{OH}{|}}{C}} \!\!-\!\! O\!-\!H \longrightarrow -\overset{|}{C}\!\!=\!\!O \;+\; H_2O$$

$$\underset{-\overset{|}{\underset{|}{C}}\!-\!O\!-\!H}{\overset{OH}{|}} \longrightarrow -\overset{|}{C}\!\!=\!\!O \;+\; H_2O$$

Let us apply this procedure to several examples here while we remember that for every C–C bond that is broken 1 mol of HIO_4 is consumed.

(a)
$$\begin{array}{c} CH_3 \\ H-\overset{|}{C}-OH \\ \text{------}\!|\!\text{------} \\ H-\overset{|}{C}-OH \\ CH_3 \end{array} + HIO_4 \longrightarrow \left\{ \begin{array}{c} CH_3 \\ H-\overset{|}{\underset{\underset{OH}{|}}{C}}-O-H \\ + \\ \overset{OH}{\underset{\underset{CH_3}{|}}{C}} \\ H-C-O-H \end{array} \right\} \xrightarrow{-2\,H_2O} 2\,CH_3\overset{O}{\overset{\|}{C}}-H$$

(b)
$$\begin{array}{c} H \\ H-\overset{|}{C}-OH \\ \text{------}\!|\!\text{------} \\ H-\overset{|}{C}-OH \\ \text{------}\!|\!\text{------} \\ H-\overset{|}{C}-OH \\ CH_3 \end{array} + 2\,HIO_4 \longrightarrow \left\{ \begin{array}{c} H \\ H-\overset{|}{\underset{\underset{OH}{|}}{C}}-O-H \\ + \\ O-H \\ H-\overset{|}{\underset{\underset{OH}{|}}{C}}-O-H \\ + \\ \overset{OH}{\underset{\underset{CH_3}{|}}{C}} \\ H-C-O-H \end{array} \right\} \xrightarrow{-3\,H_2O} \begin{array}{c} H \\ H-C=O \\ + \\ O \\ H-\overset{\|}{C}-OH \\ + \\ H-C=O \\ CH_3 \end{array}$$

(c)
$$\begin{array}{c} H \\ H-\overset{|}{C}-OH \\ \text{------}\!|\!\text{------} \\ H-\overset{|}{C}-OH \\ H-\overset{|}{C}-OCH_3 \\ OCH_3 \end{array} + HIO_4 \longrightarrow \left\{ \begin{array}{c} H \\ H-\overset{|}{C}-OH \\ OH \\ + \\ OH \\ H-\overset{|}{C}-OH \\ H-\overset{|}{C}-OCH_3 \\ OCH_3 \end{array} \right\} \xrightarrow{-2\,H_2O} \begin{array}{c} H \\ H-C=O \\ + \\ O \\ H-\overset{\|}{C} \\ H-\overset{|}{C}-OCH_3 \\ OCH_3 \end{array}$$

(d)

$$\begin{array}{c} \text{H} \\ \text{H}-\overset{|}{\underset{|}{\text{C}}}-\text{OH} \\ \text{------------} \\ \text{H}-\overset{|}{\underset{|}{\text{C}}}-\text{OH} \\ \text{------------} \\ \overset{|}{\underset{|}{\text{C}}}=\text{O} \\ \text{CH}_3 \end{array} + 2\,\text{HIO}_4 \longrightarrow$$

$$\left\{\begin{array}{c} \text{H} \\ \text{H}-\overset{|}{\underset{|}{\text{C}}}-\text{OH} \\ \text{OH} \\ + \\ \text{OH} \\ \text{H}-\overset{|}{\underset{|}{\text{C}}}-\text{OH} \\ \text{OH} \\ + \\ \text{OH} \\ \overset{|}{\underset{|}{\text{C}}}=\text{O} \\ \text{CH}_3 \end{array}\right\} \xrightarrow{-2\,\text{H}_2\text{O}}$$

$$\begin{array}{c} \text{H} \\ \text{H}-\text{C}=\text{O} \\ + \\ \overset{\text{O}}{\overset{\|}{\text{H}-\text{C}-\text{OH}}} \\ + \\ \overset{\text{O}}{\overset{\|}{\text{CH}_3\text{COH}}} \end{array}$$

(e)

$$\begin{array}{c} \text{CH}_3 \\ \overset{|}{\underset{|}{\text{C}}}=\text{O} \\ \text{------------} \\ \text{H}-\overset{|}{\underset{|}{\text{C}}}-\text{OH} \\ \text{------------} \\ \overset{|}{\underset{|}{\text{C}}}=\text{O} \\ \text{CH}_3 \end{array} + 2\,\text{HIO}_4 \longrightarrow$$

$$\left\{\begin{array}{c} \text{CH}_3 \\ \overset{|}{\underset{|}{\text{C}}}=\text{O} \\ \text{OH} \\ + \\ \text{OH} \\ \text{H}-\overset{|}{\underset{|}{\text{C}}}-\text{OH} \\ \text{OH} \\ + \\ \text{OH} \\ \overset{|}{\underset{|}{\text{C}}}=\text{O} \\ \text{CH}_3 \end{array}\right\} \xrightarrow{-2\,\text{H}_2\text{O}}$$

$$\begin{array}{c} \overset{\text{O}}{\overset{\|}{2\,\text{CH}_3\text{COH}}} \\ + \\ \overset{\text{O}}{\overset{\|}{\text{HCOH}}} \end{array}$$

(f)

$$\begin{array}{c} \text{H}_2\text{C} \qquad \overset{\text{H}}{\underset{|}{\text{C}}}-\text{OH} \\ \text{H}_2\text{C} \qquad\qquad \text{------------} \\ \text{H}_2\text{C} \qquad \underset{|}{\overset{|}{\text{C}}}-\text{OH} \\ \qquad\qquad \text{H} \end{array} + \text{HIO}_4 \longrightarrow$$

$$\begin{array}{c} \text{H}_2\text{C}\quad\overset{\text{H}}{\underset{|}{\text{C}}}-\text{OH} \\ \qquad\quad \text{OH} \\ \\ \qquad\quad \text{OH} \\ \text{H}_2\text{C}-\overset{|}{\underset{|}{\text{C}}}-\text{OH} \\ \qquad\quad \text{H} \end{array} \xrightarrow{-2\,\text{H}_2\text{O}}$$

$$\overset{\text{O}}{\overset{\|}{\text{HCCH}_2\text{CH}_2\text{CH}_2\text{CH}}}\overset{\text{O}}{}$$

(g)

$$\begin{array}{c} \text{H} \\ \text{H}-\overset{|}{\underset{|}{\text{C}}}-\text{OH} \\ \text{------------} \\ \text{H}_3\text{C}-\overset{|}{\underset{|}{\text{C}}}-\text{OH} \\ \text{CH}_3 \end{array} + \text{HIO}_4 \longrightarrow$$

$$\left\{\begin{array}{c} \text{H} \\ \text{H}-\overset{|}{\underset{|}{\text{C}}}-\text{OH} \\ \text{OH} \\ + \\ \text{OH} \\ \text{H}_3\text{C}-\overset{|}{\underset{|}{\text{C}}}-\text{OH} \\ \text{CH}_3 \end{array}\right\} \xrightarrow{-2\,\text{H}_2\text{O}}$$

$$\begin{array}{c} \text{H} \\ \text{H}-\text{C}=\text{O} \\ + \\ \text{H}_3\text{C}-\text{C}=\text{O} \\ \text{CH}_3 \end{array}$$

(h)

$$
\begin{array}{c}
\overset{\displaystyle O}{\underset{\displaystyle \|}{\text{H}-\text{C}}} \\
\text{H}-\text{C}-\text{OH} \\
\text{H}-\text{C}-\text{OH} \\
\text{H}-\text{C}-\text{OH} \\
\underset{\displaystyle \text{H}}{\text{H}-\text{C}-\text{OH}}
\end{array}
\quad + \ 3\,\text{HIO}_4 \longrightarrow
$$

D-Erythrose

$$
\left\{
\begin{array}{c}
\overset{O}{\overset{\|}{\text{H}-\text{C}-\text{OH}}} \\
+ \ \ \text{OH} \\
\text{H}-\text{C}-\text{OH} \\
\text{OH} \\
+ \\
\text{OH} \\
\text{H}-\text{C}-\text{OH} \\
\text{OH} \\
+ \\
\text{OH} \\
\underset{\text{H}}{\text{H}-\text{C}-\text{OH}}
\end{array}
\right\}
\xrightarrow{-3\,\text{H}_2\text{O}}
\begin{array}{c}
\overset{O}{\overset{\|}{3\ \text{HCOH}}} \\
+ \\
\overset{O}{\underset{\|}{\text{HCH}}}
\end{array}
$$

22.10 Oxidation of an aldohexose and a ketohexose would each require 5 mol of HIO_4 but would give different results.

$$
\begin{array}{c}
\text{CHO} \\
\text{CHOH} \\
\text{CHOH} \\
\text{CHOH} \\
\text{CHOH} \\
\text{CH}_2\text{OH}
\end{array}
\quad + \ 5\,\text{HIO}_4 \longrightarrow
\begin{array}{c}
\text{HCO}_2\text{H} \\
+ \\
\text{HCO}_2\text{H} \\
+ \\
\text{HCO}_2\text{H} \\
+ \\
\text{HCO}_2\text{H} \\
+ \\
\text{HCO}_2\text{H} \\
+ \\
\text{HCHO}
\end{array}
\quad (5\ \text{HCO}_2\text{H}\ +\ \text{HCHO})
$$

Aldohexose

$$
\begin{array}{c}
\text{CH}_2\text{OH} \\
\text{C}=\text{O} \\
\text{CHOH} \\
\text{CHOH} \\
\text{CHOH} \\
\text{CH}_2\text{OH}
\end{array}
\quad + \ 5\,\text{HIO}_4 \longrightarrow
\begin{array}{c}
\text{HCHO} \\
+ \\
\text{CO}_2 \\
+ \\
\text{HCO}_2\text{H} \\
+ \\
\text{HCO}_2\text{H} \\
+ \\
\text{HCO}_2\text{H} \\
+ \\
\text{HCHO}
\end{array}
\quad (3\ \text{HCO}_2\text{H}\ +\ 2\ \text{HCHO}\ +\ \text{CO}_2)
$$

Ketohexose

22.11 (a) Yes, D-glucitol would be optically active; only those alditols whose molecules possess a plane of symmetry would be optically inactive.

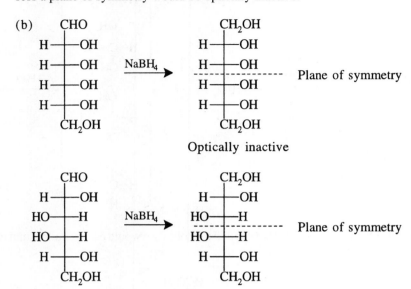

(b)

CHO
H——OH
H——OH
H——OH
H——OH
CH₂OH

NaBH₄ →

CH₂OH
H——OH
H——OH
H——OH
H——OH
CH₂OH

------ Plane of symmetry

Optically inactive

CHO
H——OH
HO——H
HO——H
H——OH
CH₂OH

NaBH₄ →

CH₂OH
H——OH
HO——H
HO——H
H——OH
CH₂OH

------ Plane of symmetry

Optically inactive

22.12 (a)

CH₂OH
C=O
HO——H
H——OH
H——OH
CH₂OH

C₆H₅NHNH₂ →

CH=NNHC₆H₅
C=NNHC₆H₅
HO——H
H——OH
H——OH
CH₂OH

(b) This experiment shows that D-glucose and D-fructose have the same configurations at C3, C4, and C5.

22.13 (a)

CHO
HO——H
HO——H
CH₂OH

L-Erythrose

CHO
H——OH
HO——H
CH₂OH

L-Threose

(b) L-Glyceraldehyde

CHO
HO——H
CH₂OH

22.14 (a)

$$
\begin{array}{c}
\text{CHO} \\
\text{H}\!-\!\!-\!\text{OH} \\
\text{H}\!-\!\!-\!\text{OH} \\
\text{CH}_2\text{OH}
\end{array}
$$

D-(–)-Erythrose

HCN

	Epimeric cyanohydrins (separated)	
$\begin{array}{c}\text{CN}\\ \text{H}\!-\!\!-\!\text{OH}\\ \text{H}\!-\!\!-\!\text{OH}\\ \text{H}\!-\!\!-\!\text{OH}\\ \text{CH}_2\text{OH}\end{array}$		$\begin{array}{c}\text{CN}\\ \text{HO}\!-\!\!-\!\text{H}\\ \text{H}\!-\!\!-\!\text{OH}\\ \text{H}\!-\!\!-\!\text{OH}\\ \text{CH}_2\text{OH}\end{array}$

(1) Ba(OH)$_2$
(2) H$_3$O$^+$

(1) Ba(OH$_2$)
(2) H$_3$O$^+$

	Epimeric aldonic acids	
$\begin{array}{c}\overset{\text{O}}{\overset{\|}{\text{C}}}\!-\!\text{OH}\\ \text{H}\!-\!\!-\!\text{OH}\\ \text{H}\!-\!\!-\!\text{OH}\\ \text{H}\!-\!\!-\!\text{OH}\\ \text{CH}_2\text{OH}\end{array}$		$\begin{array}{c}\overset{\text{O}}{\overset{\|}{\text{C}}}\!-\!\text{OH}\\ \text{HO}\!-\!\!-\!\text{H}\\ \text{H}\!-\!\!-\!\text{OH}\\ \text{H}\!-\!\!-\!\text{OH}\\ \text{CH}_2\text{OH}\end{array}$

–H$_2$O ⇅

Epimeric γ-aldonolactones

–H$_2$O ⇅

Na-Hg, H$_2$O
pH 3-5

Na-Hg, H$_2$O
pH 3-5

$\begin{array}{c}\overset{\text{O}}{\overset{\|\|}{\text{C}}}\!-\!\text{H}\\ \text{H}\!-\!\!-\!\text{OH}\\ \text{H}\!-\!\!-\!\text{OH}\\ \text{H}\!-\!\!-\!\text{OH}\\ \text{CH}_2\text{OH}\end{array}$		$\begin{array}{c}\overset{\text{O}}{\overset{\|\|}{\text{C}}}\!-\!\text{H}\\ \text{HO}\!-\!\!-\!\text{H}\\ \text{H}\!-\!\!-\!\text{OH}\\ \text{H}\!-\!\!-\!\text{OH}\\ \text{CH}_2\text{OH}\end{array}$

(b)

$$
\begin{array}{c}
\overset{\displaystyle O}{\overset{\displaystyle \|}{C}}\text{-H} \\
\text{H}-\!\!\!-\text{OH} \\
\text{H}-\!\!\!-\text{OH} \\
\text{H}-\!\!\!-\text{OH} \\
\text{CH}_2\text{OH}
\end{array}
\qquad \xrightarrow{\text{HNO}_3} \qquad
\begin{array}{c}
\overset{\displaystyle O}{\overset{\displaystyle \|}{C}}\text{-OH} \\
\text{H}-\!\!\!-\text{OH} \\
\text{H}-\!\!\!-\text{OH} \\
\text{H}-\!\!\!-\text{OH} \\
\overset{\displaystyle }{C}\text{-OH} \\
\overset{\displaystyle \|}{\displaystyle O}
\end{array}
$$

D-(–)-Ribose

Optically inactive

$$
\begin{array}{c}
\overset{\displaystyle O}{\overset{\displaystyle \|}{C}}\text{-H} \\
\text{HO}-\!\!\!-\text{H} \\
\text{H}-\!\!\!-\text{OH} \\
\text{H}-\!\!\!-\text{OH} \\
\text{CH}_2\text{OH}
\end{array}
\qquad \xrightarrow{\text{HNO}_3} \qquad
\begin{array}{c}
\overset{\displaystyle O}{\overset{\displaystyle \|}{C}}\text{-OH} \\
\text{HO}-\!\!\!-\text{H} \\
\text{H}-\!\!\!-\text{OH} \\
\text{H}-\!\!\!-\text{OH} \\
\overset{\displaystyle }{C}\text{-OH} \\
\overset{\displaystyle \|}{\displaystyle O}
\end{array}
$$

D-(–)-Arabinose

Optically active

22.15 A Kiliani-Fischer synthesis starting with D-(−)-threose would yield **I** and **II**.

$$
\begin{array}{c}
\text{CHO} \\
\text{H}-\!\!\!-\text{OH} \\
\text{HO}-\!\!\!-\text{H} \\
\text{H}-\!\!\!-\text{OH} \\
\text{CH}_2\text{OH}
\end{array}
\qquad\qquad
\begin{array}{c}
\text{CHO} \\
\text{HO}-\!\!\!-\text{H} \\
\text{HO}-\!\!\!-\text{H} \\
\text{H}-\!\!\!-\text{OH} \\
\text{CH}_2\text{OH}
\end{array}
$$

I   **II**

D-(+)-Xylose   D-(–)-Lyxose

I must be D-(+)-xylose because, when oxidized by nitric acid, it yields an optically inactive aldaric acid:

$$
\textbf{I} \quad \xrightarrow{\text{HNO}_3} \quad
\begin{array}{c}
\text{CO}_2\text{H} \\
\text{H}-\!\!\!-\text{OH} \\
\text{HO}-\!\!\!-\text{H} \\
\text{H}-\!\!\!-\text{OH} \\
\text{CO}_2\text{H}
\end{array}
$$

Optically inactive

II must be D-(−)-lyxose because, when oxidized by nitric acid, it yields an optically active aldaric acid:

$$
\begin{array}{c}
CO_2H \\
HO \!-\!\!|\!-\! H \\
HO \!-\!\!|\!-\! H \\
H \!-\!\!|\!-\! OH \\
CO_2H
\end{array}
$$

II $\xrightarrow{\text{HNO}_3}$

Optically active

22.16

$$
\begin{array}{c}
CHO \\
HO\!-\!\!|\!-\!H \\
HO\!-\!\!|\!-\!H \\
HO\!-\!\!|\!-\!H \\
CH_2OH
\end{array}
\qquad
\begin{array}{c}
CHO \\
H\!-\!\!|\!-\!OH \\
HO\!-\!\!|\!-\!H \\
HO\!-\!\!|\!-\!H \\
CH_2OH
\end{array}
\qquad
\begin{array}{c}
CHO \\
HO\!-\!\!|\!-\!H \\
H\!-\!\!|\!-\!OH \\
HO\!-\!\!|\!-\!H \\
CH_2OH
\end{array}
\qquad
\begin{array}{c}
CHO \\
H\!-\!\!|\!-\!OH \\
H\!-\!\!|\!-\!OH \\
HO\!-\!\!|\!-\!H \\
CH_2OH
\end{array}
$$

L-(+)-Ribose   L-(+)-Arabinose   L-(−)-Xylose   L-(+)-Lyxose

22.17 Since D-(+)-galactose yields an optically inactive aldaric acid, it must have either structure **III** or structure **IV**.

$$
\begin{array}{c}
CHO \\
H\!-\!\!|\!-\!OH \\
H\!-\!\!|\!-\!OH \\
H\!-\!\!|\!-\!OH \\
H\!-\!\!|\!-\!OH \\
CH_2OH
\end{array}
\xrightarrow{\text{HNO}_3}
\begin{array}{c}
CO_2H \\
H\!-\!\!|\!-\!OH \\
H\!-\!\!|\!-\!OH \\
H\!-\!\!|\!-\!OH \\
H\!-\!\!|\!-\!OH \\
CO_2H
\end{array}
\qquad
\begin{array}{c}
CHO \\
H\!-\!\!|\!-\!OH \\
HO\!-\!\!|\!-\!H \\
HO\!-\!\!|\!-\!H \\
H\!-\!\!|\!-\!OH \\
CH_2OH
\end{array}
\xrightarrow{\text{HNO}_3}
\begin{array}{c}
CO_2H \\
H\!-\!\!|\!-\!OH \\
HO\!-\!\!|\!-\!H \\
HO\!-\!\!|\!-\!H \\
H\!-\!\!|\!-\!OH \\
CO_2H
\end{array}
$$

III   Optically inactive   **IV**   Optically inactive

A Ruff degradation beginning with **III** would yield D-(−)-ribose

III $\xrightarrow[\text{H}_2\text{O}]{\text{Br}_2}$ $\xrightarrow[\text{Fe}_2(\text{SO}_4)_3]{\text{H}_2\text{O}_2}$

$$
\begin{array}{c}
CHO \\
H\!-\!\!|\!-\!OH \\
H\!-\!\!|\!-\!OH \\
H\!-\!\!|\!-\!OH \\
CH_2OH
\end{array}
$$

D-(−)-Ribose

A Ruff degradation beginning with **IV** would yield D-(−)-lyxose: thus, D-(+)-galactose must have structure **IV**.

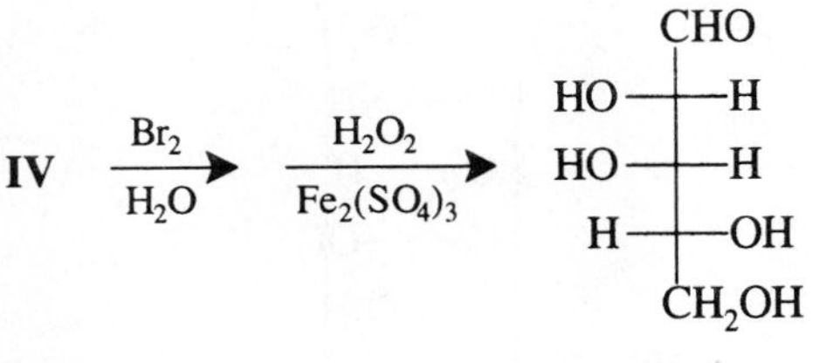

$$\textbf{IV} \xrightarrow[\text{H}_2\text{O}]{\text{Br}_2} \xrightarrow[\text{Fe}_2(\text{SO}_4)_3]{\text{H}_2\text{O}_2}$$

CHO
HO——H
HO——H
H——OH
CH$_2$OH

D-(−)-Lyxose

22.18 D-(+)-glucose, as shown here.

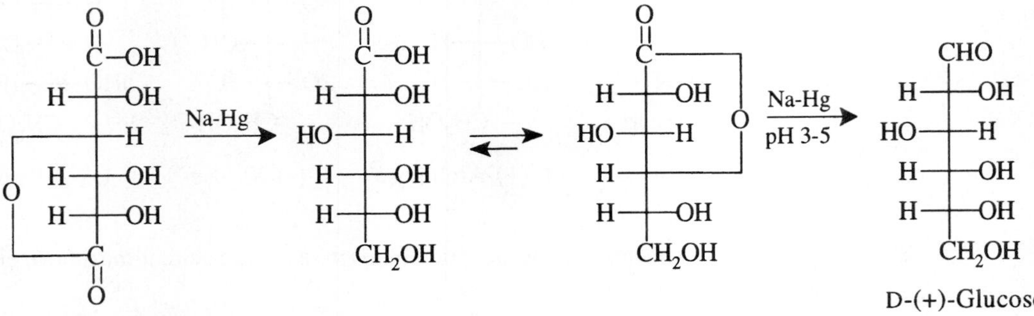

The other γ-lactone
of D-glucaric acid

$\xrightarrow{\text{Na-Hg}}$

$\xrightleftharpoons{}$

$\xrightarrow[\text{pH 3-5}]{\text{Na-Hg}}$

CHO
H——OH
HO——H
H——OH
H——OH
CH$_2$OH

D-(+)-Glucose

22.19

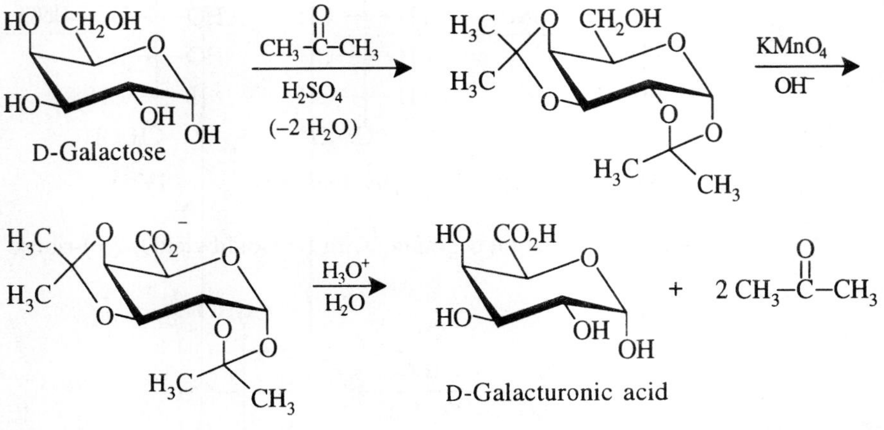

D-Galactose $\xrightarrow[\text{H}_2\text{SO}_4 \;(-2\,\text{H}_2\text{O})]{\text{CH}_3-\overset{\text{O}}{\overset{\|}{\text{C}}}-\text{CH}_3}$ $\xrightarrow[\text{OH}^-]{\text{KMnO}_4}$

$\xrightarrow[\text{H}_2\text{O}]{\text{H}_3\text{O}^+}$ D-Galacturonic acid $+$ $2\,\text{CH}_3-\overset{\text{O}}{\overset{\|}{\text{C}}}-\text{CH}_3$

22.20 (a)

CHO
CHOH
CHOH
CHOH
CH$_2$OH

(b)

CH$_2$OH
C=O
CHOH
CHOH
CHOH
CH$_2$OH

(c)

CHO
(CHOH)$_n$
HO—C—H
CH$_2$OH

or

CH$_2$OH
C=O
(CHOH)$_n$
HO—C—H
CH$_2$OH

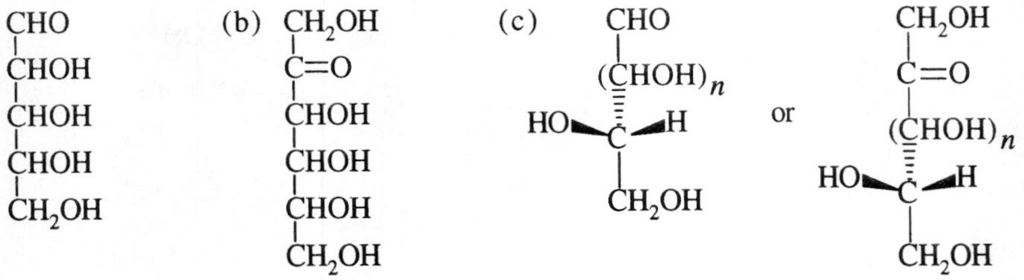

(d)
$$\begin{array}{l} \text{CHOR} \\ (\text{CHOH})_n \\ \text{CH} \\ \text{CH}_2\text{OH} \end{array} \Big] \text{O}$$

(e)
$$\begin{array}{l} \text{CO}_2\text{H} \\ (\text{CHOH})_n \\ \text{CH}_2\text{OH} \end{array}$$

(f)
$$\begin{array}{l} \text{CO}_2\text{H} \\ (\text{CHOH})_n \\ \text{CO}_2\text{H} \end{array}$$

(g)
$$\begin{array}{l} \text{C}=\text{O} \\ (\text{CHOH})_n \\ \text{CH} \\ \text{CH}_2\text{OH} \end{array} \Big] \text{O}$$

(h)
$$\begin{array}{l} \text{CHOH} \\ \text{CHOH} \\ \text{CHOH} \\ \text{CHOH} \\ \text{CH} \\ \text{CH}_2\text{OH} \end{array} \Big] \text{O} \quad \text{or}$$

CH$_2$OH—CH—O—CHOH / CHOH—CHOH

(i)
$$\begin{array}{l} \text{CHOH} \\ \text{CHOH} \\ \text{CHOH} \\ \text{CH} \\ \text{CHOH} \\ \text{CH}_2\text{OH} \end{array} \Big] \text{O} \quad \text{or}$$

CH$_2$OH / CHOH / CH—O—CHOH / CHOH—CHOH

(j) Any sugar that has a free aldehyde or ketone group or one that exists as a cyclic hemiacetal. The following are examples:

$$\begin{array}{l} \text{CHO} \\ (\text{CHOH})_n \\ \text{CHOH} \\ \text{CH}_2\text{OH} \end{array} \rightleftarrows \begin{array}{l} \text{OH} \\ \text{CH} \\ (\text{CHOH})_n \\ \text{CH} \\ \text{CH}_2\text{OH} \end{array} \Big] \text{O} \quad \text{or} \quad \begin{array}{l} \text{CH}_2\text{OH} \\ \text{C}=\text{O} \\ (\text{CHOH})_n \\ \text{CHOH} \\ \text{CH}_2\text{OH} \end{array} \rightleftarrows \begin{array}{l} \text{CH}_2\text{OH} \\ \text{C}-\text{OH} \\ (\text{CHOH})_n \\ \text{CH} \\ \text{CH}_2\text{OH} \end{array} \Big] \text{O}$$

(k)
CH$_2$OH / CH—O—CHOR / CHOH—CHOH

(l)
CH$_2$OH / CHOH / CH—O—CHOR / CHOH—CHOH

(m)   Any two aldoses that differ only in configuration at C2. (See also Section 22.8 for a broader definition.) D-Erythrose and D-threose are examples.

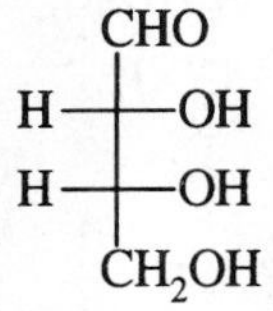

CHO
H——OH
H——OH
CH$_2$OH
D-Erythrose

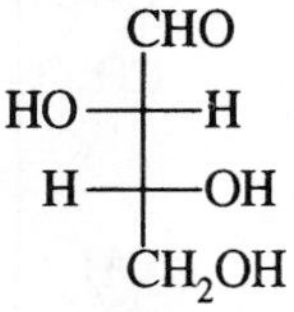

CHO
HO——H
H——OH
CH$_2$OH
D-Threose

(n)   Cyclic sugars that differ only in the configuration of C1. The following are examples:

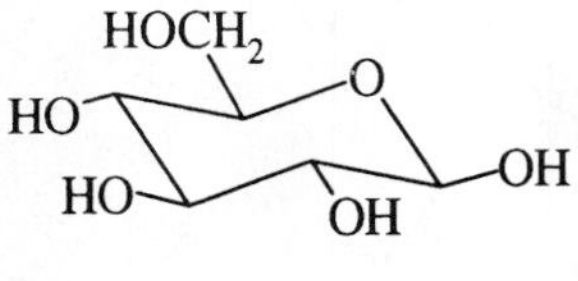

HOCH$_2$ O
HO
HO   OH   OH

and

HOCH$_2$ O
HO
HO   OH   OH

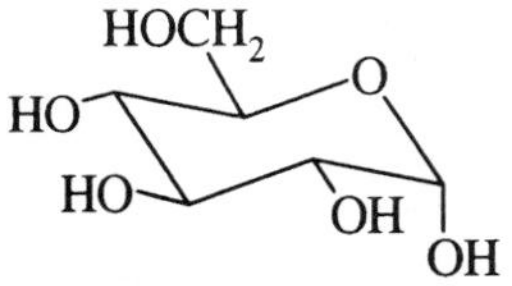

(o)   CH=NNHC$_6$H$_5$
C=NNHC$_6$H$_5$
(CHOH)$_n$
CH$_2$OH

(p)   Maltose is an example:

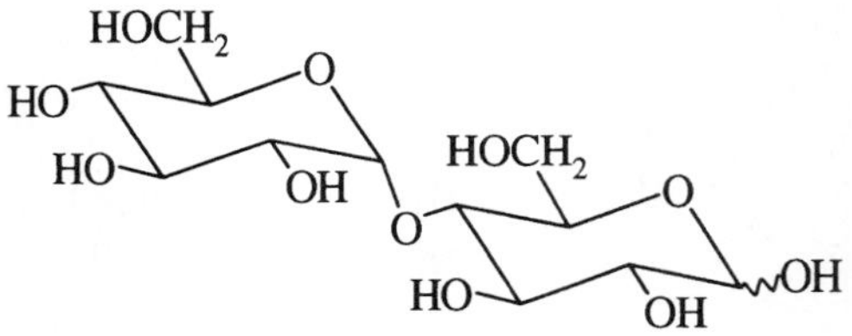

HOCH$_2$ O
HO
HO   OH   O   HOCH$_2$ O
HO   OH   OH

(q)   Amylose is an example:

HOCH$_2$ O
OH
HO   O   HOCH$_2$ O
OH
OH   O   HOCH$_2$ O
OH   OH
OH $_n$ OH

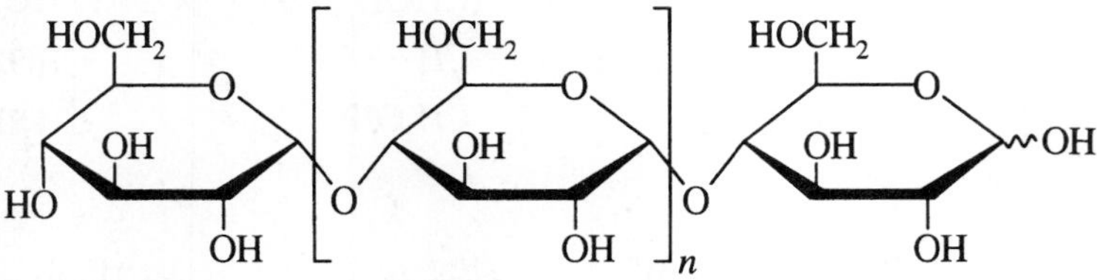

(r)   Any sugar in which all potential carbonyl groups are present as acetals (i.e., as glycosides). Sucrose (Section 22.12A) is an example of a nonreducing disaccharide; the methyl D-glucopyranosides (Section 22.4) are examples of nonreducing monosaccharides.

22.21   (a)

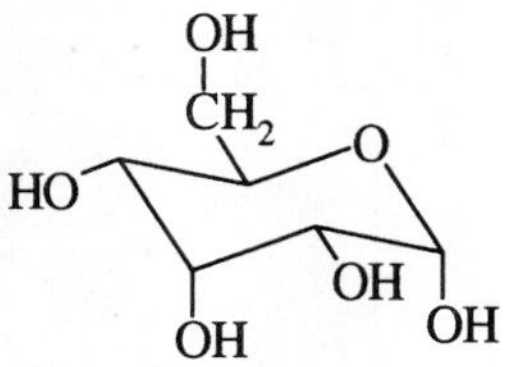

OH
CH$_2$ O
HO
OH   OH
OH

(b)

OH
CH$_2$ O
HO
OH   OCH$_3$
OH

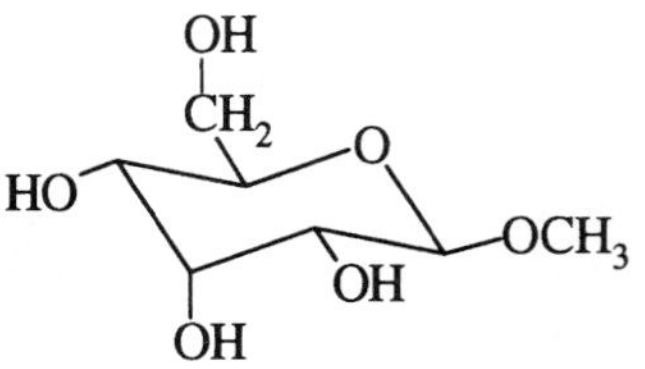

(c)

A methyl ribofuranoside would consume only 1 mol of HIO_4; a methyl ribopyranoside would consume 2 mol of HIO_4 and would also produce 1 mol of formic acid.

22.23 One anomer of D-mannose is dextrorotatory ($[\alpha]_D^{25} = +29.3°$); the other is levorotatory ($[\alpha]_D^{25} = -17.0°$).

22.24 The microorganism selectively oxidizes the –CHOH group of D-glucitol that corresponds to C5 of D-glucose.

D-Glucose $\xrightarrow{\text{H}_2 / \text{Ni}}$ D-Glucitol $\xrightarrow[\text{suboxydans}]{\text{O}_2 / \textit{Acetobacter}}$ L-Sorbose

22.25 L-Gulose and L-idose would yield the same phenylosazone as L-sorbose.

22.26

22.27 A is D-altrose, **B** is D-talose, **C** is D-galactose.

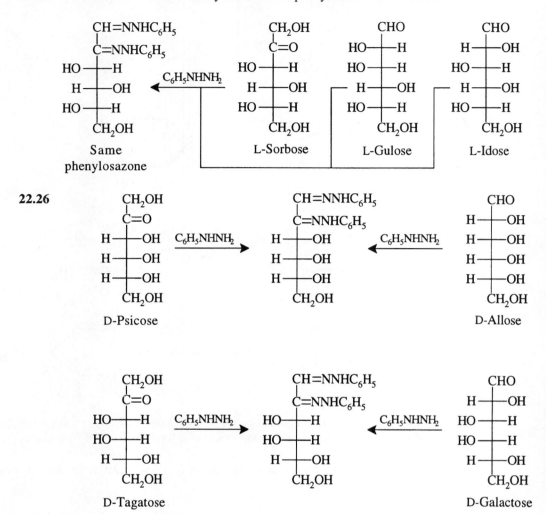

```
CH=NNHC6H5                                          CH=NNHC6H5
C=NNHC6H5                                           C=NNHC6H5
H——OH                                          HO——H
H——OH          <Different phenylosazones>      HO——H
H——OH                                          H——OH
CH2OH                                               CH2OH
```

```
CHO                  CH=NNHC6H5                    CHO
H——OH               C=NNHC6H5                  HO——H
HO——H   C6H5NHNH2→  HO——H    ←C6H5NHNH2   HO——H
HO——H               HO——H                   HO——H
H——OH               H——OH                   H——OH
CH2OH                CH2OH                       CH2OH

D-Galactose      Same phenylosazone           D-Talose
   C                                              B
```

```
   | H2, Ni                                      H2, Ni |
   ↓                                                    ↓

CH2OH                                           CH2OH
H——OH                                      HO——H
HO——H        <Different alditols>          HO——H
HO——H                                      HO——H
H——OH                                      H——OH
CH2OH                                           CH2OH
```

(*Note*: If we had designated D-talose as **A**, and D-altrose as **B**, then **C** is D-allose).

22.28

```
    O
    ||
    CH                              CH2OH
H——OH                          H——OH
HO——H      NaBH4        HO——H
H——OH      or H2/Pt→    H——OH
    CH2OH                          CH2OH

D-Xylose                        D-Xylitol
```

22.29

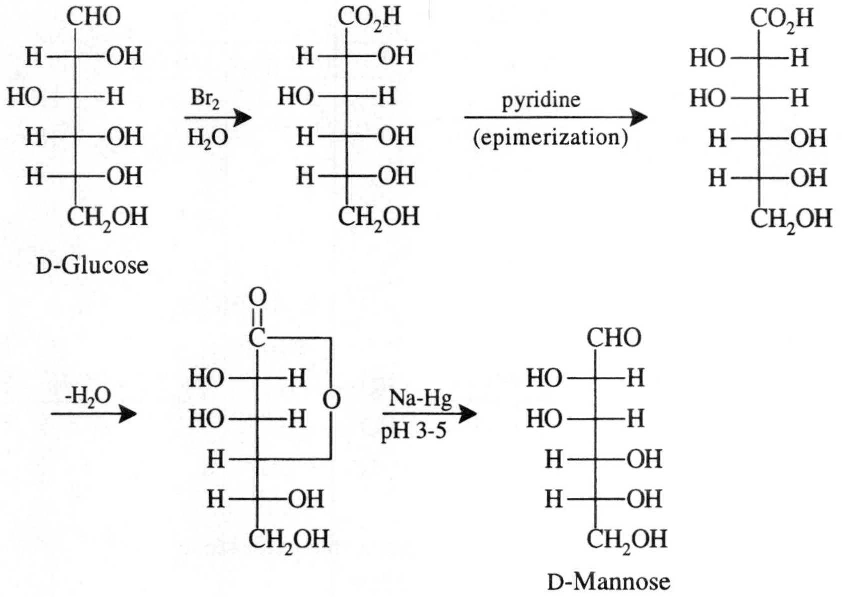

D-Glucose

D-Mannose

22.30 The conformation of D-idopyranose with four equatorial –OH groups and an axial –CH_2OH group is more stable than the one with four axial –OH groups and an equatorial –CH_2OH group.

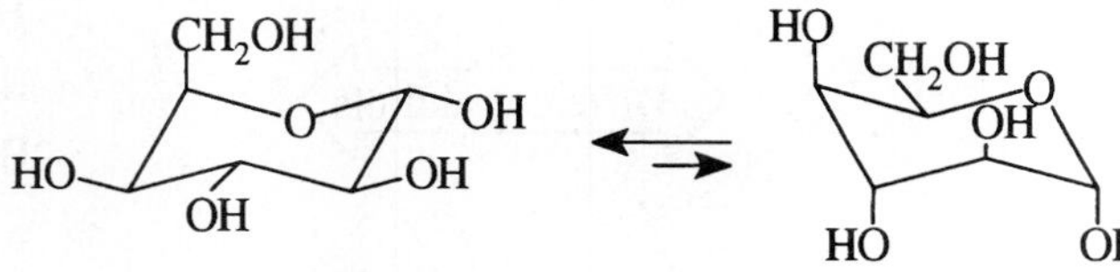

More stable

4 Equatorial –OH groups
1 Axial –CH_2OH group

Less stable

4 Axial –OH groups
1 Equatorial –CH_2OH group

22.31 (a)   The anhydro sugar is formed when the axial –CH_2OH group reacts with C1 to form a cyclic acetal.

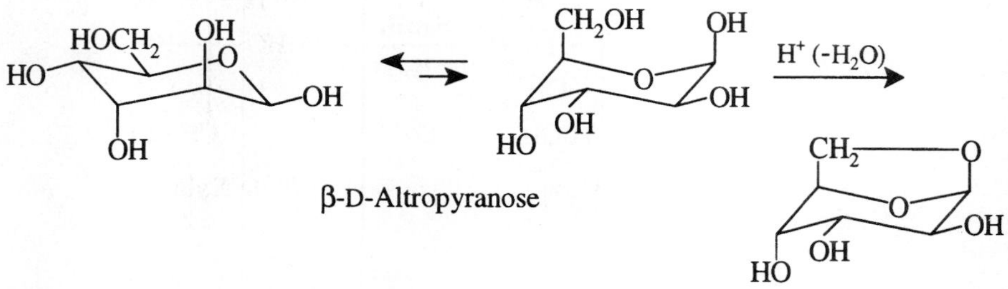

β-D-Altropyranose

Anhydro sugar

Because the anhydro sugar is an acetal (i.e., an internal glycoside), it is a nonreducing sugar.

Methylation followed by acid hydrolysis converts the anhydro sugar to 2,3,4-tri-O-methyl-D-altrose:

Anhydro-β-D-altropyranose

2,3,4-Tri-O-methyl-D-altrose

(b)   Formation of an anhydro sugar requires that the monosaccharide adopt a chair conformation with the –CH_2OH group axial. With β-D-altropyranose this requires that two –OH groups be axial as well. With β-D-glucopyranose, however, it requires that all four –OH groups become axial and thus that the molecule adopt a very unstable conformation:

Highly unstable conformation

β-D-Glucopyranose

Anhydro-β-D-glucopyranose

22.32 1.   The molecular formula and the results of acid hydrolysis show that lactose is a disaccharide composed of D-glucose and D-galactose. The fact that lactose is hydrolyzed by a *β-galactosidase* indicates that galactose is present as a glycoside and that the glycosidic linkage is beta to the galactose ring.

2.   That lactose is a reducing sugar, forms a phenylosazone, and undergoes mutarotation indicates that one ring (presumably that of D-glucose) is present as a hemiacetal and thus is capable of existing to a limited extent as an aldehyde.

3.   This experiment confirms that the D-glucose unit is present as a cyclic hemiacetal and that the D-galactose unit is present as a cyclic glycoside.

4.   That 2,3,4,6-tetra-*O*-methyl-D-galactose is obtained in this experiment indicates (by virtue of the free –OH at C5) that the galactose ring of lactose is present as a pyranoside. That the methylated gluconic acid obtained from this experiment has a free –OH group at C4 indicates that the C4 oxygen atom of the glucose unit is connected in a glycosidic linkage to the galactose unit.

　　　　Now only the size of the glucose ring remains in question, and the answer to this is provided by experiment 5.

5.   That methylation of lactose and subsequent hydrolysis gives 2,3,6-tri-*O*-methyl-D-glucose—that it gives a methylated glucose derivative with a free –OH at C4 and C5—demonstrates that the glucose ring is present as a pyranose. (We know already that the oxygen at C4 is connected in a glycosidic linkage to the galactose unit; thus, a free –OH at C5 indicates that the C5 oxygen atom is a part of the hemiacetal group of the glucose unit and that the ring is six membered.)

22.33

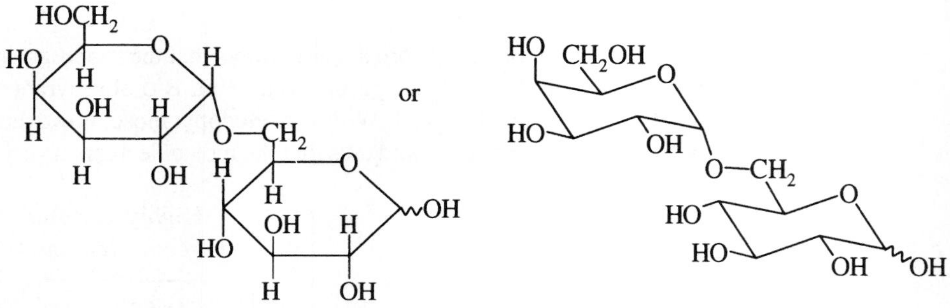

6-*O*-(α-D-Galactopyranosyl)-D-glucopyranose

　　　　We arrive at this conclusion from the data given:

1.   That melibiose is a reducing sugar and that it undergoes mutarotation and forms a phenylosazone indicate that one monosaccharide is present as a cyclic hemiacetal.

2.   That acid hydrolysis gives D-galatose and D-glucose indicates that melibiose is a disaccharide composed of one D-galactose unit and one D-glucose unit. That melibiose is hydrolyzed by an α-galactosidase suggests that melibiose is an α-D-galactosyl-D-glucose.

3.   Oxidation of melibiose to melibionic acid and subsequent hydrolysis to give D-galactose and D-gluconic acid confirms that the glucose unit is present as a cyclic hemiacetal and that the galactose unit is present as a glycoside. (Had the reverse been true, this experiment would have yielded D-glucose and D-galactonic acid.)

　　　　Methylation and hydrolysis of melibionic acid produces 2,3,4,6-tetra-*O*-methyl-D-galactose and 2,3,4,5-tetra-*O*-methyl-D-gluconic acid. Formation of the first product—a galactose derivative with a free –OH at C5—demonstrates that the galactose ring is six membered; formation of the second product—a gluconic acid derivative with a free –OH at C6—demonstrates that the oxygen at C6 of the glucose unit is joined in a glycosidic linkage to the galactose unit.

4.   That methylation and hydrolysis of melibiose gives a glucose derivative (2,3,4-tri-*O*-methyl-D-glucose) with free –OH groups at C5 and C6 shows that the glucose ring is also six membered. Melibiose is, therefore, 6-*O*-(α-D-galactopyranosyl-D-glucopyranose.

22.34 Trehalose has the following structure:

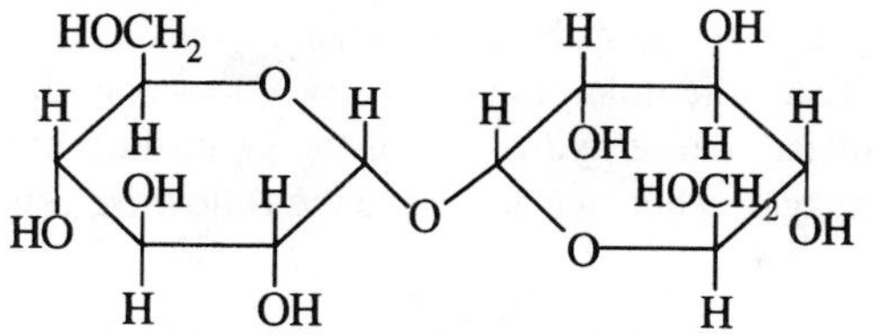

α-D-Glucopyranosyl-α-D-glucopyranoside

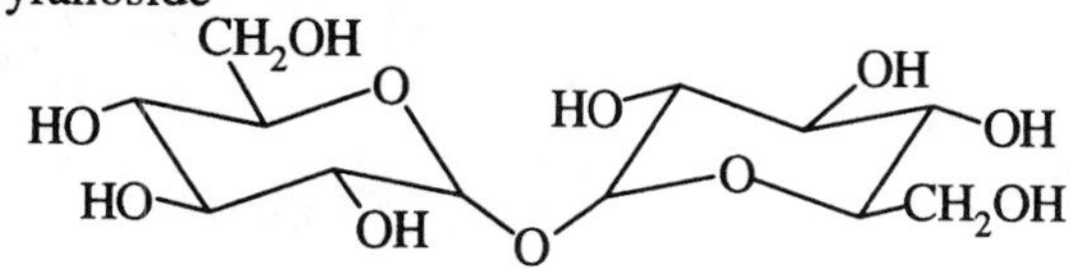

or

We arrive at this structure in the following way:

1. Acid hydrolysis shows that trehalose is a disaccharide consisting only of D-glucose units.

2. Hydrolysis by α-glucosidases and not by β-glucosidases shows that the glycosidic linkages are alpha.

3. That trehalose is a nonreducing sugar, that it does not form a phenylosazone, and that it does not react with bromine water indicate that no hemiacetal groups are present. This means that C1 of one glucose unit and C1 of the other must be joined in a glycosidic linkage. Fact 2 (just cited) indicates that this linkage is alpha to each ring.

4. That methylation of trehalose followed by hydrolysis yields only 2,3,4,6-tetra-*O*-methyl-D-glucose demonstrates that both rings are six membered.

22.35 (a) Tollens' reagent or Benedict's reagent will give a positive test with D-glucose but will give no reaction with D-glucitol.

(b) D-Glucaric acid will give an acidic aqueous solution that can be detected with blue litmus paper. D-Glucitol will give a neutral aqueous solution.

(c) D-Glucose will be oxidized by bromine water and the red brown color of bromine will disappear. D-Fructose will not be oxidized by bromine water since it does not contain an aldehyde group.

(d) Nitric acid oxidation will produce an *optically active* aldaric acid from D-glucose but an *optically inactive* aldaric acid will result from D-galactose.

(e) Maltose is a reducing sugar and will give a positive test with Tollens' or Benedict's solution. Sucrose is a nonreducing sugar and will not react.

(f) Maltose will give a positive Tollens' or Benedict's test; maltonic acid will not.

(g) 2,3,4,6-Tetra-*O*-methyl-β-D-glucopyranose will give a positive test with Tollens' or Benedict's solution; methyl β-D-glucopyranoside will not.

(h) Periodic acid will react with methyl α-D-ribofuranoside because it has hydroxyl groups on adjacent carbons. Methyl 2-deoxy-α-D-ribofuranoside will not react.

22.36 That the Schardinger dextrins are nonreducing shows that they have no free aldehyde or hemiacetal groups. This lack of reaction strongly suggests the presence of a *cyclic* structure. That methylation and subsequent hydrolysis yields only 2,3,6-tri-*O*-methyl-D-glucose indicates that the glycosidic linkages all involve C1 of one glucose unit and C4 of the next. That α-glucosidases cause hydrolysis of the glycosidic linkages indicates that they are α-glycosidic linkages. Thus, we are led to the following general structure.

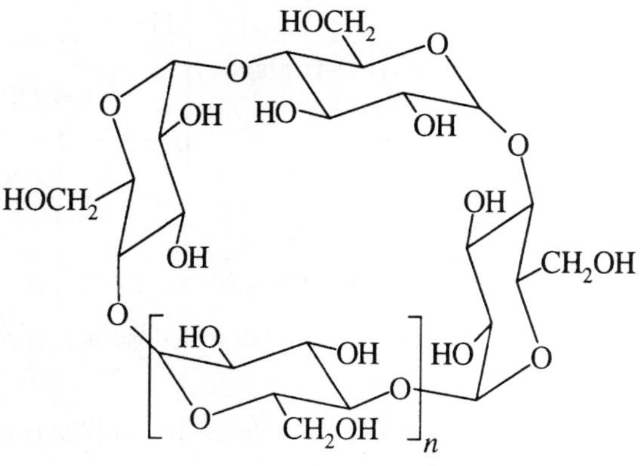

$n = 3, 4,$ or 5

Note: Schardinger dextrins are extremely interesting compounds. They are able to form complexes with a wide variety of compounds by incorporating these compounds in the cavity in the middle of the cyclic dextrin structure. Complex formation takes place, however, only when the cyclic dextrin and the guest molecule are the right size. Anthracene molecules, for example, will fit into the cavity of a cyclic dextrin with eight glucose units but will not fit into one with seven. For more information about these fascinating compounds, see R. J. Bergeron, "Cycloamyloses," *J. Chem. Educ.,* **1977,** *54,* 204.

22.37 Isomaltose has the following structure:

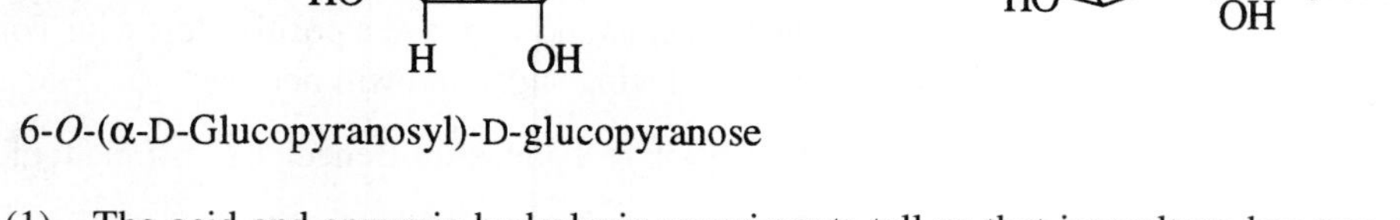

6-*O*-(α-D-Glucopyranosyl)-D-glucopyranose

(1) The acid and enzymic hydrolysis experiments tell us that isomaltose has two glucose units linked by an α linkage.

(2) That isomaltose is a reducing sugar indicates that one glucose unit is present as a cyclic hemiacetal.

(3) Methylation of isomaltonic acid followed by hydrolysis gives us information about the size of the nonreducing pyranoside ring and about its point of attachment to the reducing ring. The formation of the first product (2,3,4,6-tetra-*O*-methyl-D-glucose)—a compound with an –OH at C5—tells us that the nonreducing ring is present as a pyranoside. The formation of 2,3,4,5-tetra-*O*-methyl-D-gluconic acid—a compound with an –OH at C6—shows that the nonreducing ring is linked to C6 of the reducing ring.

(4) Methylation of maltose itself tells the size of the reducing ring. That 2,3,4-tri-*O*-methyl-D-glucose is formed shows that the reducing ring is also six membered; we know this because of the free –OH at C5.

22.38 Stachyose has the following structure:

HOCH₂
HO
H
OH H
H
H HO
D-Galactose
Hydrolysis here by an α-galactosidase yields D-galactose and raffinose

O CH₂
HO
H
OH H
H HO
D-Galactose
Hydrolysis here by an α-galactosidase yields sucrose

O CH₂
H
OH H
HO
H OH
D-Glucose

H HOCH₂
O
H HO
CH₂OH
OH H
D-Fructose

Sucrose

Raffinose

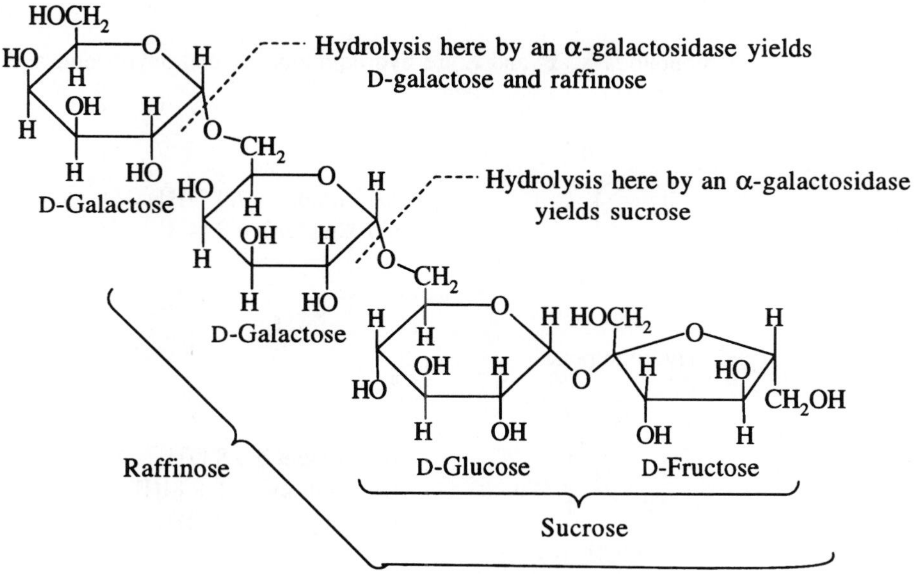

Raffinose has the following structure:

Melibiose

HOCH₂
HO
H
OH H
H
H HO
D-Galactose
Hydrolysis here by an α-galactosidase yields D-galactose and sucrose

O CH₂
H
OH H
HO
H OH
D-Glucose

Hydrolysis here by an invertase yields melibiose and fructose

CH₂OH
O
H HO
CH₂OH
OH H
D-Fructose

Sucrose

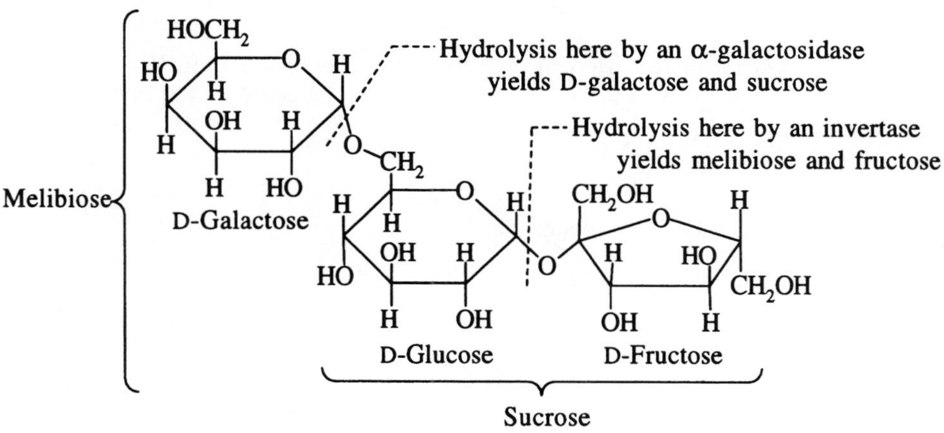

The enzymic hydrolyses (as just indicated) give the basic structure of stachyose and raffinose. The only remaining question is the ring size of the first galactose unit of stachyose. That methylation of stachyose and subsequent hydrolysis yields 2,3,4,6-tetra-*O*-methyl-D-galactose establishes that this ring is a pyranoside.

22.39 Arbutin has the following structure:

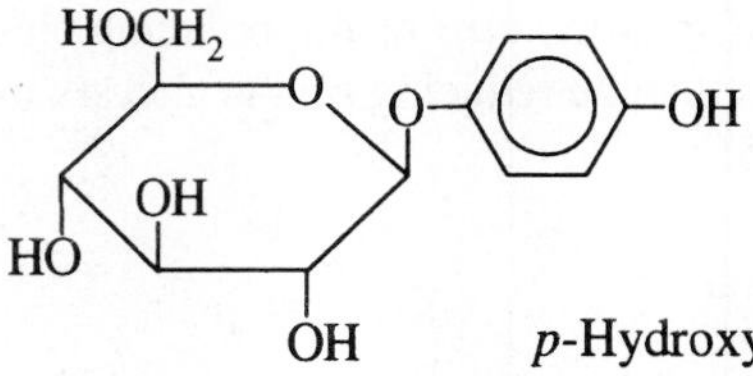

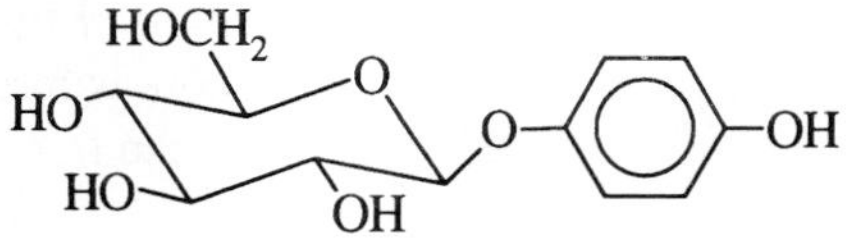

p-Hydroxyphenyl-β-D-glucopyranoside

Compounds **X**, **Y**, and **Z** are hydroquinone, *p*-methoxyphenol, and *p*-dimethoxybenzene, respectively.

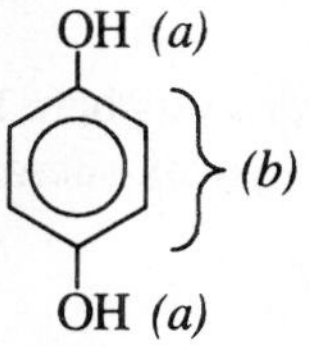

(*a*) Singlet δ 7.9 [2H]
(*b*) Singlet δ 6.8 [4H]

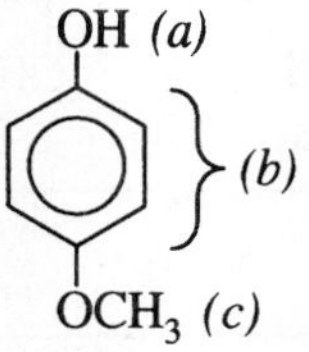

X

Hydroquinone

(*a*) Singlet δ 4.8 [1H]
(*b*) Multiplet δ 6.8 [4H]
(*c*) Singlet δ 3.9 [3H]

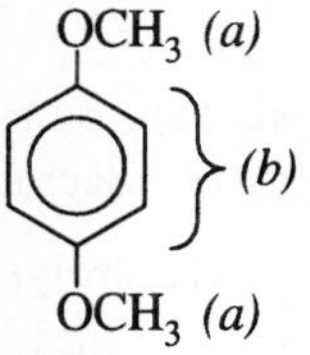

Y

p-Methoxyphenol

(*a*) Singlet δ 3.75 [6H]
(*b*) Singlet δ 6.8 [4H]

Z

p-Dimethoxybenzene

The reactions that take place are the following:

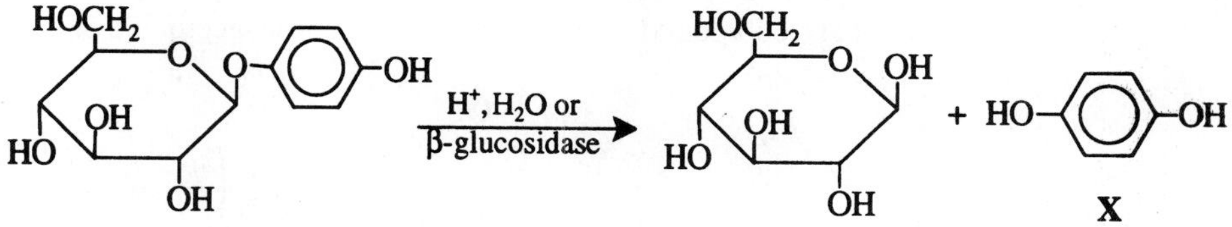

D-Glucose   Hydroquinone

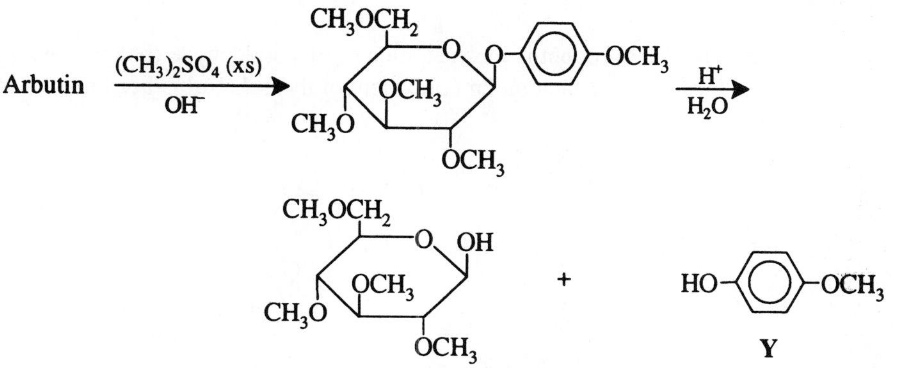

2,3,4,6-Tetra-*O*-methyl-D-glucose   *p*-Methoxyphenol

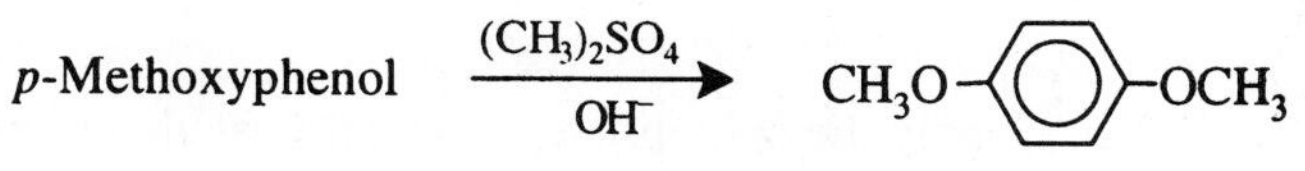

p-Methoxyphenol $\xrightarrow[\text{OH}^-]{\text{(CH}_3)_2\text{SO}_4}$ CH$_3$O—⟨benzene⟩—OCH$_3$

Z

p-Dimethoxybenzene

22.40 Aldotetrose **B** must be D-threose because the alditol derived from it (D-threitol) is optically active (the alditol from D-erythrose, the other possible D-aldotetrose, would be meso). Due to rotational symmetry, however, the alditol from **B** (D-threitol) would produce only two ^{13}C NMR signals. Compounds **A-F** are thus in the family of aldoses stemming from D-threose. Since reduction of aldopentose **A** produces an optically inactive alditol, **A** must be D-xylose. The two diastereomeric aldohexoses **C** and **D** produced from **A** by a Kiliani-Fischer synthesis must therefore be D-idose and D-gulose, respectively. **E** and **F** are the alditols derived from **C** and **D**, respectively. Alditol **E** would produce only three ^{13}C NMR signals due to rotational symmetry while **F** would produce six signals.

22.41 There are four closely spaced upfield alkyl signals in the ^{13}C NMR spectrum (δ26.5, δ25.6, δ24.9, δ24.2), corresponding to the four methyls of the two acetonide protecting groups. (The compound is, therefore, the 1,2,5,6-bis-acetonide of mannofuranose, below.)

22.42 The final product is the acetonide of glyceraldehyde (below); two molar equivalents are formed from each molar equivalent of the 1,2,5,6-bis-acetonide of mannitol.

QUIZ

22.1 Supply the appropriate structural formula or complete the partial formula for each of the following:

(a)	(b)	(c)	(d)
	CHO –C– –C– –C– –C– CH$_2$OH	CHO –C– –C– –C– CH$_2$OH	
A ketotetrose	A D-sugar	An L-sugar	An aldose

(e)	(f)	(g)	(h)
CHO H——OH H——OH HO——H H——OH CH₂OH D-Gulose	α-D-Gulo- pyranose	β-D-Gulo- pyranose	The compound that gives the same osazone as D-gulose	The compound that gives the same aldaric acid as D-gulose

22.2 Which of the following monosaccharides yields an optically inactive alditol on NaBH₄ reduction?

| CHO
HO——H
HO——H
H——OH
H——OH
CH₂OH

A | CHO
HO——H
H——OH
HO——H
H——OH
CH₂OH

B | CHO
H——OH
HO——H
HO——H
H——OH
CH₂OH

C | CHO
HO——H
HO——H
HO——H
H——OH
CH₂OH

D | Answer: [ ] |

22.3 Give the structural formula of the monosaccharide that you could use as starting material in the Kiliani-Fischer synthesis of the following compound:

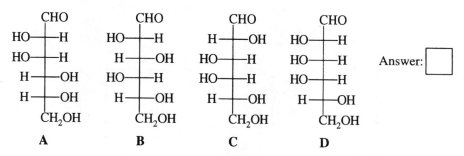

$$\xrightarrow[\text{synthesis}]{\text{Kiliani-Fischer}}$$

CHO
H——OH
HO——H
H——OH
CH₂OH
+ epimer

22.4   The D-aldopentose, (a), is oxidized to an aldaric acid, (b), which is optically active. Compound (a) undergoes a Ruff degradation to form an aldotetrose, (c), which undergoes oxidation to an optically inactive aldaric acid, (d). Supply the reagents for these transformations and the structural formulas of (a), (b), (c), and (d).

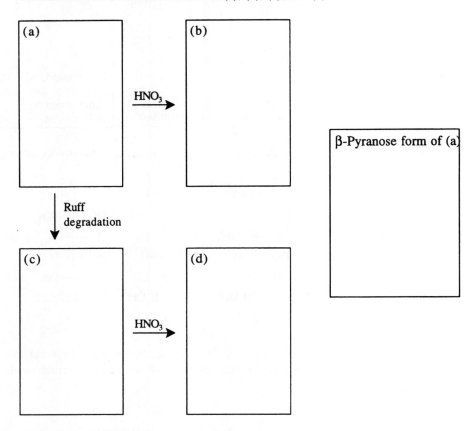

22.5   Give the structural formula of the β-pyranose form of (a) in the space just given.

22.6   Complete the following skeletal formulas and statements by filling in the blanks and circling the words that make the statements true.

The Haworth and conformational formulas of the β-cyclic hemiacetal

This cyclic hemiacetal is   (c)   reducing, nonreducing;   on reaction with Br_2/H_2O

it gives an optically   (d)   active, inactive   (e)   aldaric, aldonic   acid. On reaction

with dilute HNO_3 it gives an optically   (f)   active, inactive   (g)   aldaric, aldonic

acid. Reaction of the cyclic hemiacetal with   (h)   converts it into an

optically   (i)   active, inactive   alditol.

22.7   Outline chemical tests that would allow you to distinguish between:

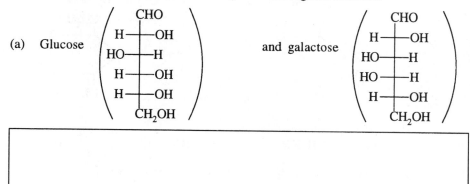

(a)   Glucose   and galactose

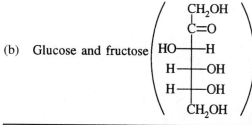

(b)   Glucose and fructose

22.8   Hydrolysis of (+)-sucrose (ordinary table sugar) yields

   (a)   D-glucose

   (b)   D-mannose

   (c)   D-fructose

   (d)   D-galactose

   (e)   More than one of the above.

22.9   Select the reagent needed to perform the following transformation:

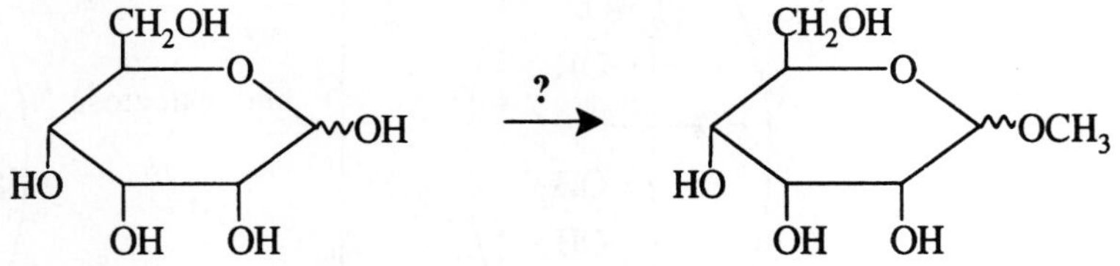

(a) CH_3OH, KOH (b) $(CH_3\overset{O}{\overset{\|}{C}})_2O$ (c) $(CH_3)_2SO_4$, OH^-

(d) CH_3OH, HCl (e) CH_3OCH_3, HCl

23 LIPIDS

SOLUTIONS TO PROBLEMS

23.1 (a) There are two sets of enantiomers, giving a total of four stereoisomers:

erythro   *threo*

(b)

(±)-*threo*-9,10-Dibromohexadecanoic acids

Formation of a bromonium ion at the other face of palmitoleic acid gives a result such that the *threo* enantiomers are the only products formed (obtained as a racemic modification).

The designations *erythro* and *threo* come from the names of the sugars called *erythrose* and *threose* (Section 22.9A).

23.2

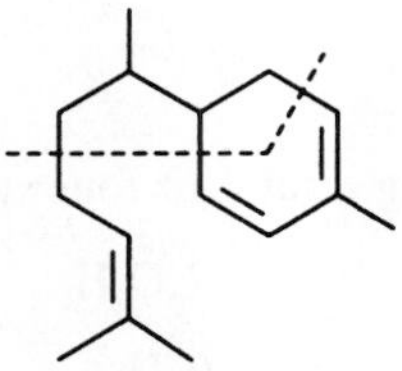

Zingiberene	β-Selinene	Caryophyllene
(a sesquiterpene)	(a sesquiterpene)	(a sesquiterpene)

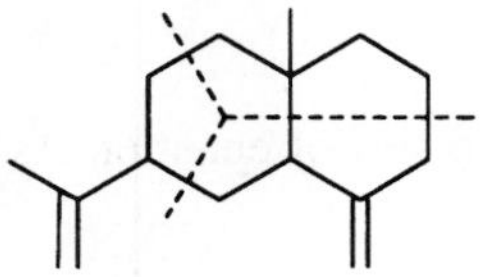

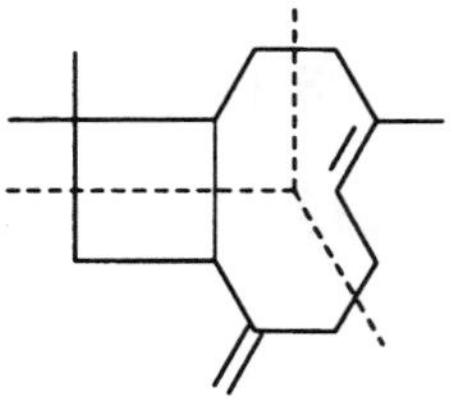

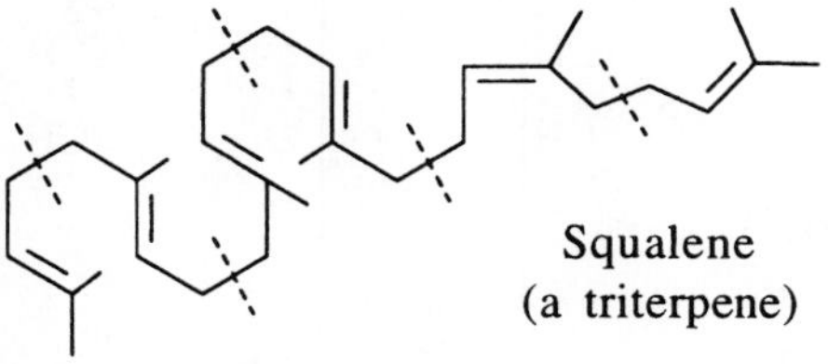

Squalene
(a triterpene)

23.3 (a)

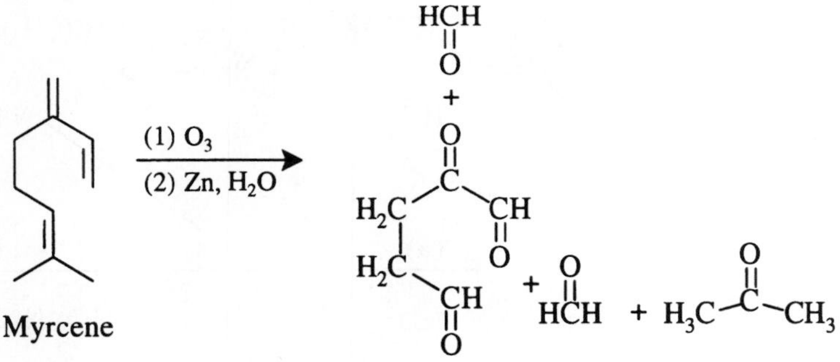

$\dfrac{(1)\ O_3}{(2)\ Zn,\ H_2O}$

Myrcene

(b)

$\dfrac{(1)\ O_3}{(2)\ Zn,\ H_2O}$

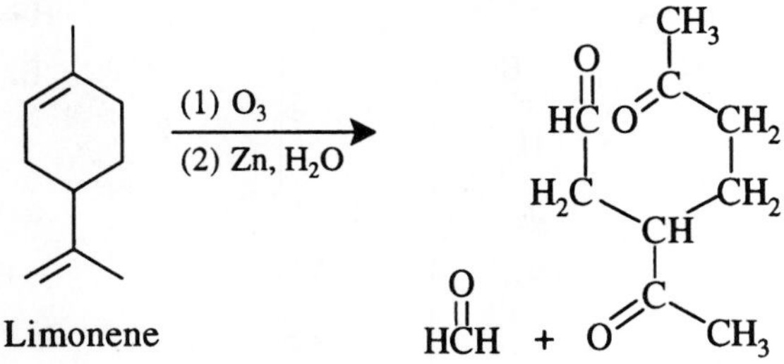

Limonene

(c) **α-Farnesene**
(see Section 23.3)

$\dfrac{(1)\ O_3}{(2)\ Zn,\ H_2O}$

$CH_3\overset{O}{\overset{\|}{C}}CH_3$ + $H\overset{O}{\overset{\|}{C}}CH_2CH_2\overset{O}{\overset{\|}{C}}CH_3$

+ $H\overset{O}{\overset{\|}{C}}CH_2\overset{O}{\overset{\|}{C}}H$ + $H\overset{O}{\overset{\|}{C}}-\overset{O}{\overset{\|}{C}}CH_3$ + $H\overset{O}{\overset{\|}{C}}H$

(d) **Geraniol**
(see Section 23.3)

$\xrightarrow[\text{(2) Zn, H}_2\text{O}]{\text{(1) O}_3}$

$$CH_3\overset{O}{\underset{}{C}}CH_3 \;+\; H\overset{O}{\underset{}{C}}CH_2CH_2\overset{O}{\underset{}{C}}CH_3$$

$$+\; H\overset{O}{\underset{}{C}}CH_2OH$$

(e) **Squalene**
(see Section 23.3)

$\xrightarrow[\text{(2) Zn, H}_2\text{O}]{\text{(1) O}_3}$

$$2\,CH_3\overset{O}{\underset{}{C}}CH_3 \;+\; H\overset{O}{\underset{}{C}}CH_2CH_2\overset{O}{\underset{}{C}}H$$

$$+\; 4\,CH_3\overset{O}{\underset{}{C}}CH_2CH_2\overset{O}{\underset{}{C}}H$$

23.4 (a)

+ CO$_2$

(+ further oxidation
products)

(b)

(c)

(+ rearranged products)

(d)

23.5 Br$_2$ in CCl$_4$ or KMnO$_4$ in H$_2$O at room temperature. Either reagent would give a positive result with geraniol and a negative result with menthol.

23.6

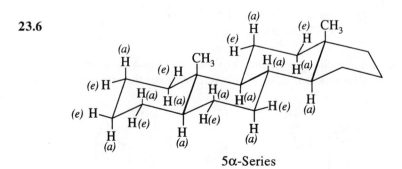

5α-Series

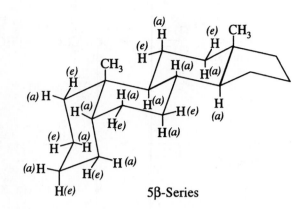

5β-Series

23.7 (a)

3α-Hydroxy-5α-androstan-17-one
(androsterone)

(b)

17α-Ethynyl-17β-hydroxy-5(10)-estren-3-one
(norethynodrel)

23.8

Absolute configuration of cholesterol
(5-cholesten-3β-ol)

23.9 Estrone and estradiol are *phenols* and thus are soluble in aqueous sodium hydroxide.
Extraction with aqueous sodium hydroxide separates the estrogens from the androgens.

23.10 (a)

Cholesterol $\xrightarrow[\text{CCl}_4]{\text{Br}_2}$ 5α,6β-Dibromocholestan-3β-ol

(b)

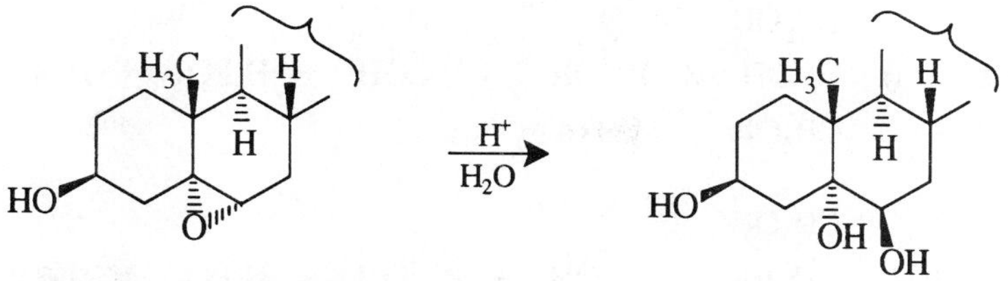

5α,6α-Epoxycholestan-3β-ol
(prepared by epoxidation
of cholesterol; cf. Section 23.4G)

Cholestan-3β,5α,6β-triol

(c)

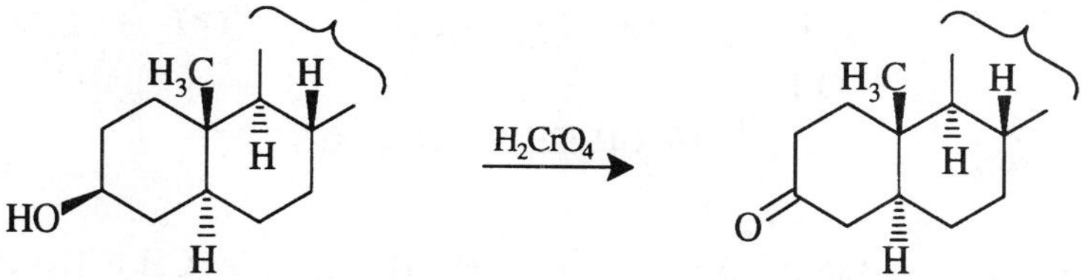

5α–Cholestan-3β-ol
(prepared by hydrogenation
of cholesterol; cf. Section 23.4G)

5α–Cholestan-3-one

(d)

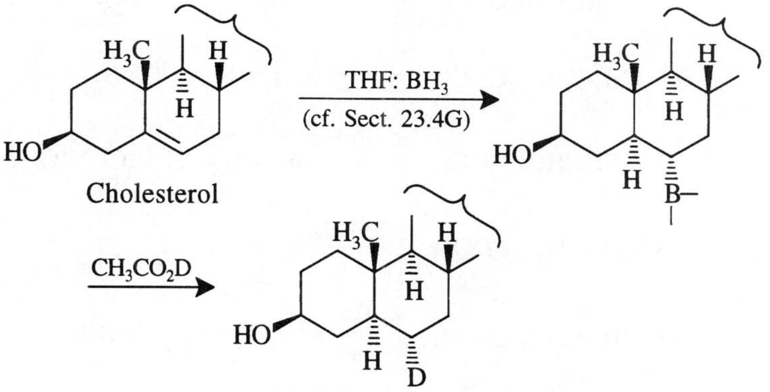

Cholesterol

6α-Deuterio-5α-cholestan-3β-ol

(e)

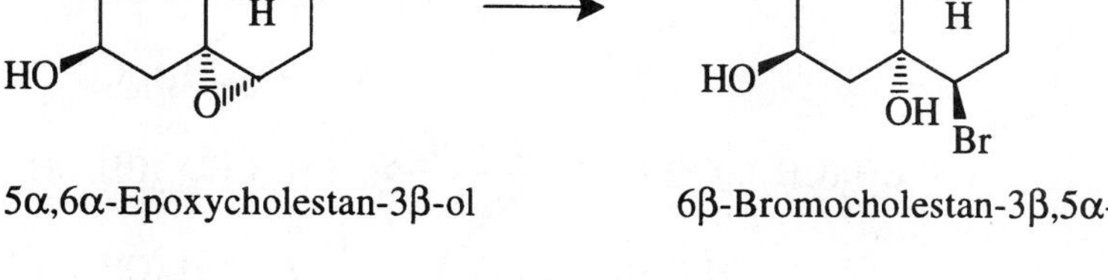

5α,6α-Epoxycholestan-3β-ol

6β-Bromocholestan-3β,5α-diol

23.11 (a) $\begin{array}{l}\text{CH}_2\text{OH}\\\text{CHOH}\\\text{CH}_2\text{OH}\end{array}$ + RCOH + R'COH + H_3PO_4 + $\text{HOCH}_2\text{CH}_2\overset{+}{\text{N}}(\text{CH}_3)_3 \text{ X}^-$

(b) $\begin{array}{l}\text{CH}_2\text{OH}\\\text{CHOH}\\\text{CH}_2\text{OH}\end{array}$ + RCOH + R'COH + H_3PO_4 + $\text{HOCH}_2\text{CH}_2\text{NH}_2$

(c) $\begin{array}{l}\text{CH}_2\text{OH}\\\text{CHOH}\\\text{CH}_2\text{OH}\end{array}$ + $\text{CH}_3(\text{CH}_2)_n\text{CH}_2\text{CH}$ + R'COH + H_3PO_4

   + $\text{HOCH}_2\text{CH}_2\overset{+}{\text{N}}(\text{CH}_3)_3 \text{ X}^-$

23.12 (a) $\text{CH}_3(\text{CH}_2)_{16}\text{CO}_2\text{H}$ + $\text{C}_2\text{H}_5\text{OH}$ $\underset{}{\overset{\text{H}^+}{\rightleftharpoons}}$ $\text{CH}_3(\text{CH}_2)_{16}\text{CO}_2\text{C}_2\text{H}_5$ + H_3O

   $\text{CH}_3(\text{CH}_2)_{16}\text{CO}_2\text{H}$ $\xrightarrow{\text{SOCl}_2}$ $\text{CH}_3(\text{CH}_2)_{16}\text{COCl}$ $\xrightarrow{\text{C}_2\text{H}_5\text{OH}}$ $\text{CH}_3(\text{CH}_2)_{16}\text{CO}_2\text{C}_2\text{H}_5$

(b) $\text{CH}_3(\text{CH}_2)_{16}\text{COCl}$ $\xrightarrow{(\text{CH}_3)_3\text{COH}}$ $\text{CH}_3(\text{CH}_2)_{16}\text{CO}_2\text{C}(\text{CH}_3)_3$

(c) $\text{CH}_3(\text{CH}_2)_{16}\text{COCl}$ $\xrightarrow{\text{NH}_3}$ $\text{CH}_3(\text{CH}_2)_{16}\text{CONH}_2$

(d) $\text{CH}_3(\text{CH}_2)_{16}\text{COCl}$ $\xrightarrow{(\text{CH}_3)_2\text{NH}}$ $\text{CH}_3(\text{CH}_2)_{16}\text{CON}(\text{CH}_3)_2$

(e) $\text{CH}_3(\text{CH}_2)_{16}\text{CONH}_2$ $\xrightarrow{\text{LiAlH}_4}$ $\text{CH}_3(\text{CH}_2)_{16}\text{CH}_2\text{NH}_2$

(f) $\text{CH}_3(\text{CH}_2)_{16}\text{CONH}_2$ $\xrightarrow[\text{OH}^-]{\text{Br}_2}$ $\text{CH}_3(\text{CH}_2)_{15}\text{CH}_2\text{NH}_2$

(g) $\text{CH}_3(\text{CH}_2)_{16}\text{COCl}$ $\xrightarrow{\text{LiAlH[OC(CH}_3)_3]_3}$ $\text{CH}_3(\text{CH}_2)_{16}\text{CHO}$

(h) $\text{CH}_3(\text{CH}_2)_{16}\text{CO}_2\text{C}_2\text{H}_5$ $\xrightarrow{\text{H}_2}_{\text{Ni}}$ $\text{CH}_3(\text{CH}_2)_{16}\text{CH}_2\text{OH}$

   $\text{CH}_3(\text{CH}_2)_{16}\text{COCl}$

   $\text{CH}_3(\text{CH}_2)_{16}\text{CO}_2\text{CH}_2(\text{CH}_2)_{16}\text{CH}_3$

(i) $\text{CH}_3(\text{CH}_2)_{16}\text{CO}_2\text{H}$ $\xrightarrow[\text{(2) H}_2\text{O}]{\text{(1) LiAlH}_4}$ $\text{CH}_3(\text{CH}_2)_{16}\text{CH}_2\text{OH}$

   $\text{CH}_3(\text{CH}_2)_{16}\text{CO}_2\text{C}_2\text{H}_5$ $\xrightarrow{\text{H}_2}_{\text{Ni}}$ $\text{CH}_3(\text{CH}_2)_{16}\text{CH}_2\text{OH}$

(j) $\text{CH}_3(\text{CH}_2)_{16}\text{COCl}$ + $(\text{CH}_3)_2\text{CuLi}$ $\longrightarrow$ $\text{CH}_3(\text{CH}_2)_{16}\text{COCH}_3$

(k) $CH_3(CH_2)_{16}CH_2OH \xrightarrow{PBr_3} CH_3(CH_2)_{16}CH_2Br$

(l) $CH_3(CH_2)_{16}CH_2Br \xrightarrow[\text{(2)H}^+\text{, H}_2\text{O, heat}]{\text{(1) NaCN}} CH_3(CH_2)_{16}CH_2CO_2H$

23.13 (a) $CH_3(CH_2)_{11}CH_2CO_2H \xrightarrow{Br_2, P} CH_3(CH_2)_{11}\underset{\overset{|}{Br}}{C}HCO_2H$

(b) $CH_3(CH_2)_{11}\underset{\overset{|}{Br}}{C}HCO_2H \xrightarrow[\text{(2) H}_3\text{O}^+]{\text{(1) OH}^-\text{, heat}} CH_3(CH_2)_{11}\underset{\overset{|}{OH}}{C}HCO_2H$

(c) $CH_3(CH_2)_{11}\underset{\overset{|}{Br}}{C}HCO_2H \xrightarrow[\text{(2) H}^+]{\text{(1) NaCN}} CH_3(CH_2)_{11}\underset{\overset{|}{CN}}{C}HCO_2H$

(d) $CH_3(CH_2)_{11}\underset{\overset{|}{Br}}{C}HCO_2H \xrightarrow[\text{(2) H}^+]{\text{(1) NH}_3 \text{ (excess)}} CH_3(CH_2)_{11}\underset{\overset{|}{NH_2}}{C}HCO_2H$

or $CH_3(CH_2)_{11}\underset{\overset{|}{NH_3^+}}{C}HCO_2^-$

23.14 (a) $CH_3(CH_2)_5CH{=}CH(CH_2)_7CO_2H \xrightarrow{Br_2} CH_3(CH_2)_5CHBrCHBr(CH_2)_7CO_2H$

(b) $CH_3(CH_2)_5CH{=}CH(CH_2)_7CO_2H \xrightarrow[\text{Ni}]{H_2} CH_3(CH_2)_{14}CO_2H$

(c) $CH_3(CH_2)_5CH{=}CH(CH_2)_7CO_2H \xrightarrow{KMnO_4}$
$CH_3(CH_2)_5CHOHCHOH(CH_2)_7CO_2H$

(d) $CH_3(CH_2)_5CH{=}CH(CH_2)_7CO_2H \xrightarrow{HCl} CH_3(CH_2)_5CH_2CHCl(CH_2)_7CO_2H$
$+$
$CH_3(CH_2)_5CHClCH_2(CH_2)_7CO_2H$

23.15 Elaidic acid is *trans*-9-octadecenoic acid:

$$CH_3(CH_2)_7 \underset{H}{\overset{}{C}}{=}\underset{(CH_2)_7CO_2H}{\overset{H}{C}}$$

It is formed by the isomerization of oleic acid.

23.16 (a)

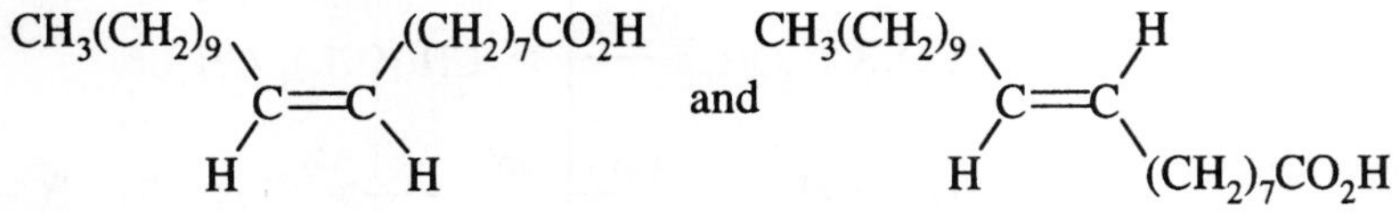

and

(b)   Infrared spectroscopy

(c)   A peak in the 675–730-cm^{-1} region would indicate that the double bond is cis; a peak in the 960–975-cm^{-1} region would indicate that it is trans.

23.17   A reverse Diels-Alder reaction takes place.

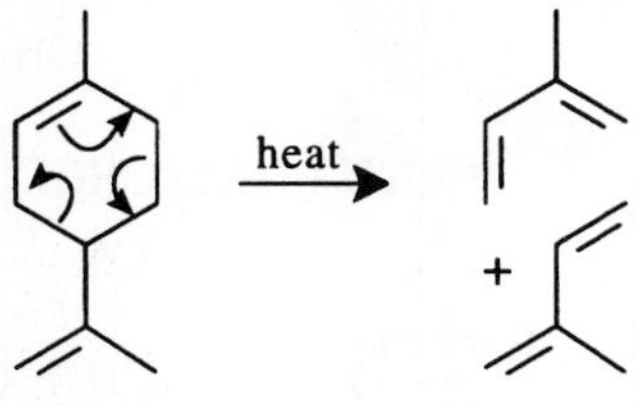

23.18

α-Phellandrene   β-Phellandrene

Note: On permanganate oxidation, the =CH$_2$ group of β-phellandrene is converted to CO$_2$ and thus is not detected in the reaction.

23.19 $CH_3(CH_2)_5C{\equiv}CH$ + $NaNH_2$ $\xrightarrow{\text{liq. NH}_3}$ $CH_3(CH_2)_5C{\equiv}CNa$

A

$\xrightarrow{ICH_2(CH_2)_7CH_2Cl}$ $CH_3(CH_2)_5C{\equiv}CCH_2(CH_2)_7CH_2Cl$ $\xrightarrow{NaCN}$

B

$CH_3(CH_2)_5C{\equiv}CCH_2(CH_2)_7CH_2CN$ $\xrightarrow[H_2O]{KOH}$ $CH_3(CH_2)_5C{\equiv}CCH_2(CH_2)_7CH_2CO_2K$

C   **D**

$\xrightarrow{H_3O^+}$ $CH_3(CH_2)_5C{\equiv}CCH_2(CH_2)_7CH_2CO_2H$ $\xrightarrow[Pd-BaSO_4]{H_2}$

E

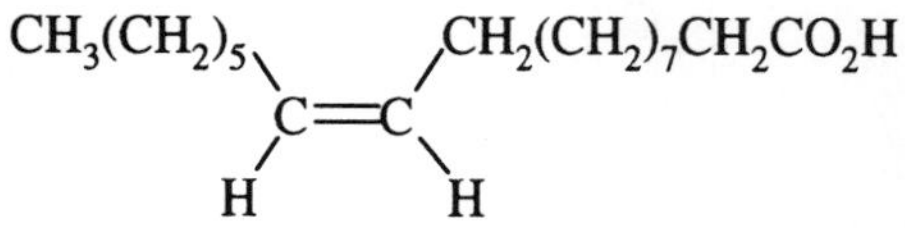

Vaccenic acid

23.20 $FCH_2(CH_2)_6CH_2Br$ + $HC{\equiv}CNa$ $\longrightarrow$ $FCH_2(CH_2)_6CH_2C{\equiv}CH$

$$\mathbf{F}$$

$$\xrightarrow[\text{(2) I(CH}_2\text{)}_7\text{Cl}]{\text{(1) NaNH}_2}$$ $FCH_2(CH_2)_6CH_2C{\equiv}C(CH_2)_7Cl$ $\xrightarrow{\text{NaCN}}$

$$\mathbf{G}$$

$FCH_2(CH_2)_6CH_2C{\equiv}C(CH_2)_7CN$ $\xrightarrow[\text{(2) H}^+]{\text{(1) KOH}}$ $FCH_2(CH_2)_6CH_2C{\equiv}C(CH_2)_7CO_2H$

$$\mathbf{H} \qquad\qquad\qquad\qquad\qquad\qquad\qquad\qquad \mathbf{I}$$

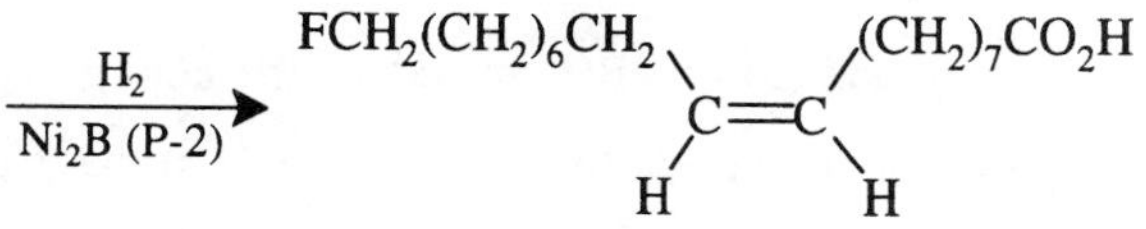

23.21

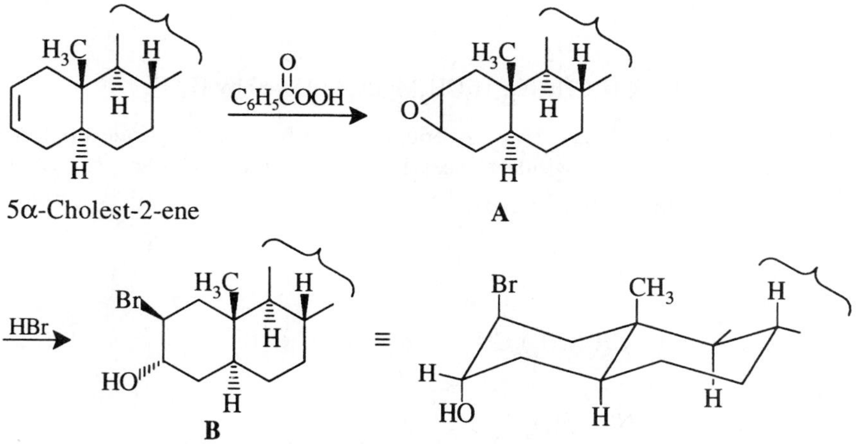

5α-Cholest-2-ene

A

Here we find that epoxidation takes place at the less hindered α face (cf. Section 23.4G). Ring opening by HBr takes place in an anti fashion to give a product with diaxial substituents.

23.22   (a) $CH_2{=}CH{-}CH{=}CH_2$

(b) OH^- (Removal of the α hydrogen atom allows isomerization to the more stable compound with a trans ring junction.)

(c) $LiAlH_4$

(d) H_3O^+ and heat. (Hydrolysis of the enol ether is followed by dehydration of one alcohol group.)

(e) $HCO_2C_2H_5$, C_2H_5ONa

(f) OsO_4, then $NaHSO_3$

(g) $CH_3\overset{\overset{\displaystyle O}{\|}}{C}CH_3$, H^+

(h) H_2, Pd catalyst

(i) H_3O^+, H_2O

(j) HIO_4

(k) Base and heat. (This reaction is an aldol condensation.)

(l) and (m) Na_2CrO_4, CH_3CO_2H to oxidize the aldehyde to an acid, followed by esterification.

(n) H_2 and Pt. (Hydrogen addition takes place from the less hindered α face of the molecule.)

(o), (p), and (q) $NaBH_4$ to reduce the keto group; OH^-, H_2O to hydrolyze the ester; and acetic anhydride to esterify the OH at the 3-position.

(r) and (s) $SOCl_2$ to make the acid chloride, followed by treatment with $(CH_3)_2Cd$.

(t) $CH_3\overset{\underset{\displaystyle CH_3}{|}}{C}HCH_2CH_2CH_2MgBr$, followed by H_3O^+.

(u), (v), and (w) Acetic acid and heat to dehydrate the tertiary alcohol; followed by acetic anhydride to acetylate the secondary alcohol; followed by H_2, Pt to hydrogenate the double bond.

23.23 (a) $CH_3(CH_2)_4\overset{\underset{\displaystyle }{\|}}{\overset{O}{C}}H$ (b) C_4H_9Li (c)   Br

(d) $NC(CH_2)_6$ ... H ... H ... NO_2 (e) Michael addition using a basic catalyst.

23.24 First: an elimination takes place,

$$R_3\overset{+}{N}CH_2CH_2\overset{O}{\overset{\|}{C}}CH_2CH_3 \;+\; NH_2^- \longrightarrow CH_2{=}CH\overset{O}{\overset{\|}{C}}CH_2CH_3 \;+\; R_3N \;+\; NH_3$$

Then a conjugate addition occurs, followed by an aldol addition:

QUIZ

23.1 Write an appropriate formula in each box.

(a)	(b)

A naturally occurring fatty acid   A soap

(c)	(d)

A solid fat   An oil

(e)	(f)

A synthetic detergent   5α-Estran-17-one

23.2 Give a reagent that would distinguish between each of the following:

(a)   Pregnane and 20-pregnanone

(b)   Stearic acid and oleic acid

(c)   17α-Ethynyl-1,3,5(10)-estratriene-3,17β-diol (ethynylestradiol) and 1,3,5(10)-estratriene-3,17β-diol (estradiol)

23.3 What product would be obtained by catalytic hydrogenation of 4-androstrene?

23.4 Supply the missing compounds:

$CH_3(CH_2)_4CH_2Br$ $\xrightarrow{HC\equiv CNa}$ (a)

$\xrightarrow{NaNH_2}$ (b) $\xrightarrow{ICH_2(CH_2)_5CH_2Cl}$

(c)

(d)

$\longrightarrow CH_3(CH_2)_5C\equiv C(CH_2)_7CN$

$\xrightarrow[\text{(2) } H_3O^+]{\text{(1) KOH, H}_2O, \text{ heat}}$ (e)

(f)

$\longrightarrow$ palmitoleic acid

23.5 The following compound is a:

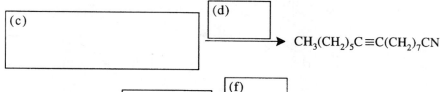

(a)   Monoterpene   (b)   Sesquiterpene   (c)   Diterpene

(d)   Triterpene   (e)   Tetraterpene

23.6 Mark off the isoprene units in the previous compound.

23.7 Which is a systematic name for the steroid shown here?

(a)   5α-Androstan-3α-ol

(b)   5β-Androstan-3β-ol

(c)   5α-Pregnan-3α-ol

(d)   5β-Pregnan-3β-ol

(e)   5α-Estran-3α-ol

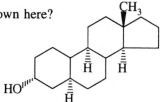

SOLUTIONS TO PROBLEM

P.1

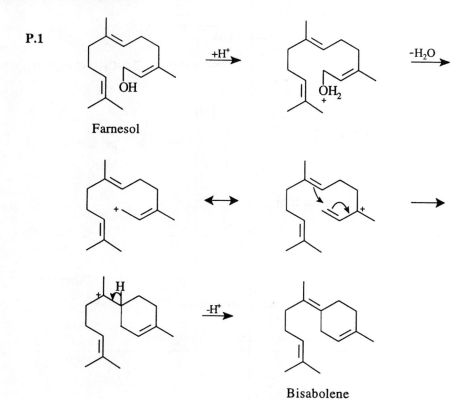

24

AMINO ACIDS AND PROTEINS

SOLUTIONS TO PROBLEMS

24.1 (a) $HO_2CCH_2CH_2CHCO_2H$
$\overset{|}{NH_3^+}$

(b) $^-O_2CCH_2CH_2CHCO_2^-$
$\overset{|}{NH_2}$

(c) $HO_2CCH_2CH_2CHCO_2^-$ predominates at the isoelectric point rather than
$\overset{|}{NH_3^+}$

$^-O_2CCH_2CH_2CHCO_2H$ because of the acid-strengthening inductive effect of the
$\overset{|}{NH_3^+}$ α-ammonio group.

(d) Since glutamic acid is a dicarboxylic acid, acid must be added (i.e., the pH must be made lower) to suppress the ionization of the second carboxyl group and thus achieve the isoelectric point. Glutamine, with only one carboxyl group, is similar to glycine or phenylalanine and has its isoelectric point at a higher pH.

24.2 The conjugate acid is highly stabilized by resonance.

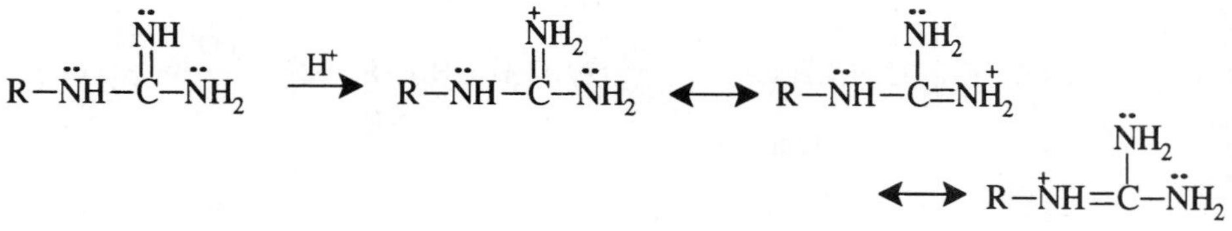

24.3 (a)

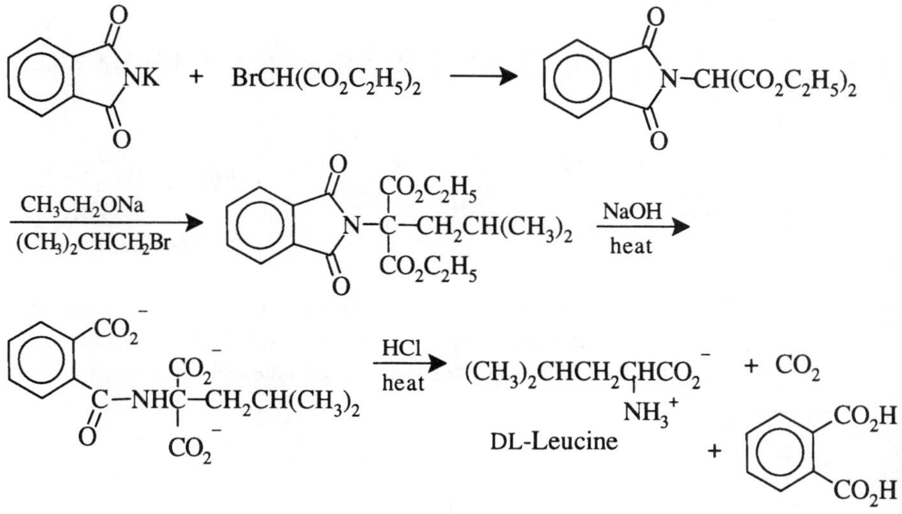

493

(b)

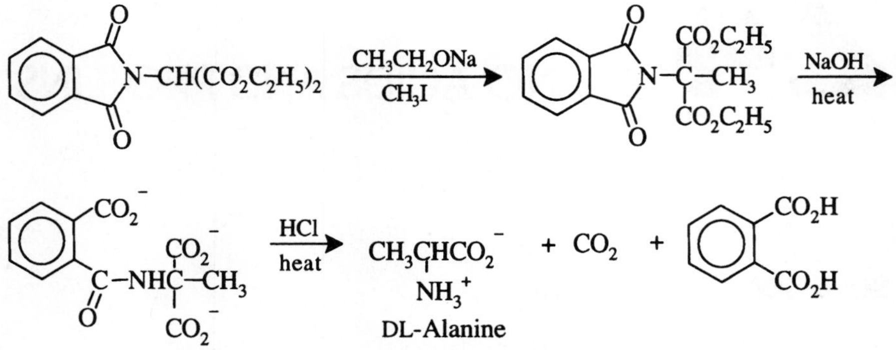

(c)

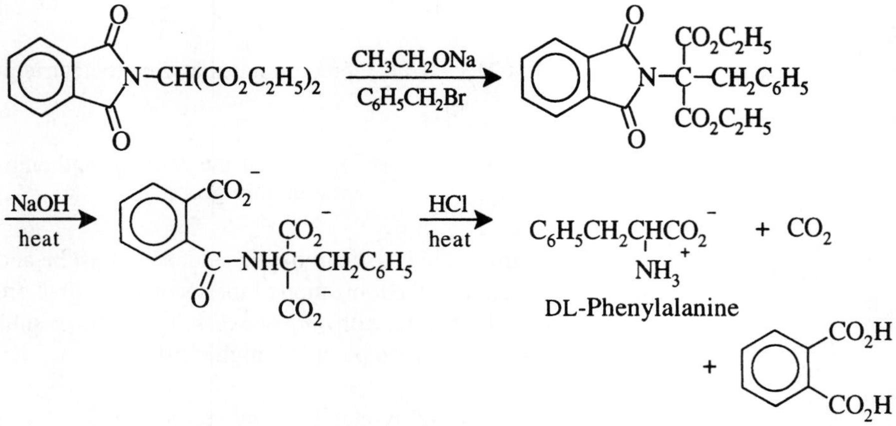

24.4 (a)

$$C_6H_5CH_2\overset{O}{\overset{\|}{C}}H \xrightarrow[HCN]{NH_3} C_6H_5CH_2\underset{NH_2}{\underset{|}{C}}HC{\equiv}N \xrightarrow{H_3O^+} C_6H_5CH_2\underset{NH_3^+}{\underset{|}{C}}HCO_2^-$$

Phenylacetaldehyde → DL-Phenylalanine

(b)

$$CH_3SH + CH_2{=}CH{-}\overset{O}{\overset{\|}{C}}H \xrightarrow{base} CH_3SCH_2CH_2\overset{O}{\overset{\|}{C}}H \xrightarrow[HCN]{NH_3}$$

$$CH_3SCH_2CH_2\underset{NH_2}{\underset{|}{C}}HC{\equiv}N \xrightarrow{H_3O^+} CH_3SCH_2CH_2\underset{NH_3^+}{\underset{|}{C}}HCO_2^-$$

DL-Methionine

24.5 Because of the presence of an electron-withdrawing 2,4-dinitrophenyl group, the labeled amino acid is relatively nonbasic and is, therefore, insoluble in dilute aqueous acid. The other amino acids (those that are not labeled) dissolve in dilute aqueous acid.

24.6 (a) $\overset{+}{H_3}NCHCONHCHCONHCH_2CO_2^-$ $\xrightarrow{\text{HCO}_3^-}$

   $\underset{\underset{CH_3}{|}}{CHCH_3}$ $\underset{}{CH_3}$ Val•Ala•Gly

$O_2N-\underset{\underset{NO_2}{}}{\bigcirc}-NHCHCONHCHCONHCH_2CO_2^-$ $\xrightarrow[\text{heat}]{H_3O^+}$

   $\underset{\underset{CH_3}{|}}{CHCH_3}$ CH_3

$O_2N-\underset{\underset{NO_2}{}}{\bigcirc}-NHCHCO_2H$ $+$ $CH_3\underset{\underset{NH_3^+}{|}}{CHCO_2^-}$ $+$ $\overset{+}{H_3}NCH_2CO_2^-$

   $\underset{\underset{CH_3}{|}}{CHCH_3}$ Alanine   Glycine

   Labeled valine
   (separate and identify)

(b) $O_2N-\underset{\underset{NO_2}{}}{\bigcirc}-NHCHCO_2H$ $+$ $O_2N-\underset{\underset{NO_2}{}}{\bigcirc}-NHCH_2CH_2CH_2CH_2\underset{\underset{NH_3^+}{}}{CHCO_2^-}$

   $\underset{\underset{CH_3}{|}}{CHCH_3}$

   α-Labeled valine   ε-Labeled lysine

   $+$ $\overset{+}{H_3}NCH_2CO_2^-$
   Glycine

24.7 $\bigcirc-N=C=S$ $+$ $H_2\ddot{N}-\underset{\underset{\underset{\underset{CH_3}{|}}{S}}{\overset{|}{\underset{CH_2}{|}}}}{CHCO}-\underset{\underset{\underset{CH_3}{|}}{CH_2}}{\overset{|}{\underset{CHCH_3}{|}}}NHCO-\underset{\underset{\underset{\underset{\underset{NH_2}{|}}{C=NH}}{\overset{|}{NH}}}{\overset{|}{\underset{CH_2}{|}}}}{NHCHCO_2^-}$ $\xrightarrow{OH^-}$

Phenyl isothiocyanate   Met-Ile-Arg

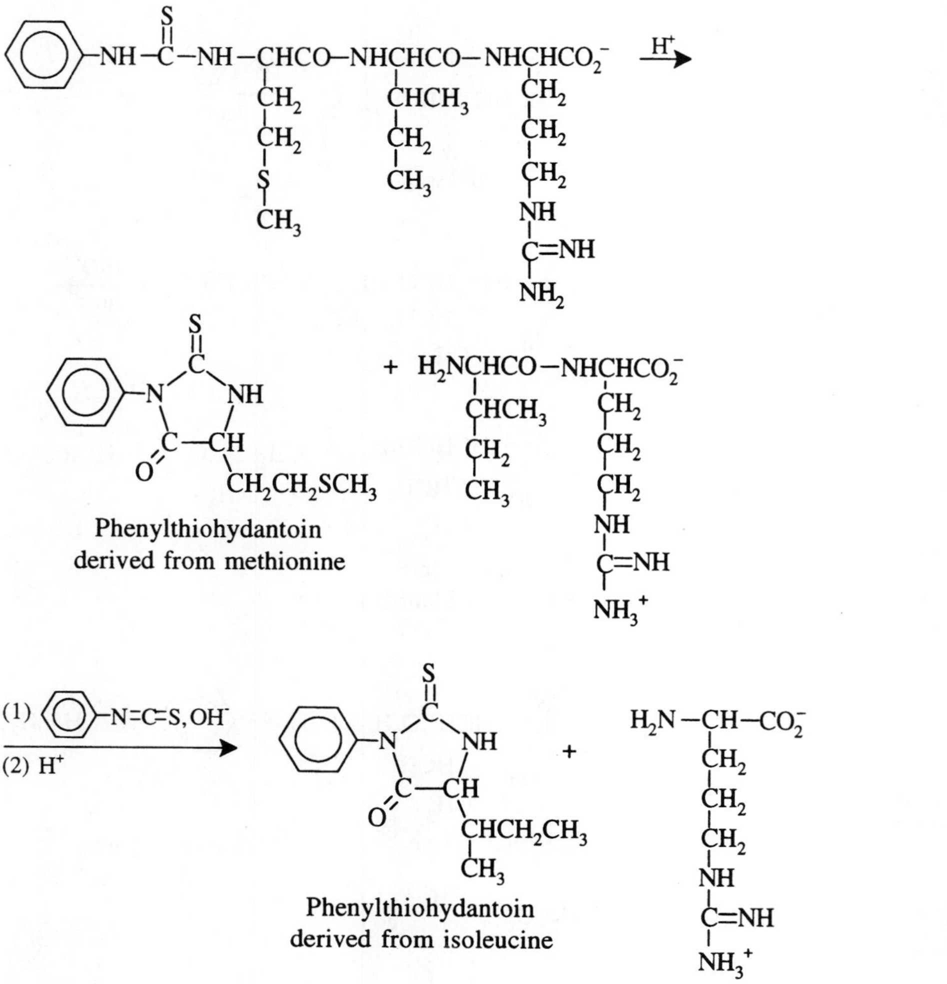

$$\text{C}_6\text{H}_5\text{—NH—C(=S)—NH—CHCO—NHCHCO—NHCHCO}_2^- \xrightarrow{\text{H}^+}$$

(structure with side chains: CH₂CH₂SCH₃ on first residue, CHCH₃CH₂CH₃ on second residue, CH₂CH₂CH₂NHC(=NH)NH₂ on third residue)

Phenylthiohydantoin
derived from methionine

$$+ \quad \text{H}_2\text{NCHCO—NHCHCO}_2^-$$

(with side chains CHCH₃CH₂CH₃ and CH₂CH₂CH₂NHC(=NH)NH₃⁺)

$$\xrightarrow[\text{(2) H}^+]{\text{(1) C}_6\text{H}_5\text{—N=C=S, OH}^-}$$

Phenylthiohydantoin
derived from isoleucine

$$+ \quad \text{H}_2\text{N—CH—CO}_2^-$$

(with side chain CH₂CH₂CH₂NHC(=NH)NH₃⁺)

24.8 (a) Two structures are possible with the sequence Glu•Cys•Gly. Glutamic acid may be linked to cysteine through its α-carboxyl group,

$$\text{HO}_2\text{CCH}_2\text{CH}_2\text{CHCO—NHCHCO—NHCH}_2\text{CO}_2^-$$
$$\qquad\qquad\quad |\text{NH}_3^+ \qquad\quad |\text{CH}_2\text{SH}$$

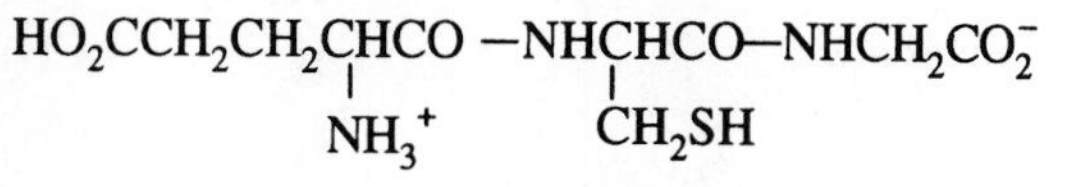

or through its γ-carboxyl group,

$$\text{H}_3\overset{+}{\text{N}}\text{CHCH}_2\text{CH}_2\text{CO—NHCHCO—NHCH}_2\text{CO}_2^-$$
$$\quad |\text{CO}_2^- \qquad\qquad\quad |\text{CH}_2\text{SH}$$

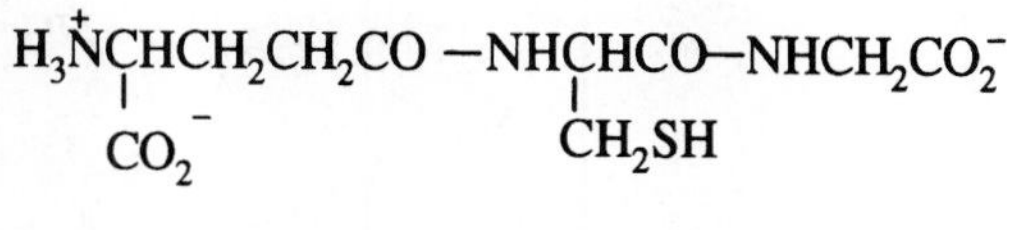

(b) This result shows that the second structure is correct, that in glutathione the γ-carboxyl group is linked to cysteine.

24.9   We look for points of overlap to determine the amino acid sequence in each case.

(a)

Ser • Thr

Thr • Hyp

Pro • Ser

Pro • Ser • Thr • Hyp

(b)

Ala • Cys

Cys • Arg

Arg • Val

Leu • Ala

Leu • Ala • Cys • Arg • Val

24.10   Sodium in liquid ammonia brings about reductive cleavage of the disulfide linkage of oxytocin to two thiol groups; then air oxidizes the two thiol groups back to a disulfide linkage:

$$
\begin{array}{c}
\text{R} \\
| \\
\text{CH}_2 \\
| \\
\text{S} \\
\diagdown \\
\quad\quad \text{S} \\
| \\
\text{CH}_2 \\
| \\
\text{R}
\end{array}
\xrightarrow[\text{NH}_3]{\text{Na}}
\begin{array}{c}
\text{R} \\
| \\
\text{CH}_2 \\
| \\
\text{SH} \\
\\
\text{SH} \\
| \\
\text{CH}_2 \\
| \\
\text{R}
\end{array}
\xrightarrow{\text{O}_2}
\begin{array}{c}
\text{R} \\
| \\
\text{CH}_2 \\
| \\
\text{S} \\
\diagdown \\
\quad\quad \text{S} \\
| \\
\text{CH}_2 \\
| \\
\text{R}
\end{array}
$$

24.11

$$\overset{+}{\text{H}_3}\text{NCH}_2\text{CO}_2^{-} \; + \; (\text{CH}_3)_3\text{COCN}_3 \xrightarrow[25°\text{C}]{\text{OH}^-} \xrightarrow{\text{H}_3\text{O}^+}$$

Glycine   *tert*-Butoxy-
                        carbonyl azide

$$(\text{CH}_3)_3\text{C}-\overset{\text{O}}{\overset{||}{\text{OC}}}\text{NHCH}_2\text{CO}_2\text{H} \xrightarrow[(2)\ \text{ClCO}_2\text{C}_2\text{H}_5]{(1)\ (\text{C}_2\text{H}_5)_3\text{N}}$$

Boc-Gly

$$(\text{CH}_3)_3\text{CO}\overset{\text{O}}{\overset{||}{\text{C}}}\text{NHCH}_2\overset{\text{O}}{\overset{||}{\text{C}}}\overset{\text{O}}{\overset{||}{\text{OC}}}\text{OC}_2\text{H}_5 \xrightarrow[-\text{CO}_2,\ -\text{C}_2\text{H}_5\text{OH}]{}$$

Mixed anhydride

$$\overset{+}{\text{H}_3}\text{NCHCO}_2^{-}$$
$$|$$
$$\text{CHCH}_3$$
$$|$$
$$\text{CH}_3$$
Valine

$$(\text{CH}_3)_3\text{CO}\overset{\text{O}}{\overset{||}{\text{C}}}\text{NHCH}_2\overset{\text{O}}{\overset{||}{\text{C}}}\text{NHCHCO}_2\text{H} \xrightarrow[(2)\ \text{ClCO}_2\text{C}_2\text{H}_5]{(1)\ (\text{C}_2\text{H}_5)_3\text{N}}$$
$$|$$
$$\text{CHCH}_3$$
Boc-Gly•Val $$\text{CH}_3$$

$$\underset{\text{Mixed anhydride}}{(CH_3)_3CO\overset{O}{\overset{\|}{C}}NHCH_2\overset{O}{\overset{\|}{C}}NHCH\underset{\underset{CH_3}{\overset{|}{CHCH_3}}}{\overset{O}{\overset{\|}{C}}}\overset{O}{\overset{\|}{C}}OC_2H_5} \xrightarrow[\text{Alanine}]{\overset{\overset{+}{H_3N}CHCO_2^-}{\overset{|}{CH_3}}}$$

$$\underset{\text{Boc-Gly•Val•Ala}}{(CH_3)_3CO\overset{O}{\overset{\|}{C}}NHCH_2\overset{O}{\overset{\|}{C}}NHCH\underset{\underset{CH_3}{\overset{|}{CHCH_3}}}{\overset{O}{\overset{\|}{C}}}NH\underset{CH_3}{\overset{|}{C}}HCO_2H} \xrightarrow[\substack{CH_3CO_2H \\ 25°C}]{CF_3CO_2H}$$

$$(CH_3)_2C=CH_2 \;+\; CO_2 \;+\; \underset{\underset{CH_3}{\overset{|}{CHCH_3}\;\; CH_3}}{\overset{+}{H_3N}CH_2\overset{O}{\overset{\|}{C}}NHCH\overset{O}{\overset{\|}{C}}NHCHCO_2^-}$$
$$\text{Gly•Val•Ala}$$

24.12 (a) $\;2\;\underset{\substack{\text{Benzyl chloro-}\\\text{formate}}}{C_6H_5CH_2O\overset{O}{\overset{\|}{C}}Cl} \;+\; \underset{\substack{\text{Lysine}}}{H_2NCH_2CH_2CH_2CH_2\underset{NH_2}{\overset{|}{C}}HCO_2^-} \xrightarrow[25°C]{OH^-}$

$$C_6H_5CH_2O\overset{O}{\overset{\|}{C}}NHCH_2CH_2CH_2CH_2\underset{\underset{C_6H_5CH_2O\overset{}{C}=O}{\overset{|}{NH}}}{\overset{|}{C}}HCO_2H \xrightarrow[\text{(2)}ClCO_2C_2H_5]{\text{(1)}(C_2H_5)_3N}$$

$$C_6H_5CH_2O\overset{O}{\overset{\|}{C}}NHCH_2CH_2CH_2CH_2\underset{\underset{C_6H_5CH_2O\overset{}{C}=O}{\overset{|}{NH}}}{\overset{|}{C}}H\overset{O}{\overset{\|}{C}}-O-\overset{O}{\overset{\|}{C}}-OC_2H_5 \xrightarrow[-CO_2,\,-C_2H_5OH]{\underset{\underset{CH_3\;\;NH_3^+}{}}{CH_3CH_2CH-CHCO_2^-}}$$

$$C_6H_5CH_2O\overset{O}{\overset{\|}{C}}NHCH_2CH_2CH_2CH_2\underset{\underset{C_6H_5CH_2O\overset{}{C}=O}{\overset{|}{NH}}}{\overset{|}{C}}H\overset{O}{\overset{\|}{C}}NH\underset{\underset{\underset{CH_3}{\overset{|}{CH_2}}}{\overset{|}{CHCH_3}}}{\overset{}{C}}HCO_2^- \xrightarrow[\substack{CH_3CO_2H \\ cold}]{HBr}$$

$$2\,C_6H_5CH_2Br \;+\; 2\,CO_2 \;+\; \overset{+}{H_3N}CH_2CH_2CH_2CH_2\underset{NH_2}{\overset{|}{C}}H\overset{O}{\overset{\|}{C}}NH\underset{\underset{\underset{CH_3}{\overset{|}{CH_2}}}{\overset{|}{CHCH_3}}}{\overset{}{C}}HCO_2^-$$
$$\text{Lys•Ile}$$

(b) $3 \text{ C}_6\text{H}_5\text{CH}_2\text{OCCl}$ $+$ H$_2$NCNHCH$_2$CH$_2$CH$_2$CHCO$_2^-$ $\xrightarrow[25°C]{OH^-}$

$\xrightarrow[\text{(2) ClCO}_2\text{C}_2\text{H}_5]{\text{(1) (C}_2\text{H}_5)_3\text{N}}$

$\xrightarrow[-\text{CO}_2, -\text{C}_2\text{H}_5\text{OH}]{\begin{array}{c}\text{CH}_3\text{CHCO}_2^- \\ \text{NH}_3^+\end{array}}$

$\xrightarrow[\substack{\text{CH}_3\text{CO}_2\text{H} \\ \text{cold}}]{\text{HBr}}$

$3 \text{ C}_6\text{H}_5\text{CH}_2\text{Br}$ $+$ 3 CO_2 $+$ H$_3$NCNHCH$_2$CH$_2$CH$_2$CHCONHCHCO$_2^-$

Arg•Ala

24.13 The weakness of the benzyl-oxygen bond allows these groups to be removed by catalytic hydrogenolysis.

24.14 (a) An electrophilic substitution reaction:

$+ \text{ CH}_3\text{OCH}_2\text{Cl} \xrightarrow{\text{BF}_3}$ ... $+ \text{ CH}_3\text{OH}$

(b) The linkage between the resin and the polypeptide is a benzylic ester. It is cleaved by HBr in CF$_3$CO$_2$H at room temperature because the carbocation that is formed initially is the relatively stable, benzylic cation.

24.15

⟨ ⟩—CH₂Cl + HO$\overset{\text{O}}{\overset{\|}{\text{C}}}$$\overset{\text{O}}{\underset{\text{CH}_3}{\text{CHNH}\overset{\|}{\text{C}}}}$OC(CH₃)₃   1. Add Boc-Ala.

↓ base

⟨ ⟩—CH₂O$\overset{\text{O}}{\overset{\|}{\text{C}}}$$\underset{\text{CH}_3}{\text{CHNH}}\overset{\text{O}}{\overset{\|}{\text{C}}}$OC(CH₃)₃   2. Purify by washing.

↓ CF₃CO₂H, CH₂Cl₂   3. Remove protecting group.

⟨ ⟩—CH₂O$\overset{\text{O}}{\overset{\|}{\text{C}}}$$\underset{\text{CH}_3}{\text{CHNH}_2}$ 4. Purify by washing.

| HO$\overset{\text{O}}{\overset{\|}{\text{C}}}$$\underset{\text{CH}_2\text{C}_6\text{H}_5}{\text{CHNH}}\overset{\text{O}}{\overset{\|}{\text{C}}}$OC(CH₃)₃   5. Add Boc-Phe.
and
dicyclohexylcarbodiimide
↓

⟨ ⟩—CH₂O$\overset{\text{O}}{\overset{\|}{\text{C}}}$$\underset{\text{CH}_3}{\text{CHNH}}\overset{\text{O}}{\overset{\|}{\text{C}}}$$\underset{\underset{\text{C}_6\text{H}_5}{\text{CH}_2}}{\text{CHNH}}\overset{\text{O}}{\overset{\|}{\text{C}}}$OC(CH₃)₃   6. Purify by washing.

↓ CF₃CO₂H, CH₂Cl₂   7. Remove protecting group.

⟨ ⟩—CH₂O$\overset{\text{O}}{\overset{\|}{\text{C}}}$$\underset{\text{CH}_3}{\text{CHNH}}\overset{\text{O}}{\overset{\|}{\text{C}}}$$\underset{\underset{\text{C}_6\text{H}_5}{\text{CH}_2}}{\text{CHNH}_2}$ 8. Purify by washing.

| HO$\overset{\text{O}}{\overset{\|}{\text{C}}}$$\underset{\underset{\underset{\text{O=COC(CH}_3)_3}{\text{NH}}}{\text{|}}}{\text{CHCH}_2\text{CH}_2\text{CH}_2\text{CH}_2\text{NH}}\overset{\text{O}}{\overset{\|}{\text{C}}}$OC(CH₃)₃   9. Add protected Lys.
and
dicyclohexylcarbodiimide
↓

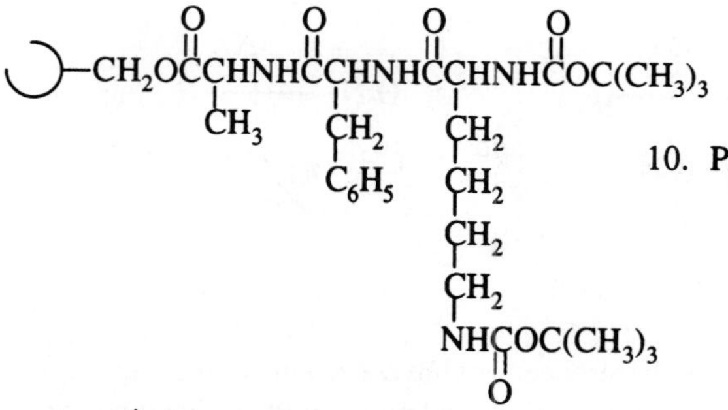

10. Purify by washing.

CF$_3$CO$_2$H, CH$_2$Cl$_2$ 11. Remove protecting groups.

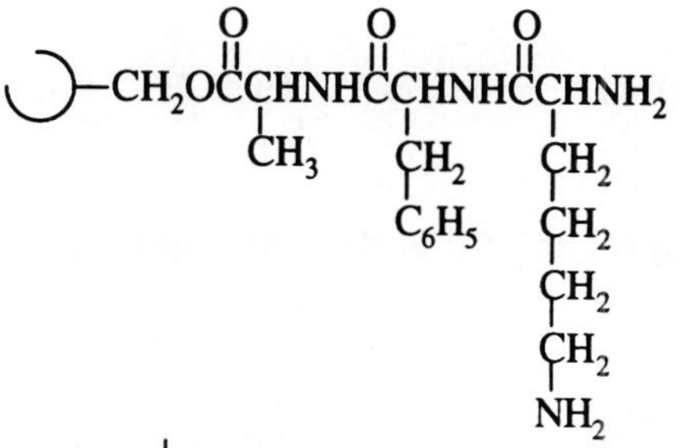

12. Purify by washing.

HBr, CF$_3$CO$_2$H   13. Detach tripeptide.

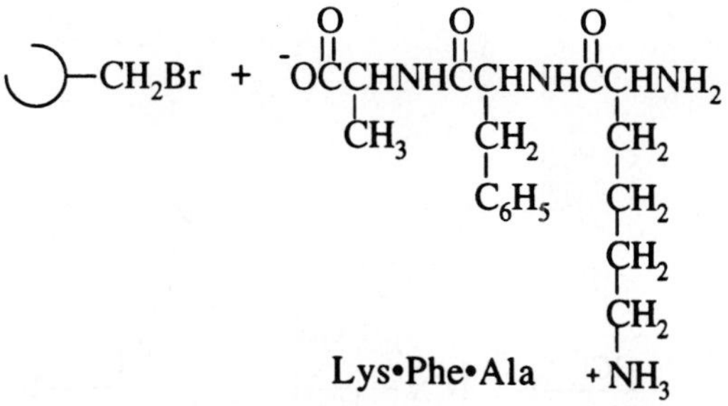

14. Isolate product.

Lys•Phe•Ala   + NH$_3$

24.16 (a) Isoleucine, threonine, hydroxyproline, and cystine.

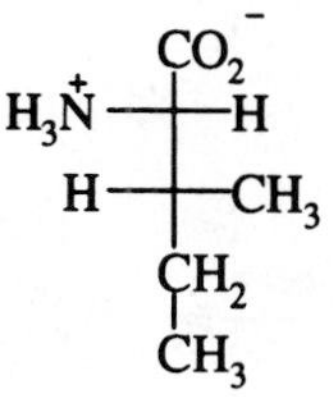

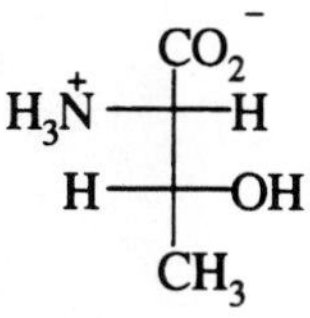

(b)

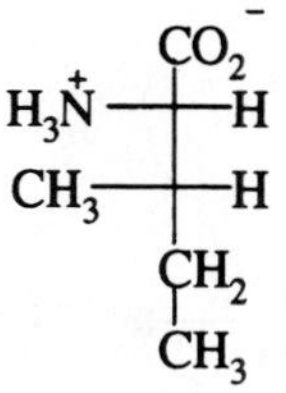

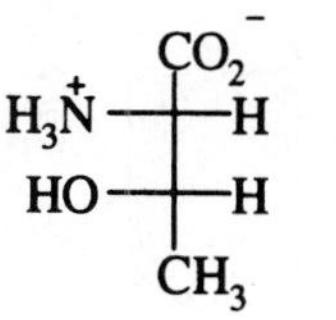

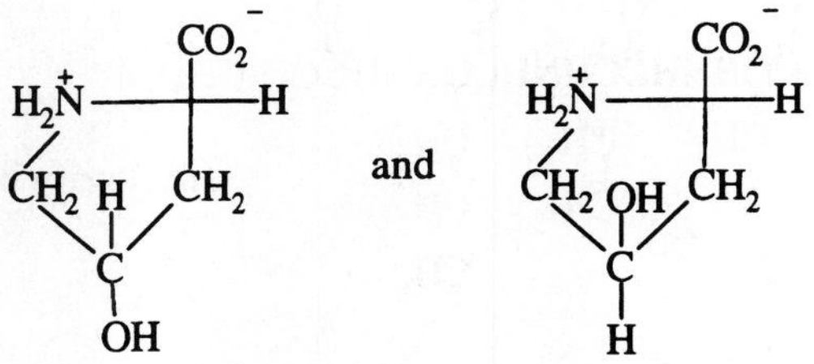

(With cystine, both stereocenters are α-carbon atoms; thus, according to the problem, both must have the L-configuration, and no isomers of this type can be written.)

(c)   Diastereomers

24.17 (a)   Alanine

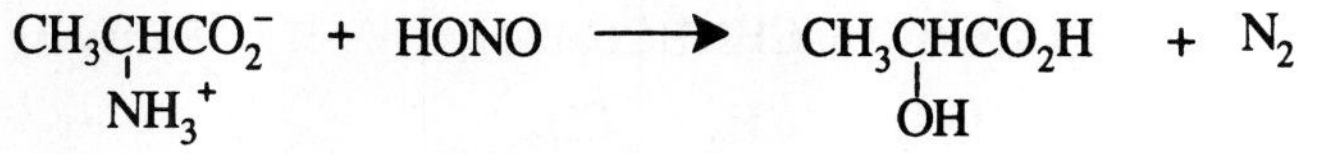

$$CH_3\underset{\underset{NH_3^+}{|}}{C}HCO_2^- + HONO \longrightarrow CH_3\underset{\underset{OH}{|}}{C}HCO_2H + N_2$$

(b)   Proline and hydroxyproline. All of the other amino acids have at least one primary amino group.

(c) 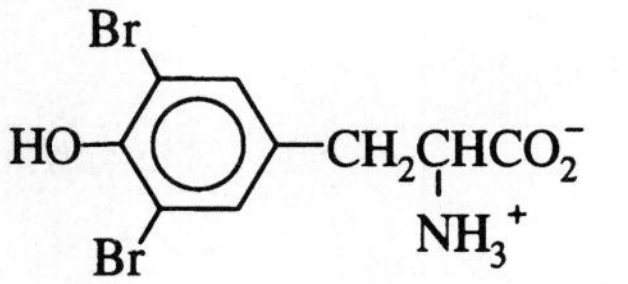   (d)

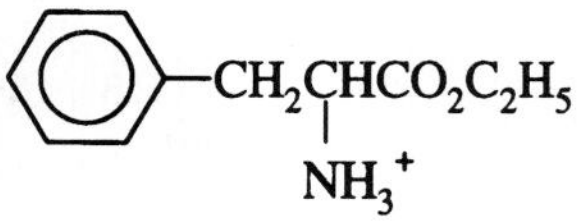

(e)

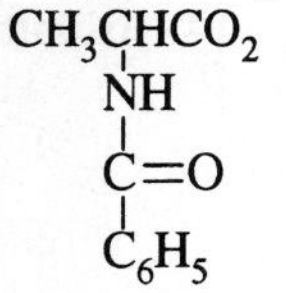

24.18 (a)

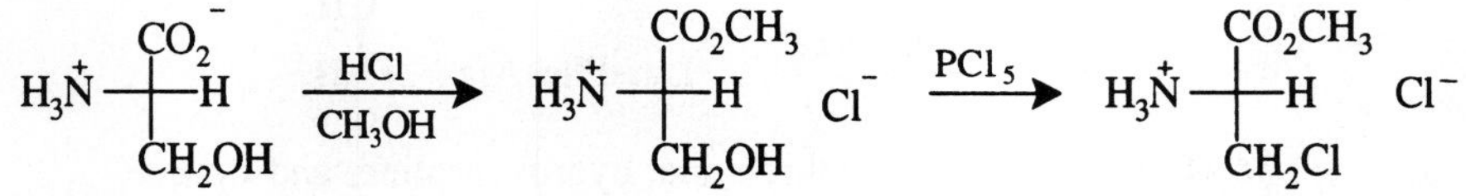

(−)-Serine   A   B
              $(C_4H_{10}ClNO_3)$ $(C_4H_9Cl_2NO_2)$

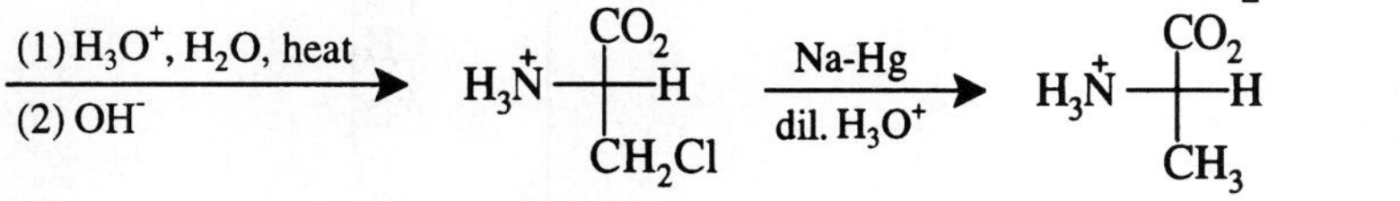

C   L-(+)-Alanine
$(C_3H_6ClNO_2)$

(b) $B \xrightarrow{OH^-}$

$$H_2N-\overset{\displaystyle CO_2CH_3}{\underset{\displaystyle CH_2Cl}{\overset{|}{\underset{|}{C}}}}-H \xrightarrow{NaSH} H_2N-\overset{\displaystyle CO_2CH_3}{\underset{\displaystyle CH_2SH}{\overset{|}{\underset{|}{C}}}}-H$$

D

$(C_4H_8ClNO_2)$

E

$(C_4H_9NO_2S)$

$$\xrightarrow[\text{(2) OH}^-]{\text{(1) H}_3O^+, \text{H}_2O, \text{heat}} \quad H_3\overset{+}{N}-\overset{\displaystyle CO_2^-}{\underset{\displaystyle CH_2SH}{\overset{|}{\underset{|}{C}}}}-H$$

L-(+)-Cysteine

(c) $H_3\overset{+}{N}-\overset{\displaystyle CO_2^-}{\underset{\displaystyle \underset{O}{\overset{|}{C}}H_2\underset{\parallel}{C}NH_2}{\overset{|}{\underset{|}{C}}}}-H$

$\xrightarrow[\substack{\text{Hofmann}\\\text{rearrangement}}]{\text{NaOBr, OH}^-}$

$H_2N-\overset{\displaystyle CO_2^-}{\underset{\displaystyle CH_2NH_2}{\overset{|}{\underset{|}{C}}}}-H$

$\xleftarrow{NH_3}$

$H_3\overset{+}{N}-\overset{\displaystyle CO_2^-}{\underset{\displaystyle CH_2Cl}{\overset{|}{\underset{|}{C}}}}-H$

L-(−)-Asparagine

F

$(C_3H_7N_2O_2)$

C

[from part (a)]

24.19 (a) $CH_3\overset{O}{\overset{\parallel}{C}}NHCH(CO_2C_2H_5)_2 + CH_2{=}CH{-}C{\equiv}N \xrightarrow[\substack{C_2H_5OH\\(95\%\ \text{yield})}]{NaOC_2H_5}$

$$CH_3\overset{O}{\overset{\parallel}{C}}NH-\overset{\displaystyle CO_2C_2H_5}{\underset{\displaystyle CO_2C_2H_5}{\overset{|}{\underset{|}{C}}}}-CH_2CH_2C{\equiv}N \xrightarrow[\substack{\text{reflux 6 h}\\(66\%\ \text{yield})}]{\text{concd. HCl}} HO_2CCH_2CH_2\underset{\displaystyle NH_3^+}{\overset{|}{C}}HCO_2^-$$

G

DL-Glutamic acid

$$+ CH_3CO_2H + 2 C_2H_5OH + NH_4^+ + CO_2$$

(b) $CH_3\overset{O}{\overset{\parallel}{C}}NH-\overset{\displaystyle CO_2C_2H_5}{\underset{\displaystyle CO_2C_2H_5}{\overset{|}{\underset{|}{C}}}}-CH_2CH_2C{\equiv}N \xrightarrow[\substack{68°C, 1000\ \text{psi}\\(90\%\ \text{yield})}]{H_2,\ Ni} \left[CH_3\overset{O}{\overset{\parallel}{C}}NH-\overset{\displaystyle CO_2C_2H_5}{\underset{\displaystyle CO_2C_2H_5}{\overset{|}{\underset{|}{C}}}}-CH_2CH_2CH_2NH_2\right]$

$$\xrightarrow{-C_2H_5OH} \quad \underset{\displaystyle CH_3\underset{\parallel O}{\overset{}{C}}NH}{\overset{\displaystyle C_2H_5O_2C}{}}{\overset{\diagup}{\underset{\diagdown}{C}}}\underset{\displaystyle \underset{O}{\overset{\parallel}{C}}{-}NH}{\overset{\displaystyle CH_2CH_2}{\overset{\diagup}{\underset{\diagdown}{\ }}}}CH_2 \xrightarrow[\substack{\text{reflux 4 h}\\(97\%\ \text{yield})}]{\text{concd. HCl}} H_3\overset{+}{N}CH_2CH_2CH_2\underset{\displaystyle NH_3^+}{\overset{|}{C}}HCO_2^- \quad \underset{\displaystyle Cl^-}{}$$

DL-Ornithine hydrochloride

H

$$+ CH_3CO_2H + C_2H_5OH + CO_2$$

24.20 We look for points of overlap:

```
                           Phe  •  Ser
                 Pro  •  Gly  •  Phe
         Pro  •   Pro                  Ser  •  Pro  •  Phe
   Arg •  Pro                                    Phe    •  Arg
   ─────────────────────────────────────────────────────────────
   Arg •  Pro  •  Pro  •  Gly  •  Phe  •  Ser  •  Pro  •  Phe   •  Arg
```

Bradykinin

24.21 1.   This experiment shows that valine is the N-terminal amino acid and that valine is attached to leucine. (Lysine labeled at the ϵ-amino group is to be expected if lysine is not the N-terminal amino acid and if it is linked in the polypeptide through its α-amino group.)

2.   This experiment shows that alanine is the C-terminal amino acid and that it is linked to glutamic acid.

At this point, then, we have the following information about the structure of the heptapeptide.

Val • Leu (Ala, Lys, Phe) Glu • Ala

The sequence here is
unknown.

3. (a)   This experiment shows that the dipeptide, **A,** is

Leu • Lys

(b)   The carboxypeptidase reaction shows that the C-terminal amino acid of the tripeptide, **B,** is glutamic acid; the DNP labeling experiment shows that the N-terminal amino acid is phenylalanine. Thus, the tripeptide **B** is

Phe • Ala • Glu

Putting these pieces together in the only way possible, we arrive at the following amino acid sequence for the heptapeptide.

```
Val  •  Leu
         Leu  •  Lys
                  Phe  •  Ala  •  Glu
                                  Glu  •  Ala
   ──────────────────────────────────────────────
Val  •  Leu  •  Lys  •  Phe  •  Ala  •  Glu  •  Ala
```

24.22 At pH 2-3 the γ-carboxyl groups of polyglutamic acid are uncharged. (They are present as $-CO_2H$ groups.) At pH 5 the γ-carboxyl groups ionize and become negatively charged. (They become γ-CO_2^- groups.) The repulsive forces between these negatively charged groups cause an unwinding of the α helix and the formation of a random coil.

24.23 The observation that the ^{1}H NMR spectrum taken at room temperature shows two different signals for the methyl groups suggests that they are in different environments. This would be true if rotation about the carbon-nitrogen bond was not taking place.

$$\delta\, 8.05\ \ H \diagdown \underset{O}{\overset{\diagup}{C}}=\!\!=N\diagup^{CH_3\ \ \delta\, 2.95}_{\diagdown CH_3\ \ \delta\, 2.80}$$

We assign the δ 2.80 signal to the methyl group that is on the same side as the electronegative oxygen atom.

    The fact that the methyl signals appear as doublets (and that the formyl proton signal is a multiplet) indicates that long-range coupling is taking place between the methyl protons and the formyl proton.

    That the two doublets are not simply the result of spin-spin coupling is indicated by the observation that the distance that separates one doublet from the other changes when the applied magnetic field strength is lowered. [*Remember!* The magnitude of a chemical shift is proportional to the strength of the applied magnetic field, while the magnitude of a coupling constant is not.]

    That raising the temperature (to 111°C) causes the doublets to coalesce into a single signal indicates that at higher temperatures the molecules have enough energy to surmount the energy barrier of the carbon-nitrogen bond. Above 111°C, rotation is taking place so rapidly that the spectrometer is unable to discriminate between the two methyl groups.

QUIZ

24.1 Write the structural formula of the principal ionic species present in aqueous solutions at pH 2, 7, and 12 of isoleucine (2-amino-3-methylpentanoic acid).

At pH = 2	At pH = 7	At pH = 12
(a)	(b)	(c)

24.2 A hexapeptide gave the following products:

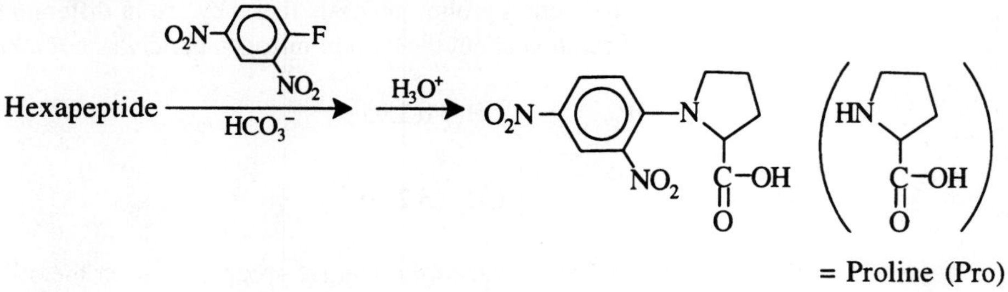

Hexapeptide $\xrightarrow[\text{HCO}_3^-]{\text{O}_2\text{N}-\langle\text{C}_6\text{H}_3(\text{NO}_2)\rangle-\text{F}}$ $\xrightarrow{\text{H}_3\text{O}^+}$ $\text{O}_2\text{N}-\langle\text{C}_6\text{H}_3(\text{NO}_2)\rangle-\text{N}\langle\text{Pro}\rangle\text{C(=O)-OH}$ $\left(\text{HN}\langle\text{Pro}\rangle\text{C(=O)-OH}\right)$ = Proline (Pro)

Hexapeptide $\xrightarrow{3\,N\text{ HCl, }100°\text{C}}$ 2 Gly, 1 Leu, 1 Phe, 1 Pro, 1 Tyr

Hexapeptide $\xrightarrow{1\,N\text{ HCl, }80°\text{C}}$ Phe•Gly•Tyr + Gly•Phe•Gly + Pro•Leu•Gly + Leu•Gly•Phe

The structure of the hexapeptide (using abbreviations such as Gly•Leu•etc.) is

25

NUCLEIC ACIDS AND PROTEIN SYNTHESIS

SOLUTIONS TO PROBLEMS

25.1 Adenine:

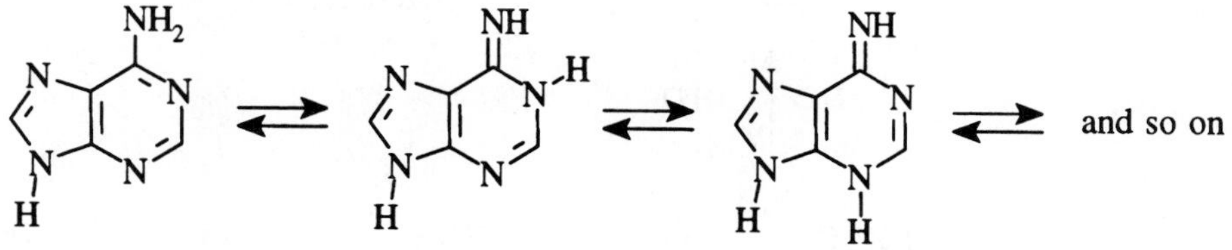

Guanine:

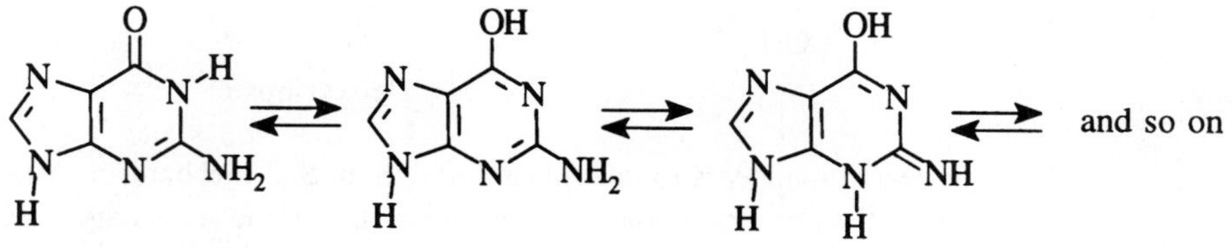

Cytosine:

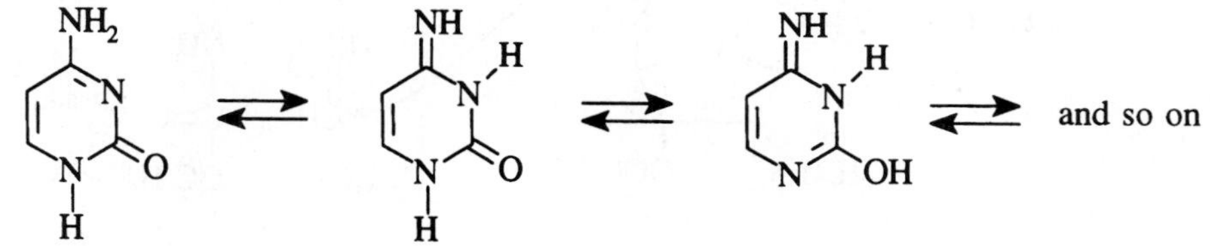

Thymine (R=CH$_3$) or Uracil (R=H):

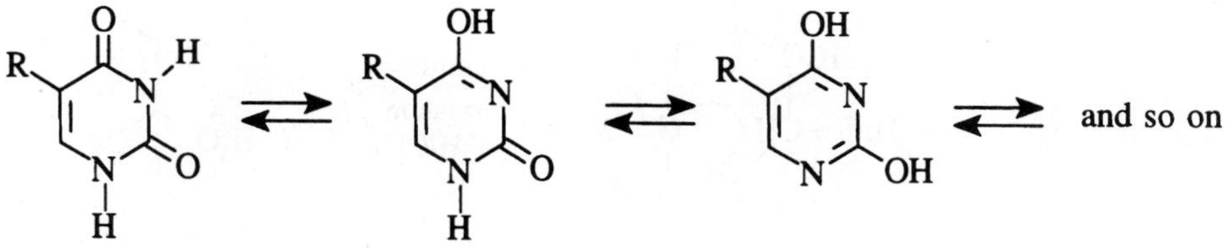

25.2 (a) The nucleosides have an *N*-glycosidic linkage that (like an *O*-glycosidic linkage) is rapidly hydrolyzed by aqueous acid but is one that is stable in aqueous base.

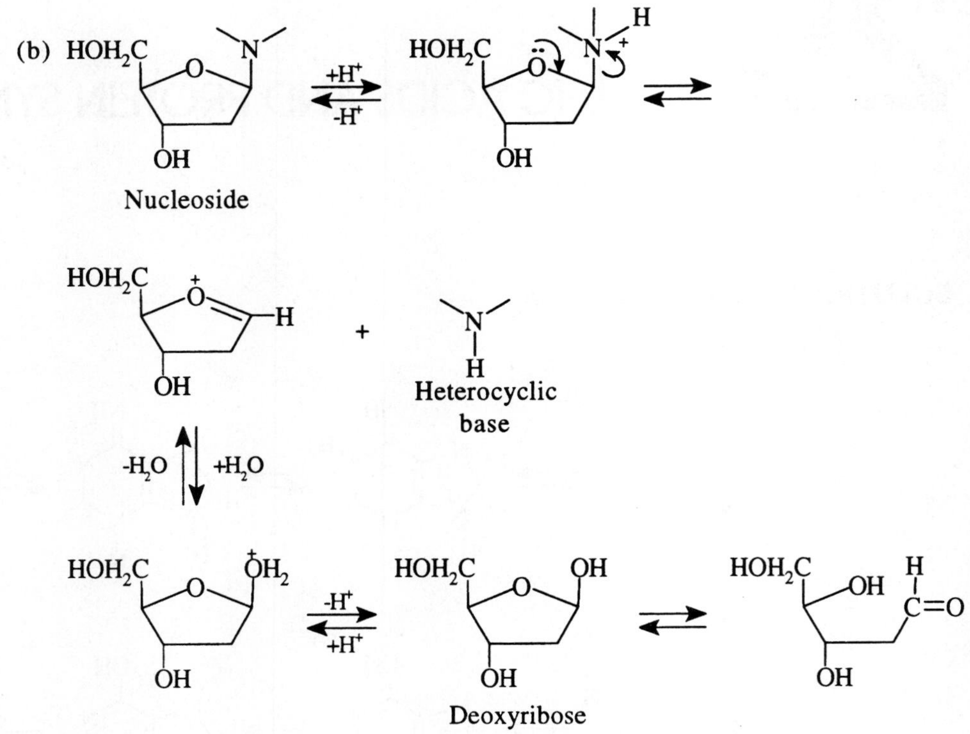
(b) HOH2C
OH
Nucleoside
+H+
-H+
HOH2C
H
OH
HOH2C
H
+
N
H
Heterocyclic base
OH
-H2O
+H2O
HOH2C
OH2
OH
-H+
+H+
HOH2C
OH
OH
Deoxyribose
HOH2C
OH
H
C=O
OH

25.3 The reaction appears to take place through an S_N2 mechanism. Attack occurs preferentially at the primary 5′-carbon atom rather than at the secondary 3′-carbon atom.

25.4

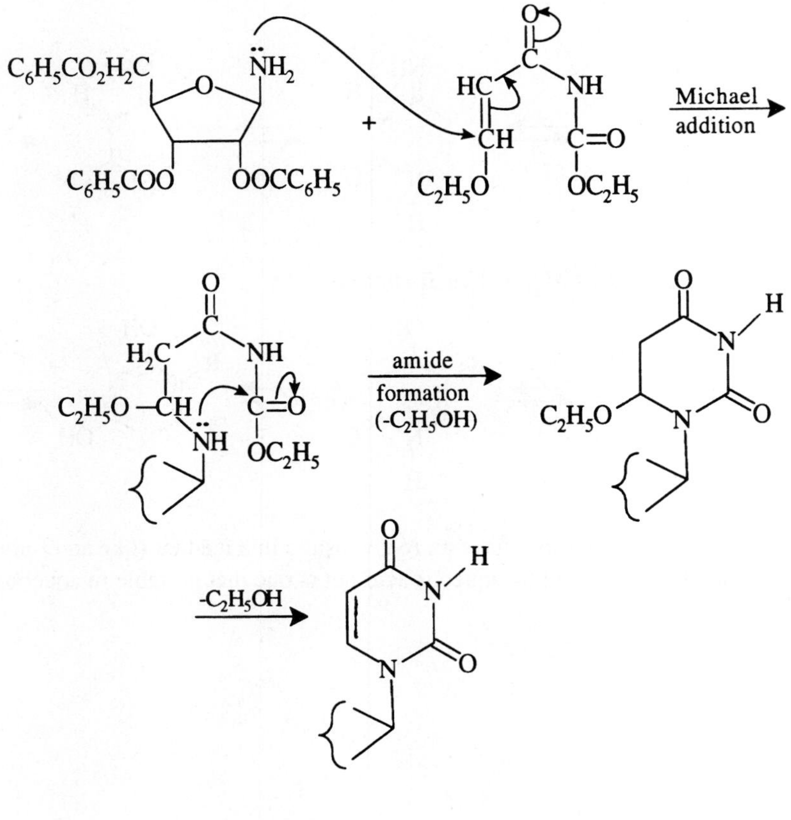
C6H5CO2H2C
NH2
C6H5COO
OOCC6H5
+
HC
CH
C2H5O
NH
C=O
OC2H5
Michael addition
H2C
NH
C2H5O
CH
C=O
NH
OC2H5
amide formation
(-C2H5OH)
C2H5O
H
-C2H5OH
H

25.5 (a) The isopropylidene group is part of a cyclic acetal and is thus susceptible to hydrolysis by mild acid.

(b) It can be installed by treating the nucleoside with acetone and a trace of acid and by simultaneously removing the water that is produced.

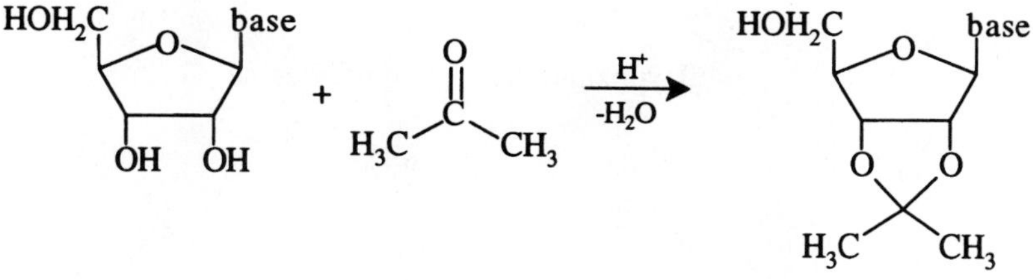

25.6

(a) $6 \times 10^9 \text{ base pairs} \times \dfrac{34 \text{ Å}}{10 \text{ base pairs}} \times \dfrac{10^{-10} \text{ m}}{\text{Å}} \cong 2 \text{ m}$

(b) $6 \times 10^{-12} \dfrac{\text{g}}{\text{ovum}} \times 3 \times 10^9 \text{ ova} = 1.8 \times 10^{-2} \text{ g}$

25.7

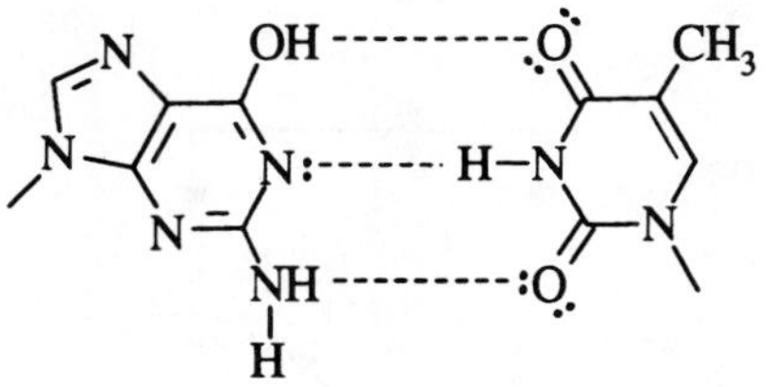

Lactim form   Thymine
of guanine

(b) Thymine would pair with adenine and thus adenine would be introduced into the complementary strand where guanine should occur.

25.8 (a) A diazonium salt and a heterocyclic analog of a phenol.

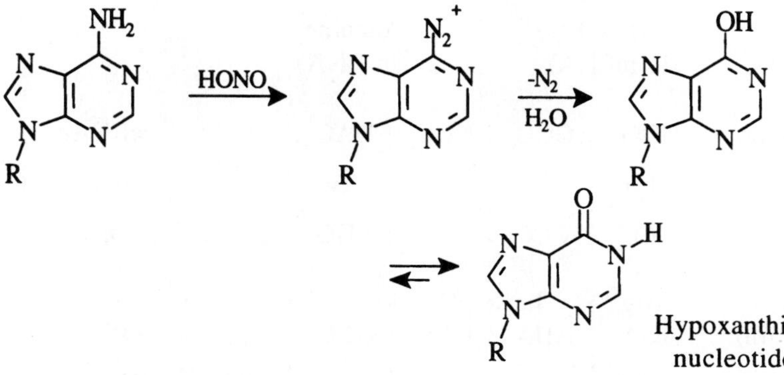

Hypoxanthine
nucleotide

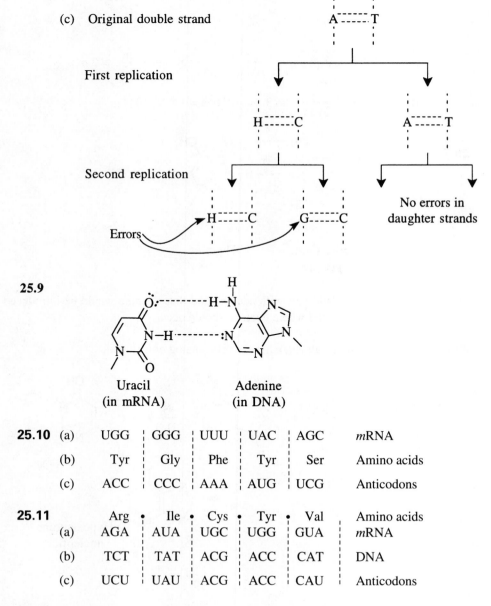

(b)

Hypoxanthine   Cytosine

(c)   Original double strand

First replication

Second replication

Errors

No errors in
daughter strands

25.9

Uracil
(in mRNA)

Adenine
(in DNA)

25.10 (a)   UGG ¦ GGG ¦ UUU ¦ UAC ¦ AGC   *m*RNA

(b)   Tyr ¦ Gly ¦ Phe ¦ Tyr ¦ Ser   Amino acids

(c)   ACC ¦ CCC ¦ AAA ¦ AUG ¦ UCG   Anticodons

25.11   Arg • Ile • Cys • Tyr • Val   Amino acids

(a)   AGA ¦ AUA ¦ UGC ¦ UGG ¦ GUA   *m*RNA

(b)   TCT ¦ TAT ¦ ACG ¦ ACC ¦ CAT   DNA

(c)   UCU ¦ UAU ¦ ACG ¦ ACC ¦ CAU   Anticodons

25.12 A change from C–T–T to C–A–T or a change from C–T–C to C–A–C.

A

APPENDIX
Empirical and Molecular Formulas

In Section 1.2B, we discussed briefly the pioneering work of Berzelius, Dumas, Liebig, and Cannizzaro in devising methods for determining the formulas of organic compounds. Although the experimental procedures for these analyses have been refined, the basic methods for determining the elemental composition of an organic compound today are not substantially different from those used in the nineteenth century. A carefully weighed quantity of the compound to be analyzed is oxidized completely to carbon dioxide and water. The weights of carbon dioxide and water are carefully measured and used to find the percentages of carbon and hydrogen in the compound. The percentage of nitrogen is usually determined by measuring the volume of nitrogen (N_2) produced in a separate procedure.

Special techniques for determining the percentage composition of other elements typically found in organic compounds have also been developed, but the direct determination of the percentage of oxygen is difficult. However, if the percentage composition of all the other elements is known, then the percentage of oxygen can be determined by difference. The following examples will illustrate how these calculations can be carried out.

EXAMPLE A

A new organic compound is found to have the following elemental analysis.

Carbon	67.95%
Hydrogen	5.69
Nitrogen	26.20
Total:	99.84%

Since the total of these percentages is very close to 100% (within experimental error), we can assume that no other element is present. For the purpose of our calculation it is convenient to assume that we have a 100-g sample. If we did, it would contain the following:

67.95 g of carbon
5.69 g of hydrogen
26.20 g of nitrogen

In other words, we use percentages *by weight* to give us the ratios *by weight* of the elements in the substance. To write a formula for the substance, however, we need *ratios by moles*.

We now divide each of these weight-ratio numbers by the atomic weight of the particular element and obtain the number of moles of each element, respectively, in 100 g of the compound. This operation gives us the ratios *by moles* of the elements in the substance:

$$C \quad \frac{67.95 \text{ g}}{12.01 \text{ g mol}^{-1}} = 5.66 \text{ mol}$$

$$H \quad \frac{5.69 \text{ g}}{1.008 \text{ g mol}^{-1}} = 5.64 \text{ mol}$$

$$N \quad \frac{26.20 \text{ g}}{14.01 \text{ g mol}^{-1}} = 1.87 \text{ mol}$$

One possible formula for the compound, therefore, is $C_{5.66}H_{5.64}N_{1.87}$.

By convention, however, we use *whole* numbers in formulas. Therefore, we convert these fractional numbers of moles to whole numbers by dividing each by 1.87, the smallest number.

$$C \quad \frac{5.66}{1.87} = 3.03 \text{ which is } \sim 3$$

$$H \quad \frac{5.64}{1.87} = 3.02 \text{ which is } \sim 3$$

$$N \quad \frac{1.87}{1.87} = 1.00$$

Thus, within experimental error, the ratios by moles are 3 C to 3 H to 1 N, and C_3H_3N is the *empirical formula*. By empirical formula, we mean the formula in which the subscripts are the smallest integers that give the ratio of atoms in the compound. In contrast, a *molecular* formula discloses the complete composition of one molecule. The molecular formula of this particular compound could be C_3H_3N or some whole number multiple of C_3H_3N; that is, $C_6H_6N_2$, $C_9H_9N_3$, $C_{12}H_{12}N_4$, and so on. If, in a separate determination, we find that the molecular weight of the compound is 108 ± 3, we can be certain that the *molecular formula* of the compound is $C_6H_6N_2$.

FORMULA	MOLECULAR WEIGHT
C_3H_3N	53.06
$C_6H_6N_2$	106.13 (which is within the range 108 ± 3)
$C_9H_9N_3$	159.19
$C_{12}H_{12}N_4$	212.26

The most accurate method for determining molecular weights is by mass spectrometry. A variety of other methods based on freezing point depression, boiling point elevation, osmotic pressure, and vapor density can also be used to determine molecular weights.

EXAMPLE B

Histidine, an amino acid isolated from protein, has the following elemental analysis:

Carbon	46.38%
Hydrogen	5.90
Nitrogen	27.01
Total:	79.29
Difference	20.71 (assumed to be oxygen)
	100.00%

Since no elements, other than carbon, hydrogen, and nitrogen, are found to be present in histidine, the difference is assumed to be oxygen. Again, we assume a 100-g sample and divide the weight of each element by its gram-atomic weight. This gives us the ratio of moles (A).

$$
\begin{array}{cccc}
 & (A) & (B) & (C) \\
C & \dfrac{46.38}{12.01} = 3.86 & \dfrac{3.86}{1.29} = 2.99 \times 2 = 5.98 \sim 6 \text{ carbon atoms} \\[2ex]
H & \dfrac{5.90}{1.008} = 5.85 & \dfrac{5.85}{1.29} = 4.53 \times 2 = 9.06 \sim 9 \text{ hydrogen atoms} \\[2ex]
N & \dfrac{27.01}{14.01} = 1.93 & \dfrac{1.93}{1.29} = 1.50 \times 2 = 3.00 = 3 \text{ nitrogen atoms} \\[2ex]
O & \dfrac{20.71}{16.00} = 1.29 & \dfrac{1.29}{1.29} = 1.00 \times 2 = 2.00 = 2 \text{ oxygen atoms}
\end{array}
$$

Dividing each of the moles (A) by the smallest of them does not give a set of numbers (B) that is close to a set of whole numbers. Multiplying each of the numbers in column (B) by 2 does, however, as seen in column (C). The empirical formula of histidine is, therefore, $C_6H_9N_3O_2$.

In a separate determination, the molecular weight of histidine was found to be 158 ± 5. The empirical formula weight of $C_6H_9N_3O_2$ (155.15) is within this range; thus, the molecular formula for histidine is the same as the empirical formula.

PROBLEMS

A.1 What is the empirical formula of each of the following compounds?

(a) Hydrazine, N_2H_4 (d) Nicotine, $C_{10}H_{14}N_2$
(b) Benzene, C_6H_6 (e) Cyclodecane, $C_{10}H_{20}$
(c) Dioxane, $C_4H_8O_2$ (f) Acetylene, C_2H_2

A.2 The empirical formulas and molecular weights of several compounds are given next. In each case, calculate the molecular formula for the compound.

EMPIRICAL FORMULA	MOLECULAR WEIGHT
(a) CH_2O	179 ± 5
(b) CHN	80 ± 5
(c) CCl_2	410 ± 10

A.3 The widely used antibiotic, penicillin G, gave the following elemental analysis: C, 57.45%; H, 5.40%; N, 8.45%; S, 9.61%. The molecular weight of penicillin G is 330 ± 10. Assume that no other elements except oxygen are present and calculate the empirical and molecular formulas for penicillin G.

ADDITIONAL PROBLEMS

A.4 Calculate the percentage composition of each of the following compounds.
(a) $C_6H_{12}O_6$
(b) $CH_3CH_2NO_2$
(c) $CH_3CH_2CBr_3$

A.5 An organometallic compound called *ferrocene* contains 30.02% iron. What is the minimum molecular weight of ferrocene?

A.6 A gaseous compound gave the following analysis: C, 40.04%; H, 6.69%. At standard temperature and pressure, 1.00 g of the gas occupied a volume of 746 mL. What is the molecular formula of the compound?

A.7 A gaseous hydrocarbon has a density of 1.251 g L^{-1} at standard temperature and pressure. When subjected to complete combustion, a 1.000-L sample of the hydrocarbon gave 3.926 g of carbon dioxide and 1.608 g of water. What is the molecular formula for the hydrocarbon?

A.8 Nicotinamide, a vitamin that prevents the occurrence of pellagra, gave the following analysis: C, 59.10%; H, 4.92%; N, 22.91%. The molecular weight of nicotinamide was shown in a separate determination to be 120 ± 5. What is the molecular formula for nicotinamide?

A.9 The antibiotic chloramphenicol gave the following analysis: C, 40.88%; H, 3.74%; Cl, 21.95%; N, 8.67%. The molecular weight was found to be 300 ± 30. What is the molecular formula for chloramphenicol?

SOLUTIONS TO PROBLEMS OF APPENDIX A

A.1   (a) NH_2 (b) CH   (c) C_2H_4O (d) C_5H_7N (e) CH_2 (f) CH

A.2

EMPIRICAL FORMULA	EMPIRICAL FORMULA WEIGHT	$\left(\dfrac{\text{MOLECULAR WEIGHT}}{\text{EMP. FORM. WT.}}\right)$	MOLECULAR FORMULA
(a) CH_2O	30	$\dfrac{179}{30} \cong 6$	$C_6H_{12}O_6$
(b) CHN	27	$\dfrac{80}{27} \cong 3$	$C_3H_3N_3$
(c) CCl_2	83	$\dfrac{410}{83} \cong 5$	C_5Cl_{10}

A.3   If we assume that we have a 100-g sample, the amounts of the elements are

	WEIGHT	Moles (A)	B
C	57.45	$\dfrac{57.45}{12.01} = 4.78$	$\dfrac{4.78}{0.300} = 15.9 = 16$
H	5.40	$\dfrac{5.40}{1.008} = 5.36$	$\dfrac{5.36}{0.300} = 17.9 = 18$
N	8.45	$\dfrac{8.45}{14.01} = 0.603$	$\dfrac{0.603}{0.300} = 2.01 = 2$
S	9.61	$\dfrac{9.61}{32.06} = 0.300$	$\dfrac{0.300}{0.300} = 1.00 = 1$
O*	$\dfrac{19.09}{100.00}$	$\dfrac{19.09}{16.00} = 1.19$	$\dfrac{1.19}{0.300} = 3.97 = 4$

(* by difference from 100)

The empirical formula is thus $C_{16}H_{18}N_2SO_4$. The empirical formula weight (334.4) is within the range given for the molecular weight (330 $\pm$ 10). Thus, the molecular formula for penicillin G is the same as the empirical formula.

A.4   (a)   To calculate the percentage composition from the molecular formula, first determine the weight of each element in 1 mol of the compound. For $C_6H_{12}O_6$,

C_6 = 6 × 12.01 = 72.06 $\dfrac{72.06}{180.2}$ = 0.400 = 40.0%

H_{12} = 12 × 1.008 = 12.10 $\dfrac{12.10}{180.2}$ = 0.0671 = 6.7%

O_6 = 6 × 16.00 = 96.00 $\dfrac{96.00}{180.2}$ = 0.533 = 53.3%

MW   180.16

(MW = molecular weight)

Then determine the percentage of each element using the formula

$$\text{Percentage of A} = \frac{\text{Weight of A}}{\text{Molecular Weight}} \times 100$$

(b) $C_2 = 2 \times 12.01 = 24.02$ $\dfrac{24.02}{75.07} = 0.320 = 32.0\%$

$H_5 = 5 \times 1.008 = 5.04$ $\dfrac{5.04}{75.07} = 0.067 = 6.7\%$

$N = 1 \times 14.01 = 14.01$ $\dfrac{14.01}{75.07} = 0.187 = 18.7\%$

$O_2 = 2 \times 16.00 = \underline{32.00}$ $\dfrac{32.00}{75.07} = 0.426 = 42.6\%$

Total $= 75.07$

(c) $C_3 = 3 \times 12.01 = 36.03$ $\dfrac{36.03}{280.77} = 0.128 = 12.8\%$

$H_5 = 5 \times 1.008 = 5.04$ $\dfrac{5.04}{280.77} = 0.018 = 1.8\%$

$Br_3 = 3 \times 79.90 = \underline{239.70}$ $\dfrac{239.70}{280.77} = 0.854 = 85.4\%$

Total $= 280.77$

A.5 If the compound contains iron, each molecule must contain at least one atom of iron, and 1 mol of the compound must contain at least 55.85 g of iron. Therefore,

$$\text{MW of ferrocene} = 55.85 \, \frac{\text{g of Fe}}{\text{mol}} \times \frac{1.000 \text{ g}}{0.3002 \text{ g of Fe}}$$

$$= 186.0 \, \frac{\text{g}}{\text{mol}}$$

A.6 First, we must determine the empirical formula. Assuming that the difference between the percentages given and 100% is due to oxygen, we calculate:

C   40.04 $\dfrac{40.04}{12.01} = 3.33$ $\dfrac{3.33}{3.33} = 1$

H   6.69 $\dfrac{6.69}{1.008} = 6.64$ $\dfrac{6.64}{3.33} \cong 2$

O $\dfrac{53.27}{100.00}$ $\dfrac{53.27}{16.00} = 3.33$ $\dfrac{3.33}{3.33} = 1$

The empirical formula is thus CH_2O.

To determine the molecular formula, we must first determine the molecular weight. At standard temperature and pressure, the volume of 1 mol of an ideal gas is 22.4 L. Assuming ideal behavior,

$$\frac{1.00 \text{ g}}{0.746 \text{ L}} = \frac{MW}{22.4 \text{ L}} \quad \text{where MW = molecular weight}$$

$$MW = \frac{(1.00)(22.4)}{0.746} = 30.0 \text{ g}$$

The empirical formula weight (30.0) equals the molecular weight; thus, the molecular formula is the same as the empirical formula.

A.7 As in Problem A.6, the molecular weight is found by the equation

$$\frac{1.251 \text{ g}}{1.00 \text{ L}} = \frac{MW}{22.4 \text{ L}}$$

$$MW = (1.251)(22.4)$$
$$MW = 28.02$$

To determine the empirical formula, we must determine the amount of carbon in 3.926 g of carbon dioxide, and the amount of hydrogen in 1.608 g of water.

$$C \quad \left(3.926 \text{ g } CO_2\right) \left(\frac{12.01 \text{ g C}}{44.01 \text{ g } CO_2}\right) = 1.071 \text{ g carbon}$$

$$H \quad \left(1.608 \text{ g } H_2O\right) \left(\frac{2.016 \text{ g H}}{18.016 \text{ g } H_2O}\right) = \underline{0.180 \text{ g hydrogen}}$$
$$1.251 \text{ g sample}$$

The weight of C and H in a 1.251-g sample is 1.251 g. Therefore, there are no other elements present.

To determine the empirical formula, we proceed as in Problem A.6 except that the sample size is 1.251 g instead of 100 g.

$$C \quad \frac{1.071}{12.01} = 0.0892 \qquad \frac{0.0892}{0.0892} = 1$$

$$H \quad \frac{0.180}{1.008} = 0.179 \qquad \frac{0.179}{0.0892} = 2$$

The empirical formula is thus CH_2. The empirical formula weight (14) is one-half the molecular weight. Thus, the molecular formula is C_2H_4.

A.8 Use the procedure of Problem A.3.

$$C \quad 59.10 \qquad \frac{59.10}{12.01} = 4.92 \qquad \frac{4.92}{0.817} = 6.02 \cong 6$$

$$H \quad 4.92 \quad \frac{4.92}{1.008} = 4.88 \quad \frac{4.88}{0.817} = 5.97 \cong 6$$

$$N \quad 22.91 \quad \frac{22.91}{14.01} = 1.64 \quad \frac{1.64}{0.817} = 2$$

$$O \quad \frac{13.07}{100.00} \quad \frac{13.07}{16.00} = 0.817 \quad \frac{0.817}{0.817} = 1$$

The empirical formula is thus $C_6H_6N_2O$. The empirical formula weight is 122.13, which is equal to the molecular weight within experimental error. The molecular formula is thus the same as the empirical formula.

A.9

$$C \quad 40.88 \quad \frac{40.88}{12.01} = 3.40 \quad \frac{3.40}{0.619} = 5.5 \quad 5.5 \times 2 = 11$$

$$H \quad 3.74 \quad \frac{3.74}{1.008} = 3.71 \quad \frac{3.71}{0.619} = 6 \quad 6 \times 2 = 12$$

$$Cl \quad 21.95 \quad \frac{21.95}{35.45} = 0.619 \quad \frac{0.619}{0.619} = 1 \quad 1 \times 2 = 2$$

$$N \quad 8.67 \quad \frac{8.67}{14.01} = 0.619 \quad \frac{0.619}{0.619} = 1 \quad 1 \times 2 = 2$$

$$O \quad \frac{24.76}{100.00} \quad \frac{24.76}{16.00} = 1.55 \quad \frac{1.55}{0.619} = 2.5 \quad 2.5 \times 2 = 5$$

The empirical formula is thus $C_{11}H_{12}Cl_2N_2O_5$. The empirical formula weight (323) is equal to the molecular weight; therefore, the molecular formula is the same as the empirical formula.

B
APPENDIX
Answers to Quizzes

CHAPTER 1

1.1 (d), **1.2** (d), **1.3** (e), **1.4** (d), **1.5** (c), **1.6**

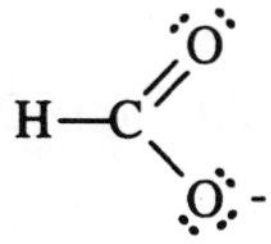

1.7

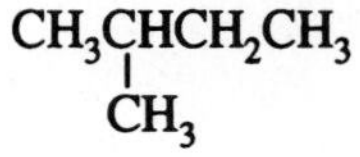

CH$_3$CHCH$_2$CH$_3$ and   CH$_3$—C—CH$_3$
with CH$_3$ substituent

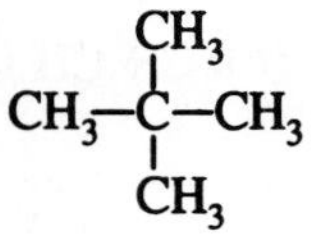

1.8

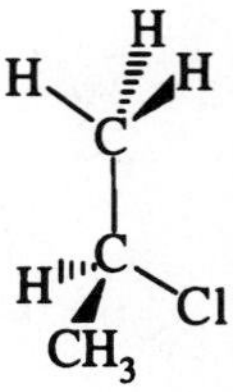

1.9 (a) sp^2 (b) sp^3 (c) 0   (d) trigonal planar   (e) 0

1.10 (a) +1   (b) 0   (c) −1

1.11

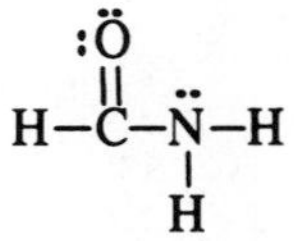

1.12

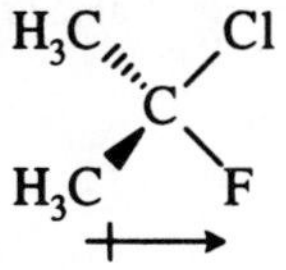

CHAPTER 2

2.1 (e)   **2.2** (a)   **2.3** (e)

2.4 (a) 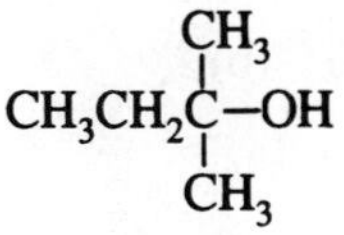 CH$_3$CH$_2$C—OH   (b) CH$_3$C—N   (c)

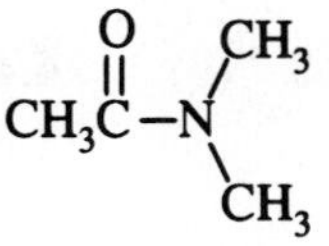

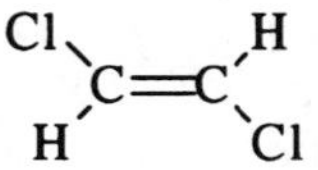

(d) $H-\overset{O}{\overset{\|}{C}}-O-CH_3$ (e) 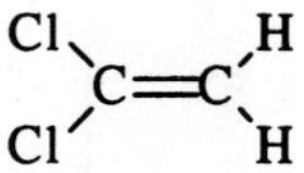 $\overset{Cl}{\underset{Cl}{>}}C=C\overset{H}{\underset{H}{<}}$ (f) $CH_3OCH_2CH_3$

(g) 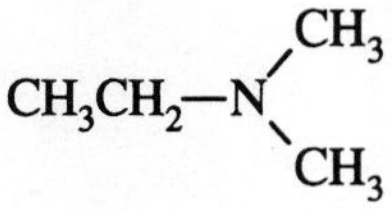 $CH_3CH_2-N\overset{CH_3}{\underset{CH_3}{<}}$

2.5   □   or   ▷—

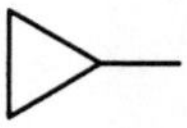

2.6 (a) $CH_3CH_2CH_2OH$ (b) 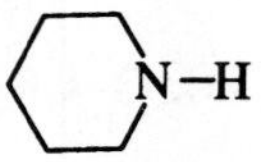 ⬡$N-H$ (c) $CH_3CH_2\overset{O}{\overset{\|}{C}}OH$

(d) $CH_3OCH_2CH_2OH$ (e) $CH_3CH_2\overset{O}{\overset{\|}{C}}NHCH_3$

2.7 (a) Isopropyl phenyl ether

(b) Ethylmethylphenylamine

(c) Isopropylamine

CHAPTER 3

3.1 (a)   **3.2** (c)   **3.3** (b)   **3.4** (e)   **3.5** (b)

3.6 (b)

3.7 $H_2SO_4 + NaF \longrightarrow NaHSO_4 + HF$

3.8 $(CH_3)_2NH$

3.9 (a) CH_3CH_2Li (b) D_2O

3.10 (a) CH_3Li (b) $CH_3\overset{CH_3}{\underset{|}{C}}HCH_2OH$ (c) $CH_3\overset{CH_3}{\underset{|}{C}}HCH_2OLi$

CHAPTER 4

4.1 (c)   **4.2** (c)   **4.3** (b)   **4.4** (a)   **4.5** (b)

4.6 (a)   **4.7** (a)

4.8 (a) 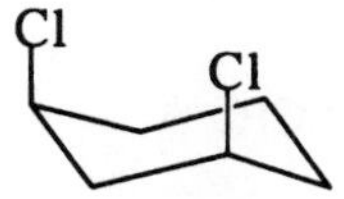 (b) 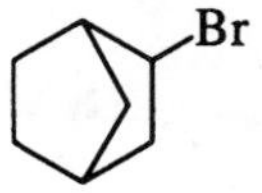 or

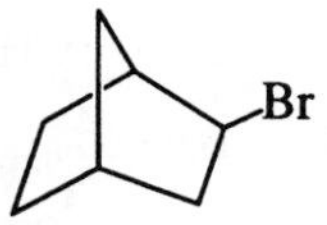

(c)

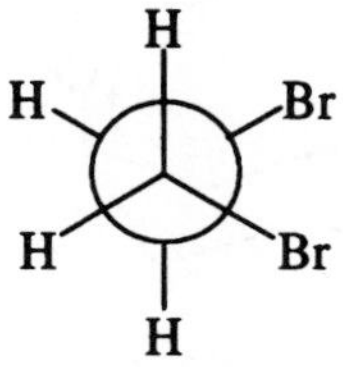

4.9 (a) H_2, Pt or H_2, Ni

(b) Zn, H^+

4.10 (a) $CH_3\overset{\displaystyle CH_3}{\underset{\displaystyle |}{C}}HCH_2CH_2Br$ (b) $CH_3\overset{\displaystyle CH_3}{\underset{\displaystyle |}{C}}HCH_2CH_2Li$

(c) $(CH_3\overset{\displaystyle CH_3}{\underset{\displaystyle |}{C}}HCH_2CH_2)_2CuLi$

4.11

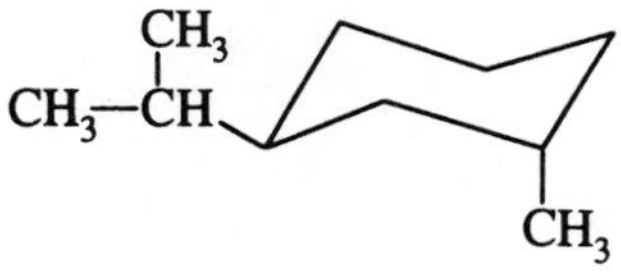

CHAPTER 5

5.1 (a) **5.2** (b) **5.3** (b) **5.4** (e) **5.5** (b)

5.6 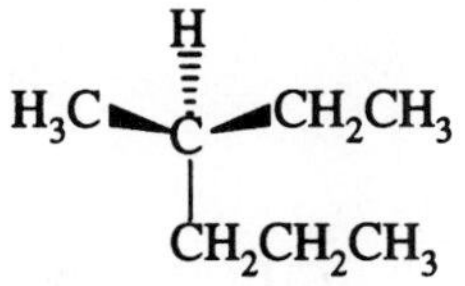 and

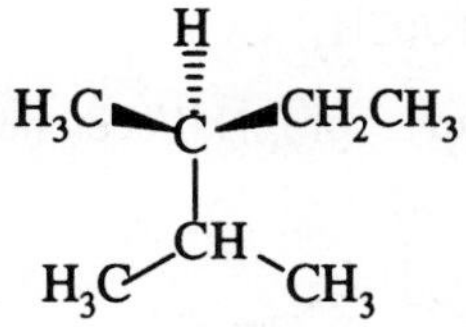

5.7 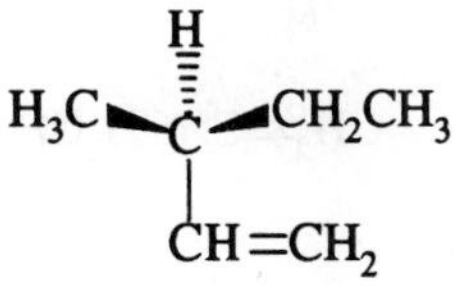 **5.8**

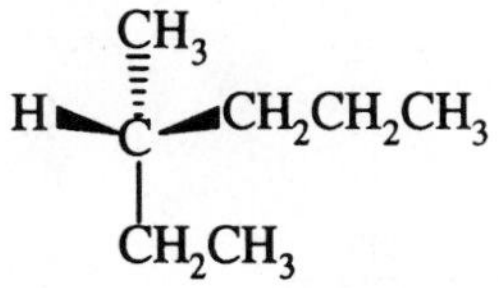

CHAPTER 6

6.1 (b) **6.2** (b) **6.3** (a)

6.4 $CH_3CH_2CH_2CH_2Br$ > $CH_3\underset{\displaystyle \underset{|}{CH_3}}{C}HCH_2Br$ > $CH_3CH_2\overset{\displaystyle CH_3}{\underset{\displaystyle |}{C}}HBr$ > $CH_3\overset{\displaystyle CH_3}{\underset{\displaystyle \underset{|}{CH_3}}{\underset{|}{C}}}-Br$

6.5   A = Br—C⟨CH$_2$CH$_3$⟩$_{\cdots}$H , CH$_3$ B = CH$_3$CH$_2$CH=CH$_2$

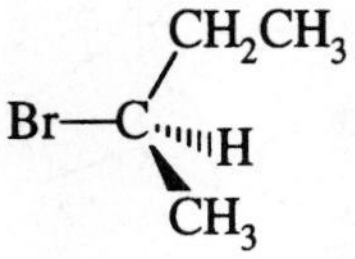

C = H—C⟨CH$_3$⟩CH$_2$OCH$_3$, CH$_2$CH$_3$ D = H$_3$C—⟨cyclohexane with CN⟩

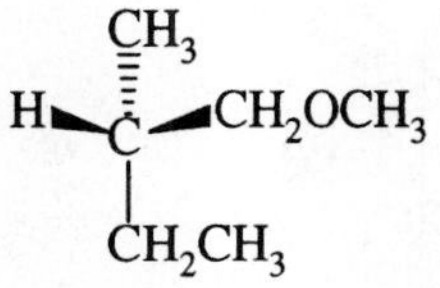

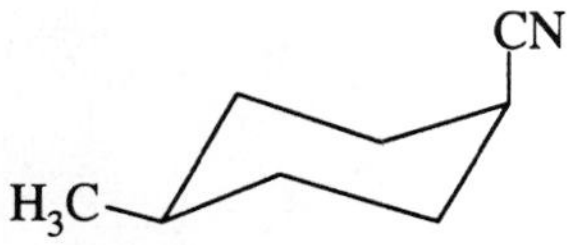

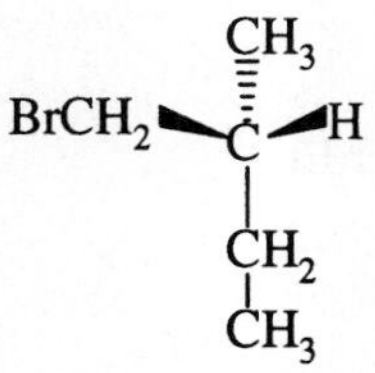

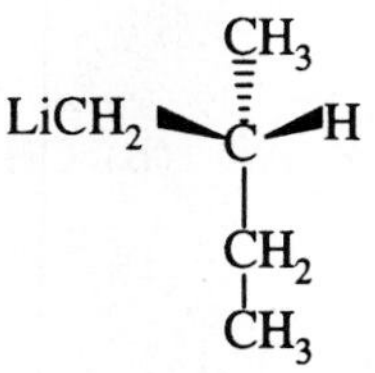

6.6   (b)

6.7   A = BrCH$_2$—C⟨CH$_3$⟩H , CH$_2$, CH$_3$ B = LiCH$_2$—C⟨CH$_3$⟩H , CH$_2$, CH$_3$ C = LiCu$\left(\text{CH}_2\text{—C}\begin{smallmatrix}\text{CH}_3\\ \text{H}\\ \text{CH}_2\\ \text{CH}_3\end{smallmatrix}\right)_2$

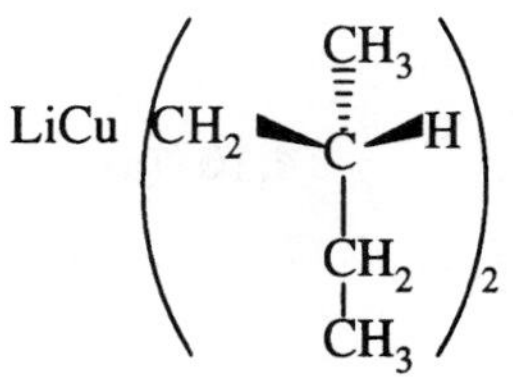

CHAPTER 7

7.1   (c)   **7.2**   (d)   **7.3**   (a)

7.4   (a)   Li, C$_2$H$_5$NH$_2$, −78°C, then NH$_4$Cl

(b)   H$_2$/Ni$_2$B(P–2)   or   H$_2$/Pd/CaCO$_3$ (Lindlar's catalyst)

(c)   H$_2$/Ni   or   H$_2$/Pt using at least 2 molar equivalents of H$_2$

(d)   C$_2$H$_5$ONa/C$_2$H$_5$OH

(e)   (CH$_3$)$_3$COK/(CH$_3$)$_3$COH

(f)   Zn/CH$_3$CO$_2$H   or   NaI/acetone

7.5   CH$_3$C(CH$_3$)=CHCH$_3$ > ⟨H$_3$C, H / C=C / H, CH$_2$CH$_3$⟩ > ⟨H$_3$C, H / C=C / H, CH$_2$CH$_3$⟩ >

CH$_2$=CHCH$_2$CH$_2$CH$_3$

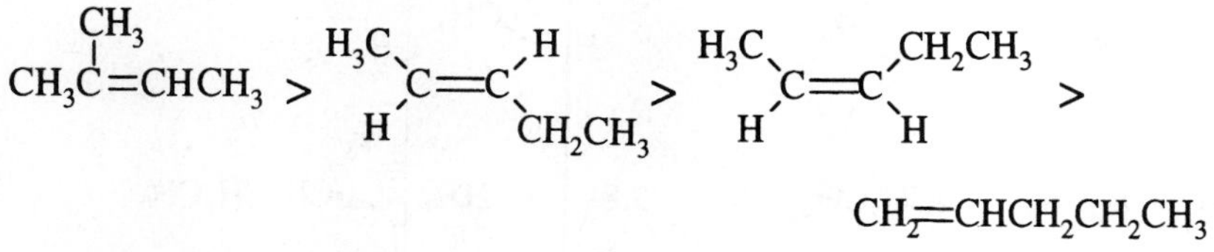

7.6   (a) 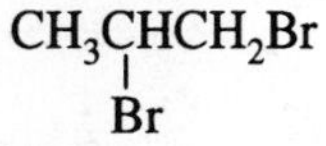   CH$_3$CHCH$_2$Br , Br   (b)   CH$_3$C≡CNa   (c)   CH$_3$C≡CH

(d)   CH$_3$C≡CNa   (e)   CH$_3$CH$_2$Br

CHAPTER 8

8.1 (e) **8.2** (c) **8.3** (e) **8.4** (a) **8.5** (d) **8.6** (c)

8.7 (e) **8.8** (b)

CHAPTER 9

9.1 (d) **9.2** (b) **9.3** (c) **9.4** (b) **9.5** (c)

9.6 $CH_3CH_2\overset{\overset{\displaystyle CH_3}{|}}{\underset{.}{C}}CH_3$

9.7 Six

9.8 (d)

CHAPTER 10

10.1 (d) **10.2** (a) **10.3** (e)

10.4 **A** = $C_6H_5CH_2ONa$ **B** = $C_6H_5CH_2OCH_2CH_2OH$

C = $C_6H_5CH_2OCH_2CH_2OSO_2CH_3$ **D** = $C_6H_5CH_2OCH_2CH_2OCH_2CH_3$

CHAPTER 11

11.1 (b) **11.2** (a)

11.3 **A** = $CH_3C\equiv CLi$ or $CH_3C\equiv CMgBr$

B = NaH

C = CH_3I

11.4 **A** = $(CH_3)_3CCH_2CH_2OH$

B = PCC/CH_2Cl_2

C = $(CH_3)_3CCH_2\overset{\overset{\displaystyle O}{||}}{C}H$

D = C_6H_5MgBr

11.5 **A** = $CH_3\underset{\underset{\displaystyle CH_3}{|}}{C}HCH_2\overset{\overset{\displaystyle O}{||}}{C}OR$ or $CH_3\underset{\underset{\displaystyle CH_3}{|}}{C}HCH_2\overset{\overset{\displaystyle O}{||}}{C}CH_3$

CHAPTER 12

12.1 (d)   **12.2** (c)   **12.3** (c)   **12.4** (c)   **12.5** (b)

CHAPTER 13

13.1 (a) $CH_3\underset{\underset{Br}{|}}{\overset{\overset{CH_3}{|}}{C}}CH_3$ (b) $BrCH_2\underset{\underset{Br}{|}}{\overset{\overset{CH_3}{|}}{C}}CH_2Br$ (c) $CH_2{=}CCH_2\underset{\underset{CH_3}{|}}{\overset{\overset{CH_3}{|}}{C}}CH_3$

(d) $-CH_2\overset{\overset{O}{\|}}{C}CH_3$ (e) $CH_3CH_2C{\equiv}CCH_2NO_2$

13.2 (c)   **13.3** (a)

CHAPTER 14

14.1 (e)   **14.2** (a)   **14.3** (b)   **14.4** (b)

14.5 $-CH_2Cl$ **14.6** Azulene

CHAPTER 15

15.1 (a)   **15.2** (a)   **15.3** (b)

15.4 (a) $A = SO_3/H_2SO_4$ $B = $

$C = H_2O, H_2SO_4$, heat $D = $

(b) $A = SOCl_2$ or PCl_5 $B = $ $+$ $AlCl_3$

$C = Zn(Hg), HCl$, reflux $D = Br_2/FeBr_3$

CHAPTER 16

16.1 (d)   **16.2** (b)   **16.3** (b)

16.4 (a)   A = —CH$_2$Br   B = NaCN

C   (1) DIBAL-H, hexane, -78°C,   (2) H$_2$O

(b) A = PCC, CH$_2$Cl$_2$ B = HOCH$_2$CH$_2$OH, H$^+$ C = H$_2$O, H$_3$O$^+$

(c) A = (C$_6$H$_5$)$_3$P   B = CH$_3$CH$_2$CH$_2$–$\overset{+}{P}$(C$_6$H$_5$)$_3$Br$^-$ C =

(d) A = (CH$_3$)$_2$CuLi   B = HCN   C = (1) LiAlH$_4$
                                                                  (2) H$_2$O

CHAPTER 17

17.1 (a) CH$_3$CH$_2$$\overset{\text{OH}}{\underset{\overset{|}{\text{CH}_3}}{\text{CH}}}$$\overset{\text{O}}{\overset{\|}{\text{CHCH}}}$ (b) CH$_3$CH$_2$$\overset{\text{OH}}{\underset{\overset{|}{\text{CH}_3}}{\text{CH}}}$CHCH$_2$OH   (c) CH$_3CH_2CH=$$\overset{\text{O}}{\overset{\|}{\underset{\overset{|}{\text{CH}_3}}{\text{C}}}}$CH

(d) LiAlH$_4$ (e) H$_2$, Ni   (f) CH$_3$OH (excess), H$^+$

(g) CH$_3$CH$_2$CH$_2$$\underset{\overset{|}{\text{CH}_3}}{\text{CH}}$CH(OCH$_3$)$_2$ (h) CH$_3$CH$_2$CH$_2$$\underset{\overset{|}{\text{CH}_3}}{\text{CH}}$$\overset{\text{O}}{\overset{\|}{\text{CH}}}$

(i) (1) CH$_3$CHBrCO$_2$CH$_2$CH$_3$, Zn   (2) H$_3$O$^+$

17.2 (a) $\overset{\text{O}}{\overset{\|}{\text{C}}}CH_3$ (b) $\overset{\text{O}}{\overset{\|}{\text{C}}}$–CH=CH (c) HCN

17.3 (a)   (b)   (c) (CH$_3$)$_2$CuLi

(d)   (e) Zn(Hg)/HCl

17.4 (e)   **17.5** (a)

CHAPTER 18

18.1 (b)   **18.2** (d)   **18.3** (d)

18.4 **A** = 3-Chlorobutanoic acid

B = Methyl 4-nitrobenzoate

C = *N*-Methylaniline

18.5 (a) **A =** (1) KMnO$_4$, OH⁻, heat   **B =** SOCl$_2$ or PCl$_5$
   (2) H$_3$O⁺

C =   **D =**

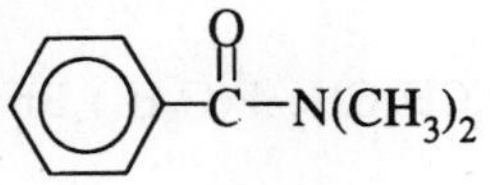

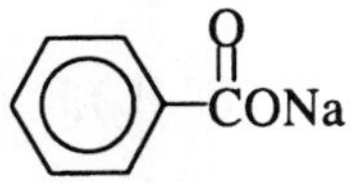

E 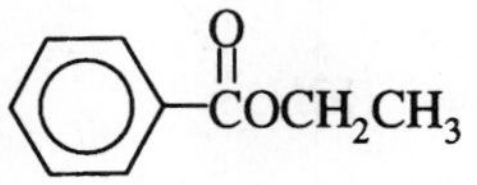 =   **F =**

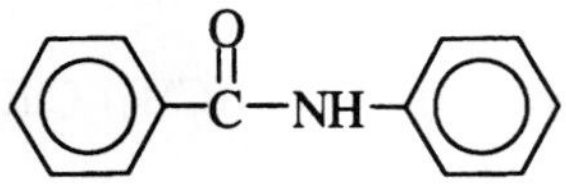

   (b) **A 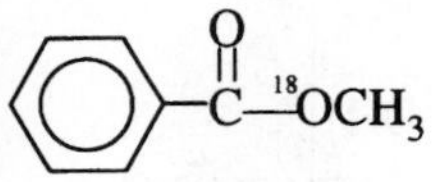 =**   **B =** CH$_3$—^{18}OH

C =

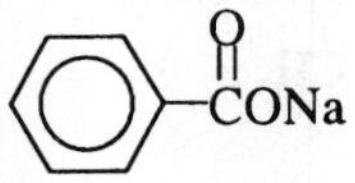

   (c) **A 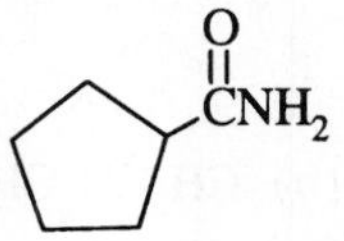 =**   **B 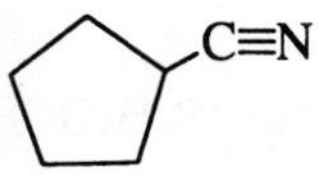 =**   **C =**

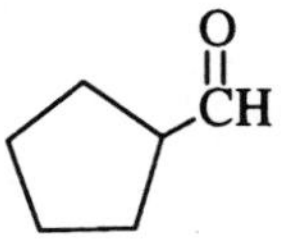

CHAPTER 19

19.1 (c)   **19.2** (e)   **19.3** (b)

19.4 (a) **A =** CH$_3$CH$_2$CH (COEt)(COEt)   **B =** CH$_3$CH$_2$C⁻ K⁺ (COEt)(COEt)

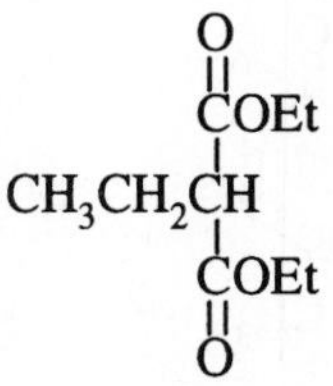

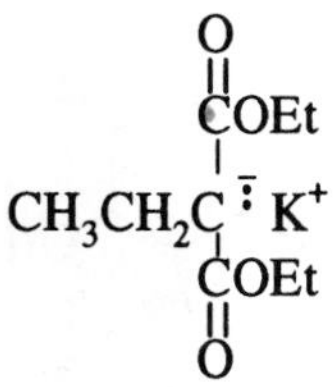

$$C = \quad CH_3CH_2\underset{\underset{O}{\overset{\Vert}{\underset{COEt}{|}}}}{\overset{\overset{O}{\Vert}}{\overset{COEt}{|}}}{C}-CH_2CH_3 \qquad\qquad D = \quad CH_3CH_2\underset{\underset{O}{\overset{\Vert}{\underset{COH}{|}}}}{\overset{\overset{O}{\Vert}}{\overset{COH}{|}}}{C}-CH_2CH_3$$

$$E = \quad \underset{CH_3CH_2}{\overset{CH_3CH_2}{\diagdown\diagup}}CH\overset{O}{\overset{\Vert}{C}}OH$$

(b) A =

B = (phenyl)–CH–CH–COEt with CH(CO₂Et)(EtO₂C) and COEt groups

C = (phenyl)–CH–CH–COH with CH(CO₂H)(HO₂C) and COH groups

(c) A = (tetrahydropyran-2-one enolate) Li⁺   B = CH_3CH_2I

(d) A = pyrrolidine (N–H)   B = $CH_2{=}CHCH_2Br$

C = pyrrolidinium cyclohexene derivative with $CH_2CH{=}CH_2$, Br^-

CHAPTER 20

20.1 (d)   **20.2** (e)

20.3 (a) (2)   (b) (4)   (c) (3)

20.4 (a) **A** = HNO_3/H_2SO_4 **B** = H_3C—⟨⟩—NO_2 **C** = $NaNO_2$, HCl

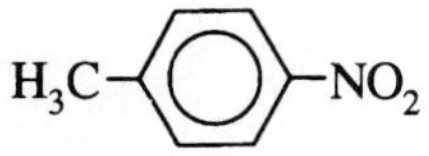

D = CuCN   **E** = $LiAlH_4$ **F** = $(CH_3)_2N$—⟨⟩

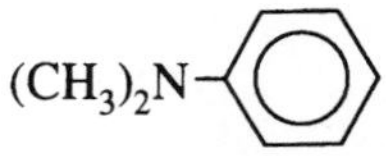

(b) **A** = NaN_3 **B** 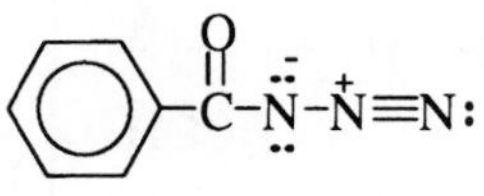 = ⟨⟩—C(=O)—N̈—N⁺≡N̈:   **C** = ⟨⟩—NH_2

D = Br—⟨⟩(Br)(Br)—NH_2 **E** = H_3PO_2

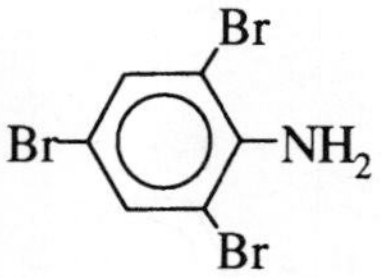

20.5 (a) (2)   (b) (2)   (c) (1)

CHAPTER 21

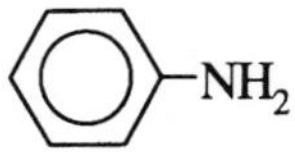

21.1 (a)   **21.2** (b)   **21.3** (b)   **21.4** (e)

21.5 (a) **A** = Br—⟨⟩—OH   **B** = KNH_2, NH_3, -33°C

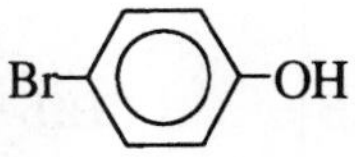

21.6 Br—⟨⟩—OH + CH_3Br

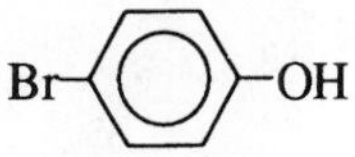

21.7 (a) (1)   (b) (1)

CHAPTER 22

22.1 (a) 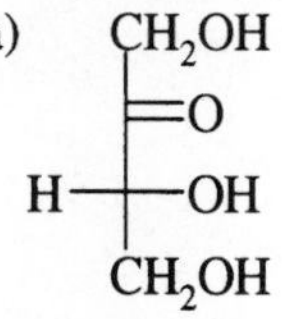
```
    CH₂OH
    ‖
    O
H——OH
    CH₂OH
```

(b) 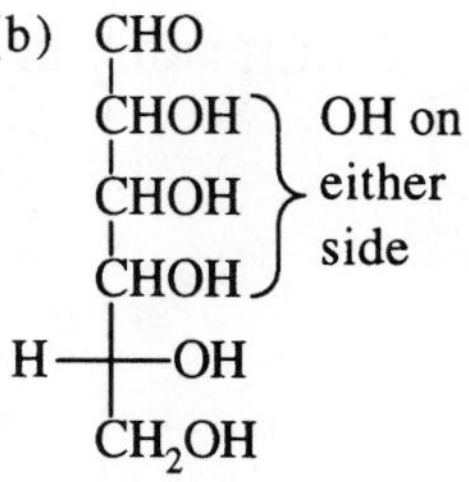
```
   CHO
   CHOH ⎫ OH on
   CHOH ⎬ either
   CHOH ⎭ side
H——OH
   CH₂OH
```

(c) 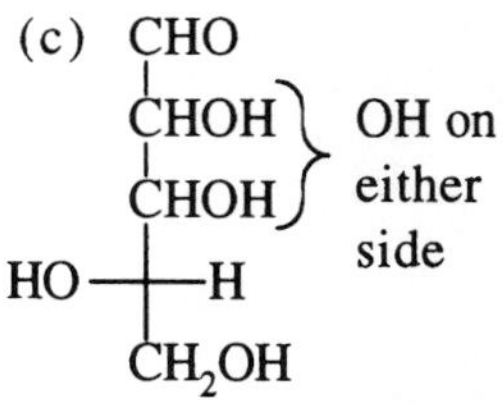
```
    CHO
    CHOH ⎫ OH on
    CHOH ⎬ either
HO——H  ⎭ side
    CH₂OH
```

(d)
```
   CHO
   (CHOH)ₙ
   CH₂OH
   n = 1,2,3...
```

(e) 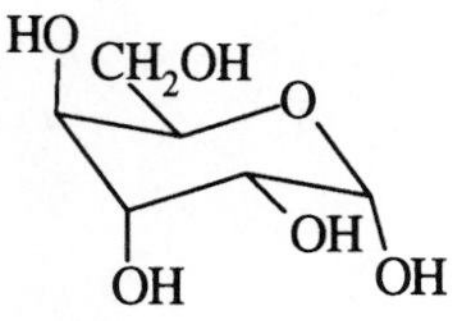 [pyranose ring structure]

(f) 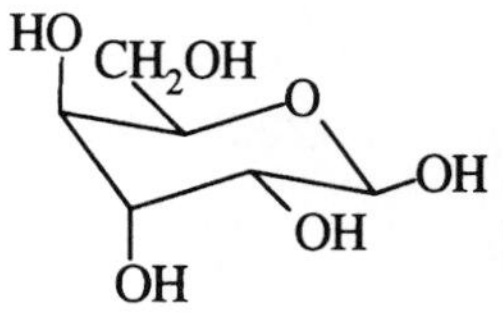 [pyranose ring structure]

(g)
```
        CHO
  HO ———— H
   H ———— OH
  HO ———— H
   H ———— OH
       CH₂OH
```

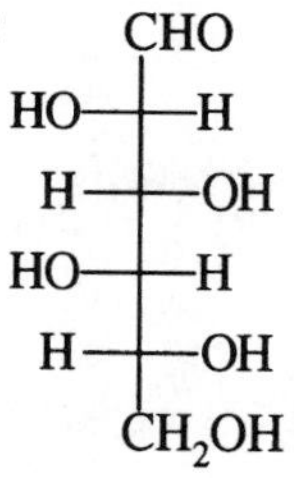

(h)
```
        CHO
  HO ———— H
   H ———— OH
  HO ———— H
  HO ———— H
       CH₂OH
```

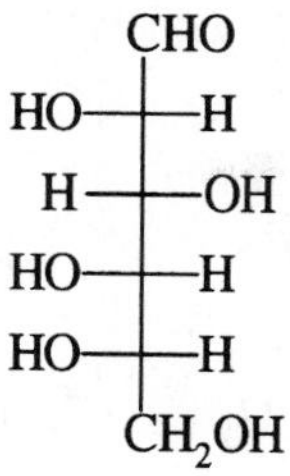

22.2   C

22.3
```
       CHO
  HO ——— H
   H ——— OH
      CH₂OH
```

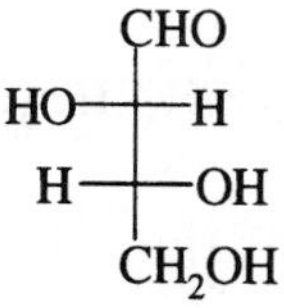

22.4   (a) 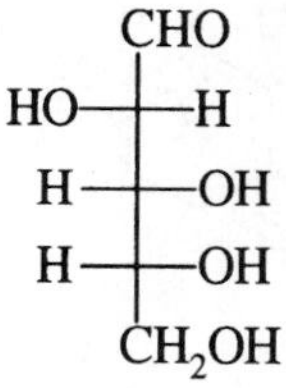
```
        CHO
  HO ———— H
   H ———— OH
   H ———— OH
       CH₂OH
```
(b) 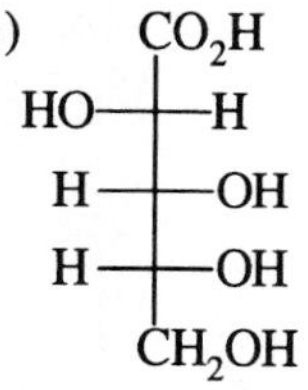
```
        CO₂H
  HO ———— H
   H ———— OH
   H ———— OH
       CH₂OH
```
(c) 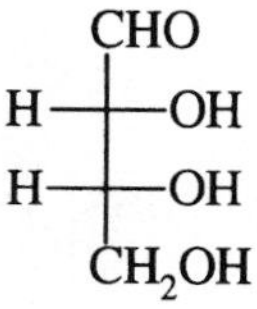
```
       CHO
   H ——— OH
   H ——— OH
      CH₂OH
```
(d)
```
        CO₂H
   H ———— OH
   H ———— OH
        CO₂H
```

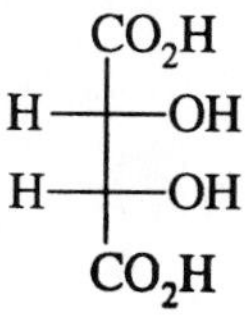

22.5

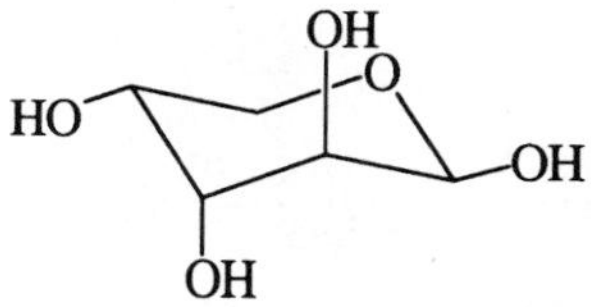

22.6   (a)

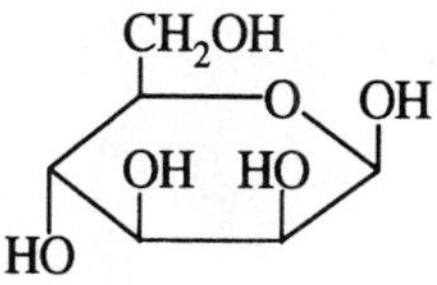

(b)

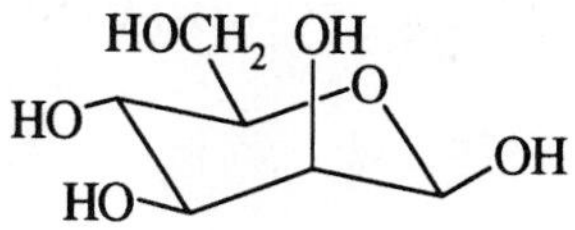

(c)   Reducing

(d)   Active   (e)   Aldonic   (f)   Active   (g)   Aldaric   (h)   NaBH₄

(i)   Active

22.7   (a)   Galactose $\xrightarrow{\text{NaBH}_4}$ optically *inactive* alditol

(b)   HIO₄ oxidation $\longrightarrow$ different products:

Fructose $\longrightarrow$ 2 mol HĊH + CO₂ + 3 HĊ–OH

Glucose $\longrightarrow$ 1 mol HĊH + 5 HĊ–OH

22.8   (e)   **22.9**   (d)

CHAPTER 23

23.1   (a) $CH_3(CH_2)_{12}CO_2H$

(b) $CH_3(CH_2)_{12}\overset{\displaystyle O}{\overset{\|}{C}}-ONa$

(c) $CH_2O\overset{\displaystyle O}{\overset{\|}{C}}(CH_2)_{12}CH_3$
$\ \ \ |$
    $CHO\overset{\displaystyle O}{\overset{\|}{C}}(CH_2)_{12}CH_3$
$\ \ \ |$
    $CH_2O\overset{\displaystyle O}{\overset{\|}{C}}(CH_2)_{12}CH_3$

(d) $CH_2O\overset{\displaystyle O}{\overset{\|}{C}}(CH_2)_7CH=CH(CH_2)_5CH_3$
$\ \ \ |$
    $CHO\overset{\displaystyle O}{\overset{\|}{C}}(CH_2)_7CH=CH(CH_2)_5CH_3$
$\ \ \ |$
    $CH_2O\overset{\displaystyle O}{\overset{\|}{C}}(CH_2)_7CH=CH(CH_2)_5CH_3$

(e) $CH_3(CH_2)_{13}SO_3Na$

(f)

23.2   (a) I_2/OH^- (iodoform test)   (b) Br_2/CCl_4 (c) $Ag(NH_3)_2OH$

23.3 5α-Androstane

23.4   (a) $CH_3(CH_2)_4CH_2C{\equiv}CH$ (b) $CH_3(CH_2)_5C{\equiv}CNa$

(c) $CH_3(CH_2)_5C{\equiv}C(CH_2)_6CH_2Cl$ (d)   KCN

(e) $CH_3(CH_2)_5C{\equiv}C(CH_2)_7CO_2H$ (f) H_2/Pd

23.5   Sesquiterpene

23.6 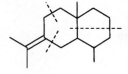

23.7   (e)

CHAPTER 24

24.1 (a)

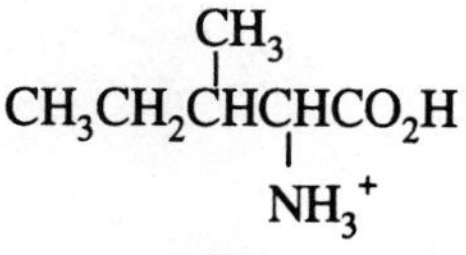

$$\underset{\underset{NH_3^+}{|}}{CH_3CH_2CHCHCO_2H}$$
$$\overset{CH_3}{|}$$

(b)

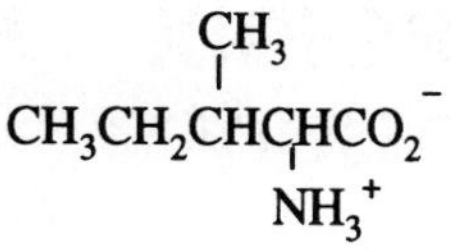

$$\underset{\underset{NH_3^+}{|}}{CH_3CH_2CHCHCO_2^-}$$
$$\overset{CH_3}{|}$$

(c)

$$\underset{\underset{NH_2}{|}}{CH_3CH_2CHCHCO_2^-}$$
$$\overset{CH_3}{|}$$

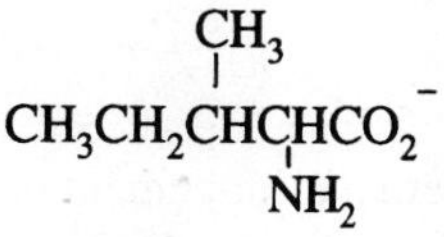

24.2 Pro·Leu·Gly·Phe·Gly·Tyr

C

APPENDIX
Glossary

Absolute configuration (Section 5.14A): The actual arrangement of groups in a molecule. The absolute configuration of a molecule can be determined by X-ray analysis or by relating the configuration of a molecule, using reactions of known stereochemistry, to another molecule whose absolute configuration is known.

Absorption spectrum (Section 13.2): A plot of the wavelength of a region of the spectrum versus the absorbance (A) at each wavelength (λ). The absorbance at a particular wavelength (A_λ) is defined by the equation $A_\lambda = \log(I_R I_S)$, where I_R is the intensity of the reference beam and I_S is the intensity of the sample beam.

Acetal (Section 16.7C): A functional group, consisting of a carbon bonded to alkoxy groups [i.e., $RCH(OR')_2$ or $R_2C(OR')_2$], derived by adding two molar equivalents of an alcohol to an aldehyde or ketone.

Acetylene (Sections 7.1 and 7.3A): A common name for ethyne. Also used as a general name for alkynes.

Acetylenic hydrogen atom (Section 7.3): A hydrogen atom attached to a carbon atom that is bonded to another carbon atom by a triple bond.

Achiral molecule (Section 5.2): A molecule that is superposable on its mirror image. Achiral molecules lack handedness and are incapable of existing as a pair of enantiomers.

Acid strength (Section 3.5): The strength of an acid is related to its acidity constant, K_a, or to its pK_a. The larger the value of its K_a or the smaller the value of its pK_a, the stronger is the acid.

Acidity constant (Section 3.5A): An equilibrium constant related to the strength of an acid. For the reaction,

$$HA \; + \; H_2O \; \rightleftharpoons \; H_3O^+ \; + \; A^-$$

$$K_a \; = \; \frac{[H_3O^+][A^-]}{[HA]}$$

Activating group (Section 15.10): A group which present on a benzene ring causes the ring to be more reactive than benzene itself.

Activation energy, E_{act} (see Energy of activation)

Acyl group (Section 15.7): The general name for groups with the structure RCO— or ArCO—.

Acyl halide (Section 15.7): Also called **acid halides.** A general name for compounds with the structure RCOX or ArCOX.

Acylation (Section 15.7): The introduction of an acyl group into a molecule.

Acylium ion (Section 15.7): The resonance-stabilized cation:

$$R-\overset{+}{C}=\overset{..}{O}: \longleftrightarrow R-C\equiv\overset{+}{O}:$$

Addition polymer (Special Topic C): A polymer that results from a stepwise addition of monomers to a chain (usually through a chain reaction) with no loss of other atoms or molecules in the process. Also known as a **chain-growth polymer.**

Addition reaction (Section 8.1): A reaction that results in an increase in the number of groups attached to a pair of atoms joined by a double or triple bond. An addition reaction is the opposite of an elimination reaction.

Aglycone (Section 22.4): The alcohol obtained by hydrolysis of a glycoside.

Aldaric acid (Section 22.6C): An α,ω-dicarboxylic acid that results from oxidation of the aldehyde group and the terminal 1° alcohol group of an aldose.

Alditol (Section 22.7): The alcohol that results from the reduction of the aldehyde or keto group of an aldose or ketose.

Aldonic acid (Section 22.6B): A monocarboxylic acid that results from oxidation of the aldehyde group of an aldose.

Aliphatic compound (Section 14.1): A nonaromatic compound such as an alkane, cycloalkane, alkene, or alkyne.

Alkaloid (Special Topic K): A naturally occurring basic compound that contains an amino group. Most alkaloids have profound physiological effects.

Alkylation (Section 15.6): The introduction of an alkyl group into a molecule.

Allyl group (Section 7.2): The $CH_2=CHCH_2$- group.

Allylic substituent (Section 12.2): A substituent on a carbon atom adjacent to a carbon-carbon double bond.

Ambident nucleophile (Special Topic H.1A): A nucleophile that is capable of reacting at two different nucleophilic sites.

Angle strain (Section 4.9) The increased potential energy of a molecule (usually a cyclic one) caused by deformation of a bond angle away from its lowest energy value.

Annulene (Section 14.7A): Monocyclic hydrocarbons that can be represented by structures having alternating single and double bonds. The ring size of an annulene is represented by a number in brackets; e.g., benzene is [6]annulene and cyclooctatetraene is [8]annulene.

Anomer (Section 22.2C): A term used in carbohydrate chemistry. Anomers are diastereomers that differ only in configuration at the acetal or hemiacetal carbon of a sugar in its cyclic form.

Anti addition (Section 7.6A): An addition that places the parts of the adding reagent on opposite faces of the reactant.

Anti conformation (Section 4.7): The anti conformation of butane, for example, has the methyl groups at an angle of 180° to each other:

Anti
conformation
of butane

Antiaromatic compound (Section 14.7D): A cyclic conjugated system whose π electron energy is greater than that of the corresponding open-chain compound.

Antibonding molecular orbital (Antibonding MO) (Section 1.12): A molecular orbital whose energy is higher than that of the isolated atomic orbitals from which it is constructed. Electrons in an antibonding molecular orbital destabilize the bond between the atoms that the orbital encompasses.

Anticodon (Section 25.5C): A sequence of three bases on transfer RNA (tRNA) that associates with a codon of messenger RNA (mRNA).

Aprotic solvent (Section 6.15C): A solvent whose molecules do not have a hydrogen atom attached to a strongly electronegative element (such as oxygen). For most purposes, this means that an aprotic solvent is one whose molecules lack an -OH group.

Arene (Section 15.1): A general name for an aromatic hydrocarbon.

Aromatic compound (Sections 14.1–14.7 and 14.11): A cyclic conjugated unsaturated molecule or ion that is stabilized by π electron delocalization. Aromatic compounds are characterized by having large resonance energies, by reacting by substitution rather than addition, and by deshielding of protons exterior to the ring in their ^{1}H NMR spectra caused by the presence of an induced ring current.

Aryl group (Section 15.1): The general name for a group obtained (on paper) by the removal of a hydrogen from a ring position of an aromatic hydrocarbon. Abbreviated Ar—.

Aryl halide (Section 6.1): An organic halide in which the halogen atom is attached to an aromatic ring, such as a benzene ring.

Atactic polymer (Special Topic C): A polymer in which the configuration at the stereocenters along the chain is random.

Atomic orbital (AO) (Section 1.11): A volume of space about the nucleus of an atom where there is a high probability of finding an electron. An atomic orbital can be described mathematically by its **wave function.** Atomic orbitals have characteristic quantum numbers. The **principal quantum number,** n, is related to the energy of the electron in an atomic orbital and can have the values 1, 2, 3, . . . The **azimuthal quantum number,** l, determines the angular momentum of the electron that results from its motion around the nucleus, and can have the values 0, 1, 2, . . . , $(n - 1)$. The **magnetic quantum number,** m, determines the orientation in space of the angular momentum, and can have values from $+l$ to $-l$. The **spin quantum number,** s, measures the intrinsic angular momentum of an electron and can have the values of $+1/2$ and $-1/2$ only.

Aufbau principle (Section 1.11): A principle that guides us in assigning electrons to orbitals of an atom or molecule in its lowest energy state or **ground state.** The aufbau principle states that electrons are added so that orbitals of lowest energy are filled first.

Autoxidation (Section 9.11B): The reaction of an organic compound with oxygen to form a hydroperoxide.

Axial bond (Section 4.11): The six bonds of a cyclohexane ring (below) that are perpendicular to the general plane of the ring, and that alternate up and down around the ring.

Base peak (Special Topic F.2): The most intense peak in a mass spectrum.

Base strength (Section 3.5): The strength of a base is inversely related to the strength of its conjugate acid; the weaker the conjugate acid, the stronger is the base. In other words, if the conjugate acid has a large pK_a, the base will be strong.

Benzenoid aromatic compound (Section 14.8A): An aromatic compound whose molecules have one or more benzene rings.

Benzyl group (Section 2.7): The $C_6H_5CH_2-$ group.

Benzylic substituent (Section 15.12A): Refers to a substituent on a carbon atom adjacent to a benzene ring.

Benzyne (Section 21.11B): An unstable, highly reactive intermediate consisting of a benzene ring with an additional bond resulting from sideways overlap of sp^2 orbitals on adjacent atoms of the ring.

Betaine (Section 16.10): An electrically neutral molecule that has nonadjacent cationic and anionic sites and that does not possess a hydrogen atom bonded to the cationic site.

Bimolecular reaction (Section 6.7): A reaction whose rate-determining step involves two initially separate species.

Boat conformation (Section 4.10): A conformation of cyclohexane (below) that resembles a boat and that has eclipsed bonds along its two sides. It is of higher energy than the chair conformation.

Bond angle (Section 2.6A): The angle between two bonds originating at the same atom.

Bond dissociation energy (see Homolytic bond dissociation energy).

Bond length (Section 2.6A): The equilibrium distance between two bonded atoms or groups.

Bond-line formula (Section 1.20D): A formula that shows the carbon skeleton of a molecule with lines. The number of hydrogen atoms necessary to fulfill each carbon's valence is assumed to be present but not written in. Other atoms (e.g., O, Cl, N) are written in.

Bonding molecular orbital (Bonding MO) (Section 1.12) The energy of a bonding molecular orbital is lower than the energy of the isolated atomic orbitals from which it arises. When electrons occupy a bonding molecular orbital, they help hold together the atoms that the molecular orbital encompasses.

Bromination (Section 8.6 and 9.3): A reaction in which bromine atoms are introduced into a molecule.

Bromohydrin (Section 8.8): A bromo alcohol.

Bromonium ion (Section 8.6A): An ion containing a positive bromine atom bonded to two carbon atoms.

Brønsted-Lowry theory of acids and bases (Section 3.2A): An acid is a substance that can donate (or lose) a proton; a base is a substance that can accept (or remove) a proton. The **conjugate acid** of a base is the molecule or ion that forms when a base accepts a proton. **The conjugate base** of an acid is the molecule or ion that forms when an acid loses its proton.

Carbanion (Section 3.3): A chemical species in which a carbon atom bears a formal negative charge.

Carbene (Special Topic D): An uncharged species in which a carbon atom is divalent. The species, CH_2, called methylene, is a carbene.

Carbenoid (Special Topic D): A carbene-like species. A species such as the reagent formed when diiodomethane reacts with a zinc-copper couple. This reagent, called the Simmons-Smith reagent, reacts with alkenes to add methylene to the double bond in a stereospecific way.

Carbocation (Section 3.3): A chemical species in which a trivalent carbon atom bears a formal positive charge.

Carbohydrate (Section 22.1A): A group of naturally occurring compounds that are usually defined as polyhydroxyaldehydes or polyhydroxyketones, or substances that undergo hydrolysis to yield such compounds. In actuality, the aldehyde and ketone groups of carbohydrates are often present as hemiacetals and acetals. The name comes from the fact that many carbohydrates possess the empirical formula $C_x(H_2O)_y$.

Carbonyl group (Section 16.1): A functional group consisting of a carbon atom doubly bonded to an oxygen atom. The carbonyl group is found in aldehydes, ketones, esters, anhydrides, amides, acyl halides, and so on. Collectively, these compounds are referred to as carbonyl compounds.

Cascade polymer (Special Topic I.5): A polymer produced from a multifunctional central core by progressively adding layers of repeating units.

CFC (see Freon)

Chain reaction (Section 9.4): A reaction that proceeds by a sequential, stepwise mechanism, in which each step generates the reactive intermediate that causes the next step to occur. Chain reactions have **chain initiating steps, chain propagating steps,** and **chain terminating steps.**

Chair conformation (Section 4.10): The all-staggered conformation of cyclohexane (below) that has no angle strain or torsional strain and is, therefore, the lowest energy conformation.

Chemical shift, δ (Section 13.7): The position in an NMR spectrum, relative to a reference compound, at which a nucleus absorbs. The reference compound most often used is tetramethylsilane (TMS), and its absorption point is arbitrarily designated zero. The chemical shift of a given nucleus is proportional to the strength of the magnetic field of the spectrometer. The chemical shift in delta units, δ, is determined by dividing the observed shift from TMS in hertz multiplied by 10^6 by the operating frequency of the spectrometer in hertz.

Chiral molecule (Section 5.2): A molecule that is not superposable on its mirror image. Chiral molecules have handedness and are capable of existing as a pair of enantiomers.

Chirality (Section 5.2): The property of having handedness.

Chlorination (Section 9.3): A reaction in which chlorine atoms are introduced into a molecule.

Chlorohydrin (Section 8.8): A chloro alcohol.

Cis-trans isomers (Section 2.4B and 4.12): Diastereomers that differ in their stereochemistry at adjacent atoms of a double bond or on different atoms of a ring.

Codon (Section 25.5C): A sequence of three bases on messenger RNA (mRNA) that contains the genetic information for one amino acid. The codon associates, by hydrogen bonding, with an anticodon of a transfer RNA (tRNA) which carries the particular amino acid for protein synthesis on the ribosome.

Condensation polymer (Special Topic I): A polymer produced when bifunctional monomers (or potentially bifunctional monomers) react with each other through the intermolecular elimination of water or an alcohol. Polyesters, polyamides, and polyurethanes are all condensation polymers. Also called **step-growth polymers.**

Condensation reaction (Section 17.5A): A reaction in which molecules become joined through the intermolecular elimination of water or an alcohol.

Configuration (Section 5.6): The particular arrangement of atoms (or groups) in space that is characteristic of a given stereoisomer.

Conformation (Section 4.6): A particular temporary orientation of a molecule that results from rotations about its single bonds.

Conformational analysis (Section 4.6) An analysis of the energy changes that a molecule undergoes as its groups undergo rotation (sometimes only partial) about the single bonds that join them.

Conformer (Section 4.6): A particular staggered conformation of a molecule.

Conjugate acid (Section 3.2A): The molecule or ion that forms when a base accepts a proton.

Conjugate addition (Section 17.9): A form of nucleophilic addition to an α,β-unsaturated carbonyl compound in which the nucleophile adds to the β-carbon.

Conjugate base (Section 3.2A): The molecule or ion that forms when an acid loses its proton.

Conjugated system (Section 12.1): Molecules or ions that have an extended π system. A conjugated system has a *p* orbital on an atom adjacent to a multiple bond; the *p* orbital may be that of another multiple bond, or that of a radical, carbocation, or carbanion.

Connectivity (Section 1.3): The sequence, or order, in which the atoms of a molecule are attached to each other.

Constitutional isomers (Section 1.3A): Compounds that have the same molecular formula but differ in their connectivity (i.e., molecules that have the same molecular formula but have their atoms connected in a different way).

Copolymer (Special Topic C): A polymer synthesized by polymerizing two different monomers.

Coupling constant, J$_{ab}$ (Section 13.9): The separation in frequency units (hertz) of the peaks of a multiplet caused by spin-spin coupling between atoms *a* and *b*.

Covalent bond (Section 1.4B): The type of bond that results when atoms share electrons.

Cracking (Section 4.1C): A process used in the petroleum industry for breaking down the molecules of larger alkanes into smaller ones. Cracking may be accomplished with heat (thermal cracking) or with a catalyst (catalytic cracking).

Crown ether (Section 10.22A): Cyclic polyethers that have the ability to form complexes with metal ions. Crown ethers are named as *x*-crown-*y* where *x* is the total number of atoms in the ring and *y* is the number of oxygen atoms in the ring.

Cyanohydrin (Section 16.9): A functional group consisting of a carbon atom bonded to a cyano group and to a hydroxyl group, i.e., $RHC(OH)(CN)$ or $R_2C(OH)(CN)$, derived by adding HCN to an aldehyde or ketone.

Cycloaddition (Section 12.10): A reaction, like the Diels-Alder reaction, in which two connected groups add to the end of a π system to generate a new ring.

D and L designations (Section 22.2B): A method for designating the configuration of monosaccharides and other similar compounds in which the reference compound is (+)- or (−)- glyceraldehyde. According to this system, (+)-glyceraldehyde is designated D-(+)-glyceraldehyde and (−)-glyceraldehyde is designated L-(−)-glyceraldehyde. Therefore, a monosaccharide whose highest numbered stereocenter has the same general configuration as D-(+)-glyceraldehyde is designated a D-sugar; one whose highest numbered stereocenter has the same general configuration as L-(+)-glyceraldehyde is designated an L-sugar.

Deactivating group (Section 15.10): A group which present on a benzene ring causes the ring to be less reactive than benzene itself.

Debromination (Section 7.16): The elimination of two atoms of bromine from a vic-dibromide, or, more generally, the loss of bromine from a molecule.

Debye unit (Section 1.18): The unit in which dipole moments are stated. One debye, D, equals 1×10^{-18} esu cm.

Decarboxylation (Section 18.11): A reaction whereby a carboxylic acid loses CO_2.

Degenerate orbitals (Section 1.11): Orbitals of equal energy. For example, the three $2p$ orbitals are degenerate.

Dehydration reaction (Section 7.13): An elimination that involves the loss of a molecule of water from the substrate.

Dehydrohalogenation (Section 6.17): An elimination reaction that results in the loss of HX from adjacent carbons of the substrate and the formation of a π bond.

Delocalization (Section 6.13B): The dispersal of electrons (or of electical charge). Delocalization of charge always stabilizes a system.

Dextrorotatory (Section 5.7B): A compound that rotates plane-polarized light in a clockwise direction.

Diastereomers (Section 5.1): Stereoisomers that are not mirror images of each other.

Diastereotopic hydrogens (or ligands) (Section 13.8B): If replacement of each of two hydrogens (or ligands) by the same group yields compounds that are diastereomers, the two hydrogen atoms (or ligands) are said to be diastereotopic.

Dielectric constant (Section 6.15D): A measure of a solvent's ability to insulate opposite charges from each other. The dielectric constant of a solvent roughly measures its polarity. Solvents with high dielectric constants are better solvents for ions than solvents with low dielectric constants.

Dienophile (Section 12.10): The diene-seeking component of a Diels-Alder reaction.

Dipolar ion (Section 24.2C): The charge-separated form of an amino acid that results from the transfer of a proton from a carboxyl group to a basic group.

Dipole moment, μ (Section 1.18): A physical property associated with a polar molecule that can be measured experimentally. It is defined as the product of the charge in electrostatic units (esu) and the distance that separates them in centimeters: $\mu = e \times d$.

Dipole-dipole interaction (Section 2.16B): An interaction between molecules having permanent dipole moments.

Disaccharide (Section 22.1A): A carbohydrate that, on a molecular basis, undergoes hydrolytic cleavage to yield two molecules of a monosaccharide.

E-Z system (Section 7.2A): A system for designating the stereochemistry of alkene diastereomers based on the priorities of groups in the Cahn-Ingold-Prelog convention.

E1 reaction (Section 6.19): A unimolecular elimination in which, in a slow, rate-determining step, a leaving group departs from the substrate to form a carbocation. The carbocation then in a fast step loses a proton with the resulting formation of a π bond.

E2 reaction (Section 6.18): A bimolecular 1,2-elimination in which, in a single step, a base removes a proton and a leaving group departs from the substrate, resulting in the formation of a π bond.

Eclipsed conformation (Section 4.6): A temporary orientation of groups around two atoms joined by a single bond such that the groups directly oppose each other.

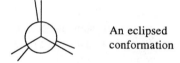

An eclipsed
conformation

Electromagnetic spectrum (Section 13.1): The full range of energies propagated by wave fluctuations in electromagnetic field.

Electronegativity (Section 1.4A): A measure of the ability of an atom to attract electrons it is sharing with another, and thereby polarize the bond.

Electrophile (Sections 3.3 and 8.1): A Lewis acid, an electron-pair acceptor, an electron-seeking reagent.

Electrophoresis (Section 25.6A): A technique for separating charged molecules based on their different mobilities in an electric field.

Elimination (Section 6.17): A reaction that results in the loss of two groups from the substrate and the formation of a π bond. The most common elimination is a 1,2-elimination or β elimination, in which the two groups are lost from adjacent atoms.

Empirical formula (Section 1.2B): a formula that expresses the relative proportions of atoms in a molecule as smallest whole numbers.

Enantiomeric excess or enantiomeric purity (Section 5.8B): A percentage calculated for a mixture of enantiomers by dividing the moles of one enantiomer minus the moles of the other enantiomer by the moles of both enantiomers and multiplying by 100. The enantiomeric excess equals the percent optical purity.

Enantiomers (Section 5.1): Stereoisomers that are mirror images of each other.

Enantiotopic hydrogens (or ligands) (Section 13.8B): If replacement of each of two hydrogens (or ligands) by the same group yields compounds that are enantiomers, the two hydrogen atoms (or ligands) are said to be enantiotopic.

Endergonic reaction (Section 6.9): A reaction that proceeds with a positive free energy change.

Endothermic reaction (Section 1.9A): A reaction that absorbs heat. For an endothermic reaction $\Delta H°$ is positive.

Energy (Section 1.9): Energy is the capacity to do work.

Energy of activation, E_{act} (Section 9.5B): A measure of the difference in potential energy between the reactants and the transition state of a reaction. It is related to, but not the same as, the free energy of activation, $\Delta G^{\ddagger}$.

Enolate ion (Section 17.1): The delocalized anion formed when an enol loses its hydroxylic proton or when the aldehyde or ketone that is in equilibrium with the enol loses an α proton.

Enthalpy change (Section 1.9A and 3.8): Also called the heat of reaction. The standard enthalpy change, $\Delta H°$, is the change in enthalpy after a system in its standard state has undergone a transformation to another system, also in its standard state. For a reaction, $\Delta H°$ is a measure of the difference in the total bond energy of the reactants and products. It is one way of expressing the change in potential energy of molecules as they undergo reaction. The enthalpy change is related to the free-energy change, $\Delta G°$, and to the entropy change, $\Delta S°$, through the expression: $\Delta H° = \Delta G° + T\Delta S°$.

Entropy change (Section 3.8): The standard entropy change, $\Delta S°$, is the change in entropy between two systems in their standard states. Entropy changes have to do with changes in the relative order of a system. The more random a system is, the greater is its entropy. When a system becomes more disorderly, its entropy change is positive.

Epoxide (Section 10.18): An oxirane. A three-membered ring containing one oxygen atom.

Equatorial bond (Section 4.11): The six bonds of a cyclohexane ring (below) that lie generally around the "equator" of the molecule.

Equilibrium constant (Section 3.5A): A constant that expresses the position of an equilibrium. The equilibrium constant is calculated by multiplying the molar concentrations of the products together and then dividing this number by the number obtained by multiplying together the molar concentrations of the reactants.

Equilibrium control (see Thermodynamic control)

Essential oil (Section 23.3): A volatile odoriferous compound obtained by steam distillation of plant material.

Exergonic reaction (Section 6.9): A reaction that proceeds with a negative free-energy change.

Exothermic reaction (Section 1.9A): A reaction that evolves heat. For an exothermic reaction, $\Delta H°$ is negative.

Fat (Section 23.2): A triacylglycerol. The triester of glycerol with carboxylic acids.

Fatty acid (Section 23.2): A long-chained carboxylic acid (usually with an even number of carbon atoms) that is isolated by the hydrolysis of a fat.

Fischer projection formula (Sections 5.12 and 22.2C): A two-dimensional formula for representing the configuration of a chiral molecule. By convention, Fischer projection formulas are written with the main carbon chain extending from top to bottom with all groups eclipsed. Vertical lines represent bonds that project behind the plane of the page (or that lie in it). Horizontal lines represent bonds that project out of the plane of the page.

Fischer
projection

Wedge-dashed wedge
formula

Fluorination (Section 9.3): A reaction in which fluorine atoms are introduced into a molecule.

Formal charge (Section 1.7): The difference between the number of electrons assigned to an atom in a molecule and the number of electrons it has in its outer shell in its elemental state. Formal charge can be calculated using the formula: $F = Z - S/2 - U$, where F is the formal charge, Z is the group number of the atom (i.e., the number of electrons the atom has in its outer shell in its elemental state), S equals the number of electrons the atom is sharing with another atom, and U is the number of unshared electrons the atom possesses.

Free energy of activation, $\Delta G^{\ddagger}$ (Section 6.9): The difference in free energy between the transition state and the reactants.

Free-energy change (Section 3.8) The standard free-energy change, $\Delta G°$, is the change in free energy between two systems in their standard states. At constant temperature, $\Delta G° = \Delta H° - T\Delta S° = -RTlnK_{eq}$, where $\Delta H°$ is the standard enthalpy change, $\Delta S°$ is the standard entropy change, and K_{eq} is the equilibrium constant. A negative value of $\Delta G°$ for a reaction means that the formation of products will be favored when the reaction reaches equilibrium.

Freon (Section 9.11C): A chlorofluorocarbon, or CFC.

Frequency (abbreviated ν)(Section 13.1): The number of full cycles of a wave that pass a given point in each second.

Functional group (Section 2.8): The particular group of atoms in a molecule that primarily determines how the molecule reacts.

Functional group interconversion (Section 6.16): A process that converts one functional group into another.

Furanose (Section 22.2C): A sugar in which the cyclic acetal or hemiacetal ring is five-membered.

Gauche conformation (Section 4.7): A gauche conformation of butane, for example, has the methyl groups at an angle of 60° to each other:

A gauche
conformation
of butane

Geminal (gem) substituents (Section 7.16): Substituents that are on the same atom.

Glycol (Section 4.3F): A diol.

Glycoside (Section 22.4): A cyclic mixed acetal of a sugar with an alcohol.

Grignard reagent (Section 11.6B): An organomagnesium halide, usually written RMgX.

Ground state (Section 1.12): The lowest electronic energy state of an atom or molecule.

Halogenation (Section 9.3): A reaction in which a halogen atom is introduced into a molecule.

Halohydrin (Section 8.8): A halo alcohol.

Halonium ion (Section 8.6A): An ion containing a positive halogen atom bonded to two carbon atoms.

Hammond-Leffler postulate (Section 6.15A): A postulate stating that the structure and geometry of the transition state of a given step will show a greater resemblance to the reactants or products of that step depending on which is closer to the transition state in energy. This means that the transition state of an endothermic step will resemble the products of that step more than the reactants, while the transition state of an exothermic step will resemble the reactants of that step more than the products.

Heat of combustion (Section 4.8): The standard enthalpy change for the complete combustion of one mole of a compound.

Heat of hydrogenation (Section 7.9A): The standard enthalpy change that accompanies the hydrogenation of one mole of a compound to form a particular product.

Heisenberg uncertainty principle (Section 1.12): A fundamental principle that states that neither the position nor momentum of an electron (or of any object) can be exactly measured simultaneously.

Hemiacetal (Section 16.7B): A functional group, consisting of a carbon atom bonded to an alkoxy group and to a hydroxyl group [i.e., $RCH(OH)OR'$ or $R_2C(OH)(OR')$]. Hemiacetals are synthesized by adding one molar equivalent of an alcohol to an aldehyde or a ketone.

Hemiketal (Section 16.7B): Also called a hemiacetal. A functional group, consisting of a carbon atom bonded to an alkoxy group and to a hydroxyl group [i.e., $R_2C(OH)(OR')$]. Hemiketals are synthesized by adding one molar equivalent of an alcohol to a ketone.

Hertz (abbreviated Hz) (Section 13.1): Now used instead of the equivalent cycles per second as a measure of the frequency of a wave.

Heterocyclic compound (Section 14.9): A compound whose molecules have a ring containing an element other than carbon.

Heterolysis (Section 3.1B): The cleavage of a covalent bond so that one fragment departs with both of the electrons of the covalent bond that joined them. Heterolysis of a bond normally produces positive and negative ions.

Hofmann rule (Section 7.12B): When an elimination yields the alkene with the less substituted double bond, it is said to follow the Hofmann rule.

HOMO (Section 13.2): The highest occupied molecular orbital.

Homologous series (Section 4.5): A series of compounds where each member differs from the next member by a constant unit.

Homolysis (Section 3.1B): The cleavage of a covalent bond so that each fragment departs with one of the electrons of the covalent bond that joined them.

Homolytic bond dissociation energy, DH^o (Section 9.2): The enthalpy change that accompanies the homolytic cleavage of a covalent bond.

Hückel's rule (Section 14.7): A rule stating that planar monocyclic rings with $(4n + 2)$ delocalized π electrons (i.e., with 2, 6, 10, 14, . . ., delocalized π electrons) will be aromatic.

Hund's rule (Section 1.11): A rule used in applying the **aufbau principle.** When orbitals are of equal energy (i.e., when they are **degenerate**), electrons are added to each orbital with their spins unpaired, until each degenerate orbital contains one electron. Then electrons are added to the orbitals so that the spins are paired.

Hybridization of atomic orbitals (Section 1.13): A mathematical (and theoretical) mixing of two or more atomic orbitals to give the same number of new orbitals, called **hybrid orbitals,** each of which has some of the character of the original atomic orbitals.

Hydration (Section 8.5): The addition of water to a molecule, such as the addition of water to an alkene to form an alcohol.

Hydroboration (Section 10.6): The addition of a boron hydride (either BH_3 or an alkylborane) to a multiple bond.

Hydrogen bond (Section 2.16C): A strong dipole-dipole interaction ($1 - 9$ kcal mol^{-1}) that occurs between hydrogen atoms bonded to small strongly electronegative atoms (O, N, or F) and the nonbonding electron pairs on other such electronegative atoms.

Hydrogenation (Sections 4.15A and 7.5): The addition of hydrogen to a molecule, usually to a multiple bond in the molecule. Hydrogenation is often accomplished through the use of a metal catalyst such as platinum, palladium, rhodium, or ruthenium.

Hydrophilic group (Section 2.16E): A polar group that seeks an aqueous environment.

Hydrophobic group (also called a lipophilic group) (Sections 2.16E and 10.22): A nonpolar group that avoids an aqueous surrounding and seeks a nonpolar environment.

Hydroxylation (Section 8.9): The addition of hydroxyl groups to each carbon of a double bond.

Index of hydrogen deficiency (Section 7.8): The index of hydrogen deficiency (or IHD) equals the number of pairs of hydrogen atoms that must be subtracted from the molecular formula of the corresponding alkane to give the molecular formula of the compound under consideration.

Inductive effect (Sections 3.7B and 15.11B): An intrinsic electron attracting or releasing effect that results from a nearby dipole in the molecule and that is transmitted through space and through the bonds of a molecule.

Infrared (IR) spectroscopy (Section 13.3): A type of optical spectroscopy that measures the absorption of infrared radiation. Infrared spectroscopy provides structural information about functional groups present in the compound being analyzed.

Intermediate (Sections 3.1A, 6.11, and 6.12): A transient species that exists between reactants and products in a state corresponding to an energy minimum on a potential energy diagram.

Iodination (Section 9.3): A reaction in which iodine atoms are introduced into a molecule.

Ion (Section 1.4): A chemical species that bears an electrical charge.

Ion-dipole interaction (Section 2.16E): The interaction of an ion with a permanent dipole. Such interactions (resulting in solvation) occur between ions and the molecules of polar solvents.

Ionic bond (Section 1.4A): A bond formed by the transfer of electrons from one atom to another, resulting in the creation of oppositely charged ions.

Ionic reaction (Section 3.1B): A reaction involving ions as reactants, intermediates, or products. Ionic reactions occur through the heterolyis of covalent bonds.

Isoelectric point (Section 24.2C): The pH at which the number of positive and negative charges on an amino acid or protein are equal.

Isomers (Sections 1.3A and 5.1): Different molecules that have the same molecular formula.

Isoprene unit (Section 23.3): A name for the following structural unit found in all terpenes.

Isotactic polymer (Special Topic C): A polymer in which the configuration at each stereocenter along the chain is the same.

Kekulé structure (Sections 1.3 and 14.4): A structure in which lines are used to represent bonds. The Kekulé structure for benzene is a hexagon of carbon atoms with alternating single and double bonds around the ring and with one hydrogen atom attached to each carbon.

Ketal (Section 16.7C): Properly called an acetal. A functional group, consisting of a carbon bonded to alkoxy groups [i.e., $R_2C(OR')_2$], derived by adding two molar equivalents of an alcohol to a ketone.

Kinetic control (Sections 6.6 and 12.9A): A principle stating that when the ratio of products of a reaction is determined by relative rates of reaction, the most abundant product will be the one that is formed fastest.

Kinetic energy (Section 1.9): Energy that results from the motion of an object. Kinetic energy (KE) = $1/2\ mv^2$, where m is the mass of the object and v is its velocity.

Kinetics (Section 6.6): A term that refers to rates of reactions.

Lactam (Section 18.8I): A cyclic amide.

Lactone (Section 18.7C): A cyclic ester.

Leaving group (or nucleofuge) (Section 6.5): The substituent with an unshared electron pair that departs from the substrate in a nucleophilic substitution reaction.

Leveling effect of a solvent (Section 3.13): An effect that restricts the use of certain solvents with strong acids and bases. In principle, no acid stronger than the conjugate acid of a particular solvent can exist to an appreciable extent in that solvent, and no base stronger than the conjugate base of the solvent can exist to an appreciable extent in that solvent.

Levorotatory (Section 5.7B): A compound that rotates plane-polarized light in a counterclockwise direction.

Lewis structure (or electron-dot structure) (Section 1.5): A representation of a molecule showing electron pairs as a pair of dots or as a dash.

Lewis theory of acids and bases (Section 3.2B): An acid is an electron pair acceptor, and a base is an electron pair donor.

Lipid (Section 23.1): A substance of biological origin that is soluble in nonpolar solvents. Lipids include fatty acids, triacylglycerols (fats and oils), steroids, prostaglandins, terpenes and terpenoids, and waxes.

Lipophilic group (or hydrophopic group) (Sections 2.16E and 10.22): A nonpolar group that avoids an aqueous surrounding and seeks a nonpolar environment.

LUMO (Section 13.2): The lowest unoccupied molecular orbital.

Macromolecule (Section 9.10): A very large molecule.

Magnetic resonance imaging (MRI) (Section 13.12): A technique based on NMR spectroscopy that is used in medicine.

Markovnikov's rule (Section 8.2): A rule for predicting the regiochemistry of electrophilic additions to alkenes and alkynes that can be stated in various ways. As originally stated (in 1870) by Vladimir Markovnikov, the rule provides that "if an unsymmetrical alkene combines with a hydrogen halide, the halide ion adds to the carbon with the fewer hydrogen atoms." More commonly, the rule has been stated in reverse: that in the addition of HX to an alkene or alkyne, the hydrogen atom adds to the carbon atom that already has the greater number of hydrogen atoms. A modern expression of Markovnikov's rule is: *In the ionic addition of an unsymmetrical reagent to a multiple bond, the positive portion of the reagent (the electrophile) attaches itself to a carbon atom of the alkene or alkyne in the way that leads to the formation of the more stable intermediate carbocation.*

Mass spectrometry (Special Topic F): A technique that is useful in structure elucidation and is based on fragmenting a molecule in a magnetic field, then instrumentally determining the mass/charge ratio and relative amounts of the fragments that result.

Mechanism (see Reaction mechanism)

Meso compound (Section 5.11A): An optically inactive compound whose molecules are achiral, even though they contain tetrahedral atoms with four different attached groups.

Mesylate (Section 10.10): A methanesulfonate ester.

Methylene (Special Topic D): The carbene with the formula CH_2.

Methylene group (Section 4.5): The $-CH_2-$ group.

Micelle (Section 23.2C): A spherical cluster of ions in aqueous solution (such as those from a soap) in which the nonpolar groups are in the interior and the ionic (or polar) groups are at the surface.

Molar absorptivity (abbreviated ϵ (Section 13.2): A proportionality constant that relates the observed absorbance (*A*) at a particular wavelength (λ) to the molar concentration of the sample and the length (*l*) (in centimeters) of the path of the light beam through the sample cell:.

$$\epsilon = A/C \times l$$

Molecular formula (Section 1.2B): A formula that gives the total number of each kind of atom in a molecule. The molecular formula is a whole number multiple of the empirical formula. For example, the molecular formula for benzene is C_6H_6; the empirical formula is CH.

Molecular ion (Special Topic F.1): The cation produced in a mass spectrometer when one electron is dislodged from the parent molecule.

Molecular orbital (MO) (Section 1.12): An orbital that encompass more than one atom of a molecule. When atomic orbitals combine to form molecular orbitals, the number of molecular orbitals that result always equals the number of atomic orbitals that combine.

Molecularity (Section 6.8) The number of species involved in a single step of a reaction (usually the rate-determining step).

Monomer (Section 9.10): The simple starting compound from which a polymer is made. For example, the polymer polyethylene is made from the monomer, ethylene.

Monosaccharide (Section 22.1A): The simplest class of carbohydrate, one that does not undergo hydrolytic cleavage to a simpler carbohydrate.

Mutarotation (Section 22.3): The spontaneous change that takes place in the optical rotation of α-and β-anomers of a sugar when they are dissolved in water. The optical rotations of the sugars change until they reach the same value.

Neighboring group participation (Special Topic B): The effect on the course or rate of a reaction brought about by another group near the functional group undergoing reaction.

Newman projection formula (Section 4.6): A means of representing the spatial relationships of groups attached to two atoms of a molecule. In writing a Newman projection formula, we imagine ourselves viewing the molecule from one end directly along the bond axis joining the two atoms. Bonds that are attached to the front atom are shown as radiating from the center of a circle; those attached to the rear atom are shown as radiating from the edge of the circle:

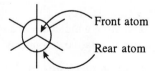

Nitrogen rule (Special Topic F): A rule that states that if the mass of the molecular ion in a mass spectrum is an even number, the parent compound contains an even number of nitrogen atoms, and conversely.

Node (Section 1.10): A place where a wave function or ψ function is equal to zero. The greater the number of nodes in an orbital, the greater is the energy of the orbital.

Nonbenzenoid aromatic compound (Section 14.8B): An aromatic compound, such as azulene, that does not contain benzene rings.

Nuclear magnetic resonance (NMR) spectroscopy (Section 13.4): A spectroscopic method for measuring the absorption of radiofrequency radiation by certain nuclei when the nuclei are in a strong magnetic field. The most important NMR spectra for organic chemists are ^{1}H NMR spectra and ^{13}C NMR spectra. These two types of spectra provide structural information about the carbon framework of the molecule, and about the number of hydrogen atoms attached to each carbon atom.

Nucleic acids (Sections 25.1 and 25.2): Biological polymers of nucleotides. DNA and RNA are, respectively, nucleic acids that preserve and transcribe hereditary information within cells.

Nucleophile (Section 3.3): A Lewis base, an electron pair donor that seeks a positive center in a molecule.

Nucleophilic substitution reaction (Section 6.3): A reaction initiated by a nucleophile (a species with an unshared electron pair) in which the nucleophile reacts with a substrate to replace a substituent (called the leaving group) that departs with an unshared electron pair.

Nucleophilicity (Section 6.15B): The relative reactivity of a nucleophile in an S_N2 reaction as measured by relative rates of reaction.

Nucleoside (Section 25.2): A five-carbon monosaccharide bonded at the 1′-position to a purine or pyrimidine.

Nucleotide (Section 25.2): A five-carbon monosaccharide bonded at the 1′-position to a purine or pyrimidine and at the 3′- or 5′-position to a phosphate group.

Olefin (Section 7.1): An old name for an alkene.

Optical purity (Section 5.8B): A percentage calculated for a mixture of enantiomers by dividing the observed specific rotation for the mixture by the specific rotation of the pure enantiomer and multiplying by 100. The optical purity equals the enantiomeric purity or enantiomeric excess.

Optically active compound (Section 5.7): A compound that rotates the plane of polarization of plane polarized light.

Orbital (Section 1.11): A volume of space in which there is a high probability of finding an electron. Orbitals are described mathematically by wave functions, and each orbital has a characteristic energy. An orbital can hold two electrons when their spins are paired.

Organometallic compound (Section 11.5): A compound that contains a carbon-metal bond.

Oxidation (Section 11.2): A reaction that increases the oxidation state of atoms in a molecule or ion. For an organic substrate, oxidation usually involves increasing its oxygen content or decreasing its hydrogen content. Oxidation also accompanies any reaction in which a less electronegative substitutent is replaced by a more electronegative one.

Oxonium ion (Section 3.11): A chemical species with an oxygen atom that bears a formal positive charge.

Oxymercuration (Section 10.5): The addition of -OH and $-HgO_2CR$ to a multiple bond.

Ozonolysis (Section 8.10A): The cleavage of a multiple bond that makes use of the reagent O_3, called ozone. The reaction leads to the formation of a cyclic compound called an **ozonide,** which is then reduced to carbonyl compounds by treatment with zinc and water.

***p* orbitals** (Section 1.11): A set of three degenerate (equal energy) atomic orbitals shaped like two tangent spheres with a nodal plane at the nucleus. For *p* orbitals, principal quantum number, *n* (see **atomic orbital**), is 2, the azimuthal quantum number, *l*, = 1 and the magnetic quantum numbers, *m,* are +1, 0, or − 1.

Paraffin (Section 4.14): An old name for an alkane.

Pauli exclusion principle (Section 1.11): A principle that states that no two electrons of an atom or molecule may have the same set of four quantum numbers. It means that only two electrons can occupy the same orbital, and then only when their spin quantum numbers are opposite. When this is true, we say that the spins of the electrons are paired.

Periplanar (Section 7.12C): A conformation in which vicinal groups lie in the same plane.

Peroxide (Section 9.1): A compound with an oxygen-oxygen single bond.

Peroxyacid (Section 10.18): An acid with the general formula, RCO_3H, containing an oxygen-oxygen single bond.

Phase sign (Section 1.10): A sign, either + or $-$, that is characteristic of all equations that describe the amplitudes of waves.

Phase transfer catalysts (Section 10.22): A reagent that transports an ion from an aqueous phase into a nonpolar phase where reaction takes place more rapidly. Tetraalkylammonium ions and crown ethers are phase transfer catalysts.

Phospholipid (Section 23.6): Compound that is structurally derived from **phosphatidic acid.** Phosphatidic acids are derivatives of glycerol in which two hydroxyl groups are joined to fatty acids, and one terminal hydroxyl group is joined in an ester linkage to phosphoric acid. In a phospholipid, the phosphate group of the phosphatidic acid is joined in ester linkage to a nitrogen-containing compound such as choline, 2-aminoethanol, or L-serine.

Pi (π) bond (Section 2.4): A bond formed when electrons occupy a bonding π molecular orbital (i.e., the lower energy molecular orbital that results from overlap of parallel p orbitals on adjacent atoms).

Pi (π) molecular orbital (Section 2.4): A molecular orbital formed when parallel p orbitals on adjacent atoms overlap. *Pi* orbitals may be **bonding** (p lobes of the same phase sign overlap) or **antibonding** (p orbitals of opposite phase sign overlap).

pK_a (Section 3.5): The pK_a is the negative logarithm of the acidity constant, K_a.

Plane of symmetry (Section 5.5): An imaginary plane that bisects a molecule in a way such that the two halves of the molecule are mirror images of each other. Any molecule with a plane of symmetry will be achiral.

Plane polarized light (Section 5.7A): Ordinary light in which the oscillations of the electrical field occur only in one plane.

Polar covalent bond (Section 1.18): A covalent bond in which the electrons are not equally shared because of differing electronegativities of the bonded atoms.

Polar molecule (Section 1.19): A molecule with a dipole moment.

Polarimeter (Section 5.7B): A device used for measuring optical activity.

Polarizability (Section 6.15C): The susceptibility of the electron cloud of an uncharged molecule to distortion by the influence of an electric charge.

Polymer (Section 9.10): A large molecule made up of many repeating subunits. For example, the polymer polyethylene is made up of the repeating subunit, $-(CH_2CH_2)-$.

Polysaccharide (Section 22.1A): A carbohydrate that, on a molecular basis, undergoes hydrolytic cleavage to yield many molecules of a monosaccharide.

Potential energy (Section 1.9): Potential energy is stored energy; it exists when attractive or repulsive forces exist between objects.

Potential energy diagram (Section 6.9): A plot of potential energy changes that take place during a reaction versus the reaction coordinate. It displays potential energy changes as a function of changes in bond orders and distances as reactants proceed through the transition state to become products.

Primary carbon (Section 2.10): A carbon atom that has only one other carbon atom attached to it.

Primary structure (Section 24.5): The covalent structure of a polypeptide or protein. This structure is determined, in large part, by determining the sequence of amino acids in the protein.

Protecting group (Section 16.7D): A group that is introduced into a molecule to protect a sensitive group from reaction while a reaction is carried out at some other location in the molecule. Later, the protecting group is removed.

Protein (Section 24.1): A large biological polymer of α-amino acids joined by amide linkages.

Protic solvent (Sections 3.10 and 6.15C): A solvent whose molecules have a hydrogen atom attached to a strongly electronegative element such as oxygen or nitrogen. Molecules of a protic solvent can, therefore, form hydrogen bonds to unshared electron pairs of oxygen or nitrogen atoms of solute molecules or ions, thereby stabilizing them. Water, methanol, ethanol, formic acid, and acetic acid are typical protic solvents.

Proton decoupling (Section 13.11B): An electronic technique used in ^{13}C NMR spectroscopy that allows decoupling of spin-spin interactions between ^{13}C nuclei and ^1H nuclei. In spectra obtained in this mode of operation, all carbon resonances appear as singlets.

Proton off-resonance decoupling (Section 13.11D): An electronic technique used in ^{13}C NMR spectroscopy that allows one-bond couplings between ^{13}C nuclei and ^1H nuclei. In spectra obtained in this mode of operation, CH_3 groups appear as quartets, CH_2 groups appear as triplets, CH groups appear as doublets, and carbon atoms with no attached hydrogen atoms appear as singlets.

Psi function (or ψ function) (Section 1.10): A mathematical expression derived from **quantum mechanics** corresponding to an energy state for an electron. The square of the ψ function, ψ^2, gives the probability of finding the electron in a particular location in space.

Pyranose (Section 22.2C): A sugar in which the cyclic acetal or hemiacetal ring is six-membered.

R (Section 2.8A): A symbol used to designate an alkyl group. Oftentimes, it is taken to symbolize any organic group.

***R-S* system** (Section 5.6): A method for designating the configuration of tetrahedral stereocenters.

Racemic form (racemate or racemic mixture) (Section 5.8A): An equimolar mixture of enantiomers. A racemic form is optically inactive.

Racemization (Section 6.14A): A reaction that transforms an optically active compound into a racemic form is said to proceed with racemization. Racemization will take place whenever a reaction causes chiral molecules to be converted to an achiral intermediate.

Radical (or free radical) (Section 3.1B): An uncharged chemical species that contains an unpaired electron.

Radical reaction (Section 9.1): A reaction involving radicals. Homolysis of covalent bonds occur in radical reactions.

Radicofunctional nomenclature (Section 4.3E): A system for naming compounds that uses two or more words to describe the compound. The final word corresponds to the functional group present; the preceding words, usually listed in alphabetical order, describe the remainder of the molecule. Examples are: methyl alcohol, ethyl methyl ether, and ethyl bromide.

Rate control (see Kinetic control)

Rate-determining step (Section 6.11A): If a reaction takes place in a series of steps, and if the first step is intrinsically slower than all of the others, then the rate of the overall reaction will be the same as (will be determined by) the rate of this slow step.

Reaction coordinate (Section 6.9): The abscissa in a potential energy diagram that represents the progress of the reaction. It represents the changes in bond orders and bond distances that must take place as reactants are converted to products.

Reaction mechanism (Section 3.1): A step-by-step description of the events that are postulated to take place at the molecular level as reactants are converted to products. A mechanism will include a description of all intermediates and transition states. Any mechanism proposed for a reaction must be consistent with all experimental data obtained for the reaction.

Rearrangement (or 1,2-shift) (Section 7.15): A reaction that results in a product with a different carbon skeleton from the reactant. The type of rearrangement called a 1,2-shift inolves the migration of an organic group (with its electrons) from one atom to the atom next to it.

Reducing sugar (Section 22.6A): Sugar that reduces Tollens' or Benedict's reagents. All sugars that contain hemiacetal or hemiketal groups (and, therefore, are in equilibrium with aldehydes or α-hydroxyketones) are reducing sugars. Sugars in which only acetal or ketal groups are present are nonreducing sugars.

Reduction (Section 11.2): A reaction that lowers the oxidation state of atoms in a molecule or ion. Reduction of an organic compound usually involves increasing its hydrogen content or decreasing its oxygen content. Reduction also accompanies any reaction that results in replacement of a more electronegative substituent by a less electronegative one.

Regioselective reaction (Section 8.2C): A reaction that yields only one (or a predominance of one) constitutional isomer as the product when two or more constitutional isomers are possible products.

Relative configuration (Section 5.14A): The relationship between the configurations of two chiral molecules. Molecules are said to have the same relative configuration when similar or identical groups in each occupy the same position in space. The configurations of molecules can be related to each other through reactions of known stereochemistry, for example, through a reaction that causes no bonds to a stereocenter to be broken.

Resolution (Sections 5.15 and 20.3E): The process by which the enantiomers of a racemic form are separated.

Resonance effect (Sections 3.9 and 15.11B): An effect by which a substituent exerts either an electron-releasing or withdrawing effect through the pi system of the molecule.

Resonance energy (Section 14.5): An energy of stabilization that represents the difference in energy between the actual compound and that calculated for a single resonance structure. The resonance energy arises from delocalization of electrons in a conjugated system.

Resonance structures (or resonance contributors) (Sections 1.8 and 12.5): Lewis structures that differ from one another only in the position of their electrons. A single resonance structure will not adequately represent a molecule. The molecule is better represented as a **hybrid** of all of the resonance structures.

Retrosynthetic analysis (Section 4.16): A method for planning syntheses that involves reasoning backward from the target molecule through various levels of precursors and thus finally to the starting materials.

Ring flip (Sections 4.10 and 4.11): The change in a cyclohexane ring (resulting from partial bond rotations) that converts one ring conformation to another. A ring flip converts any equatorial substituent to an axial substituent and vice versa.

Ring strain (Section 4.8): The increased potential energy of the cyclic form of a molecule (usually measured by heats of combustion) when compared to its acyclic form.

s **orbital** (Section 1.11): A spherical atomic orbital. For *s* orbitals the azimuthal quantum number, l,=0 (see **Atomic orbital**).

Saponification (Section 18.7B): Base-promoted hydrolysis of an ester.

Saturated compound (Section 2.2): A compound that does not contain any multiple bonds.

Secondary carbon (Section 2.10): A carbon atom that has two other carbon atoms attached to it.

Secondary structure (Section 24.8): The local conformation of a polypeptide backbone. These local conformations are specified in terms of regular folding patterns such as pleated sheets, α-helices, and turns.

Shielding and deshielding (Section 13.6): Effects observed in NMR spectra caused by the circulation of sigma and pi electrons within the molecule. Shielding causes signals to appear at higher magnetic fields (upfield); deshielding causes signals to appear at lower magnetic fields (downfield).

Sigma σ **bond** (Section 1.13): A single bond. A bond formed when electrons occupy the bonding σ orbital formed by the end-on overlap of atomic orbitals (or hybrid orbitals) on adjacent atoms. In a sigma bond, the electron density has circular symmetry when viewed along the bond axis.

Sigma σ **orbital** (Section 1.13): An orbital formed by end-on overlap of orbitals (or lobes of orbitals) on adjacent atoms. Sigma orbitals may be **bonding** (orbitals or lobes of the same phase sign overlap); or **antibonding** (orbitals or lobes of opposite phase sign overlap).

S_N1 **reaction** (Sections 6.11 and 6.15): Literally, substitution nucleophilic unimolecular. A multistep nucleophilic substitution in which the leaving group departs in a unimolecular step prior to the attack of the nucleophile. The rate equation is first order in substrate but zero order in the attacking nucleophile.

S_N2 reaction (Sections 6.7 and 6.8): Literally, substitution nucleophilic bimolecular. A bimolecular nucleophilic substitution reaction that takes place in a single step in which a nucleophile attacks a carbon bearing a leaving group from the backside, causing an inversion of configuration at this carbon and displacement of the leaving group.

Solvent effect (Sections 6.15C and D): An effect on relative rates of reaction caused by the solvent. For example, the use of a polar solvent will increase the rate of reaction of an alkyl halide in an S_N1 reaction.

Solvolysis (Section 6.14B): Literally, cleavage by the solvent. A nucleophilic substitution reaction in which the nucleophile is a molecule of the solvent.

***sp* orbital** (Section 1.15): A hybrid orbital that is derived by mathematically combining one *s* atomic orbital and one *p* atomic orbital. Two *sp* hybrid orbitals are obtained by this process, and they are oriented in opposite directions with an angle of 180° between them.

***sp²* orbital** (Section 1.14): A hybrid orbital that is derived by mathematically combining one *s* atomic orbital and two *p* atomic orbitals. Three *sp²* hybrid orbitals are obtained by this process, and they are directed toward the corners of an equilateral triangle with angles of 120° between them.

***sp³* orbital** (Section 1.13): A hybrid orbital that is derived by mathematically combining one *s* atomic orbital and three *p* atomic orbitals. Four *sp³* hybrid orbitals are obtained by this process, and they are directed toward the corners of a regular tetrahedron with angles of 109.5° between them.

Specific rotation (Section 5.7C): A physical constant calculated from the observed rotation of a compound using the following equation:

$$[\alpha]_D = \frac{\alpha}{c \; x \; l}$$

where α is the observed rotation, using the D line of a sodium lamp, c is the concentration of the solution in grams per milliliter, or the density of a neat liquid in g mL^{-1}, and l is the length of the tube in decimeters.

Spin decoupling (Section 13.10): An effect that causes signal splitting not to be observed.

Spin-spin coupling (Section 13.9): An effect observed in NMR spectra. Spin-spin couplings result in a signal appearing as a multiplet (i.e., doublet, triplet, quartet, etc.) and are caused by magnetic couplings of the nucleus being observed with nuclei of nearby atoms.

Staggered conformation (Section 4.6): A temporary orientation of groups around two atoms joined by a single bond such that the bonds of the back atom exactly bisect the angles formed by the bonds of the front atom in a Newman projection formula:

A staggered conformation

Stereocenter (Section 5.2): An atom bearing groups of such nature that an interchange of any two groups will produce a stereoisomer.

Stereochemistry (Section 5.4): Chemical studies that take into account the spatial aspects of molecules.

Stereoisomers (Sections 2.4B, 5.1, and 5.2): Compounds wth the same molecular formula that differ *only* in the arrangement of their atoms in space. Stereoisomers have the same connectivity, and, therefore are not constitutional isomers. Stereoisomers are classified further as being either enantiomers or diastereomers.

Stereospecific reaction (Section 8.7A): A reaction in which a particular stereoisomeric form of the reactant reacts in such a way that it leads to a specific stereoisomeric form of the product.

Steric effect (Section 6.15A): An effect on relative reaction rates caused by the space-filling properties of those parts of a molecule attached at or near the reacting site.

Steric hindrance (Section 6.15A): An effect on relative reaction rates caused when the spatial arrangement of atoms or groups at or near the reacting site hinders or retards a reaction.

Steroid (Section 23.4): Steroids are lipids that are derived from the following perhydrocyclopentanophenanthrene ring system:

Structural formula (Sections 1.3 and 1.20): A formula that shows how the atoms of a molecule are attached to each other.

Substituent effect (Sections 3.7B and 15.10): An effect on the rate of reaction (or on the equilibrium constant) caused by the replacement of a hydrogen atom by another atom or group. Substituent effects include those effects caused by the size of the atom or group, called steric effects, and those effects caused by the ability of the group to release or withdaw electrons, called electronic effects. Electronic effects are further classified as being inductive effects or resonance effects.

Substitution reaction (Sections 6.3 and 9.3): A reaction in which one group replaces another in a molecule.

Substitutive nomenclature (Section 4.3F): A system for naming compounds in which each atom or group, called a substituent, is cited as a prefix or suffix to a parent compound. In the IUPAC system, only one group may be cited as a suffix. Locants (usually numbers) are used to tell where the group occurs.

Substrate (Section 6.3): The molecule or ion that undergoes reaction.

Sugar (Section 22.1A): A carbohydrate.

Syn addition (Section 7.6A): An addition that places both parts of the adding reagent on the same face of the reactant.

Syndiotactic polymer (Special Topic C): A polymer in which the configurations at the stereocenters along the chain alternate regularly, (R), (S), (R), (S), etc.

Synthon (Section 8.16): The fragments that result (on paper) from the disconnection of a bond. The actual reagent that will, in a synthetic step, provide the synthon is called the **synthetic equivalent.**

Tautomers (Section 17.2): Constitutional isomers that are easily interconverted. Keto and enol tautomers, for example, are rapidly interconverted in the presence of acids and bases.

Terpene (Section 23.3): A lipid that has a structure that can be derived on paper by linking isoprene units.

Tertiary carbon (Section 2.10): A carbon atom that has three other carbon atoms attached to it.

Tertiary structure (Section 24.8B): The three-dimensional shape of a protein that arises from foldings of its polypeptide chains superposed on its α helixes and pleated sheets.

Thermodynamic control (Sections 6.6 and 12.9A): A principle stating that the ratio of products of a reaction that reaches equilibrium is determined by the relative stabilities of the products (as measured by their standard free energies, $\Delta G°$). The most abundant product will be the one that is the most stable.

Torsional barrier (Section 4.6): The barrier to rotation of groups joined by a single bond caused by repulsions between the aligned electron pairs in the eclipsed form.

Torsional strain (Section 4.9): The strain associated with an eclipsed conformation of a molecule; it is caused by repulsions between the aligned electron pairs of the eclipsed bonds.

Tosylate (Section 10.10): A p-toluenesulfonate ester.

Transition state (Sections 6.8 and 6.9): A state on a free energy diagram corresponding to an energy maximum (i.e., characterized by having higher free energy than immediately adjacent states). The transition state is also a term used to refer to the species that occurs at this state of maximum free energy; another term used for this species is *the activated complex.*

Unimolecular reaction (Section 6.11): A reaction whose rate-determining step involves only one species.

Unsaturated compound (Section 2.2): A compound that contains multiple bonds.

van der Waals force (or London force) (Sections 2.16D and 4.7): A weak force that acts between nonpolar molecules or between parts of the same molecule. Bringing two groups (or molecules) together first results in an attractive force between them because a temporary unsymmetrical distribution of electrons in one group induces an opposite polarity in the other. When groups are brought closer than their *van der Waals radii,* the force between them becomes repulsive because their electron clouds begin to interpenetrate each other.

Vicinal (vic) substituents (Section 7.16): Substituents that are on adjacent atoms.

Vinyl group (Section 6.1): The CH_2=CH- group.

Vinylic halide (Section 6.1): An organic halide in which the halogen atom is attached to a carbon atom of a double bond.

Vinylic substituent (Section 6.1): A substituent on a carbon atom that participates in a carbon-carbon double bond.

Visible-ultraviolet (visible-UV)spectroscopy (Section 13.2): A type of optical spectroscopy that measures the absorption of light in the visible and ultraviolet regions of the spectrum. Visible-UV spectra primarily provide structural information about the kind and extent of conjugation of multiple bonds in the compound being analyzed.

Wave function (or ψ function) (Section 1.10): A mathematical expression derived from **quantum mechanics** corresponding to an energy state for an electron, i.e., for an orbital. The square of the ψ function, ψ^2, gives the probability of finding the electron in a particular place in space.

Wavelength (abbreviated λ) (Section 13.1): The distance between consecutive crests (or troughs) of a wave.

Wavenumber (abbreviated $\bar{\nu}$) (Section 13.3): A way to express the frequency of a wave. The wavenumber is the number of waves per centimeter, expressed as cm^{-1}.

Ylide (Section 16.10): An electrically neutral molecule that has a negative carbon with an unshared electron pair adjacent to a positive heteroatom.

Zaitsev's rule (Section 7.12A): A rule stating that an elimination will give as the major product the most stable alkene (i.e., the alkene with the most highly substituted double bond).

Zwitterion Another name for a dipolar ion (**see also Dipolar ion**).

D

APPENDIX
Molecular Model Set Exercises

The exercises in this appendix are designed to help you gain an understanding of the three-dimensional nature of molecules. You are encouraged to perform these exercises with a model set as described.

These exercises should be performed as part of the study of the chapters shown below.

Chapter in Text	Accompanying Exercises
4	1, 3, 4, 5, 6, 8, 10, 11, 12, 14, 15, 16, 17, 18, 20, 21
5	2, 7, 9, 13, 24, 25, 26, 27
7	9, 19, 22, 28
2	31
3	23
22	29
24	30
12	31

The following molecular model set exercises were developed by Ronald Starkey for use with the Theta Molecular Model Set (J. Wiley & Sons, Inc.).

Refer to the instruction booklet that accompanies the model set for details of molecular model assembly.

EXERCISE 1 (Chapter 4)

Assemble a molecular model of methane, CH_4. Note that the hydrogen atoms describe the apexes of a regular tetrahedron with the carbon atom at the center of the tetrahedron. Demonstrate by attempted superposition that two models of methane are identical.

Replace any one hydrogen atom on each of the two methane models with a halogen (a green atom-center in the Theta Molecular Model Set) to form two molecules of CH_3X. Are the two structures identical? Does it make a difference which of the four hydrogen atoms on a methane molecule you replace? How many different configurations of CH_3X are possible?

Repeat the same considerations for two disubstituted methanes with two identical substituents (CH_2X_2), and then with two different substituents (CH_2XY). Two shades of green atom-centers could be used for the two different substituents.

Methane, CH_4

EXERCISE 2 (Chapter 5)

Construct a model of a trisubstituted methane molecule (CHXYZ). Four different colored atom-centers (red, blue, yellow, and white) are attached to a central tetrahedral black carbon atom center. Note that the carbon now has four different substituents. Compare this model with a second model of CHXYZ. Are the two structures identical (superposable)?

Interchange any two substituents on one of the carbon atoms. Are the two CHXYZ molecules identical now? Does the fact that interchange of any two substituents on the carbon interconverts the stereoisomers indicate that there are only two possible configurations of a tetrahedral carbon atom?

Compare the two models that were not identical. What is the relationship between them? Do they have a mirror-image relationship? That is, are they related as an object and its mirror image?

EXERCISE 3 (Chapter 4)

Make a model of ethane, CH_3CH_3. Does each of the carbon atoms retain a tetrahedral configuration? Can the carbon atoms be rotated with respect to each other without breaking the carbon-carbon bond?

Rotate about the carbon-carbon bond until the carbon-hydrogen bonds of one carbon atom are aligned with those of the other carbon atom. This is the eclipsed conformation. When the C–H bond of one carbon atom bisects the H–C–H angle of the other carbon atom the conformation is called staggered. Remember, conformations are arrangements of atoms in a molecule that can be interconverted by bond rotations.

In which of the two conformations of ethane you made are the hydrogen atoms of one carbon closer to those of the other carbon?

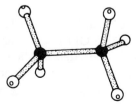

Ethane, CH_3CH_3

EXERCISE 4 (Chapter 4)

Prepare a second model of ethane. Replace one hydrogen, any one, on each ethane model with a substituent such as a halogen (a green atom-center), to form two models of CH_3CH_2X. Are the structures identical? If not, can they be made identical by rotation about the C–C bond? With one of the models, demonstrate that there are three equivalent staggered conformations (see Exercise 3) of CH_3CH_2X. How many equivalent eclipsed conformations are possible?

EXERCISE 5 (Chapter 4)

Assemble a model of a 1,2-disubstituted ethane molecule, CH_2XCH_2X. Note how the orientation of and the distance between the X groups changes with rotation about the carbon-carbon bond. The arrangement in which the X substituents are at maximum separation is the *anti*-staggered conformation. The other staggered conformations are called *gauche*. How many *gauche* conformations are possible? Are they energetically equivalent? Are they identical?

EXERCISE 6 (Chapter 4)

Construct two models of butane, $CH_3CH_2CH_2CH_3$. Note that the structures can be viewed as dimethyl substituted ethanes. Show that rotations about the C2,C3 bond of butane produce eclipsed, *anti*-staggered, and *gauche*-staggered conformations. Measure the distance between C1 and C4 in the conformations just mentioned. The scale of the Theta Molecular Model Set is: 3 cm in a model corresponds to approximately $1.0 \text{Å} (0.1 \text{ nm})$ on a molecular scale. In which eclipsed conformation are the C1 and C4 atoms closest to each other? How many eclipsed conformations are possible?

EXERCISE 7 (Chapter 5)

Using two models of butane, verify that the two hydrogen atoms on C2 are not stereochemically equivalent. Replacement of one hydrogen leads to a product that is not identical to that obtained by replacement of the other C2 hydrogen atom. Both replacement products have the same condensed formula, $CH_3CHXCH_2CH_3$. What is the relationship of the two products?

EXERCISE 8 (Chapter 4)

Make a model of hexane, $CH_3CH_2CH_2CH_2CH_2CH_3$. Extend the six-carbon chain as far as it will go. This puts C1 and C6 at maximum separation. Notice that this *straight-chain* structure maintains the tetrahedral bond angles at each carbon atom and therefore the carbon chain adopts a zigzag arrangement. Does this extended chain adopt staggered or eclipsed conformations of the hydrogen atoms? How could you describe the relationship of C1 and C4?

EXERCISE 9 (Chapter 7)

Prepare models of the four isomeric butenes, C_4H_8. Note that the restricted rotation about the double bond is responsible for the cis-trans stereoisomerisim. Verify this by observing that breaking the π bond of *cis*-2-butene allows rotation and thus conversion to *trans*-2-butene. Is any of the four isomeric butenes chiral (nonsuperposable with its mirror image)? Indicate pairs of butene isomers that are structural (constitutional) isomers. Indicate pairs that are diastereomers. How does the distance between the C1 and C4 atoms in *trans*-2-butene compare with that of the *anti* conformation of butane? Compare the C1 to C4 distance in *cis*-2-butene with that in the conformation of butane in which the methyls are eclipsed.

| 1-Butene | *cis*-2-Butene | *trans*-2-Butene | 2-Methylpropene |

EXERCISE 10 (Chapter 4)

Make a model of cyclopropane. The Theta Molecular Model Set requires the use of 1.5-cm flexible tubing for the carbon-carbon bonds of the cyclopropane ring. The flexible tubes illustrate quite well the "bent-bond" nature of the ring bonds. It should be apparent that the ring carbon atoms must be coplanar. What is the relationship of the hydrogen atoms on adjacent carbon atoms? Are they staggered, eclipsed, or skewed?

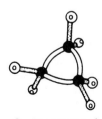

Cyclopropane, Δ

EXERCISE 11 (Chapter 4)

A model of cyclobutane can be assembled in a conformation that has the four carbon atoms coplanar. For this exercise, the rigid 2.0-cm tubes of the Theta Molecular Model Set should be used for the carbon-carbon bonds of the ring. How many eclipsed hydrogen atoms are there in the conformation? Torsional strain (strain due to deviations from an eclipsed conformation) can be relieved at the expense of increased angle strain by a slight folding of the ring. The deviation of one ring carbon from the plane of the other three carbon atoms is about 25°. This folding compresses the C–C–C bond angle to about 88°. Rotate the ring carbon bonds of the planar conformation to obtain the folded conformation. Are the hydrogen atoms on adjacent carbon atoms eclipsed or skewed? Considering both structural and stereoisomeric forms, how many dimethylcyclobutane structures are possible? Do deviations of the ring from planarity have to be considered when determining the number of possible dimethyl structures?

Cyclobutane, □

EXERCISE 12 (Chapter 4)

Cyclopentane is a more flexible ring system than cyclobutane or cyclopropane. A model of cyclopentane in a conformation with all the ring carbon atoms coplanar exhibits minimal deviation of the C–C–C bond angles from the normal tetrahedral bond angle. How many eclipsed hydrogen interactions are there in this planar conformation? If one of the ring carbon atoms is pushed slightly above (or below) the plane of the other carbon atoms, a model of the envelope conformation is obtained. Does the envelope conformation relieve some of the torsional strain? How many eclipsed hydrogen interactions are there in the envelope conformation?

Cyclopentene

EXERCISE 13 (Chapter 5)

Make a model of 1,2-dimethylcyclopentane. How many stereoisomers are possible for this compound? Identify each of the possible structures as either cis or trans. Is it apparent that cis-trans isomerism is possible in this compound because of restricted rotation? Are any of the stereoisomers chiral? What are the relationships of the 1,2-dimethylcyclopentane stereoisomers?

EXERCISE 14 (Chapter 4)

Assemble the six-membered ring compound cyclohexane. Is the ring flat or puckered? Place the ring in a chair conformation and then in a boat conformation. Demonstrate that the chair and boat are indeed conformations of cyclohexane—that is, they may be interconverted by rotations about the carbon-carbon bonds of the ring.

Chair form   Boat form

Note that in the chair conformation carbon atoms 2, 3, 5, and 6 are in the same plane and carbon atoms 1 and 4 are below and above the plane, respectively. In the boat conformation, carbon atoms 1 and 4 are both above (they could also both be below) the plane described by carbon atoms 2, 3, 5, and 6. Is it apparent why the boat is sometimes associated with the flexible form? Are the hydrogen atoms in the chair conformation staggered or eclipsed? Are any hydrogen atoms eclipsed in the boat conformation? Do carbon atoms 1 and 4 have an *anti* or *gauche* relationship in the chair conformation? (*Hint:* Look down the C2, C3 bond).

A twist conformation of cyclohexane may be obtained by slightly twisting carbon atoms 2 and 5 of the boat conformation as shown.

Boat form   Twist form

Note that the C2, C3 and the C5, C6 sigma bonds no longer retain their parallel orientation in the twist conformation. If the ring system is twisted too far, another boat conformation results. Compare the nonbonded (van der Waals repulsion) interactions and the torsional strain present in the boat, twist, and chair conformations of cyclohexane. Is it apparent why the relative order of thermodynamic stabilities is chair > twist > boat?

EXERCISE 15 (Chapter 4)

Construct a model of methylcyclohexane. How many chair conformations are possible? How does the orientation of the methyl group change in each chair conformation?

Identify carbon atoms in the chair conformation of methylcyclohexane that have intramolecular interactions corresponding to those found in the *gauche* and *anti*conformations of butane. Which of the chair conformations has the greatest number of *gauche* interactions? How many more? If we assume, as in the case for butane, that the *anti* interaction is 0.9 kcal mol^{-1} more favorable than *gauche,* then what is the relative stability of the two chair conformations of methylcyclohexane? *Hint:* Identify the relative number of *gauche* interactions in the two conformations.

EXERCISE 16 (Chapter 4)

Compare models of the chair conformations of monosubstituted cyclohexanes in which the substituent alkyl groups are methyl, ethyl, isopropyl, and *tert*-butyl.

Rationalize the relative stability of axial and equatorial conformations of the alkyl group given in the table for each compound. The chair conformation with the alkyl group equatorial is more stable by the amount shown.

ALKYL GROUP	$\Delta G°$ (kcal mol^{-1}) EQUATORIAL $\leftrightarrows$ AXIAL
CH_3	1.7
CH_2CH_3	1.8
$CH(CH_3)_2$	2.2
$C(CH_3)_3$	5.0 (approximate)

EXERCISE 17 (Chapter 4)

Make a model of 1,2-dimethylcyclohexane. Answer the questions posed in Exercise 13 with regard to 1,2-dimethylcyclohexane.

EXERCISE 18 (Chapter 4)

Compare models of the neutral and charged molecules shown next. Identify the structures that are isoelectronic, that is, those that have the same electronic structure. How do those structures that are isoelectronic compare in their molecular geometry?

CH_3CH_3	CH_3NH_2	CH_3OH
$CH_3CH_2^-$	$CH_3NH_3^+$	$CH_3OH_2^+$
CH_3NH^-		

EXERCISE 19 (Chapter 7)

Prepare a model of cyclohexene. Note that chair and boat conformations are no longer possible, as carbon atoms 1, 2, 3, and 6 lie in a plane. Are cis and trans stereoisomers possible for the double bond? Attempt to assemble a model of *trans*-cyclohexene. Can it be done? Are cis and trans stereoisomers possible for 2,3-dimethylcyclohexene? For 3,4-dimethylcyclohexene?

Cyclohexene

Assemble a model of *trans*-cyclooctene. Observe the twisting of the π-bond system. Would you expect the cis stereoisomer to be more stable than *trans*-cyclooctene? Is *cis*-cyclooctene chiral? Is *trans*-cyclooctene chiral?

EXERCISE 20 (Chapter 4)

Construct models of *cis*-decalin (*cis*-bicyclo[4.4.0]decane) and *trans*-decalin. Observe how it is possible to convert one conformation of *cis*-decalin in which both rings are in chair conformations to another all-chair conformation. This interconversion is not possible in the case of the *trans*-decalin isomer. Suggest a reason for the difference in the behavior of the cis and trans isomers. *Hint:* What would happen to carbon atoms 7 and 10 of *trans*-decalin if the other ring (indicated by carbon atoms numbered 1–6) is converted to the alternative chair conformation. Is the situation the same for *cis*-decalin?

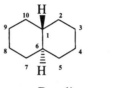

trans-Decalin   *cis*-Decalin

EXERCISE 21 (Chapter 4)

Assemble a model of norbornane (bicyclo[2.2.1]heptane). Observe the two cyclopentane ring systems in the molecule. The structure may also be viewed as having a methylene (CH_2) bridge between carbon atoms 1 and 4 of cyclohexane. Describe the conformation of the cyclohexane ring system in norbornane. How many eclipsing interactions are present?

Norbornane

Using a model of twistane, identify the cyclohexane ring systems held in twist conformations. In adamantane, find the chair conformation cyclohexane systems. How many are present? Evaluate the torsional and angle strain in adamantane. Which of the three compounds in this exercise are chiral?

Twistane   Adamantane

EXERCISE 22 (Chapter 7)

An hypothesis known as Bredt's Rule states that a double bond to a bridgehead of a small-ring bridged bicyclic compound is not possible. The basis of this rule can be seen if you attempt to make a model of bicyclo[2.2.1]hept-1-ene, **A.** One approach to the assembly of this model is to try to bridge the number 1 and number 4 carbon atoms of cyclohexene with a methylene (CH$_2$) unit. Compare this bridging with the ease of installing a CH$_2$ bridge between the 1 and 4 carbon atoms of cyclohexane to form a model of norbornane (see Exercise 21). Explain the differences in ease of assembly of these two models.

A   **B**

Bridgehead double bonds can be accommodated in larger ring-bridged bicyclic compounds such as bicyclo[3.2.2]non-1-ene, **B.** Although this compound has been prepared in the laboratory, it is an extremely reactive alkene. The Theta Molecular Model Set model of bicyclo [3.2.2] non-1-ene clearly shows the strained (twisted) double bond system.

EXERCISE 23 (Chapter 14)

Not all cyclic structures with alternating double and single bonds are aromatic. Cyclooctatetraene shows none of the aromatic characteristics of benzene. From examination of molecular models of cyclooctatetraene and benzene, explain why there is π-electron delocalization in benzene but not in cyclooctatetraene. *Hint:* Can the carbon atoms of the eight-membered ring readily adopt a planar arrangement?

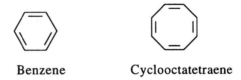

Benzene   Cyclooctatetraene

Note that benzene can be represented in several different ways with the Theta Molecular Model Set. In this exercise, the Kekulé representation with alternating double and single bonds is appropriate. Alternative representations of benzene are shown in the model set instruction booklet.

EXERCISE 24 (Chapter 5)

Consider the CH$_3$CHXCHYCH$_3$ system. A butane that has at C2 and C3 different shades of green atom-centers is representative. Assemble all possible stereoisomers of this structure. How many are there? Indicate the relationship among them. Are they all chiral?

Repeat the analysis with the CH$_3$CHXCHXCH$_3$ system. The green atom-centers are suitable for representation of the X substituent.

EXERCISE 25 (Chapter 5)

The $CH_3CHXCHXCH_3$ molecule can exist as the stereoisomers shown here. In the eclipsed conformation (meso) shown on the left, the molecule has a plane of symmetry that bisects the C2, C3 bond. This is a more energetic conformation than any of the three staggered conformations, but it is the only conformation of this configurational stereoisomer that has a plane of symmetry. Can you consider a molecule achiral if only one conformation, and in this case not even the most stable conformation, has a plane of symmetry? Are any of the staggered conformations achiral (superposable on their mirror image)? Make a model of the staggered conformation shown here and make another model that is the mirror image of it. Are these two structures different conformations of the same configurational stereoisomer (e.g., are they conformers that can be interconverted by bond rotations), or are they configurational stereoisomers? Based on your answer to the last question, suggest an explanation for the fact that the molecule is not optically active.

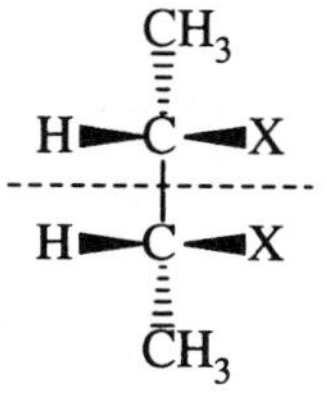

E 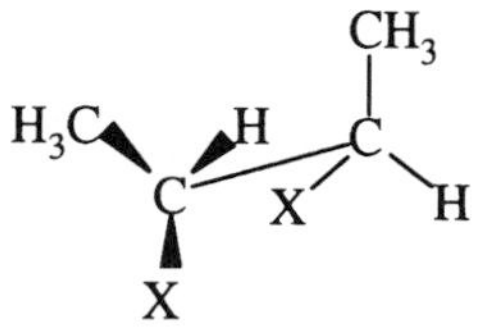   S

EXERCISE 26 (Chapter 5)

Not all molecular chirality is a result of a center of chirality, such as CHXYZ. Cumulated dienes (1,2-dienes or allenes) are capable of generating molecular chirality.

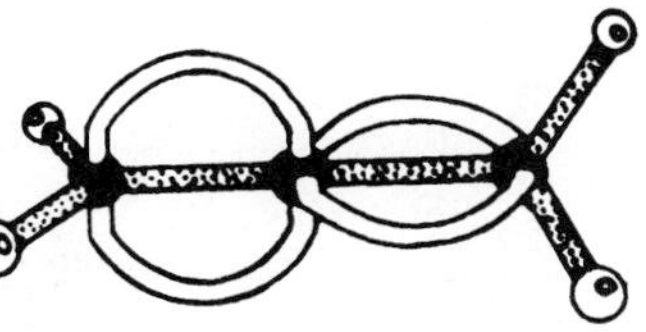

1,2-Propadiene (allene), $H_2C=C=CH_2$

Identify, using models, which of the following cumulated dienes are chiral.

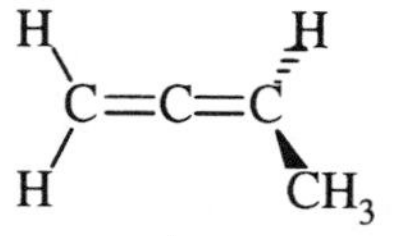

A 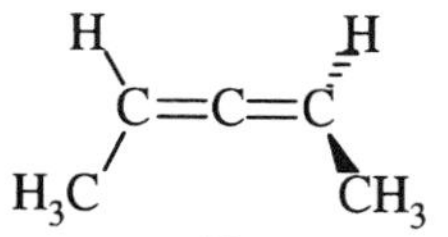   B 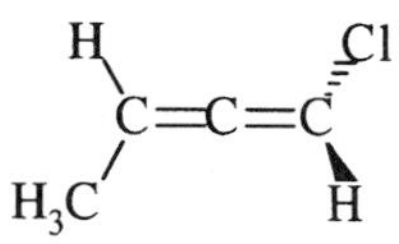   C

Are the following compounds chiral? How are they structurally related to cumulated dienes?

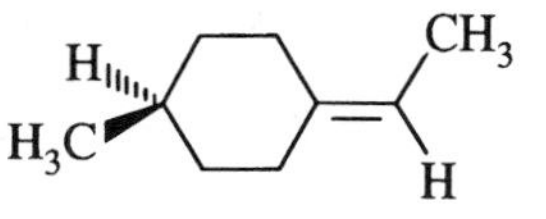

D 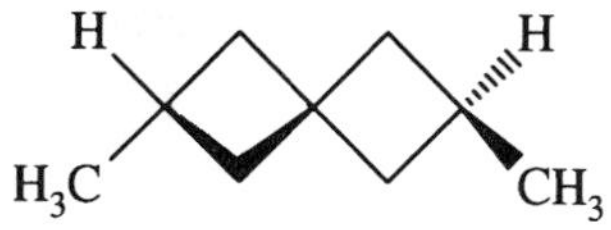   E

Is the cumulated triene **F** chiral? Explain the presence or absence of molecular chirality. More than one stereoisomer is possible for triene **F.** What are the structures, and what is the relationship between those structures?

H\C=C=C=C/CH₃, H₃C/ , \H

F

EXERCISE 27 (Chapters 5 and 14)

Substituted biphenyl systems can produce molecular chirality if the rotation about the bond connecting the two rings is restricted. Which of the three biphenyl compounds indicated here are chiral and would be expected to be optically active?

a   f

b   e

J. a = f = CH₃   K. a = b = CH₃   L. a = f = CH₃

b = e = N(CH₃)₃⁺   e = f = N(CH₃)₃⁺   b = e = H

EXERCISE 28 (Chapter 8)

Assemble a model of ethyne (acetylene). The linear geometry of the molecule should be readily apparent. Note that the Theta Molecular Model Set depicts the σ and both the π bonds of the triple bond system. Based on attempts to assemble cycloalkynes, predict the smallest cycloalkyne that is stable.

Ethyne, HC≡CH

EXERCISE 29 (Chapter 22)

Construct a model of β-D-glucopyranose. Note that in one of the chair conformations all the hydroxyl groups and the CH₂OH group are in an equatorial orientation. Convert the structure of β-D-glucopyranose to α-D-glucopyranose, to β-D-mannopyranose, and to β-D-galactopyranose. Indicate the number of large ring substituents (OH or CH₂OH) that are axial in the more favorable chair conformation of each of these sugars. Is it reasonable that the β-anomer is more stable than the α-anomer of D-glucopyranose?

Make a model of β-L-glucopyranose. What is the relationship between the D and L configurations? Which is more stable?

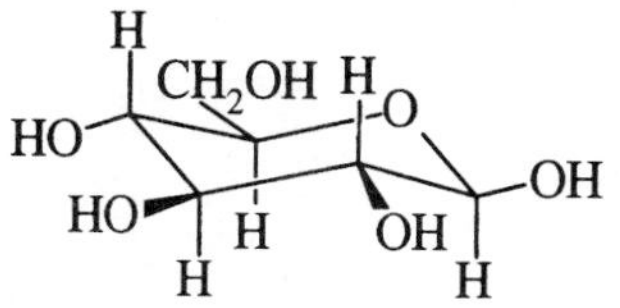

β-D-Glucopyranose

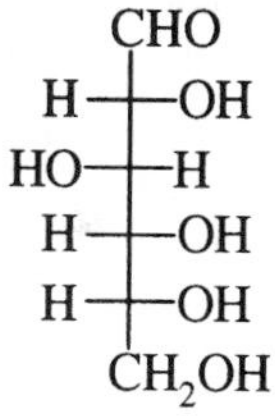

D-(+)-Glucose 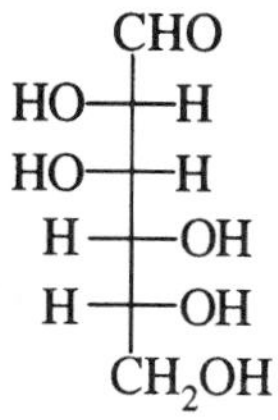   D-(+)-Mannose 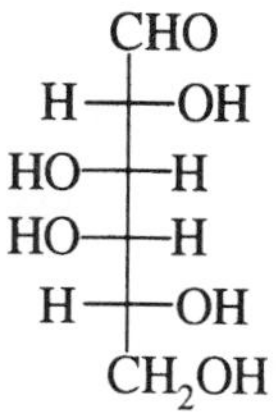   D-(+)-Galactose

EXERCISE 30 (Chapter 24)

Assemble a model of tripeptide **A** shown here. If the model is made according to the representation of a peptide shown in the Theta Molecular Model Set, you will be able to observe the restricted rotation of the C–N bond in the amide linkage. Note the planarity of the six atoms associated with the amide portions of the molecule. Which bonds along the peptide chain are free to rotate? The amide linkage can either be cisoid or transoid. How does the length (from the N-terminal nitrogen atom to the C-terminal carbon atom) of the tripeptide chain that is transoid compare with one that is cisoid? Which is more "linear"? Convert a model of tripeptide **A** in the transoid arrangement to a model of tripeptide **B**. Which tripeptide has a longer chain?

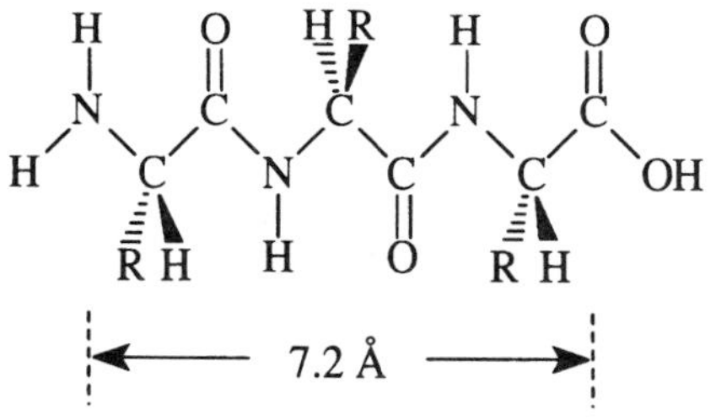

$\longleftarrow$ 7.2 Å $\longrightarrow$

Tripeptide **A** $R = CH_3$ (L-Alanine)
Tripeptide **B** $R = CH_2OH$ (L-Serine)

EXERCISE 31 (Chapter 12)

Make models of the π molecular orbitals for the following compounds. Use the phase representation of each contributing atomic orbital shown in the Theta Molecular Model Set instruction booklet. Compare each model with π molecular orbital diagrams shown in the textbook.

(a) π_1 and π_2 of ethene (CH_2=CH_2)

(b) π_1 thru π_4 of 1,3-butadiene (CH_2=CH–CH=CH_2)

(c) π_1, π_2 and π_3 of the allyl (propenyl) radical (CH_2=CH–$\dot{C}H_2$)

EXERCISE 32 (Special Topic O)

Explain the observed stereochemistry of the pericyclic reactions shown here. The course of the reactions are controlled by orbital symmetry.

(a) An electrocyclic reaction.

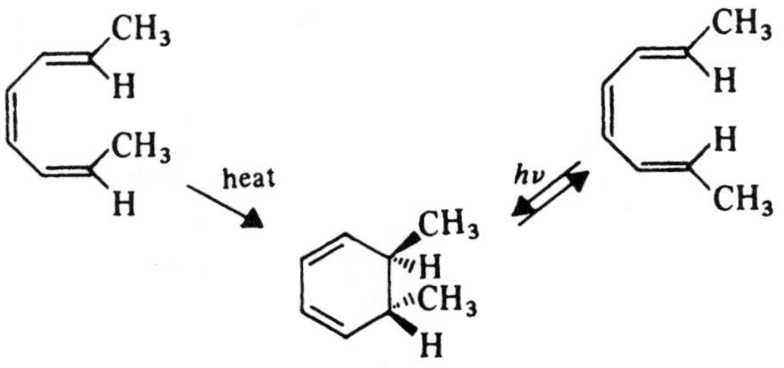

(b) (4+2) Cycloaddition reactions.

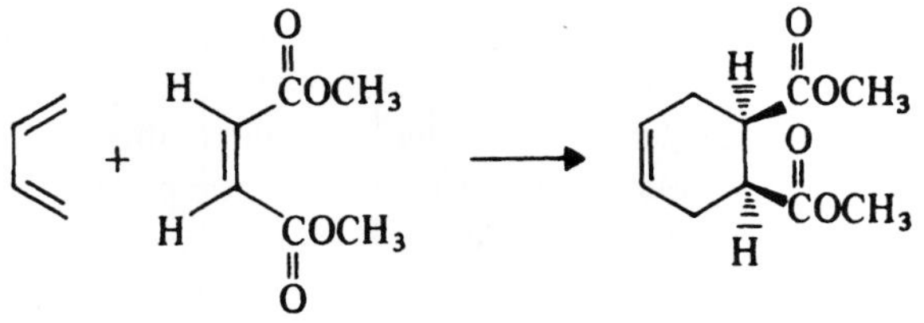

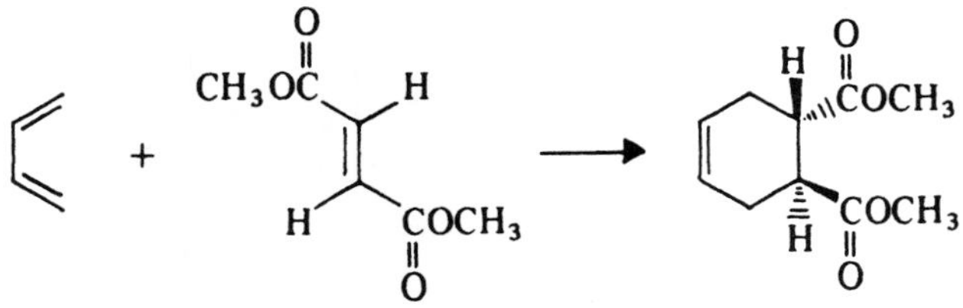

EXERCISE 33

The Theta Molecular Model Set is well suited for the assembly of many fairly complex natural products. Several interesting representative natural product structures, suitable for your model-making pleasure, are shown here.

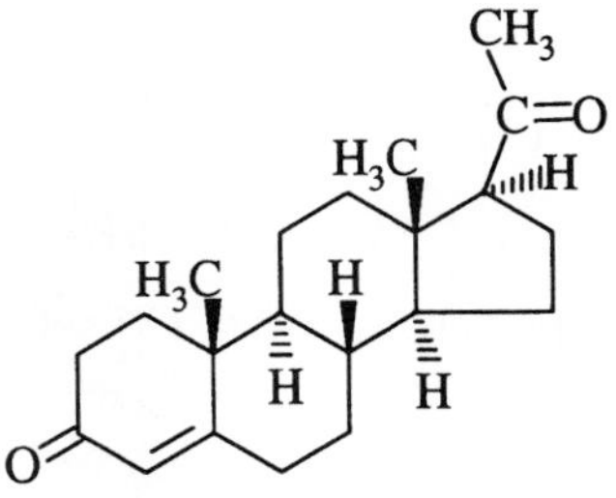

Progesterone 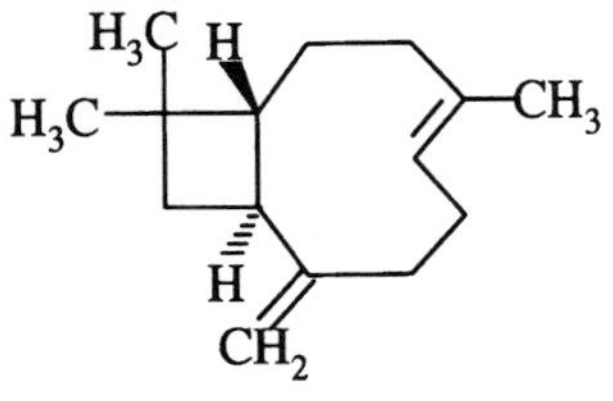   Caryophyllene 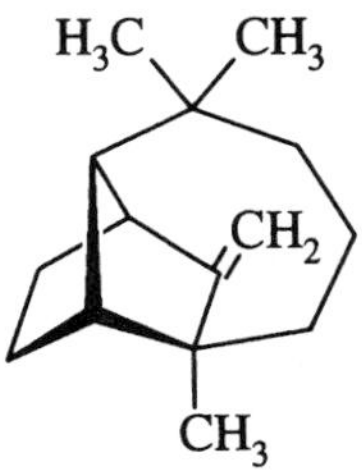   Longifolene

Morphine   Strychnine

MOLECULAR MODEL SET EXERCISES SOLUTIONS

Solution 1   Replacement of any hydrogen atom of methane leads to the same monosubstituted product CH_3X. Therefore, there is only one configuration of a monosubstituted methane. There is only one possible configuration for a disubstituted methane of either the CH_2X_2 or CH_2XY type.

Solution 2   Interchange of any two substituents converts the configuration of a tetrahedral stereocenter to that of its enantiomer. There are only two possible configurations. If the models are not identical, they will have a mirror-image relationship.

Solution 3   The tetrahedral carbon atoms may be rotated without breaking the carbon-carbon bond. There is no change in the carbon-carbon bond orbital overlap during rotation. The eclipsed conformation places the hydrogen atoms closer together than they are in the staggered conformation.

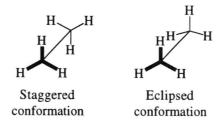

Staggered   Eclipsed
conformation   conformation

Solution 4   All monosubstituted ethanes (CH_3CH_2X) may be made into identical structures by rotations about the C–C bond. The following structures are three energetically equivalent staggered conformations.

The three equivalent eclipsed conformations are

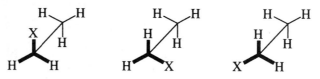

Solution 5   The two *gauche* conformations are energetically equivalent, but not identical (superposable) since they are conformational enantiomers. They bear a mirror-image relationship and are interconvertible by rotation about the carbon-carbon bond.

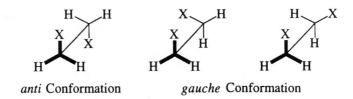

anti Conformation   *gauche* Conformation

Solution 6   There are three eclipsed conformations. The methyl groups (C1 and C4) are closest together in the methyl-methyl eclipsed conformation. The carbon-carbon internuclear distances between C1 and C4 are shown in the following table. The number of conformations of each type, the model distances, and the corresponding molecular distances in angstroms (Å) are shown.

| | | DISTANCES | |
CONFORMATION	NUMBER	(cm)	(Å)
Eclipsed (CH_3, CH_3)	1	7.4	2.5
Gauche	2	8.5	2.8
Eclipsed (H, CH_3)	2	10.0	3.3
Anti	1	11.0	3.7

Solution 7   The enantiomers formed from replacement of the C2 hydrogen atoms of butane are

$$
\begin{array}{cc}
CH_3 & CH_3 \\
H-C-X & X-C-H \\
CH_2CH_3 & CH_2CH_3
\end{array}
$$

Solution 8   The extended chain assumes a staggered arrangement. The relationship of C1 and C4 is *anti*.

$$
\begin{array}{ccc}
HH & HH & HH \\
H\diagdown C & C & C \\
C & C & C \diagup H \\
HH & HH & HH
\end{array}
$$

Solution 9 None of the isomeric butenes is chiral. They all have a plane of symmetry. All the isomeric butenes are related as constitutional (or structural) isomers except *cis*-2-butene and *trans*-2-butene, which are diastereomers.

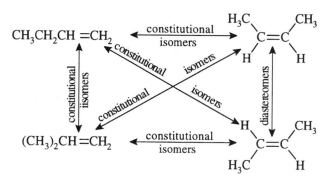

Molecular Model Set C1 to C4 distances in centimeters:

COMPOUND	DISTANCES (cm)
cis-1-Butene	6.0
trans-2-Butene	11.0
Butane (*gauche*)	8.5
Butane (*anti*)	11.0

Solution 10 The hydrogen atoms are all eclipsed in cyclopropane.

Solution 11 All the hydrogen atoms are eclipsed in the planar conformation of cyclobutane. The folded ring system has skewed hydrogen interactions. There are six possible isomers of dimethylcyclobutane. Since the ring is not held in one particular folded conformation, deviations of the ring planarity need not be considered in determining the number of possible dimethyl structures.

Solution 12 In the planar conformation of cyclopentane, all five methylene pairs of hydrogen atoms are eclipsed. That produces 10 eclipsed hydrogen interactions. Some torsional strain is relieved in the envelope conformation since there are only six eclipsed hydrogen interactions.

Solution 13 The three configurational stereoisomers of 1,2-dimethylcyclopentane are shown here. Both trans stereoisomers are chiral, while the cis configuration is an achiral meso compound.

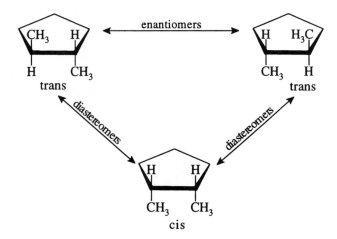

Solution 14 The puckered ring of the chair and the boat conformation may be interconverted by rotation about the carbon-carbon bonds. The chair is more rigid than the boat conformation. All hydrogen atoms in the chair conformation have a staggered arrangement. In the boat conformation, there are eclipsed relationships between the hydrogen atoms on C2 and C3, and also between those on C5 and C6. Carbon atoms that are 1,4 to each other in the chair conformation have a *gauche* relationship. An evaluation of the three conformations confirms the relative stability: chair > twist > boat. The boat conformation has considerable eclipsing strain and nonbonded (van der Waals repulsion) interactions, the twist conformation has slight eclipsing strain, and the chair conformation has a minimum of eclipsing and nonbonded interactions.

Solution 15 Interconversion of the two chair conformations of methylcyclohexane changes the methyl group from an axial to a less crowded equatorial orientation, or the methyl that is equatorial to the more crowded axial position.

Axial methyl   Equatorial methyl

The conformation with the axial methyl group has two *gauche* (1,3 diaxial) interactions that are not present in the equatorial methyl conformation. These *gauche* interactions are axial methyl to C3 and axial methyl to C5. The methyl to C3 and methyl to C5 relationships with methyl groups in an equatorial orientation are anti.

Solution 16 The $\Delta G°$ value reflects the relative energies of the two chair conformations for each structure. The crowding of the alkyl group in an axial orientation becomes greater as the bulk of the group increases. The increased size of the substituent has little effect on the steric interactions of the conformation that has the alkyl group equatorial. The *gauche* (1,3-diaxial) interactions are responsible for the increased strain for the axial conformation. Since the ethyl and isopropyl groups can rotate to minimize the nonbonded interactions, their effective size is less than their actual size. The *tert*-butyl group cannot relieve the steric interactions by rotation and thus has a considerably greater difference in potential energy between the axial and equatorial conformations.

Solution 17 All four stereoisomers of 1,2-dimethylcyclohexane are chiral. The *cis*-1,2-dimethylcyclohexane conformations have equal energy and are readily interconverted, as shown here.

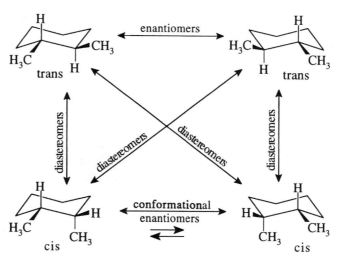

Solution 18 The structures that are isoelectronic have the same geometry. Isoelectronic structures are

$$CH_3CH_3 \quad \text{and} \quad CH_3NH_3{}^+$$

$$CH_3NH_2 \quad CH_3CH_2{}^- \quad \text{and} \quad CH_3OH_2{}^+$$

Structure CH_3NH^- would be isoelectronic to CH_3OH.

Solution 19 Cis-trans stereoisomers are possible only for 3,4-dimethylcyclohexene. The ring size and geometry of the double bond prohibit a trans configuration of the double bond. Two configurational isomers (they are enantiomers) are possible for 2,3-dimethylcyclohexene.

cis-Cyclooctene is more stable because it has less strain than the *trans*-cyclooctene structure. The relative stability of cycloalkene stereoisomers in rings larger than cyclodecane generally favors trans. The *trans*-cyclooctene structure is chiral.

trans-Cyclooctene
(one enantiomer)

Solution 20 The ring fusion in *trans*-decalin is equatorial, equatorial. That is, one ring is attached to the other as 1,2-diequatorial substituents would be. Interconversion of the chair conformations of one ring (carbon atoms 1 through 6) in *trans*-decalin would require the other ring to adopt a 1,2-diaxial orientation. Carbon atoms 7 and 10 would both become axial substituents to the other ring. The four carbon atoms of the *substituent* ring (carbon atoms 7 through 10) cannot bridge the diaxial distance. In *cis*-decalin both conformations have an axial, equatorial ring fusion. Four carbon atoms can easily bridge the axial, equatorial distance.

Solution 21 The cyclohexane ring in norbornane is held in a boat conformation, and therefore has four hydrogen eclipsing interactions. All the six-membered ring systems in twistane are in twist conformations. All four of the six-membered ring systems in adamantane are chair conformations.

Solution 22 Bridging the 1 and 4 carbon atoms of cyclohexane is relatively easy since in the boat conformation the flagpole hydrogen atoms (on C1 and C4) are fairly close and their C–H bonds are directed toward one another. With cyclohexene, the geometry of the double bond and its inability to rotate freely make it impossible to bridge the C1, C4 distance with a single methylene group. Note, however, that a cyclohexene ring can accommodate a methylene bridge between C3 and C6. This bridged bicyclic system (bicyclo[2.2.1] hept-2-ene) does not have a bridgehead double bond.

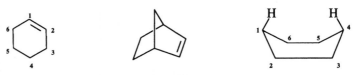

Bicyclo[2.2.1]hept-2-ene

Solution 23 The 120° geometry of the double bond is ideal for incorporation into a planar six-membered ring, as the internal angle of a regular hexagon is 120°. Cyclooctatetraene cannot adopt a planar ring system without considerable angle strain. The eight-membered ring adopts a "tub" conformation that minimizes angle strain and does not allow significant *p*-orbital overlap other than that of the four double bonds in the system. Thus, the cyclooctatetraene has four isolated double bonds and is not a delocalized π-electron system.

Cyclooctatetraene (tub conformation)

Solution 24 In the $CH_3CHXCHYCH_3$ system, there are four stereoisomers, all of which are chiral.

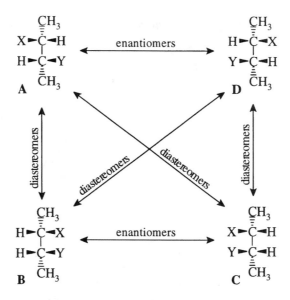

In the $CH_3CHXCHXCH_3$ system, there are three stereoisomers, two of which are chiral. The third stereoisomer (**G**) (shown on page 575) is an achiral meso structure.

Solution 25 If at least one conformation of a molecule in which free rotation is possible has a plane of symmetry, the molecule is achiral. For a molecule with the configurations specified, there are two achiral conformations. The eclipsed conformation **E** shown in the exercise and staggered conformation **F.**

A model of **F** is identical with its mirror image. It is achiral, although it does not have a plane of symmetry, due to the presence of a *center* of symmetry that is located between C2 and C3. A center of symmetry, like a plane of symmetry, is a reflection symmetry element. A center of symmetry involves reflection through a point; a plane of symmetry requires reflection about a plane. A model of the mirror image of **S** (structure **T**) is not identical to **S,** but is a conformational enantiomer of **S.** They can be made identical by rotation about the C2, C3 bond. Since **S** and **T** are conformational enantiomers, each will be present in equal amounts in a solution of this configurational stereoisomer. Both conformation **S** and conformation **T** are chiral and therefore should rotate the plane of plane polarized light. Since they are enantiomeric, the rotations of light will be equal in magnitude but *opposite* in direction. The net result is a racemic form of conformational enantiomers, and thus optically inactive. A similar argument can be made for any other chiral conformation and this configuration of $CH_3CHXCHXCH_3$.

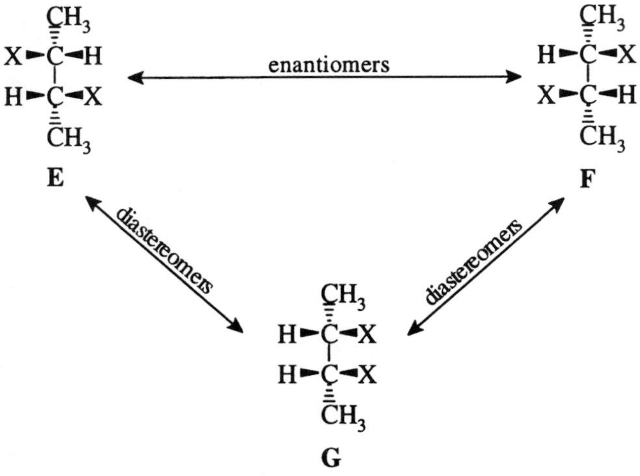

Solution 26 Structures B and C are chiral. Structure A has a plane of symmetry and is therefore achiral. Compounds D and E are both chiral. The relative orientation of the terminal groups in D and E is perpendicular, as is the case in the cumulated dienes.

Cumulated triene F is achiral. It has a plane of symmetry passing through all six carbon atoms. Structure F has a trans configuration. The cis diastereomer is the only other possible stereoisomer.

Solution 27 Structure **J** can be isolated as a chiral stereoisomer because of the large steric barrier to rotation about the bond connecting the rings. Biphenyl **K** has a plane of symmetry and is therefore achiral. The symmetry plane of **K** is shown here. Any chiral conformation of **L** can easily be converted to its enantiomer by rotation. It is only when a ≠ b and f ≠ e and rotation is restricted by bulky groups that chiral (optically active) stereoisomers can be isolated.

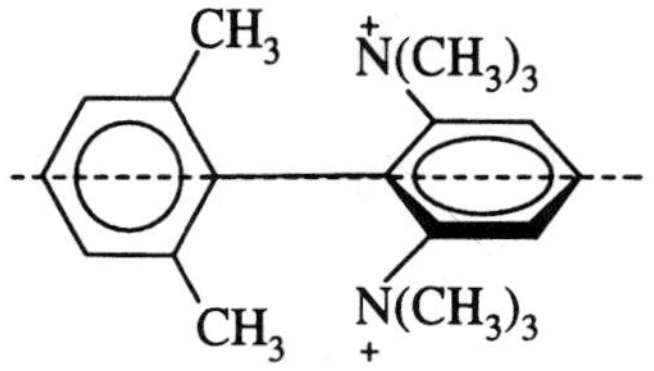

A plane of symmetry

Solution 28 The smallest stable cycloalkyne is the nine-membered ring cyclononyne. A model of this alkyne can easily be assembled with the Theta Molecular Model Set.

Solution 29 As shown here, the alternative chair conformation of β-D-glucopyranose has all large substituents in an axial orientation. The structures α-D-glucopyranose, β-D-mannopyranose, and β-D-galactopyranose all have one large axial substitutent in the most favorable conformation. β-L-Glucopyranose is the enantiomer (mirror image) of β-D-glucopyranose. Enantiomers are of equal thermodynamic stability.

β-D-Glucopyranose

α-D-Glucopyranose

β-D-Galactopyranose

β-D-Mannopyranose

β-L-Glucopyranose

Solution 30 The peptide chain bonds not free to rotate are indicated by the bold lines in the structure shown here. The transoid arrangement produces a more linear tripeptide chain. The length of the tripeptide chain does not change if you change the substituent **R** groups.

Solution 31 The models of the π molecular orbitals for ethene are shown here. A representation of these orbitals can be found in the text on page 58.

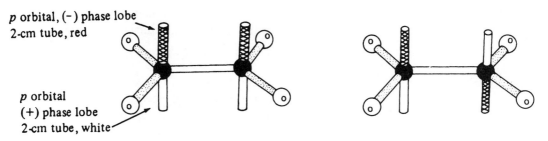

p orbital, (−) phase lobe
2-cm tube, red

p orbital
(+) phase lobe
2-cm tube, white

Ethene, π bonding molecular orbital   Ethene, π* antibonding molecular orbital

The π molecular orbitals for 1,3-butadiene are shown in the text on page 514. A model of one of the π molecular orbitals of 1,3-butadiene is shown in the model set instruction booklet. The phases of the contributing atomic orbitals to the molecular orbitals of the allyl radical can be found in the text on page 502. The π molecular orbital of the allyl radical has a node at C2. This can be illustrated with the Theta Molecular Model Set by not placing red or white p-orbital tubes on the C2 atom center prongs. The absence of tubes indicates an orbital phase of zero.

Solution 32 The complete solution to this exercise is given in the text Sections O.2B and O.3B. The orbitals involved are shown below.

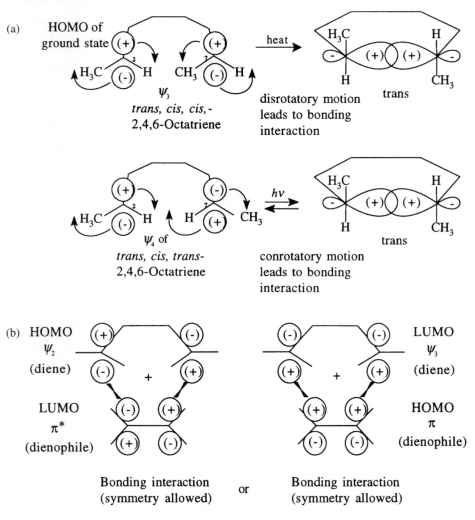

(a)   HOMO of ground state

ψ_3

trans, cis, cis,-
2,4,6-Octatriene

heat

disrotatory motion leads to bonding interaction

trans

ψ_4 of

trans, cis, trans-
2,4,6-Octatriene

hv

conrotatory motion leads to bonding interaction

trans

(b) HOMO
ψ_2
(diene)

LUMO
π^*
(dienophile)

Bonding interaction (symmetry allowed)

or

LUMO
ψ_3
(diene)

HOMO
π
(dienophile)

Bonding interaction (symmetry allowed)